Handbook of 3D Integration

Edited by
Philip Garrou, Christopher Bower
and Peter Ramm

Further Titles of Interest

Ban P. Wong

Nano-CMOS Design for Manufacturability: Robust Circuit and Physical Design for Technology Nodes

2008
ISBN: 978-0-470-11280-9

Volker Saile, Ulrike Wallrabe, Osamu Tabata, Jan G. Korvink (Eds.)

LIGA and its Applications

Advanced Micro & Nanosystems Volume 7

2008
ISBN: 978-3-527-31698-4

Osamu Tabata, Toshiyuki Tsuchiya (Eds.)

Reliability of MEMS

Advanced Micro & Nanosystems Volume 6

2008
ISBN: 978-3-527-31494-2

Georges Hadziioannou, George G. Malliaras (Eds.)

Semiconducting Polymers

Chemistry, Physics and Engineering (2 Volumes)

2nd edition, 2007
ISBN: 978-3-527-31271-9

Volkan Kursun, Eby G. Friedman

Multi-voltage CMOS Circuit Design

2006
ISBN: 978-0-470-01023-5

Hagen Klauk (Ed.)

Organic Electronics

Materials, Manufacturing and Applications

2006
ISBN: 978-3-527-31264-1

Handbook of 3D Integration

Technology and Applications of 3D Integrated Circuits

Volume 1

Edited by
Philip Garrou, Christopher Bower and Peter Ramm

WILEY-VCH Verlag GmbH & Co. KGaA

The Editors

Dr. Philip Garrou
Microelectronic Consultants of North Carolina
3021 Cornwallis Road
Research Triangle Park, NC 27709-2889
USA

Dr. Christopher Bower
Semprius, Inc.
2530 Meridian Parkway
Durham, NC 27713
USA

Dr. Peter Ramm
Fraunhofer IZM
Hansastraße 27d
80686 München
Germany

Cover Description:
The artistic illustration explains the principle of 3D integration: Stacking and vertical interconnecting of device layers. The magnified portion depicts a cross section of a 3D integrated test chip which was fabricated at Fraunhofer Munich. The wafer-level 3D integration technology is based on bonding of thinned device substrates and a metallization process providing a high vertical wiring density. This technology presents a solution to performance and reliability limitations due to the wiring of microelectronic devices and enables completely new ultra-miniaturized products.

Library of Congress Card No.: applied for

British Library Cataloguing-in-Publication Data
A catalogue record for this book is available from the British Library.

Bibliographic information published by the Deutsche Nationalbibliothek
Die Deutsche Nationalbibliothek lists this publication in the Deutsche Nationalbibliografie; detailed bibliographic data are available in the Internet at http://dnb.d-nb.de.

Typesetting Thomson Digital, Noida, India
Printing Strauss GmbH, Mörlenbach
Binding Litges & Dopf Buchbinderei GmbH, Heppenheim

Printed in the Federal Republic of Germany
Printed on acid-free paper

ISBN: 978-3-527-32034-9

Contents

Volume 1

Handbook of 3D Integration: Technology and Applications of 3D Integrated Circuits.
Edited by Philip Garrou, Christopher Bower and Peter Ramm

ISBN: 978-3-527-32034-9

Preface

Many of readers of this book will doubtless see the title "Handbook of 3D Integration" and expect the two volumes to cover all manner of devices and packages that can be argued to be 3D. The editors are aware that stacked packages and stacked dies without through semiconductor vias (TSVs) are 3D structures, but we contend that these structures can ultimately be catalogued under "3D Packaging". The focus of this book is the technology and applications of "3D Integration," which we classify as the vertical integration of thinned and bonded silicon integrated circuits with vertical electrical interconnects between the IC layers. Most of the vertical interconnects discussed here will be through silicon vias (TSVs).

The editors are aware that the terminology "3D Integration" lacks specificity. Why not a handbook of "3D ICs" or "3D Silicon Integration" or "Vertical System Integration"? While each of these has pros and cons, we feel that the simple term "3D Integration" has already been accepted by many researchers, and it is very unlikely that new terminology will take hold.

These volumes intend to provide engineers and scientists with a timely and fairly comprehensive overview of the field. Although our goal was to be as complete as possible, there will clearly be some technical areas that are not covered. We do not cover monolithic growth approaches to 3D integration and we do not cover 3D integration of heterogeneous materials. Perhaps, as they mature, these technologies will be included in a future edition. The book is organized into five parts:

- Part I covers the processing technology for TSVs. This section includes chapters on deep reactive ion etching of TSVs, laser-drilled TSVs, TSV sidewall insulation, Cu electroplating and chemical vapor deposition of Cu and W.

- Part II covers wafer thinning and bonding technology. Included are chapters on thinning and singulation of silicon wafers, techniques and tools for wafer alignment and bonding, polymer bonding and intermetallic bonding.

- Part III covers the various integration processes being pursued across the globe. This section begins with a chapter that surveys commercial activity in 3D integration. Next, there is a grouping of chapters from universities and institutes that have distinct approaches to 3D integration. This group includes chapters from

Handbook of 3D Integration: Technology and Applications of 3D Integrated Circuits.
Edited by Philip Garrou, Christopher Bower and Peter Ramm

ISBN: 978-3-527-32034-9

Fraunhofer IZM, University of Arkansas, Japan's ASET consortium, CEA-LETI, MIT Lincoln Labs, IMEC, MIT and RPI. This part ends with a grouping of chapters from 3D integration start-up companies, including chapters from Tezzaron, Ziptronix and Zycube.

- Part IV covers design, performance and thermal management for 3D integration. Design for 3D is covered in chapters from NC State University, Fraunhofer IIS, Lincoln Labs and University of Minnesota. This part includes a chapter from Intel on testing of 3D circuits and a chapter from IBM on thermal management in 3D ICs.
- Part V contains chapters on specific applications of 3D integration. Individual chapter topics are on 3D microprocessors, 3D memory, sensor arrays, power devices and wireless sensor systems.

We would like to acknowledge all the authors of each chapter. It is the individual author contributions that made this work possible. We would also like to thank the authors for providing reviews of chapters. The editors are deeply grateful for the time and effort you each put into your chapters. We also wish to express our gratitude to those at Wiley-VCH who have been a great help in keeping us on schedule.

We hope that this book will serve as a valuable resource for practitioners of 3D integration. Based on the progress made over the last few years, we anticipate that the next decade will be a very exciting time to be working in this area.

November 2007

Phil Garrou, RTP, North Carolina
Chris Bower, RTP, North Carolina
Peter Ramm, Munich, Germany

List of Contributors

Sitaram Arkalgud
SEMATECH
2706 Montopolis Boulevard
Austin, TX 78741
USA

Brian Aull
Massachusetts Institute of Technology
Lincoln Laboratory
244 Wood Street
Lexington, MA 02420-9108
USA

Rozalia Beica
Semitool, Inc.
655 West Reserve Drive
Kalispell, MT 59901
USA

Robert Berger
Massachusetts Institute of Technology
Lincoln Laboratory
244 Wood Street
Lexington, MA 02420-9108
USA

Eric Beyne
IMEC
Kapeldreef 75
3001 Leuven
Belgium

Michiel A. Blauw
Eindhoven University of Technology
PO Box 513
5600 MB Eindhoven
The Netherlands

Christopher Bower
Semprius, Inc.
2530 Meridian Parkway
Durham, NC 27713
USA

Thomas Brunschwiler
IBM Zurich Research Laboratory
Advanced Thermal Packaging
Säumerstrasse 4
8803 Rüschlikon
Switzerland

Susan Burkett
University of Arkansas
Department of Electrical Engineering
3217 Bell Engineering Center
Fayetteville, AR 72701
USA

James Burns
Massachusetts Institute of Technology
Lincoln Laboratory
244 Wood Street
Lexington, MA 02420-9108
USA

Handbook of 3D Integration: Technology and Applications of 3D Integrated Circuits.
Edited by Philip Garrou, Christopher Bower and Peter Ramm

ISBN: 978-3-527-32034-9

Tim S. Cale
Rensselaer Polytechnic Institute
Mailstop CII-6015/CIE
110 8th Street
Troy, NY 12180-3590
USA

S. M. Chang
Industrial Technology Research
Institute of Taiwan
195 Chung Hsing Road
Chutung, Hsinchu
Taiwan 310, ROC

Barbara Charlet
CEA-LETI, MINATEC
Département Integration Hétérogene
Silicium
17, rue des Martyrs
38054 Grenoble Cedex 9
France

Nisha Checka
Massachusetts Institute of Technology
Lincoln Laboratory
244 Wood Street
Lexington, MA 02420-9108
USA

Chang-Lee Chen
Massachusetts Institute of Technology
Lincoln Laboratory
244 Wood Street
Lexington, MA 02420-9108
USA

Chenson Chen
Massachusetts Institute of Technology
Lincoln Laboratory
244 Wood Street
Lexington, MA 02420-9108
USA

Wouter Dekkers
NXP-TSMC Research Center
High Tech Campus 4
Mailbox WAG02
5656 AE Eindhoven
The Netherlands

Marc de Samber
Philips Applied Technologies
High Tech Campus 7
5656 AE Eindhoven
The Netherlands

Léa Di Cioccio
CEA-LETI, MINATEC
Département Integration Hétérogene
Silicium
17, rue des Martyrs
38054 Grenoble Cedex 9
France

R. Ecke
TU Chemnitz
Zentrum für Mikrotechnologien
Reichenhainer Straße 70
09126 Chemnitz
Germany

Günter Elst
Fraunhofer IIS
Design Automation Division
Zeunerstraße 38
01069 Dresden
Germany

Paul Enquist
Ziptronix
800 Perimeter Park, Suite B
Morrisville, NC 27560
USA

Andy Fan
Massachusetts Institute of Technology
Department of Electrical Engineering
77 Massachusetts Avenue
Cambridge, MA 02139
USA

Paul D. Franzon
North Carolina State University
Monteith GRC 443
ECE, Box 7914
Raleigh, NC 27695
USA

Philip Garrou
Microelectronic Consultants of North
Carolina
3021 Cornwallis Road
Research Triangle Park, NC 27709-2889
USA

Pascale Gouker
Massachusetts Institute of Technology
Lincoln Laboratory
244 Wood Street
Lexington, MA 02420-9108
USA

Ronald J. Gutmann
Rensselaer Polytechnic Institute
Mailstop CII-6015/CIE
110 8th Street
Troy, NY 12180-3590
USA

David Henry
CEA-LETI, MINATEC
Département Integration Hétérogene
Silicium
17, rue des Martyrs
38054 Grenoble Cedex 9
France

Arne Heittmann
Qimonda AG
Gustav-Heinemann-Ring 212
81739 Munich
Germany

Thomas Herndl
Infineon Technologies
Operngasse 20b/32
1010 Vienna
Austria

David Heyes
NXP Semiconductors
Bramhall Moove Lane
Stockpat, Cheshire SK7 5B
UK

Thierry HILT
CEA-LETI
17, avenue des Martyrs
38054 Grenoble Cedex
France

Adrian Ionescu
Ecole Polytechnique Fédérale de
Lausanne
Institute of Microelectronics and
Microsystems
Electronics Laboratory
1015 Lausanne
Switzerland

Jean-Pierre Joly
CEA-LITEN, INES
Département des Technologies Solaires
50, avenue du Lac Léman
73377 Le Bourget du Lac
France

Craig Keast
Massachusetts Institute of Technology
Lincoln Laboratory
244 Wood Street
Lexington, MA 02420-9108
USA

Ervin (W. M. M.) Kessels
Eindhoven University of Technology
PO Box 513
5600 MB Eindhoven
The Netherlands

Armin Klumpp
Fraunhofer IZM
Hansastraße 27d
80686 Munich
Germany

Jeffrey Knecht
Massachusetts Institute of Technology
Lincoln Laboratory
244 Wood Street
Lexington, MA 02420-9108
USA

Werner Kröninger
Infineon Technologies AG
Postfach 10 09 44
93009 Regensburg
Germany

Yann Lamy
NXP-TSMC Research Center
High Tech Campus 4
Mailbox WAG02
5656 AE Eindhoven
The Netherlands

Patrick Leduc
CEA-LETI, MINATEC
Département Integration Hétérogene
Silicium
17, rue des Martyrs
38054 Grenoble Cedex 9
France

Paul Lindner
EV Group
Erich Thallner GmbH
DI Erich Thallner Straße 1
4782 St.Florian/Inn
Austria

W. C. Lo
Industrial Technology Research
Institute of Taiwan
195 Chung Hsing Road
Chutung, Hsinchu
Taiwan 310, ROC

James Jian-Qiang Lu
Rensselaer Polytechnic Institute
Mailstop CII-6015/CIE
110 8th Street
Troy, NY 12180-3590
USA

T. M. Mak
Intel Corporation
2200 Mission College Blvd., SC 12-604
Sauta Clara, CA 95052-8119
USA

Thorsten Matthias
EV Group
7700 South River Parkway
Tempe, AZ 85284
USA

Bruno Michel
IBM Zurich Research Laboratory
Advanced Thermal Packaging
Säumerstrasse 4
8803 Rüschlikon
Switzerland

Patrick Morrow
Intel Corporation
Mail Stop: RA3-252
5200 N.E. Elam Young Parkway
Hillsboro, OR 97124-6467
USA

Makoto Motoyoshi
ZyCube Co. Ltd.
ZyCube Sendai Lab.
519-1176 Aoba Aramaki, Aoba-ku,
Sendai-shi, Miyagi
985-0845 Japan

Sriram Muthukumar
Intel Corporation
Mail Stop: CH4-109
5000 W Chandler Blvd
Chandler, AZ 85226
USA

Pierre Nicole
THALES systèmes aéroportés
2 Avenue Gay Lussac
78851 Elancourt Cedex
France

Stefan Pargfrieder
EV Group
Erich Thallner GmbH
DI Erich Thallner Straße 1
4782 St. Florian/Inn
Austria

Robert Patti
Tezzaron Semiconductor Corp.
1415 Bond Street
Naperville, IL 60563
USA

Ulrich Ramacher
Infineon Technologies AG
Am Campeon 1-12
85579 Neubiberg
Germany

Peter Ramm
Fraunhofer IZM
Hansastraße 27d
80686 Munich
Germany

Rafael Reif
Massachusetts Institute of Technology
Department of Electrical Engineering
77 Massachusetts Avenue
Cambridge, MA 02139
USA

Thomas L. Ritzdorf
Semitool, Inc.
655 West Reserve Drive
Kalispell, MT 59901
USA

Fred Roozeboom
NXP-TSMC Research Center
High Tech Campus 4
Mailbox WAG02
5656 AE Eindhoven
The Netherlands

Mihai Sanduleanu
Philips Applied Technologies
High Technology Campus 7
5656 AE Eindhoven
The Netherlands

Sachin S. Sapatnekar
University of Minnesota
Department of Electrical
and Computer Engineering
200 Union Street
Minneapolis, MN 55455
USA

Anton Sauer
Fraunhofer IZM
Hansastraße 27d
80686 Munich
Germany

Leonard Schaper
University of Arkansas
Department of Electrical Engineering
3217 Bell Engineering Center
Fayetteville, AR 72701
USA

Peter Schneider
Fraunhofer IIS
Design Automation Division
Zeunerstraße 38
01069 Dresden
Germany

Stefan E. Schulz
TU Chemnitz
Zentrum für Mikrotechnologien
Reichenhainer Straße 70
09126 Chemnitz
Germany

Charles Sharbano
Semitool, Inc.
655 West Reserve Drive
Kalispell, MT 59901
USA

Herbert Shea
Ecole Polytechnique Fédérale de
Lausanne
Institute of Microelectronics and
Microsystems
Electronics Laboratory
1015 Lausanne
Switzerland

Antonio Soares
Massachusetts Institute of Technology
Lincoln Laboratory
244 Wood Street
Lexington, MA 02420-9108
USA

Vyshnavi Suntharalingam
Massachusetts Institute of Technology
Lincoln Laboratory
244 Wood Street
Lexington, MA 02420-9108
USA

Kenji Takahashi
Toshiba Corp.
1 Komukai Toshiba-cho, Saiwai-ku,
Kawasaki-shi, Kanagawa
212-8583 Japan

Maaike Taklo
SINTEF ICT
Microsystems and Nanotechnology
Gaustadalléen 23
0373 Oslo
Norway

Chuan Seng Tan
Nanyang Technological University
School of Electrical and Electronic
Engineering
50 Nanyang Avenue
Singapore 639798
Singapore

Kazumasa Tanida
Toshiba Corp.
1 Komu Kai Toshibacho, Saiwai-Ku
Kawasaki-shi, Kanagawa
212-8583 Japan

Mark E. Tuttle
Micron Technology, Inc.
Mail Stop 1-717
8000 S. Federal Way
Boise, ID 83707-0006
USA

Brian Tyrrell
Massachusetts Institute of Technology
Lincoln Laboratory
244 Wood Street
Lexington, MA 02420-9108
USA

Eric (F.) van den Heuvel
Philips Applied Technologies
High Technology Campus 7
5656 AE Eindhoven
The Netherlands

Emile van der Drift
Delft University of Technology
PO Box 5053
2600 GB Delft
The Netherlands

Richard (M. C. M.) van de Sanden
Eindhoven University of Technology
PO Box 513
5600 MB Eindhoven
The Netherlands

Eric van Grunsven
Philips Applied Technologies
High Tech Campus 7
5656 AE Eindhoven
The Netherlands

Co Van Veen
Philips Applied Technologies
High Technology Campus 7
5656 AE Eindhoven
The Netherlands

Jan F. Verhoeven
Philips Applied Technologies
High Technology Campus 7
5656 AE Eindhoven
The Netherlands

Susan Vitkavage
Lockheed Martin
5600 Sand Lake Road
Orlando, FL 32819
USA

Keith Warner
Massachusetts Institute of Technology
Lincoln Laboratory
244 Wood Street
Lexington, MA 02420-9108
USA

Josef Weber
Fraunhofer IZM
Hansastraße 27d
80686 Munich
Germany

Werner Weber
Infineon Technologies AG
Am Campeon 1–12
85579 Neubiberg
Germany

Bruce Wheeler
Massachusetts Institute of Technology
Lincoln Laboratory
244 Wood Street
Lexington, MA 02420-9108
USA

Robert Wieland
Fraunhofer IZM
Hansastraße 27d
80686 Munich
Germany

Markus Wimplinger
EV Group
7700 South River Parkway
Tempe, AZ 85284
USA

Jürgen M. Wolf
Fraunhofer IZM
Gustav-Meyer-Allee 25
13355 Berlin
Germany

Bernhard Wunderle
Fraunhofer IZM
Gustav-Meyer-Allee 25
13355 Berlin
Germany

Peter Wyatt
Massachusetts Institute of Technology
Lincoln Laboratory
244 Wood Street
Lexington, MA 02420-9108
USA

Donna Yost
Massachusetts Institute of Technology
Lincoln Laboratory
244 Wood Street
Lexington, MA 02420-9108
USA

1
Introduction to 3D Integration

Philip Garrou

1.1
Introduction

Wafer level 3-dimensional (3D) integration is an emerging, system level integration architecture wherein multiple strata (layers) of planar devices are stacked and interconnected using through silicon (or other semiconductor material) vias (TSV) in the Z direction as shown in Figure 1.1.

The technical and market drivers for such a new architecture are discussed in Chapter 2. Several process sequences have been developed to fabricate such stacks, which are discussed in Chapter 3. All of them depend on the following enabling technologies.

- TSV formation – realization of electrically isolated connections through the silicon substrate. The diameter of the TSV is dependent on the degree of access needed to an individual strata, which differs with application area.
- Thinning of the strata – usually to below 50 μm in memory stacks, 25 μm for CMOS silicon circuits and to below 5 μm for SOI circuits.
- Alignment and bonding – either as die to wafer or wafer to wafer. Several technologies are available (Chapter 3).

The main attribute of these stacked structures are the z axis interconnects which are usually called "through silicon vias" (TSV) but are also described as "through wafer vias" (TWV) and/or "through wafer interconnect" (TWI).

Conceptually, 3D can alleviate interconnect delay problems, while reducing chip area. When the large number of the long interconnects needed in 2D structures are replaced by short vertical interconnects this greatly enhances the performance of logic circuits. For instance, logic gates on a critical path can be placed very close to each other by stacking them and interconnecting in the z direction. Circuits with different voltage requirements and or performance requirements can also be put on different layers [1].

Handbook of 3D Integration: Technology and Applications of 3D Integrated Circuits.
Edited by Philip Garrou, Christopher Bower and Peter Ramm

ISBN: 978-3-527-32034-9

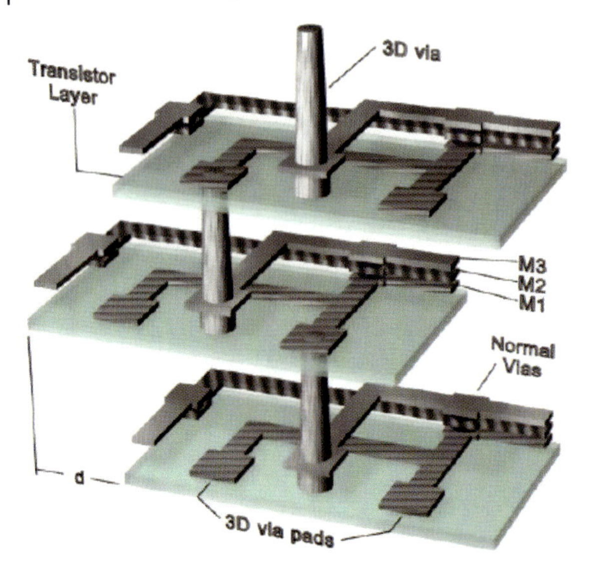

Figure 1.1 3D wafer level stacking using through silicon interconnect vias. Figure courtesy of Professor Duncan Elliott, Department of Electrical and Computer Engineering, University of Alberta.

Figure 1.2 depicts memory to logic interconnections created using a current 2D interconnection scheme, a system on chip (SOC) solution and a 3D integration solution using through silicon vias (TSV).

SOC, system-on-chip, refers to the integration of nearly all aspects of a system design on a single chip. Such chips are often mixed signal and/or mixed technology designs, which include embedded DRAM, logic, analog, RF, and so on. While this technology at first glance looks appealing, integration of such disparate technologies on a single chip dramatically increases the chip area and increases long global interconnect, which can lead to significant signal transmission delays. Since it takes

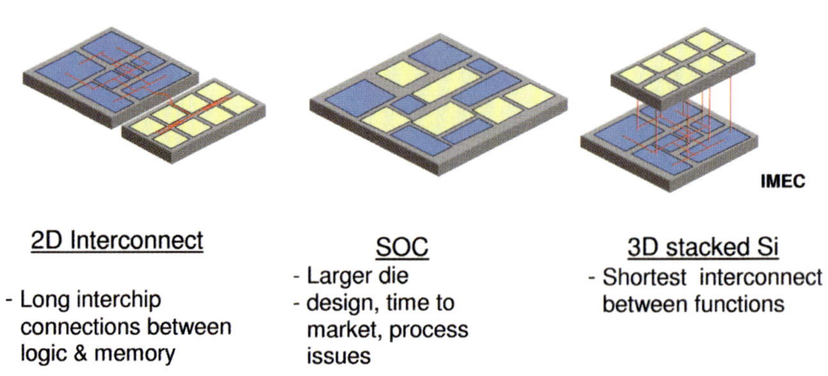

Figure 1.2 2D vs SOC vs 3D. Figure courtesy Professor Eric Beyne, MEC.

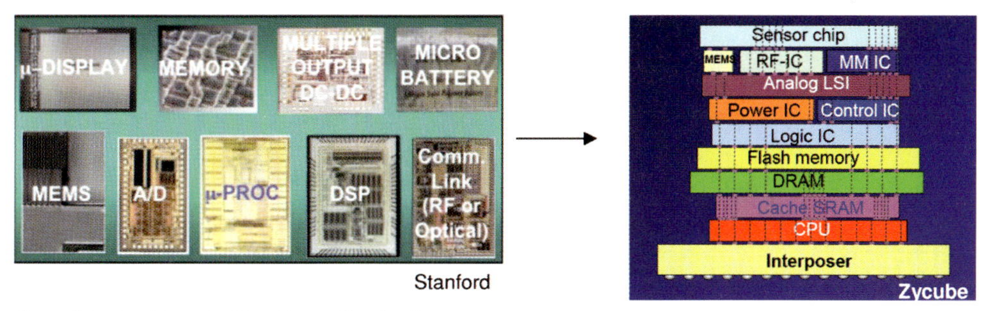

Figure 1.3 The "Holy Grail" – 3D stacking for "heterogeneous integration." Figures courtesy of Stanford University, CA, and Zycube, Japan.

different process technologies to produce these different functions, the complexity of materials and process issues is significant.

In 3D architecture, device fabrication is achieved by the production of full wafers of a specific function, that is, embedded processors, DSPs, SRAM, DRAM, and so on. These are then thinned, aligned and vertically interconnected (chip to wafer or wafer to wafer) to create a functional device. Thus, the 3D concept allows integration of otherwise incompatible technologies, and offers significant advantages in performance, functionality, and form factor. In some sectors this has become known as "heterogeneous integration." This is shown pictorially in Figure 1.3. Other technologies that could be conceivably included in the stack include antenna, sensors, power management and power storage devices.

Such technology requires both a common die size and a common interconnection scheme. We will see in Chapter 3 that the interconnecting vias can be created by fabrication in the IC foundry (FEOL) or fabrication by the assembly and packaging house after the chip is finished. If done by the latter, open areas must be left in cells, or between cells, to accommodate these interconnecting vias. While some Si real estate is consumed by such post chip TSV fabrication, a high interwafer interconnect density can be achieved with a minor area penalty.

Having shorter signal paths between die make it possible to improve the system's performance by permitting the system to run faster, it also wastes less power. Wire length is directly related to power usage, and keeping wire lengths short helps keep power use down. As we shall see later (Chapter 33) one of the concerns about using stacked-die is heat removal, but the use of TSV reduces the overall wire length, which reduces heat generation, somewhat.

1.2 Historical Evolution of Stacked Wafer Concepts

GE started to investigate the possibilities of forming electrical interconnections through semiconductor wafers for NASA in 1981 [2].

During 1986–1990 Akasaka [3] and Hayashi [4] laid out the basic concepts for and proposed technologies for 3D ICs. Later, Hayashi [5] proposed fabrication of separate devices in separate wafers, reduction in the thickness of the wafers, providing front and back leads and connecting the thinned die to each other. This was dubbed CUBIC (CUmulatively Bonded IC) and a two active layer device was fabricated in a top to bottom fashion and tested.

1.3 3D Packaging vs 3D Integration

Over the past few years, die (chip) stacking has emerged as a significant packaging option. Integrating chips vertically in a single package multiplies the amount of silicon that can be crammed in a given package footprint, conserving hand held device real estate. At the same time, it enables shorter chip to chip routing, which speeds communication between them. Another benefit is the simplification of board assembly because there are fewer components to be placed on the board.

Initial applications consisted of two-chip memory combinations such as flash and SRAM and flash plus flash. Today, chip stacking has been extended beyond memories to logic and analog ICs in packages that may also contain surface-mount passives. In addition, chip stacking has evolved to include three or four die stacks and side-by-side combinations of stacked and unstacked die within a package. The die are typically mounted to a substrate, which is bumped to create either a chip scale package (CSP) or ball grid array (BGA) as the final package.

Though chip stacking began with mounting smaller dies onto larger ones to enable wirebonding of both, packaging vendors have developed techniques for stacking same-size die or for stacking a larger die on top of a smaller one such as placing a spacer (a dummy piece of silicon) between the two. The spacer lifts the top die just enough to allow wirebonding to the bottom die. While standard wirebonding might have a loop height of 150–175 μm, die stacking could require loop heights under 100 μm. Figure 1.4 shows typical wire bonding in such 3D stacked structures.

Figure 1.4 Wire bonded chips stacked in 3D package.

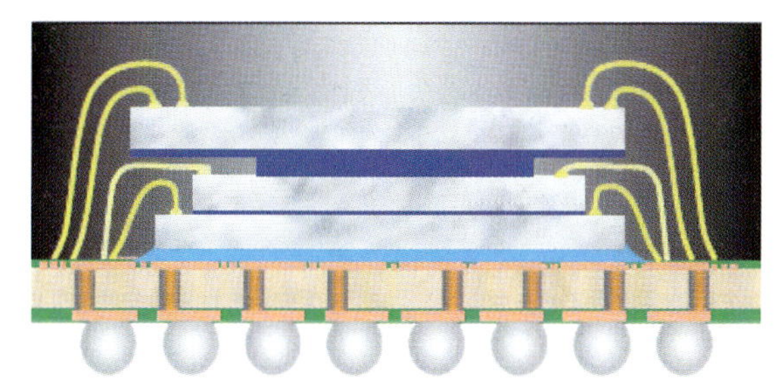

Figure 1.5 3D stacked die BGA package.

Such variations have helped expand the number of stacked-die package options, creating whole portfolios of what vendors commonly call 3-D packages. A variety of these 3D BGA packages are now in high-volume production (e.g., Figure 1.5).

The number of dies that can be stacked depends on the required thickness of the final package and the thickness of each layer (substrate, die, spacers, and BGA ball diameter) within the package. Typical ball diameters range from 0.75 mm for 1.27-mm pitch down to 0.2 mm for 0.35-mm pitch. Package height of 1.4 mm was the standard for stacked-chip packages in portable applications. Demand has recently shifted to 1.2- and 1.0-mm high packages, and even 0.8 mm is a possibility. It is currently possible to build three- and four-die stacks in 1.4-mm packages.

Another 3D packaging alternative is called package stacking or PoP (package on package). While package stacking increases material costs per package and overall package height, it provides higher yields per stacked device, which lowers cost. Package stacking needs thin, flat, high-temperature, moisture-resistant packages to handle the multiple reflows and rework associated with SMT. Vendors like Amkor have been developing processes to stack CSPs and BGAs (Figure 1.6).

Digital camera and cell phone applications are currently stacking two packages for logic + memory architectures. High density DRAM and Flash memory modules are stacking up to four packages high, with this capability demonstrated to eight high stacks.

3D packaging technology does:

- Thin die to save weight and volume.
- Stack die to save x–y space (wire bonding).

3D packaging technology does not:

- Minimize interconnect or enhance electrical performance (C and L parasitics).

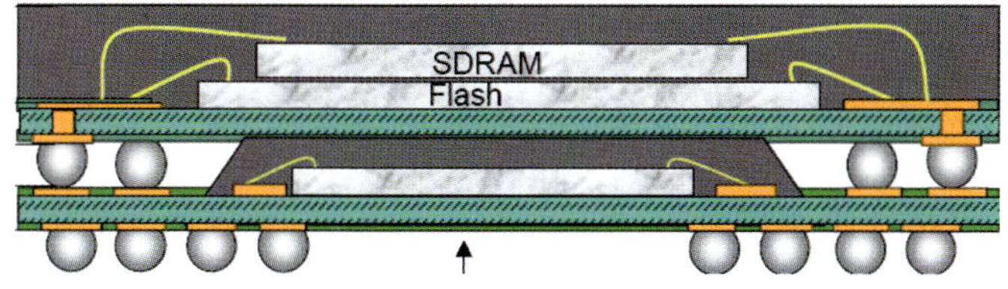

Figure 1.6 Typical Amkor PoP.

1.4
Non-TSV 3D Stacking Technologies

1.4.1
Irvine Sensors

Irvine sensors started delivering stacked Si memory from the IBM Burlington factory in 1992. In their first generation process, gold rerouting metallization was added to bring all signals to an outer die edge, and then the wafer was diced. The die were stacked, and the stack lapped to expose the ends of the gold rerouting metal. Bus metallization was deposited on the side of the stack, interconnecting the dice and a ceramic top cap substrate was added, which allowed signals into and out of the stack [6, 7]. Limitations to this technology included:

- All dice must be the same size, limiting the stack to a single die type.
- Frequent die shrinks required substantial retooling.
- The trend in commercial wafers is for street widths to shrink, which made the process more and more difficult.

Their newer technology, known as neo-stack, addresses previous limitations. In the Neo-stacking approach KGD are bumped using a gold wire bonder. A new wafer, or "Neo-wafer," is constructed using many of the bumped dice in a potting compound matrix. A standard Neo-die size, slightly larger than the largest die in the stack, is used for all dice in the stack. This feature allows the stack to be heterogeneous. Blank silicon is added to open areas on layers where smaller dice are used to enhance thermal conduction between layers. The Neo-wafer is metallized and thinned before dicing into individual Neo-die. Other die types are similarly fabricated into Neo-die of the same dimension. All of the necessary dice are then laminated into a single stack, with all signals to be interconnected brought out to two sides of the stack. On the top of the stack is a cap, with metallization on both sides, connected through vias. Metallization is added to the two sides of the stack to complete the interconnection between dice, bringing all input/output signals to the cap chip.

Figure 1.7 shows a cross section of the "Neostack" is. A Flash neo-stack and its composite layers are shown in Figure 1.8.

1.4.2
UTCS (Ultrathin Chip Stacking) IMEC, CNRS, U. Barcelona

Similar, though not identical technology has been proposed by IMEC, CNRS, U. Barcelona [8–10]. This technology named Ultra Thin Chip Stacking (UTCS) can be fabricated by the following sequence: chips are thinned down to 10 μm, interspaced with BCB dielectric layers and the vertical interconnection is achieved with metallized vias. The final stack is significantly thinner than the individual silicon chip.

Figure 1.7 Irvine sensors neo-stack wafer concept [7].

BCB, the adhesive and planarization layer, has poor thermal conductivity, which degrades the heat extraction efficiency through the vertical path. Heat extraction is vastly improved by the use of copper grids or full metal plates to remove heat from the thinned chips.

Figure 1.9 shows schematically how the chips are interconnected. The detailed procedure is shown in Figure 1.10.

1.4.3 Fujitsu

In the summer of 2002 Fujitsu introduced its similar CS Module based on wafer-thinning, chip stacking and re-distribution technologies [11]. The technology stacks chips into two layers and redistributes signal circuitry between them. Figure 1.11

Figure 1.8 Flash neo-stack and its composite layers.

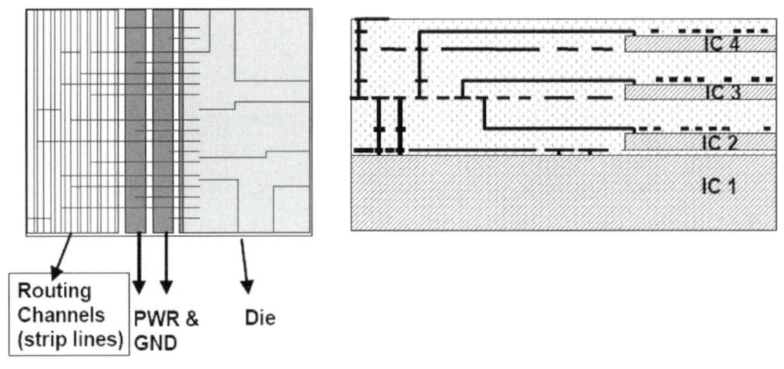

Figure 1.9 UTCS 3D routing in thin film dielectric [8].

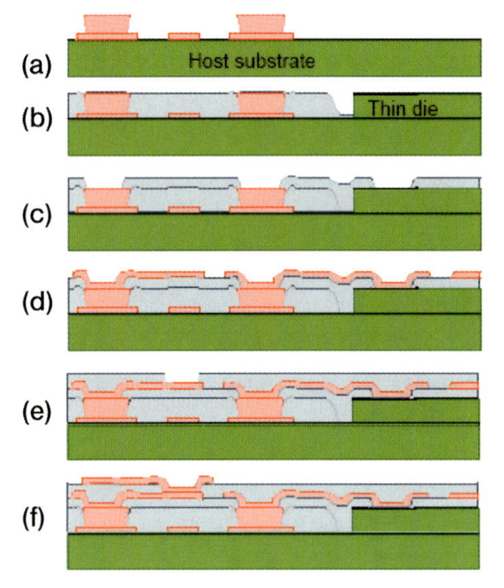

Figure 1.10 UTCS process sequence: (a) Patterning of the first interconnection level and growth of studs; (b) transfer of the thin die, deposition of a thick photo-BCB layer, opening of the cavity around the chip and vias on studs; (c) deposition of BCB planarization layer, opening of contacts and dry etching to remove BCB residues; (d) patterning of the second metal layer; (e) deposition of an insulating and planarization BCB layer; (f) patterning of the contact metal layer for pad definition. [8].

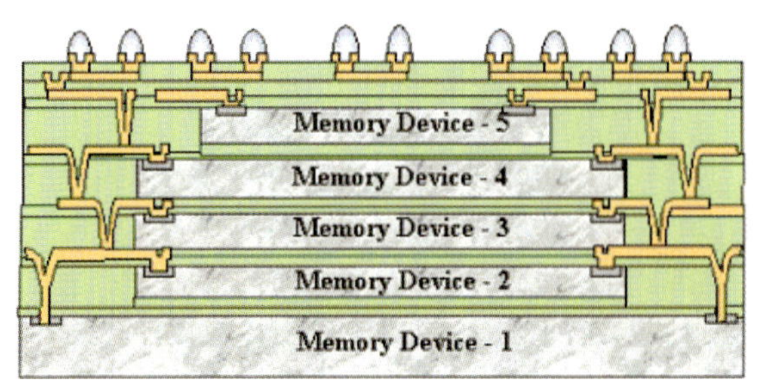

Figure 1.11 Fujitsu chip scale module memory stack [11].

shows a schematic of a five-layer, high-capacity memory product with four chips stacked on top of the base-level memory device.

1.4.4
Fraunhofer/IZM

Fraunhofer/IZM have a similar technology they call "Chip in Polymer" [12]. It is based on embedding of ultrathin chips into build up layers on a printed circuit board.

Working with Nokia, Philips, AT&S Datacon and IMEC under a European STREP (specific targeted research project) they have attempted to determine whether such processes are suitable for manufacturing [13]. Figure 1.12 shows the process sequence they developed.

The structures consist of a double layer core fabricated from high Tg FR4. The die are bonded to the top surface of the core and then high Tg RCC (resin coated copper) is bonded to both sides of the laminate substrate. The RCC has a Cu thickness of 5 μm and a dielectric thickness of 70 μm. Vias are laser drilled and then plated after typical desmear and electroless seeding.

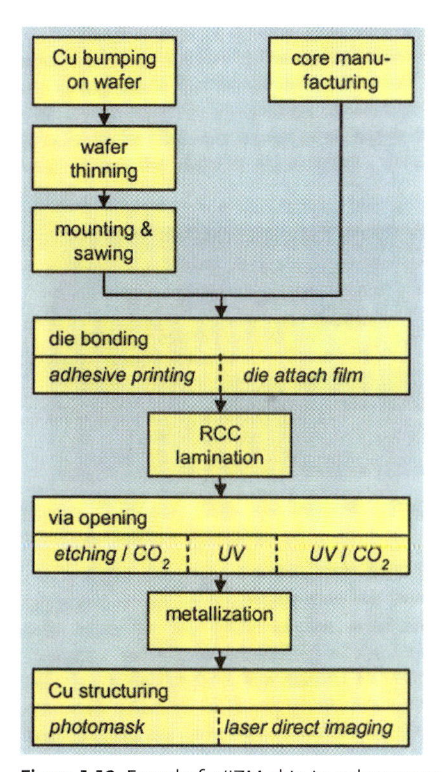

Figure 1.12 Fraunhofer/IZM chip in polymer process flow [13].

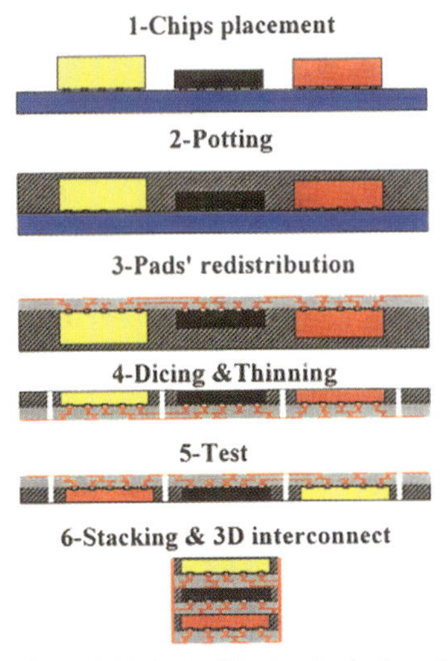

Figure 1.13 Leti – 3D plus "re-built wafer" technology [14].

1.4.5
3D Plus/Leti [14]

Leti and 3D Plus have proposed the 3D structure shown in Figure 1.13 which they call "re-built wafer." In this approach chips and passives of various sizes are imbedded into a resin matrix (active side down). The pads are redistributed, the

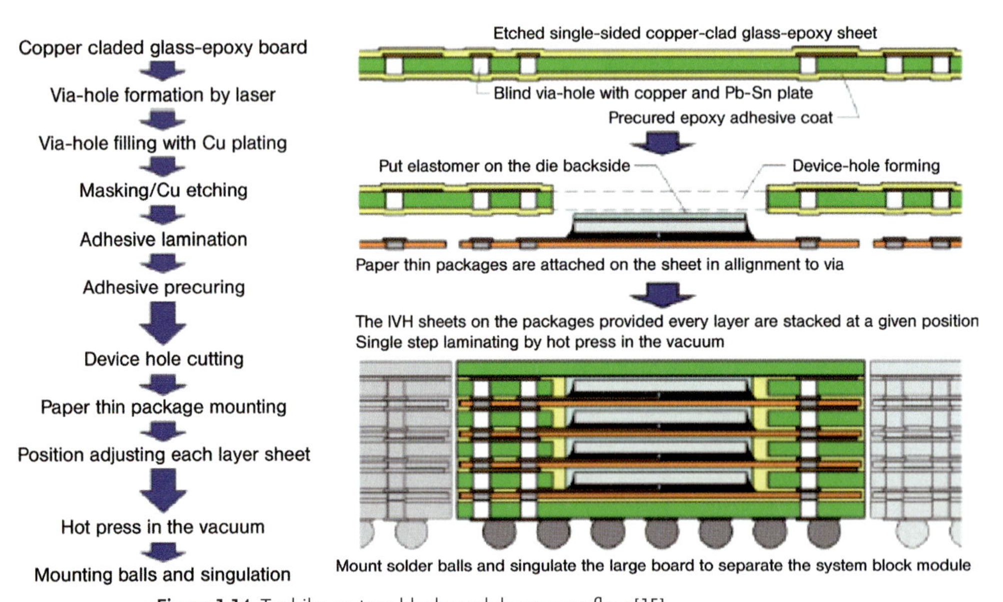

Figure 1.14 Toshiba system block module process flow [15].

imbedded substrate is thinned and the devices are re diced to equal size. After testing, the devices are stacked and connected using 3D Plus edge connecting technique. Two processes were studied for the redistribution: BCB/Cu and laminated film/Cu. The BCB/Cu process was reportedly complicated by "substrate" warpage after thinning.

1.4.6
Toshiba System Block Module [15]

The fabrication process proposed by Toshiba for their "System Block Module" is shown in Figure 1.14 [15].

References

1 Banerjee, K., Souri, S., Kapur, P. and Saraswat, K. (2001) 3D ICs: a novel chip design for improving deep sub-micron interconnect performance and subsystems-on-chip integration. *Proceedings of IEEE*, **89**, 602.

2 Anthony, T. (1981) Forming electrical interconnection through semiconductor wafers. *Journal of Applied Physics*, **52**, 5340.

3 Akasaka, Y. and Nishimura, T. (1986) Concept and basic technologies for 3D IC structure. *IEDM Technical Digest*, 488.

4 Kunio, T., Oyama, K., Hayashi, Y. and Morimote, M. (1989) 3D Ics having 4 stacked active device layers. *IEDM Technical Digest*, 837.

5 Takahashi, S., Hayashi, Y., Kunio, T. and Endo, N. (1992) Characteristics of thin film devices for a stacked type MCM. *Proceedings IEEE MCMC*, 159.

6 Bertin, C., Perlman, D. and Shanken, S. (1993) Evaluation of a 3-D memory cube system. *IEEE Transactions CHMT*, **16**, 1006.

7 Gann, K. (1998) High density packaging of flash memory. Proceedings IEEE Int. Non Volatile Memory Technology Conference, p. 96.

8 Pinal, S. Tassleei, J. Lepinois, F. *et al.* (2001) Ultra thin chip verticle interconnect technique. *Proceedings IMAPS Europe*, p. 42.

9 Pinel, S. *et al.* (2002) Thermal modeling and management in ultrathin chip stack technology. *IEEE Transactions CPMT*, **25**, 244.

10 European Patent UTCS EP 992011061.

11 Fujitsu press release July 15th 2002 at Semicon West.

12 Reichl, H., Ostermann, A., Weiland, R. and Ramm, P. (2003) The 3rd dimension in microelectronic packaging. Proceedings 14th European Micro & Packaging Conf. Friedrichshafen GR, p. 1.

13 Ostmann, A. *et al.* (2005) Technology for embedding active die. Proceedings European Microelectronic Packaging Conference, Brugge BE, p. 101.

14 Souriau, J.C, Lignier, O., Charrier, M. and Poupon, G. (2005) Wafer level processing of 3D system in package for Rf and data applications. Proceedings Electronic Component and Technology Conference, p. 356.

15 Imoto, T. *et al.* (2001) Development of 3-dimensional module package, system module block. Proceedings 51st Elect. Component Technology Conference, Orlando, p. 552.

2
Drivers for 3D Integration

Philip Garrou, Susan Vitkavage, and Sitaram Arkalgud

2.1
Introduction

For a technology to be used in mainstream microelectronic applications, several drivers must simultaneously demonstrate significant benefits when compared to the existing alternatives. In the semiconductor industry, these drivers typically are:

- Better electrical performance
- Lower power consumption and noise
- Form factor improvement
- Lower cost
- More functionality

In this chapter we make the case that such requirements will be satisfied in the near future by 3D integration technology using through silicon vias (TSVs).

2.2
Electrical Performance

Is 3D integration technology positioned to become a paradigm shift for the semiconductor industry, which has, up to now, been dominated by shrinking gate dimensions to improve gate switching delay?

Semiconductor manufacturers have been shrinking the transistor size in integrated circuits (ICs) to improve chip performance for several decades. This has resulted in increased speed and device density, both of which are described by Moore's Law, which states that chip complexity (i.e., transistor count or performance) will double every 24 months. Performance in a semiconductor chip is driven by several factors. At the device level, gate and interconnect delays must be considered. At the die and system level, bandwidth and latency must be considered.

Handbook of 3D Integration: Technology and Applications of 3D Integrated Circuits.
Edited by Philip Garrou, Christopher Bower and Peter Ramm

ISBN: 978-3-527-32034-9

It is becoming apparent that 3D integration technology (described in Chapter 1) will be required to overcome the roadblocks currently being encountered in the traditional device shrinking technology approach and allow the industry to stay on the traditional productivity curve as the industry approaches the 32 nm node.

While many IC manufacturers of logic like IBM, Toshiba, Sony and NEC have demonstrated that performance increases are possible in the 45 nm generation, it is not clear that further performance gains will be achievable at the 32 nm generation. Toshiba's Advanced Logic Dept has noted "... we just cannot tell whether there will be any significance to dropping to the 32 nm generation for logic" [1]

Emma of IBM has recently shown IBM data that agree with this conclusion [2].

2.2.1 Signal Seed

The speed of an electrical signal in an IC is governed by two components – the transistor gate delay, which is the switching time of an individual transistor, known as transistor gate delay, and the *RC* delay, or signal propagation time between transistors, *RC* delay (where *R* is the metal wire resistance, *C* is the interlevel dielectric capacitance). This is shown in Equation 2.1:

$$RC \text{ delay} = 2\rho\varepsilon(4L^2/P^2 + L/T^2), \quad (2.1)$$

where ρ is the metal resistivity, ε is the permittivity of the dielectric (ILD), *L* is the line length, *P* is the metal pitch and *T* is the metal thickness.

For sub-micron technology, the *RC* delay becomes the dominant factor over gate delay. While shrinking gate dimensions and decreasing operating voltage improves device performance, the performance of the interconnect wires are degraded as technology nodes continue to evolve. Smaller cross section wire dimensions have increased resistance, and tighter pitches can raise the capacitance, resulting in an overall increase in *RC* delay. The interconnect delay will start to dominate because the combined gate and interconnect delay are increasing with each new technology node (Figure 2.1) [3].

The outlook for continued performance enhancements through reduction of RC delay is not promising. In a seminal paper in the Proceedings of IEEE in 2001 there were predictions that chip interconnect threatens to "... decelerate or halt the historical progression of the semiconductor industry ..." and proposals that 3D integration of circuits "... should be rigorously explored to help alleviate interconnect delay and density problems ... and reduce chip area" [4].

As the interconnect cross section continues to shrink, line resistivity and capacitance become a major problem even for Cu lines. This increase in resistivity is caused by the electron surface scattering effect, which depends on the interconnect temperature and on the copper–barrier interface quality (Figure 2.2). For future process generations, the conductor cross-sectional dimensions will become smaller than the mean free path of electrons in bulk copper.

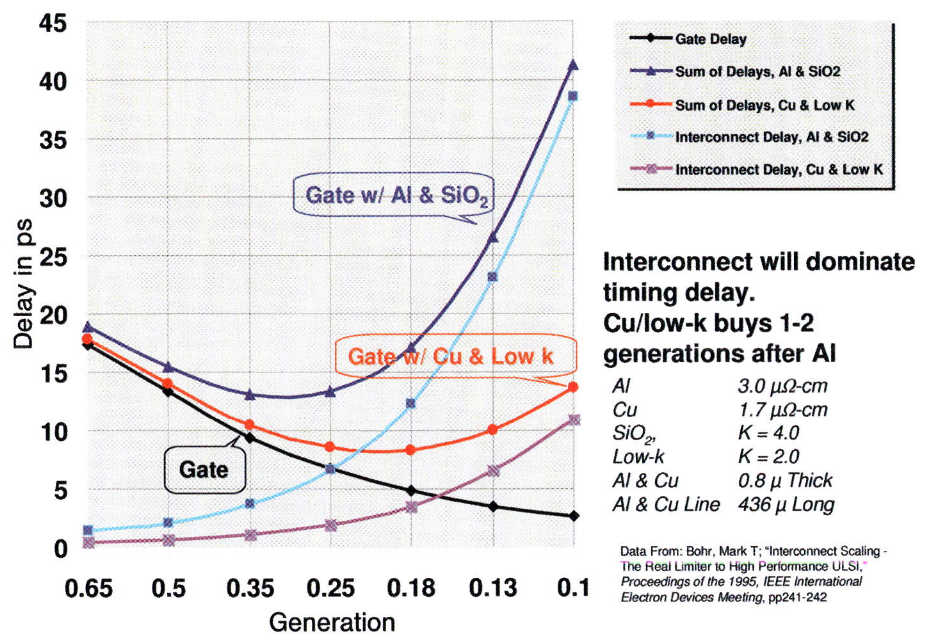

Figure 2.1 Interconnect delay versus IC node [3].

Also, the required barrier layer thickness (required to prevent migration of copper ions into the silicon substrate) does not scale and, therefore, becomes a significant fraction of the allowable interconnect cross-sectional area, which increases the effective resistivity [6].

In addition, integration of low-k dielectrics has not been as easy as initially predicted. Low-k dielectrics have incorporated fluorine, carbon and porosity to achieve ever decreasing k_{bulk} values. As a consequence, the mechanical properties

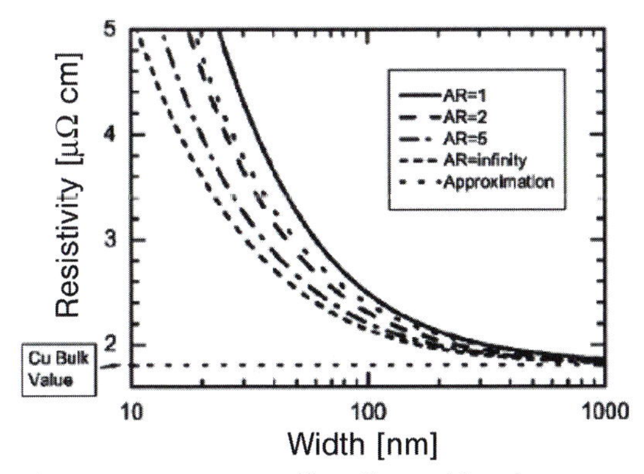

Figure 2.2 Cu resistivity: effect of line width scaling [5].

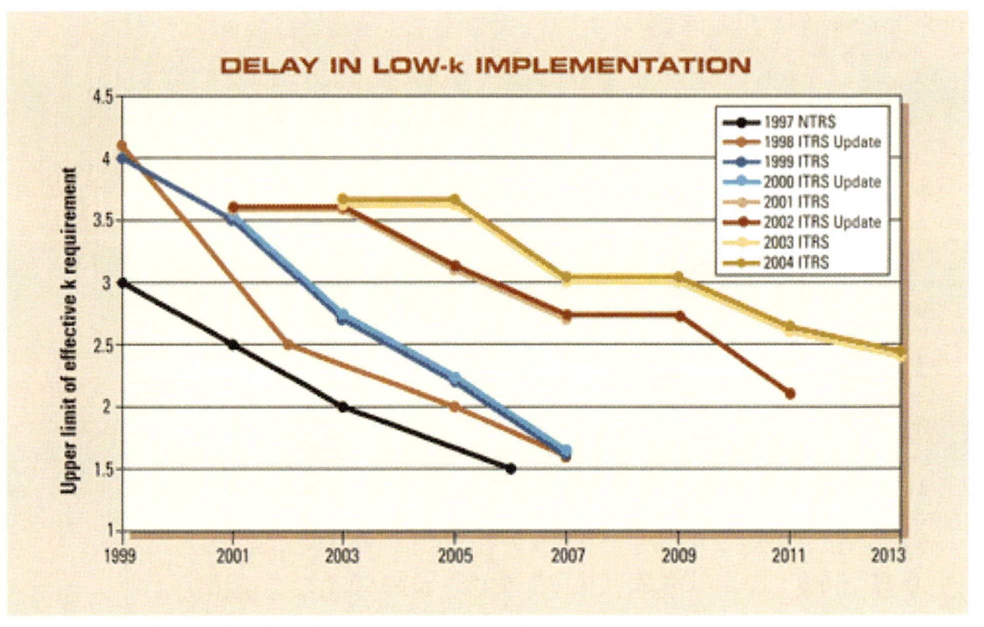

Figure 2.3 Delay in low-k implementation [7].

and the reliability of the dielectrics have degraded considerably and the dielectric is more susceptible to damage and moisture uptake.

Delays in the IRTS roadmap due to the delays in implementing low-k dielectrics have been well documented [7] and are shown in Figure 2.3.

Many trade press and technical articles in the last few years have documented the issues such as CTE and fracture toughness that have arisen when trying to integrate low-k dielectrics [8–11]. Most processes have stayed with generic FSG through the 65 nm node. Spin-on organic dielectrics like SiLK have been all but abandoned by the industry and CVD materials like Coral and Black Diamond have had significant problems integrating into production lines at TSMC and UMC, respectively. It has been said that the significant delays and technical impediments discovered while implementing low k dielectrics have created a crisis for on chip interconnect [3, 12].

SEMATECH representatives have noted. "... at the 45 nm node, ultra low-k dielectrics are so fragile and sustain so much damage from standard processing that the issues associated with incorporating them can essentially negate the advantages provided by these ultra low-k materials." SEMATECH added "... while it may be technically possible to solve these processing issues, and even to drive interconnect materials below 2.5 k-effective, doing so is not likely to be economical" [13].

To summarize the interconnect issues:

1. While Cu is one of the lowest resistivity metals, migration problems require Cu trace encapsulation in highly resistive refractory materials.

2. To achieve dielectric constants below ~2.5, porosity must be incorporated into the material, which weakens it and causes major processing and packaging issues.
3. Scaling deteriorates the performance of interconnects, and that deterioration will (already has) become a significant limiter in overall circuit performance.

While the industry will continue to use copper in combination with some form of barrier and low-k material, beyond the 65 nm node, no material set has been identified that will offset the impact of scaling [14].

Such results have generated many discussions concerning the end of device scaling as we know it and has led to a search for solutions beyond the perceived limits of silicon devices [4].

The industry k-effective (k,eff) historical performance can be compared to the impact of converting into 3D as shown in Figure 2.3. The 3D comparison was obtained by converting the normalized curve predicting the effect of adding 3D layers on *RC* delay [4] into an effective dielectric constant. Each successive point in Figure 2.4(b) assumes an additional layer is stacked. As seen in the curve, the *RC* delay improvement for 3D integration will flatten out beyond a certain number of stacked layers [3].

2.2.2 Memory Latency

Memory access is increasing at a much slower rate than processor speeds. This processor–memory performance gap is shown in Figure 2.5.

The slower speed of the memory causes the processor to stall, waiting on the memory to access data. Cache has been used to help interface processors with slower main memory. Most designs use multiple levels of cache. Cache pipelines require multiple memory levels to move data from main memory to the processor and vice versa; these pipelines reduce the impact of the slower memory on processor performance.

Conventional 2D architectures for processors have the processor and cache in the same plane. L0 and L1 caches are present on the same die as the processor, whereas the L2 cache can be on a separate die. Interconnection wires that connect the

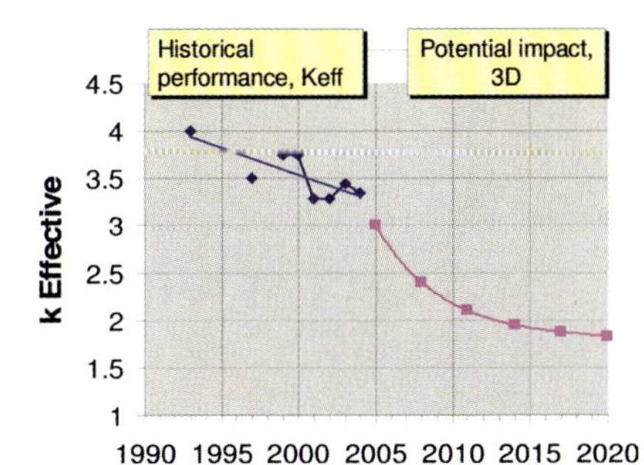

Figure 2.4 Impact of adding 3D layers compared to history of k effective reduction [3].

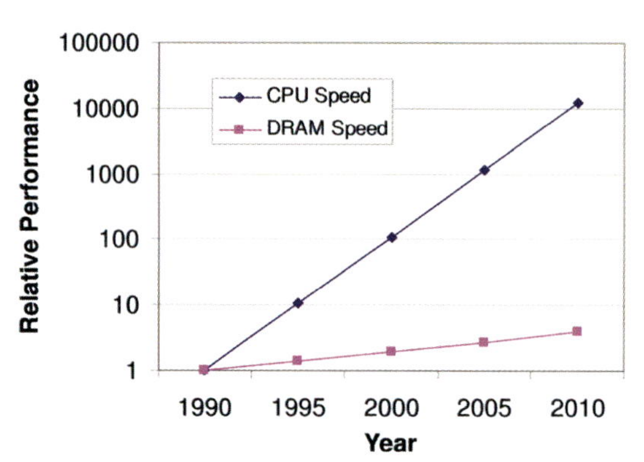

Figure 2.5 Relative speed improvements for CPU versus DRAM over time [15].

processor and cache are long, especially with L2 on a separate die. This causes multiple clock cycles to pass before data moves from one end to another.

In initial 3D processor designs, the L0 and L1 cache could be on the same level as the processor. Higher levels of cache, such as L2, would stack on top of the processor wafer. The L2 cache itself could be formed on multiple wafer layers. In later 3D architectures, complete repartitioning of the processor die to take full advantage of 3D architecture would be expected.

For example, Putswammy and Loh of Ga Tech have reported on a repartitioning of the Intel "Core 2" processor to better understand the potential impact of 3D on latency. The results are shown in Figure 2.6 and Table 2.1. The reduction of latency by repartitioning the cache and cores is significant.

It is generally assumed that the use of 3D architecture may well be the only way to avoid memory latency issues for future generations of multicore microprocessors (see later processor discussion)

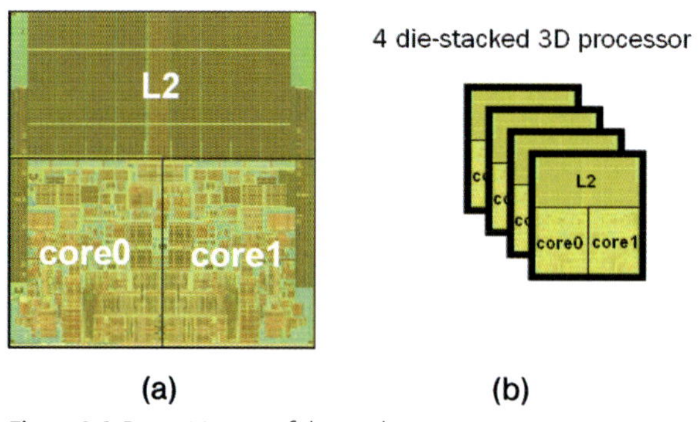

Figure 2.6 Repartitioning of the Intel "Core 2" Processor: (a) current processor; (b) repartition of core and cache onto four levels [16].

Table 2.1 Impact of partitioning on latency for Intel Core 2 Processor [16].

Name	% Latency reduction
Scheduler	32
ALU + bypass	36
Reorder buffer	52
Register file	53
L1 cache	31
L2 cache	51
Register alias table	36

2.3 Power Consumption and Noise

3D integration provides a smaller wire-length distribution, with the largest effect associated with the longest paths. Shorter wires, in turn, will decrease the average load capacitance and resistance and decrease the number of repeaters needed for long wires. Since interconnect wires with their supporting repeaters consume a significant portion of total active power, the reduced average interconnect length in 3D (compared with 2D counterparts) will improve power consumption.

Along with a reduction in the longest global interconnects by $1/L^{1/2}$ (where $L =$ the number of layers in the 3D stack) and a reduction in total wiring required for a given circuit comes a reduction in energy dissipation that varies roughly as the square root of the number of layers [17]. Such wire reduction will lower power consumption since interconnect consumes about one half of a chip's power.

2.3.1 Noise

The shorter interconnects and consequent reduction of load capacitance in 3D ICs will reduce the noise due to simultaneous switching events. The shorter wires will also have lower wire-to-wire capacitance, resulting in less noise coupling between signal lines.

2.4 Form Factor

One of the most influential requirements in today's marketplace is to include as much low cost memory in as small a space as possible. This is primarily driven by the fact that processor speeds have risen much more quickly than memory access, resulting in a need by the microprocessor to rapidly access large amounts of system memory. Stacking memory using 3D TSVs alleviates this bottleneck while producing maximum memory access speed and absolute minimum size package.

The size of a memory chip is typically defined by optimum yield, memory density, cell size, and chip efficiency. Typically, high volume, high density memory technologies use the most aggressive lithography available and rely on utilizing the next generation of lithography to "shrink" to the next node to stay on the productivity curve. When lithography does not yield the productivity gains, one of the few options to increase density and remain on the productivity curve at a given lithography generation is to stack chips on each other [1]. This has been demonstrated recently on several conventional memory technologies.

2.4.1 Non-Volatile Memory Technology: Flash

Industry experts are pessimistic about shrinking design rules for NAND flash beyond the 32 nm generation because "... memory cells will be so small that operation will be unstable" and "... the big problem is not the transistors but the increase in delay" [18].

In 2006 Samsung announced the development of a small footprint, 16 GB high density memory that uses through-silicon vias to stack eight 2 GB NAND flash die (Figure 2.7). According to Samsung, the stacked device showed a 15% smaller footprint and is 30% thinner than an equivalent wire-bonded MCP solution.

Samsung indicated that such technology has tremendous potential for cell phones (due to tight form factor and large storage requirements because space is so limited) but is also being developed for "next-generation computing systems in 2010 and beyond." [19]

ASE recently announced that memory stacks with TSV technology would "... come into its own in the 2008–2011 timeframe when 45 nm process technology becomes mainstream." They conclude that memory stacks will be in "... SiPs as thin as 1.0–1.2 mm (20% thinner than todays' packages), which will allow the manufacture of thinner mobile phones" [20].

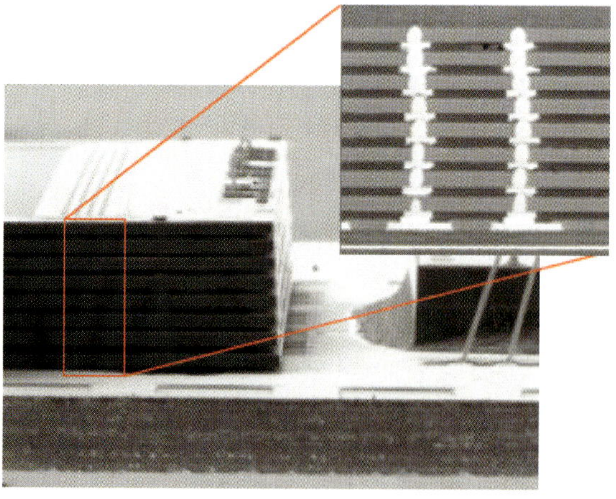

Figure 2.7 Samsung 16 GbBIT NAND Flash, eight chip stack with TSV [19].

2.4.1.1 **FLASH or DRAM for "Osmium"?**

In 2006 Micron announced "Osmium" technology [21] which will reportedly stack thinned die and interconnect them with TSV created through the current peripheral die pads. They did not announce process details or whether it will be used on their flash or DRAM product lines or both.

2.4.2 Volatile Memory Technology: SRAM and DRAM

3D technologies developed by Samsung, Micron, NEC and Tezzaron are all pointed towards targeting cell phones and other portable devices having enough RAM to run high-definition video and other 3D graphics applications in the near future.

Samsung Electronics announced that it has developed the first all-DRAM stacked memory package using "through silicon via" (TSV) technology. Prototypes using their wafer-level-processed stacked package (WSP) have consisted of four 512 megabit (Mb) DDR2 DRAM for a combined 2 gigabits (Gb) of high density memory and a 4 GB (gigabyte) DIMM stack made up of TSV-processed 2 Gb DRAMs. Samsung has said it was developing the process for "next-generation computing systems in 2010 and beyond" [22].

NEC, working with Oki and memory manufacturer Elpida, has developed 3D stacking technology using poly-silicon TSVs. Their motivation for stacking DRAM is shown in Figure 2.8 [23] They predict that a 3× increase in memory density can be achieved by using TSV technology by 2010 at the 32 nm node.

2.4.3 CMOS Image Sensors

One of the first applications to make use of 3D stacking/TSV will be CMOS image sensor chips for cell phones. CMOS sensing chips must mount face up. Minimal

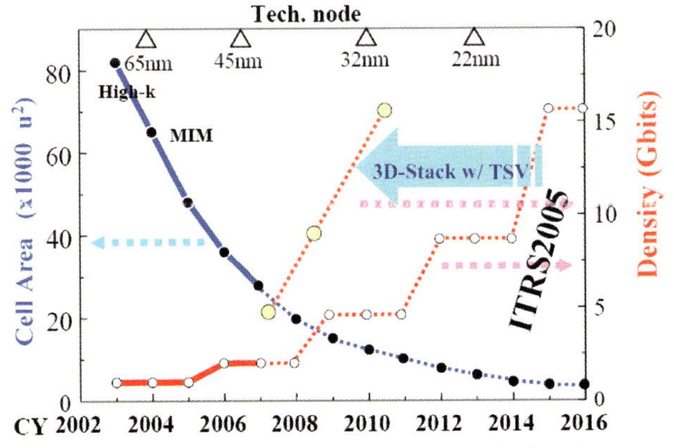

Figure 2.8 Elpida motivation for DRAM with TSV technology [23].

packaged device size is obtained by creating TSV and connecting directly to the back side of the chip. Tessera, Schott Glass, Fujikura, Sanyo, Toshiba, Zycube and others are developing and commercializing such TSV "packages." For further detail see Chapter 15.

2.5 Lower Cost

Cost is a primary force behind the acceptance of any new technology into mainstream production. While technologies may find applications based on performance improvements alone, they tend to remain in specialized, niche applications until their cost can be driven down to an acceptable price point for the industry. Insuring that 3D integration processes are stripped of unnecessary cost will be an important factor to watch and will determine the speed with which this technology is broadly implemented.

2.6 Application Based Drivers

We now look at various device types and how 3D integration could effect them.

2.6.1 Microprocessors

3D Integration architectures for microprocessors includes both "logic + memory" stacking and "logic + logic" stacking. It is likely that we will first see cache memory bonded to processor chips and later see full repartitioning of the processor chip.

One of the biggest performance bottlenecks in today's microprocessors is the access time between the CPU and the main memory. To reduce the bottleneck, memory caches are used to interface the processor to the main memory. Because of this, the processor area is becoming a very small fraction of the total die size. The tremendous bandwidth needed to avoid latency issues in the multicore processor systems of the future can likely only be addressed by TSV. By stacking memory directly on top of a massively multicore processor and connect directly to the memory chip, Intel claims that they can get transfer rates between the processor and memory of up to a terabyte per second. For further detail on microprocessor applications see Chapters 15 and 34.

2.6.2 Memory

Memory will be another early adopter of 3D integration technology. All of the major global memory suppliers are actively working on 3D technologies and determining the appropriate insertion point in their product lines. As we have detailed, there is worry about memory performance past the 32 nm node using current technologies and memory is most in need of the form factor advantages of 3D technology for its incorporation into portable consumer devices. For further details see Chapters 15 and 35.

2.6.3 Sensors

One area that will likely see early implementation in 3D is focal plane array image sensors. Active pixel focal plane architectures are well suited for 3D interconnection because the signal integration, amplification, and readout can be placed close to the photodetection elements. Current 2D solutions are unable to handle the data transfer rates needed with today's high speed imaging applications. This change from sequential signal processing to per-pixel parallel signal processing will significantly improve real time imaging, which is not achievable today. These applications are covered thoroughly in Chapter 37.

2.6.4 Fields Programmable Gate Arrays (FPGAs)

FPGAs (Fields Programmable Gate Arrays) consist of large arrays of simple, programmable logic elements with a hierarchy of programmable interconnect. FPGAs have always had problems with wire delays. 3D integration can improve FPGAs by removing the programmable interconnect from the logic block layer and placing it on another tier in the stack, thus reducing the interconnect delay. For more detail on FPGAs see Chapter 15.

References

1 Ooishi, M. (April 2007) Vertical stacking to redefine chip design. *Nikkei Electronics Asia*, 20.

2 Emma, P. (July 2007) Technology scaling after Moore's law. SEMATECH "The 3D Buzz: Making TSVs Real" session, SEMICON West, San Francisco CA.

3 Vitkavage, S. and Monnig, K. (June 2005) 3D interconnects and the IRTS roadmap. Proceedings 3D Architectures for Semiconductor Integration and Packaging Conference, Phoenix AZ.

4 Davis, J., Venkatesan, R., Kaloyeros, A. *et al.* (2001) Interconnect limits on gigascale integration (GSI) in the 21st century. *Proceedings of IEEE*, **89**, 305.

5 Steinhogl, W., Schindler, G., Steinlesberger, G. and Engelhardt, M. (2002) Size-dependent resistivity of metallic wires in the mesoscopic range. *Physical Review B*, **66**, 75414.

6 Meindl, J. (May/June 2003) Interconnect opportunities for gigascale integration. *IEEE Micro*, **23**, 28.

7 Braun, A. (1st May 2005) Low-k bursts into the Mainstream . . . incrementally. *Semiconductor International*.

8 Peters, I. (1st Jan 2003) Industry confronts sub-100 nm challenges. *Semiconductor International.*

9 Lammers, D. (21st April 2003) Worries Dull SiLK's Sheen at IBM Micro. *EE Times.*

10 Cataldo, and Lammers, D. (March 17th 2003) Altera Pounces as Xilinx becomes latest to abandon low-K. *EE Times.*

11 Goldstein, H. (December 2003) SiLK Slips: IBM Follows Industry Trend and Chucks Spin-on Insulator. *IEEE Spectrum,* **40**, 14.

12 Garrou, P. (June 2005) 3D Integration: A Status Report. Proceedings 3D Architectures for Semiconductor Integration and Packaging, Phoenix AZ.

13 Pfeifer, K. (October 2004) Sematech Low-k Symposium, San Diego CA.

14 Chambra, N., Monnig, K., Augar, R. *et al.* (Feb. 2002) Interconnect challenges and strategic solutions. *Future Fab International,* **12**.

15 Banerjee, K., Souri, S., Kapur, P. and Saraswat, K. (2001) 3D ICs: A novel chip design for improving interconnect performance and system on chip integration. *Proceedings of IEEE,* **89**, 602.

16 Puttaswamy, K. and Loh, G. (2007) Thermal herding: microarchitecture techniques for controlling hotspots in high-performance 3D-integrated processors. Proceedings IEEE 13th International Symposium, **10**, 193.

17 Joyner, J. and Meindl, J.D. (2002) Opportunities for reduced power dissipation using 3D integration. Proceedings IEEE International Interconnect Technology Conference, 148.

18 Patti, R. (June 2005) FaStack technology: 3D transition to manufacturing. Proceedings 3D Architectures for Semiconductor Integration and Packaging, Tempe AZ.

19 Lee, K. (November 2006) Next generation package technology for higher performance and smaller systems. 3D Architectures for Semiconductor Integration and Packaging Conference, Burlingame CA.

20 Tsuda, K. (24th September 2007) 3D Interconnect Coming in Thin Phones, *Semiconductor International.*

21 Davis, J. (3rd August 2006) Micron takes wraps off packaging innovation, *Semiconductor International.*

22 Samsung International (23rd April 2007) Samsung develops new, highly efficient stacking process for DRAM. *Semiconductor International.*

23 Ikeda, H. (May/June 2007) 3D stacked DRAM using TSV, Plenary Session Electronics Components and Technology Conference, Reno, Nevada.

3
Overview of 3D Integration Process Technology

Philip Garrou and Christopher Bower

3.1
3D Integration Terminology

In general, the 3D process sequences discussed in this book share three common technologies: (i) through silicon via (TSV) formation; (ii) IC wafer thinning and (iii) aligned wafer or die bonding. Later in this chapter we attempt to classify the various approaches to 3D integration according to the sequence in which these operations are performed, but first we will set some basic definitions.

3.1.1
Through Silicon Vias (TSVs)

TSVs can be categorized by when they are fabricated relative to the IC fabrication process. The major classifications include the following:

1. TSVs fabricated during the IC fabrication process:
   a. Front-end-of-line (FEOL) TSVs are fabricated before the IC wiring processes occur.
   b. Back-end-of-line (BEOL) TSVs are made at the IC foundry during the metal wiring processes.
2. TSVs fabricated following the complete IC fabrication (also referred to as a Post-BEOL TSVs in this chapter).

3.1.1.1 FEOL TSVs
In general, the term front-end-of-line (FEOL) refers to all of the IC process steps preceding the first wiring metal level. The back-end-of-line (BEOL) begins with the first wiring metal level in the IC. It is possible to fabricate TSVs as part of the FEOL process. The conducting material in the FEOL TSV must be doped polysilicon to

Handbook of 3D Integration: Technology and Applications of 3D Integrated Circuits.
Edited by Philip Garrou, Christopher Bower and Peter Ramm

ISBN: 978-3-527-32034-9

achieve thermal and material compatibility with the subsequent device processing. These polysilicon TSVs are analogous to deep trench polysilicon technology [1]. A major drawback of such FEOL TSVs is the high resistivity of polysilicon compared to metals. However, they can be made conductive enough for many applications and multiple groups are currently developing this technology. CEA Leti (Chapter 19) [2] NEC (Chapter 15) [3] and Zycube (Chapter 26) [4] have described processes for FEOL polysilicon TSVs.

3.1.1.2 BEOL TSVs

TSVs fabricated in the BEOL can consist of either tungsten (W) or copper (Cu). In general the TSV formation occurs early in the BEOL process to insure the TSV will not occupy valuable interconnect routing real estate. Tezzaron describes a W BEOL process in Chapter 24 and IMEC describes a BEOL copper TSV technology they call "copper nails" in Chapter 21. In both the FEOL and BEOL cases the TSV must be designed into the IC wiring. Figure 3.1 depicts FEOL and BEOL TSV processes.

3.1.1.3 Post-BEOL TSVs

Another option is to fabricate the TSV after the IC fabrication is complete. To fabricate post-BEOL TSVs the ICs must still be specifically designed for 3D integration. For TSVs introduced from the front side of the wafer, an exclusion zone must be present in the IC wiring levels. A major advantage of this approach is that the IC wafers can be fabricated at foundries that might not yet offer TSV processes. This could be of particular importance for applications requiring heterogeneous device technologies (e.g., analog, digital, RF, high voltage, etc.) from various foundries. The various options for Post-BEOL TSVs are discussed later in the chapter.

3.1.1.4 "Vias First" and "Vias Last"

The terms "vias first" and "vias last" are used to describe when the TSV fabrication process occurs relative to the other 3D processes of wafer thinning and aligned bonding. The term "vias first" is used to describe processes where the TSV is introduced into the IC wafer prior to that wafer being bonded to the 3D IC stack. The

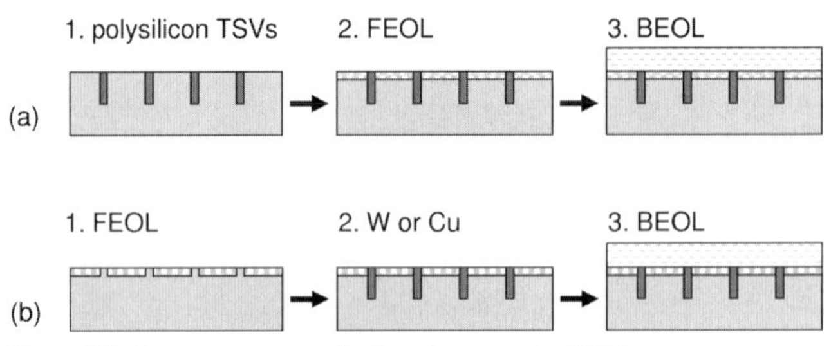

Figure 3.1 Process sequence for foundry-generated TSVs: (a) FEOL polysilicon TSV and (b) BEOL W or Cu TSV.

term "vias last" has been used to describe processes where the TSV is fabricated in the IC wafer after it is thinned and attached to the 3D IC stack.

3.1.1.5 **IC Wafers not Designed for 3D Integration**

It is possible to envision strategies for making TSV through IC wafers that were not specifically designed for 3D integration, when redesign is not a viable option. The ASET consortium in Japan (see Chapter 18) studied this option thoroughly and concluded that TSVs could be fabricated in the empty regions between the bonding pads and the dicing streets [5]. One can fabricate the TSV right through the peripheral bond pads; however, the presence of support pillars under the pads (to resist cracking during wire bonding of fragile low-k ILD based chips), sometimes precludes this option [6]. This approach will only be viable for applications that only require TSV along the perimeter of the die.

3.1.2
Wafer Thinning

It becomes very difficult to process wafers thinned below 100 microns and the challenge increases with wafer diameter. Presently, most 3D IC processes are aiming for individual IC tiers considerably thinner than 100 microns. For this reason wafers are often mounted on temporary handle wafers, also called carrier wafers, for thinning and backside processing. The IC wafer must be mounted "face-down" onto the handle wafer, and in general must be bonded in a "face-up" configuration onto the 3D IC stack. The other handling option is to bond the IC wafer directly to the 3D IC stack. In this case the wafer must be bonded to the 3D stack in a "face-down" configuration. Figure 3.2 shows these two options for IC wafer thinning.

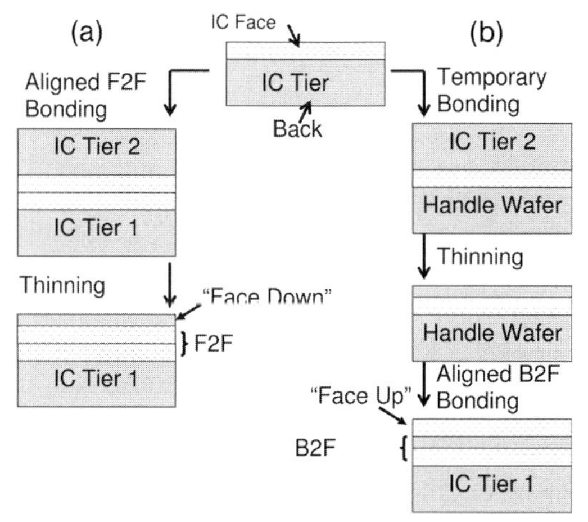

Figure 3.2 Direct (a) and intermediate handle (b) approaches for the fabrication of ultrathin IC stacks.

3.1.3
Aligned Wafer/IC Bonding

3.1.3.1 Wafer-to-Wafer versus Die-to-Wafer

The technique for building 3D ICs is based on wafer-to-wafer (W2W) and die-to-wafer (D2W) bonding and interconnect technology. Some die-to-die stacking is done, but mainly for prototyping to save costs.

There are several problems associated with stacking wafers but the most critical one is yield. Wafer-to-wafer stacking is most practical for high yielding individual wafer layers, like memory, which all contain the same size die. For instance a 90% yield on two individual wafers, when stacked, results in a 81% yield assuming *no* yield loss due to the stacking process. Another potential challenge associated with stacked memory wafers is how to implement speed sorting.

Die-to-wafer bonding is best suited for lower yielding wafer layers and/or die that are not the same size. Assembly time is an issue when done die-by-die since this offers no wafer scale economics. W2W and D2W issues are discussed further in Chapter 12.

3.1.3.2 Face-to-Face and Back-to-Face Bonding

Face-to-face (F2F) bonding may be performed either with or without TSV present in the IC wafer and is typically done without a handle wafer. Once the wafers are bonded, if TSVs are present, the transferred wafer is thinned down to expose the TSVs. If TSVs are not present, the wafer is thinned and TSVs are fabricated by etching from the “back side” of the IC wafer. F2F bonding can only occur between IC Tier 1 and IC Tier 2 in a 3D stack because the face of the transferred wafer will not be available for the next wafer in the stack.

Back-to-face (B2F) bonding may be done either with or without TSV present in the wafer (or die) and is done with a handle wafer. Once the front side is bonded to the handle wafer the wafer, in all cases, is thinned. If TSV are present backside processing is performed to create backside pads and the wafers are bonded. If TSV are to be fabricated post BEOL two options exist: (i) The wafer can be backside etched to form the TSV and subsequently backside processed to form pads for bonding or (ii) the wafer can be bonded to the stack by polymer bonding and subsequently the TSV created from the front of the wafer, through the polymer to the face of the wafer underneath.

Another B2F option is to first create the TSV from the front of the wafer through the exclusion zones followed by mounting to the handle, thinning, backside processing and bonding.

3.2
Processing Sequences

This section gives the reader an appreciation of the multitude of process sequences currently being proposed or developed for 3D integration. Where possible, we point readers to detailed material found within other chapters.

As previously stated, 3D integration sequences can be classified by combinations of the three major process technologies; (i) TSV fabrication, (ii) wafer thinning and (iii) aligned wafer or die bonding. In this attempt at categorization each of the major technologies can be further subdivided as follows:

1. TSV fabrication:
   a. TSV fabricated during IC fabrication.
      i. FEOL TSV
      ii. BEOL TSV
   b. TSV fabricated after IC fabrication (called post-BEOL TSV).
      i. Before bonding (vias first);
      ii. after bonding (vias last)
2. Wafer thinning:
   a. Thinning on a handle wafer,
   b. thinning after bonding to the 3D IC stack.
3. Aligned wafer or die bonding:
   a. Metal bonding (multiple methods, all create electrical interconnect between layers).
      i. Direct Cu/Cu, Au/Au, and so on,
      ii. eutectic CuSn, and so on,
      iii. hybrid SiO_2/metal.
   b. Direct bonding (e.g., SiO_2–SiO_2).
   c. Adhesive bonding.

Table 3.1 shows a compilation of process sequences used for fabricating 3D IC stacks.

Figure 3.3 is a schematic depiction of processes B and D from Table 3.1. Here, the TSVs are fabricated by an IC foundry that offers either a polysilicon FEOL TSV or a BEOL Cu or W TSV. The availability of these processes will be a near-term limitation of this approach, especially for applications that require separate device technologies. For example, an application requiring an analog IC process for Tier 3 and a digital IC process for Tier 2 would require both IC processes to offer the TSV technology. Here, the tiers are bonded to the 3D IC stack in a "face-down" configuration. Tier 1 and Tier 2 will have a "face-to-face" bonding arrangement, while the remaining levels will be oriented "back-to-face." This approach requires bonding methods that form inter-layer electrical interconnections during the bond process. Here, the generic term "metal bonding" is used to describe the various approaches to achieve electrical interconnections during the bonding process. These approaches include direct bonding (e.g., Cu-to-Cu), eutectic bonding (CuSn), microbumps and hybrid methods that include dielectric and metal bonding. Although not illustrated in the figure, the electrical interconnection from the bond interface to the TSV will be accomplished using the BEOL wiring levels in the IC. The IC is thinned after being bonded to the 3D

Table 3.1 Process sequences for 3D integration[a,b].

Process	Figure	IC Wafer	Step #1	Step #2	Step #3	Examples
A	Figure 3.4	FEOL TSV (vias first)	Wafer thinning (on handle)	"Face-up" bond (metal bonding)		NEC [3] CEA-LETI [2]
B	Figure 3.3	FEOL TSV (vias first)	"Face-down" bond (metal bonding)	Wafer thinning (on 3D stack)		Ziptronix (Chapter 25)
C	Figure 3.4	BEOL TSV (vias first)	Wafer thinning (on handle)	"Face-up" bond (metal bonding)		IMEC (Chapter 21)
D	Figure 3.3	BEOL TSV (vias first)	"Face-down" bond (metal bonding)	Wafer thinning (on 3D stack)		Tezzaron (Chapter 24)
E	Figure 3.5	No TSV	TSV from front (vias first)	"Face-down" bond (metal bonding)	Wafer thinning	Tezzaron (Chapter 24) (on 3D stack)
F	Figure 3.6	No TSV	TSV from front (vias first)	Wafer thinning (on handle)	"Face-up" bond (metal bonding)	Fraunhofer Munich, (Chapter 16) Arkansas (Chapter 17)
G	Figure 3.7	No TSV	"Face-down" bond (all methods)	Wafer thinning (on 3D stack)	TSV from back (vias last)	Intel (Chapter 34) Lincoln Labs (Chapter 20)
H	Figure 3.8	No TSV	Wafer thinning (on handle)	"face-up" bond (all methods)	TSV from front (vias last)	RTI (Chapter 36)
I	Figure 3.9	No TSV	Wafer thinning (on handle)	TSV from back (vias first)	"face-up" Bond (metal bonding)	IMEC (Chapter 21) Zycube (Chapter 26) Sanyo (Chapter 15)

[a] Metal bonding = Cu-to-Cu, CuSn, microbumps, and so on.
[b] All methods = can use metal, direct oxide-oxide or adhesive bonding.

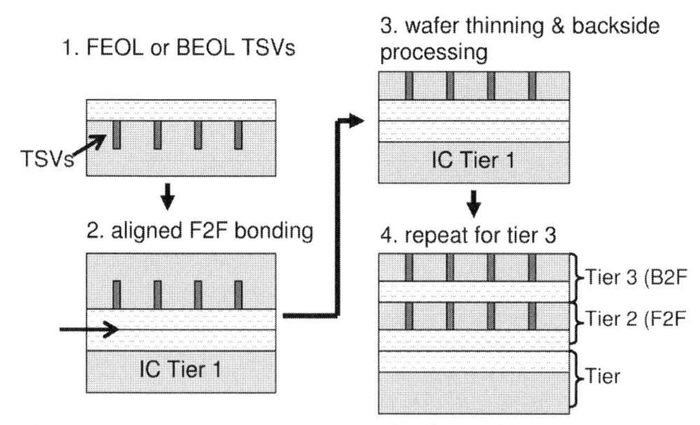

Figure 3.3 3D process sequences B and D from Table 3.1.

IC stack. One advantage of this process is that a handle wafer is not required. However, if a problem occurs during the thinning process, a 3D stack of multiple wafers could be yielded compared to loss of a single wafer when using the handle wafer approach. Other advantages of this process are that the TSVs can serve as an etch stop for the thinning and the exposed TSVs can also serve as alignment marks. This process sequence has been developed by Tezzaron; details of their process are in Chapter 24.

Figure 3.4 illustrates the process sequences A and C from Table 3.1. These 3D sequences utilize wafers that have the TSVs added at the IC foundry (either FEOL or BEOL TSVs). The wafer is temporarily bonded to a handle wafer and thinned down to the TSVs. While attached to the handle wafer, additional backside processing such as passivation and redistribution metal layers or bumps can be added. The thinned IC is bonded "face-up" onto the 3D IC stack. In this case, all of the tiers will be arranged "back-to-front". The bonding can be either "wafer-to-wafer" or "die-to-wafer." These approaches require a metal (interconnect forming) bonding method. This approach is being developed by NEC [3] and CEA-LETI (Chapter 19) with polysilicon FEOL TSVs.

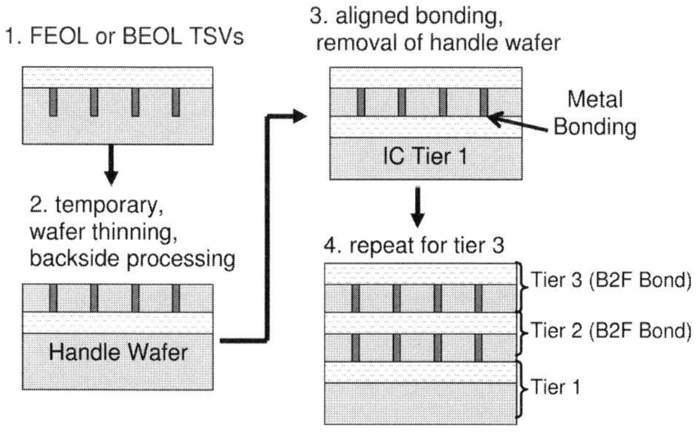

Figure 3.4 3D process sequences A and C from Table 3.1.

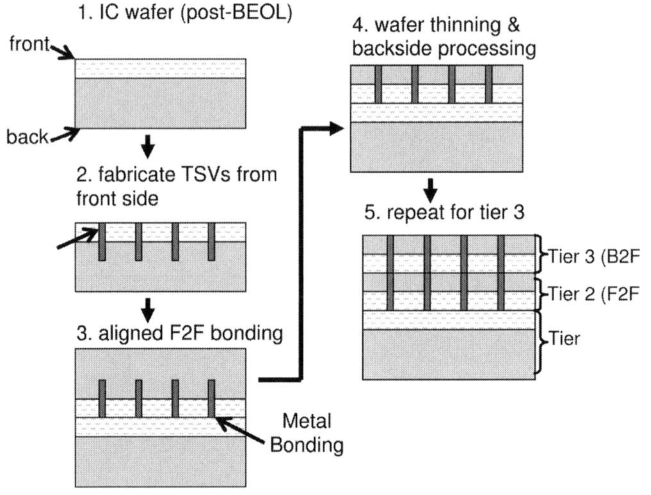

Figure 3.5 3D process sequence E from Table 3.1.

Figure 3.5 illustrates process sequence E from Table 3.1. This approach to 3D integration starts with bulk IC wafers that do not have TSVs added during the IC fabrication. One major advantage of post-BEOLTSVs is that the starting wafers could be fabricated using any available IC process (i.e., does not require an IC process that offers FEOL or BEOLTSVs). Here, the TSVs are fabricated from the front side of the IC wafer. One major disadvantage of this approach is the exclusion zone in the wiring levels to accommodate the TSV. Also, this process is inherently more complex than FEOL and BEOL approaches because the TSV process must etch through the thick dielectrics on the wafer surface. In this sequence, the full thickness wafers are bonded face down onto the 3D stack and then thinned to expose the backsides of the TSVs. A metal bonding process is required. An example is the Tezzaron SuperVia technology discussed in Chapter 24.

Figure 3.6 illustrates process sequence F from Table 3.1. The TSVs are fabricated from the front side of the wafer. The disadvantages from sequence E apply to this approach also. The wafer is thinned on a handle wafer and bonded face up onto the 3D IC stack. A metal bonding process would be used in this case.

Figure 3.7 shows an example of process sequence G from Table 3.1. This process begins with wafers that do not contain TSVs. First, the IC wafer is bonded face-down onto the 3D IC stack. Next, the wafer is thinned. The TSVs are fabricated from the back side of IC Tier 2. The TSVs will be designed to land on a metal pad within the IC wiring levels. Intel describe a 3D microprocessor in Chapter 34 that uses a similar approach.

Figure 3.8 illustrates process sequence H from Table 3.1. The process starts with fully fabricated IC wafers that do not contain foundry provided TSVs. The wafer is temporarily bonded to a handle wafer and thinned. The thinned IC is then bonded "face-up" onto the 3D IC stack. This approach can potentially utilize any of the bonding technologies. Fraunhofer (Chapter 16) and RTI (Chapter 36) have described

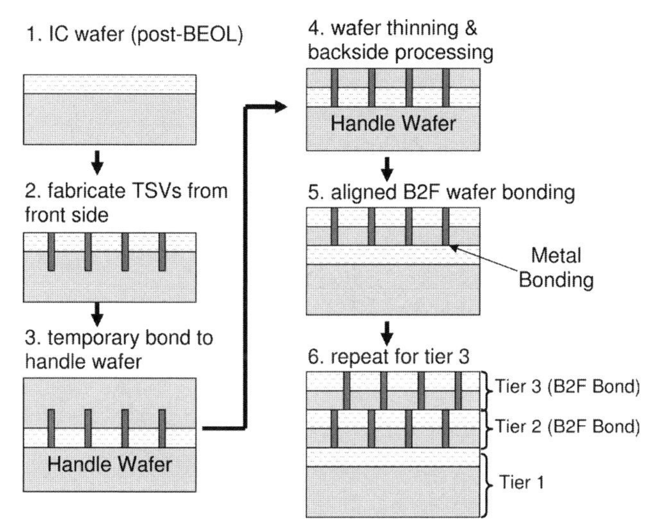

Figure 3.6 3D process sequence F from Table 3.1.

processes that use thermocompression bonding using polymer adhesives. After bonding the handle wafer is removed and the TSVs are fabricated through the bonded tier. In this case, the TSV will traverse the BEOL interlevel dielectrics and the device silicon. The design must include an exclusion zone in the wiring metals to accommodate the TSVs.

Figure 3.9 illustrates process I from Table 3.1. Here, the post-BEOL IC wafer is attached to a handle wafer and thinned. Next, TSVs are fabricated from the backside of the wafer still attached to the handle wafer. The IC is then bonded onto the 3D stack using metal bond methods. Such a backside via approach is used for CMOS image sensor processes (see Chapter 15).

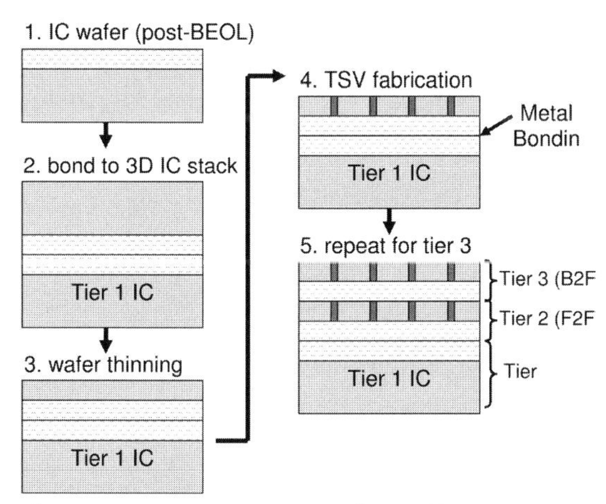

Figure 3.7 3D process sequence G from Table 3.1.

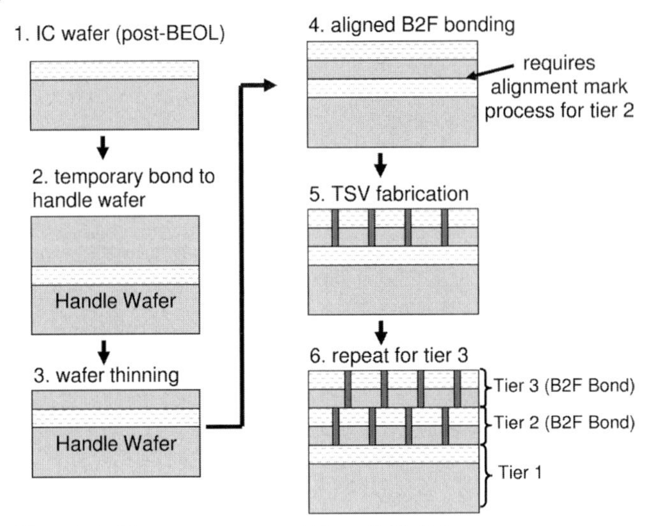

Figure 3.8 3D process sequence H from Table 3.1.

3.3 Technologies for 3D Integration

3.3.1 TSV Formation

3.3.1.1 Deep Reactive Ion Etching (DRIE) for TSV

The breakthrough for TSV technology came in the mid-1990s, with the development of what has become known as deep reactive ion etching (DRIE) or the "Bosch

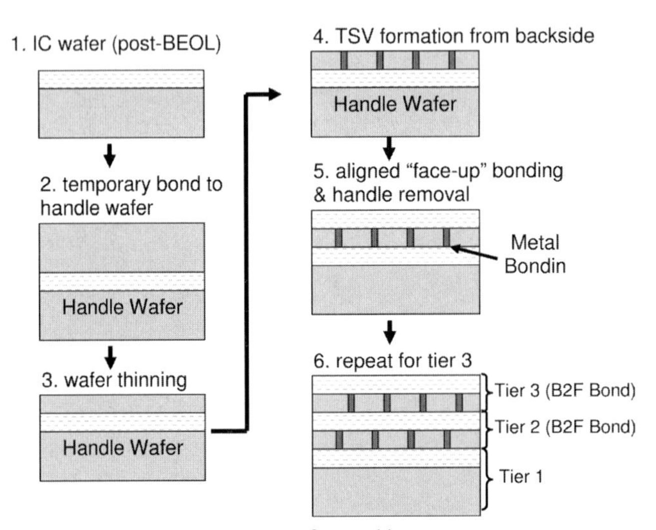

Figure 3.9 3D process sequence I from Table 3.1.

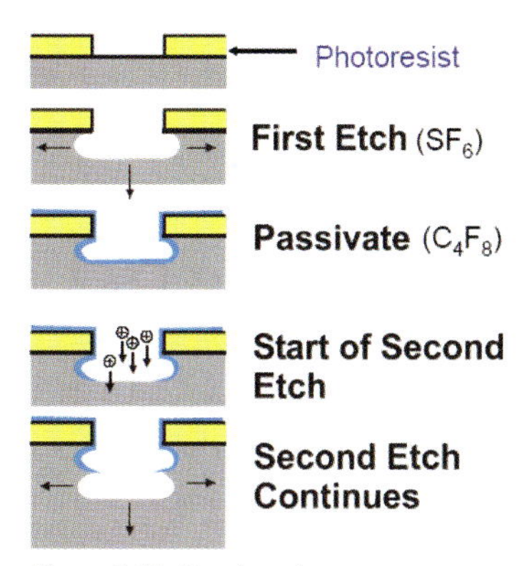

Figure 3.10 Bosch etch.

Process." This plasma etch technique uses SF_6 to rapidly etch the Si, and C_4F_8 to generate CF_2 to passivate the sidewalls of the via during this anisotropic etch. This process is capable of high selectivity, and very vertical side walls. Bosch etching is depicted in Figure 3.10.

Chapter 4 gives complete details of DRIE Bosch etching.

3.3.1.2 **Laser TSV Formation**

UV lasers can also be used to fabricate TSV. Reports indicate that UV laser machined features can be within 2 microns to an active device without producing device degradation [7]. Laser pulse and scan rate play a significant role in determining the throughput and quality of the resulting laser machined feature.

Figure 3.11 shows 10 μm vias that have been drilled to a depth of 70 microns for an aspect ratio of 7 : 1. The natural sidewall taper angle for laser vias is reported to be 85°, which makes them suitable for sputter metallization [8].

Surface finish inside the via is a strong function of "drilling" rate. Rapid drilling producing a rougher sidewall. Figure 3.12 shows the drilling rate for various diameter vias in Si wafers [8].

It is likely that 10 μm vias will meet many of today's needs. Laser proponents claim that they can reduce the via diameters down to 1 um [8], in which case this technology will move itself into mainstream 3D via generation technologies.

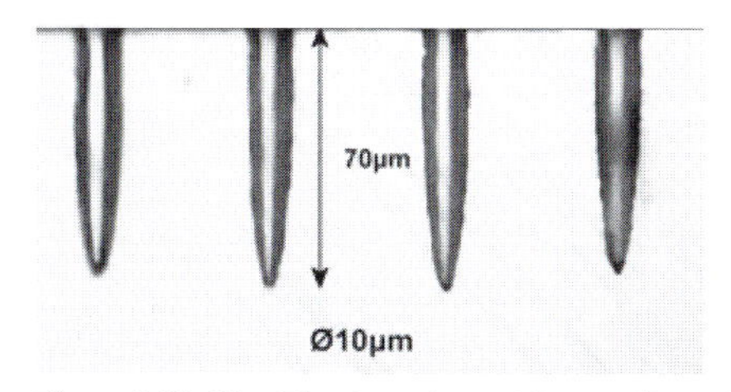

Figure 3.11 10 × 70 micron Laser Vias in silicon [8].

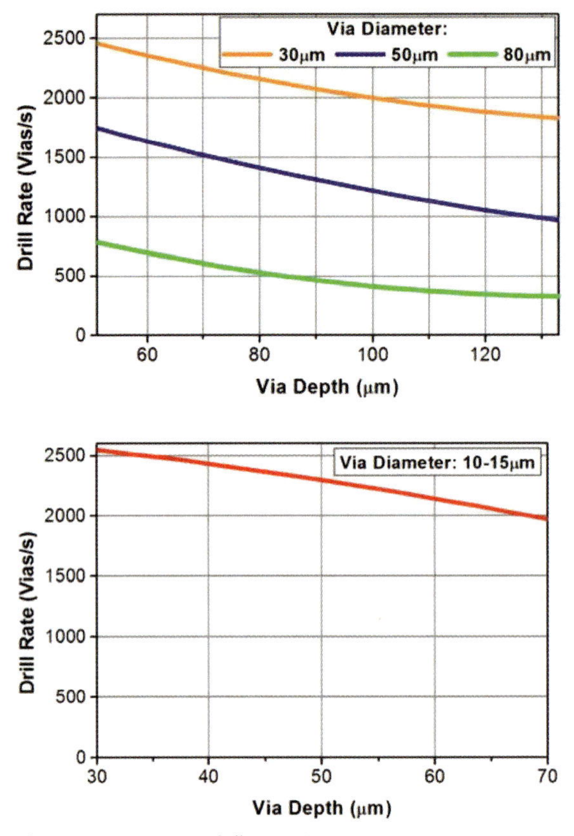

Figure 3.12 Laser via "drill rate" [8].

Chapter 5 gives further information on laser vias.

Via Fill Via filling includes lining of the deep vias with an inorganic or organic insulator, deposition of a diffusion/adhesion layer and subsequent conductor fill (Cu, W or pSi). Whether W, Cu or poly silicon, the vias need to be completely filled, that is, they cannot have voids that would serve to trap processing chemicals.

3.3.1.3 CVD SiO_2 Insulator

In general, SiO_2 insulator is deposited by CVD. The use of TEOS (tetraethylorthosilicate) as a silicon source for PECVD silicon dioxide became common in the 1980s [9]. Silicon dioxide films deposited with TEOS (300 °C) generally show higher conformality than so-called "low temperature oxides, (LTO)" deposited from silane and oxygen. Silane-based SiO_2 processes are generally used in Damascene processing since the vias are shallow and the higher conformality is not needed. PECVD TEOS is used when conformal filling of high aspect ratio vias is needed [10]. Non-plasma based TEOS films require high temperature anneals for densification. Chapter 6 gives further information on SiO_2 as insulation.

3.3.1.4 **Organic Insulators**

Although many polymeric materials have been used in conjunction with thin film processing [11], very few can be deposited conformally. The use of parylene as a conformal organic insulator has been reported by several groups [12, 13] and is detailed further in Chapter 7.

3.3.1.5 **Diffusion Barrier/Adhesion Layer**

Filling of high aspect ratio vias with electroplated copper requires smooth and continuous seed layer. It is essential to prevent copper from entering device silicon and forming deep level traps or diffusing into the insulator between the copper lines and decreasing its insulation resistance. Copper electromigration has been reviewed in detail [14].

For Cu vias a TiN adhesion/barrier layer is generally deposited before deposition of a seed layer of copper. The TiN barrier layer and the copper seed layer can be deposited by sputtering; however, for high aspect ratios (i.e., >4) conventional PVD DC magnetron techniques are inadequate because of insufficient step coverage, especially that of Cu on the sidewalls of the high aspect ratio vias. Further information on diffusion, barrier and seed layers can be found in Chapters 8 and 9.

Ionized metal plasma (IMP) based PVD technology enables more conformal deposition of Cu seed layer on the sidewalls as well as the bottom of via holes than conventional sputtering. IMP provides superior step coverage because of the directionality of the deposited atoms and utilization of ion bombardment to sputter material from the bottom of the via to the sidewalls, thus yielding continuous and conformal barrier and seed layers [15]. Recently, IMP sputtering, has been examined to fill such high AR 3D vias. [16, 17]

As shown in Figure 3.13, conventional sputtering deposits a conformal Cu seed on the sidewalls of a $7.5 \times 60\,\mu m$ via no more than $10\,\mu m$ deep whereas using IMP

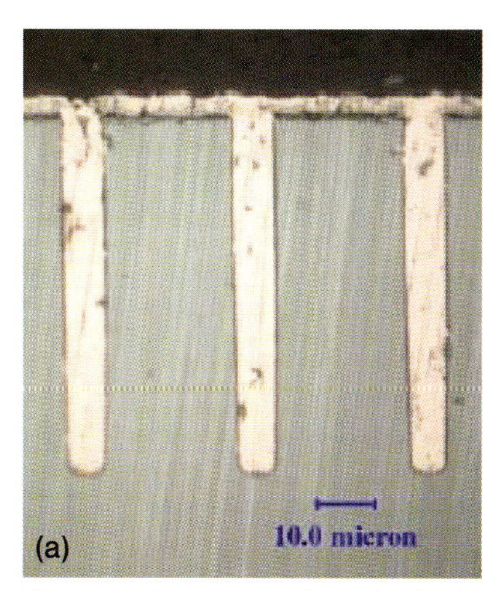

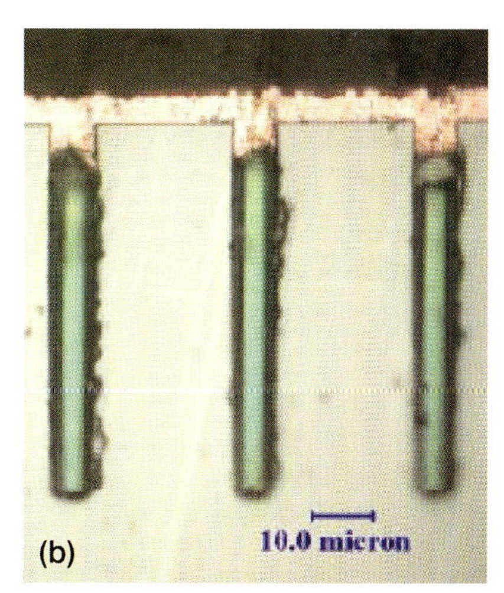

Figure 3.13 Cross section of Vias plated with Cu after Cu seed deposition by (a) IMP sputtering and (b) normal sputtering of $7.5 \times 60\,\mu m$ vias [16].

sputtering results in complete via fill for the same dimension via. Even this technique has its limits, however, and it fails to completely coat the inside of a via of dimensions $5.7 \times 52\,\mu m$ [16].

3.3.1.6 TiN Barrier Layer by MOCVD

Conformal coatings of TiN in deep vias are achieved by MOCVD of precursors such as tetrakis(diethylamido)titanium (TDEAT) [18]. The deposition temperature for this precursor is reported to be $\sim$350 °C. Reports from the ASET consortium in Japan, however, indicate that TiN can be deposited at temperatures as low as 170 °C with rates of $5\,nm\,min^{-1}$, if the concentration of NH_3 in the feed stream is increased [19].

3.3.1.7 Copper Seed by MOCVD

Another option is trench filling with MOCVD Cu. The MOCVD precursor is usually (hfac)Cu(DMB) for Cu. The impact of precursor composition on copper morphology and deposition rate with and without ethyl iodide catalyst has been studied [20]. It is thought that iodine enhances surface diffusion and increases Cu nucleation density, leading to lateral growth and low surface roughness films.

3.3.1.8 Tungsten Metallization

Tungsten metallization is usually deposited by standard CVD processing and requires annealing at $\sim$450 °C. Further information on W CVD can be found in Chapter 9.

3.3.1.9 Copper Metallization

Copper is usually deposited by electrop onto a seed layer. To prevent voids in the copper plugs, copper plating for vias with aspect ratios >2 is performed by reverse pulse plating. In this technique Cu is not continuously deposited, but rather the current is applied in short pulses to give the plating chemicals adequate time to refresh concentrations at the surface. The reverse current pulses remove Cu from the thicker deposited regions [21].

Chapters by ASET (Chapter 18) and Semitool (Chapter 8) give further details on Cu plating for 3D integration.

3.3.2 Temporary Bonding to Carrier Wafer

Handling methodology for the thinned wafer is one of the most critical technologies in realizing 3D chip stack with TSV. The carrier wafer (also called handling wafer) is used as well.

- Support the substrate during the thinning process.
- Transfer layers to the 3D stack.

Silicon or glass substrates can be used as wafer carrier. The temporary bonding adhesive properties must include:

- Level/protect the topography of the device wafer.
  - 2 um TTV (total thickness variation).
- Strong enough to withstand grind and polish.
- Resistant to polishing chemicals.
  - Thermally stable enough to withstand backside processing.
  - Resistant to chemicals and backside processing conditions.
- Easily debonded.
- Residue free.

3.3.3 Thinning

Thinning is performed in two steps. First, a coarse backgrinding or lapping of the wafer, followed by one or more of the following: plasma dry etching (SF_6), wet etching (KOH, TMAH), and/or CMP.

Commercially available backgrinding systems use a two-step process including a coarse grind ($5\,\mu m\,s^{-1}$) and a fine grind ($<1\,\mu m\,s^{-1}$). The finer grind is necessary to remove most of the damage layer created by the coarse grinding step and reduce the surface roughness.

Backgrinding causes physical damage to the wafer, including scratches, crystal defects and stress. The amount of damage produced in the silicon depends on several operating parameters such as grit size, wheel speed and coolant flow.

A layer at least 5–10 μm thick is normally observed, damaged by micro cracks due to the grinding process. The next several microns down contains crystal dislocations that can cause degradation of electrical properties.

Such defects and surface roughness must be removed by plasma dry etching (SF_6), wet etching (KOH, TMAH), and/or CMP.

3.3.3.1 Plasma Etching

Plasma etching of silicon gives far less damage than grinding of silicon. Plasma thinning can achieve very good thickness homogeneity. Drawbacks include the slow etch rate and costly equipment. More recently, atmospheric downstream plasma (ADP) has been proposed [22]. ADP operates at ambient pressure, the reactant gas is CF_4 and etch rates of $20\,\mu m\,min^{-1}$ at 2% uniformity are reportedly achieved. A surface roughness of 0.3 nm has been measured by wafers thinned by ADP.

3.3.3.2 Wet Etching

Due to its crystalline plane dependent etching, wet etching by KOH or TMAH is limited to aspect ratios of ~0.7, so it cannot be used for deep via etching. Since the etch rate of a silicon wafer in 25% THAH solution at 80 °C is $\sim 40\,\mu m\,h^{-1}$ [26] it would take ~11 hours to etch 450 μm, limiting the usefulness of this solution as a bulk thinning process. Thus, a more aggressive etching solution is required. A typical wet etching process uses a mixture of HF, HNO_3 and HOAc. The thinning is conducted as a batch process in a tank system.

A wet chemical spin process has also been proposed by SEZ [23]. The front surface of the wafer is protected by additional layers or by special chucks that allow wet processing without further protective layers. A common etch rate for such spin etching is $10\,\mu m\,min^{-1}$.

3.3.3.3 CMP

CMP is usually performed with a 0.3 μm silica slurry at pH 10. After thinning and polishing are complete the wafers are cleaned in a solution of $NH_4OH : H_2O_2 : H_2O$ to remove debris.

Bonding and thinning are covered thoroughly in Chapter 10.

3.3.3.4 Impact of Thinning on Electrical Characteristics

Pinel et al. have studied the impact of wafer thinning on the electrical characteristics of MOSFET transistors. They conclude that under normal low power-dissipation levels thinning has no noticeable impact on the fundamental performance of the devices [24].

Takahashi of NEC has shown that for Si layer thickness of >300 nm (after thinning) the electrical characteristics of the circuits appear stable [25].

Ramm and coworkers have reported that a via exclusion zone of 15 μm from a transistor appears to be sufficient to insure that the electricals are not negatively affected by the stack or through vias [26].

3.3.4 Alignment/Bonding

3.3.4.1 Wafer Alignment

EVG (Austria), Suss Micotech (Munich), AML (UK) and Ayumi (Japan) are the largest suppliers of wafer bond equipment. See Chapter 12 for detailed coverage on alignment and bonding. In general, bonders are usually configured with the aligners. Both the Suss and EVG bonders accept wafers for various bonding techniques. Both systems maintain very accurate control of temperature (ramp, uniformity), atmosphere (vacuum or process gas) and contact force. Current wafer alignment limits (best case now ±1–2 μm) in turn limit 3D TSV stacking technology to global interconnect.

3.3.4.2 Wafer Bonding Options

Wafer bonding techniques used for 3D integration include:

- Silicon dioxide (SiO_2) fusion bonding,
- metal (Cu) fusion bonding,
- metal eutectic bonding (Cu/Sn),
- bumping (Pb/Sn, Au, In),
- polymer adhesive bonding.

some of which are shown in Figure 3.14.

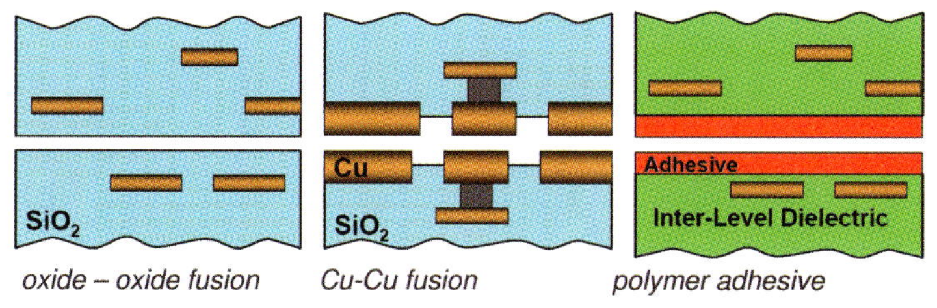

Figure 3.14 Wafer bonding approaches for 3D integration. (Courtesy of RPI.)

Silicon Fusion Bonding Silicon fusion bonding uses temperature and pressure to join a highly polished silicon device wafer and a silicon donor wafer [27]. This is usually carried out on SiO_2 surfaces.

Requirements for wafer fusion bonding are:

- Smoothness (microscopic),
- flatness (macroscopic),
- cleanliness,
- surface chemistry.

A key drawback to fusion bonding is the high-temperature anneal required to form the bond. Plasma processing has been used to reduce annealing temperatures from ~1000 °C down to 200–300 °C [27, 28]. Others have used a $H_2SO_4/H_2O_2/$ plasma pre-clean [29].

The plasma alters the surface of the wafer, making it hydrophilic. Silicon fusion bonding occurs in three steps: (i) cleaning and/or plasma treatment, leaving a hydrophilic surface with a specific chemistry and contact angle; (ii) particle removal combined with surface reactivation and bonding; and (iii) a high-temperature anneal in a standard vertical furnace. Figure 3.15 shows the relationship between anneal temperature and bonding energy for silicon fusion bonding.

Bonding voids are created by particles or protrusions from the wafer surface or trapped air. These voids can be observed after surface contact is made and are not changed during annealing. It is widely reported that a RMS roughness of <1.0 nm and a wafer bow of <4 μm (4″ wafer) is needed for successful bonding. Si fusion bonding is discussed further in Chapter 25.

Copper–Copper Bonding Direct copper bonding involves bonding conditions of 350 or 400 °C for 30 min followed by a nitrogen anneal of 350 °C for 60 min or 400 °C for 30 min, respectively, to produce an excellent quality bond. Highly polished Cu surfaces similar to that needed for silicon are required [30]. Once again, care must be taken to prevent small voids due to particles or air entrapment during the bonding process. Figure 3.16 shows a Cu–Cu bonded interface. Copper–copper bonding is discussed further in Chapter 22.

Copper–Tin Eutectic Bonding Eutectic bonds can also be formed by the interaction of the Cu layers on two wafers with an intermediate tin layer (Figure 3.17). A Cu-Sn

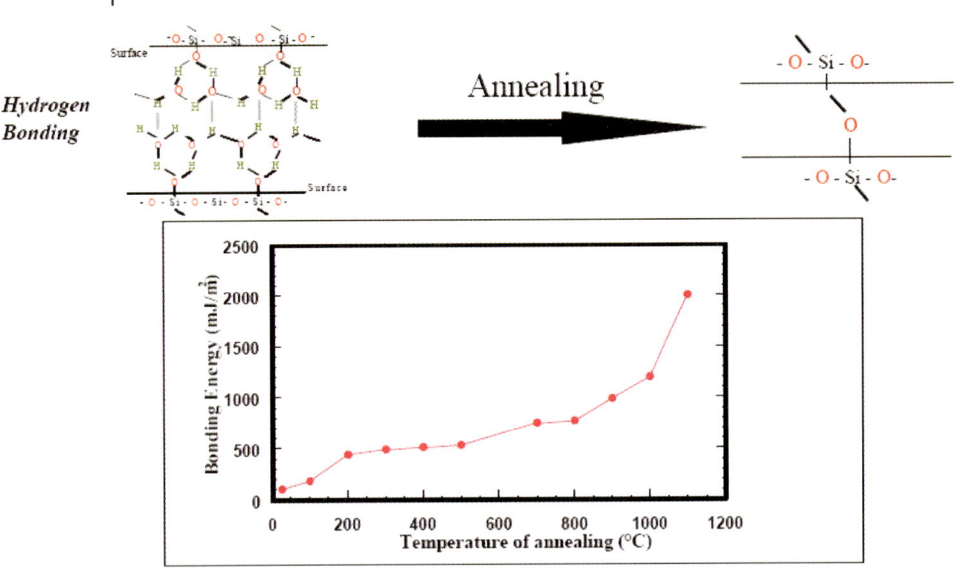

Figure 3.15 Chemistry of low temperature silicon fusion bonding [27].

eutectic forms when the tin is melted. Cu-Sn solid–liquid mixing occurs at temperatures slightly above the eutectic point with high contact force (80 N). See Chapter 14.

Polymer Adhesive Bonding Adhesive bonding uses polymers to deposit a planarizing material between two wafers. Such materials can be cured at low temperature to provide a low-stress wafer stack. Figure 3.18 shows a via etched through a BCB bonded interface.

It is difficult to maintain wafer alignment precision during adhesive wafer bonding. There is a claim that the 2–5 μm alignment tolerances available on today's wafer alignment tools drop to 10–15 μm when attempting adhesive wafer bonding due to shear forces exerted on the polymer when the wafers are pushed together [31]. To preserve the initial alignment accuracy, a frictional surface at the wafer edge

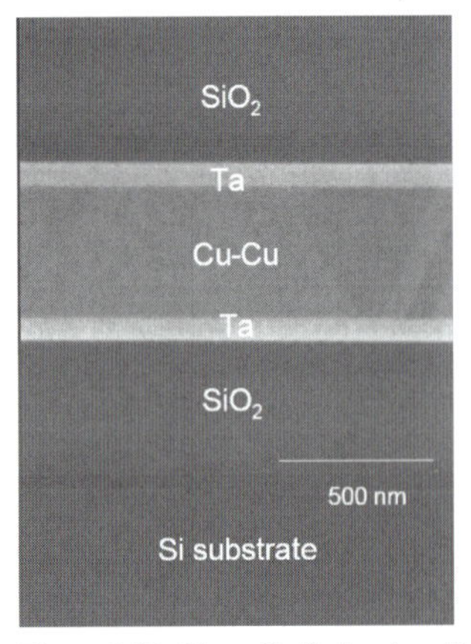

Figure 3.16 Direct Cu fusion bonding. (Courtesy of Reif – MIT.)

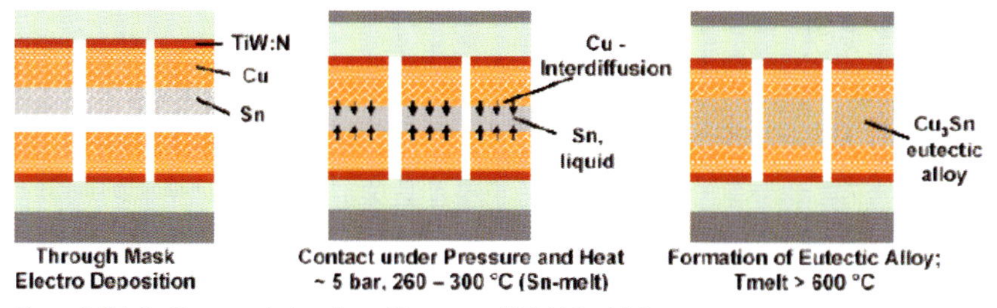

Figure 3.17 Cu/Sn eutectic bonding. (Courtesy of IZM-Munich.)

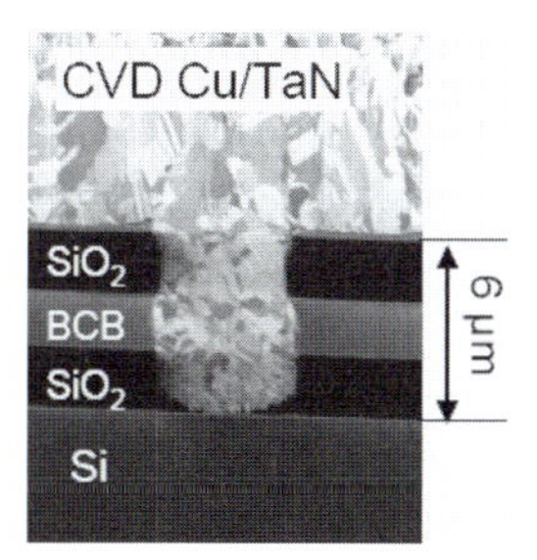

Figure 3.18 Cu Via etched through BCB wafer bonding layer. (Courtesy of RPI.)

(a 8 mm wide, 1.2 μm thick Al ring) is incorporated to prevent the wafers from shifting when pressure is applied. See Chapter 13 for a complete discussion.

References

1 Roozeboom, F. *et al.* (2006) Passive and heterogeneous integration towards a Si based system in package concept. *Thin Solid Films*, **504**, 391.

2 Henry, D., Baillin, X., Lapras, V., Vaudaine, M.H. *et al.* (2007) Via first technology development based on high aspect ratio trenches filled with doped polysilicon. 57th ECTC, Reno, Nevada, p. 830.

3 Mitsuhashi, T. *et al.* (2007) Development of 3D processing process technology for stacked memory. MRS Symposium Proceedings, Enabling Technologies for 3D Integration (eds C. Bower, P. Garrou, P. Ramm and K. Takahashi), **970**, p. 155.

4 Koyanagi, M. *et al.* (2006) 3D integration technology based on wafer bonding with vertical buried interconnect. *IEEE Transactions on Electron Devices*, **53**, 2799.

5 Takahashi, K., Taguchi, Y., Tomisaka, M. *et al.* (2004) Process integration of 3D chip stack with vertical interconnection, Proceedings of the 54th Electronic Components and Technology Conference (ECTC 2004), Las Vegas, NV, p. 601–609.

6 Garrou, P. (Oct 2006) 3D Integration moving forward. *Semiconductor International*, p. SP 12.

7 Toftness, R., Boyle, A. and Gillen, D. (2005) Laser technology for wafer dicing and microvia drilling for next generation wafers. *Proceedings SPIE*, **5713**, 54.

8 Rodin, A. (2007) High Throughput Laser Via and Dicing Process. Proceed. Peaks in Packaging, Whitefish MT.

9 Chin, B. and Van de Ven, E. (1988) Plasma TEOS for interlayer dielectric applications. *Solid State Technology*, **31**, 119.

10 Cote, D. *et al.* (1999) Plasma assisted CVD of dielectric thin films for ULSI semiconductor circuits. *IBM Journal of Research*, **43**, p. 5–38.

11 Garrou, P. *et al.* (1997) Polymers in packaging. in *Microelectronics Packaging*

Handbook (eds Tummala, Rymaszewski and Klopfenstein), Chapman & Hall, New York.

12 Gobert, J. *et al.* (1997) IC compatible fabrication of through-wafer conductive vias. *Proceedings SPIE*, **3223**, 17.

13 (a) Sabuncuoglu, D., Pham, N., Majeed, B. *et al.* (2007) Sloped through wafer vias for 3D wafer level packaging, Proceedings of the 57th Electronic Components and Technology Conference (ECTC, 2007), Reno, NV, p. 643.
(b) Jang, D.M. *et al.* (2007) Development and Evaluation of 3D SiP with Vertically Interconnected TSV,Proceedings of the 57th Electronic Components and Technology Conference (ECTC 2007), Reno, NV, p. 847.

14 Ogawa, E., Lee, K., Blaschke, V. and Ho, P. (2002) Electromigration reliability issues in dual damascene Cu interconnections. *IEEE Transactions on Reliability*, **51**, 403.

15 Hashim, I., Pavate, V., Ding, P. *et al.* IMP Ta/Cu Seed Layer Technology for High Aspect Ratio Via Fill. Proc. SPIE-Int. Soc. Opt. Eng., Volume 3508, Multilevel Interconnect Technology II, (eds M. Graf, D. Patel, and Klopfenstein).

16 Cho, B. and Lee, W. (2007) Filling of very fine via holes for 3D SiP by using ionized metal plasma sputtering and electroplating. International Conference Electronic Packaging, Tokyo, Japan.

17 Jang, D.M. *et al.* (2007) Development and Evaluation of 3D SiP with Vertically Interconnected TSV, ECTC, 847.

18 Ko, Y., Seo, B., Park, D. *et al.* (2002) Additive vapor effect on the conformal coverage of a high aspect ratio trench using MOCVD copper metallization. *Semiconductor Science and Technology*, **17**, 978.

19 Koide, T. and Sekiguchi, A. (2003) Formation of copper feed-through electrodes using CVD. *Proceedings MES*, 404.

20 Zhang, M. *et al.* (1999) Optimization of copper CVD film properties using precursor of Cu(hfac)(tmvs) with variations of additive content. *Proceedings IITC*, 170.

21 Kenny, S. and Matejat, K. (Feb. 21 2001) HDI production using pulse plating with insoluble anodes. *CircuiTree*.

22 Siniaguine, O. (1998) Atmospheric downstream plasma etching of Si wafers. Proceedings International Elect. Manuf. Tech. Symposium, p. 139.

23 Hendrix, M., Drews, S. and Hurd, T. (2000) Advances of wet chemical spin processing for wafer thinning and packaging applications. Proceed. International Elect. Manuf. Tech. Symp., p. 229.

24 Pinel, S., Lepinos, F., Cazarre, A. *et al.* (2002) Impact of ultra-thinning on DC characteristics of MOSFET devices. *European Physical Journal*, **17**, 41.

25 Takahashi, S., Hayashi, Y., Kunio, T. and Endo, N. (1992) Characteristics of thin film devices for a stacked-type MCM. IEEE Multi Chip Module Conference, p. 159.

26 Ramm, P. (Oct 2004) Vertical system integration technologies. Adv. Metals Conference Workshop on 3D Integration of Semiconductor Devices.

27 Pasquariello, D. (2001) Plasma assisted low temperature semiconductor wafer bonding, Dissertation, Uppsala University, Sweden.

28 Zucker, O., Langheinrich, W. and Kulozik, M. (1993) Application of oxygen plasma processing to silicon direct bonding. *Sensors & Actuators A*, **6**, 227.

29 Kurahashi, T., Onada, M. and Hatton, T. (1991) Sensors utilizing Si wafer direct bonding at low temperature. Proceed. 2nd International Symp Micro Machine and Human Science, Nagoya, p. 173.

30 Rief, R., Tan, C.S., Fan, A. *et al.* (April 2004) Technology and applications of 3D integration enabled by bonding. 3D Architectures for 3D Semiconductor Integration and Packaging Conference, Burlingame CA.

31 Niklaus, F., Enoksson, P., Kalvesten, E. and Stemme, G. (2003) A method to maintain wafer bonding alignment precision during adhesive wafer bonding. *Sensors & Actuators A*, **107**, 273.

I
Through Silicon Via Fabrication

Handbook of 3D Integration: Technology and Applications of 3D Integrated Circuits.
Edited by Philip Garrou, Christopher Bower and Peter Ramm

ISBN: 978-3-527-32034-9

4
Deep Reactive Ion Etching of Through Silicon Vias

Fred Roozeboom, Michiel A. Blauw, Yann Lamy, Eric van Grunsven, Wouter Dekkers, Jan F. Verhoeven, Eric(F.) van den Heuvel, Emile van der Drift, Erwin (W.M.M.) Kessels, and Richard (M.C.M.) van de Sanden

4.1
Introduction

4.1.1
Deep Reactive Ion Etching as Breakthrough Enabling Through-Wafer Interconnects

In recent years the conventional scaling in CMOS[1)] (Complementary Metal Oxide Semiconductor) microfabrication has reached its limitations. Revolutionary solutions are underway using novel gate stack material combinations, nanodevice fabrication and new 3D designs to continue the scaling of the computing and data storage power on a chip.

A technology revolution with similar or even larger impact than regular IC mass manufacturing has recently taken place by opening up the third dimension in silicon. Originally, plasma etching had just been used for thin-film patterning but it became more popular in deep trench capacitors and trench isolation applications. Today, Reactive Ion Etching (RIE) based on fluorine plasma chemistry is the technique of choice for the development of microelectromechanical systems (MEMS). The technique was shown to be a viable alternative with reasonable etch rates and selectivity to hard masks with nearly vertical sidewalls being independent of crystal orientation, as opposed to wet etching techniques based on potassium hydroxide or tetramethylammonium hydroxide (TMAH). In the mid-1990s Deep Reactive Ion Etching (DRIE) was introduced by Bosch [1] and commercialized by several equipment manufacturers.

New applications based on, or adopted from, Microsystems Technology (MST), or MEMS [2], could be developed in silicon technologies to add more functionality in other domains than microelectronic, such as the mechanical, acoustic, fluidic, photonic, biomedical and so on. These new applications include sensors and

1) A glossary of abbreviations used is listed in Appendix A.

Handbook of 3D Integration: Technology and Applications of 3D Integrated Circuits.
Edited by Philip Garrou, Christopher Bower and Peter Ramm

ISBN: 978-3-527-32034-9

actuators, and are usually developed on separate chips for rapidly growing markets of, for example, accelerometers and gyroscopes [3], micromirror-based projectors, inkjet printers, and so on. New markets are emerging in wireless communication [4] and in medical and health care [5]. Characteristic for these new products is the use of silicon DRIE, at first to realize the high aspect ratio features in the individual chips and later to create the *through-silicon interconnects*, needed to accomplish 3D die stacking in System-in-Package (SiP) devices. It is only in the 2005 edition of the International Technology Roadmap for Semiconductors [6] that this heterogeneous integration was recognized as a fully established and emerging technology in advanced packaging technology. Stacked-dies packaging technology was till then a subchapter in the previous roadmaps. Today, the mass fabrication of the first SiP devices containing stacked dies is a fact for flash memory devices or imaging sensors [7, 8] where thermal issues are not a limiting factor.

4.1.2
State of the Art and Basic Principles in DRIE

Today, modern DRIE tools contain an inductively coupled plasma (ICP) source connected to a diffusion chamber. An intense plasma with ion concentrations of 10^{10}–10^{11} cm^{-3} is generated by a high-power RF coil antenna around a remote ceramic enclosure and then diffused into a larger chamber. The substrate is located within this diffusion chamber and usually clamped on a bipolar electrostatic chuck that can be cooled by backside helium flow and/or liquid nitrogen cooling. The chuck can be biased by a low-power LF or RF source such that ions from the plasma can be accelerated to the substrate with independent control. Gases enter at the top of the plasma chamber and are pumped off by a high conductance pump assembly at the bottom of the diffusion chamber (Figure 4.1).

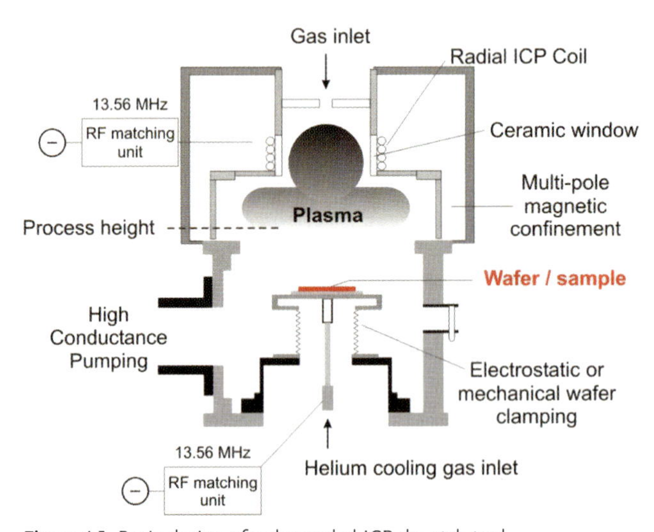

Figure 4.1 Basic design of a decoupled ICP dry-etch tool.

4.1.3
Bosch Process

As mentioned earlier, the Bosch process is the mainstay in silicon micromachining, and is also referred to as deep reactive ion etching, time-domain multiplexed etching or switched etching process. Originally, the Bosch process was based on alternating cycles of Si-etching with SF_6 or NF_3 in Ar to form gaseous SiF_x etch products, and passivation with CHF_3 or CF_4 (later also C_4F_8) in Ar to form a protecting fluorocarbon polymer deposit on the sidewalls and bottom of the feature [1] (Figure 4.2). During each etch step a bias voltage is applied to the substrate chuck. This causes a directional physical ion bombardment from the plasma onto the substrate that breaks down the polymer at the bottom part of the feature.

Yet, the etching remains in fact isotropic. Without interruption it would proceed mainly by the non-directional neutral species (F-containing radicals). To minimize this lateral etching component the etch steps are quickly interrupted by the next wall passivation step. Typical etch or passivation cycle lengths are 1–10 s with 0.1–1 μm

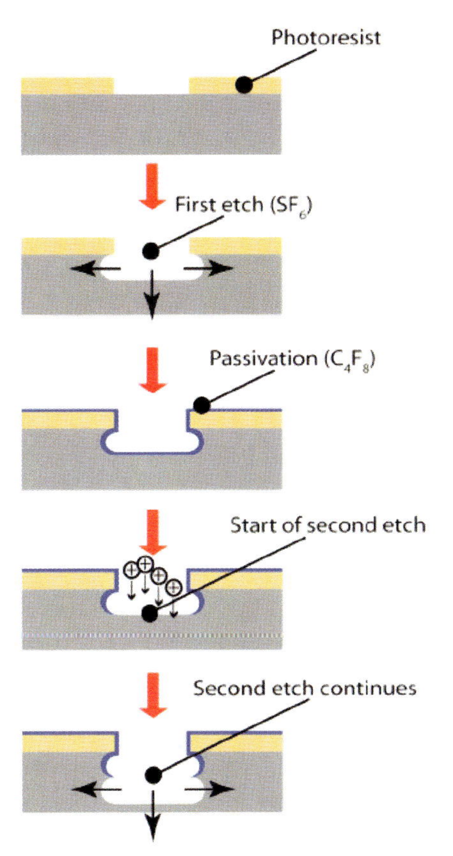

Figure 4.2 Basic steps in the Bosch process.

etched depth per cycle. The process enables dry-etching of deep vertical microstructures in silicon with relatively high etch rates and selectivities (up to ~200 : 1) against a hard oxide and photoresist mask material. This basic Bosch process has been licensed to several equipment manufacturers. Many have optimized the process further and often refer to it with their own trademarks.

By definition, the more traditional continuous etching methods rely upon simultaneous etching and sidewall passivation. Often, different chemistry is used (e.g., C_4F_8) or oxygen is blended into the SF_6 to facilitate the formation of the passivating layer [9]. Another way to achieve anisotropic profiles is to suppress spontaneous isotropic etching by keeping the wafer at cryogenic temperatures (see Section 4.3.3.1). The Bosch process as a room temperature process is preferred over cryogenic processing since the latter is more sensitive to temperature variations. Also, cracking of the mask materials at lower temperatures has been reported.

Of course the Bosch process is not perfect. Several characteristic features have been listed [10, 11] as non-ideal, such as the initial mask undercut and scallops on the sidewall, aspect ratio dependent etching (ARDE) rate, notching at dielectric interfaces and sidewall roughness (mouse bites and striations). These features are illustrated in Figure 4.3, and explained later in Section 4.3. Methods to suppress these effects are discussed in Section 4.4.

4.1.4
Alternatives for Via Hole Creation

DRIE is the most popular method to make through-wafer interconnects, especially for high interconnect densities (>1000 mm^{-2}) with etch depths of often 30–100 μm or more. Alternative via hole formation techniques are laser drilling powder blasting and wet-chemical via hole etching. Wet etching of ultrafine high aspect ratio via hole arrays with small pitch is dealt with below.

4.1.4.1 Wet Etching of High Aspect Ratio Pore Arrays

The first work on vertical pore wet-etching was published by Theunissen [12]. He studied spontaneous macropore formation during anodic dissolution of lightly doped n^--type silicon in aqueous HF and explained the photo-electrochemical etching process by a depletion of holes in the porous region due to space charge effects. The self-adjusting pore etching mechanism has been further investigated by Lehmann *et al.* [13, 14]. For the pore initiation they defined a regular pattern of micro-indentations along the slow-etching {1 1 1} crystallographic planes in small Si(100) samples by pre-etching with hot KOH.

The anisotropic wet-etching is based on the preferential anodic dissolution of Si in the etch pit regions where the holes are collected more efficiently due to the enhanced electrical field in the space charge layer (Figure 4.4). When the rate-determining step for the dissolution reaction:

$$Si + 2H^+ + 6F^- + 2h^+ \rightarrow [SiF_6]^{2-} + H_2\uparrow$$

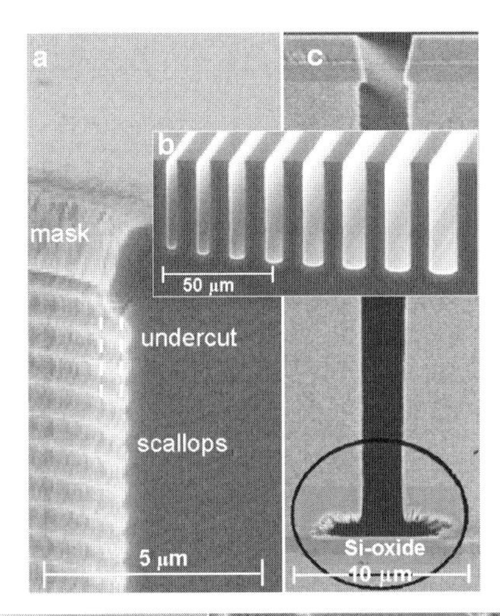

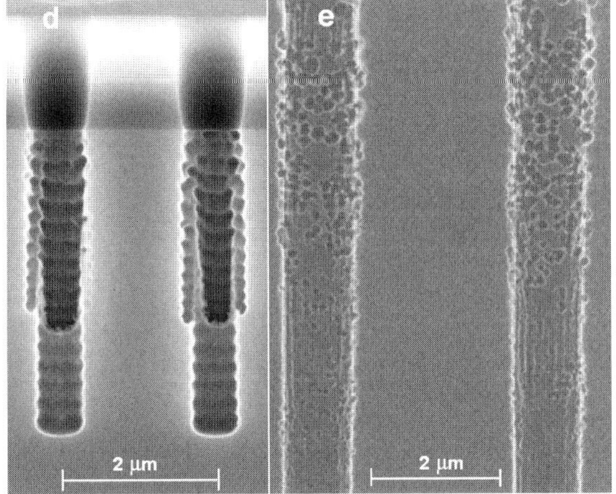

Figure 4.3 Non-ideal characteristics of the Bosch process: (a) mask undercut and scallops, and (b) aspect ratio dependent etching, (c) notching at dielectric interfaces, (d) striations and (e) mouse bites.

is controlled by the number of holes (h^+) generated by white light illumination of the wafer backside, the pore walls become depleted of the minority carriers (holes) that drive the dissolution, and are thus passivated. By proper tuning of the light intensity the hole transport and the diffusion of F^- ions in the liquid can be balanced. In that case the electrical field lines are pointing at the pore tip. The trajectories of the holes are drawn in Figure 4.4, which illustrates how the holes (minority carriers) find their way through the space charge layer by "current crowding" to the pore tip. In this zone they locally outnumber the electrons (intrinsic majority carriers) and find their way to the pore tips. There they loosen the atomic Si–Si bonds to dissolve the Si-atoms at the bottom, according to the reaction given above.

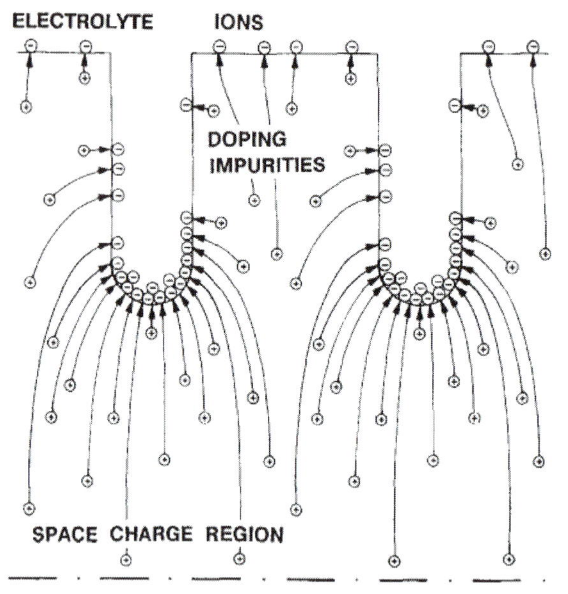

Figure 4.4 Electrical field line distribution in the space charge layer around pores during anodic dissolution of silicon. From Reference [13].

In our wet etching experiments [15] standard g-line contact photolithography was used to apply a Si_3N_4 mask with a pseudo-hexagonal array of circular openings with 1.5 μm diameter and 3.5 μm spacing onto lightly n-type (phosphorus) doped (10 Ω cm) 150-mm silicon (100) wafers. This pattern was used for pre-etching the {111}-oriented micro-indentations with hot KOH.

The experimental set-up used is shown in Figure 4.5. The wafer is placed in a polypropylene holder containing a K_2SO_4 electrolyte, used for a uniform anodic

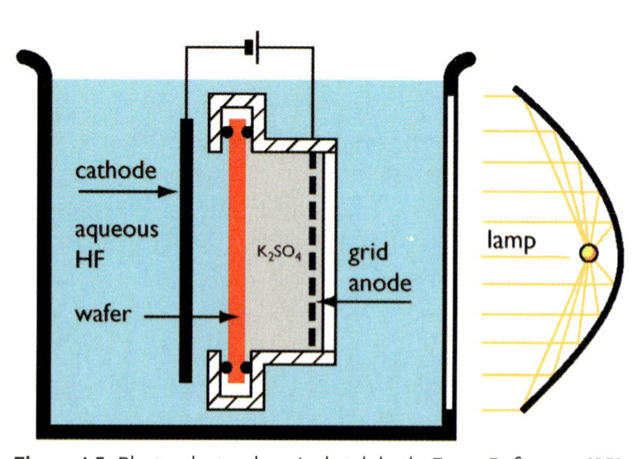

Figure 4.5 Photo-electrochemical etch bath. From Reference [15].

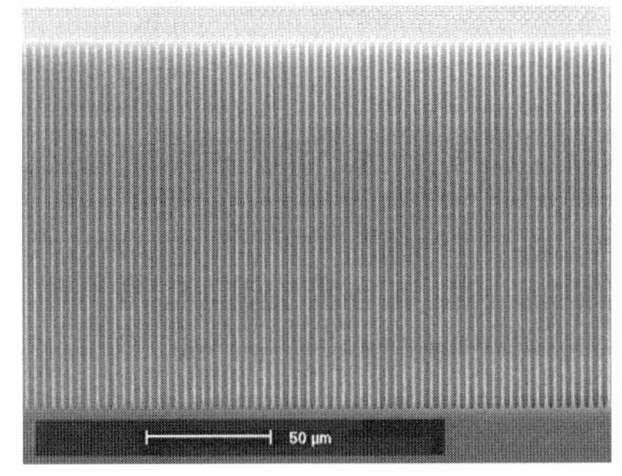

Figure 4.6 SEM image of a wet-etched pore array in n^- Si. From Roozeboom *et al.* [15].

contact at the wafer backside. The contact is made through a platinum grid anode placed in the electrolyte. The holder is placed in an aqueous HF solution with the wafer front side facing the Pt cathode. A tungsten-halogen lamp illuminates the wafer backside through transparent polycarbonate windows in the wafer holder and the etch bath.

The pore diameter is controlled by the anode current, which in turn is controlled by the light intensity. In practice the current is monitored and used to adjust the lamp power in an automated cycle. Typical etch conditions were 7.5 V bias and 0.7 A, using a 1.45 *M* HF/4.62 *M* ethanol solution, which is circulated through a thermostat by a pneumatic Teflon pump. Figure 4.6 gives a typical result and shows the excellent depth (~150 µm) and diameter (~2 µm) uniformity obtained with this etch technique. The etch rate at 30 °C was typically 0.6 µm min^{-1}, but we also reached 4.0 µm min^{-1} when using higher HF-concentrations in combination with higher light intensities. This work has been published elsewhere, along with details on the mechanism and kinetics of this process [16, 17].

We obtained uniform pore arrays as deep as 400 µm. An advantage of the wet-etch process is that it could be carried out as an inexpensive multi-wafer process, as opposed to the single-wafer DRIE process. Moreover, it was confirmed that one can also wet-etch lightly p-type [18] silicon (B-doped, 10 Ω cm) without wafer backside irradiation (holes being the majority carriers in p-type Si).

However, we should also list a few limitations here:

(1) Single vias cannot be wet-etched. Owing to the physical phenomenon of the space charge layer formation around the pore, wet etching is limited to arrays of pores with a certain pitch. The outer pores will show poorly fillable lateral branching.

(2) The dimensions of the pore diameter and the pore pitch (i.e., space charge layer) are limited and coupled to the substrate dopant level. Figure 4.7 shows the range of stable diameters, only from 0.5 to ~10 µm, and the corresponding substrate dopant levels (0.1 to 40 Ω cm).

Both limitations may impose physical restrictions on the lay-out and design of the functional devices on the substrate.

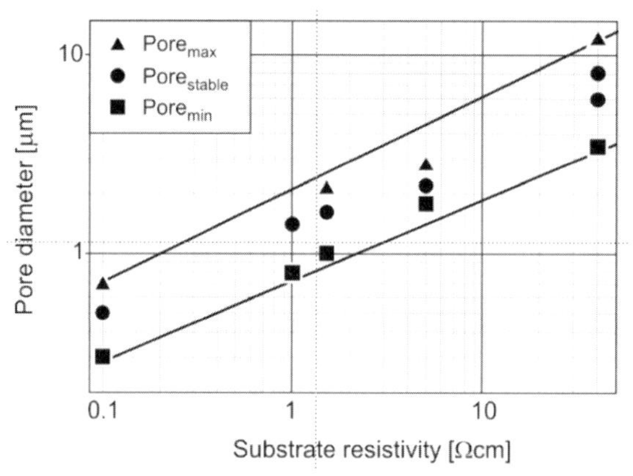

Figure 4.7 Stability range for wet-etched pore diameters in n^- Si vs substrate dopant level. From Lehmann *et al.* [19].

4.2
DRIE Equipment and Characterization

Traditionally, silicon etching was performed in capacitively coupled plasma (CCP) reactors. Figure 4.8 shows a schematic drawing of such a reactor. However, a trend has set in to inductively coupled plasma (ICP) reactors to improve etch rate and profile control. Much higher ion and radical densities can be obtained in ICP reactors while the ion energy can be controlled independently with a separate power source in the downstream region of the plasma source.

4.2.1
High-Density Plasma Reactors

Silicon etching has evolved considerably due to the everlasting demands for higher etch rates, higher selectivity and higher anisotropy. These targets can be obtained

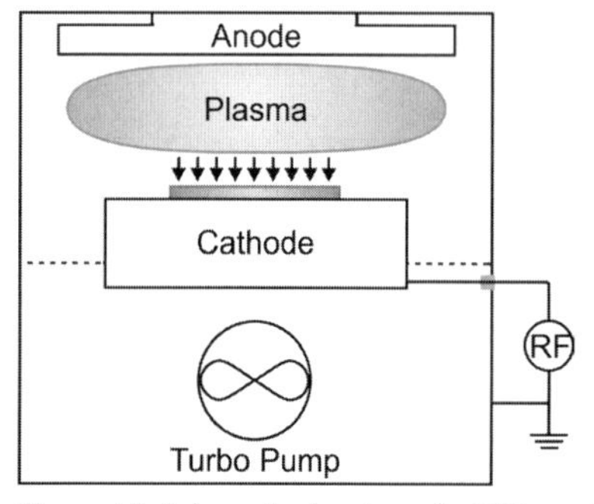

Figure 4.8 Schematic drawing of a CCP reactor with a single power source for plasma generation and substrate biasing.

almost without compromise in high-density plasma (HDP) etching systems such as ICP, electron cyclotron resonance (ECR) plasma, and expanding thermal plasma (ETP) reactors.

Due to the high density of radicals and ions the etch rate is several orders of magnitude higher than in CCP systems. As a consequence the gas flow, the source power and the pumping speed are also higher, with the aim of keeping the supply of radicals in proportion with the consumption. In addition, the concentration of etch products in the gas phase is low due to the low residence time.

Mask erosion and plasma-induced damage of the etched materials are strongly reduced if the ion energy is reduced. In RIE systems a higher plasma density is always coupled to a higher bias voltage. However, in HDP etching systems the plasma density and the ion energy can be more or less independently controlled by separate power sources for plasma generation and substrate biasing. As a result, a high plasma density and a low ion energy can be obtained simultaneously.

4.2.1.1 **Inductively Coupled Plasma**

Figure 4.9 gives a schematic drawing of an ICP reactor. Inductive coupling is obtained with a coil around a ceramic tube or with a spiral loop on a ceramic plate. The RF power is delivered by an impedance matching network. A strong rise of the plasma density is observed above a certain threshold of RF power. Above this threshold, the ion density is of the order of $10^{10}\,cm^{-3}$. Below the threshold, the plasma density is comparable to RIE systems due to capacitive coupling.

An Alcatel MET reactor and an STS Multiplex reactor were used for most of the research described here. Both ICP etching systems typically have a 13.56 MHz RF

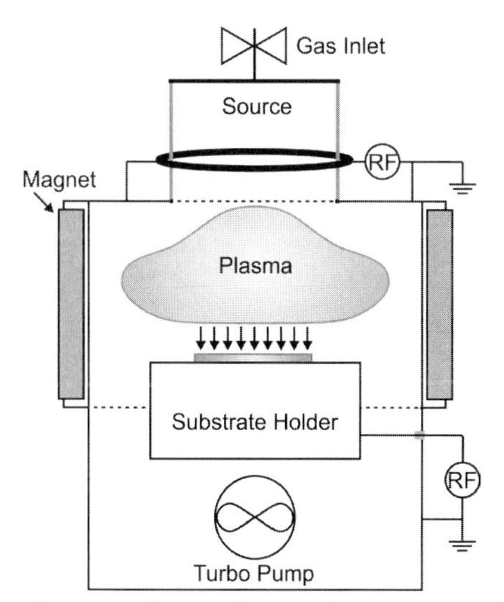

Figure 4.9 Schematic drawing of an ICP reactor with independent control of plasma generation and substrate biasing.

source of the order of 1000 W for plasma generation and a 13.56 MHz RF power supply of the order of 100 W for independent substrate biasing.

A vacuum is obtained with turbomolecular pumps with a pump speed of $\sim$1000 l s^{-1}. The reactor chamber is accessible through a load-lock system. With full pumping speed the residence time is about 0.2 s. The gases mostly used are SF_6, C_4F_8, O_2 and Ar. The reactor pressure is measured with a Baratron capacitance manometer and can be regulated with an adjustable butterfly valve. Normally, the total gas flow is in the range 50–500 sccm and the pressure is in the range 0.5–5 Pa. For the pulsing of gases in the time-multiplexed etching process the reactor is equipped with fast-response mass flow controllers and short gas lines, reducing dead volumes. The substrate holder temperature can be regulated in the range -150 to $+25$ °C, with fluctuations smaller than ± 1 °C, by liquid nitrogen cooling in combination with resistive heating. The low substrate temperature regime is necessary for the cryogenic etching process, for which the optimal substrate temperature is approximately -125 °C. Accurate control of the substrate temperature is important, because a change of ± 5 °C already leads to a significant degradation of the anisotropy of the cryogenic etching process.

4.2.1.2 **Electron Cyclotron Resonance Plasma**

Another type of plasma generation is based on the interaction of microwave radiation with a plasma. ECR plasma generation occurs only if the mean free path of electrons is large enough to gain sufficient energy for the dissociation and ionization of molecules. To increase the power transfer efficiency the electrons are confined by a magnetic field near the antennas. In this magnetic field ($\sim$87.5 mT in our system) the electrons follow a circular trajectory with the electron cyclotron frequency, thus exposing them longer to the oscillating electromagnetic field. The movement of the electrons is in phase with the electromagnetic field and leads to resonant energy transfer. In practice the pressure is $\sim$0.1 to 0.5 Pa. In an ICP reactor the radical fluxes are one order of magnitude higher due to the higher pressure. This makes the ICP reactor more suitable for high etch rate applications. For this reason the ECR reactor is not used too often in DRIE.

However, the ECR reactor type is ideal for fundamental etching studies because of the excellent control of ion and radical fluxes. This is discussed for Aspect Ratio Independent Etching in Section 4.3.2.3. For that study an Alcatel RCE 200 distributed electron cyclotron resonance (DECR) reactor was used with 14 antennas to obtain a better uniformity. A schematic drawing of the reactor is shown in Figure 4.10.

The microwave power from a 2000 W, 2.45 GHz magnetron tube is coupled into a distributor and the reflected power is reduced to nearly zero with a manual stub tuner. The chamber pumping, the wafer handling capabilities, gas flow and substrate temperature control and the substrate biasing are equally versatile as for the ICP reactor. However, the optimal substrate temperature for the cryogenic etching process in the DECR reactor is approximately -95 °C. An important modification to this system consists of a 15 cm high quartz cylinder placed on the clamping ring around the wafer. Originally, it was intended to eliminate contamination from the

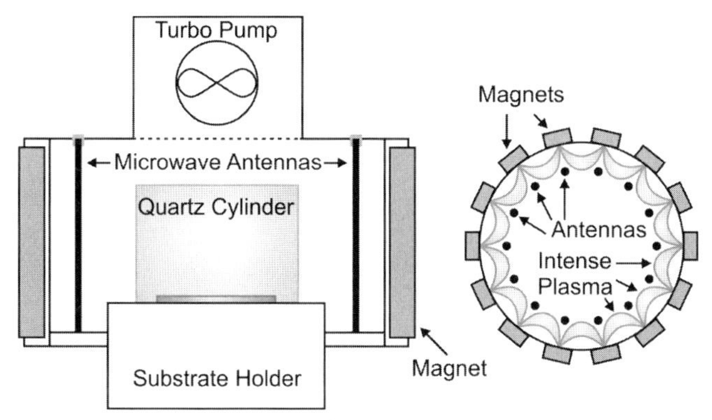

Figure 4.10 Schematic drawing of a DECR plasma reactor. Left-hand side: the quartz cylinder blocking the direct path between source and substrate. Right-hand side: magnet and antenna positions with most intense plasma regions.

antennas and the reactor wall. However, it effectively results in a spatial separation of the plasma generation and the substrate biasing because the direct path between the source region and the wafer is blocked. Thus, an extremely low ion-to-radical flux ratio is obtained at the wafer level, because electron–ion recombination is much faster than radical–radical recombination. The ion flux is reduced by two orders of magnitude due to the quartz cylinder. Without the quartz cylinder the ion flux in a DECR reactor is comparable to that in an ICP reactor.

4.2.1.3 Expanding Thermal Plasma

Figure 4.11 gives a schematic overview of an ETP reactor consisting of a plasma source and a reactor chamber. An Ar plasma is generated in a DC-current, wall-stabilized cascaded arc. The arc consists of three cathodes and four cascaded plates that are electrically insulated and stacked on top of an anode plate. The plasma source current that flows through a 4 mm diameter plasma channel is 75 A, unless stated otherwise. A continuous Ar flow of 50 ml s^{-1} is ionized up to an ionization degree of ~5–10% at a plasma source pressure of 36 kPa [21]. The sub-atmospheric, thermal plasma with a temperature around 1 eV expands supersonically through a nozzle in the anode plate into the reactor chamber. This leads to a stationary shock front at a certain distance from the nozzle because the reactor chamber is maintained at a three orders of magnitude lower pressure. The reactor pressure is controlled between 25 and 81 Pa with a valve in front of the roots pump. After supersonic expansion, the pure Ar plasma yields an electron and ion density in the reactor chamber on the order of 10^{13} cm^{-3}. The electron temperature decreases to approximately 0.3 eV. No additional power is coupled into the expanding plasma in the downstream region [22]. As a result, the plasma chemistry in the reactor chamber is ion-driven rather than electron-induced.

The process gases SF_6, O_2, and C_4F_8 are injected through a 10 cm diameter ring with eight symmetrically placed holes ~5 cm downstream from the nozzle of the

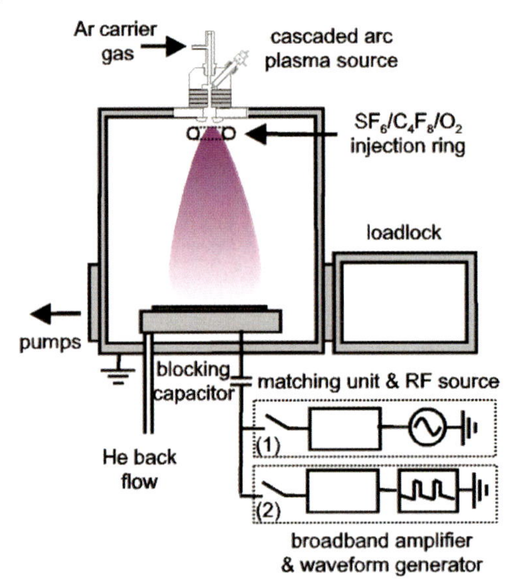

Figure 4.11 Schematic drawing of the ETP reactor as it is used for deep anisotropic Si etching. The two methods to generate energetic ion impact at the substrate are indicated: (1) RF substrate bias and (2) pulse-shaped substrate bias. From Reference [20].

plasma source. Charge transfer from the Ar ions to the process gas molecules that are injected in the reactor chamber is followed by dissociative recombination of the formed molecular ions and electrons. This leads to fragmentation of the process gas molecules and consumption of Ar ions and electrons [23]. As the plasma source has a remote character, the highest ion and radical densities in the reactor chamber are obtained if the highest amount of Ar ions is produced in the plasma source. The distance between the plasma source and the substrate is variable and can be set to its maximum of 60 cm by default. The reactor is equipped with a load-lock to allow substrate transfer without breaking the vacuum in the reactor chamber. The temperature of the substrate holder is controlled by a combination of liquid N_2 cooling and electrical heating. Good thermal contact between the substrate and the substrate holder is ensured by a He backside flow.

Energetic ion impact is generated in two distinct ways. First, a 13.56 MHz sinusoidal RF substrate bias source is used, which is coupled to the substrate holder through an impedance matching circuit. Second, inspired by the original work of Wang and Wendt [24] we used a pulse-shaped low-frequency (typically 400 kHz) substrate bias source. The pulse-shaped bias consists of a positive voltage spike followed by a slow linear negative voltage ramp. It is generated by an arbitrary waveform generator connected in series with a broadband amplifier. The output of the broadband amplifier is connected to the substrate holder with a ceramic capacitor.

In principle, a pulse-shaped substrate bias results in a mono-energetic ion energy distribution, whereas an RF substrate bias can lead to a bimodal ion energy distribution, especially in the case of high-density plasmas. The improved control of the ion energy leads to a higher etch selectivity because the etching threshold of the mask material is usually higher than that of Si [25]. Furthermore, an RF substrate bias creates a bright glow around the substrate holder in the ETP reactor and a capacitively coupled plasma can be obtained. This enhancement of the electron temperature and electron and ion density in this region of the reactor is related to the very low electron temperature in the expanding thermal plasma without any substrate bias. No such bright glow is visible with a pulse-shaped substrate bias. This is an indication that additional plasma formation is minimal low in this case. Since the spread in ion energy is related to electron temperature, the directionality of the ions is better. Thus, a better anisotropy can be obtained by pulse-shaped biasing.

4.2.2 Plasma Chemistry

Chlorine- and bromine-based plasmas are frequently used in CMOS processing to obtain perfect anisotropy. Lateral etching is minimal for these plasmas because chlorine and bromine do not etch silicon spontaneously at room temperature. For the same reason, the etch rate is relatively low. However, the required etch depth is relatively small and the etch rate is thus less critical. In contrast, the etch rate is the limiting factor in DRIE of silicon. For this application, fluorine-based plasmas are able to provide substantially higher etch rates.

The reaction of fluorine atoms with silicon is spontaneous and exothermic. Therefore, a component that passivates the silicon surface is added to the plasma to suppress lateral etching. Energetic ion impact is needed to sputter the component in the direction of the substrate surface because the surface passivation blocks the etching in all directions. Both continuous and time-multiplexed etching modes have been developed to achieve deep anisotropic etching.

4.2.2.1 Cryogenic and Room Temperature Plasma Etching

It is not possible to prevent lateral etching in high-density SF_6 plasmas by lowering the substrate temperature because of the high fluorine radical partial pressure [26]. The addition of O_2 is necessary to obtain anisotropic profiles [27]. Surface passivation with O_2 is especially effective below $-80\,^{\circ}C$, which can be reached with liquid N_2 cooling. The cryogenic etching process is named after the need for such liquid N_2 cooling. A change of 10% of the O_2 flow rate, which is approximately 1% of the total gas flow rate, already leads to a significant degradation of the anisotropy. The SiO_xF_y passivation layer is removed by the ion bombardment, so that etching proceeds in the direction of the ion flux, that is, in the vertical direction as desired. The combined effect of continuous processing, a low temperature and O_2 addition leads to a high etch rate and a high selectivity. The need for liquid N_2 cooling and etching artifacts such as crystal orientation dependent etching make the cryogenic etching less popular for industrial processes.

The addition of a fluorocarbon gas instead of O_2 to high-density SF_6 plasmas results in the deposition of a thin polymer layer at room temperature. The fluorocarbon polymer passivation layer blocks the spontaneous chemical reaction of fluorine radicals and silicon. Sputtering removes the polymer layer so that the etching proceeds in the direction of the ion impact. Mixtures of SF_6 and fluorocarbon gases such as CHF_3 have been used for etching. However, the optimal plasma conditions for a high etch rate and a high degree of anisotropy are contradictory. The fraction of fluorine radicals should be as high as possible for high etch rates but it should be much lower for the deposition of a qualitatively good passivation layer.

4.2.3 Plasma Diagnostics and Surface Analysis

The plasma is a complex medium and the ion and radical densities are difficult to predict for a given reactor. Therefore, plasma diagnostics are needed to control the ion and radical densities. A quantitative measurement of the ion and radical fluxes also makes it possible to model the reaction mechanism. A well-known model for ion-induced etching is the ion–neutral synergy model [28]. It takes into account only two species (ions and radicals). For etching with sidewall passivation, an extended ion–neutral synergy model has been developed that includes three species (ions, etching radicals and passivating radicals). It shows that ion-limited and radical-limited etching regimes exist in deep silicon etching. The ion flux can be measured with a Langmuir probe system and the radical densities can be measured with optical emission spectroscopy (OES). More background information can be found elsewhere [29].

The profiles of the etched features can be inspected with cross-sectional SEM. Further surface analysis is typically performed with TEM, AFM, ellipsometry, and XPS to investigate the surface reaction mechanism.

4.2.3.1 Langmuir Probe

Several plasma parameters such as plasma potential, floating potential, electron and ion density, and electron temperature can be measured with a Langmuir probe. The plasma potential and ion density are the most important parameters since the ion impact on the substrate depends on these data. Furthermore, plasma sheath properties such as the plasma sheath thickness and the ion transit time can be calculated with these data using the Child–Langmuir law.

If the plasma sheath is much thinner than the ion mean free path, the plasma sheath is collisionless and the ion angle distribution (IAD) is narrow. A collisionless plasma sheath leads to a better anisotropy. If the ion transit time is much larger than the RF period, the ions experience an average electric field in the plasma sheath and the ion energy distribution (IED) is narrow. The ion energy is usually taken as equal to the difference of plasma potential and DC bias voltage, assuming a narrow IED.

4.2.3.2 Optical Emission Spectroscopy

The characteristic color of emitted light is one of the most striking properties of a plasma. Each plasma has a distinct color spectrum, which is a fingerprint of the plasma composition. With optical emission spectroscopy (OES) the light intensity is measured as a function of the wavelength. The electron temperature of the plasma is often several eV, with the exception of the ETP. Electrons in the high-energy tail of the electron distribution can excite the plasma species. The excitation energy depends on the type of plasma species, but normally it is in the range 6–15 eV.

The intensity is not only proportional to the plasma species density but also depends on the electron density and the electron temperature. A technique named actinometry has to be deployed for a quantitative measurement of the plasma species density by OES. A small amount of an inert gas, the actinometer gas, is added to the plasma such that it does not influence the plasma chemistry. The plasma species density is proportional to the emission intensity ratio of the actinometer gas and the plasma species and to the density of the actinometer gas [30, 31].

4.2.3.3 Ellipsometry

To investigate the reaction mechanism it is useful to measure the reaction layer thickness during processing. Contamination with atmospheric species and evaporation of volatile reaction products are avoided with *in situ* ellipsometry. The measurement of the reaction layer thickness during cryogenic etching is highly complicated because of surface roughness, which is easily formed. Surface roughness shows up as an increase of the measured layer thickness. This obscures the thickness of the thin reaction layer.

4.2.3.4 X-ray Photoelectron Spectroscopy

X-ray photoelectron spectroscopy (XPS) is a technique for quantitative analysis of the chemical composition of a substrate surface. The sample is irradiated with X-rays and emitted core electrons are detected. The characteristic photoelectron energy spectrum is a fingerprint of the elemental composition. The area under a peak of a certain element is proportional to its concentration and its photoionization cross-section. In a high-resolution mode the chemical bonding state can also be observed due to a small shift of the photoelectron energy. The probing depth is defined by the inelastic mean free path of photoelectrons in a solid, which is of the order of a few nanometers. This makes XPS a true surface analysis technique.

4.2.3.5 Microscopy

Optical and Scanning Electron Microscopy (SEM) are the mainstay techniques in sample inspection. Samples are cleaved along a crystal plane perpendicular to the surface to obtain the cross-section of the structures. For more details on the cleaving of samples see Section 4.4.5.

For a quick measurement of the etch depth on wide structures, a surface profiler can also be used. Transmission electron microscopy (TEM) and atomic force microscopy (AFM) are used less frequently for advanced surface analysis. With TEM, the cross-sectional topography and the crystal structure of a sample can be investigated. With AFM the surface topography of a structure can be measured to determine its roughness.

4.3
DRIE Processing

This section discusses the practical situations and limitations in DRIE processing. These include mask-related issues, high aspect ratio features and sidewall control.

4.3.1
Mask Issues

Here we discuss the preparation and patterning of hard masks and some important issues on the sidewall roughness of via holes that can be related to mask properties during processing.

The DRIE etching reported here was done on 150-mm wafers in an STS Multiplex ICP etch tool, using SF_6/10% O_2 in etch cycles and C_4F_8 in passivation cycles (see Appendix B for recipe examples). We designed mask patterns with via openings ranging from 1 μm up to 100 μm diameter. Open areas in the masks used ranged typically from a few % to as high as 40%.

4.3.1.1 Mask Preparation and Patterning

The patterns with the features (usually circular and elongated holes with race-track cross-section) are applied by standard photolithography, that is, g-line contact printing or i-line stepper. Normally, silicon dioxide (~1 μm thick thermally grown SiO_2, LPCVD-TEOS or PECVD) and/or photoresist (PR, ~2 to 3 μm thick) are the masking materials of choice. The etch selectivities are roughly Si : PR : oxide = 200 : 2.5 : 1. Note that the mask material deposition methods (e.g., spin coating) normally give very good layer thickness uniformity. However, despite all hardware optimizations for silicon etch rate uniformity (such as balanced inductive coil drives, spatial ion discrimination and collimator improvements) photoresist erosion is reported to increase by 20–30% from center to edge across 150-mm diameter wafers [11]. Thus, one should always apply enough mask thickness to tolerate the mask consumption across the entire wafer during the entire process. Often one applies a stacked mask layer (PR and oxide). This is easiest done by simply leaving the photoresist layer needed to pre-etch the openings in an oxide hard mask after this pre-etch process.

4.3.1.2 Mask Undercut and Scallops

A characteristic inherent to the alternation of etch and passivation pulses in the Bosch process is the initial mask undercut and the corrugation ("scalloping") of the sidewall, especially for longer etch pulses; (Figure 4.3a). This is because the etching with SF_6 remains in fact semi-isotropic and proceeds mainly by the non-directional neutral species (radicals).

A special case of mask undercutting that can happen is shown in Figure 4.12. It has been referred to as "mouse-bitten collars" [32]. The figure illustrates that the balance of etching and (de)passivation can be fatally disturbed during an etch process that starts seemingly well. One can see that after a critical time of 40–60 min etching the

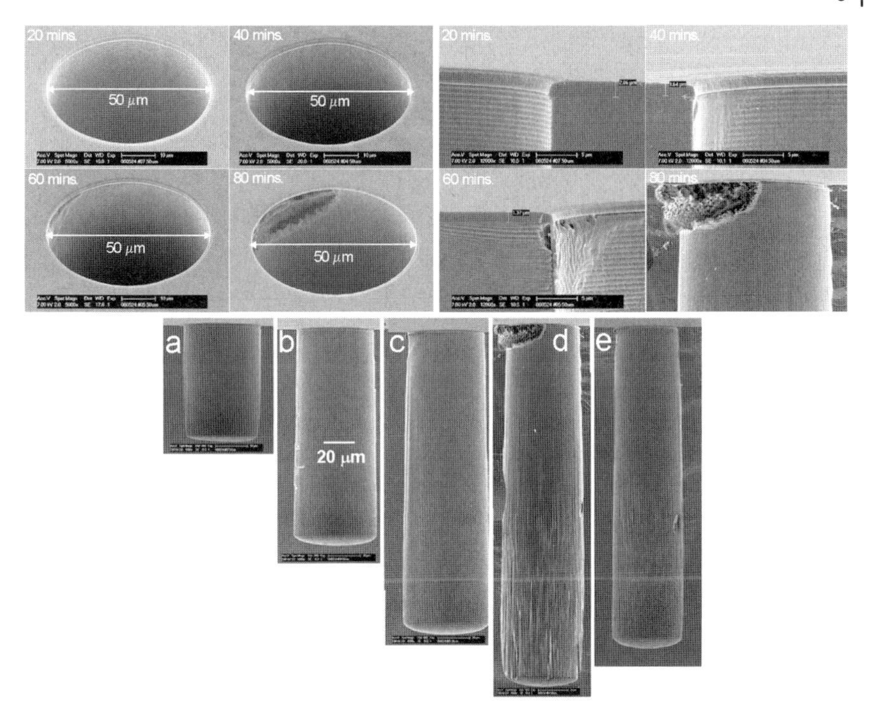

Figure 4.12 Different stages in the creation of a mouse-bitten collar at a via top. Shown is the development of a nominally 50 μm diameter via hole after (a) 20, (b) 40, (c) 60 and (d) 80 min. of etching with recipe A. Subsequent depths are: 75, 147, 206 and 258 μm, corresponding with decreasing average etch rates of 3.7, 3.6, 3.4 and 3.2 μm min^{-1}. For comparison a via etched for 80 min with the less aggressive recipe B is shown in (e) with 232 μm depth (2.9 μm min^{-1}). Recipes A and B are listed in Appendix B.

polymer layer right under the mask is no longer homogeneously sticking or deposited. It is possible that the stress balance in the dissimilar layers (eroding mask and passivation layer) is disturbed by the continuously eroding mask layer and may cause cracking. Consequently, radicals can diffuse laterally through these cracks and cause an isotropic etch profile behind the degrading polymer layer. At this weak spot a mouse-bitten collar is formed after ~80 min of etching. This is also the stage where shadow effects caused by the degrading polymer give further rise to marked vertical striations in the lower via part. Figure 4.12e shows how a less aggressive etch recipe with only a 10% lower etch rate can be used with virtually no mouse-bitten collar and no striations.

4.3.1.3 **Striations and Mouse Bites**

The oxide mask patterns are usually pre-etched in a $CHF_3/CF_4/Ar$ plasma or in buffered HF solution. This is a critical step in the control of via sidewall smoothness. The masking oxide should be removed completely, that is, no thin oxide mask bottom should be left behind after pre-etching of the patterns. Also, the slope of the

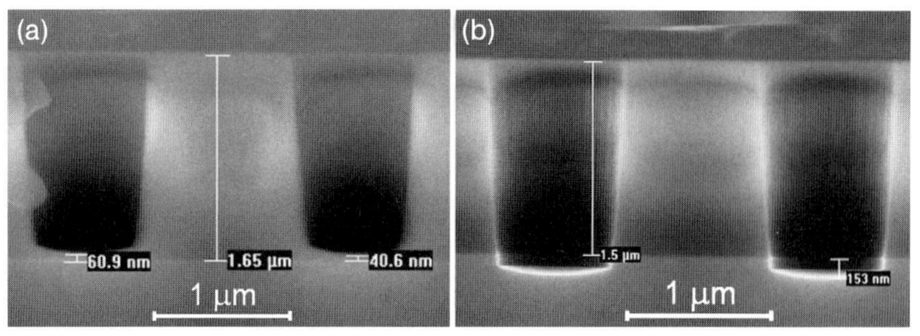

Figure 4.13 SEM images of (a) an insufficiently and (b) a properly opened oxide mask layer.

pre-etched holes in the mask layer should be steeper than approximately 80°. Figures 4.13 and 4.14 illustrate what happens if these conditions are not fulfilled. Figure 4.13 shows an incompletely and a correctly pre-etched oxide mask layer. During the subsequent silicon etching step the narrow inner part of the insufficiently etched oxide mask will break through first and the remaining thin annular oxide mask acts initially as a masking layer for the narrow channel shape ("spike" or "striation") initially etched in the silicon, see Figure 4.14a.

Next, the thinnest mask parts are eroded in a rather irregular way, and as soon as the mask effectively retracts a new part of the silicon is exposed to the ion bombardment. This is manifested as vertical striations carved out by the bombarding

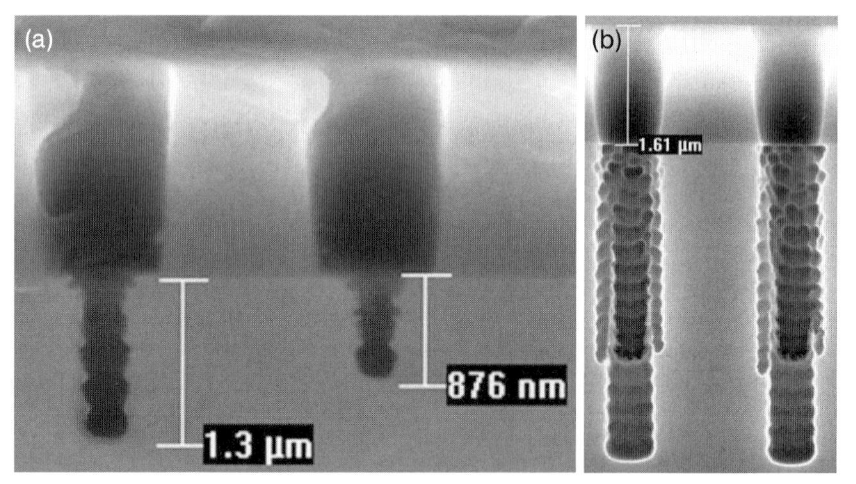

Figure 4.14 Stages upon extended "Bosch" etching through an insufficiently pre-etched oxide mask layer at the bottom of openings with nominal 1 μm diameter: (a) initial break-through and (b) after complete consumption of the annular oxide mask bottom.

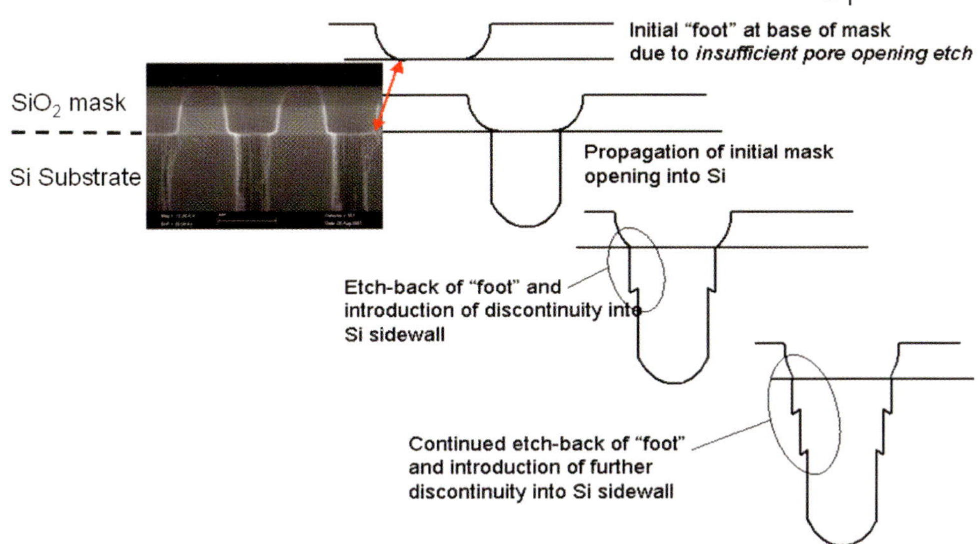

Figure 4.15 Schematic view of striation formation due to retracting residual mask shadowing.

ion particles through the vertical passivation layer and further etched laterally by neutral particles (Figure 4.14b). Figure 4.15 illustrates the entire process.

If the wall passivating Teflon polymer is too thin or too poorly sticking onto the silicon to protect the sidewall, the chemical etching in lateral direction by neutral fluorine particles will be able to create the cavities shown in Figure 4.16, often referred to as "mouse bites".

Especially in via profiles with too large scallops this can also take place at the concave surfaces right below the scallops since these are shadowed by the overhanging scallops and are more sensitive to chemical-induced etching by laterally diffusing radicals (see the "key holes" in Figure 4.17).

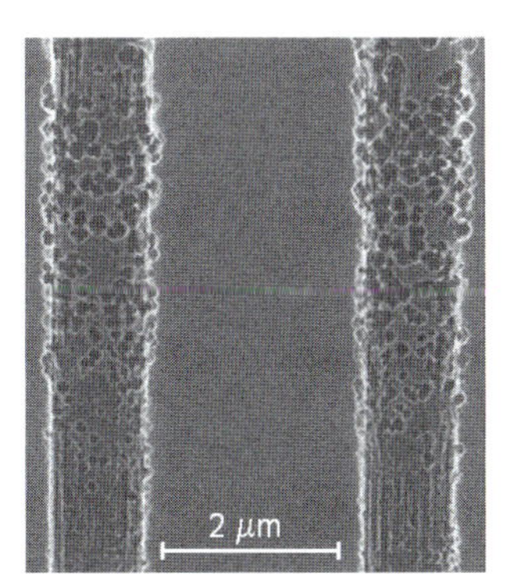

Figure 4.16 SEM image of mouse bites formed along the via sidewalls.

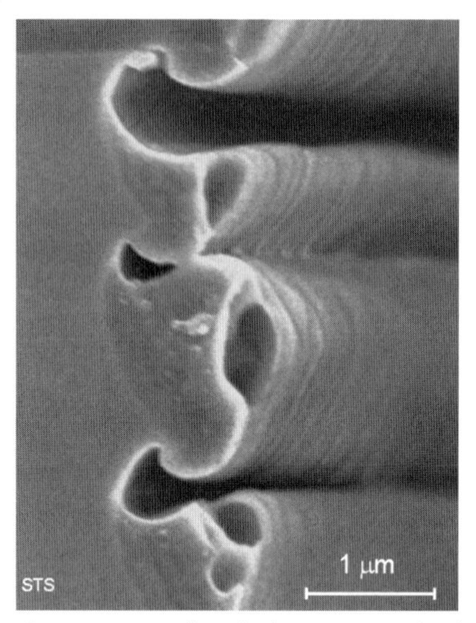

Figure 4.17 Laterally etched cavities immediately below large "shadowing" scallops. Courtesy from STS.

Mouse bites can pose serious problems in the via filling processes. An example is the formation of pinholes in insulating, barrier and seeding layers, which can result in leakage currents. Thus, mouse-bitten sidewalls should be smoothened first. Some smoothening techniques are described in Section 4.4.2.

4.3.2
High Aspect Ratio Features

Aspect ratio dependent etching (ARDE, "etch rate decreasing with etch time") and the related RIE lag ("lower etch rate in smaller features") is a normal phenomenon appearing in most etching processes [33]. Especially under aggressive conditions, the etch rate is often limited by the transport of etching species to and of reaction products from the base of the structure. This effect can be reduced by several approaches but typically at the expense of etch rate, selectivity or profile control [29].

4.3.2.1 **Ion-Limited and Radical-Limited Regimes**

With increasing ion-to-radical flux ratio, the etching shifts from an ion-limited regime to a radical-limited regime if the etching is described by the ion–neutral synergy model. In the ion-limited regime, the etch rate is proportional to the ion flux and it is unaffected by the radical flux. In the radical-limited regime it is the other way around.

In high aspect ratio structures, the ion flux and the radical flux decrease as a function of the aspect ratio because of sidewall collisions. A decrease of either flux has repercussions on the etch rate. It can be assumed that the ion flux is well-collimated whereas the radical flux has an isotropic distribution. As a result, the decrease of the ion flux is much smaller than the decrease of the radical flux in high-aspect-ratio structures. Therefore, ARDE is the most prominent in a radical-limited regime.

4.3.2.2 **Knudsen Transport**

Knowledge of the transport mechanisms of ions and radicals in high aspect ratio structures is needed to clarify ARDE. However, it is difficult to separate the effects of ions and radicals on the etch rate in an anisotropic silicon etching experiment. To circumvent this problem, in a dedicated experiment, silicon lines were fabricated on a Silicon-on-Insulator (SOI) wafer with their orientation parallel to the substrate surface. They are covered by silicon dioxide to exclude the ion flux while the fluorine atoms reach the silicon from one of the sides. As a result, the radical transport mechanism can be investigated separately. In this experiment, the etching is fully chemical because the etching is performed at +25 °C. A hollow trapezoid-shaped silicon dioxide tube remains on the substrate when the silicon is removed from within.

In Figure 4.18 the etch rate is plotted as a function of the aspect ratio of the remaining silicon dioxide tube. The transport of radicals is dominated by sidewall collisions if the mean free path is much larger than the feature dimensions. It is called Knudsen transport if the scattering of the radicals on the sidewalls is diffuse. The Knudsen model is fitted to the data, showing a good correlation [34]. The fitted parameters are the initial etch rate and the reaction probability, which is 0.47 for fluorine. The etch rate obtained from etching a planar silicon substrate is also plotted (♦, zero aspect ratio). It shows that the ion impact has a negligible influence

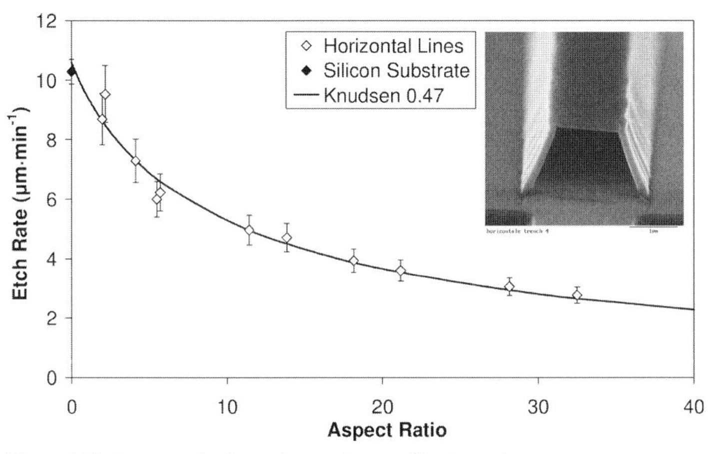

Figure 4.18 Aspect ratio dependent etch rate of horizontal silicon lines that are parallel to the substrate surface. These are covered with silicon dioxide to exclude the ion flux. It shows that the radical transport is limited by Knudsen transport. Inset: trapezoid-shaped silicon dioxide cover with ~1 µm top width, remaining on the substrate after etching. From Reference [29].

on the etch rate for these process conditions because it coincides exactly with the fitted curve.

4.3.2.3 Aspect Ratio Dependent Etching and Aspect Ratio Independent Etching

The etch rate is reduced in the cryogenic etching process if the temperature is lowered and if oxygen is added to the fluorine plasma. Ion impact is needed to continue etching perpendicular to the substrate surface. In the ECR reactor the ion flux was manipulated by placing a quartz cylinder around the substrate, which blocks the direct path between the source region and the substrate (Section 4.2.1.2). Effectively, a remote plasma source is created. The density of the radicals is only slightly influenced because the lifetime of the radicals is much larger.

Figure 4.19 shows the etch rate as a function of the aspect ratio for a high ion flux and for a low ion flux caused by inserting the quartz cylinder. Although the initial etch rate is higher, the etch rate decreases strongly with aspect ratio in the case of a high ion flux. For a low ion flux the etch rate is essentially constant. The process can be tuned from ARDE to Aspect Ratio Independent Etching (ARIE) by adjusting the ion flux. This behavior can be understood by assuming that a transition takes place from a radical-limited regime to an ion-limited regime. However, the transition is accompanied by a reduction of the etch rate because the reaction probability of the etching species and the passivating species is reduced in the ion-limited regime.

The chemically-enhanced ion-neutral synergy model, which was developed to describe the cryogenic etching process, fits well to the data [35]. The model takes into account three species: ions, etching species, and passivating species, in contrast

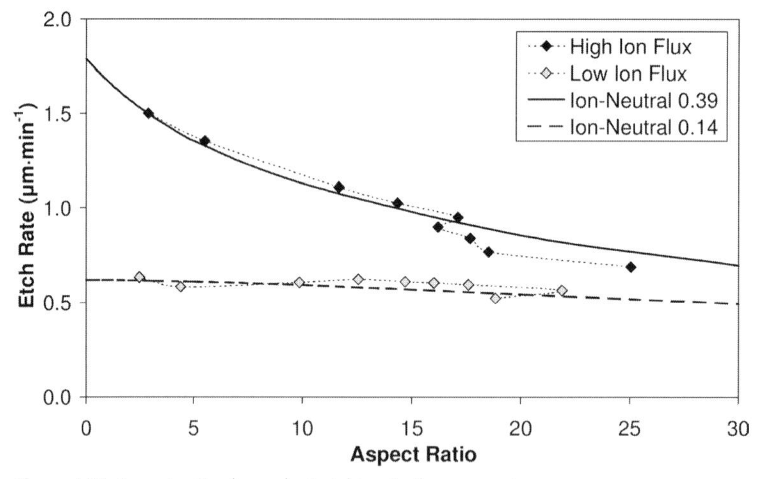

Figure 4.19 Aspect ratio dependent etching in the cryogenic etching process can be turned into aspect ratio independent etching by reducing the ion flux at the expense of a reduced etch rate. Results have been fitted by a chemically enhanced ion-neutral synergy model, with fitted reaction probabilities of 0.39 and 0.14. From Reference [29].

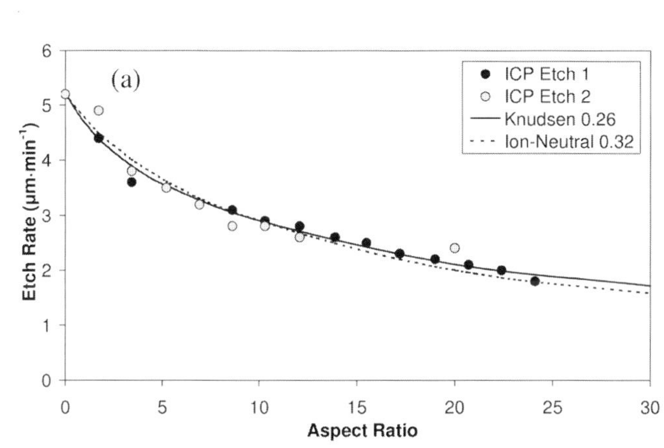

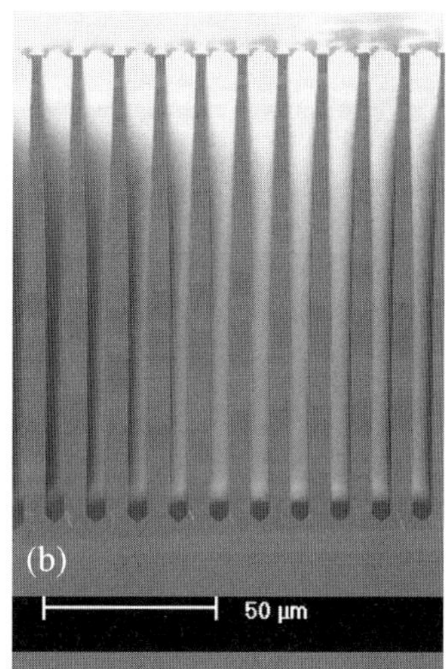

Figure 4.20 Results obtained by cryogenic etching process in an ICP reactor. (a) ARDE is largely caused by Knudsen transport of fluorine radicals. (b) Anisotropic trench profiles. From Reference [29].

to the standard ion–neutral synergy model, which only includes ions and etching species. The name of the model arises from the fact that the interaction between the ions and the passivating species follows ion–neutral synergy behavior and that the etch rate is enhanced by the chemical reaction of the etching species. In the model, it is assumed that the fluxes of the etching species and the passivating species are determined by Knudsen transport and that the ion flux is not aspect ratio dependent.

ARDE was also obtained with the cryogenic etching process in the ICP reactor. The etch rate was measured *in situ* with laser interferometry. It is shown in Figure 4.20 as a function of aspect ratio together with the final trench profile. Both the Knudsen transport model and the chemically enhanced ion–neutral synergy model are fitted to the data, which shows that the decrease of the etch rate is mainly caused by the decrease of the fluorine flux.

4.3.2.4 **Depassivation in the Time-Multiplexed Process**

In the time-multiplexed process, trench profiles tend to taper positively, leading to an etch stop when the sidewalls converge. This prevents the aspect ratio exceeding a certain value. Figure 4.21 shows the etch rate as a function of the aspect ratio for three different ion fluxes. The passivation was adjusted to the ion flux as well to maintain the right balance between etching and passivation. The decrease of the etch rate is largely caused by the Knudsen transport of the radicals [36, 37]. The remarkable turning point in the curves, which defines the maximal achievable aspect ratio, shifts to higher aspect ratios for higher ion fluxes. An increase of the ion flux thus leads to

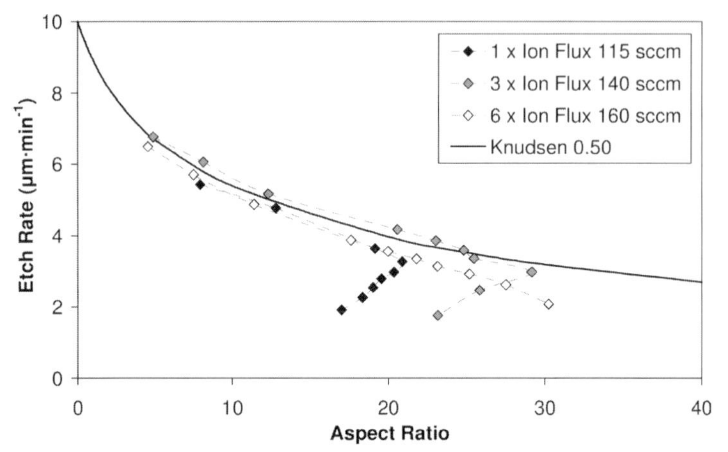

Figure 4.21 Knudsen transport of radicals largely determines the aspect ratio dependent etching for the time-multiplexed process. The maximum obtainable aspect ratio increases with the ion flux. From Reference [29].

better profile control. This is shown in Figure 4.22, which gives the trench profiles for a 1.5 μm wide mask aperture. The figure clearly shows that a ~6× higher ion flux leads to a better anisotropy.

A similar improvement of the maximal achievable aspect ratio is obtained for higher ion energies and lower reactor pressures. In general, higher ion-to-radical flux ratios and better collimation of the ion flux yield better results. The trend is caused by the increased removal of the polymer passivation layer from the base of the trench for these process conditions.

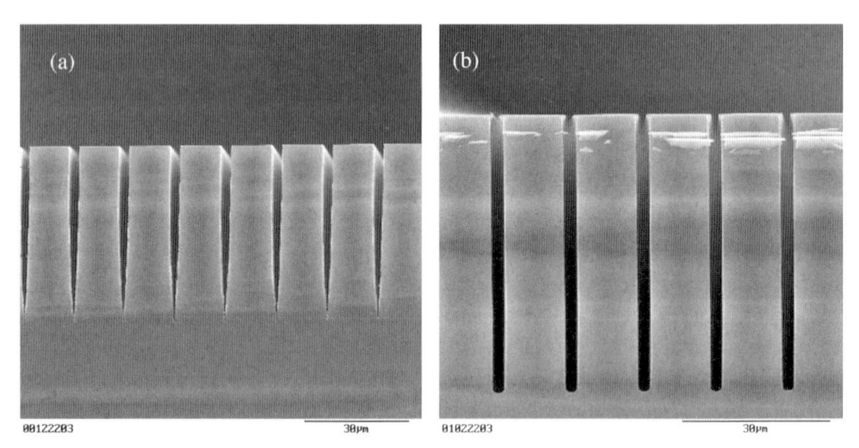

Figure 4.22 An increased ion-to-radical flux ratio improves the anisotropy. Trenches obtained with (a) a regular Bosch process and (b) 6× higher ion flux (see Figure 4.21). Marker is 30 μm. From Reference [29].

4.3.2.5 Aspect Ratio Dependent Etching in the Triple-Pulse Process

In the standard time-multiplexed etching process, the etching and the passivation are optimized independently. However, it was shown that depassivation also plays an important role in the optimization of the etching process. Therefore, all three subprocesses in deep anisotropic silicon etching (etching, passivation and depassivation) are optimized independently in the so-called triple pulse process by adding a third pulse to the standard time-multiplexed etching process [38]. This extra plasma pulse consists of a low-pressure oxygen plasma, which efficiently removes the polymer passivation layer from the base of the feature.

The sidewall scallops are more pronounced in the triple pulse process because of the efficient removal of the polymer passivation layer from the base of the trench. For the same reason, the etching starts immediately upon switching to the etching pulse. In one experiment, the etch rate was calculated from the distance between two consecutive scallops. The instantaneous etch rate is plotted in Figure 4.23 as a function of the aspect ratio for a trench with a 1.5 μm wide mask aperture. The decrease of the etch rate is well described by Knudsen transport of the radicals. This is expected because of the purely chemical character of the etching pulse.

Efficient removal of the polymer passivation layer may also be a solution to avoid the mouse-bitten collars caused by a too thick polymer passivation layer (Section 4.3.1.2).

4.3.3 Sidewall Passivation, Depassivation and Profile Control

Anisotropic profiles are obtained under the appropriate process conditions. Finding the right balance between etching and passivation is crucial. However, shifting of the balance to either side can be used to fine tune the profile, e.g., to change the sidewall

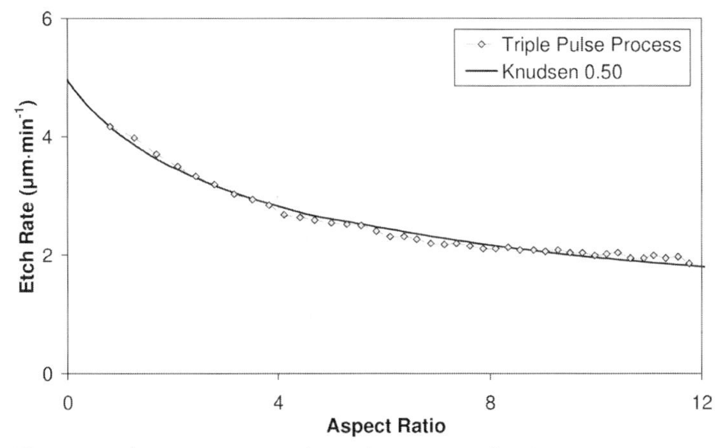

Figure 4.23 The instantaneous etch rate decreases as a function of the aspect ratio. The decrease is caused by depletion of radicals, which is described by the Knudsen transport model. Data are derived from the distance between the sidewall ripples. From Reference [29].

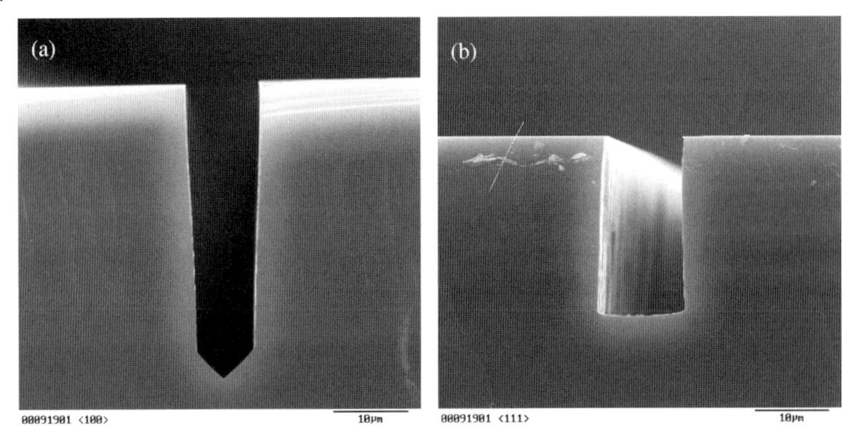

Figure 4.24 Trench profiles etched by cryogenic etching. (a) Si(100) wafer showing crystal orientation dependent etching at the base of the trench. (b) In a Si(111) wafer no crystal orientation dependent etching is visible and the sidewall passivation is reduced. Marker is 10 μm. From Reference [29].

taper from negative to positive. Different approaches are used for the various etching processes.

4.3.3.1 Cryogenic Etching

In cryogenic etching, the lateral etch is minimized by modifying the oxygen flow rate and the substrate temperature. The etch rate and the sidewall taper can then be adjusted by modifying the ion flux and the ion energy. Another characteristic of the cryogenic etching process is that the etch rate is crystal orientation dependent. Trench profiles that were etched simultaneously in a Si(100) wafer and a Si(111) wafer are shown in Figure 4.24. In the Si(100) wafer a facet is formed at the base of the trench whereas no facet is visible in the Si(111) wafer. The etch rate of the Si(111) crystal planes is lower than that of the Si(100) crystal planes since the former planes are more densely occupied by atoms than the latter. Therefore, the etch front in the Si(100) wafer will advance up to the Si(111) crystal planes, which have the lowest etch rate. The exposed planes are not exactly the Si(111) planes, which have an angle of 54.74° with the substrate (100) surface. An explanation could be that the etch yield depends on the ion angle. If the etch yield increases with increasing angle of incidence the crystal planes with the lowest etch rate have a lower angle towards the substrate surface.

4.3.3.2 Room Temperature Etching

For the room temperature time-multiplexed etching process the sidewall tapering has been shown to be strongly passivation-induced. The sidewall taper can even shift from slightly negative to positive. Figure 4.25 shows some trench profile examples for increasing C_4F_8 passivation gas flow.

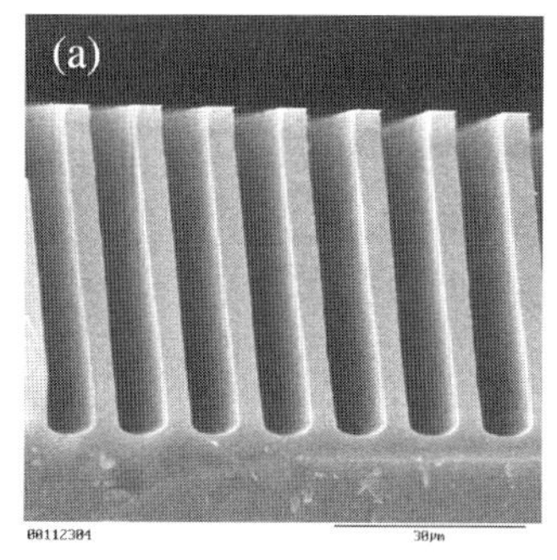

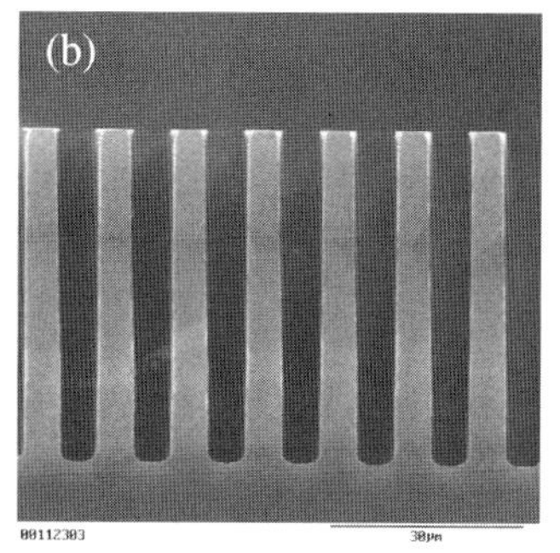

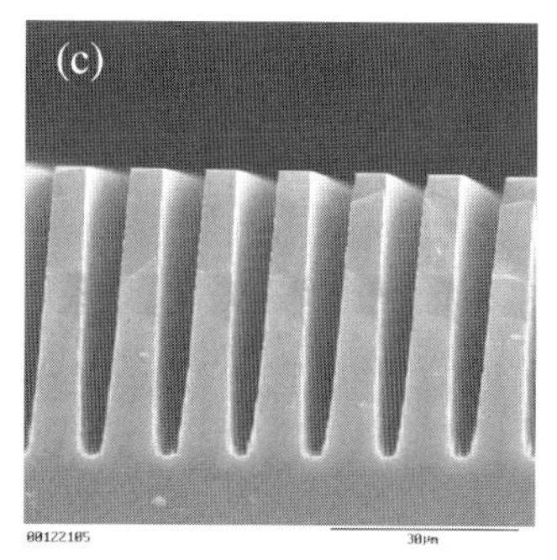

Figure 4.25 The sidewall taper of the trench profile can change from slightly negative to positive for an increasing C_4F_8 flow: (a) 95, (b) 115 and (c) 135 sccm. Note: images (a) and (c) are not taken in full front view but the negative and positive tapering within the trenches can clearly be seen. Marker is 30 μm. From Reference [29].

4.3.3.3 O_2 Plasma Triple-Pulse Process

The fluorocarbon polymer layer is efficiently removed in the triple pulse process due to the low-pressure oxygen plasma. The depassivation rate can be 3× higher than in the standard time-multiplexed process. It leads to a well-defined etching pulse but no improvement of the maximally obtainable aspect ratio was achieved. This is caused by increased sidewall erosion during the depassivation pulse. For the same reason, a negative sidewall taper is obtained if the sidewall erosion is not sufficiently compensated during the passivation pulse. An example is shown in Figure 4.26.

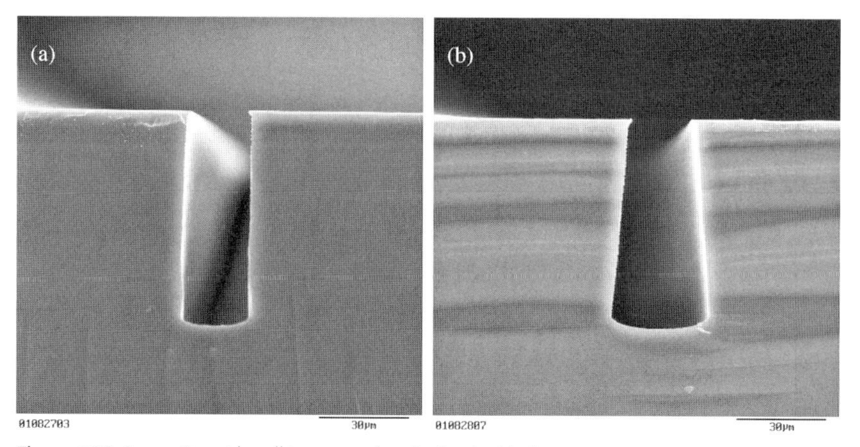

Figure 4.26 A negative sidewall taper can be obtained with the triple pulse process by increasing the depassivation pulse time from (a) 2 to (b) 4 s. Marker is 30 μm. From Reference [29].

4.3.3.4 **Expanding Thermal Plasma Etching**

Anisotropic via profiles are obtained with the ETP reactor using both the cryogenic etching process and the time-multiplexed etching process [22]. The results are very similar to those obtained with the ICP reactor although the plasma chemistry is completely different. The etch rate in the ETP reactor increases strongly with increasing pressure, increasing source current, and decreasing source-to-wafer distance. Lateral etching is minimal as a result of good substrate temperature control. Figure 4.27 shows anisotropic via profiles for three different process conditions using the time-multiplexed etching process and RF substrate bias. For the initial conditions, the etch rate was 5.9 $\mu m\, min^{-1}$ and the selectivity towards the silicon dioxide mask was 127. The etch rate increases to 10.0 $\mu m\, min^{-1}$ upon doubling the reactor pressure. An increase of the source current from 75 to 90 A leads to an etch rate of 7.7 $\mu m\, min^{-1}$ but at the expense of selectivity. Preliminary experiments with the ion-energy-specific, pulse-shaped substrate bias showed that the selectivity can be increased more than 3× while maintaining high etch rates and anisotropic profiles.

4.3.3.5 **Sidewall Taper Control**

Traditionally, people aim at perfectly straight through-wafer vias for a large scope of applications, including 3D integration. However, the use of high aspect ratio vias not only complicates DRIE reaction rates by limiting Knudsen diffusion. For the same reasons it also limits subsequent via filling processes once the vias have been etched. Recently, it has been recognized that easy access for lining and filling of via interconnect holes can be greatly facilitated by etching vias with a tapered shape [39]. As described previously, the right balance between etching and passivation is of great importance regarding the undercut, mouse bites or striation effects. Below, we

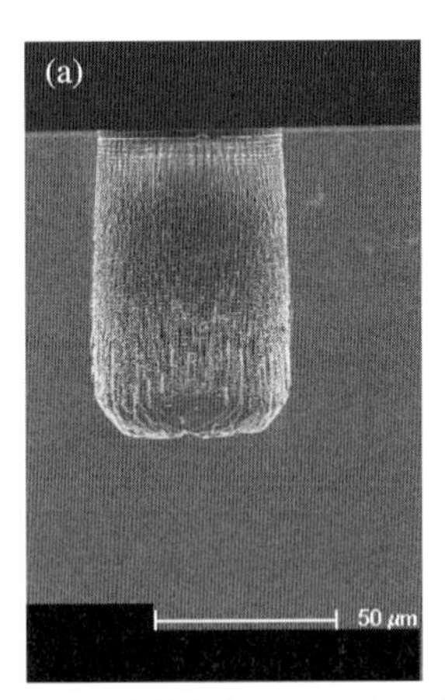

Figure 4.27 Anisotropic via profiles obtained with a time-multiplexed process in the ETP reactor. (a) Initial conditions, (b) higher reactor pressure and (c) higher source current. From Reference [20].

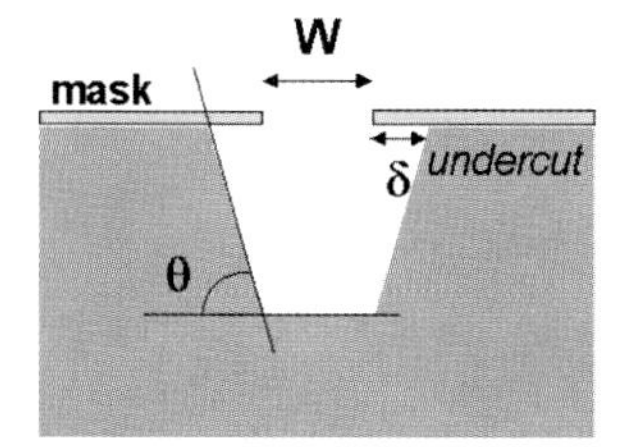

Figure 4.28 Scheme of tapered via parameters. *W* is the nominal width of the mask opening, δ is the undercut, and θ the slope angle resulting from the balance of etching and passivation.

discuss how the balance of etching and passivation can also be used to directly control the slope of via walls. This leads to a new type of vias, as illustrated in Figure 4.28.

Modification of the Bosch Process Tapered vias can be obtained in a straightforward way by modifying the ratio of passivation and etching cycle times. By continuously ramping up the passivation time (or ramping down the etch time), a tapered via shape can be obtained [39–41]. The etch cycle becomes increasingly shorter and the effect of lateral isotropic etching inside the via is progressively reduced. Therefore, the via will be slightly tapered.

Another way to obtain tapered vias with the Bosch process is to insert an additional isotropic etch. This extra etch step is responsible for the tapering of the vias. The slope is controlled by the cycle time of this third step, which is ramped down during the process [42]. Via shapes resulting from this approach are given in Figure 4.29 together with the corresponding process parameters. The figure shows that the slope angle θ can be tuned in a range from 60 to 85°. The etch rate is similar to those obtained with a classical Bosch process. Moreover, this third step smoothens the scallops due to the sequential process.

Continuous Processing The balance of etching and passivation is even more important in the continuous etching mode. This was the first mode used in the early days of RIE processing before the introduction of the Bosch process. Continuous processing has been extensively studied, especially for the CMOS interconnect metal vias (with depths ranging from a few hundred nm to a few μm), but is now also scaled up and adapted for through-wafer via etching. Passivation and etch gases are injected together in the plasma chamber and the competition between these gases results in a tapered via. The etching is still performed by fluorine-containing plasma (SF_6, CHF_3, etc.), whereas the passivation is performed by carbon-containing gas and oxygen. Extensive literature is available on conventional RIE etching and several models such as the fluorine-to-carbon model [43] or the black-silicon method [44] describe the competition on their respective effects and their influence on the slope angle. Modern deep etch tools can now provide very high-density plasmas. This makes continuous mode processing competitive again for through-wafer vias, especially for tapered vias. In this case, slope control can be carried out in several ways:

	Anisotropic etch step	Anisotropic passivation step	Subsequent isotropic etch
Time	9 s	7 s	t = 30-60 s
Step overlap	1 s	0.5 s	N/A
Gas flow	150 sccm SF_6	100 sccm C_4F_8	100 sccm SF_6
Pressure	APC[a] fixed: 60°	APC fixed: 60°	15 mTorr
Coil power	600 W	600 W	600 W
Platen power	12 W	0 W	0 W

[a] Automatic pressure control (APC) deactivated and exhaust valve set at a fixed angle

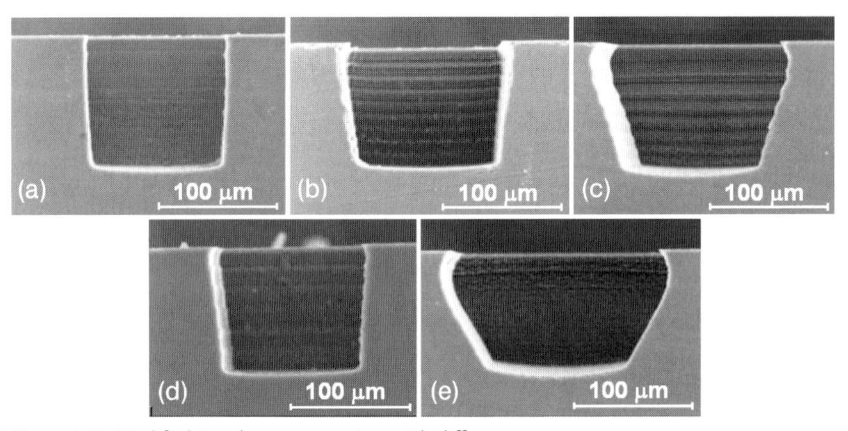

Figure 4.29 Modified Bosch process recipes with different number of cycles followed by an extra isotropic etch and corresponding tapered vias obtained. (a) 87°, 13 Bosch cycles only, (b) 82°, 13 Bosch cycles, 30 s isotropic etch, (c) 72°, 13 Bosch cycles, 60 s isotropic etch, (d) 85°, 21 Bosch cycles, 30 s isotropic etch and (e) 60°, 5 Bosch cycles, 60 s isotropic etch. After Roxhed *et al.* [42].

- Slope control with carbon-containing gas: Also in the continuous process, the balance of etching and passivation can be controlled by the amount of passivation gas, as reported for C_4F_8 [32] (Figure 4.30). An increase in passivation gas increases the passivation locally and reduces the inclination of the wall. Again, other parameters like pressure or platen power can also play a role in the global shape of the via. The etch rate reported was about ~3.5 μm min^{-1}.

- Slope control with oxygen concentration: By changing the O_2-fraction in the total gas flow one can change the local oxidation on the vias walls. This results in a passivation of the silicon surface with the growth of an $Si_xO_yF_z$ passivation film [45]. This modifies the lateral etch rate and the inclination of the wall (Figure 4.31). The pressure can also be a relevant parameter to tune the balance between the etching and the passivation. However, the etching conditions are highly dependent on the RIE tool. Note also that the selectivity and the etch rate will both decrease dramatically with O_2 plasma concentration.

Figure 4.30 Example of a tapered via obtained with a continuous process using SF_6 etching gas (260 sccm) competing with C_4F_8 passivation gas (150 sccm). Top width is ~100 μm. From Reference [32].

- Control of the slope angle with temperature: In general, temperature is a very important parameter controlling the passivation. Indeed, the deposition and removal of SiO_xF_y in continuous RIE mode are very often kinetically limited. As described in Section 4.3.3.1, low temperatures (<0 °C) lead to a better passivation

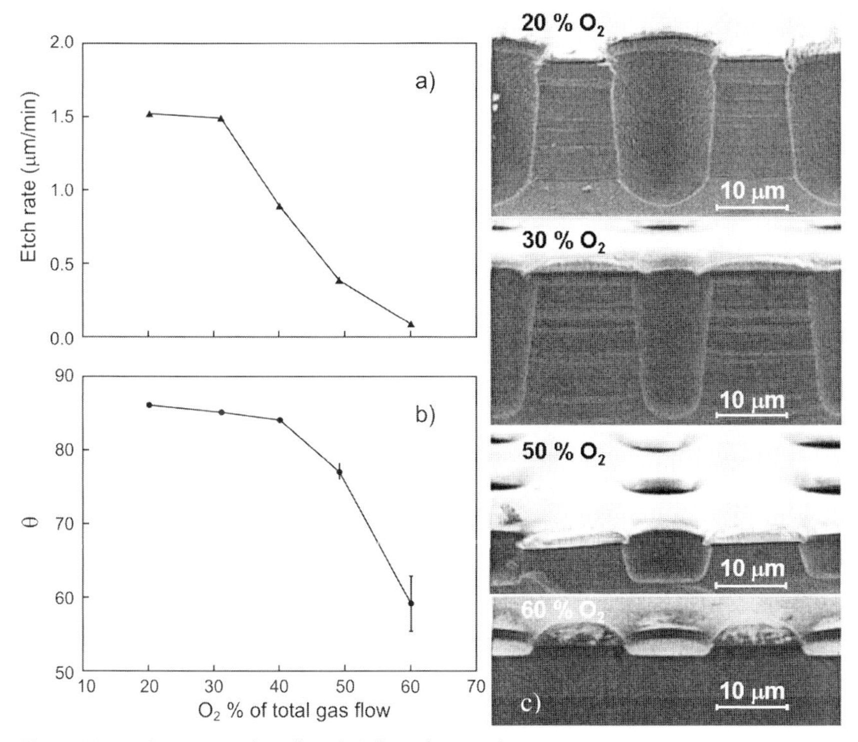

Figure 4.31 Etch rate (a), side wall angle θ (b) and (c) via shape as a function of O_2-fraction of total gas flow. After Figueroa *et al.* [45].

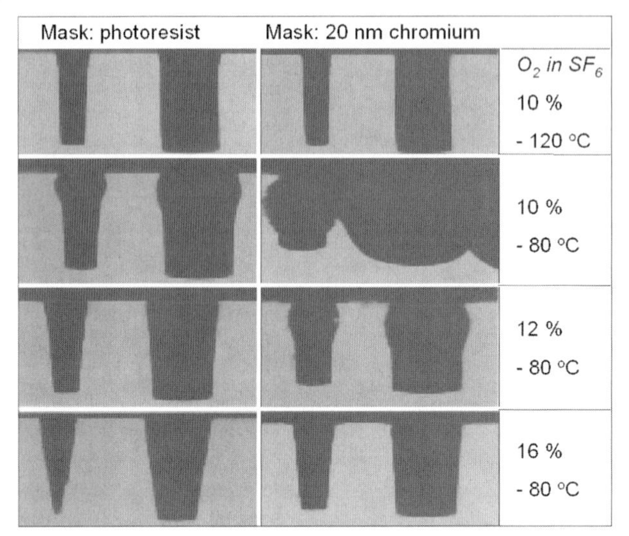

Figure 4.32 Example of tapered vias obtained with a cryogenic continuous process for different mask materials (Olin 907-12 resist and chromium), O_2-fractions in SF_6 and process temperatures. Nominal trench widths are 4 and 8 μm. After de Boer *et al.* [46].

than high (or room) temperature processing. Modern cryogenic deep reactive ion etching allows etching in a very wide temperature range (−120 to 20 °C). It is possible to enhance or decrease the passivation with the wafer temperature [46]. As a general rule of thumb, for a given via shape, the lower the temperature the lower the required O_2 content, all other parameters remaining constant. Depending on the cryogenic DRIE recipe a wide variety of shapes can be obtained for vias or trenches (Figure 4.32). One should also notice the different shapes achieved upon using different mask materials.

- Combining Bosch and continuous processing: Some authors tried to combine Bosch and continuous processing modes sequentially to take advantages of the two techniques [47]. Although it shows interesting results regarding the via shape and etch rate, the basic principles of passivation remain the same in each of the two modes. However, we note that aspect ratios as high as 6 and etch rates of $\sim$3.5 μm min^{-1} can be obtained for tapered vias. This could imply a step further in pushing the limits for through-wafer vias and 3D integration.

4.4 Practical Solutions in Via Etching

This section briefly considers a few recent developments and practical measures that can greatly enhance the quality and quality control of the via shape and dimensions.

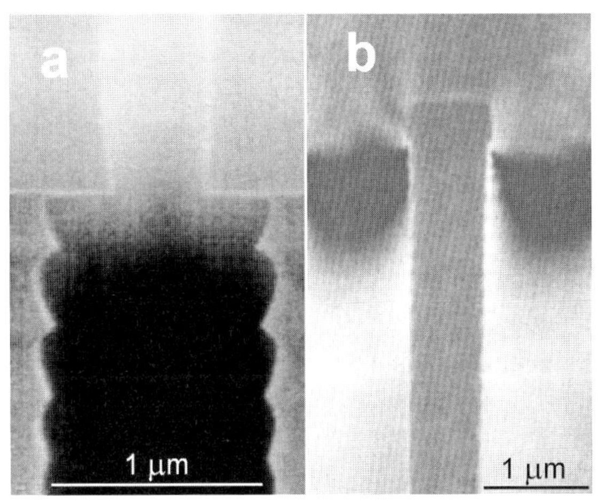

Figure 4.33 Vias etched without (a) and with (b) advanced parameter ramping, resulting in reduced mask undercut and scalloping. Courtesy of STS.

4.4.1 Undercut and Scallop Reduction

The newer generations of dry-etch tools contain advanced gas switching valves with actuation times below 20 ms, mass flow controllers with fast response and internally purged gas injectors. These additions in combination with more efficient etch chamber design and decoupled plasma sources enable faster ramping and switching. Figure 4.33 illustrates how the initial undercut can be reduced from typically 0.3 to 0.1 μm and the scalloping from ~0.2 μm to 40 nm.

4.4.2 Sidewall Roughness Minimization

The roughness of the feature walls is mainly defined by the scallops and mouse bites as described in Section 4.3, where we discussed a few process and mask related measures to smoothen the via surfaces. Here we describe briefly several post-etch methods, both wet- and dry-chemical, for further smoothening.

4.4.2.1 Wet-Etch Smoothening

Via sidewalls can have several μm of roughness that can be removed relatively easily with wet-etch solutions. A classical, rather aggressive one is an isotropically etching mixture of HF, HNO_3 and acetic acid. Its high etch rate of 13 μm min^{-1} at room temperature and low selectivity towards SiO_2, Al and photoresist make it less suitable as a controllable 3D integration process [32].

A typical polysilicon etch (mixture of HF, HNO_3 and H_2O) has a much milder etch rate of 0.35 μm min^{-1} and good selectivity towards SiO_2, but a poor selectivity towards all.

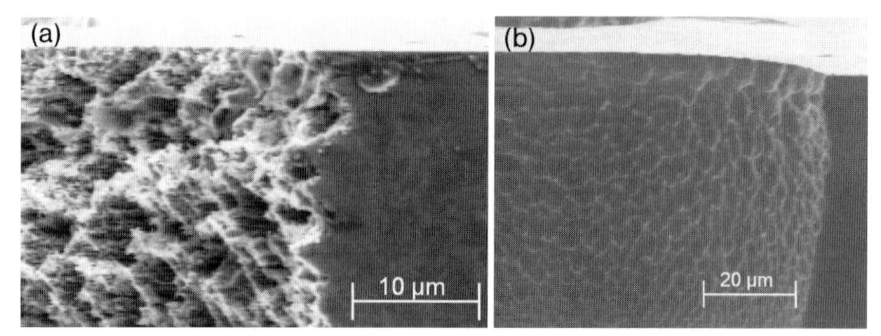

Figure 4.34 Smoothening effect of isotropic dry-etching with SF_6. (a) A non-optimum sidewall and (b) after 10 min of etching the sidewall. After Reference [32].

The last smoothening wet-etchant is a hot KOH solution. It is known as an anisotropic etchant with high selectivity towards SiO_2. However, it is not considered as fully compatible with silicon processing.

4.4.2.2 Dry-Etch Smoothening

Smoothening by dry-etching with SF_6 has also been reported. Figure 4.34 shows a typical example of how a very rough sidewall is improved after 10 min of dry-etching in pure SF_6 plasma at low power. In this case the smoothening was carried out in the same dry-etch tool directly after the via etching [32].

4.4.3 Loading Effects

The etch rate can be slowed in several ways, that is, on a local scale (feature size or chip size) and on a global scale (wafer size). Depending on the supply of etchant species and the exposed silicon area this can have an adverse impact on the etch depth uniformity of the feature size across a die or a wafer, respectively. The reduction of the etch rate with increasing exposed area is referred to as the "loading" or pattern density effect. The loading phenomenon was first modeled by Mogab [48] and can be explained by the depletion of the etchant by the substrate material. This is most pronounced for high-speed etching where chemical etching is the dominant reaction mechanism. Depending on the etch recipe and the reactor configuration the Bosch process with its highly chemical component can be quite sensitive to loading effects. More details were published by Kiihamäki *et al.* [10, 49]. They designed clever nested ring test structures and chip lay-outs to distinguish between RIE lag and loading effects, and to determine feature-scale, chip-scale and wafer-scale loading effects.

The SEM image in Figure 4.35 illustrates a feature-scale microloading effect on a trench array that we etched with an aggressive Bosch etch recipe. The etched structure was a dense array of trenches that we used in high-density trench capacitors [50].

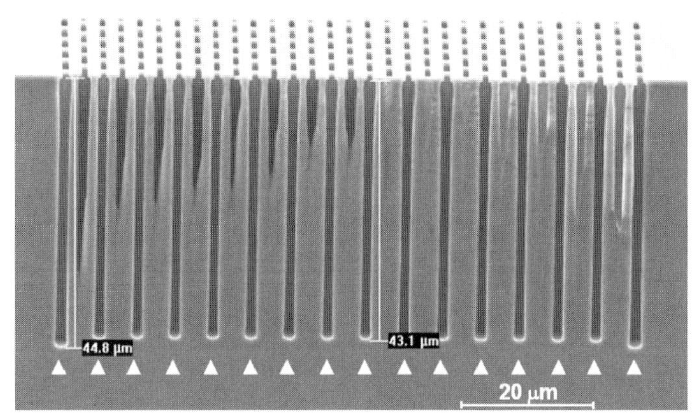

Figure 4.35 Feature-size loading effect on a trench array, causing deeper etching of the outer trenches. Trench width is 1.0 µm, length 10 µm, longitudinal pitch 11.5 µm and transversal pitch 3 µm. Note that only the trenches indicated by open triangles were cleaved correctly and considered here.

In the center of the array the relative exposed area of silicon to be etched is high. In contrast, at the edge of the array the areal density of open silicon is much lower and, consequently, the depletion of reactants in this region will be less. This will cause a higher average etch rate and, therefore, deeper features.

The effect as described above is revealed very locally on the wafer and depends strongly on the pattern density in the mask layout. Therefore, this effect is often referred to as "feature-size loading." The same effect, reduction of the etch rate, can also be observed when the overall open silicon area differs on a wafer scale. Patterns can influence the etch rates of adjacent patterns at distances up to a few mm. In that case it is referred as "wafer-scale loading". The loading effect is further determined by the type of equipment and the etching recipes. One needs to be aware of this effect to develop predictable etching processes.

We characterized this effect for an in-house process, using a "Bosch" etch recipe in an STS Multiplex tool. Arrays of different feature sizes and shapes were etched, varying from round to elongated trenches (Figure 4.36). The density of this pattern was designed per wafer from 5 to 10, 20, and 40% open silicon area. After etching the wafers were inspected by means of cross-sectional SEM to derive the average etch rates of the different features.

These average etch rates are plotted in Figure 4.37 as a function of the aspect ratio. The graphs represent the results for four pattern densities. For each pattern density the four lines show the effect of the size and elongation of the structures. This effect is mainly caused by Knudsen diffusion and is very similar for 5 to 20% exposed area (Figures 4.37a–c). Figure 4.37d with 40% exposed area deviates. Whereas the lines for 5, 10, and 20% loading converge to an initial etch rate of $\sim$4 $\mu m\,min^{-1}$, those in the graph for 40% loading converge to an initial etch rate of $\sim$3 $\mu m\,min^{-1}$. The lines are similar, thus the effect of different geometry of the structures is similar, yet the average etch rate is lower. This is caused by "wafer-scale loading" and it becomes

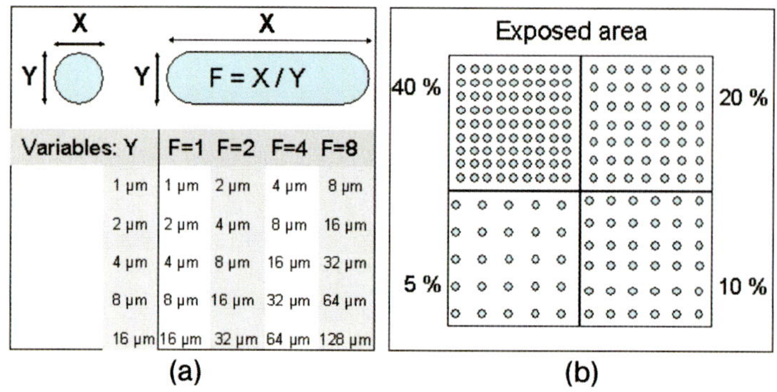

Figure 4.36 Sketch of basic mask design for different feature shapes and sizes (a) and exposed areas (b). See also Figure 4.37.

significant (for this specific process) for mask lay-outs with an exposed area higher than 20%.

Such characterization enables us to predict the etch rates of different structures and pattern densities. This not only accelerates the realization of new designs and concepts but also gives an opportunity to use these differences in etch rates to form complex 3D structures.

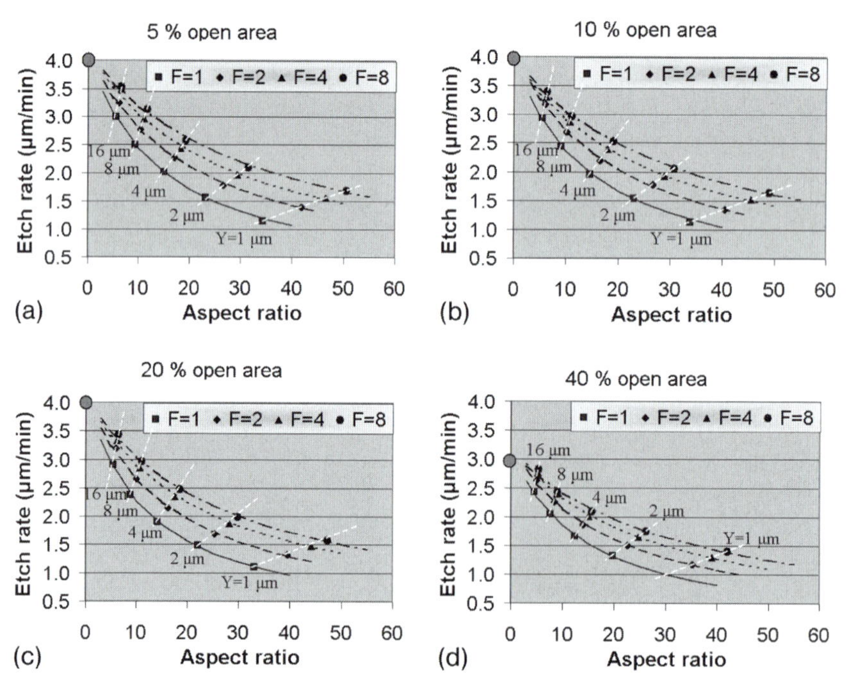

Figure 4.37 Etch rates for different features and exposed areas. For symbols and dimensions see Figure 4.36.

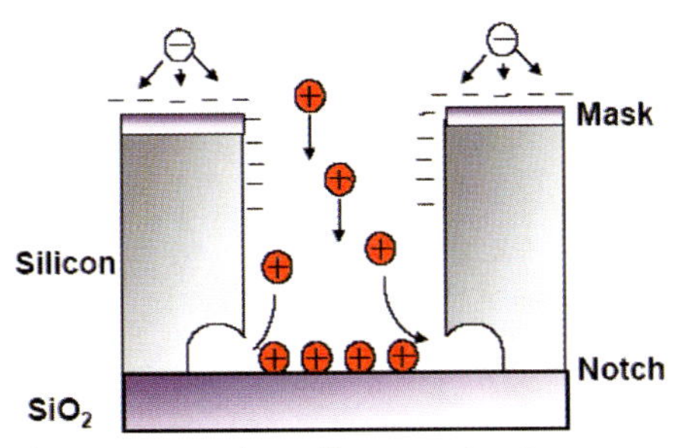

Figure 4.38 Notching effect occurring due to accumulation of charge at the dielectric bottom of high aspect ratio vias, causing deflection of ions to the sidewalls.

4.4.4
Notching at Dielectric Interfaces

An increasingly popular substrate in microelectronics and MEMS manufacturing is Silicon-on-Insulator, SOI. It offers an ideal sharp interface of Si with silicon dioxide that can be used as a very selective RIE-etch stop layer and as a sacrificial layer after RIE. Figure 4.38 illustrates what happens at the bottom of the via hole during conventional RIE etching using continuous biasing at 13.56 MHz. Upon the slightest overetch the dielectric bottom will become charged. Consequently, the incoming ions will be re-directed to the sides and they will attack the sidewalls at the via bottom [51]. This phenomenon has been greatly overcome or eliminated by using a low-frequency bias pulse of 380 kHz [52]. Typically, notching is reported to occur only for aspect ratios above 1 (Figure 4.39).

4.4.5
Inspection of Via Structures

The mainstay in the inspection of via shapes and dimensions is Scanning Electron Microscopy (SEM) imaging of cross-sections obtained by cleaving. For small pitch, small diameter via arrays this is a relatively easy practice. Figure 4.40 shows an example for blind vias etched through a mask with nominally 1.5 μm diameter holes and 2.5 μm pitch. For these fine dimensions the cleaved intersect is automatically an ideal half-circle, across the line A-A in the horizontal plane (Figure 4.41). However, for crude vias of $\geq$20 μm diameter and $\geq$50 μm pitch this is no longer the case. Often

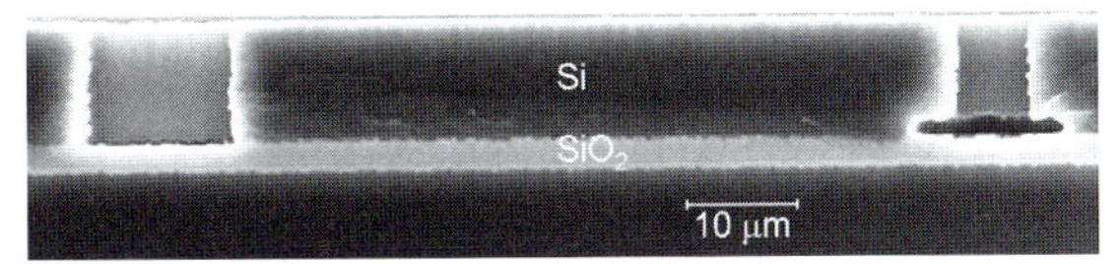

Figure 4.39 SEM image of notching of vias with an aspect ratio higher than unity. After Reference [10].

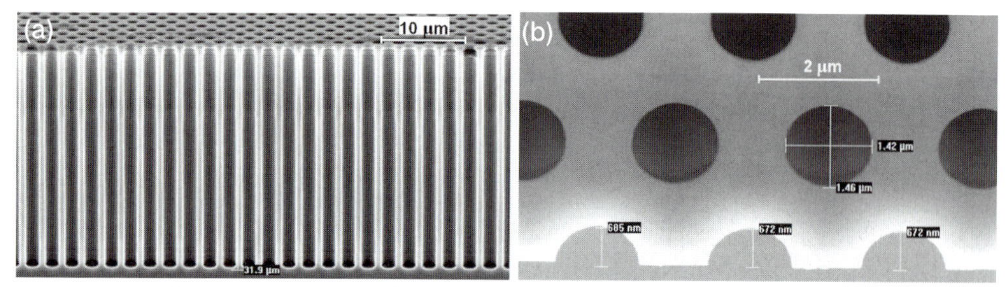

Figure 4.40 SEM images of 32 µm deep via holes etched through a mask with nominally 1.5 µm diameter openings and 2.5 µm pitch; (a) side view (from Reference [50] and (b) top view of cleaved specimen.

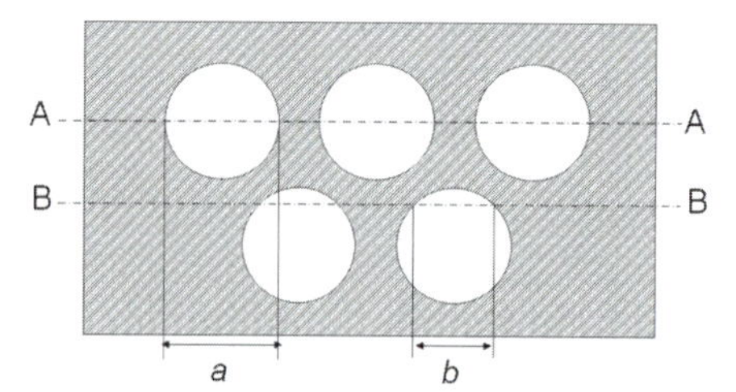

Figure 4.41 Illustration of a perfect intersection (line A-A), yielding the correct via diameter *a*, and of an artifact intersection B-B, yielding a too small via size *b*.

specimens cleave along non-perfect intersections, for example along line B-B in Figure 4.41. As a consequence, this artifact might give unnecessary errors in the measured data for via dimensions and profiles.

To exclude these errors we utilize laser scribing of a pre-cleavage line connecting the exact via centers. The laser is operated to approach the via hole until a distance of ~15 µm (i.e., the spot size of our YAG laser) to the via circumference (Figure 4.42). Thus the laser creates a deep vertical groove by ablation of the Si. This groove will impose the scribed direction upon cleavage of the silicon substrate.

This method can also be used for the inspection of Cu-filled vias (Figure 4.43). Here one can clearly distinguish the laser-ablated areas on the cleaved plane.

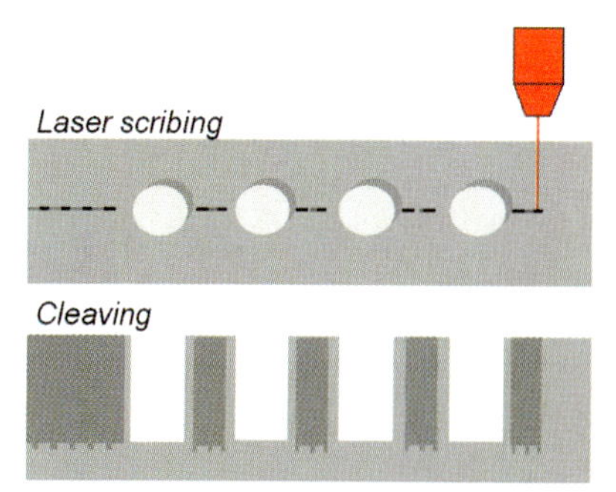

Figure 4.42 Illustration of laser scribing for unequivocal inspection of via dimensions and profile.

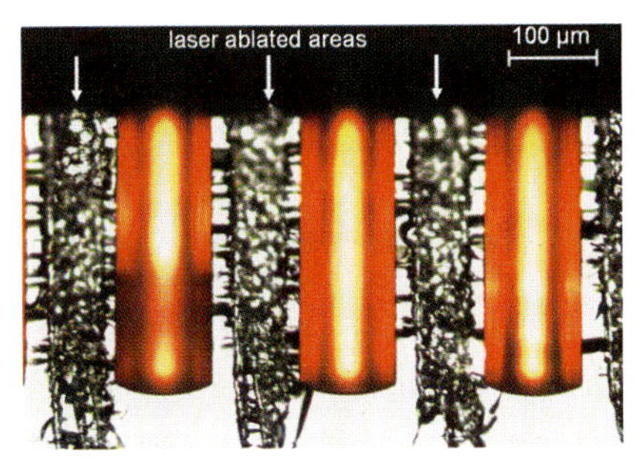

Figure 4.43 Optical micrograph of blind Cu-filled via holes. Cross-section obtained by laser scribing and subsequent cleaving.

4.4.6
In Situ Trench Depth Measurement

Reproducibility of trench size and shape is essential in modern through-wafer via technology. Earlier in this chapter we described how to inspect and improve via features such as sidewall smoothness and profile. Currently, SEM is used as a post-process inspection and process optimization technique. By its nature this is an expensive, labor intensive, invasive and *ex situ* technique used on cleaved samples.

Etch depth control is no problem for SOI wafers. These contain a buried silicon dioxide layer acting as a stop layer that terminates the etching when one uses conventional *in situ* diagnostics such as Optical Emission Spectroscopy. However, for regular silicon wafers the via depth is still typically controlled by plainly programming the process time. In that case one obtains a wafer-to-wafer accuracy of typically 4%. This is owing to non-constant factors such as the condition of the etch chamber and the loading effects due to the different exposed silicon areas generated either intentionally (mask design) or non-intentionally (mask pre-etching).

Several non-invasive methods – preferably optical, *in situ* real-time diagnostics – are being assessed to measure and control the via depth. This is done in an attempt to develop fast, inexpensive techniques that can be used for on-line measurement of via depths in production, as an alternative to elaborate SEM inspection. The methods are all optical (interferometry, and spectroscopic reflectometry and ellipsometry). All methods have their limitations and there seems to be no generic method that covers the wide span of via diameters ranging from a few µm to ~100 µm and depths from 10 to 200 µm for all via densities (~10 mm^{-2} to >1000 mm^{-2}).

One method being developed is interferometric endpoint detection (IEPD) [53]. The IEPD unit is mounted on a viewport of the etch chamber and uses a white light source. The emitted light, with a beam size of ~15 mm, reaches both etched and non-etched areas of the silicon wafer where it is reflected differently. The reflected light is collected and fed through optical fibers into a detector that monitors the oscillating light intensity at a pre-selected wavelength. The optical signal is maximized for output intensity by selecting the wavelength λ that

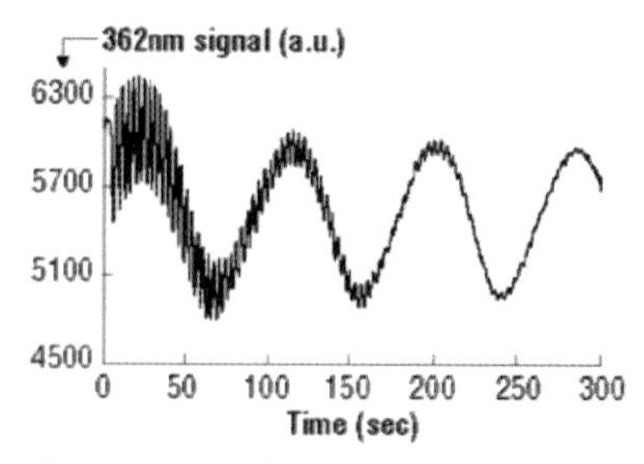

Figure 4.44 Interferometric endpoint detection example of etching a 15 μm deep trench. The ratio of oscillation periods (100 : 3) equals the Si : SiO_2 etch selectivity. From Reference [53].

is optimal for a given set of mask reflectivity, plasma intensity and via depth. Figure 4.44 shows a typical output signal. It is composed of two superimposed oscillations: a low-frequency response with a ~100 s period due to the silicon dioxide mask erosion and a high-frequency oscillation with a ~3 s period representing the actual silicon via etching. The latter oscillation frequency f can then be converted into the etch rate $S = \lambda \cdot f \cdot 60/2$.

The IEPD method is limited, as is clearly illustrated by Figure 4.44. The amplitude of the fine fringes, representing the silicon etching, is reduced rapidly during the trench etching until the point that no fringes can be resolved to derive the silicon etch rate. The trench bottom is then too deep (~10 μm in this case) and too rounded to reflect the incident light into the detector. For larger via depths one has to rely on model-based computing of the etch times based on the measured early etch stages.

The limited reflection of incident light by the via bottoms is a setback for most broadband optical monitoring methods. This might be overcome if one uses two lasers with a small spot size (~20–30 μm) compared to the via size. Pore depth measurement is then based on interferometry where one laser is focused on the surface of the wafer and the other spot is pointed at the via or etched feature. The via depth is measured by the phase shift of the two lasers.

Better results may be obtained using ellipsometry as compared to reflectometry. This is probably because in ellipsometry one does not measure intensity but, rather, a ratio of intensities. This generally causes the precision of ellipsometric measurements to be superior to those of reflectometry.

4.5
Concluding Remarks

In this chapter we have reviewed fluorine-based high-density plasma etching of deep anisotropic microstructures to form via interconnect holes in silicon. The first section introduced the state of the art in DRIE and alternative wet etching of high aspect ratio pores. The next section covered different types of plasma reactors, the plasma chemistry used and the diagnostics and analysis techniques. These techniques have proven useful and necessary in studying the particle fluxes and their interactions with the substrate.

A section on processing then followed, including mask preparation and mask-related issues, transport phenomena of plasma particles in high aspect ratio features and sidewall control. We ended with a section on practical solutions in DRIE to optimize and characterize feature dimensions and smoothness and to better control the DRIE processes.

Acknowledgments

The authors acknowledge the work on via technology done by our partners in the national Senter-Novem projects *EXSTC* and *INNOVia* as well as the European projects *"MICROSPECT"* (FP-5) and *"e-CUBES"* (FP-6). The co-operation and technical data and updates given by colleagues from Surface Technology Systems, UK, are gratefully acknowledged.

The research of Erwin (W.M.M.) Kessels has been made possible by a fellowship of the Royal Netherlands Academy of Arts and Sciences (KNAW).

Part of the experimental RIE work was performed by internship students A. Alami-Idrissi and P.J.W. van Lankvelt.

Appendix A: Glossary of Abbreviations

AFM	Atomic Force Microscopy
ARDE	Aspect Ratio Dependent Etching
ARIE	Aspect Ratio Independent Etching
CCP	Capacitively Coupled Plasma
CMOS	Complementary Metal Oxide Semiconductor
DECR	Distributed Electron Cyclotron Resonance
DRIE	Deep Reactive Ion Etching
ECR	Electron Cyclotron Resonance
ETP	Expanding Thermal Plasma
HDP	High Density Plasma
IAD	Ion Angle Distribution
ICP	Inductively Coupled Plasma
IED	Ion Energy Distribution
IEPD	Interferometric Endpoint Detection
LF	Low Frequency
LPCVD	Low Pressure Chemical Vapor Deposition
MEMS	Microelectromechanical System

MST	Microsystem Technology
OES	Optical Emission Spectroscopy
PECVD	Plasma Enhanced Chemical Vapor Deposition
PR	Photoresist
RF	Radio Frequency
RIE	Reactive Ion Etching
sccm	standard cubic centimeter
SEM	Scanning Electron Microscopy
SiP	System-in-Package
SOI	Silicon-on-Insulator
TEM	Transmission Electron Microscopy
TEOS	Tetraethyl orthosilicate
TMAH	Tetramethylammonium Hydroxide
XPS	X-ray Photoelectron Spectroscopy

Appendix B: Examples of DRIE Recipes

DRIE recipes for ~20–50 μm diameter vias in an STS Multiplex tool.

Recipe A	Etch	Passivation
Target	Vias with 20–50 μm diameter (see Figure 4.12a–d)	
Cycle time (s)	7.0	5.0
Gas flow (sccm)		
C_4F_8	50 (ramp–0.4 sccm min^{-1})	120
SF_6	260	0
O_2	26	0
RF-coil power (W)	2500	800
RF-platen power (W)	20	0
Pressure (mTorr)	94	94
Temperature (°C)	10	10
Total time (min)	80	

Recipe B	Etch	Passivation
Target	Vias with 20–50 μm diameter (see Figure 4.12e)	
Cycle time (s)	7.0	5.0
Gas flow (sccm)		
C_4F_8	50	120
SF_6	260	0
O_2	26	0
RF-coil power (W)	2500	800
RF-platen power (W)	5	0
Pressure (mTorr)	94	94
Temperature (°C)	10	10
Total time (min)	80	

References

1 Laermer, F. and Schilp, A. (March 12, 1996) Method for anisotropic plasma etching of substrates, US Patent 5,498,312 and Laermer, F. and Schilp, A. (March 26 1996) Method of anisotropically etching silicon US Patent 5,501,893.

2 Madou, M.J. (2002) *Fundamentals of Microfabrication: The Science of Miniaturization*, 2nd edn. CRC Press, Boca Raton, USA.

3 Funk, K., Emmerich, H., Schilp, A. *et al.* (1999) A surface micromachined silicon gyroscope using a thick polysilicon layer. Proceedings 12th IEEE Conf. on Micro-Electromechanical Systems, 1999, (MEMS'), Orlando, p. 57–60.

4 Roozeboom, F., Kemmeren, A.L.A.M., Verhoeven, J.F.C. *et al.* (2006) Passive and heterogeneous integration towards a silicon-based System-in-Package concept. *Thin Solid Films*, **504**, 391–396.

5 Aarts, E. and Marzano, S. (2003) *The New Everyday; Views on Ambient Intelligence*, 010 Publishers, Rotterdam, The Netherlands.

6 ITRS Roadmap 2005 edition, Assembly and Packaging; www.itrs.net/Links/2005ITRS/Home2005.htm.

7 Garrou, P.E. and Vardaman, E.J. (March 2006) 3D Integration at the Wafer Level, TechSearch International Report.

8 Bower, C.A., Garrou, P.E., Ramm, P. and Takahashi, K. (eds) (2007) Enabling Technologies for 3-D Integration, *Material Research Society Symposium Proceedings*. 970, Materials Research Society, Warrendale, Pennsylvania.

9 Legtenberg, R., Jansen, H., de Boer, M. and Elwenspoek, M. (1995) Anisotropic reactive ion etching of silicon using SF_6/O_2/CHF_3 gas mixtures. *Journal of the Electrochemical Society*, **142**, 2020–2027.

10 Kiihamäki, J. (2005) Fabrication of SOI Micromechanical Devices, PhD thesis, VTT, Finland, http://www.vtt.fi/inf/pdf/publications/2005/P559.pdf. (accessed on September 2007).

11 Laermer, F. and Urban, A. (2003) Challenges, developments and applications of silicon deep reactive ion etching. *Microelectronic Engineering*, **67–68**, 349–355.

12 Theunissen, M.J.J. (1972) Etch channel formation during anodic dissolution of *n*-type silicon in aqueous hydrofluoric acid. *Journal of the Electrochemical Society*, **119**, 351–360.

13 Lehmann, V. and Föll, H. (1990) Formation mechanism and properties of electrochemically etched trenches in *n*-type silicon. *Journal of the Electrochemical Society*, **137**, 653–659.

14 Lehmann, V. (1993) The physics of macropore formation in low doped n-type silicon. *Journal of the Electrochemical Society*, **140**, 2836–2843.

15 Roozeboom, F., Elfrink, R.J.G., Rijks, T.G.S.M. *et al.* (2001) High-density, low-loss MOS capacitors for integrated RF decoupling. *International Journal of Microcircuits and Electronic Packaging*, **24**, 182–196.

16 van den Meerakker, J.E.A.M., Elfrink, R.J.G., Roozeboom, F. and Verhoeven, J.F.C.M. (2000) Etching of deep macropores in 6 in. Si wafers. *Journal of the Electrochemical Society*, **147**, 2757–2761.

17 van den Meerakker, J.E.A.M. and Mellier, M.R.L. (2001) Kinetic and diffusional aspects of the dissolution of Si in HF solutions. *Journal of the Electrochemical Society*, **148**, G166–G171.

18 van den Meerakker, J.E.A.M., Elfrink, R.J.G., Weeda, W.M. and Roozeboom, F. (2003) Anodic silicon etching; the formation of uniform arrays of macropores or nanowires. *Physica Status Solidi A -Applied Research*, **197**, 57–66 and references therein.

19 Lehmann, V. and Grüning, U. (1997) The limits of macropore array fabrication. *Thin Solid Films*, **297**, 13–17.

20 Blauw, M.A., van Lankvelt, P.J.W., Roozeboom, F. *et al.* (2007) High-rate anisotropic silicon etching with the expanding thermal plasma technique. *Electrochemical and Solid-State Letters*, **10**, H309–H312.

21 van Hest, M.F.A.M., Haartsen, J.R., van Weert, M.H.M. *et al.* (2003) Analysis of the expanding thermal argon–oxygen plasma gas phase. *Plasma Sources Science and Technology*, **12**, 539–553.

22 van de Sanden, M.C.M., de Regt, J.M. and Schram, D.C. (1993) Recombination of argon in an expanding plasma jet. *Physical Review E*, **47**, 2792–2797.

23 van de Sanden, M.C.M., Severens, R.J., Kessels, W.M.M. *et al.* (1998) Plasma chemistry aspects of a-Si:H deposition using an expanding thermal plasma. *Journal of Applied Physics*, **84**, 2426–2435.

24 Wang, S.B. and Wendt, A.E. (2000) Control of ion energy distribution at substrates during plasma processing. *Journal of Applied Physics*, **88**, 643–646.

25 Wang, S.B. and Wendt, A.E. (2001) Ion bombardment energy and SiO_2/Si fluorocarbon plasma etch selectivity. *Journal of Vacuum Science & Technology A*, **19**, 2425–2432.

26 Puech, M. and Maquin, P. (1996) Low temperature etching of Si and photoresist in high density plasmas. *Applied Surface Science*, **100–101**, 579–582.

27 Dussart, R., Boufnichel, M., Marcos, G. *et al.* (2004) Passivation mechanisms in cryogenic SF_6/O_2 etching process. *Journal of Micromechanics and Microengineering*, **14**, 190–196.

28 Coburn, J.W. and Winters, H.F. (1979) Ion- and electron-assisted gas-surface chemistry – An important effect in plasma etching. *Journal of Applied Physics*, **50**, 3189–3196.

29 Blauw, M.A. (2004) Deep anisotropic dry etching of silicon microstructures by high-density plasmas, PhD thesis, TU Delft, The Netherlands, and references therein. www.library.tudelft.nl/dissertations/diss_html_2004/as_blauw_2004. (accessed on September 2007).

30 Coburn, J.W. and Chen, M. (1980) Optical emission spectroscopy of reactive plasmas: A method for correlating emission intensities to reactive particle density. *Journal of Applied Physics*, **51**, 3134–3136.

31 Donnelly, V.M., Flamm, D.L., Dautremont-Smith, W.C. and Werder, D.J. (1984) Anisotropic etching of SiO_2 in low-frequency CF_4/O_2 and NF_3/Ar plasmas. *Journal of Applied Physics*, **55**, 242–252.

32 Tezcan, D.S., de Munck, K., Pham, N. *et al.* (2006) Development of vertical and tapered via etch for 3D through wafer interconnect technology. Electronics Packaging Technology Conference.

33 Gottscho, R.A., Jurgensen, C.W. and Vitkavage, D.J. (1992) Microscopic uniformity in plasma etching. *Journal of Vacuum Science & Technology B*, **10**, 2133–2147.

34 Coburn, J.W. and Winters, H.F. (1989) Conductance considerations in the reactive ion etching of high aspect ratio features. *Applied Physics Letters*, **55**, 2730–2732.

35 Blauw, M.A., van der Drift, E., Marcos, G. and Rhallabi, A. (2003) Modeling of fluorine-based high-density plasma etching of anisotropic silicon trenches with oxygen sidewall passivation. *Journal of Applied Physics*, **94**, 6311–6318.

36 Kiihamäki, J. (2000) Deceleration of silicon etch rate at high aspect ratios. *Journal of Vacuum Science & Technology A*, **18**, 1385–1389.

37 Lai, S.L., Johnson, D. and Westerman, R. (2006) Aspect ratio dependent etching lag reduction in deep silicon etch processes. *Journal of Vacuum Science & Technology A*, **24**, 1283–1288.

38 Blauw, M.A., Craciun, G., Sloof, W.G. *et al.* (2002) Advanced time-multiplexed plasma etching of high aspect ratio silicon structures. *Journal of Vacuum Science & Technology B*, **20**, 3106–3110.

39 Spiesshoefer, S., Rahman, Z., Vangara, G. *et al.* (2005) Process integration for through-silicon vias. *Journal of Vacuum Science & Technology A*, **23**, 824–829.

40 Ayon, A.A., Bayt, R.L. and Breuer, K.S. (2001) Deep reactive ion etching: a promising technology for micro- and nanosatellites. *Smart Materials & Structures*, **10**, 1135–1144.

41 Yeom, J., Wu, Y. and Shannon, M.A. (2003) Critical aspect ratio dependence in deep reactive ion etching of silicon, Proceedings IEEE 12th International Conference on Solid State Sensors, Actuators, Microsystems, 1631–1634.

42 Roxhed, N., Griss, P. and Stemme, G. (2007) A method for tapered deep reactive ion etching using a modified Bosch process. *Journal of Micromechanics and Microengineering*, **17**, 1087–1092.

43 Coburn, J.W. and Winters, H.F. (1979) Plasma etching: a discussion of mechanisms. *Journal of Vacuum Science & Technology*, **16**, 391–403.

44 Jansen, H., de Boer, M., Legtenberg, R. and Elwenspoek, M. (1995) The black silicon method: a universal method for determining the parameter setting of a fluorine-based reactive ion etcher in deep silicon trench etching with profile control. *Journal of Micromechanics and Microengineering*, **5**, 115–120.

45 Figueroa, R.F., Spiesshoefer, S., Burkett, S.L. and Schaper, L. (2005) Control of sidewall slope in silicon vias using SF_6/O_2 plasma etching in a conventional reactive ion etching tool. *Journal of Vacuum Science & Technology B*, **23**, 2226–2231.

46 de Boer, M.J., Gardeniers, J.G.E., Jansen, H.V. *et al.* (2002) Guidelines for etching silicon MEMS structures using fluorine high-density plasma at cryogenic temperatures. *Journal of Microelectromechanical Systems*, **11**, 385–401.

47 Nagarajan, R., Ebin, L., Dayong, L. *et al.* (2006) Development of a novel deep silicon tapered via etch process for through-silicon interconnection in 3-D integrated systems Electronic Components and Technology Conference, pp. 383–387.

48 Mogab, C.J. (1977) The loading effect in plasma etching. *Journal of the Electrochemical Society*, **124**, 1262–1268.

49 Karttunen, J., Kiihamäki, J. and Franssila, S. (2000) Loading effects in deep silicon etching. *Proceedings of SPIE*, **4174**, 90–97.

50 Roozeboom, F., Kemmeren, A.L.A.M., Verhoeven, J.F.C. *et al.* (2005) More than 'Moore': towards passive and system-in-package integration. *Electrochemical Society Symposium Proceedings*, **2005–8**, 16–31.

51 Hwang, G.S. and Giapis, K.P. (1997) On the origin of the notching effect during etching in uniform high density plasmas. *Journal of Vacuum Science & Technology B*, **15**, 70–87.

52 Hopkins, J., Johnston, I.R., Bhardwaj, J.K. *et al.* (Feb. 13 2001) Method and apparatus for etching a substrate US Patent 6,187,685.

53 Thomas, D. (2007) Maximizing power device yield with *in situ* trench depth measurement. *Solid State Technology*, **50** (4), 48–60.

5
Laser Ablation

Wei-Chung Lo and S.M. Chang

5.1
Introduction

High-density 3D stacked LSI technologies have been developed extensively in the past few years. In particular, 3D stacked LSI technology with vertical connections has been studied as one of the most promising packaging technologies, because vertical interconnections provide a very short electrical path and low signal loss. There are two major methods to make the through silicon via (TSV), deep reactive ion etching (DRIE) and laser drilling. This chapter focuses on the status and results related to integration of 3D chip stacking based on laser-drilled technology.

Laser drilling has proven workable for forming TSVs and blind vias [3]. The flexibility of the laser process makes it possible to control via depth, diameter and sidewall slope. Indeed, some companies have announced laser-based TSV in their product application that will be commercialized in the near feature. Samsung's industry-first 3D memory using TSV is a 16 Gbit memory solution that stacks eight 2 Gbit NAND chips. Samsung's wafer-level processed stack package (WSP) introduces a simplified process for the TSV. Instead of using a conventional dry etching method, a tiny laser drills the TSV. This reduces production cost significantly as it eliminates the typical photolithography-related processes required for mask-layer patterning and also shortens the dry-etching process needed to penetrate through a multilayer structure. Samsung will apply its WSP technology to the production of NAND-based memory cards for mobile applications and other consumer electronics. Later, the company will extend the technique to high-performance system-in-package (SiP) and high-capacity DRAM stack packages used in servers that require fast data processing [4].

Toshiba's CMOS image sensor with vertical interconnections at low cost also used laser technology. Toshiba's technology, being based on the printed-circuit-board fabrication process, has no need for expensive wafer fabrication technologies such as RIE, CVD and CMP. They fabricated vertical interconnections by, first, via drilling with laser ablation through silicon chip, followed by dielectric film lamination,

Handbook of 3D Integration: Technology and Applications of 3D Integrated Circuits.
Edited by Philip Garrou, Christopher Bower and Peter Ramm

ISBN: 978-3-527-32034-9

a second via drilling by laser ablation on dielectric film and pattern plating of Cu [5]. This low cost process for TSV making was published at the same time by ITRI [6].

5.2
Laser Technology for 3D Packaging

5.2.1
Advantages

To achieve high density through via interconnects in the die or substrate for 3D application, high aspect ratio via hole diameter to hole length is required. One way to achieve this rapidly and economically is to use laser-drilled techniques. Making TSV using a laser tool has several benefits: (i) "One Step" process for TSV, hence avoiding complex and costly photolithography steps; (ii) dramatic reduction in capital costs, operating costs, factory floor space utilization, and engineering staff costs for the same operation; (iii) no tooling, no mask sets, and purely CAD based; (iv) allows for fast design changes, migration from prototype phase to volume production recipes in minutes, without the costs of expensive mask sets [7].

5.2.2
Disadvantages

Although laser-based TSV technology offers so many benefits there are some disadvantages compared to DRIE or other methods. One is the thermal effect generated by a laser beam punched to substrate with high energy. The zone of thermal effect could affect or damage the performance of devices if the laser-drilled holes too close to the active area of devices. A second disadvantage is the Si debris splashed from the melted Si wafer, which is not easily removed by traditional cleaning process. However, several methods have been developed to remove or reduce debris. XSiL laser systems will coat a sacrificial layer to protect against debris. After laser drilling, a system integrated washer can remove the coating layer and ensure clean wafers [3]. A third disadvantage is that the via profile of sidewall is not straight and smooth but, rather, is slightly saw-toothed. Finally, the laser tool is not suitable for application in high I/O products due to the present limitation of laser beam size and accuracy of total system position. Throughput of holing by laser is also highly related to hole size and number. To date, holes smaller than 10 micron in diameter can not be achieved by commercial laser machines.

5.3
For Si Substrate

5.3.1
Difficulties

Some previous studies were conducted at three different laser wavelengths to determine which wavelength would minimize damage to the die or substrate.

The drilling of Si with a focus laser beam is a three-dimensional heat flow problem. The duration of pulse determines the period of time for heating the material to a vaporization point. Different wavelengths will result in different degrees of Si debris. The studies showed that through holes processed with a laser wavelength of 266 nm exhibited the cleanest sidewall geometries with minimum residual Si debris. This is due to the absorption coefficient of this wavelength in Si being significantly greater than the fundamental and second harmonic wavelengths, thus creating a mechanism of material removal that includes vaporization and ablation [8]. To achieve maximum density of device I/O an area array of vias is desired through the active wafer or chip. Design rules would have to be developed to insure that the active area of devices is not damaged. In previous experimental trials, test wafers were constructed to determine if these effects occur when a laser creates vias near functional structures. Such studies showed that laser-created holes near transistors were highly dependent upon distance from the device. Hole as close as 90 microns to devices yielded acceptable structure characteristics [9].

5.3.2
Results

Research based on laser technology was also carried out at ITRI for 3D chip stacking [6, 10–12]. A Siemens (Model MB3205) laser machine with a 15 μm focused beam size was used with coherent UV light (355 nm). It ablates at least 155 μm-thick Si and also polymeric dielectric material. In this study, two patterns, including peripheral and area array, were produced separately, 25, 50, 75, 100, 150 and 200 μm diameter. Table 5.1 shows the difference and uniformity of the top and bottom diameters of the laser-drilled holes. The design and location of fiducial marks are important as the reference of each set of patterns. The second ablation of dielectric material cannot proceed precisely without it. Figure 5.1 illustrates the top view of silicon with different diameters (25, 50, 100 and 150 μm) fabricated by the laser drilling technique. The laser ablation afforded a straight sidewall as shown in the OM picture and the top/bottom slope ratio >0.8. To ensure stability of output power, the laser power and Q switch frequency were calibrated and adjusted in accordance a with master curve. Finally, a Q switch frequency of 30 kHz was chosen for silicon drilling and the power was 2700 mV.

Table 5.1 Uniformity of silicon laser drilling.

Dimension in diameter (μm)	X (top) (μm)	Y (top) (μm)	X (bottom) (μm)	Y (bottom) (μm)
25	25	27	24	24
50	49	52	44	45
75	74	76	67	70
100	99	101	90	97
150	148	150	142	145
200	200	198	189	192

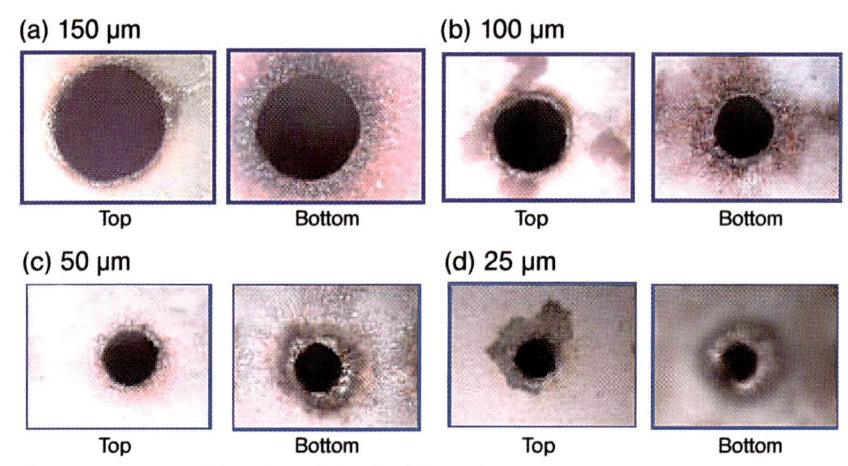

Figure 5.1 Top view of silicon laser drilling for different diameters.

Figure 5.2 shows an OM photograph of a laser-drilled wafer through hole inside a pad with polymeric insulation, 150 μm (left), 100 μm (middle) and 50 μm (right) in diameter. Table 5.2 shows the difference and uniformity of top and bottom diameters for laser drilling hole. For high accuracy laser drilling, we use fiducial marks on chip as the local alignment. The appropriate design and location of fiducial marks used as reference for each set of patterns led to the implementation of a second ablation of dielectric material.

Figure 5.3 shows dust with particle sizes of up to 6 μm around the hole. After chemical or mechanical cleaning, the dust can be removed. The Si splash also deposited around the ring pad. It is expected that the Si, along with the metal on the pad, were melted and evaporated by the laser beam. Some of the splash from the hole

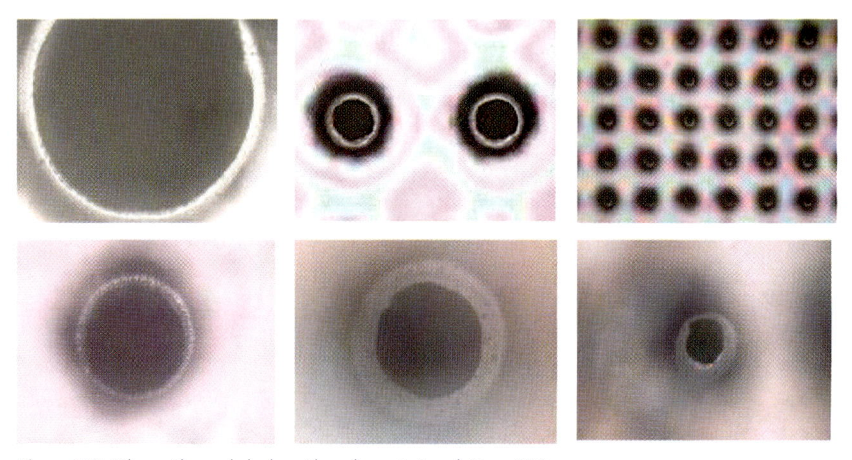

Figure 5.2 Silicon through hole with polymeric insulation, 150 (left), 100 (middle) and 50 μm in diameter (right).

Table 5.2 Uniformity of dielectric laser drilling.

Dimension in diameter (μm)	X (top) (μm)	Y (top) (μm)	X (bottom) (μm)	Y (bottom) (μm)
25	—	—	—	—
50	20	20	20	23
75	30	33	67	22
100	54	56	53	26
150	111	116	104	111
200	164	167	158	162

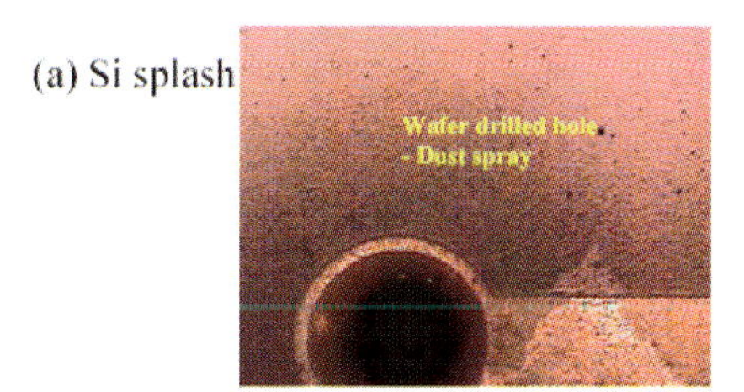

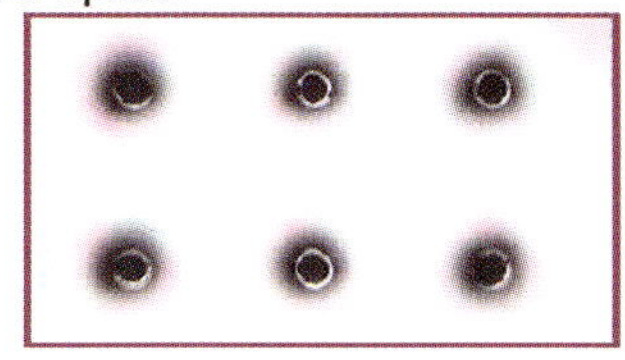

Figure 5.3 Silicon through hole by laser (a) without and (b) with polymeric insulation.

was deposited around the ring pad and is difficult to clean. The other type of damage is ablation of the metal pad by the laser beam. The high power (nearly peak power of the machine, 3.4 W) required while drilling Si wafer material might also lead to ablation of the metal layer on the ring pad. After the formation of an insulating layer by laser, the PCB compatible process is introduced into wafer level deposition of copper metal film. Copper through-hole plating was achieved with electroless copper activation. Figure 5.4 shows the deposition results.

It was found that the Ni/Au UBM, the Cu plating metal, and the graphite seed layers were deposited all over the wafer surface and the through-hole sidewall. This implies that PCB compatible processes might be the potential integrated solution to forming the silicon through-hole interconnection. Since the polymeric layer is directly fabricated into the hole, the integration will be challenging due to the CTE mismatch during thermal history. In addition, adhesion is another important part during processing and depends on the polymeric material selected. Compared to other materials used, the epoxy type of polymer clearly afforded a flattened surface at the interface of silicon and interconnect. In our research, we filled the through-hole with resin by vacuum lamination and then laser-drilled a second smaller diameter

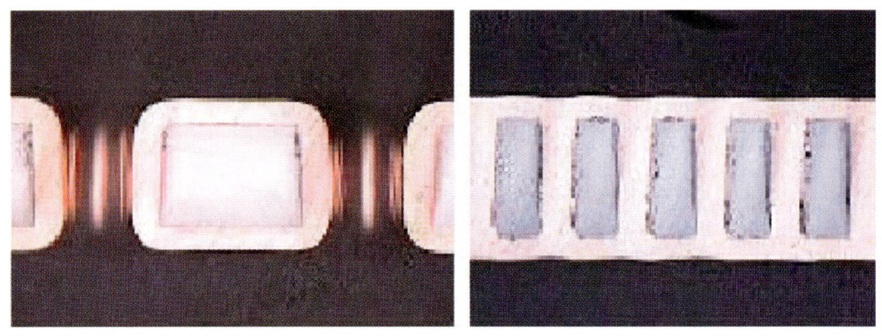

Figure 5.4 Silicon through hole with polymeric insulation, 200 μm in diameter (left), 50 μm in diameter (right).

through-hole on the resin, which helped us to obtain a sufficient thickness of isolator on the wall of the original through hole. The dielectric material we used, ABF (Ajinomoto Build-up Film), is a high resin-flow material that easily gave various diameters of resin-filled holes without voids. Figure 5.5 shows the result of a 150 μm thin wafer double-sides symmetrically laminated using 40 μm thick ABF. It illustrates that thin wafers with different diameters (50, 75, 100, 150 and 200 μm) could be filled successfully with resin by vacuum lamination. It also demonstrates that the laminating condition is pretty good with void-free, thin wafer integrity, and nice surface evenness.

OM observation results are shown in Figure 5.6. For Figure 5.6a, ten pulses were used to punch the silicon, forming a hole. The hole is circular and some dust is deposited radically around the hole. For Figure 5.6b, 100 pulses were used to punch the silicon. The hole is oval and has splashes that are several μm higher than the

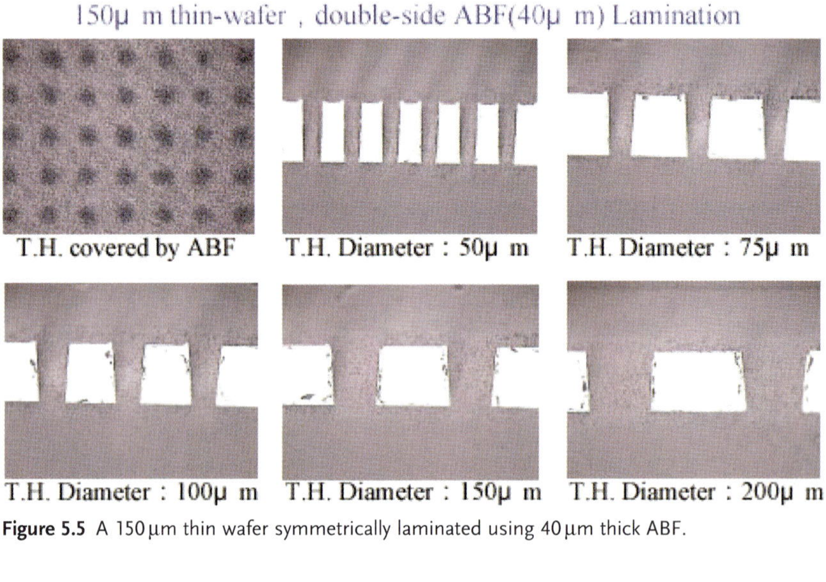

Figure 5.5 A 150 μm thin wafer symmetrically laminated using 40 μm thick ABF.

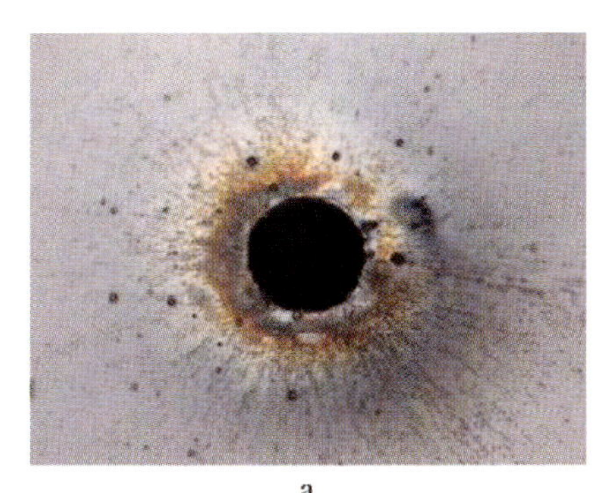

a

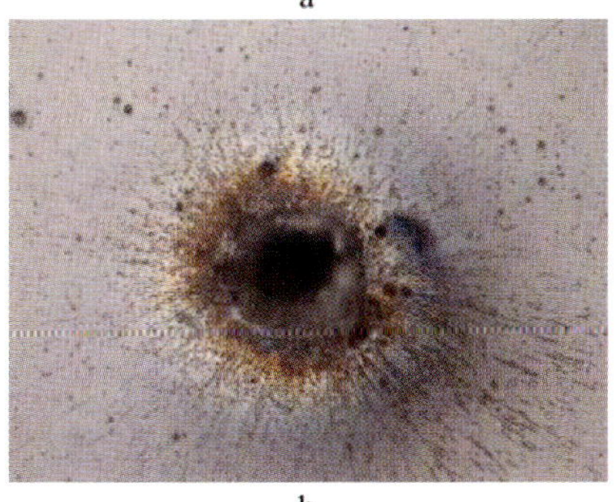

b

Figure 5.6 Top views of (a) holes drilled with 10 pulses of 90 μJ at 30 k frequency and (b) holes drilled with 100 pulses of 90 μJ at 30 k frequency.

silicon. This demonstrates plasma pressure in the hole when laser drilling the silicon. After grinding and polishing this sample, we saw both micro-pores (1–2 μm diameter) around the hole and micro-cracks (average length several μm) extending from the wall into the silicon. These features covered an area of about 10 μm from the hole edge.

Figure 5.7 is a scanning electron micrograph of an ion-polished cross-section of a laser-drilled hole. There are two clear features. One is a pore-containing deposition layer 3–5 μm thick and the other is cracks up to about 10 μm from the hole edge. The cracks characteristically change their initial radical direction and assume a peripheral course. It is concluded from this that these cracks were not caused by the laser-induced pressure surges but by the thermally induced tensile stresses in accordance with reference [13]. Figure 5.8 shows a typical transmission electron micrograph of the edge of a laser-drilled hole of a silicon wafer. The following details can be seen. (i) The re-deposition area caused by the molten material on the wall of the hole with the formation of pores and cracks. The range is 2–6 μm in this study. (ii) A plastically deformed boundary zone extending around 6 μm and containing dislocations of mainly radical alignment. These are glide dislocations of different types that commence from the wall of the hole and run into the crystal on different slip planes. (iii) The single crystal of different slip planes. These defects are formed by the

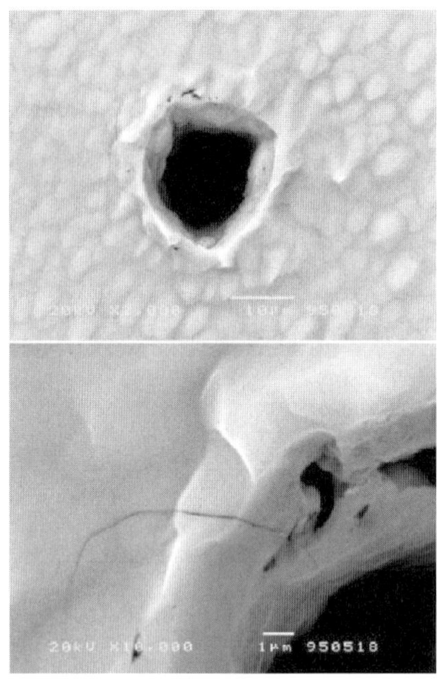

Figure 5.7 SEM pictures show micro-pores and micro-cracks around the hole.

laser-induced heat influence zone. If laser drilled silicon technology is applied in semiconductor application the transistor near the hole must be considered. Consequently, the range of the thermal effect area is an important issue that requires study.

5.4
Results for 3D Chip Stacking

Figure 5.9 shows a schematic diagram of 3D high-density BGA packaging used in an ongoing research project in EOL/ITRI. It consists of six major parts: micro-bump, micro-gap filling, wafer alignment, stacked wafer thinning, wafer bonding and 3D interconnect assembly.

The wafer/chip thickness used here was 100 μm, down to 50 μm. The results show that this structure can provide more reliable wafer stacking without voids, including for four major processes, that is, wafer thinning, direct laser drilling/patterning, insulation layer formation and PCB compatible via filling/wet etching. This provides more advantages than competing technologies not only in terms of compatibility with low cost silicon-through processes but also in the flexibility to inter-chip or inter-wafer assembly of connecting different components. Figure 5.10 shows the process flow chart of the Si-through vertical 3D interconnection.

Figure 5.11 illustrates that the results show two different wafer thicknesses with through-hole forming vertical interconnection by electroplating. Figure 5.11a shows

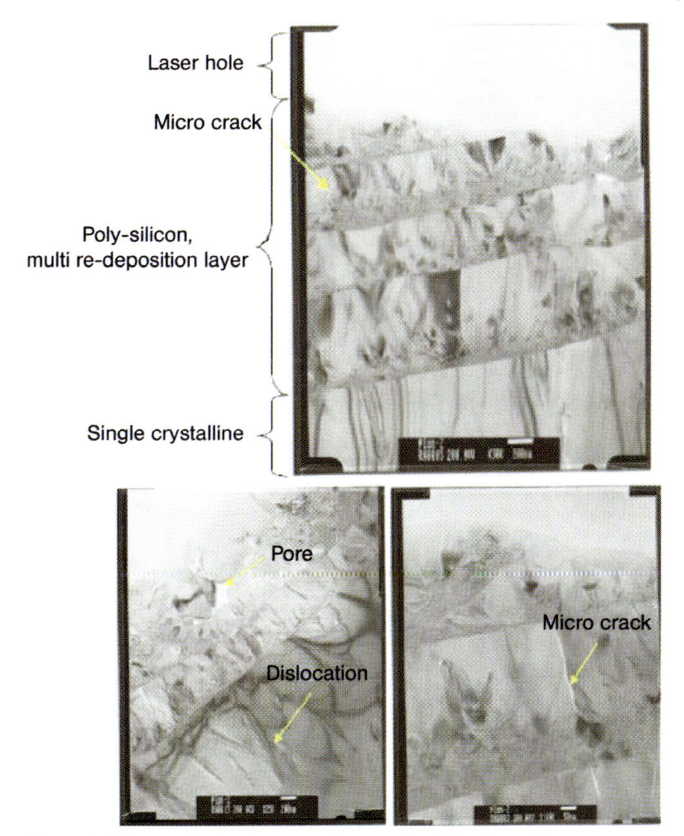

Figure 5.8 Transmission electron micrograph of the deposition layer and the influenced boundary zone of hole in a (1 0 0) silicon wafer drilled with a diode pumped solid state laser (Vanadate).

a thin wafer (100 μm thick) and Figure 5.11b shows an ultrathin wafer (50 μm thick). Both the original through-hole diameters are 100 μm with polymeric insulation about 15 μm thick, which means that the diameter of hole filled with copper is about 70 μm. Since all of the processes described above are PCB compatible, they have the potential to provide a cost-effective solution for 3D chip-stacking packaging. Compared to the Bosch process [14], we provided the polymeric material as the dielectric layer between

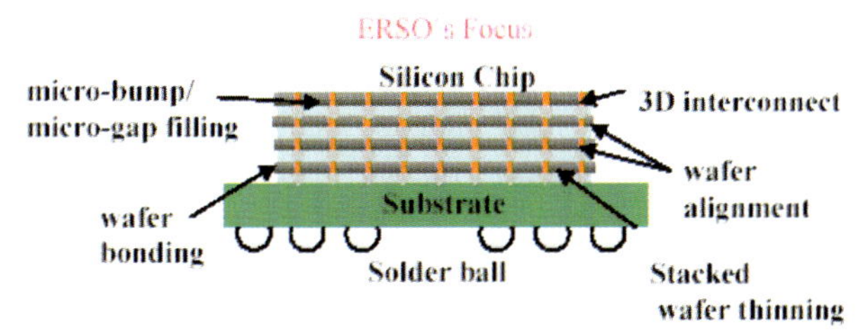

Figure 5.9 Schematic diagram of 3D stacking packaging focus in ITRI.

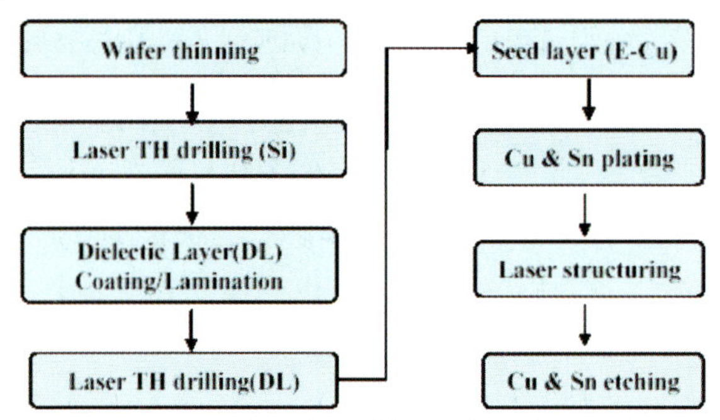

Figure 5.10 Process flow chart of the Su-through vertical 3D interconnection.

copper column and silicon sidewall. The material we chose here is an epoxy-based flowing resin commonly used in the high-density organic substrate. After the ABF lamination process, the holes are filled with ABF dielectric material as an electrical isolation wall. Local fiducial marks fabricated by previous laser process on the corner were used as reference for each set of patterns, allowing the implementation of the second ablation of dielectric material. The laser structuring process is a fast prototyping method to pattern copper trace, but it still had some disadvantages when we used it as the patterning tool. Because the polymeric dielectric film can be removed more easily than copper metal by the ablation of UV laser, it is very difficult to protect the film during copper trace patterning. This is situation is exacerbated, and may even lead to silicon damage, with thinner dielectric films. Therefore, to avoid damaging both ABF and silicon wafer, we adopted a combination of a laser structuring process and a wet etching process to pattern copper traces. After copper through-hole plating, the local fiducial marks were covered. To structure precisely, the four plated via holes are regarded as local fiducial marks. Then, 1 μm tin was plated on a 40 μm copper layer as an etching mask. After that, copper traces are patterned by laser structuring, ablating not only the tin mask but also, partially, the Cu layer.

Figure 5.12c shows OM micrograph of wafer after dicing and laser structuring process; the white region is tin and the dark red region is revealed Cu layer after

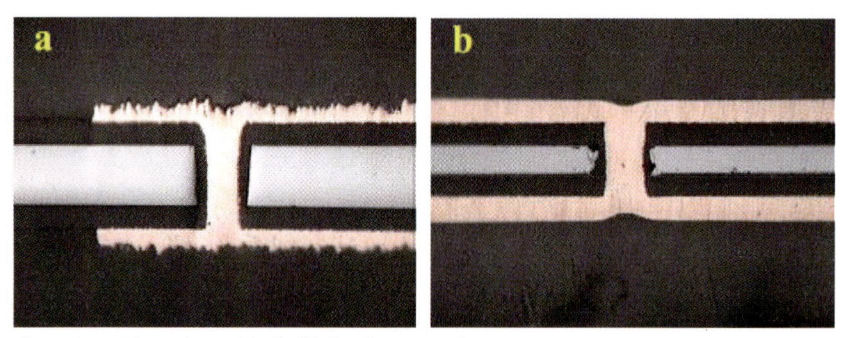

Figure 5.11 Silicon through-hole filled with copper plating, (a) 70 μm hole diameter/100 μm thick wafer and (b) 70 μm hole diameter/50 μm thick wafer.

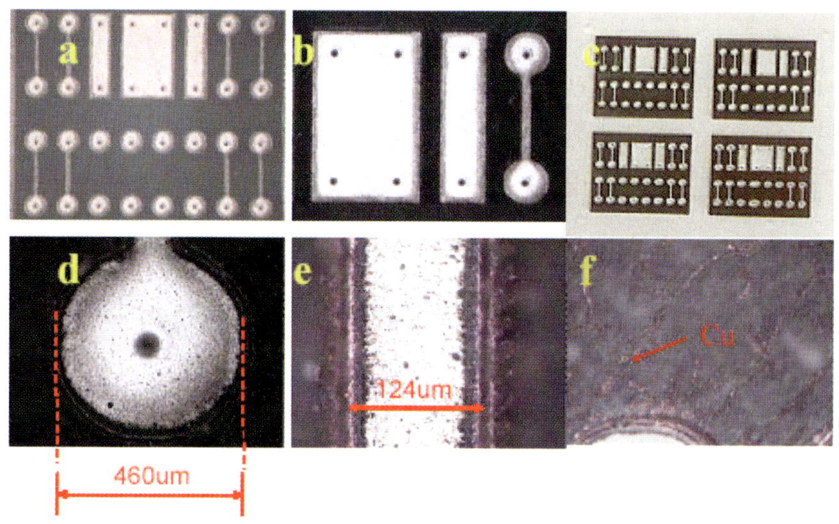

Figure 5.12 OM micrograph of test vehicles after electroplating, laser structuring and etching processes.

removal of the tin strip. Figure 5.12 indicates that the patterns of metal traces/pads can be positioned precisely by laser structuring. The conductive line is 124 μm wide and the pad is 460 μm in diameter. Final copper etching and tin strip wet etching complete trace pattern transfer from tin to the copper layer. As shown in Figure 5.12, the copper traces are patterned successfully, with 84 μm wide conductive lines and 420 μm diameter pads, without damaging the ABF and Si wafer. It takes advantage of the fact that the Cu and tin etchant does not hurt the dielectric film. A 20 μm undercut is also seen on each side of the trace. The undercut exists in the wet etching process intrinsically, and can be solved by compensating the pattern design or by violent agitation. Furthermore, compared with the lithographic method, the undercut can also be reduced by partial removal of copper during laser structuring.

Figure 5.13a shows a stacking of ten chips. The electrical resistance of this stacking module with daisy chain design is less than 2 Ω. The total electrical length inside the stacking is about 10 cm. Figure 5.13b shows the result of X-ray inspection of the inner structure of this ten-chip-stacked module. The TSV and bonding condition between chips are good [15].

5.5 Reliabilities

By putting through-hole interconnects composed of Cu and polymeric material insulation in a programmable temperature and humidity test chamber for period of time, the water absorption was determined. The test condition is 85 °C and 85% RH. The samples passed the 85%/85 °C humidity test. For components with multi-chip

Figure 5.13 (a) OM outlook and (b) X-ray inspection for a 10 chips stacking with daisy chain design.

stacking, some reliability tests such as the pre-condition test and thermal cycling test are ongoing.

5.6 The Future

3D integration offers significant advantages in performance, functionality and form factor. For decades, semiconductor manufactures have been shrinking the size of ICs to achieve the yearly increases in speed and performance described by Moore's Law. Moore's Law exists only because the RC delay has been negligible in comparison with signal propagation delay. However, for submicron technology, RC delay becomes a dominant factor. To continue and extend Moore's Law in deep submicron generation, one emerging solution is 3D integration. Replacement of the large number of long interconnects needed in 2D structures by short vertical interconnects would greatly enhance performance. A 3D wafer stacking would allow specific functions to be vertically interconnected to create a product. Such 3D integration, also referred to as "chip stacking", can be categorized as a system-in-a-package (SiP) solution [16]. SiP with chip stacking by wire bonding technology for electrical interconnects is widely used by industry, especially for mobile phone applications. SiP using TSV technology will be coming soon. Mobile phone applications are still the major driving force for this technology. Customers desire products with more functions, higher performance, smaller packaging size, and lower cost. Therefore, for initial SiP application with TSV, low cost processing technology such as laser-based has the large potential to reduce manufacturing cost.

Acknowledgment

The authors appreciate research support from the Ministry of the Economic Affairs (MOEA), Taiwan, ROC for performing related projects and from all members of 3D projects in R division/EOL/ITRI.

References

1 Ko, C.T. *et al.* (2006) Next standard packaging method for DRAM-chip-in-substrate package. IMPACT 2006, Microsystems, Packaging, Assembly Conference Taiwan, pp. 91–96.
2 Pelzer, R. *et al.* (2003) Vertical 3D interconnect through aligned wafer bonding, ICEPT Proceedings, pp. 512–517.
3 Billy Diggin, *et al.* (2007) Laser Drilling for TSVs & Thin Wafer Dicing, presentation material of EMC-3D Europe Technical Tour.
4 Website news: www.physorg.com. (accessed on October 2007).
5 Masahiro Sekiguchi, *et al.* (2006) Novel low cost integration of through chip interconnection and application to CMOS image sensor, Proceedings 56th Electron. Components and Technology Conference, pp. 1367–1374.
6 Lo, W.C. *et al.* (2006) An innovative chip-to-wafer and wafer-to-wafer stacking. Proceedings 56th Electronic Components and Technlogy Conference, pp. 409–414.
7 XSiL's presentation material.
8 Dahwey Chu and Doyle Miller, W. (1995) Laser micromachining of through via interconnects in active die for 3D multichip module. IEEE/CPMT Inationall Electronics Manufacturing Technology Symposium, pp. 120–126.
9 Lee, Rex. A., Whittaker and Dennis, R. (1991) Laser created silicon vias for stacking dies in MCMs. IEEE/CHMT '91 IEMT Symposium, pp. 262–265.
10 Lo, W.C. *et al.* (2005) Development and characterization of low cost ultrathin 3D interconnect. Proceedings 55th Electronic Components and Technlogy Conference, pp. 337–342.
11 Chen, Y.H. *et al.* (2006) Thermal effect characterization of laser-ablated silicon-through interconnect. Electronics Systemintegration Technology Conference, Dresden, Germany, pp. 594–599.
12 Lo, W.C. *et al.* (2007) 3D chip-to-chip stacking with through silicon interconnects, International Symposium on VLSI-TSA, pp. 72–73.
13 Luft, A. *et al.* (1996) A study of thermal and mechanical effects on materials induced by pulsed laser drilling. *Applied Physics A*, **63**, 93–101.
14 Ranganathan, N. *et al.* (2005) High aspect ratio through-wafer interconnect for three dimensional integrated circuits. Proceedings 55th Electronic Components and Technlogy Conference, pp. 343–348.
15 To be published, Lo, W.C. and Chang, S.M.
16 Garrou, Philip (Feb. 2005) Future ICs Go Vertical. *Semiconductor International*, SP10.

6
SiO_2

Robert Wieland

6.1
Introduction

The electrical isolation properties of silicon dioxides vary over a wide range, depending on the process technology of the deposition or thermal grow of such layers. Metal-filled Through Silicon Vias (TSVs) need sufficient electrical isolation to the surrounding bulk silicon. In principle, many thermal oxidation processes would deliver a sufficient dielectric layer for the electrical isolation of the Si-substrate to the metal within the TSVs. For a 3D integration process flow where the TSVs are formed in an early state of the CMOS fabrication flow, during or at the end of the front end processing block – which typically means there is no metallization layer present – a thermal oxidation process could be used to grow an appropriate SiO_2 layer for TSV isolation. The maximum allowed process temperatures would be in the range 700–900 °C or higher.

However, most currently used 3D integration technology flows form the desired TSVs during or at the end of the back end processing block. Thus it is no longer possible to run processes above 400 °C due to the presence of metallization layers on the device substrates.

Therefore, CVD based SiO_2 films at moderate temperatures, in the range 200–400 °C, are often used for electrical isolation of the TSVs. Because the growth of thermal oxides is well characterized in the semiconductor industry [1–4], this chapter deals mainly with CVD-based SiO_2 films for TSVs.

6.2
Dielectric CVD

CVD (chemical vapor deposition) allows the deposition of amorphous dielectric films on a substrate using gaseous phase reactions. Process gases flow through a vacuum

Handbook of 3D Integration: Technology and Applications of 3D Integrated Circuits.
Edited by Philip Garrou, Christopher Bower and Peter Ramm

ISBN: 978-3-527-32034-9

reactor and become dissociated by heat or a by plasma (PECVD: Plasma Enhanced CVD). For details of the CVD processing see References [5, 6, 10].

CVD processes allow precise control of all relevant process parameters with a high purity of the reactants and can result in a wide variety of dielectric film types. Process requirements for a SiO_2 CVD film in the 3D environment are a high deposition rate with good uniformity both within and from wafer to wafer, controlled stoichiometry and purity of the film, good adhesion and conformality and sufficient dielectric properties. Typical characteristics of CVD based SiO_2 films are film thickness, refractive index, wet etching rate in buffered HF, mechanical film stress, shrinkage (800 °C, N_2 ambient) and step coverage. The main process parameters for the adjustment of the desired film properties are the mixture of gases or vapors, gas flow, vacuum pressure, wafer temperature and the geometry of the process chamber. For plasma-enhanced processes, RF-power, RF-frequency, electrode design and electrode spacing are additional parameters.

Table 6.1 gives an overview of typical dielectric films and their properties.

The use of TEOS instead of silane (SiH_4) results in an improved conformality of the SiO_2-film [13]. TEOS-based SiO_2 films contain less hydrogen in the film than those deposited using silane; addition of O_2 minimizes the inclusion of traces of C or N in the film [14]. Uniformity of the film thickness across a wafer is typically in the 3% 1 sigma range. Figure 6.1a and b show the above-mentioned films deposited on a 90° step of approximately 800 nm, with the PETEOS resulting in improved step coverage.

LPCVD TEOS typically generates films with high conformality [14] at rather low deposition rates; however, the comparatively high process temperature, 620–690 °C, limits their use for 3D applications.

As seen from Table 6.1, a SACVD (Sub-Atmospheric CVD) film is well suited for the requirements of 3D integration in terms of process temperature and conformality. The latter is of great importance when HAR TSVs (high aspect ratio through silicon vias) are to be isolated.

Table 6.1 Properties of different SiO_2 film types as dielectric layer for 3D integration.

Film	Process temp. (°C)	Process gases	Film thickness (nm)	Stress (MPa)	Conformality (step coverage)
Thermal oxide	700–1150	O_2, H_2	5–1000	400–500	High
PECVD oxide	150–400	SiH_4, N_2O		150	Low
PECVD TEOS	250–400	$Si(OC_2H_5)_4$, O_2	50–5000	100–200	Moderate
SACVD ozone-TEOS	400	$Si(OC_2H_5)_4$, O_3	150–500	100–200	High
LPCVD TEOS	650–750	$Si(OC_2H_5)_4$	20–500	80–120	High
PSG, BPSG	400–750	$Si(OC_2H_5)_4$, PH_3, TMB	300–900		Moderate–high

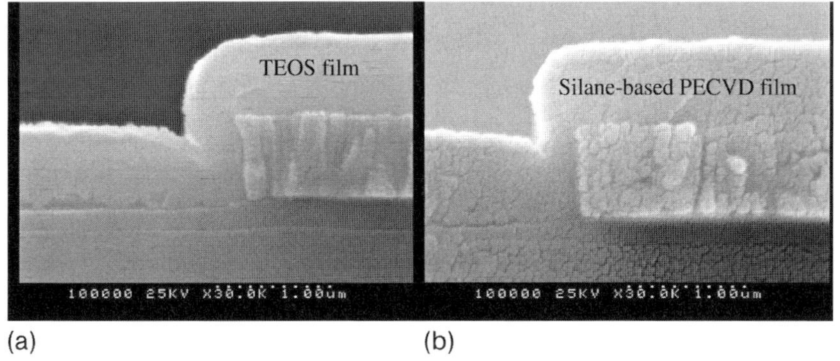

Figure 6.1 (a) Step coverage of a PETEOS based SiO_2 film; (b) step coverage of a silane-oxide based SiO_2 film.

6.2.1 Sub-Atmospheric CVD

Sub-atmospheric ozone-TEOS CVD is typically performed in a conventional PECVD single wafer chamber, which is mechanically altered to handle high chamber pressures safely. The process runs without plasma and uses a mixture of TEOS (Tetra-Ethyl-Ortho-Silicate, $Si(OC_2H_5)_4$) and ozone (O_3) in a pressure range of 100–600 Torr.

Figure 6.2 shows a cross-sectional view of a single wafer SACVD universal chamber (Applied Materials P 5000 system).

To achieve repeatable film qualities, a multi-chamber system with a vacuum load lock and an automatic robotic system for wafer exchange and wafer storage is a standard requirement in the industry (Figure 6.3).

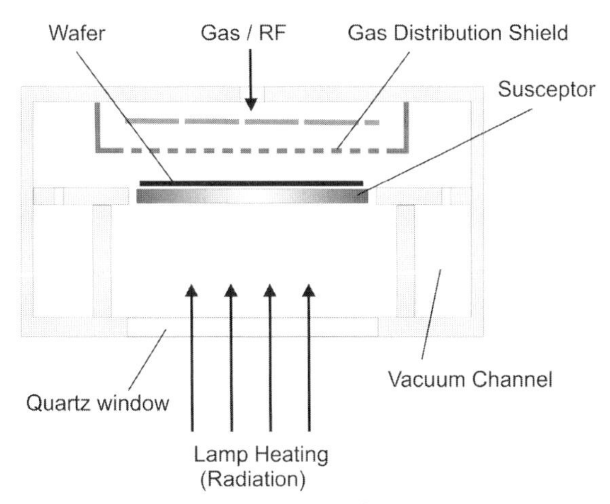

Figure 6.2 Schematic cross section of a CVD chamber (Applied Materials P 5000).

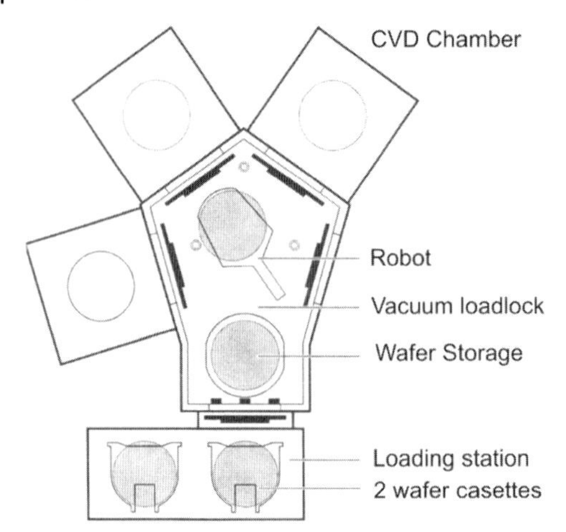

Figure 6.3 Schematic top view of a multi-chamber system with central loadlock and robotic handling system.

The ozone-activated deposition of a SACVD film occurs in a process region far from the thermodynamical equilibrium. Due to the excessive amount of ozone species at relatively high pressure near atmosphere, the film deposition rate depends mainly on the transport of oxygen atoms and reactive by-products of the thermally diffracted precursor molecules to the wafer substrate. The amount of hydroxyl groups on the wafer substrate also influences the deposition rate.

Chemically relevant processes during the deposition occur in the so-called mass transport limited process region. High process gas flow is used to secure efficient transport of precursor molecules and their reactive species to the wafer substrate. Precise control of processing temperature, chamber pressure and gas/vapor flows by closed-loop servo control allows for repeatable film results.

Wafer temperature can be controlled either by use of a servo-loop lamp heating system, where heat is transformed mainly by radiation to a susceptor, or by using an electrical resistance heating system within the susceptor. To maintain chamber pressures in the range 400–600 Torr as well as in the 2 Torr regime, a dual manometer system with high pressure/low pressure capacitance manometers and a closed loop throttle valve is a basic requirement. TEOS vapor is supplied to the process chamber either by bubbling the heated TEOS liquid with MFC-controlled He-flow or by direct injection of the TEOS vapor with a liquid pump and a vaporizer.

A stable ozone (O_3) gas flow is, typically, produced by flowing O_2 (MFC-controlled) through a water-cooled ozone generator. Typical O_2-flow is in the range 2–10 $L\,min^{-1}$; the generator power is 1–3 kW. As shown in Reference [7], an O_3 flow with O_3-concentrations of up to 250 $g\,m^{-3}$ in O_2 can be achieved by adding an MFC-controlled N_2-flow of ca. 1.0–2.0 $mL\,min^{-1}$ to the O_2-gas line prior to entering the ozonator. Table 6.2 shows the process parameter regime of a typical SACVD process.

Table 6.2 Process parameters of a SACVD O_3/TEOS deposition process.

Parameter	Value
Chamber pressure (Torr)	1–600
Susceptor temperature (°C)	280–400
Spacing shower head/susceptor (mm)	10–20
TEOS-flow (He-bubbler) (slm)	1–2
O_2-flow (slm)	3–5
O_3-concentration in O_2 ($g\,m^{-3}$)	150–250
RF-power (high frequency) (W)	0
RF-power (low frequency) (W)	200–400
NF_3 and C_2F_6 flow (chamber clean) (slm)	0.5–1.5

6.2.2 Process Sequence of O_3-Activated SACVD Deposition

A typical SACVD process starts by filling the reactor chamber and then stabilizing the chamber pressure with inert gases. Since Ar and N_2 influence the O_3 concentration, only O_2 and He are used for this purpose. Once the desired chamber pressure has been achieved and stabilized, the precursor vapor is allowed to flow into the deposition chamber. This avoids the pre-deposition of a pure TEOS seed layer on the wafer substrate. Ramping up of the precursor flow to the desired O_3/TEOS ratio allows the build-up of a conformal SiO_2-seed layer with increasing deposition rate, until the final O_3/TEOS ratio has been reached. To improve the uniformity of the deposited film, a surface pre-treatment, using a low frequency N_2 plasma step, can be run prior to the filling-up steps of the deposition chamber. It is assumed that the low frequency N_2 plasma pre-treatment reduces the non-uniformity of atomic SiO_2, SiO, SiH and SiOH layers, thus resulting in a more uniform SiO_2 film thickness.

Once the desired SiO_2-film has been deposited, the chamber pressure is slowly ramped down, to avoid any particle deposition from gas phase reactions onto the wafer substrate. After unloading the wafer to the storage place in the load lock, the deposition chamber is then automatically cleaned with a NF_3/C_2F_6 plasma cleaning step, followed by a short PECVD-O_2/TEOS deposition step. This assures equal chamber surface conditions after each wafer prior to the next SACVD process.

6.2.3 Conformal SACVD O_3 TEOS Films for 3D Integration

SACVD based SiO_2 films have already been used in conventional IC fabrication for the sub-micron filling of shallow trenches (STI – shallow trench isolation) due to their high conformal deposition behavior [12, 14, 15]. At Fraunhofer IZM, early work on TEOS/O_3 based SiO_2 layers for 3D integration was published in 1995 [8]. Electrical isolation of HAR TSVs with aspect ratios of 10 : 1 and higher requires a film conformality in the range of 50–60% to isolate sufficiently the bottom part of a metal-filled TSV.

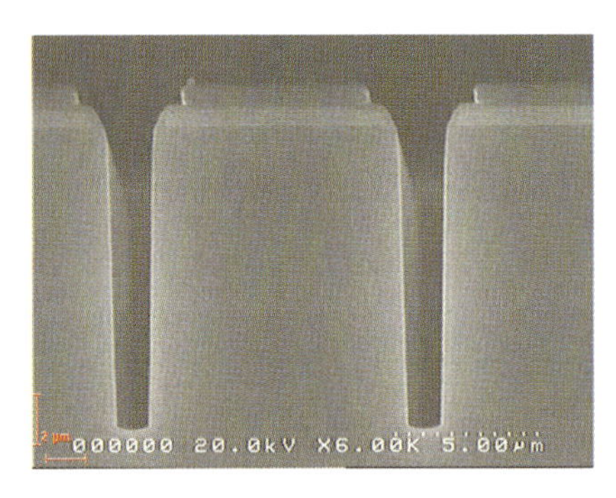

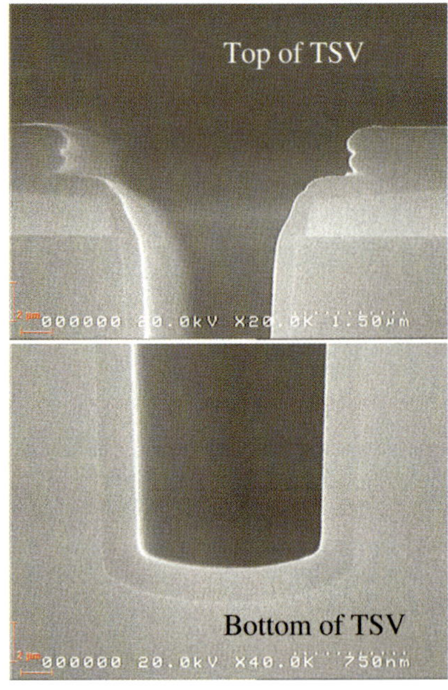

Figure 6.4 Conformality of a SACVD O_3/TEOS film deposited in a TSV with an AR = 7 : 1; taper angle of the trench hole ~89° (HBr based process).

Figure 6.4 shows a 22 µm deep TSV with a 270 nm thick SACVD O_3/TEOS layer and an aspect ratio (AR) of 7 : 1. Conformality is ca. 80%, if one compares the film thickness at the top surface with that at the bottom of the trench hole. Conformality is defined by the film thickness ratio of the deposited film on the top surface and the bottom sidewall of the trench hole. The maximum achievable conformality of the SiO_2 film depends strongly on the taper angle of the trench hole for a given aspect ratio. This assumption is also valid for the subsequently deposited metal-organic CVD layers like TiN, W or even Cu.

Depending on the type of etch process used to build the HAR trench holes, the achievable taper angle of the TSV varies in the range ~85° up to 90° or slightly higher. HAR TSVs with aspect ratios of 15 : 1 or above are usually etched with the so-called "Bosch etch process", which leaves more or less rough sidewalls with a "scallop" like form. Using a SACVD process, sidewall roughness can be reduced tremendously if the deposited SiO_2 film thickness is equal or thicker than the size of the trench sidewall scallops.

Figure 6.5 shows a 50 µm deep TSV trench with an aspect ratio of >16 : 1, etched with the Bosch process. A SiO_2 film (O_3/TEOS SACVD) with a nominal film thickness of 403 nm has been deposited afterwards. As seen in Figure 6.5b (top of TSV) and c (bottom of TSV), the resulting surface of the O_3/TEOS layer is comparably smooth. However, film conformality only reached 43%, probably due to the TSV taper angle of almost 90°.

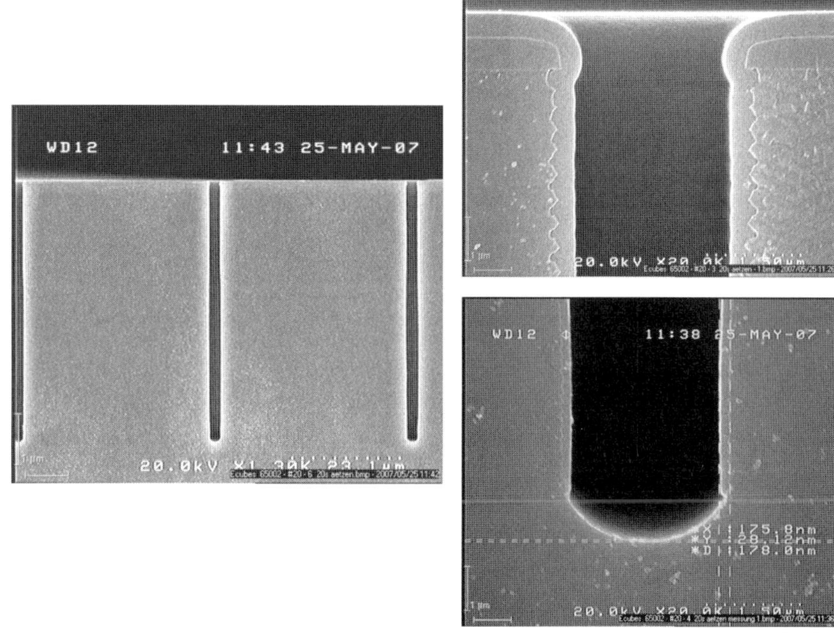

Figure 6.5 Conformality of ~43% – SACVD O_3/TEOS film deposited in a TSV with an AR >16 : 1; taper angle of the trench hole ~89.5–90 °C (Bosch process).

Besides the achievable taper angle of the etched TSVs, the aspect ratio of the formed trench holes defines the achievable O_3/TEOS conformality over a wide range. Figure 6.6 shows the aspect ratio dependent O_3/TEOS deposition conformality for trench holes of the same lateral size and taper angle (89.4°); the trench depth varied between 20 and 76 μm. By increasing the trench depth and keeping the open area constant (same lithography) as was done in the experiments of Figure 6.6, the deposition area can become several times bigger compared to a blank wafer surface at an open area ≥5%. This increase in deposition area is most likely the reason for a decreasing lateral and vertically deposited film thickness with increasing AR of the TSVs to be covered. Even an SACVD process is mass transport limited; thus the amount of reactive species decreases for a doubled or tripled surface if the TEOS net gas flow is kept constant. This study concludes that further optimization of the SACVD process to fulfill the specific 3D requirements is necessary. One way to increase film conformality for HAR TSVs is the deposition temperature: Shareef reports a noticeable effect on conformality by deposition temperature – increasing temperature increases conformality [11].

To compare SiO_2 film behavior in terms of the method of deposition/growing, the very same TSV trenches as those of Figures 6.4 and 6.5 were used to deposit/grow a

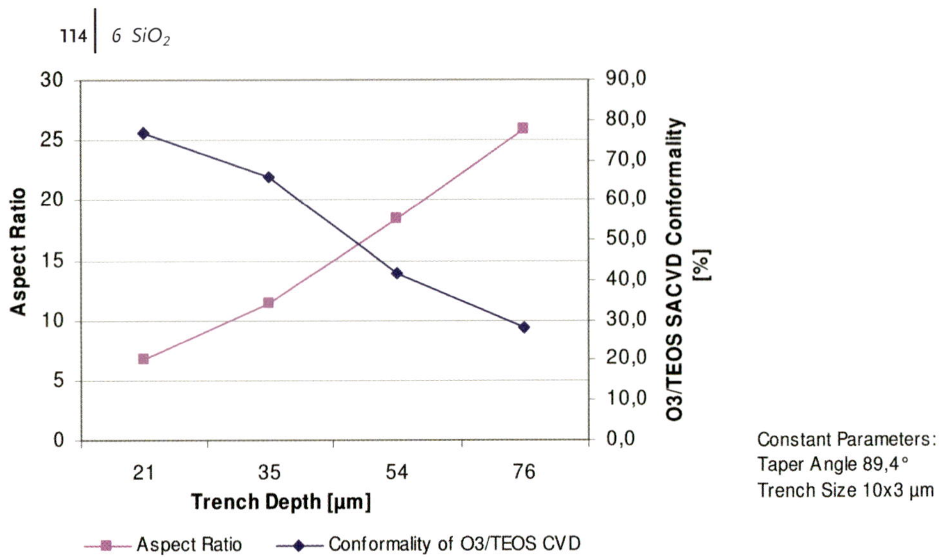

Figure 6.6 Conformality of SACVD O_3/TEOS films deposited in TSVs (3 × 10 µm) versus trench depth.

thermal oxide and a PETEOS oxide. As shown in Figure 6.7, thermally grown oxide (wet oxidation at 1000 °C) leaves the scallops from the Bosch process almost unchanged, since there is no add-on deposition in a diffusion-driven thermal oxidation process. Instead, the Si-surface of the formerly etched TSV hole is used to grow a SiO_2-layer. Conformality is better than 85%. Thermally grown oxides at temperatures in the range 800–1050 °C are grown whether the Si-surfaces are perfectly clean or not. Possible contaminations caused by the previous TSV etching process do not have a noticeable effect on the conformality of the grown oxide film thickness. Of course, any organic or inorganic contamination present can later lead to reliability issues regarding the electrical isolation quality of the TSVs.

Figure 6.8 show the conformality of a PETEOS SiO_2 film, deposited at the typical plasma-enhanced CVD pressure in the 2–4 Torr range. As expected, scallops are

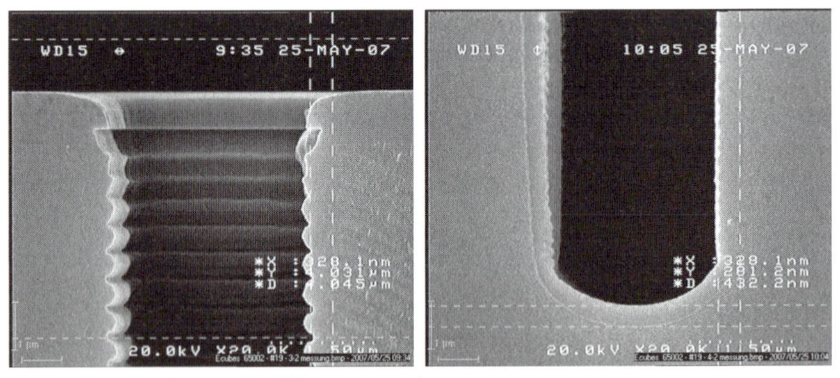

Figure 6.7 Thermally grown SiO_2 film in a TSV with an aspect ratio of >16 : 1; scallops only slightly reduced compared to the Bosch process. Conformality ~85%.

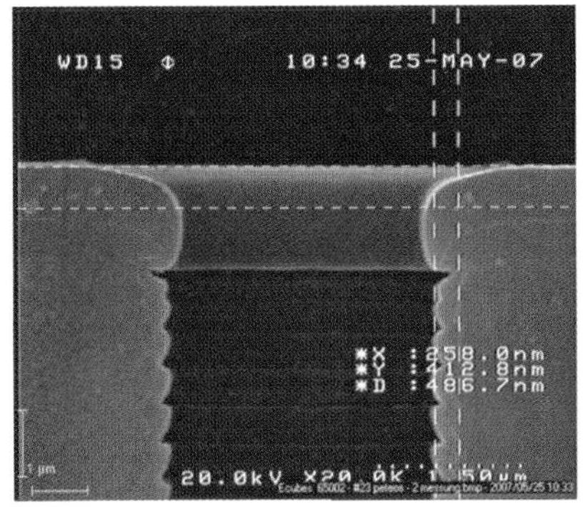

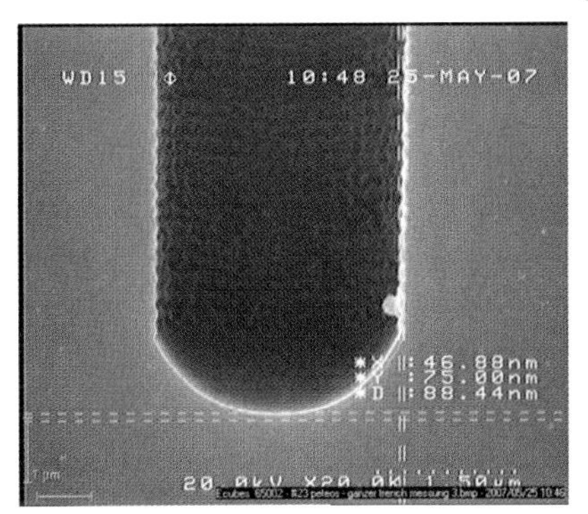

Figure 6.8 PETEOS SiO_2 film in a TSV with an aspect ratio of >16 : 1; scallops somewhat reduced compared to the Bosch process. Conformality ~11%.

somewhat reduced, but the low conformality of ~11% for the SiO_2 film is not suitable for 3D integration requirements.

Smoothening of the scallops can only be expected with the O_3/TEOS SACVD process due to its unique growing properties during deposition, as long as the scallop size stays below the absolute film thickness deposited.

The quality of the deposited SACVD SiO_2 films is also influenced by the condition of the trench surfaces prior to deposition. For example, moisture or even insufficient polymer removal from the preceding trench etch process might hinder a conformal and uniform film deposition.

6.3
Dielectric Film Properties

SiO_2-based dielectric film properties depend mainly on the reactive species used for deposition, on the deposition temperature, pressure regime and plasma enhancement. Table 6.3 shows some typically achievable film properties of different types of SiO_2-films. Regarding the reactive species, thermally grown oxide delivers by far the highest film quality, since there is only O_2 and/or H_2 involved during film growth. Within-wafer uniformity, surface roughness, wet etching rate in HF and electrical break down voltage are unsurpassed compared to CVD-films. One disadvantage of thermally grown oxide, however, is the comparably high extrinsic film stress of 400 MPa and more.

Silane-based CVD films contain hydrogen; the amount depends mainly on the deposition temperature and the N_2O/SiH_4 ratio [14]. Also, TEOS-based films are known to contain O-H groups [14] and tend to absorb H_2O over time [14]. Due to the required film conformality for 3D purposes, ozone-based SACVD films deliver acceptable results regarding within-wafer uniformity, surface roughness and extrinsic film stress.

As shown in Table 6.3, O_3/TEOS SACVD films can be deposited at relatively low temperatures with an acceptable deposition rate, thus resulting in a cost-efficient

Table 6.3 Typical dielectric film properties of different SiO_2 films.

Dielectric film type	Thermal, wet	LPCVD TEOS	PETEOS	SATEOS	Plasma-oxide
Deposition tool	Horiz. furnace	Horiz. furnace	CVD reactor	CVD reactor	CVD reactor
Deposition pressure (Torr)	760	0.25	2.8	600	2.5
Gases for deposition/ diffusion	H_2, O_2, HCl	TEOS	TEOS, O_2	TEOS, O_3	SiH_4, N_2O
Temperature (°C)	850	680	400	400	400
Within-wafer uniformity (% 1σ)	1	2.1	3	3.9	2.2
Deposition rate (nm min^{-1})	2	4	580	160	140
Mean roughness (Ra) (nm)	0.3	0.5	0.5	2	1.7
Wet Etch Rate (HF 1%) (nm min^{-1})	7	>43	16	32	40
Stress (extrinsic) (MPa)	410 (compr.)	100 (compr.)	130 (compr.)	150 (tensile)	170 (compr.)
El. Breakdown voltage (V μm^{-1})	2602	891	870	361	348

process step. The measured electrical breakdown voltage of 361 V μm^{-1} for SACVD O_3/TEOS films is ~7 times lower than for thermally grown oxide films. A minimal film thickness of ~150 nm SiO_2 is thus required for the electrical isolation of TSVs. Therefore, the breakdown voltage of an O_3/TEOS SACVD film in a HAR TSV can be estimated to be >50 V.

Electrical measurements of W-filled TSVs with a 260 nm thick SACVD-based SiO_2 have shown electrical leakage current values in the fA range [9].

6.4 3D-Specifics Regarding SiO_2 Dielectrics

6.4.1 Wafer Pre-Processing

Due to the mass transport limited deposition character of SACVD processes, ozone/ TEOS based films are quite sensitive to the surfaces on which the film has to be deposited. Therefore, appropriate wet/dry cleaning of the wafer surfaces after DT etching is mandatory for repeatable deposition results, especially concerning the film conformality in HAR trenches. One practical cleaning flow (Fraunhofer IZM Munich) is given below:

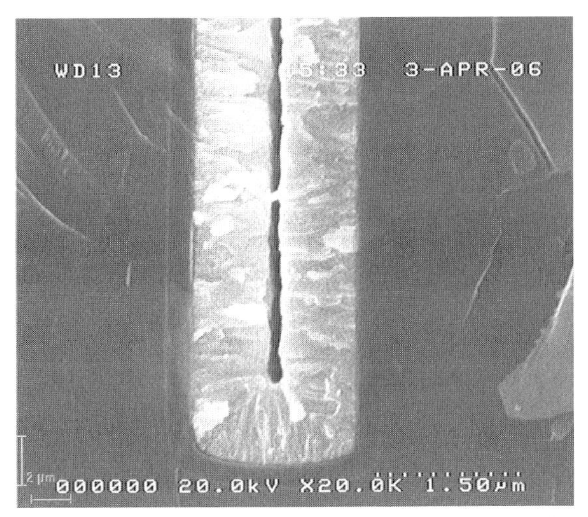

Figure 6.9 Conformal SACVD O_3/TEOS SiO_2 film in a TSV after W-fill. Aspect ratio ~8 : 1, taper angle 89° (HBr-process).

1. *O_2/H_2O ashing:* Microwave downstream plasma strip program for removal of sidewall polymers after DT etching (carbon fluoride based polymer).
2. *Chemical wet cleaning:* Removal of organic and inorganic contaminants, i.e., Caro's acid (H_2O_2 and H_2SO_4) followed by SC1/SC2 cleaning (SC1: NH_4OH and H_2O_2; SC2: HCl and H_2O_2).

Also a short O_2-plasma etch step (RIE), either *in situ* after DT etch or done afterwards, can help in removing sidewall passivates.

Quantifying the possible influence of pre-cleaning on the achievable conformality is difficult, because the surfaces of the trench holes, the trench depth, trench size, AR and taper angle all influence the deposition result. Figure 6.9 shows a W-filled TSV (AR ~8 : 1, trench depth 18 µm) with a 240 nm thick, conformal O_3/TEOS film, deposited after the above-mentioned cleaning sequence.

6.4.2
Backside Processing Requirements on SiO_2 Film Conformality in TSVs

Many 3D process flows require Si-thinning from the wafer backside until the TSVs are opened. Once this point of the thinning sequence is reached, either by Si-CMP (chemical-mechanical polishing) or by Si dry etching, the TSVs have to be electrically contacted from the back by an additional metal layer, in most cases an additional Cu or AlSiCu structure. To keep the electrical isolation intact, the former trench bottom of the TSV – seen from the front side – which in turn becomes the freshly opened "tip" of the metal-filled TSV, must have enough SiO_2 isolation to withstand the etch attack of the Si-etch back step (Figure 6.10).

This SiO_2 thickness depends mainly on two parameters: The prior deposited absolute SiO_2 film thickness and its achieved film conformality. If a maskless

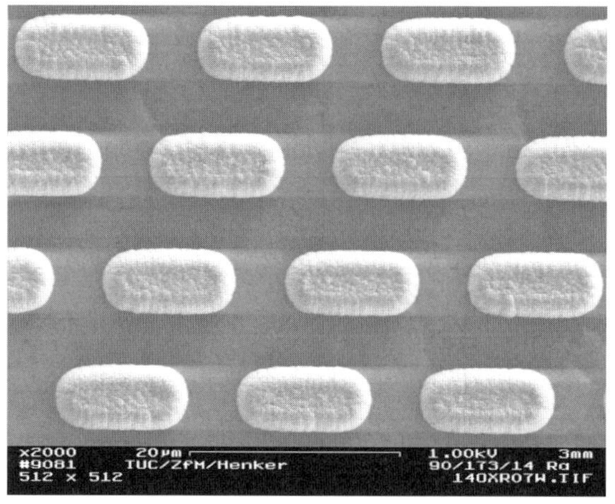

Figure 6.10 Backside view (SEM) of a Cu-filled TSVs, just opened from the back after Si-etching. O_3/TEOS film still covering the Cu-metallization. TSV depth 40 µm; nominal size 3 × 10 µm at front side.

technique is used to continue the 3D integration after "touching" the tips of the TSVs, the SiO_2 film thickness at the TSV tip should be ≥200 nm. Modern Si-dry etching processes can achieve selectivities up to 50 : 1 between Si and SiO_2. If, for example, a Si recess of ~2 µm is targeted, the estimated total SiO_2 removal will be ~40 nm, thus leaving ~160 nm SiO_2. For a backside photoresist mask, which is used to structure a previously deposited backside SiO_2 film, even thicker SiO_2 films are advisable due to known alignment accuracies of stepper lithography or mask aligners.

6.4.3
SiO_2 Film Deposition on Thinned Silicon Substrates

Once the Si-thinning has been completed, a backside dielectric film is required to fully isolate the thinned Si-substrate from the metal-filled TSVs. For this purpose, silane-based PECVD oxides at low deposition temperature are preferable. In most cases the thinned silicon substrate has to be temporarily glued onto a handling substrate, otherwise the Si-substrate could no longer be handled safely. Therefore, the glue material, often a thermoplastic or wax, which can be chemically removed later, requires a maximum processing temperature in the range of 180 °C or lower, otherwise delamination between handling wafer and thinned substrate will occur. Since PECVD of TEOS – or other precursors like HMDS – tends to show early condensation of the TEOS vapor at low deposition temperatures, some reactor types would not allow running a deposition process below 250 °C. Therefore, silane-based PECVD processes at temperatures down to 150 °C can be used instead. The achievable SiO_2 film quality is of course lower than that of standard processes. However, film thickness can well be increased above 1 µm, thus compensating for the lower electrical breakdown voltage of a low temperature film. Figure 6.11 shows a

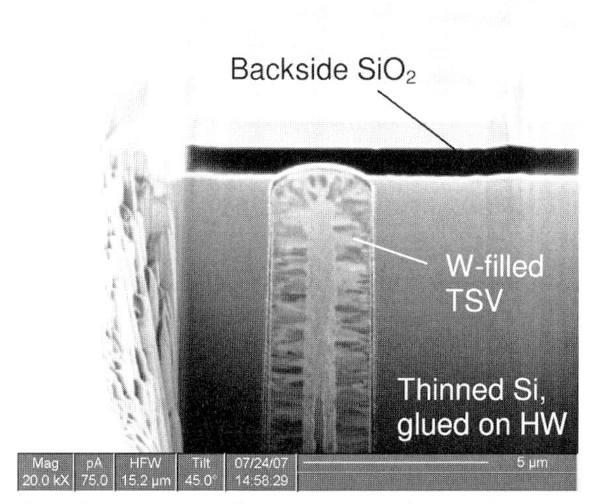

Figure 6.11 FIB image of a 50 μm thin Si substrate glued on a handling wafer, with W-filled TSV; SiO_2 CVD-film (silane-based) deposited at 150 °C on backside of handling wafer stack.

FIB-image of thinned Si with a 50 μm deep W-filled TSV, covered with 1 μm backside SiO_2 (silane-based PECVD, 150 °C deposition temperature).

6.5 Concluding Remarks

SiO_2-based dielectrics are widely used in 3D applications. Their well-known film characteristics, electrical performance and comparably easy deposition make them strong candidates for use in 3D. SACVD O_3/TEOS films are well suited for electrical isolation of HAR TSVs, if temperature limitations due to metal presence have to be considered. Film conformality might become one of the limiting factors for the electrical isolation of HAR TSVs when SACVD processes are used. However, taper angle, aspect ratio, sidewall roughness and cleanliness of the etched TSVs play important roles for the maximum achievable conformalities of these films.

References

1 Schumicki, G. and Seegebrecht, P. (1991) *Prozeßtechnologie – Fertigungsverfahren für Integrierte MOS-Schaltungen*, Springer Verlag, Berlin, 143 ff.

2 Widmann, D., Mader, H. and Friedrich, F. (1996) *Technologie hochintegrierter Schaltungen*, Springer Verlag, Berlin, 21 ff.

3 Wolf, S. and Tauber, R.N. (2000) *Silicon Processing for the VLSI Era Volume 1: Process Technology*, Lattice Press, Sunset Beach, California, 265 ff.

4 Nicollian, E.H. and Brews, J.R. (1982) *MOS (Metal Oxide Semiconductor) Physics and Technology*, Wiley, 709 ff.

5 Schumicki, G. and Seegebrecht, P. (1991) *Prozeßtechnologie – Fertigungsverfahren für Integrierte MOS-Schaltungen*, Springer Verlag, Berlin, 189 ff.

6 Widmann, D., Mader, H. and Friedrich, F. (1996) *Technologie hochintegrierter Schaltungen*, Springer Verlag, Berlin, 13 ff.

7 Graßl, T. (1998) Oberflächeninduzierte Abscheidung von Siliziumdioxid aus der Gasphase, Technical University of Munich – Physical Department, E16, p. 30–32.

8 Graßl, T., Ramm, P. *et al.* (1995) Deposition of TEOS/O_3 oxide layers for application in vertically integrated cicuit technology. Proceedings of the first International Dielectrics for VLSI/ULSI Mulitlevel Interconnection Conference, p. 382.

9 Wieland, R., Bonfert, D., Klumpp, A. *et al.* (2005) 3D Integration of CMOS transistors with ICV-SLID technology, Proceedings European Workshop Materials for Advanced Metallization MAM 2005, Microelectronic Engineering 82, pp. 529–533.

10 Bunshah, R.F. (1994) *Handbook of Deposition Technologies for Films and Coatings*, Noyes Publications, Berkshire, UK, 374 ff.

11 Shareef, I.A. *et al.* (Jul/Aug 1995) Subatmospheric chemical vapor deposition ozone/TEOS process for SiO_2 trench filling. *Journal of Vacuum Science & Technology B*, **13**, p. 1891.

12 Chatterjee, A. *et al.* (1995) A shallow trench isolation study for 0.25/0.18 μm CMOS technologies and beyond. Symposium on VLSI Technology Digest of Technical Papers, 16.3, IEEE.

13 Chin, B.L. and van de Ven, E.P. (April 1988) Plama TEOS process for interlayer dielectric applications. *Solid State Technology*, 119–122.

14 Wolf, S. and Tauber, R.N. (2000) *Silicon Processing for the VLSI Era Volume 1: Process Technology*, Lattice Press, Sunset Beach, California, 192 ff.

15 Fujino, K. *et al.* (1990) Silicon dioxide deposition by atmospheric pressure and low-temperature CVD using TEOS and ozone. *Journal of the Electrochemical Society*, **137** (9), 2883–2887.

7
Insulation – Organic Dielectrics

Philip Garrou and Christopher Bower

There are many organic insulators like benzocyclobutene (BCB) and polyimide (PI) commercially available and qualified for fabrication of microelectronic thin film structures [1]. However, such spin-coated materials cannot, in general, be deposited as conformal thin films, which is a requirement for the insulation of TSVs [2]. Here we will briefly discuss organic insulators that can be deposited in a conformal manner and review those demonstrated as sidewall insulators in TSVs.

7.1
Parylene

Parylene is the trade name for a well-known family of vapor deposited conformal polymers that have been used in a wide range of applications. Originally developed and commercialized by Union Carbide over 40 years ago [3, 4], parylene deposition equipment and materials are now available from several vendors. Figure 7.1 shows the process for deposition of parylene.

First, parylene dimer (di-*para*-xylylene) is placed into a small vacuum chamber where it is heated and vaporized. Next, the dimer vapor passes through a high-temperature zone where it is pyrolyzed into the monomeric form (*para*-xylylene). Finally, the reactive monomer enters the room temperature deposition chamber and deposits a conformal transparent polymer film [(Poly)-*para*-xylylene]. Figure 7.2 shows the molecular structures of the more well-known members of the parylene family.

Parylene-C and -D have one and two chlorine atoms, respectively, on the aromatic ring. Parylene-F (also called Parylene-AF4), with totally fluorinated methylene, groups was studied in the mid-1990s as a possible low-k dielectric [5–8]. Notably, the deposition of Parylene-F is not as simple as the traditional room temperature deposition process shown in Figure 7.1 [8]. Specialty Coating Systems is currently offering a parylene called Parylene-HT [9, 10] that has properties similar to Parylene-F. Table 7.1 summarizes Parylene properties.

Handbook of 3D Integration: Technology and Applications of 3D Integrated Circuits.
Edited by Philip Garrou, Christopher Bower and Peter Ramm

ISBN: 978-3-527-32034-9

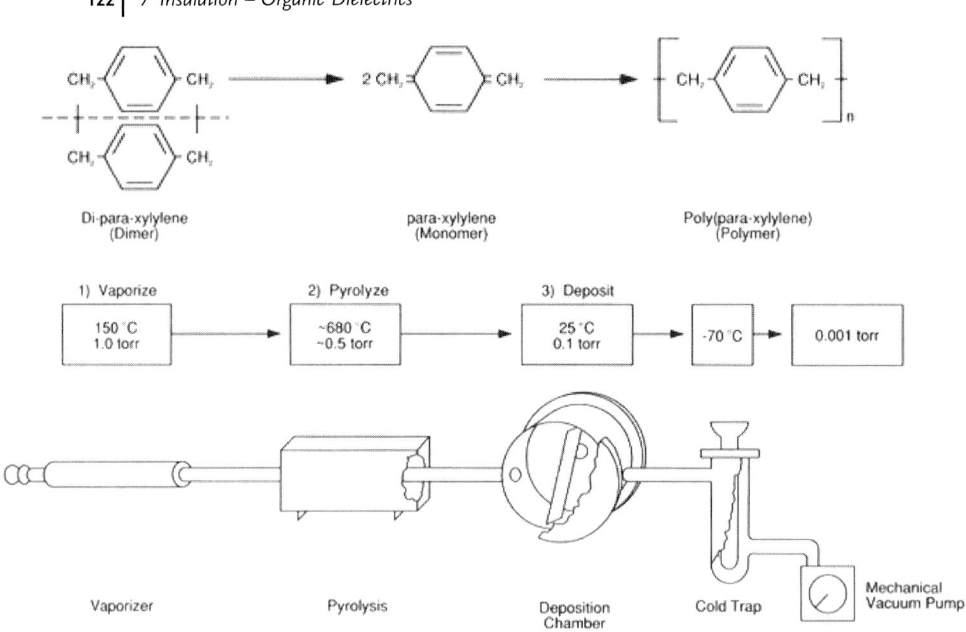

Figure 7.1 Parylene deposition process. From Reference [4].

7.1.1
Parylene in TSVs

Parylene has several attributes that make it an attractive candidate for insulation of TSVs. Parylene films are highly conformal. Burkett *et al.* reported 48% conformality in a 14 : 1 depth-to-width aspect ratio trench [11]. Parylene films are considered to be free of pinholes and exhibit low water absorption. The room temperature deposition

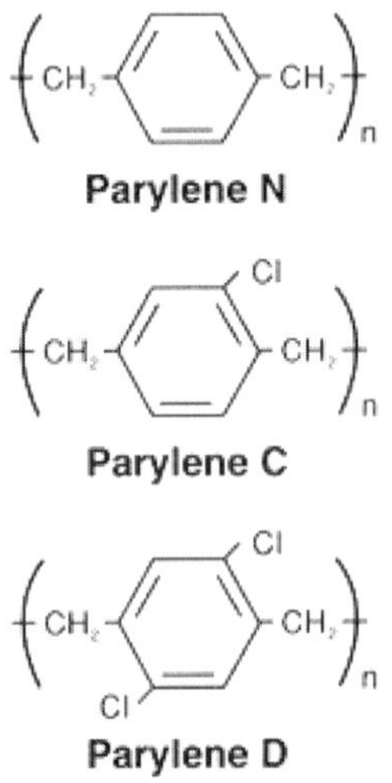

Parylene N

Parylene C

Parylene D

Figure 7.2 Molecular structures of common types of parylene. From Reference [4].

Table 7.1 Properties of Parylene-N, Parylene-C, and Parylene-F. Information compiled from References [3, 4].

	Parylene-N	Parylene-C	Parylene-F
Mechanical properties			
Tensile strength (MPa)	45	45–55	52
Elongation at break (%)	40	200	10
Modulus (GPa)	2.4	3.2	2.6
Density ($g\,cm^{-3}$)	1.110	1.28	NA
Water absorption (%) (24 h)	0.01 (0.019″)	<0.01	<0.01
Electrical Properties			
Dielectric constant (1 MHz)	2.65	2.95	2.17
Dissipation factor (1 MHz)	0.0006	0.013	0.001
Typical barrier properties			
Moisture vapor transmission[b]	1.50	0.21	NA
Typical thermal properties			
Melting temperature (°C)	410	290	NA
T_g (°C)	200–250	150	
CTE (10^{-5}/°C)	69	35 (50 after anneal)	36
Thermal conductivity, 10^{-4} ($cal\,s^{-1}$)/($cm^2\,°C\,cm^{-1}$)	3	2	NA

[a] cm3-mil/100 in2/24 h-atm (23 °C).
[b] g-mil/100 in^2/24 h, 37 °C, 90% RH; data from Cookson Specialty coatings website unless noted.

temperature makes parylene a good candidate material for CMOS compatible "TSV-last" approaches to 3D integration. Several groups have demonstrated TSVs that utilize parylene as the TSV insulation material.

The Centre Suisse d'Electronique et de Microtechnique (CSEM) demonstrated an IC-compatible process for fabricating TSVs using Parylene-C insulation [12]. In this work, 90 × 90 μm through holes were etched completely through 380 μm thick silicon wafers. A 4–5 μm thick layer of Parylene-C was deposited as the TSV insulation. Figure 7.3 shows the measured parylene uniformity within the through-wafer vias.

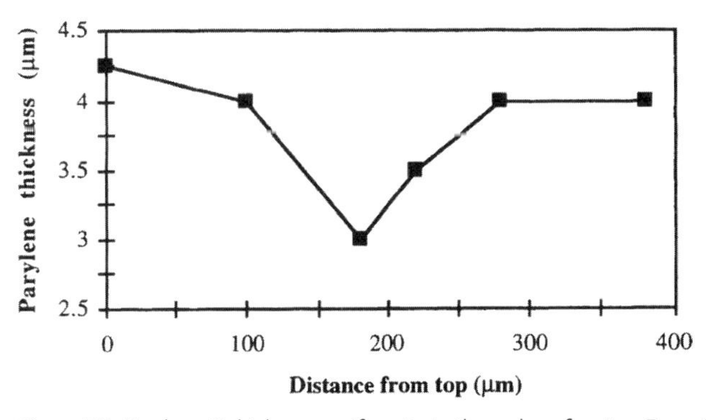

Figure 7.3 Parylene-C thickness uniformity in through-wafer vias. From Reference [12].

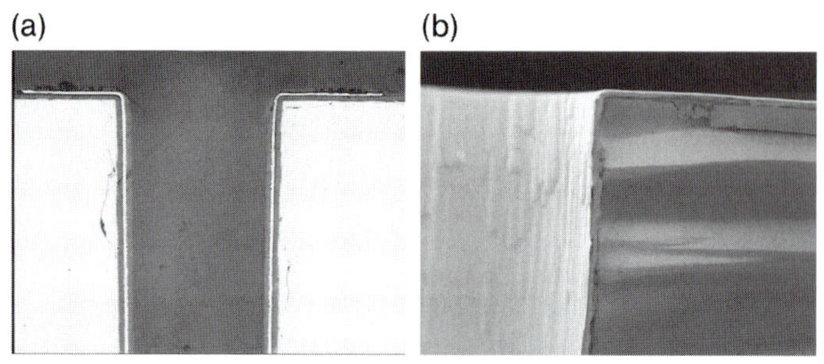

Figure 7.4 (a) Cross-sectional image of a through-wafer via coated with Parylene-C insulation, MOCVD TiN and Cu metallization, and Parylene-C encapsulation; (b) Electron micrograph showing the Parylene-C coating on the sidewall of the through-wafer via. Images courtesy of F. E. Rasmussen and O. Hansen [13, 14].

The leakage current between a 20 TSV daisy chain and the silicon substrate was found to range from 0.1 to 10 nA at 50 V bias. The dielectric breakdown was in excess of 100 V.

A group at the Technical University of Denmark has reported on a process for fabricating through-wafer vias in CMOS wafers that utilizes Parylene-C insulation (Figure 7.4) [13, 14]. They note that the wafer front and back sides and the through-holes are coated in a single deposition step. Improved metal-to-parylene adhesion was obtained when the parylene was treated with oxygen plasma prior to metal deposition. The authors consider that the oxygen plasma facilitates the formation of metal–oxygen–carbon and metal–carbon bonds.

IMEC has reported a process designed to allow TSVs to be fabricated at under 250 °C for post-CMOS compatibility [15]. For this process, a 1–2 μm layer of Parylene-N was used as the TSV insulation. Figure 7.5 shows the process flow used to fabricate the

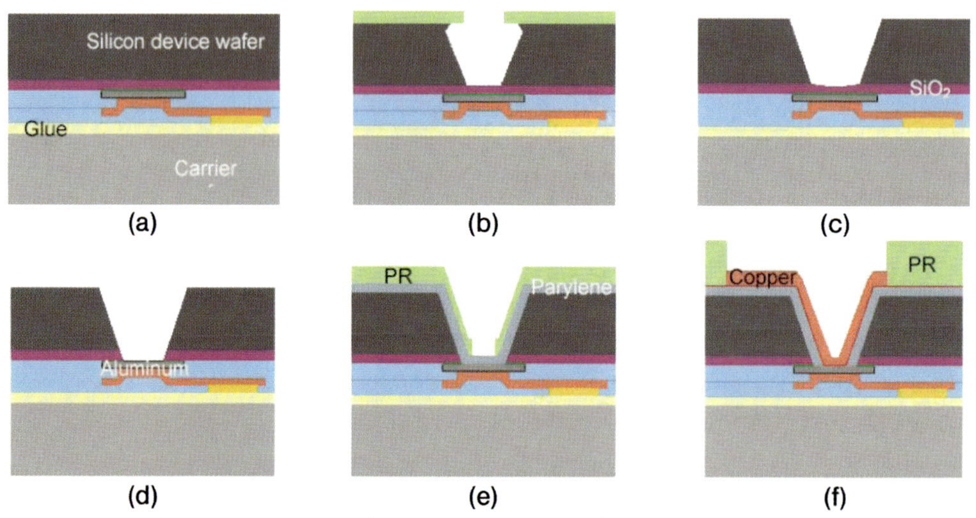

Figure 7.5 IMEC process for post-CMOS through-wafer vias that uses Parylene-N as the sidewall insulation. From Reference [15].

through-wafer vias. The wafer is first attached to a carrier wafer and thinned to 100 μm. Next, the TSV is etched from the backside of the wafer and stops on an aluminium landing pad. A special silicon RIE process was developed to generate sloped sidewalls in the vias. The parylene is deposited as the sidewall insulator. A via through the parylene at the base of the vias is achieved using a specially developed spray-coated photoresist and dry etching using an oxygen plasma. The TSV metallization consists of a sputtered Ti/Cu/Ti seed layer followed by electroplated copper.

7.1.2 Limiting Aspects of Parylene

The film stress of as-deposited Parylene-N is approximately 20 MPa (compressive). Since it is deposited on substrates at ambient temperature, no thermal stresses develop on cooling. However, thermal annealing, or subsequent thermal processing, results in the development of tensile stress. The thermal stress resulting from temperature excursions to 180 °C results in stress levels close to the yield strength of the material (55 MPa) [16, 17]. Processing temperatures resulting in stresses in excess of the tensile strength of the film can result in delamination and fracturing of the films. The tensile stress in Parylene-F after annealing at 350 °C is reported to be 19 MPa compared to a tensile strength of 55 MPa [16]. In addition, parylene N and C are known to oxidize in ambient air at temperatures in excess of 150 °C whereas Parylene-F exhibits significantly better thermal and oxidative stability.

Copper ion migration occurs in both inorganic dielectrics (i.e., SiO_2) and polymeric dielectrics to varying degrees, which is detrimental to both resistivity and breakdown voltage. Stanford University led a classic study on Cu^+ migration in dielectrics [18]. Results for Cu^+ drift are shown in Figure 7.6. The data reveal that LPCVD oxide is the worst and PECVD oxynitride is the best barrier. The polymeric materials fall in between with BCB showing excellent Cu migration resistance and Parylene-F (AF-4) also being significantly better than polyimides and SiO_2. Whether one uses an inorganic oxide or Parylene polymer, it is likely a diffusion barrier layer will be needed to prevent Cu^+ migration in TSVs.

Fluorine has been reported to react with metals like Cu, Ta and Ti [6]. In Cu damascene structures in particular, fluorine readily attacks tantalum-based barriers, leading to volatile TaF_2 formation and loss of low-k/barrier adhesion. This remains a key impediment to damascene integration of fluorine-based low-k materials with copper. It is not clear what the temperature onset of such bound fluorine reactivity is, but clearly such information would be important to any decision to use fluorinated polymeric materials, such as Parylene-F, in a 3D integration process.

7.2 Plasma-Polymerized BCB

A group at NEC has studied plasma-polymerization of BCB monomers to generate conformal BCB thin films [19–21]. The deposition tool consisted of a liquid mass flow

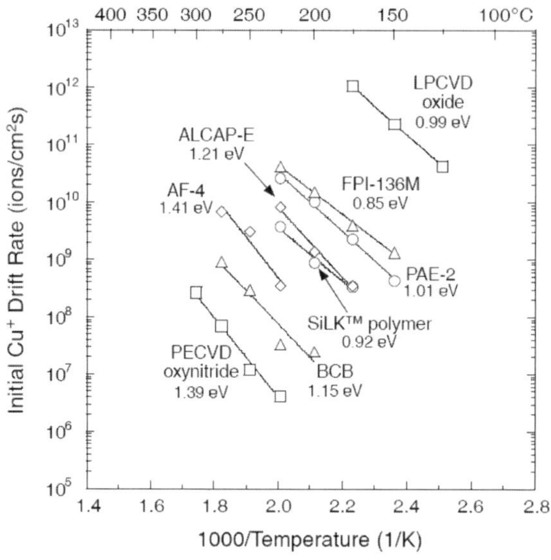

Figure 7.6 Cu^+ drift rates in various dielectrics. From Reference [18].

meter to control introduction of the monomer vapor and an RF plasma CVD reactor. The thermal stability of the BCB is reported to be significantly enhanced while other properties remain about the same (Table 7.2). The NEC data show that the BCB becomes highly conformal when deposited from the gas phase and thus may be applicable for such TSV insulation applications. Figure 7.7 illustrates the plasma-polymerization method and shows an electron micrograph of the conformal plasma-polymerized BCB film.

7.3 Spray-Coated Organic Insulators

Spray coating of a photoresist has been investigated recently as a method to generate conformal polymeric films over large topographies [22]. In this method an atomizing nozzle creates a spray of small polymer droplets that are deposited onto the surface. Several equipment manufacturers now offer wafer spray-coaters that can coat polymeric materials over considerable step topographies. Figure 7.8 shows a simple

Table 7.2 Properties of plasma-polymerized BCB films. From Reference [21].

Dielectric constant	2.7
Index of refraction (n)	1.59
Modulus	4.7 GPa
Leakage current @ 1 MV cm^{-1}	8×10^{-9} A cm^{-2}
Breakdown Field (>1 mA cm^{-2})	5 MV cm^{-1}
Thermal stability	400 °C

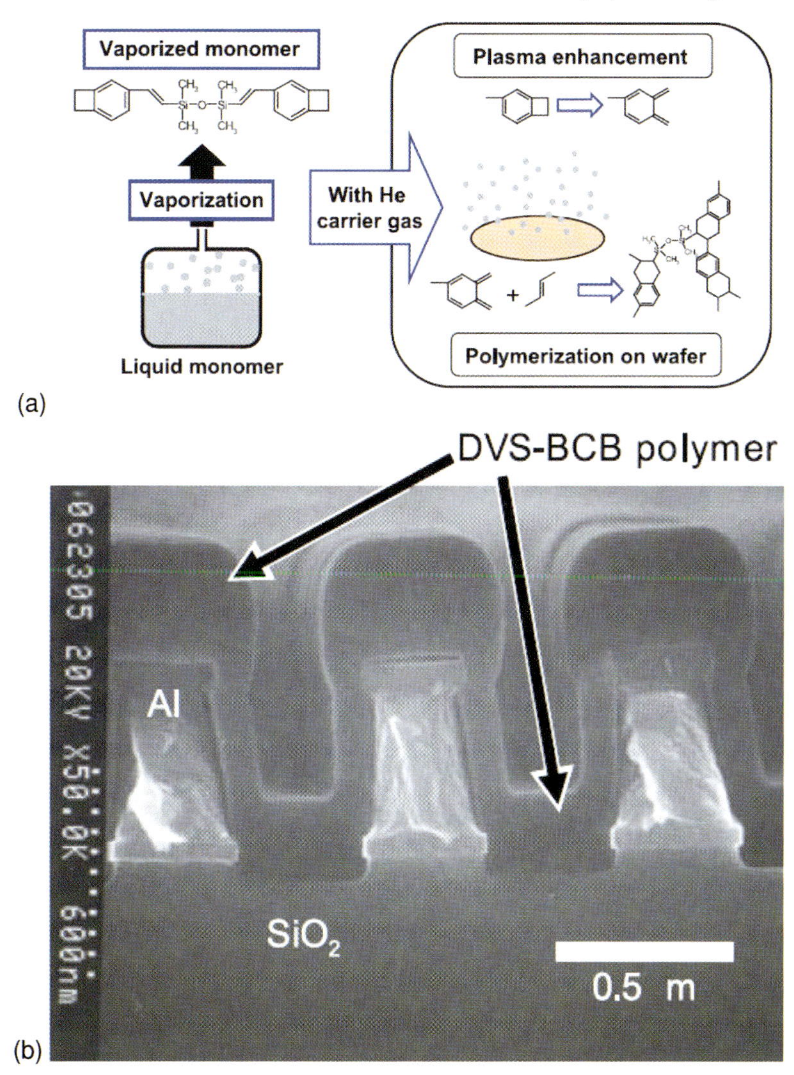

Figure 7.7 (a) Schematic illustration of the plasma-polymerization BCB process and (b) an electron micrograph showing the conformal plasma-BCB. From Reference [20].

drawing of the EV Groups EVG101 spray coater. During the coating process, the wafer is slowly rotated (30–60 rpm) while the spray nozzle is scanned across the wafer. In addition to photoresists, this method is generally applicable to all spin-on polymeric materials, including BCB and polyimide. To be spray-coated, the materials must be diluted to achieve a viscosity lower than approximately 20 cSt. Spray-coating of conformal dielectrics in high aspect ratio TSVs is still unproven. However, the use of spray-coating has already been reported in several 3D integration studies.

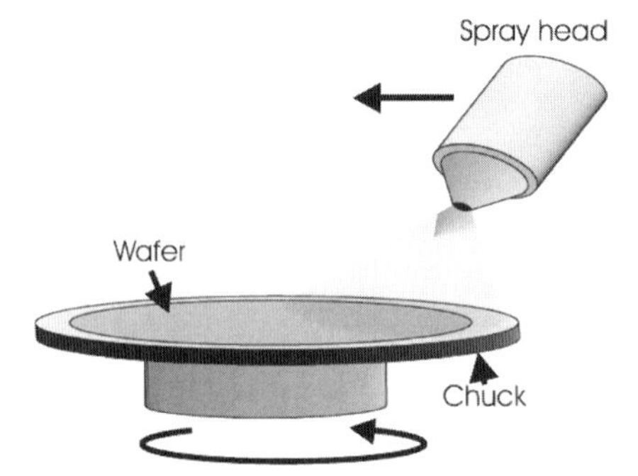

Figure 7.8 Illustration of a spray-coating system. From Reference [24].

Tezcan *et al.* [15] used spray-coated photoresist to allow for patterning at the base of a TSV. This was possible because the researchers developed a TSV with a sloped sidewall.

Schott Advanced Packaging has developed a TSV process that employs a spray-coated organic insulator [23]. This TSV has a relatively low aspect ratio and tapered sidewalls to effectively allow for this type of TSV insulation. Figure 7.9 shows an electron micrograph of the TSV with the spray-coated sidewall insulation.

The EV Group has recently announced a significant improvement in the feature topographies that can be coated by spray-coating [24, 25]. Figure 7.10 is an electron micrograph cross section of 200 μm diameter, 300 μm deep vias that have a continuous spray-coated polymer layer.

7.4 Laser-Drilled Organics

Laser ablation of silicon and organic resins has been utilized to fabricate organically insulated TSVs. Toshiba has demonstrated a low cost TSV process for vertically

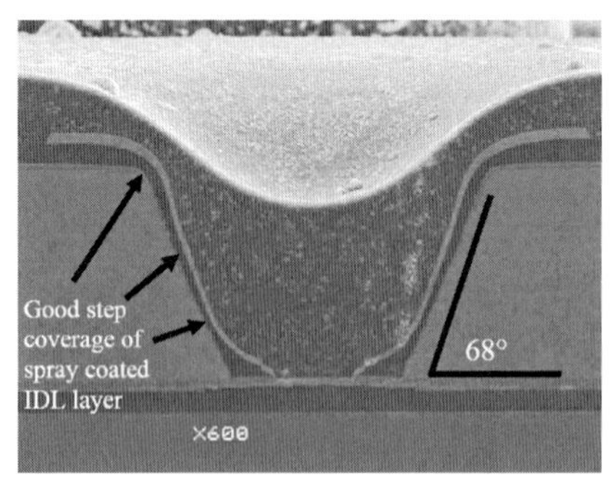

Figure 7.9 Electron micrograph of Schott Advanced Packaging's through silicon via with spray-coated insulation. From Reference [23].

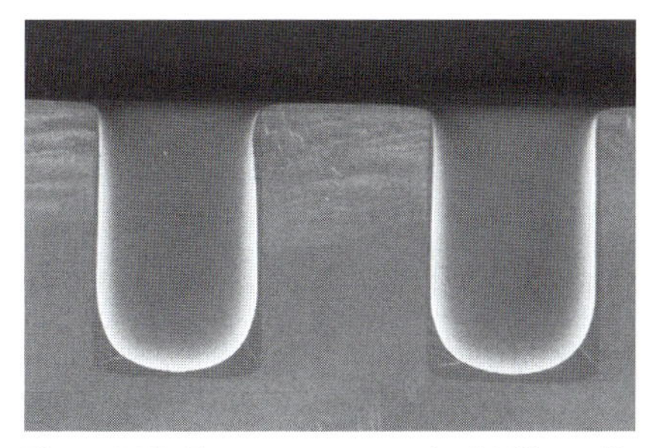

Figure 7.10 Electron micrograph of 200 μm diameter, 300 μm deep vias spray-coated with EVGroup's "nanospray" method. Figure courtesy of M. Wimplinger, EVGroup [24, 25].

integrated CMOS image sensors [26]. The process flow is shown in Figure 7.11. In this process, a YAG laser was first used to drill a via in the thinned silicon wafer. Next, an epoxy resin was vacuum laminated onto the wafer to achieve void-free filling of the vias. A noted advantage of this process is the low process temperature (180 °C max.) and the low-cost equipment required for vacuum lamination. A second via was YAG laser drilled through the dielectric resin. A process was developed to laser drill through both the silicon (1st via) and epoxy resin (2nd via) without damaging the metal landing pad. The laser-drilled via in the epoxy resin had sloped sidewalls to facilitate subsequent electroplating of the Cu interconnect. A group at ITRI has reported a through-wafer TSV process that utilizes laser-drilled vias through both the silicon and the epoxy resin [27]. The ITRI work is described in more detail in Chapter 5.

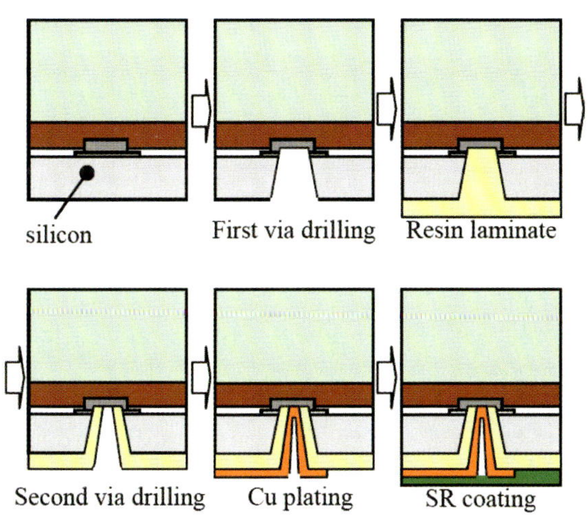

Figure 7.11 Toshiba process for low cost TSVs that utilize a laser-drilled resin laminate for the TSV insulation. From Reference [28].

7.5 Concluding Remarks

There are several potential methods to fabricate organically insulated TSVs. These range from vapor deposited polymers like the parylenes to laser-drilled epoxy resins. Owing to the low processing temperatures and potentially low cost, organics will continue to receive attention for demanding low-cost post-CMOS TSV applications.

References

1 Garrou, P. *et al.* (1997) Polymers in packaging. in *Microelectronics Packaging Handbook* (eds Tummala, Rymaszewski and Klopfenstein), Chapman & Hall, New York.

2 Garrou, P. (February 2005) Future ICs go vertical. *Semiconductor International*, p. SP10.

3 Fortin, J.B. and Lu, T.-M. (2004) *Chemical Vapor Deposition Polymerization – The Growth and Properties of Parylene Thin Films*, Kluwer Academic Publishers, Amsterdam.

4 Specialty Coating Systems Website http://www.scscoatings.com/parylene_knowledge. (accessed on October 2007).

5 You, L., Yang, G.-R., Lang, C.-I. *et al.* (1993) Vapor deposition of parylene-F by pyrolysis of dibromotetrafluoro-p-xylene. *Journal of Vacuum Science & Technology A*, **11** (6), 3047–3052.

6 Yang, G.-R., Zhao, Y.-P., Wang, B. *et al.* (1998) Chemical interactions at Ta/fluorinated polymer buried interfaces. *Applied Physics Letters*, **72**, 1846–1847.

7 You, L., Yang, G.-R., Lang, C.-I., Moore, J.A., Wu, P., McDonald, J.F., and Lu, T.-M., (1993) Vapor Deposition of Parylene-F by Pyrolysis of DiBromoTetraFluoro-p-Xylylene, *Journal of Vacuum Science and Technologie A*, **11**, 3047–3051.

8 Dolbier, W.R. and Beach, W.F. (2003) Parylene-AF4: a polymer with exceptional dielectric and thermal properties. *Journal of Fluorine Chemistry*, **122**, 97–104.

9 Kumar, R., Molin, D., Young, L. and Ke, F. (2004) New high temperature polymer thin coating for power electronics. Proceedings of the Applied Power Electronics Conference, APEC'04, Vol. 2, pp. 1247–1249.

10 http://www.scscoatings.com/parylene_knowledge/parylene-ht.aspx. (accessed on October 2007).

11 Burkett, S., Craigie, C., Qiao, X. *et al.* (2001) Processing techniques for 3-D integration techniques. *Superficies y Vacio*, **13**, 1–6.

12 Gobet, J., Thiebaud, J.-P., Crevoisier, F. and Moret, J.-M. (1997) IC compatible fabrication of through-wafer conductive vias. *Proceedings of the SPIE*, **3223**, 17–25.

13 Rasmussen, F.E., Frech, J., Heschel, M. and Hansen, O. (2003) Fabrication of high aspect ratio through-wafer vias in CMOS wafers for 3-D packaging applications. Transducers'03, Boston, MA, pp. 1659–1662.

14 Rasmussen, F.E. (2004) Electrical interconnections through CMOS wafers, Ph.D. Thesis, Technical University of Denmark.

15 Tezcan, D.S., Pham, N., Majeed, B. *et al.* (2007) Sloped through wafer vias for 3D wafer level packaging. Proceedings of the 57th Electronic Components and Technology Conference (ECTC 2007), Reno, NV, pp. 643–647.

16 Dabral, S., Van Etten, J., Zhang, X. *et al.* (1992) Stress in thermally annealed parylene films. *Journal of Electronic Materials*, **21**, 989–994.

17 Harder, T.A., Yao, T.-J., He, Q. *et al.* (2002) Residual stress in thin film Parylene-C. 15th International Conference on Micro Electro Mechanical Systems (MEMS'02), Las Vegas, NV.

18 Loke, A.L.S., Wetzel, J.T., Townsend, P.H. *et al.* (1999) Kinetics of copper drift in low-k polymer interlevel dielectics. *IEEE Transactions on Electron Devices,* **46**, 2178–2187.

19 Kawahara, J., Nakano, A., Saito, S. *et al.* (1999) High performance Cu interconnects with low-k BCB-polymers by plasma-enhnaced monomer-vapor polymerization (PE-MVP) method. *VLSI Technology Symposium Digest,* 45–46.

20 Kawahara, J., Nakano, A., Kinoshita, K. *et al.* (2003) Highly thermal-stable, plasma-polymerized BCB polymer film. *Plasma Sources Science and Technology,* **12**, S80–S88.

21 Tada, M., Ohtake, H., Kawahara, J. and Hayashi, Y. (2004) Effects of material interfaces in Cu/low-k damascene interconnects on their performance and reliability. *IEEE Transactions on Electron Devices,* **51**, 1867–1876.

22 Pham, N.P., Burghartz, J.N. and Sarro, P.M. (2005) Spray coating of photoresist for pattern transfer on high topography surfaces. *Journal of Micromechanics and Microengineering,* **15**, 691–697.

23 Shariff, D., Suthiwongsunthorn, N., Bieck, F. and Lieb, J. (2007) Via interconnections for wafer level packaging: impact of tapered via shape and via geometry on product yield and reliability. Proceedings of the 57th Electronic Components and Technology Conference (ECTC 2007), Reno, NV, pp. 858–863.

24 Wimplinger, M. (2006) New nanospray technology achieves conformal coating of extreme surface topographies. International Wafer-Level Packaging Conference, San Jose, CA.

25 http://www.evgroup.com/NanoSpray.asp. (accessed on October 2007).

26 Sekiguchi, M., Numata, H., Sato, N. *et al.* (2006) Novel low cost integration of through chip interconnection and application to CMOS image sensor. Proceedings of the 56th Electronic Components and Technology Conference (ECTC 2006), San Diego, CA, pp. 1367–1374.

27 Lo, W.-C., Chen, Y.-H., Ko, J.-D. *et al.* (2006) An innovative chip-to-wafer and wafer-to-wafer stacking. Proceedings of the 56th Electronic Components and Technology Conference (ECTC 2006), San Diego, CA, pp. 409–414.

8
Copper Plating

Tom Ritzdorf, Rozalia Beica, and Charles Sharbono

8.1
Introduction

Many groups are currently working with copper as the conductive material to provide conductors through silicon for through-silicon via (TSV) applications. These applications include a wide range of feature sizes and architectures, but many of them utilize copper electroplating to provide copper conductor deposition. The first products to employ copper-filled TSVs are flash memory devices that stack multiple chips, and CMOS image sensors, which utilize back-side contact to reduce interference of light transmission by the electrical wiring and produce more compact cameras.

Copper electrochemical deposition (ECD) processes have several characteristics that make them attractive for depositing thick layers of copper for lining or filling relatively large features. ECD processes are typically performed at close to room temperature and pressure, so the equipment is not as complex and expensive as vacuum deposition equipment. In addition, these aqueous processes provide superconformal deposition under the correct conditions [1]. This feature provides the capability to deposit more material within high aspect ratio features than at the mouth or on the top field portion of the substrate.

Copper plating has been used in the production of semiconductor devices for over a decade, and has been extensively studied [2–4]. The capability of copper ECD processes to achieve superconformal deposition in submicron features and proposed mechanisms have been extensively reported [5, 6]. While it is fairly well understood how to achieve superconformal copper deposition in these small features, the same processes cannot be utilized successfully for filling large vias that are tens of microns deep. In submicron damascene interconnect processes, the feature depths are one to two orders of magnitude less than the hydrodynamic and diffusion boundary layer thickness typical of most industrial wafer plating equipment. On the other hand, TSV applications sometimes utilize via depths that are up to an order of magnitude larger than the hydrodynamic boundary layer thickness. This difference drives how the processes interact with the plating process parameters and how the processes must be optimized differently to achieve the same result.

Handbook of 3D Integration: Technology and Applications of 3D Integrated Circuits.
Edited by Philip Garrou, Christopher Bower and Peter Ramm

ISBN: 978-3-527-32034-9

In submicron interconnect processes, the deposition rate can be high enough to lead to significant variations in the cupric ion concentration through the depth of a feature. While this variation can be high enough in some cases to impact the deposition rate at the bottom of the feature, the typical concentration difference is on the order of 20–50% of the bulk cupric ion concentration. Since the objective is usually to maximize the deposition rate in the TSV fill processes, the cupric ion concentration becomes a very important factor in determining the optimal process conditions. The other main difference in these two classes of processes is that the potential distribution in submicron features is usually fairly constant [7, 8], while the large features used in TSV processes can drive substantial potential variation through the feature depth due to the ohmic drop within the solution itself. These differences lead to differences in the optimal deposition conditions in order to achieve what appear to be similar results but which on dimensional scales are two or three orders of magnitude apart.

Alternative conductive materials for TSV applications include CVD tungsten and doped polysilicon. Both of these materials provide a better match of the coefficient of thermal expansion (CTE) with silicon, but at the expense of lower electrical and thermal conductivity than copper, as well as the use of more expensive vacuum deposition equipment. While announcements have been made regarding the use of all of these materials [9, 10], it remains to be seen which will gain more market acceptance, or which materials will fit what applications.

8.2 Copper Plating Equipment

The equipment that can be used for copper deposition can vary from simple manually operated tank systems to fully automated wafer-processing equipment. The important basic features of ECD equipment for copper plating include the ability to contain the corrosive plating chemistries and to control the hydrodynamic boundary layer thickness everywhere on the wafer, the ability to provide the optimized electrical input, and the ability to handle silicon wafers (sometimes thinned) with minimal chance of contamination or damage. In addition, it is common to provide equipment that is capable of automated chemistry composition control [11–13].

Equipment used for wafer plating typically falls into one of two categories: wet bench or single wafer fountain platers. Wet benches for wafer plating usually handle wafers that are loaded in a handling and electrical contact fixture and move the fixtures from cell to cell in a vertical orientation. In contrast, fountain plating equipment usually handles one wafer at a time and the wafers are processed in a face-down horizontal orientation. The fluid handling requirements are similar regardless of the type of equipment utilized. There must be a tank that can handle and monitor all the chemistry for the plating process, as well as pumps, filters, temperature controllers and flow controllers to control the process related fluid parameters.

The electroplating process itself is controlled by the electrical characteristics and the mass transfer characteristics of the electroplating reactor. This means that it is very important to control the fluid agitation, especially at the wafer surface where it directly impacts the hydrodynamic boundary layer thickness at the deposition surface. These agitation conditions set up the mass transfer characteristics by determining the length over which diffusion of the chemical species must occur. Typical industrial wafer plating equipment utilizes agitation schemes that provide uniform hydrodynamic boundary layer thickness, on the order of 10–50 microns.

Electrolytic deposition is controlled by the electroplating potential and applied current. Therefore, it is critical to provide adequate power supply capabilities and to design the plating reactor in a way that optimizes the global distribution of current density across the wafer. This is typically done through the use of current shields, auxiliary cathodes (current thieves), and virtual anode elements in the plating reactor. It is also common to provide pulse or pulse reverse waveform capability as part of the power supply. This enables modification of the surface concentrations locally through the use of an electrical waveform. For example, the use of "off" time in a recipe allows the concentration gradient imposed by deposition to relax, and the addition of an anodic pulse may remove a portion of a deposited film such as copper and allow for better control of the local deposition morphology. Reversing the potential on the surface is also likely to cause desorption of some or all of the organic additives on the metal surface, which will impact the deposition morphology [14, 15].

Wafer handling is a critical aspect of automated wafer electroplating equipment, as it is not trivial, and must be robust for the equipment to be industrially useful. In addition to impacting tool reliability, a poor wafer handling system can cause wafer breakage or scrap, which could quickly become the largest cost associated with processing wafers. It is also important to be able to handle thinned wafers, or wafers mounted on carriers, for some 3D interconnect processes. Even if the wafer handling system does not cause any failures, it must be done correctly to utilize efficiently the process chambers on the equipment and allow for efficient capital utilization.

Chemical composition control is sometimes considered to be an ancillary function to the electroplating tool, but it is very important in terms of maintaining a manufacturing process that is operating within its control limits and providing high wafer yields. It is common to include at least some of the functional components for the automated analysis and/or dosing of various bath constituents either on-board the plating tool or in a separately integrated chemical control system. These chemical control systems usually have some form of controlled constituent dosing, which uses a dosing model based on time or number of wafers plated. They may also make use of chemical analysis of one or more of the components of the plating bath [11, 12].

8.3
Copper Plating Processes

Copper is the most common metal plated, owing to its mechanical and electrical properties, and has been used in various applications, from plating on plastics, printed wiring boards, zinc die castings, automotive bumpers, rotogravure rolls, electro-refining and electroforming [16, 17]. In the semiconductor industry, electroplated copper has played a major role in the change from aluminium to copper [2, 18–21]. This materials change has been one of the most important changes that the semiconductor industry has experienced since its creation [16, 20, 22].

Compatibility with conventional multilayer interconnection in LSI (large scale integration) and back-end-of line (BEOL) processes, as well as superior properties like low resistivity, high conductivity and purity, makes copper one of the preferred materials for through-silicon via (TSV) applications [23–25].

The different methods that have been proposed and tested for through silicon deep vias, including lining and filling of features, are chemical vapor deposition (CVD), plasma vapor deposition (PVD), electroless and electrodeposition processes. Applicability of each of these methods is limited by feature dimensions, ease of process applicability, final deposit characteristics, process reliability and cost considerations.

Copper cannot be effectively applied by "dry" deposition techniques: PVD does not provide acceptable fill and CVD, as shown in Figure 8.1, is more suitable for small features and also requires costly, unstable and hazardous organometallics, resulting in deposits with lower purity and high resistivity, as well as poor adhesion [26, 27].

Metal deposition through wet processes (plating) involve the use of chemical solutions and, in a very simplistic way, mean "to convert ions to metallic form by providing electrons to the solution." Depending upon the carrier providing the electrons, there could be three types of plating: electroplating (electron from a power supply), electroless plating (electron from reducing agent in the solution), and immersion plating or displacement plating (electron from a base metal). Of the "wet" processes, electroplating is the process of choice since electroless plating is

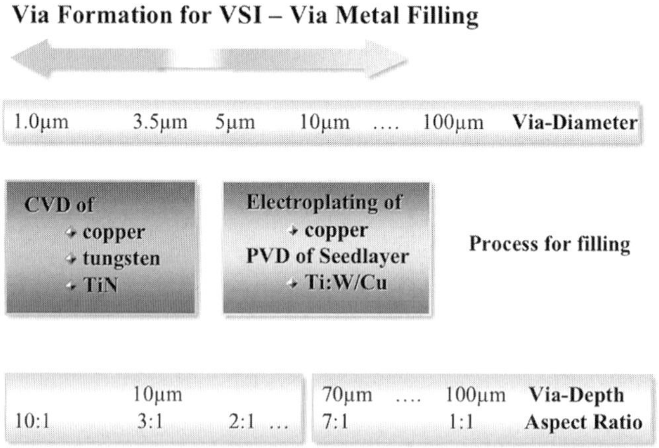

Figure 8.1 "Through via challenges" [27].

slower, involves more complex and costly chemistry and controls, and the bath requires frequent replacement. Moreover, electroplating is characterized by faster deposition rates (~1 μm min^{-1}) than electroless plating (~0.2 μm min^{-1}), CVD (~0.2 μm min^{-1}) or PVD (0.05–0.1 μm min^{-1}), and requires less costly equipment than the vacuum-based processes [28].

Because copper electrodeposition is already a well-known semiconductor process technology due to its application in copper damascene interconnects, it was believed that its transfer to filling through silicon vias would be easily adopted. However, as with any new technology at its beginning, its applicability to TSV interconnection had, and continues to have, its challenges and own learning curves on the way to becoming a fully mature technology.

The first attempts at via filling through electrodeposition started with larger features. The intent was to minimize integration issues at the expense of increased filling difficulty, but this resulted in lower throughput and increased overall processing costs. Also, when it comes to large features, especially on a thinned wafer, the coefficient of thermal expansion mismatch becomes an issue because copper expands 5–6 times more than silicon. Therefore, for larger vias, to minimize the stress concentration that can result in wafer breakage, polymer coating has been suggested as a more appropriate filling method [27–29]. Another approach considered to minimize the CTE mismatch problem is applying a thin metal lining and filling with a non-conductive polymer that can more easily accommodate thermal expansion of the copper during heat treatment.

Each method has its own limitations. However, due to cost effectiveness, popularity and wide applicability to various feature dimensions without any limitations of shape, copper electroplating processes are the technology of choice for filling through silicon vias.

There are typically three types of blind-via filling processes applicable through electrodeposition (Figure 8.2): lining, full filling and full filling with pattern formation. The lining process, applied especially in the case of larger features, is generally used for sensor applications. Photoresist patterns can be added to form directly a metal stud (Figure 8.2c) or a redistribution line (RDL) pattern above vias using a single plating process step. The stud can further be used as a mini-bump for eutectic bonding [23].

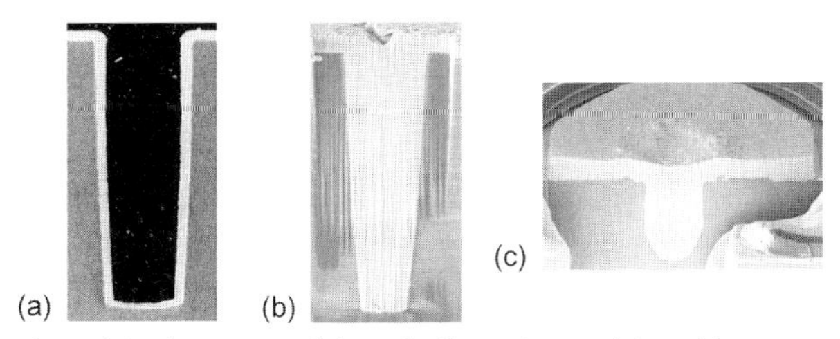

Figure 8.2 Three types of through-silicon via metal deposition: (a) lining, (b) full filling and (c) full filling with stud formation.

8.3.1
Copper Lining

The lining process is typically employed on large features and in applications such as MEMS or sensors. As the name implies, a feature is lined with a copper deposit. Typically, requirements for copper lining are a high degree of conformality and thickness of 5–15% of the feature width.

The plating process parameters affecting the copper lining process include bath composition (organic and inorganic concentrations), deposition waveform, average current density, and flow conditions. Optimization of these parameters is needed to manage two primary effects or trends that affect the deposition within the features: current crowding at the feature mouth and mass transfer limitations at the feature base. These effects are due to an increased electric field at via corners (mouth) and difference in mass transport rates between the top and bottom of the features [23]. Figure 8.3 compares two conditions where these effects have been managed (Figure 8.3a) and not managed (Figure 8.3b).

Typical methods or paths to reduce the effects of current crowding (or deposit thickness) at the feature mouth are a reduced current density, a pulse waveform, and/or proper chemical composition, which balances the suppression at the via mouth with the acceleration of deposition deeper within the feature. Mass transfer limitations at the via bottom are reduced with a high limiting current density (LCD) bath and strong agitation.

Following the deposition of copper, the remaining open portion of a feature is lined or filled with other materials. The material options are primarily polymers or materials with a CTE near that of silicon such as a conductive ceramic paste. [30, 31]. Figure 8.4 illustrates the application of a copper lining and non-conductive polymer.

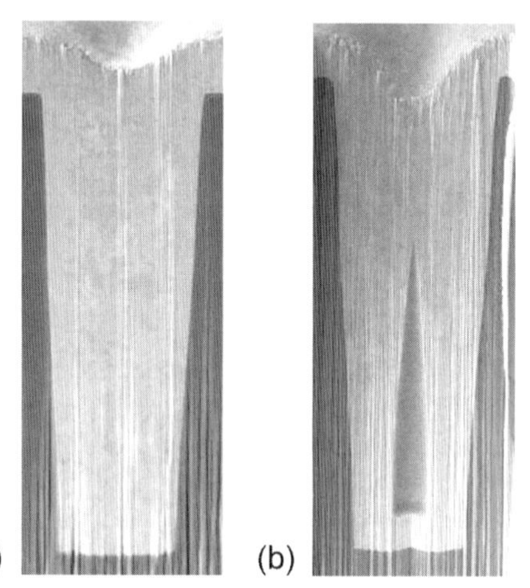

Figure 8.3 Effects of ECD process parameters on fill performance: (a) optimized conditions and (b) non-optimized conditions. Feature dimensions: 30 μm wide and 100 μm deep vias.

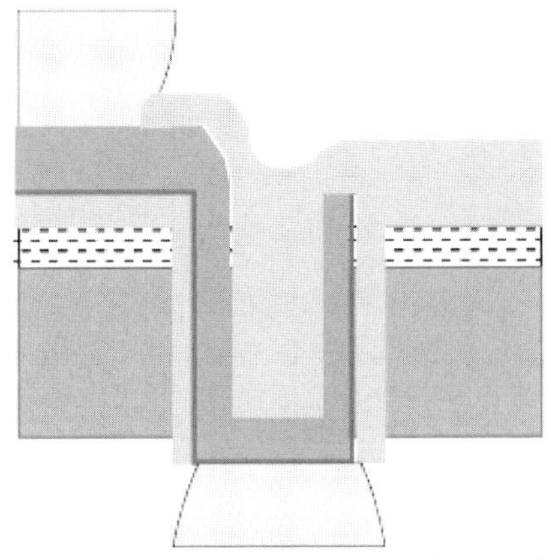

Figure 8.4 Via lined with insulating polymer and copper. Non-conductive polymer fills the interior part of the via. (Source: IMEC.)

Advantages of lining process vs. the full fill options are:

(a) decreased deposition time;
(b) elimination of CTE mismatch issue between copper and silicon.

The use of copper lining with a polymer filler functions as a buffer to mitigate the induced stress on the silicon during thermal treatment, which is especially important with thinned wafers.

8.3.2 Copper Full Fill With and Without Stud Formation

Full copper fill is employed over a wide range of feature dimensions, from near-damascene-scale features to large features used for sensor applications. The general requirement is a robust, void-free deposition within the features. The choice of adding a photodefined resist mask to allow the creation of a copper stud on top of the filled via depends on the integration scheme and the tradeoff between CMP removal of copper overburden and chemical etch of the barrier and seed layer. To adequately fill vias, a superconformal fill mechanism, where the rate of deposition is faster at the bottom of the feature than at the top, is required. Figure 8.5 shows an example of a strong superconformal fill mechanism through a fill series ($^1/_3$ fill, $^2/_3$ fill and full fill). As with the copper lining process, the copper ECD process parameters affecting the fill capability include bath composition (organic and inorganic concentrations), average current density, deposition waveform, and flow conditions.

For completely full filling, with or without a stud, the most critical barrier to copper metallization by plating has been the requirement for completely void-free deposition without any electrolyte entrapped within a reasonable process time [23, 32]. Several studies, very similar to copper damascene electrodeposition, have been performed, and to achieve void-free via filling it has been suggested that this application would also require superconformal fill (Figure 8.6) [6, 23, 25, 33, 34].

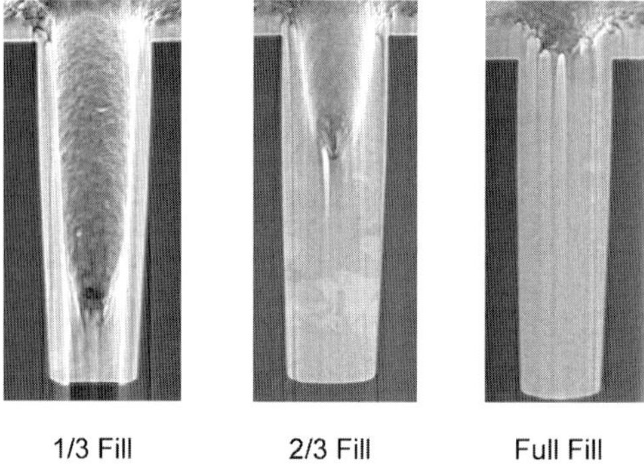

Figure 8.5 Fill profile evolution of electrodeposited copper. Feature dimensions: 5 μm wide and 25 μm deep vias.

Copper cannot be plated in its "normal" mode where the top ridges build-up first (due to enhanced transport and a higher field at the small curvature, highly accessible areas), generating a trapped void (Figure 8.7a, sub-conformal plating). Nor is it acceptable to plate conformally, by suppressing the deposition process through the use of excess additives (Figure 8.7b). The preferred process is shown in Figure 8.7c, where the copper growth rate is larger at the bottom of the via than at the top, resulting in a superconformal filling [25, 35].

Conventional copper plating systems have been tried for deep via filling as well; however, defects such as seams, voids and electrolyte inclusions can occur with conventional methods when attempting to plate features that are smaller or have higher aspect ratios, causing serious interconnect reliability problems by disrupting the ability of the electrodeposited metal to carry a coherent signal.

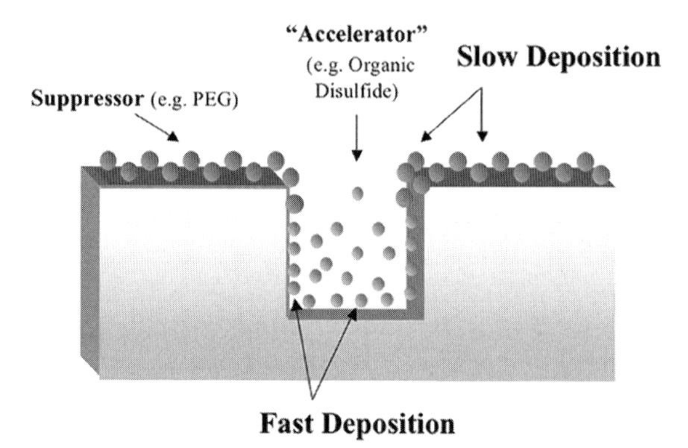

Figure 8.6 Additive distribution near and within a via. Variable adsorption leads to variable kinetics and superconformal deposition. Larger, slow diffusing, suppressor adsorbs primarily at the flat surface and along the rim, while a fast diffusing, smaller additive penetrates the via and enhances the deposition rate there [26].

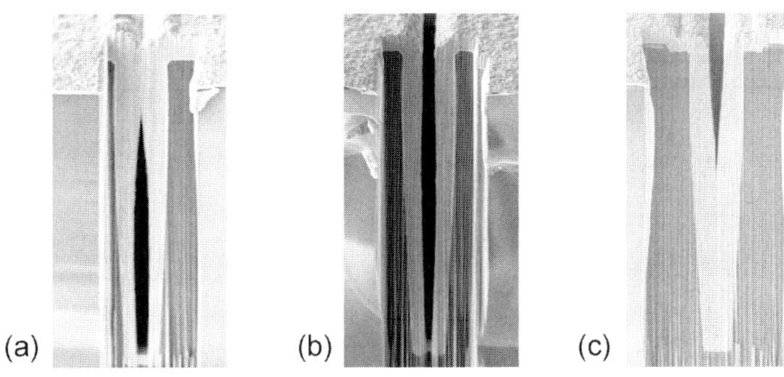

Figure 8.7 Various filling profiles: (a) sub-conformal, (b) conformal and (c) superconformal. Feature dimensions: 12 μm wide and 100 μm deep vias [37].

8.4 Factors Affecting Copper Plating

Factors that impact the ability to successfully fill TSV features with copper are:

- Via profile and smoothness
- Insulator/barrier/seed layer coverage
- Feature wetting.

8.4.1 Via Profile and Smoothness

Figure 8.8 illustrates the various via profiles that are typically obtained. The tapered via profile usually has a profile angle between 85 and 90° and generally represents the easiest profile to fill due to the ease of chemical transport into the feature and geometric leveling. The straight via profile has a sidewall angle of 90°. The "re-entrant" via profile has a smaller width at the top than in the middle of the feature and is the most difficult profile to fill due to the increased difficulty of chemical

Figure 8.8 Etched via profiles: (a) tapered, (b) straight side-wall and (c) re-entrant (negative slope). (Courtesy of Alcatel Micro Machining Systems.)

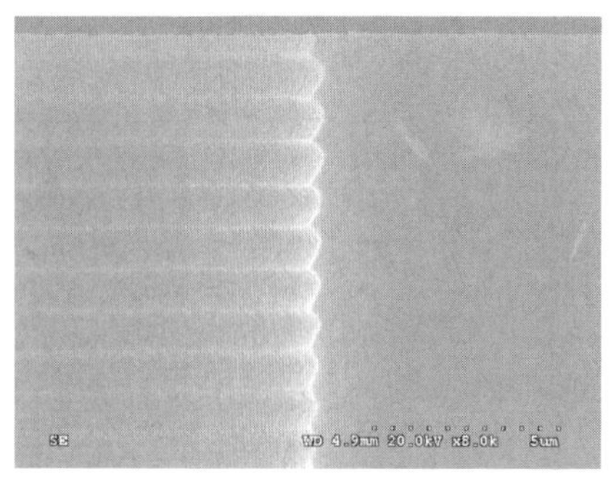

Figure 8.9 Via sidewall scalloping induced by the Bosch etch process. (Courtesy of Alcatel Micro Machining Systems.)

transfer into the feature and higher demands on the superconformal nature of the process.

The sidewall smoothness of the via is another important factor, due to its impact on the ability to provide a continuous layer of insulator/barrier/and seed. Figure 8.9 illustrates the "scalloped" profile obtained with some DRIE etch processes. The severity of the "scalloped" profile will impact the required thickness and coverage profile for a continuous layer of insulator, barrier and seed layer [27, 32, 38, 39].

8.4.2
Insulator/Barrier/Seed Layer Coverage

The general requirement for these layers is a continuous coverage over the whole feature of sufficient thickness with good interlayer adhesion. The quality of the insulator/barrier/seed layer coverage is impacted primarily by the deposition method, feature width/aspect ratio and via profile and smoothness [23, 39]. Figure 8.10

Figure 8.10 Void in via bottom due to poor seed layer coverage.

shows a bottom void in the copper plating due to insufficient copper seed layer coverage at the feature bottom.

8.4.3 Feature Wetting

For successful copper deposition, a feature must be wettable, or chemistry must be easily transferable into the features. Factors that impact the feature wetting are prewet process, feature geometry, surface condition of the seed layer, and surface tension of the copper plating chemistry [32, 39, 41].

The function of the prewet process is the removal of trapped air within the features, typically through immersion and/or impinging spray of liquid. As the feature geometry becomes more aggressive (increased aspect ratio, reduced width), a more aggressive prewet process may be needed or, in severe cases, a dilute surfactant solution is used to wet the features. The surface condition of the seed layer, such as a copper oxide layer, will also impact the feature wetting due to increased contact angle of the DI water on the seed surface. Treatments or solutions for the copper oxide layer include the use of a dilute acid solution, typically sulfuric acid, to etch away the copper oxide layer. Finally, the surface tension of the chemistry will impact the ability to wet the features. A wetting agent, or " surfactant," is typically added to the copper plating bath in small amounts to reduce the surface tension and promote good feature wetting.

Figure 8.11 illustrates the impact feature wetting can have on the deposition of copper. With poor feature wetting (Figure 8.11a), the bottom portion of the feature had a trapped air bubble present, preventing the deposition of copper. With a good prewet (Figure 8.11b), the trapped air has been removed, allowing the transfer of plating chemistry to the bottom of the feature and the subsequent deposition of copper. Allowing time for plating solution to diffuse into wetted features is also important prior to initiating electrodeposition.

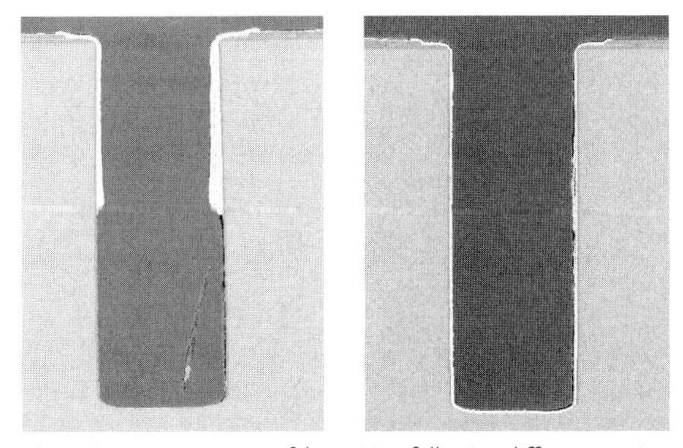

Figure 8.11 Comparison of deposition following different wetting processes: (a) poor wetting and (b) adequate wetting.

8.5 Plating Chemistries

Copper electroplated deposits can conventionally be obtained from various plating solutions. The main systems known today are sulfate, methane sulfonic or fluoroboric acid based, pyrophosphate complex ion or cyanide based. Of the plating systems that have been studied, relatively few have revealed a stage of commercial importance, and even fewer with respect to deep via filling [16].

8.5.1 Acid Copper Sulfate Chemistry

Literature publications and patents for sulfate solutions far outnumber those of all the other solutions combined, making this system the most widely and heavily used, especially for printed wiring boards and semiconductor interconnect technologies.

A common acid copper plating chemistry would be a solution containing a dissolved copper salt (copper sulfate), an acidic electrolyte (such as sulfuric acid) in an amount sufficient to impart conductivity to the bath, chloride ions to reduce anode polarization and increase the activity of certain additives, and organic additives to enhance the effectiveness and quality of plating. Table 8.1 summarizes the information concerning the major electrolyte components.

Many of today's commercially available chemistries contain three organic components designated as suppressor (or carrier), leveler and accelerator (or brightener), similar to additives commonly used for copper interconnect metallization. The suppressors are typically polyalkylene glycol type polymers with a molecular weight of around 2000. The levelers are typically alkane surfactants containing sulfonic acids and amine or amide functionalities, and the accelerators are typically sulfur derivatives of propane sulfonic acid. Table 8.2 summarizes the classes of commonly used plating additives.

Table 8.1 Major electrolyte components. Arrows indicate recent trends [26].

			Concentration	
Species	**Function**	**Polarization effect**	**Wafer plating**	**Conventional**
Copper Sulfate	Reactant	Mild accelerator	0.2–0.6 M ↓ 0.5–1.0 M	0.25 (M) 0.2–0.6 M
Sulfuric acid	Conductivity (supporting electrolyte)	Mild inhibitor	0.5–2 M (pH = 0) ↓ 0.003–0.1 M (pH = 1–3.5)	1.8 M 0.5–2 M (pH = 0)

Table 8.2 Classes of plating additives commonly used for copper metallization [26].

Species	Name	Function	Polarization	Concentration (ppm)
Chloride ion		Brightens deposit color	Mild inhibitor	40–100
Polyether (PAG polymer) (polyalkylene-glycol)	Carrier (wetting agent, leveler)	Levels (macroscopic scale) by monolayer film adsorption	Inhibitor (suppressor)	PAG: 50–500
Organic sulfide	Additive Brightner	Micro-leveler Grain refiner	Accelerator	SPS: 5–100
Nitrogen compound	Leveler dye Surfactant	Micro-leveler Grain refiner Brightner	Strong inhibitor	0–20

The presence of only accelerator in addition to the inorganic components was reported to result in a very rough and poor fill deposition profile. Addition of suppressor improved the morphology and profile of the via fill, which was enhanced even more by addition of leveler, a synergistic effect observed in copper damascene applications as well [26].

8.5.2 Methane Sulfonic Acid Chemistry

Methane sulfonic acid chemistries are similar to sulfuric acid based solutions, with the exception that higher copper solubility is achievable in methane sulfonic acid.

8.5.3 Cyanide Chemistry

Deposits produced in cyanide solutions are typically thin (<12.5 μm) and generally not suitable for deposition of relatively thick deposits. Cyanide solutions, in any industry in general, and specifically in the semiconductor industry, are finding less and less favor because of their perceived toxicity and waste treatment problems. In recent years, owing to environmental concerns, some alkaline non-cyanide systems have also been developed in attempts to replace the cyanide ones [16].

8.5.4 Other Copper Plating Chemistries

Pyrophosphate solutions, once used heavily for plating through holes on printed wiring boards, have been almost completely replaced by high-throw acid sulfate solutions.

Fluoroborate solutions have been advertised for many years as having the capability to deposit copper at very high current densities. However, present commercial usage of this type of solution is minimal, simply because other solutions such as those with sulfate ions can do the same job and are less expensive, easier to control and less susceptible to impurities. The chemical cost of acid fluoroborate electrolyte is approximately twice that of acid sulfate, and for this reason fluoroborate copper has not gained a significant share of the electronic and semiconductor industry [12].

8.6
Plating Process Requirements

When it comes to via filling processes, the key elements critical for successful manufacturability and high reliability of a TSV chip integration process are process robustness and speed control. The main requirements are excellent fill performance (void-free) and thickness uniformity (<3% across the wafer) that needs to be extended to the wafer circumference, with only a few millimeters edge exclusion. The plated copper must adhere well and the overburden be minimized to withstand subsequent CMP.

8.6.1
Suggested Mechanisms for Superconformal Deposition

To better predict the fill performance and thus achieve a more robust process, several studies have aimed at understanding the mechanism and behavior of sulfuric acid based electrolyte components. The effects of electrolyte components as well as process parameters on fill and process performance have been intensively studied.

Prior to electrochemical deposition (ECD), the via structures are lined with an insulator, a barrier and a copper seed layer. The copper layer, which is necessary for nucleation of the plated copper, is usually applied through PVD. The entire sidewall of the vias is metallized, including the bottom and the top.

Two general types of filling mechanism have been proposed for deep via filling process:

(a) Superconformal growth, which involves a reaction-diffusion mechanism involving generation and destruction of accelerant in the contrasting environments in the via bottom and near the via top [5, 6, 35, 36].

(b) Sidewall growth phenomena, which include final seam closure by a curvature-dependent growth mechanism [6]. Which growth phenomenon is dominant depends on the organic additive system, and the type of plating current modulation techniques (such as pulse plating and variation in waveforms), used for the deposition [26].

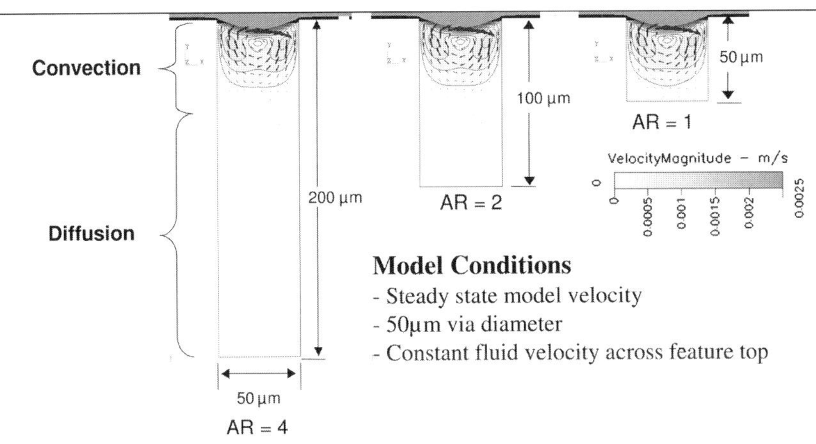

Figure 8.12 Mass transfer through high aspect ratio via. As aspect ratio (AR) of the via increases, a larger portion comes under diffusion control [40].

Unlike the current distribution on the wafer scale, which is typically controlled by the electric field, the current distribution inside the via is dominated by kinetics and mass transport. Since the plating additives are present in very small quantities (ppm ranges), their flux is always transport limited. Because flow is absent at the bottom of the via, and to a certain extent within the via (how much depends on the dimensions and aspect ratios of the vias, as shown in Figure 8.12), the copper and additives are transported inside the via, and especially to the bottom of the via, solely by diffusion. In terms of mass transfer, the availability of metallic ions would be much higher at the via mouth than at the via bottom where depletion of copper takes place due to transport limitation, adversely affecting the deposit characteristics and resulting in pinch-off at the via mouth.

In recent modeling, diffusion and convection were the two major mass flow mechanisms considered. The mass transport of cupric ion due to migration is very small, therefore it was not included in this model. When the via aspect ratio is small, the flow is more dominated by convection, which means mechanical flow strength is quite important to change overall metal ion concentration within the via. However, as the aspect ratio increases, a larger portion of the via becomes diffusion-controlled, which means that ion transfer is not substantially increased by increased forced convection and uniform ion concentration along the via depth is more difficult to achieve [40].

In terms of charge transfer, there can be a severe potential drop, which means the via mouth can have a higher charge transfer due to low resistance. The modeling results illustrated in Figure 8.13 with respect to charge transfer conditions show the potential variation along the via depth at a given current density with a uniform seed layer. The major factors affecting potential distribution include the thickness and coverage of seed layer, feature dimensions and bath conductivity.

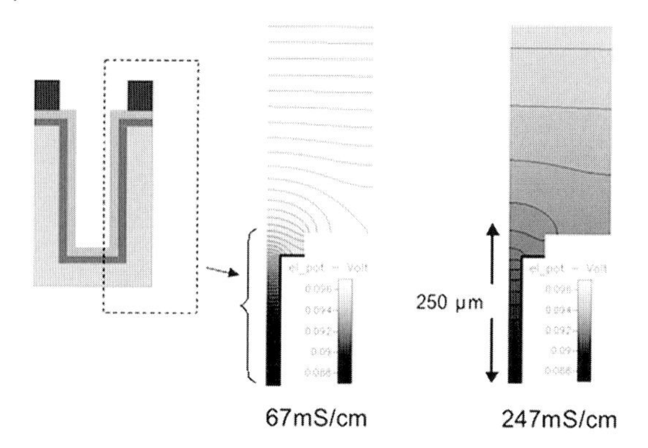

Model Parameters

- Uniform seed layer
- Constant current density

Factors

- Copper seed layer (thickness and coverage)
- Feature dimensions
- Bath conductivity

Additive adsorption/desorption behavior is dependent on local potential

Figure 8.13 Potential variation through high aspect ratio vias. Additive adsorption/desorption behavior depends on local potential [40].

The modeling results suggest that a higher conductivity bath generates a more uniform potential distribution, but still there exists a potential drop along the via, due mainly to length and depth effect [40].

Owing to higher electronic charge density and transport conditions at the top of the via, under conventional conditions, including process parameters and formulations, the vias will be closed before complete filling takes place (Figure 8.14). New electroplating compositions are required that can achieve superconformal filling and thus plate, effectively, void-free deep vias [40].

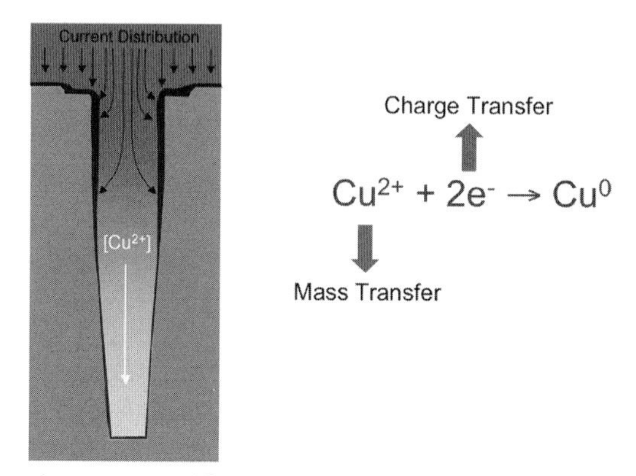

Figure 8.14 Void formation mechanism. Higher local current density and increased concentration of cupric ions near the via mouth lead to pinch-off.

8.6.2
Effect of Waveform and Current Density on Fill Performance

Pulse plating involves application of a cathodic current for a certain time interval, usually followed by a short, high-energy anodic pulse periodically interposed. The main difference between direct current (DC) and pulsed current is that with the DC plating only voltage (or current) can be controlled, while with pulse plating three parameters – on-time, off-time, and peak current density – can be varied independently. These variables create a mass transport situation and adsorption and desorption phenomena that are not otherwise possible, improving the fill performance [16].

Experimental studies confirmed that the waveform can indeed improve filling performance significantly. By applying the appropriate waveform parameters, the overhang is reduced and, thus, a superconformal fill is obtained (Figure 8.15) [40].

Void-free fill can be achieved with an appropriate combination of reactor design, chemical composition and process conditions. Due to complexity of this process and the numerous variables that could affect the final process performance, the plating mechanism for through silicon via applications is still not completely understood and continues to be investigated.

As shown in Figure 8.16, an increased copper concentration formulation can also improve the bottom fill rate due to increased mass transport of copper ions to the via bottom. The amount of the fill improvement will be highly dependent on the applied current density vs. limiting current density (LCD) and on feature dimensions and aspect ratio [38].

The probability of void formation typically increases with increasing average current density (Figure 8.17). With increasing average current density, at a fixed deposition condition and feature dimension, a significant change in deposit

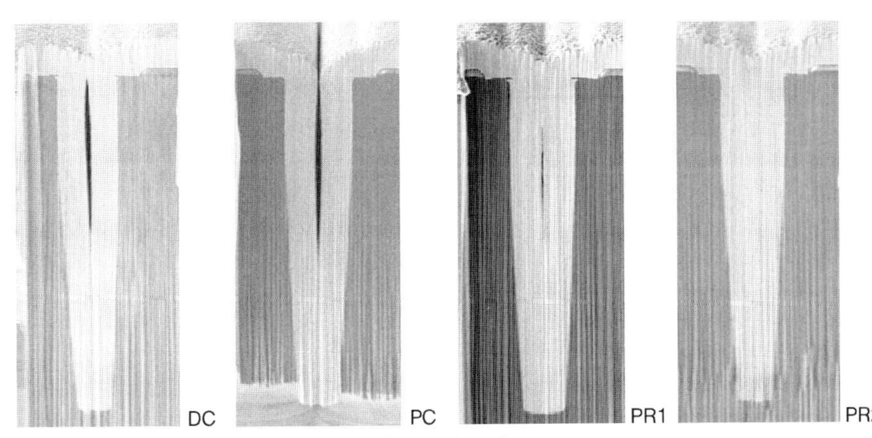

Figure 8.15 Effect of waveform on the via filling, where the difference between PR1 and PR2 is the amount of reverse current. DC: Direct Current, PC: Pulse Current, PR: Pulse Reverse. (Source: Semitool).

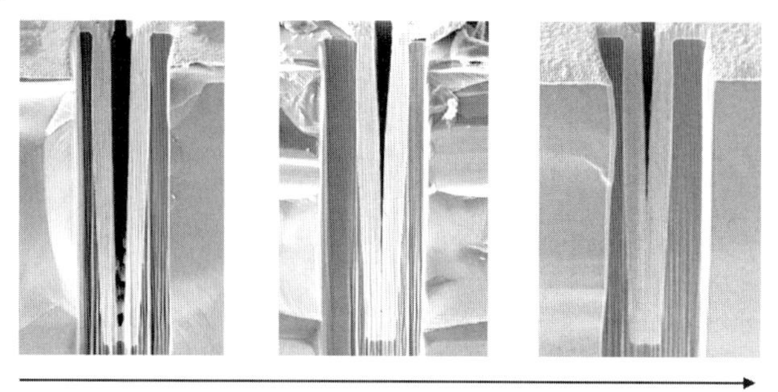

Figure 8.16 Impact of increased copper concentration. Feature dimensions: 12 μm wide and 100 μm deep vias.

profiles is observed over filling time. The profile at each deposition time shows less superconformal growth, meaning that the ratio of the thickness at feature mouth to that at the via bottom becomes larger. This is due mainly to higher current crowding at the feature mouth and more significant mass transfer limitations at the via bottom at a higher current density. This phenomenon leads to a higher probability of void formation within the via with increasing current density. Figure 8.17b clearly shows higher deposition rate near the via mouth and less deposition rate near the via bottom with increasing current density. This figure also shows that the amount of deposit within the via is less at the same amp-time due to more severe current crowding on the field area and via mouth areas than at the inside of vias [40].

8.6.3
Effect of Deposition Waveform on Fill Performance

The deposition waveform shows a comparison between direct current (Figure 8.18a) and pulse reverse (Figure 8.18b) waveform. Particularly in the case of high aspect ratio features, such as the feature in Figure 8.18, a pulse reverse waveform can yield improved fill performance over direct current due to its ability to selectively remove

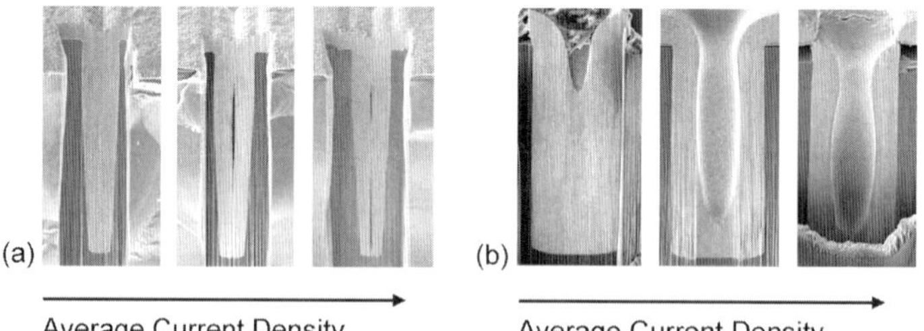

Figure 8.17 Effect of average current density on the filling profile evolution: (a) full filling (12 μm wide and 100 μm deep vias) and (b) partial filling (40 μm wide and 100 μm deep vias).

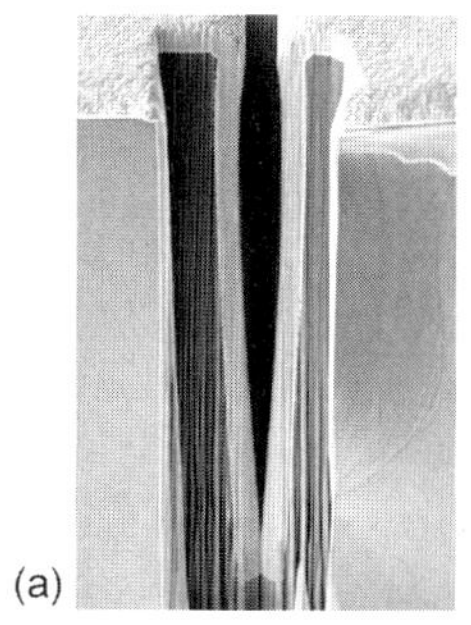

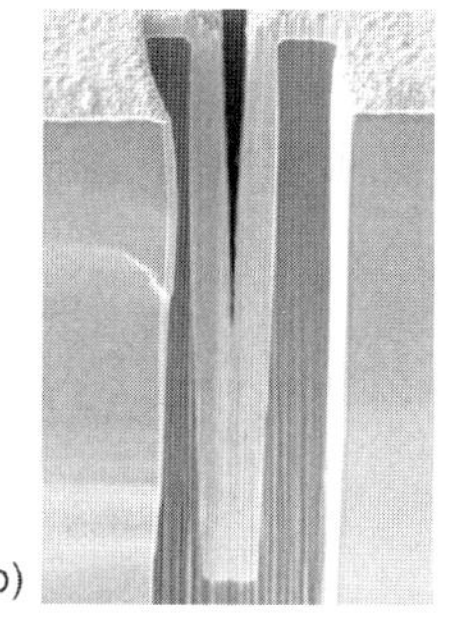

Figure 8.18 Effect of deposition waveform on the filling profile: (a) direct current and (b) pulse reverse. Feature dimensions: 12 μm wide and 100 μm deep vias.

copper from the via mouth, thus reducing the chances of via pinch off and ensuring adequate mass transport of chemistry into the feature.

Figure 8.19 illustrates the impact of increased cycle time (decreasing frequency) with a pulse reverse waveform (average current density kept constant). As shown in Figure 8.19, a longer cycle time can improve the fill performance up to a point. This beneficial trend is due to the increased reverse time or deplate time as the cycle time is increased, which reduces the likelihood of pinch off at the via mouth [40].

8.6.4
Impact of Feature Dimension on Fill Time

An important consideration for all feature dimensions is the time required for full void-free fill. As shown in Figure 8.20, fill time is reduced as the feature dimension is decreased. Thus, to maximize the tool throughput and minimize cost per wafer to process, the feature size should be reduced as much as possible and yet meet the design requirements for the feature, without exceeding the process capability of the fill or barrier and seed deposition processes.

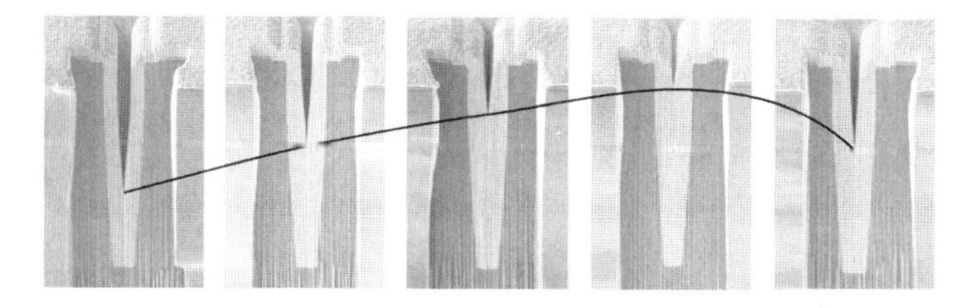

Figure 8.19 Effect of increased cycle time (lower frequency) with pulse reverse waveform on the filling profile. Feature dimensions: 12 μm wide and 100 μm deep vias.

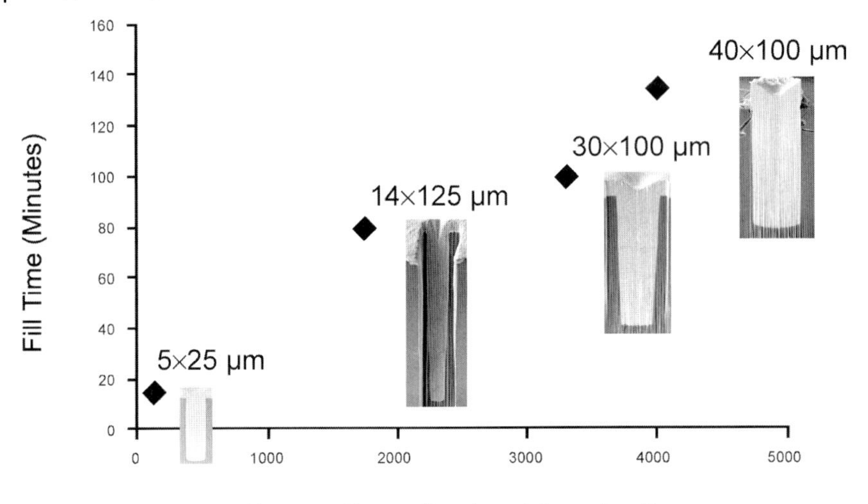

Figure 8.20 Impact of feature dimensions on fill time.

8.6.5
Impact of Feature Dimension on Overburden

Another factor to consider, specifically in the case without stud formation, is the copper deposition thickness in the field area, or the overburden. Removal of this field or bulk copper requires a chemical mechanical polish (CMP), which is an expensive process. By reducing the feature dimensions and thus the fill time, the copper

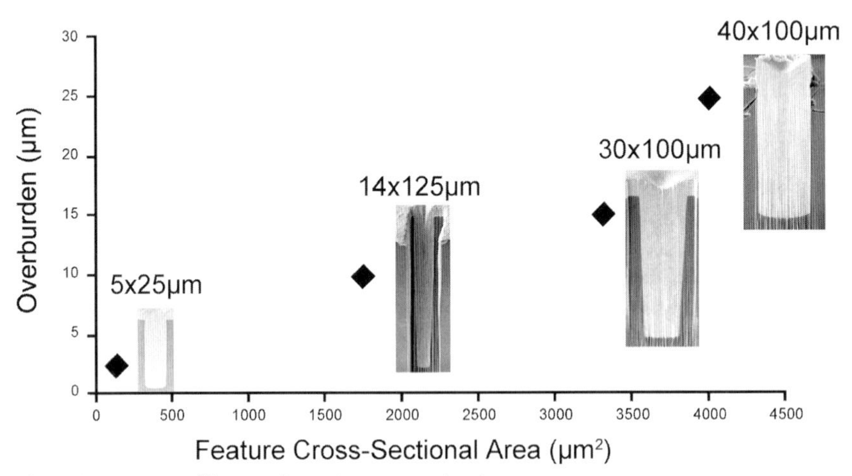

Figure 8.21 Impact of feature dimensions on overburden.

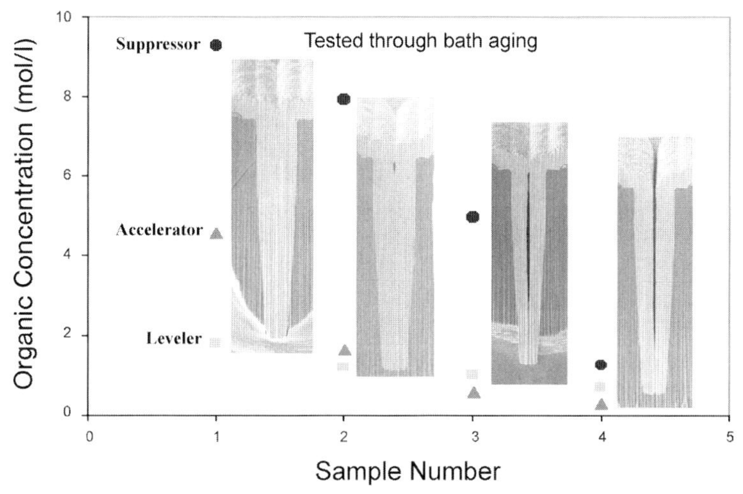

Figure 8.22 Impact of organic component consumption on fill performance.

thickness in the field region is reduced and, accordingly, the amount of copper that must be removed by CMP. Figure 8.21 illustrates the trend of decreasing overburden with a reduction in feature dimension.

8.6.6 Bath Analysis and Maintenance

Another important factor is analysis and maintenance of the bath components. All components must be maintained within specified ranges but the organic components or additives (accelerator, suppressor, and leveler) will be consumed the most rapidly with time and applied current. Figure 8.22 shows the progression from void free deposition to void formation over a short time period without dosing of the organic components.

8.7 Summary

Electrochemical deposition of copper has been proven to be a robust, industrially useful process for semiconductor chip manufacturing. It has been used for copper interconnects in the front end manufacturing process for about a decade. There is now much research targeted at adapting copper ECD processes to 3D interconnects for chip stacking (whether chip-to-chip, chip-to-wafer or wafer-to-wafer) and for creating chips with multiple layers of active devices. While there is still work to be done to provide economically attractive processes, ECD copper has been proven to be technically feasible for 3D interconnect formation.

References

1 Andricacos, P.C., Uzoh, C., Dukovic, J.O. *et al.* (1998) Damascene copper electroplating for chip interconnections. *IBM Journal of Research and Development*, **42** (5), 567–574.

2 Edelstein. D., Heidenreich, J., Goldblatt, R.*et al.* (1997) Full copper wiring in a sub-0.25 μm CMOS ULSI technology. *Proceedings IEEE IED*, 773–776.

3 Venkatesan, S., Gelatos, A.V., Misra, V. *et al.* (1997) A high performance 1.8V, 0.20 mm CMOS technology with copper metallization. *Proceedings IEEE IEDM*, 769–772.

4 Zielinski, E.M., Russell, S.W., List, R.S. *et al.* (1997) Damascene integration of copper and ultra-low-k xerogel for high performance interconnects. *Proceedings IEEE IEDM*, 936–938.

5 Vereecken, P.M., Binstead, R.A., Deligianni, H. and Andricacos, P.C. (2005) The chemistry of additives in damascene copper plating. *IBM Journal of Research and Development*, **49** (1), 3–18.

6 Moffat, T.P., Wheeler, D., Edelstein, M.D. and Josell, D. (2005) Superconformal film growth: Mechanism and quantification. *IBM Journal of Research and Development*, **49** (1), 19–36.

7 Takahashi, K.M. and Gross, M.E. (1999) Analysis of transport phenomena in electroplated copper filling of submicron vias and trenches. Advanced Metallization Conference, pp. 57–63.

8 Takahashi, K. and Gross, M. (1999) Transport phenomena that control electroplated copper filling of submicron vias and trenches. *Journal of the Electrochemical Society*, **146** (12), 4499–4503.

9 Niklaus, F., Lu, J.Q., McMahon, J.J. *et al.* (2005) Wafer level 3D integration technology platforms for ICs and MEMS, http://www.ee.kth.se/php/modules/publications/reports/2005/IR-EE-MST_2005_001.pdf. (accessed on April 2007).

10 Vardaman, J. (2007) 3-D through-silicon vias become a reality, *Semiconductor International*, http://www.semiconductor.net/article/CA6445435.html. (accessed on April 2007).

11 Ritzdorf, T. and Fulton, D. (2005) Electrochemical deposition equipment, in *New Trends in Electrochemical Technology, Volume 3, Microelectronic Packaging* (eds M. Datta, T. Osaka and J.W. Schultze), CRC, Boca Raton, pp. 495–509.

12 Taylor, T., Ritzdorf, T., Lindberg, F. *et al.* (1998) Electrolyte composition monitoring for copper interconnect applications. Electrochemical Processing in ULSI Fabrication I and Interconnect and Contact Metallization: Materials, Processes, and Reliability, ECS, Pennington, NJ, pp. 33–47.

13 Graham, L., Ritzdorf, T. and Lindberg, F. (2000) Steady-state chemical analysis of organic suppressor additives used in copper plating baths. Interconnect and Contact Metallization for ULSI, PV 99-31, ECS, Pennington, NJ, pp. 143–151.

14 Ritzdorf, T., Fulton, D. and Chen, L. (1999) Pattern-dependent surface profile evolution of electrochemically deposited copper. Proceedings of the Advanced Metallization Conference, pp. 101–107.

15 Reid, J., Gack, C. and Hearne, S.J. (2003) Cathodic depolarization effect during Cu electroplating on patterened wafers. *Electrochemical Solid-State Letter*, **6** (2), C26–C29.

16 Schlesinger, M. and Paunovic, M. (2000) *Modern Electroplating – Fourth Edition*, John Wiley & Sons, Inc., 61–81.

17 Flott, L.W. (1996) *Metal Finishing*, **94**, 55.

18 Sard, R. (1986) in *Encyclopedia of Materials Science and Engineering*, Volume 2 (ed. M.B. Bever), Wiley, New York, p. 1423.

19 Kanellos, M. (2002) Chipmakers make smooth shift to copper, http://www.news.com/2102-1001_3-273620.html. (accessed on April 2007).

20 Andricacos, P.C. (1998) Electroplated copper wiring on IC chips. *The Electrochemical Society Interface*, **7**, 23.

21 Singer, P.(November, 1997) *Semiconductor International*, 67–70.

22 Andricacos, P.C.(Spring 1998) Copper On-chip Interconnections – A Breakthrough in Electrodeposition to Make Better Chips. *Interface (Electrochemical Society)*, **8**, 32–39.

23 Kim, B. and Ritzdorf, T. (2006) High aspect ratio via filling with copper for 3D integration. SEMI Technical Symposium: Innovations in Semiconductor Manufacturing, SEMICON Korea 2006, STS, S6: Electropackage System and Interconnect Product, p. 269.

24 Lee, S.R. and Hon, R. (2006) Multi-stacked flip chips with copper plated through silicon vias and re-distribution for 3D system-in-package integration, MRS Fall Meeting, Symposium Y.

25 Nguyen, T., Boellaard, E., Pham, N.P. *et al.* (2002) *Journal of Micromechanics and Microengineering*, **12**, 395.

26 Landau, U. (2000) Copper metallization of semiconductor interconnects – issues and prospects. CMP Symposium, Abstract # 505, Electrochemical Society Meeting, Phoenix, AZ, pp. 22–27.

27 Klumpp, A., Ramm, P., Wieland, R. and Merkel, R. (2006) Integration Technologies for 3D Systems, http://www.atlas.mppmu.mpg.de/~sct/welcomeaux/activities/pixel/3DSystemIntegration_FEE2006.pdf. (accessed on April 2007).

28 Sun, J., Kondo, K., Okamura, T. *et al.* (2003) *Journal of the Electrochemical Society*, **150** (6), G355.

29 Burkett, S., Schaper, L., Rowbotham, T. *et al.* (2007) Proceedings Material Research Society Symposium, Volume 970.

30 Gonzalez, M. *et al.* (2005) Influence of dielectric materials and via geometry on the thermomechanical behavior of silicon through interconnects, Proceedings of 10th Pan Pacific Microelectronics Symposium, SMTA, Hawaii.

31 Gonzalez, O.C., Vandevelde, M., Swinnen, B. *et al.* (2007) Analysis of the induced stresses in silicon during thermocompression Cu-Cu bonding of Cu-through-vias in 3D-SIC architechture. Proceedings of 57th ECTC Conference, pp. 249–255.

32 Dory, T. (2005) Challenges in copper deep via plating. PEAKS – Wafer Level Packaging Symposium, June, Whitefish, MT.

33 Kang, S.K., Buchwalter, S.L., LaBianca, N.C. *et al.* (September 2001) Development of conductive adhesive materials for via fill applications, *IEEE Transactions on Components and Packaging Technologies*, **24**, p. 431–435.

34 Edelstein, D., Heidenreich, J., Goldblatt, R. *et al.* (1997) Full copper wiring in a Sub-0.25 μm CMOS ULSI technology, Proceedings of the IEEE International Electron Devices Meeting, pp. 773–776.

35 Kondo, K., Yonezawa, T., Mikami, D. *et al.* (2005) High-aspect-ratio copper-via-filling for three-dimensional chip stacking. reduced electrodeposition process time. *Journal of the Electrochemical Society*, **152** (11), H173–H177.

36 Barkey, D.P., Callahan, J., Keigler, A. *et al.* (2006) Studies on through-chip via filling for wafer-level 3D packaging. Proceedings of 210th ECS Meeting.

37 Kim, B., Sharbono, C., Ritzdorf, T. and Schmauch, D. (2006) Factors affecting copper filling process within high aspect ratio deep vias for 3D chip stacking. Proceedings of 56th ECTC Annual Meeting, 1, p. 838.

38 Polamreddy, S., Spiesshoefer, S., Figueroa, R. *et al.* (March 2005) Sloped sidewall DRIE process development for through silicon vias (TSVs). IMAPS Device Packaging Conference.

39 Worwag, W. and Dory, T. (2007) Copper via plating in three dimensional interconnects. Proceedings of 57th ECTC Annual Meeting.

40 Kim, B. (2006) Through-silicon-via copper deposition for vertical chip integration. Proceedings of MRS Fall Meeting.

41 Forman, B. (2007) Advances in wafer plating the next challenge: through silicon via plating. Proceedings EMC-3D SE Asia Technical Symposium, January 22–26.

9
Metallization by Chemical Vapor Deposition of W and Cu

Armin Klumpp, Robert Wieland, Ramona Ecke, and Stefan E. Schulz

9.1
Introduction

To build vertical electrical interconnects between thinned chip stacks or wafer stacks, TSVs with high aspect ratios are used in several currently presented 3D process flows [1–4]. For this purpose, the required metallization needs to be deposited after the electrical isolation of Through Silicon Vias (TSV; see Chapter 6, Dielectrics) in such a way, that reliable electrical interconnects are formed. Mostly, the TSV metallization consists of a bi- or multilayer stack of a thin diffusion barrier, adhesion layer and/or seed layer and the conductor material like tungsten (W) or copper (Cu). In general, different deposition techniques are available to realize metal deposition: electrochemical deposition (ECD, electroplating), chemical vapor deposition (CVD), electroless plating and physical vapor deposition (PVD: Sputtering). Apart from PVD all processes have the potential to fill high aspect ratio (HAR) patterns. Figure 9.1 shows the applicability of different deposition processes and filling concepts depending on the TSV diameter. The pure metal CVD approach is suited for complete TSV fill with lateral width of up to ~3 μm at high aspect ratios. It is currently unclear as to what extent the electroplating technology is able to substitute MOCVD (metal-organic chemical vapor deposition) processes for HAR TSV metallization [5, 6]. However, if aspect ratios of larger than 7 : 1 are to be filled with a conducting material, CVD is the process with the highest currently available conformality. To date, CVD of poly-Si, tungsten or copper has been used to completely fill TSVs. In the region above 3 μm lateral size, several factors limit the application of metal CVD, like high film stress for W CVD and process limitations for Cu CVD. In this case, filling can be performed by electroplating. CVD can be used to deposit a so-called seed layer only, which then acts as a base layer to grow metals by subsequent electroplating techniques. This is especially of great interest since even most advanced PVD tools are hardly able to cover vias with aspect ratios larger than 7 : 1 [7, 8]. Depending on the chosen

Handbook of 3D Integration: Technology and Applications of 3D Integrated Circuits.
Edited by Philip Garrou, Christopher Bower and Peter Ramm

ISBN: 978-3-527-32034-9

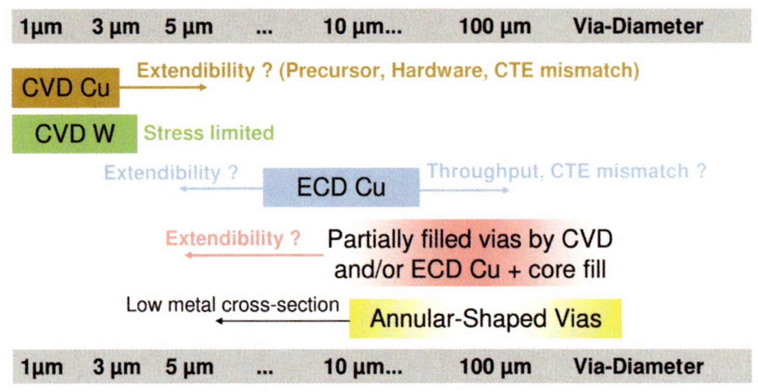

Figure 9.1 Overview of deposition processes and concepts for metallization of TSVs.

metallization scheme, a thin conducting CVD layer such as W or barrier/Cu can be used as a seed layer for the subsequent electroplating process.

TSV dimensions vary over a wide range, from ~2 up to 100 µm and beyond. An increasing demand for TSVs with HAR will probably require an increasing amount of metal CVD applications now and in the future, because the use of metal-PVD for a seed layer is strongly limited by the desired AR. For completely filled Cu TSVs with large diameters the mismatch in the coefficient of thermal expansion (CTE) of copper and silicon should be considered. This leads to vertical movement of the Cu plug in the TSV during repeated thermal cycling in further BEOL processes [9]. With the hybrid method, where the TSVs are only covered by Cu at the sidewalls and are backfilled by a composite material, the damage of dielectric layers due to thermal mismatch is prevented [9]. At the same time this has the potential to provide a low cost process that avoids the complexity of the void-free complete fill by ECD. The annular shaped TSV in contrast to cylindrical TSV is a further approach to reduce the impact of the thermal mismatch due to the reduction of metal volume in the TSV and to produce low resistance vias at the same time [9].

In this chapter we focus on the CVD of metals, that is, W and Cu, and metal nitrides (TiN). Metal-CVD or metal-organic. CVD (MOCVD) is typically carried out in a conventional CVD single wafer chamber (cold wall reactor). The deposition process is a strictly chemically based, temperature driven reaction of the decomposed precursor material on the wafer surface. Since moisture causes metal oxides, a load lock based system is required to accomplish repeatable and stable process conditions.

9.2 Commercial Precursors

The availability and choice of a precursor is one of the most delicate issues of metal deposition by CVD. The precursor choice determines the quality of the deposited films (impurities, adhesion), deposition parameters (temperature, pressure, reactants) and therefore influences the reliability of the 3D integrated TSVs. Most commonly used metals for 3D integration are, currently, tungsten and copper for TSV filling and often TiN and also TaN as barrier/seed layer films. The former three materials will be discussed in more detail.

9.2.1
TiN Precursors

Commercially available precursors for CVD of TiN are $TiCl_4$ with NH_3 addition [11, 12], TDMAT [tetrakis(dimethylamino)titanium] [13, 14] and TDEAT [tetrakis(diethylamino)titanium) [15]. With $TiCl_4$, the chemical reduction of $TiCl_4$ with NH_3 to TiN and HCl occurs instantly and requires sophisticated management of the gas/precursor flows into the reaction chamber to avoid unwanted gas phase reactions above the wafer surface, thus creating particles and TiN deposition behind the gas distribution plate. Additionally, an acceptable Cl-content in the TiN-film would require temperature treatments in the 600 °C range [16], which is far too high for backend processes like 3D integration.

Therefore, metal-organic precursors like TDMAT or TDEAT are often applied in commercially available metal-CVD reactors, for example from Applied Materials or from Novellus Systems. With TDMAT, the TiN film is thermally deposited on a heated substrate from TDMAT that is delivered from a bubbler by using a helium carrier gas. The underlying deposition reaction can be described by the following equation:

$$Ti[N(CH_3)_2]_4 \rightarrow TiN(C,N) + HN(CH_3)_2 + \text{other hydrocarbons}$$

The deposition is carried out under surface reaction limited conditions for films with high step coverage [17]. To remove impurities and to achieve a stable film, a plasma densification step is needed after the thermal decomposition of TDMAT (see Section 9.3.1).

9.2.2
Copper Precursors

There are inorganic and metalorganic precursors for the CVD of Cu. The copper halides have the disadvantage that they are solids, which causes problems of the precursor delivery. Turning to liquid sources, which are more convenient than solid ones, metalorganic precursors are attractive because of the lower precursor processing temperature. Metalorganic precursors are classified with respect to the oxidation state of the copper as Cu(II) and Cu(I) compounds. Most compounds in use are β-diketonate complexes. In a Cu(II) compound the central Cu^{2+} ion is bonded to two singly-charged β-diketonate ligands. In contrast, Cu(I) compounds contain one singly charged β-diketonate ligand that is strongly bonded to the central Cu^{+} ion and a second neutral ligand that is more weakly bonded to the Cu. The difference in the Cu–ligand bond strengths is reflected in the minor stability, lower deposition temperatures and higher growth rates typical of the Cu(I) compounds. An important characteristic of both kinds of precursors is the use of heavily fluorinated ligands such as hexafluoroacetylacetonate (hfac). Substitution of hydrogen with fluorine induces a significant increase in volatility of the complexes, but also affects negatively the adhesion of the deposited films. [18]

Major efforts have gone into the design and development of new precursors, for both Cu(II) and Cu(I) compounds. Some authors focus on the thin film deposition as conformal seed layer for the Cu electroplating [19–21].

The most widely used precursor for Cu CVD is (hfac)Cu(TMVS) (copper hexafluoroacetylacetonate trimethylvinylsilane), named CupraSelect, which is commercially available from Schumacher (Air Products). Beneficial for the industrial employment of CupraSelect is its availability in large quantities and good purity. Its vapor pressure at room temperature is 8 Pa and increases to 260 Pa at 65 °C. Nevertheless, it has insufficient thermal stability at elevated temperatures, so that a bubbling system should not be used for precursor delivery [22].

Another commercially available precursor is GigaCopper [(hfac)Cu(MHY)] (MHY = 2-methyl-1-hexene-3-yne) from BASF (formerly Merck). Joulaud *et al.* [23, 24] compared both CupraSelect and GigaCopper in regard to the deposition rate and film properties.

Both precursors decompose in the same way:

$$2Cu^{I}(hfac)L_{(g)} \rightarrow Cu^{0} + Cu^{II}(hfac)_{2(g)} + 2L \quad (L = TMVS,\ MHY)$$

The Cu deposition with GigaCopper shows no essential improvement concerning deposition rate and adhesion behavior. For both precursors only half of the Cu contained in the precursor contributes to film formation. Moreover, one reaction product in both cases is $Cu^{II}(hfac)_2$, which has a low vapor pressure as it is a solid under normal conditions. This results in extensive pump down steps between single deposition processes.

9.2.3 Tungsten Precursor

WF_6 (tungsten hexafluoride) is a well established precursor and has been used successfully for contact and via metallization in VLSI fabrication since the mid-1980s [25]. As an inorganic vapor, WF_6 is comparably easy to handle. It is highly corrosive, forming HF (hydrofluoric acid) if exposed to a humid atmosphere.

The main reaction to deposit a tungsten film is the hydrogen reduction of WF_6 (bulk deposition):

$$WF_6(\text{vapor}) + 3H_2(\text{gas}) \rightarrow W(\text{solid}) + 6HF(\text{gas})$$

As is typical for CVD-processes, depending on the amount of WF_6-flow, there is a mass-transport-limited deposition regime, with deposition rate = f(WF_6-flow) and also a surface-reaction-rate-limited regime, with deposition rate = f(temperature, H_2-flow). For HAR trench filling, the latter regime is used for W-filling [10].

9.3
Deposition Process Flow

No matter if W or Cu is chosen as the main filling material in HAR TSVs, a conformal deposition of a barrier seed layer is always the first step in the filling of the TSVs. With W, the growth on and the adhesion to silicon oxide is poor. Therefore, a thin nucleation/adhesion layer, also called "liner", has to be deposited, for example, a TiN, TiW or a tungsten silicide (WSi_x) film. For copper metallization, a diffusion barrier is required to prevent the diffusion of copper towards the bulk silicon. Copper in dielectrics can lead to increased leakage current up to dielectric breakdown. In silicon, copper has a high diffusion coefficient and a high solubility at elevated temperatures. It forms Cu-Si compounds at relatively low temperatures and introduces deep level traps in Si. This causes electronic device failures. Barrier layers should prevent copper diffusion and reaction with adjacent materials. The nitrides of refractory metals (e.g., Ti, Ta, W) are promising candidates for barrier layers based on their properties. For Cu CVD, an additional adhesion layer may be necessary between the diffusion barrier and the Cu film.

Depending on the aspect ratio of the TSVs, the above-mentioned adhesion layers can be deposited by either PVD or MOCVD. In general, chemical vapor deposition is superior to physical vapor deposition concerning conformality. With increasing aspect ratio of the TSVs to larger than 7 : 1, MOCVD of TiN is a commonly used process. Once the barrier/adhesion layer has been deposited, the actual metal CVD process can be started. It is in any case preferable to use a multi-chamber system, where the TSV metallization flow can be performed without exposing the wafer to atmosphere until all metal is deposited, thus avoiding unwanted oxidation on the barrier layer surface (Figure 9.2). To finish the metallization sequence, an etch back

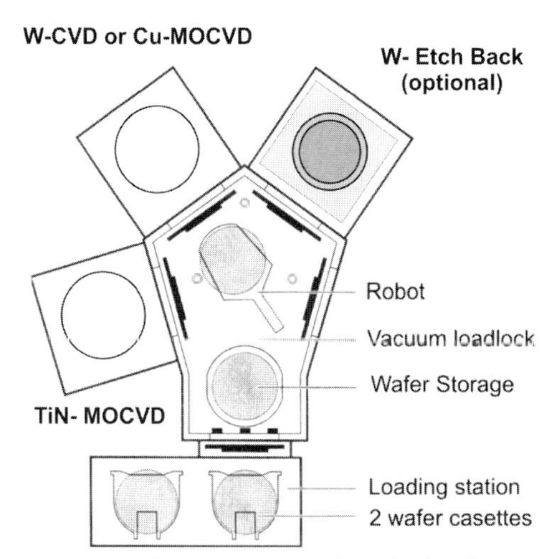

Figure 9.2 Schematic top view of a multi-chamber system with central load lock and robotic handling system for CVD of metal.

process, CMP of metal or a combination of both techniques, is required to remove the laterally deposited metal film until the TSV is electrically isolated.

9.3.1 Barrier Deposition

Presently, TiN is the most promising barrier material for application in 3D integration concerning precursor availability, deposition temperature (400 °C), good conformality and barrier properties.

For tungsten CVD, a TiN layer is used to provide an adequate nucleation and adhesion layer. For Cu filling it acts as barrier against Cu diffusion. Both metallization schemes use the same precursor and deposition process.

TiN depositions can be performed in commercially available CVD reactors with plasma capability. The precursor, TDMAT, is delivered by a bubbler system. The as-deposited film by pyrolysis of TDMAT contains substantial amounts of impurities (C,H) and is not stable to air. Therefore, a plasma densification treatment is needed after the deposition step. For CVD-W this step is also essential to promote the required adhesion for the highly stressed W films. The TiN barrier layers are produced by a multistep process consisting of alternating pyrolysis steps and plasma treatment steps. The TiN layer is plasma-densified after depositing ~5 nm TiN, resulting in a final thickness of ~2.5 to 3 nm after plasma treatment, depending on treatment time. The number of cycles determines the thickness of the entire barrier, for example, eight cycles for ~20 nm thickness. The deposition process and the TiN layer properties are described in detail in References [26, 27].

The step coverage in high aspect ratio TSVs depends on their geometry and diameter size as well as their depth. Vias with a width of 20 μm diameter and greater are uncritical with respect to the step coverage. Conversely, in slot holes 20 μm long, 1–5 μm wide, and 30 μm deep TiN films show a step coverage of 30–70% at the bottom of the via, depending on the width of the slot hole. In vias with circular cross-section the same dependence was found (Table 9.1). Higher step coverage is observed on sidewalls, particularly in the upper via range (over 100%, see Figure 9.3). Since the plasma treatment is quite directional, the densification is not very effective at surfaces parallel to the incident direction of the ions.

Table 9.1 Step coverage in TSVs with circular cross section and different sizes.

TSV size	1 μm	1.7 μm
depth	15 μm	15 μm
AR	15	8
Top	82 nm = 100%	82 nm = 100%
Upper side wall	110 nm = 163%	10 nm = 163%
Lower side wall	59 nm = 73%	63 nm = 78%
Bottom	27 nm = 33%	46 nm = 57%

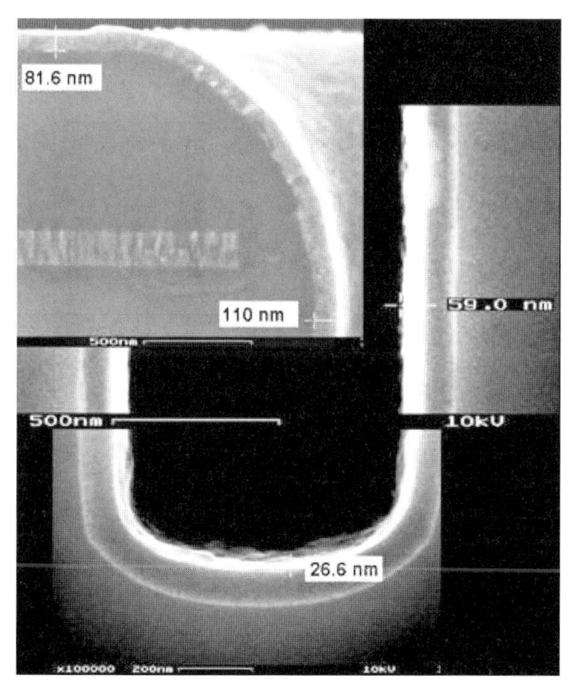

Figure 9.3 Step coverage of TiN in a TSV with AR = 15 and a width of 1 μm.

9.3.2
Adhesion Layer

Copper deposited by CVD often has insufficient adhesion to the barrier layer, especially when fluorine-containing precursors are used. Copper is a relatively precious metal, thus it is not possible to achieve good adhesion with oxide bonds. These oxide bonds, however, are the reason for the excellent adhesion of CVD TiN to SiO_2. Copper adhesion is influenced by the deposition process. For a PVD process the adhesion of copper on different layers is always sufficient. The kinetic energy of the particles forming the layer is greater than for CVD and ECD (electrochemical deposition) processes. During sputtering, adsorbates are removed and there is a forced intermixing in the monolayer range. Another reason for the insufficient adhesion of CVD copper is the chemistry used. Although very pure copper layers are available with the CupraSelect precursor, impurities are enriched at the interface. They form an amorphous interlayer 3 nm thick, consisting of fluorine, oxygen, and carbon (Figure 9.4a) [28]. This interlayer was also found by other authors [29].

Furthermore, there is a contradiction between the barrier properties and the adhesion to the copper. On the one hand, no reactions and mixing with copper are allowed to prevent the diffusion. On the other hand, a very thin layer between TiN and CVD copper can promote the adhesion. However, this layer must not negatively influence the properties of the barrier and the conductor. Such an adhesion layer is a

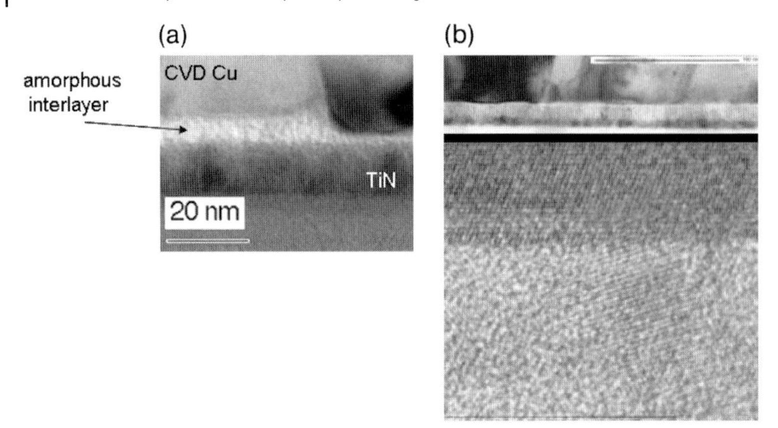

Figure 9.4 (a) XTEM of 20 nm TiN/CVD-Cu with amorphous interlayer. (b) XTEM of 5 nm TiN/10 nm PE-TiN/Cu HRTEM of the interface PE-TiN/Cu showing no interlayer.

very thin film with strong adhesion to the upper and lower layers, but in the case of Cu it is not suitable as a barrier layer. Therefore, a process leading to a Ti-rich TiN layer was developed. The free Ti can react with Cu, which strongly promotes the adhesion.

The adhesion layer is grown by PECVD with the same precursor, TDMAT. The deposition process and the layer properties are described in detail in Reference [27]. The electrical resistivity of the adhesion layer (~3–4 nm) was investigated together with a 5 nm thick standard TiN layer. The layer stack had a resistivity of 320 μΩ cm, slightly higher than 5 nm TiN (290 μΩ cm). This increase results from the higher impurity content of the PE-TiN, but it is acceptable.

The Ti-rich TiN layer shows only excellent adhesion to the copper if adhesion layer deposition and Cu CVD are carried out *in situ* with a subsequent heat treatment. The PECVD-TiN is not stable in air for long so that a vacuum break should be avoided. The complete stack, TiN barrier/TiN adhesion/CVD-Cu, shows no interlayer (Figure 9.4b) and successfully passed the subsequent step of CMP.

Normally ~ 2 nm of this adhesion layer are sufficient to obtain good adhesion of CVD-Cu. Since step coverage is reduced for plasma-enhanced CVD processes the deposition process had to be optimized and extended. By varying the spacing (distance from shower head to wafer), step coverage of 20–30% at side walls and 15–40% on the bottom can be achieved, depending on TSV size and depth ($d = 1$–5 μm, depth 30 μm). For larger sizes step coverage increases. For a reliable TSV formation, especially to achieve a good bottom contact, the adhesion layer is also essential in the structures and not only on top. Adhesion of Cu on the inner surfaces of the TSVs must be ensured during further temperature steps in processing. If the TSVs are not completely filled with Cu because of overhang at the entrance of the TSVs, or when only a Cu seed layer is used for further Cu plating to fill the via, the Cu film can be extracted. In this case, no electrical contact can be realized by this TSV (Figure 9.5).

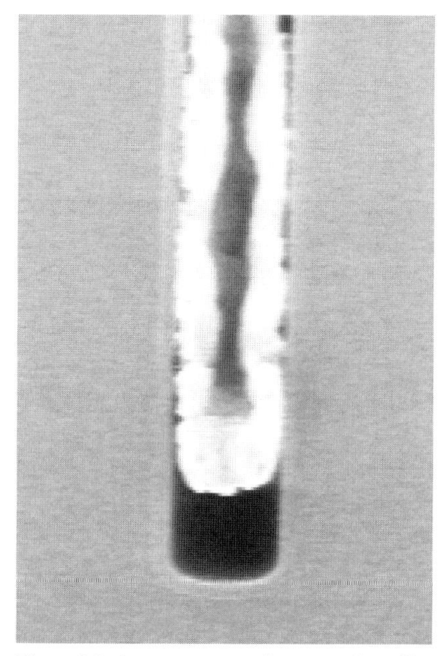

Figure 9.5 Cu extraction in the case of insufficient adhesion in TSV.

9.3.3 Copper Deposition

9.3.3.1 Deposition Equipment and Liquid Precursor Delivery Systems

The deposition equipment used was a P5000 with a lamp-heated Blanket Tungsten Chamber, but we recommend a resistively heated deposition chamber to minimize undesired Cu depositions, for example, at the heater window. A shadow ring is suggested, to prevent Cu deposition at the wafer backside and on the susceptor in either case. The results described in this chapter were obtained with the precursor (hfac)Cu(TMVS), which was evaporated using a liquid delivery system (LDS). This system is composed of a liquid flow meter for the precursor, a mass-flow controller for the carrier gas and a controlled evaporator unit from Bronkhorst. Within the system the precursor is dosed as a liquid with high accuracy and is vaporized just before the deposition. The heated stainless-steel gas line between the vaporizer and the deposition chamber should be as short as possible to avoid precursor loss due to Cu deposition on the inner surface of the line. In the worst case, precursor condensation can occur for very high precursor flow rates and long deposition times. Direct liquid injection systems (DLI) are an alternative means of precursor delivery.

9.3.3.2 Cu CVD Process Characteristics and Blanket Film Properties

The deposition rate of Cu with CupraSelect onto blanket wafers is relatively high and can reach up to 200 nm min^{-1} depending on the wafer size. This rate can be achieved with high temperature ($\geq$190 °C), high total pressure ($\geq$2 kPa) and high precursor flow. A long incubation time compared to other CVD processes is observed for film formation, which depends on deposition temperature and pressure in the chamber. With increasing temperature, as well as for lower pressure, the incubation time is shortened. The pressure influences the free adsorption sites. Their density is reduced with higher pressure. In the same manner, the barrier material and potential pre-treatments of the surface, such as H_2 plasma, impact the nucleation. The addition of water vapor during the nucleation stage leads to higher nuclei density and uniform distribution [22, 30].

The properties of the deposited Cu layers are influenced by the deposition parameters. The specific resistivity of the films depends on film thickness and grain size and is in the range 2.2–3 μΩ cm for unpolished films. The root mean square (RMS) roughness ranges from 15 to 20% of the Cu thickness. Hence, the determined layer thickness is usually too high, which in turn affects the calculation of the electrical resistivity. After polishing, the Cu layers exhibit electrical resistivity, <2 μΩ cm. The film density represents 90–96% of the bulk value. For a 350 nm thick Cu layer deposited at 200 °C the average grain size is 110 nm. The grains show a columnar structure. Impurity concentrations in the films resulting from the metal-organic precursor (Si, F, C and O) are below the detection limit of Auger electron spectroscopy (AES). However, there is an enrichment of impurities at the interface to the diffusion barrier, which heavily impacts the adhesion.

With the equipment described above, a deposition rate of 165 nm min^{-1} on 150 mm wafers at 200 °C and 2.7 kPa were achieved. However, the high deposition rate process is inapplicable for filling TSV structures because of the reduced step coverage of 65%.

9.3.3.3 MOCVD Cu Application to TSV Fill

Presently, complete filling of TSVs by Cu CVD using the described processes is only applicable for TSV sizes $\leq$2 μm after oxide passivation (Figure 9.6).

One way to achieve higher step coverage is to lower the deposition temperature. The parameters for the high-rate process are close to the transport-controlled regime. With lower temperatures (<170 °C wafer temperature) deposition takes place in the reaction-limited region. However, also with decreased temperature the deposition rate decreases drastically, to about 70 nm min^{-1}. Furthermore, the deposition surface enlargement with increasing TSV depth and TSV structure density becomes important for the deposition rate and step coverage. Therefore, the deposition parameters should be modified for each TSV layout to achieve the desired Cu thickness and step coverage. It is advised to use only one TSV size per layout. For moderate via diameters, that is, below 2.5 μm, and aspect ratios of lower or around 20 a complete TSV fill can be achieved using low deposition temperature (see Figure 9.6 for filling of 1.2 μm wide TSV with an aspect ratio of 12.5). For larger TSV diameters the filling process becomes more critical and the throughput will decrease further.

Moreover, the etch profile at the entrance of via is also important for the filling capability of the Cu CVD. A V-shaped via entrance is effective against the formation of

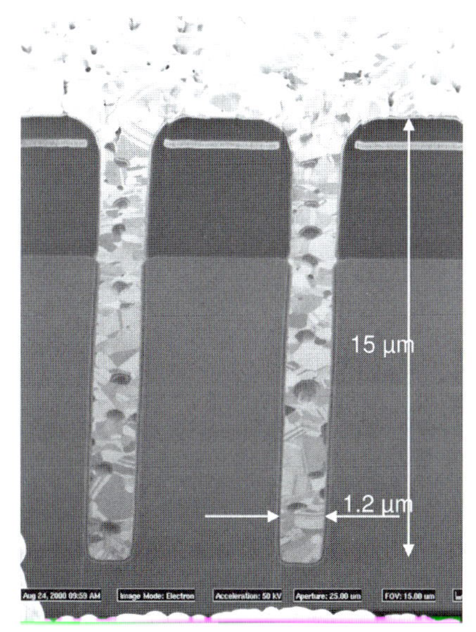

Figure 9.6 Cu-filled via holes with AR 12.5 : 1 after Cu CVD at 170 °C and anneal (focused ion beam, FIB).

overhangs due to the reduced step coverage of the deposition processes (passivation, barrier, metal). A completely V-shaped via further improves the filling capability.

9.3.3.4 **MOCVD Cu Application to Cu Seed Layer**

Depending on the application of the devices and stacks, the TSVs can have different sizes and depth. TSV sizes $\geq$2.5 µm can not be filled completely by Cu CVD in an economic manner. In this case the TSVs can be filled by electrochemical deposition. Therefore, this deposition technique requires a seed layer that should have a homogenous thickness in all structures. A thin Cu layer deposited by PVD is used as seed layer in Cu damascene applications. However, in vias with high aspect ratios the deposited Cu layer will become thinner with increasing depth and can be discontinuous at the bottom of TSV. Here, a complete fill by electrochemical deposition is impossible. Therefore, a conformal seed layer deposited by Cu CVD can be applied in HAR patterns.

The requirement of a homogenously thick Cu layer also demands modifications in the CVD process for high conformality. Reduction of the deposition temperature leads to high step coverage of Cu films in TSVs. In addition, the deposition pressure can be reduced to further improve step coverage achieved for very aggressive aspect ratios of 20 and more. This influences the deposition rate only marginally. The lower total pressure enables a longer mean free path of the reactants so that the precursor molecules can reach the bottom of the vias. The desorption of $Cu(hfac)_2$ from the surface is enhanced as well as the diffusivity in the gas phase. This is also noticed in the pump down steps after the deposition. The initial pressure before deposition can be reached in shorter time due to the lower residence time of the by-products in the

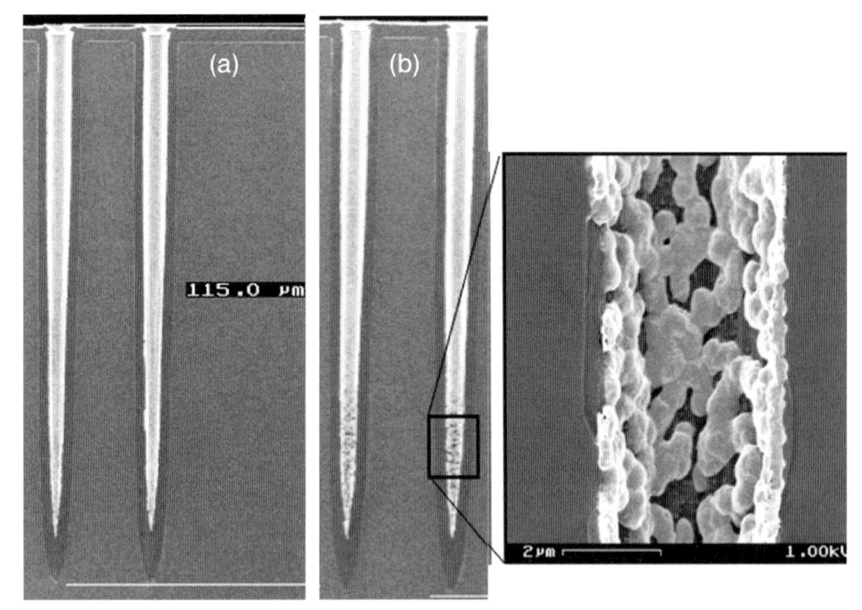

Figure 9.7 Step coverage of a Cu CVD seed layer in aggressive aspect ratios >20 : 1: (a) deposited at lower total pressure, giving a continuous film, and (b) deposited at higher total pressure – agglomeration of Cu grains due to the adhesion anneal and too low thickness achieved in the lower via region.

reactor. The application as seed layer for Cu ECD requires a homogenous Cu layer thickness. The lowest CVD Cu seed layer thickness for plating processes should be at least 120 nm to avoid agglomeration of Cu grains during the adhesion temperature anneal. As shown in Figure 9.7, the reduced step coverage leads to thinner films that agglomerate during subsequent temperature steps. Discontinuous films are unsuitable for the electroplating process to further fill the TSVs. The process with lowered total process pressure yields conformal and continuous Cu films suitable as seed layers for Cu ECD.

9.3.4 Tungsten CVD Application to TSV Fill

Conformal W-CVD processes require an appropriate barrier/seed layer film to be deposited on SiO_2-surfaces without peeling. A thin TiN barrier layer, approx. 20–30 nm thick, is sufficient for this purpose (Section 9.3.1). The chosen thickness of the TiN layer depends on the lateral size, the taper angle and the aspect ratio of the TSV to be metallized. Once the barrier film is deposited, the wafer is transported directly to the W-deposition chamber under vacuum. The TSV W-fill process is carried out in four steps:

Step 1: Heat to 430 °C, H_2-purge and W-nucleation layer deposition. Nucleation on adhesion layers like TiN is carried out by silane reduction of the WF_6 to form a thin, uniform W-layer, thus improving the film quality of the subsequent via-fill process [31].

$$3SiH_4 + 2WF_6 \rightarrow 2W + 3SiF_4 + 6H_2$$

Step 2: W-via fill process (bulk deposition) of conformal W up to ~1100 nm, which is a surface-reaction-rate-limited process [32]. The final thickness is determined by the maximum lateral TSV size to be filled. The bulk deposition step uses hydrogen reduction of the WF_6. The deposition rate mainly depends on the wafer temperature and the WF_6-flow.
Deposition rate for a typical W-Via-fill process at 430 °C is ~300 nm min^{-1}. The sheet resistance of a 1100 nm thick W-film is in the range of 95–100 mΩ cm^{-2}; the extrinsic film stress (tensile) is ~0.95 GPa. Surface roughness is ~17 nm (R_a), measured by AFM after deposition of 1100 nm W and plasma etch back to ~400 nm remaining thickness. The maximum achievable W-film thickness is mainly limited by the comparably high film stress, which reaches the GPa range once 1 μm film thickness is exceeded. Si wafers with the above-mentioned film thickness and stress can reach wafer bow levels of 300 μm and higher, which in turn can cause handling problems with some robot types. Besides this, peeling of the W-film has also been observed, depending on the subsequent dielectric layers and the adhesion of the TiN to the underlying SiO_2.

Step 3: W-interconnect process to eventually close any gaps on top of the TSV-hole; this is a transport-limited process with reduced WF_6 flow and a 20% slower deposition rate and lower step coverage. The need for an additional W-interconnect deposition step depends strongly on the shape and size of the TSVs to be filled. Tapered TSVs result in almost void-free filling; thus the main W-via fill process is sufficient. If voids are expected due to rather vertical sidewall angles of the TSVs, an additional W-interconnect step is helpful to reduce the attack of the top W-layer during W-etch back.

Step 4: Plasma chamber clean with an F-containing etching gas after every wafer. For example, NF_3 is a commonly used gas for a W cleaning process.

9.4 Complete TSV Metallization Including Filling and Etchback/CMP

9.4.1 W-CVD Metallization

After the TSVs are isolated by a dielectric layer, for example by a conformal deposition of a SACVD based ozone-TEOS film (Chapter 6), the W-CVD metallization and subsequent W etch back can be performed. Since CVD of W requires processing temperatures in the 400 °C range, the dielectric isolation film inside the TSVs must sustain the same temperature level. The complete process sequence is carried out as follows:

After the isolation of the TSVs is completed, a full *in situ* process sequence under vacuum is performed, consisting of an Ar pre-sputtering step in the W etch back chamber (RIE), multiple deposition/etch steps to achieve a 20 to 30 nm thick, plasma densified TiN seed layer in the TiN-MOCVD chamber, followed by ~1100 nm W CVD, cool down under vacuum and partial etching of the W in the W etch back chamber.

Since relatively thick W-layers are required to fill the TSV-holes, a partial blanket W etch back to a remaining film thickness <500 nm is carried out, thus avoiding any peeling or delamination of the W-layer. A conventional RIE plasma chamber with an SF_6-based etch chemistry can be used for the partial blanket W-etch step as well as for the following structured W-etch step described below. The partial etch back also helps decrease the wafer bow to a moderate level, so wafers can be handled normally into a lithography tool, that is, a wafer stepper. By use of a photoresist mask, a PR-structured W etch is done afterwards to remove the remaining W after *in situ* partial W etch back. The remaining W film is structured in such a way that the area of the original TSV plus some overlap is fully covered with photoresist (light field mask). Therefore, excessive etch attack on the W inside the TSV is avoided during W-etch. The first stop layer is the thin TiN-layer, preventing the underlying oxide being etched excessively during the W over-etching step. Finally, the remaining TiN layer is etched with chlorine chemistry, so that the TSVs are electrically isolated from each other and can be further integrated.

Figure 9.8 shows a typical W-filled TSV (20 µm depth) after the completed W-structuring sequence as a cross-section (left) and from the top (right). The achievable conformality of the W-CVD step depends mainly on the taper angle and the aspect ratio of the TSV and is >80% for a TSV with a 10 : 1 AR at 3 µm lateral size (trench width). The left W-filled TSV shows part of the formerly produced dielectric

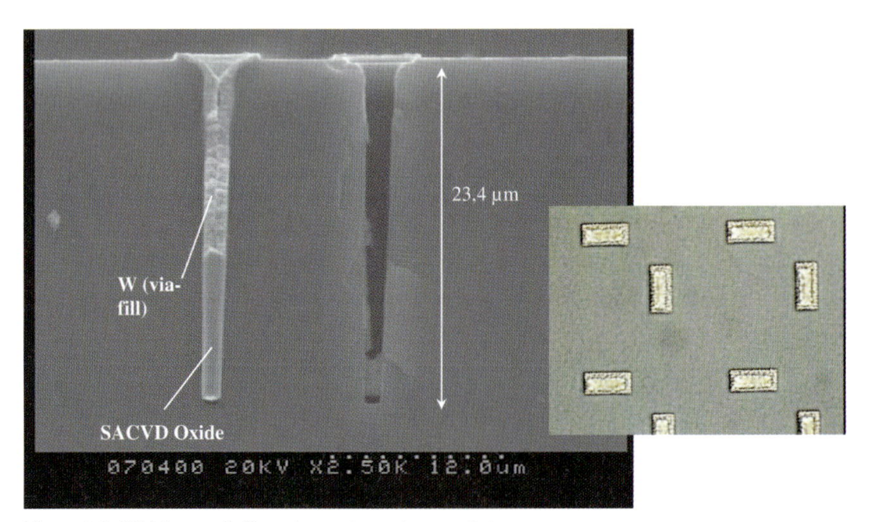

Figure 9.8 SEM-image (left) and top-view micrograph image (right) of a W-filled TSV, AR > 10 : 1, taper angle 88° (HBr-chemistry) after a structured W-etch step.

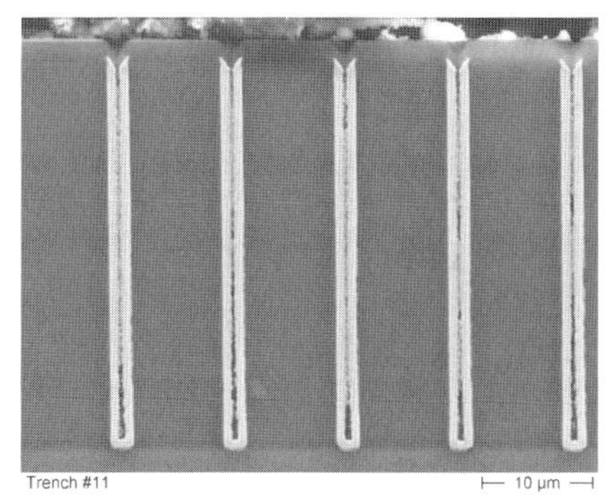

Figure 9.9 SEM-image of a W-filled TSV, AR > 15 : 1, taper angle 89.5° (SF_6/C_4F_8-chemistry), after blanket W etch back.

isolation layer (ozone-TEOS based SACVD film) and also the W-film inside of it at the upper part of the TSV. (Owing to its stiffness and hardness the W-film in the right tungsten filled plug was completely removed while cleaving the sample).

A void free W-fill result is difficult to achieve for HAR TSVs, the taper angle of the formerly fabricated TSV is the main parameter for the size of the void. Figure 9.9 shows an SEM image of a W-filled TSV with an aspect ratio >15 : 1 and a taper angle of 89.5°, resulting in a noticeable void.

Besides the above-mentioned possibility of an W etch back process, CMP of W could also be applied to polish the remaining bulk tungsten until the underlying dielectric layer is reached, thus resulting in electrically isolated TSV with negligible topography.

9.4.2
Cu CVD Metallization

In case of copper metallization for TSV fill the process sequence has been set up as follows.

After TSV isolation, TiN barrier deposition is carried out. To meet the demands of a diffusion barrier the TiN barrier should be ~20 nm thick in the lower region of TSV. With smaller sizes and higher aspect ratios the step coverage decreases. Therefore, the deposited barrier thickness should be chosen with respect to the TSV geometry. We apply TiN barrier thickness in the range 30 nm on top for larger TSV and up to 90 nm for smaller diameters for depths between 20 and 50 µm. Deposition of the adhesion layer TiN takes place in the same deposition chamber as for the barrier, but at lower temperatures. In contrast to the TiN barrier deposition this is a pure plasma enhanced process. The Cu deposition process is accomplished in situ, without vacuum break. After the complete TSV fill a thermal treatment is necessary to obtain sufficient adhesion between the adhesion layer and the Cu. This thermal treatment is carried out in one of the CVD chambers of the P5000 cluster tool. The wafer is treated in vacuum at a susceptor temperature range of 380–420 °C for 3–5 min. Without this temperature anneal the layer stack does not pass the following

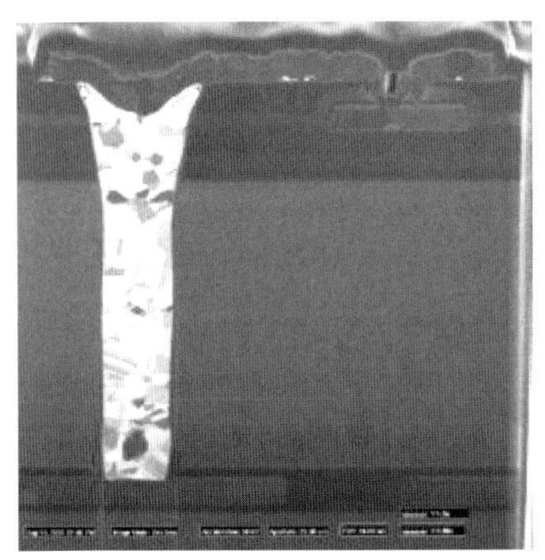

Figure 9.10 Complete CVD-Cu filled TSV after thinning and top and bottom metallization (tilted FIB cross-section).

CMP process. Chemical Mechanical Polishing removes the surplus copper as well as the TiN barrier/adhesion layer from the top of the wafer. Usually, this requires two steps. In a first step the protruding Cu is removed, while the second step removes the TiN layer. Different slurries with defined selectivity are used for these processes to achieve well-insulated, in-plane TSVs and to avoid issues like dishing and copper corrosion.

Subsequent to the CMP, insulator depositions and interconnect processing using Al or Cu damascene metallization is carried out for assembling the interconnection between the TSVs and/or the stacked chips. Figure 9.10 shows an example of a TSV completely filled by CVD-Cu after thinning and top and bottom metallization.

9.5 Conclusions

Both W CVD and Cu CVD can be used for the metallization of TSVs. Depending on the lateral TSV size, either a complete CVD-based TSV fill or the use of CVD as a seed layer with a subsequent electroplating process to complete the fill is employed. A complete fill by CVD is generally applicable for small TSV diameters in the range of up to ~3 μm, especially at high aspect ratios. This limitation to comparably small TSV sizes is due to the metal properties and also to the deposition process itself. The W fill process is stress limited. The Cu fill process is determined by efficiency factors, such as precursor costs and the relatively long processing times. New precursor developments could make the present Cu CVD process more efficient.

Both processes are highly qualified for making seed layers. The high conformality of the CVD allows for a film deposition with homogenous thickness on top and in the TSV structures. Homogenous film thickness is a major condition for equal current density distribution in electroplating processes, thus enabling void-free filling.

References

1 Temple, D., Bower, C.A. *et al.* (2007) in 2006 MRS Symposium Proceedings, volume 970, 0970-Y03-04, p. 115–117.

2 Wieland, R. *et al.* (2007) in Smart Systems Integration 2007, 649, Verlag GmbH, 978-3-8007-3009-4.

3 Ramm, R. *et al.* in AMC 2001 Proceedings, 159, 1-55899-670-2.

4 Sakuma, K. *et al.* ECTC 2007 Proceedings, 627, IEEE 1-4244-0985-3/07/

5 Kim, B. (2007) 2006 MRS Symposium Proceedings, 970, 253 0970-Y06-02, p. 259–260.

6 Burkett, S., Schaper, L. *et al.* (2007) in 2006 MRS Symposium Proceedings, 970, 0970 Y06 01, p. 261 273.

7 Wang, S.-Q. *et al.* (1996) *Journal of Vacuum Science & Technology*, **14** (3), 1846.

8 Rossnagel, S.M. and Kim, H. (2001) 2001 Proceedings IITC, pp. 3–5.

9 Tsang, C.K. *et al.* (2007) *Materials Research Society Symposium Proceedings*, **970**, p. 261–273.

10 Körner, H. and Seidel, U. (1994) in 1993 Advanced Metallization for ULSI Appl. VII, MRS, p. 513.

11 Ohshita, Y. *et al.* (1995) *Journal of Crystal Growth*, **146**, 188–192.

12 Hillmann, J. *et al.* (July 1995) *Solid State Technology*, **38**, p. 147.

13 Chang, Y.H. *et al.* (1996) *Applied Physics Letters*, **68** (18), 2580.

14 Weber, A., Gross, M.E. *et al.* (1995) *Journal of the Electrochemical Society*, **142** (6), L79.

15 Tsau, L. *et al.* (1996) in 1996 VIMC Conference, Vol. 106, p. 596–598.

16 Leutenecker, R. *et al.* (1995) *Thin Solid Films*, **270**, 621–626.

17 Paranjpe, A. and Islam Raja, M. (September/October 1995) *Journal of Vacuum Science & Technology B*, **13**, p. 2105–2114.

18 Kodas, T. and Hampden-Smith, M. (1994) *The Chemistry of Metal CVD*, VCH, Weinheim, 3-527-29071-0.

19 Kim, C.K. *et al.* (2003) Chemical Vapour Deposition XVI and EUROCVD 14. Electrochemical Society Proceedings, PV, 2003-08 (2), p. 1284.

20 Kim, K. and Yong, K. (2003) Chemical Vapour Deposition XVI and EUROCVD 14. Electrochemical Society Proceedings, PV, 2003-08 (2), 1290, 1-56677-378-4.

21 Das, M. and Shivashankar, S.A. (2007) *Applied Organometallic Chemistry*, **21**, 15–25.

22 Röber, J., Kaufmann, C. and Gessner, T. (1995) *Applied Surface Science*, **91**, 134–138.

23 Joulaud, M., Angekort, C., Doppelt, P. *et al.* (2002) *Microelectronic Engineering*, **64**, 107.

24 Joulaud, M. *et al.* (2003) Chemical Vapour Deposition XVI and EUROCVD 14. Electrochemical Society Proceedings, PV, 2003-08 (2) p. 1268. 1-56677-378-4.

25 Blewer, R.S. (1988) Tungsten and Other Refractory Metals for VLSI Applications. Proceedings of the 1988 Workshop Held October 4–6, Albuquerque, USA Materials Research Society Conference Proceedings, (ed. Carol M. Conica), p. 65–76.

26 Riedel, S., Schulz, S.E. and Gessner, T. (2000) *Microelectronic Engineering*, **50**, 533.

27 Ecke, R. *et al.* (2003) Chemical Vapour Deposition XVI and EUROCVD 14 (eds Allendorf, M., Maury, F and Teyssandier, F), Electrochemical Society Proceedings, PV, 2003-08 (2), p. 1284.

28 Riedel, S., Weiss, K., Schulz, S.E. and Gessner, T. (2000) AMC 1999 Proceedings, p. 195, 1-55899-539-0.

29 Gandikota, S. *et al.* (2000) *Microelectronic Engineering*, **50**, 547.

30 Yang, D., Hong, J., Richards, D.F. and Cale, T.S. (2002) *Journal of Vacuum Science & Technology B*, **20** (2), 495–506.

31 Körner, H. and Seidel, U. (1994) *Advanced Metallization for USLI Applications 1993* (ed. Fareau *et al.*), MRS, Pittsburgh, PA, p. 513.

32 Körner, H. *et al.* (1991) *Tungsten and other Refractory Metals for ULSI VII* (eds R. Blumenthal and G. Smith), MRS, Pittsburgh, PA, p. 369

II
Wafer Thinning and Bonding Technology

Handbook of 3D Integration: Technology and Applications of 3D Integrated Circuits.
Edited by Philip Garrou, Christopher Bower and Peter Ramm

ISBN: 978-3-527-32034-9

10
Fabrication, Processing and Singulation of Thin Wafers

Werner Kröninger

10.1
Applications for Thin Silicon Dies

Today, many applications use *thin silicon*. Why? Using thin silicon improves several key parameters of integrated circuits and some applications have been created by the ability of producing thin silicon.

The main spheres of application for silicon chips include the processors you find in your PCs, memory-chips used in nearly every electronic module, the power-devices your car is full of and the smart-labels found in access-cards, tickets and labeling applications that make our daily life easier. For each of these applications the chips must be thinned. What is the driving force of this development, why are we going thin?

For processors the main advantage of thin substrates is improved heat dissipation. Stacking of memory chips needs also thin silicon. For power devices it is reduction in electrical resistance. For smart-cards and related applications the main feature is the flexibility of thin silicon, which makes the IC-chip capable of surviving daily use. Thus, the question should be: How much bulk silicon is necessary to guarantee the function of the device? Several investigations have been conducted to clarify this matter. The answer is also product dependent; however, an enormous number of applications benefit from thinner silicon. Most of them only need some microns of silicon beyond the active layers. Metaphorically, there is only an inch of water needed beyond the bottom of a ship.

10.2
Principal Facts: Thinning and Wafer Bow

Two main factors contribute to the wafer bow (Figure 10.1). First, the final wafer thickness, which acts directly on the wafer bow. The amount of bow rises with reduction of thickness. Second, the active layers on the device side also affect the wafer bend. The level of defection depends on the number and kind of active layers,

Handbook of 3D Integration: Technology and Applications of 3D Integrated Circuits.
Edited by Philip Garrou, Christopher Bower and Peter Ramm

ISBN: 978-3-527-32034-9

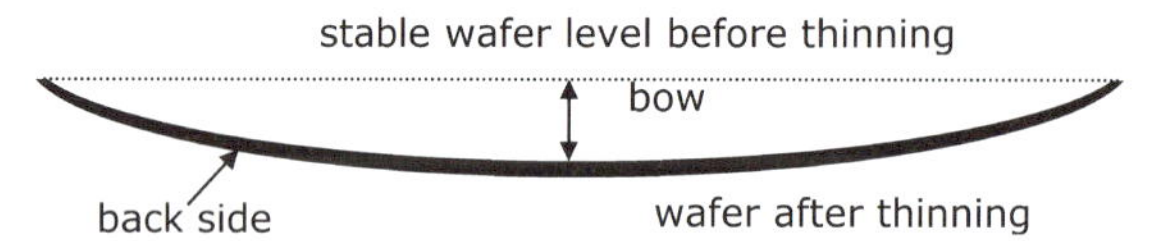

Figure 10.1 Indication of bow orientation: wafer showing bow due to tensile stress of active layers.

because many of them are under tensile stress. In general, "more layers, more bow." These layers also create topology (leading to additional mechanical tension); some of them are famous for creating tension like metal and photo imide. Chemical Mechanical Polishing (CMP) reduces topology, thus reducing tension, resulting in a reduction of bow. Chip dimensions also have an effect. The abrasive grinding process leaves a damaged layer of silicon containing silicon oxide. This layer causes the biggest part of the bow. This is why getting rid of this layer (stress-release) is an important part of thinning silicon. The final thickness of the wafer is of course the main player, the thinner the bulk the less rigid the chip becomes.

For a particular chip, having installed a proper and suitable thinning process, the wafer-bow depends exclusively on the product and the final wafer thickness. Let us look at this wafer bow along the process flow (Figure 10.2). Starting from the original thickness (725 μm for 8-inch) the amount of bow is nearly zero for most applications, reaching its maximum at the coarse grinding step. From there we try to reduce the bow by fine grinding and, especially, by stress release procedures.

10.2.1 Where Does this Phenomenon Come From?

To explain this phenomenon let us look at a diamond grain scratching into silicon.

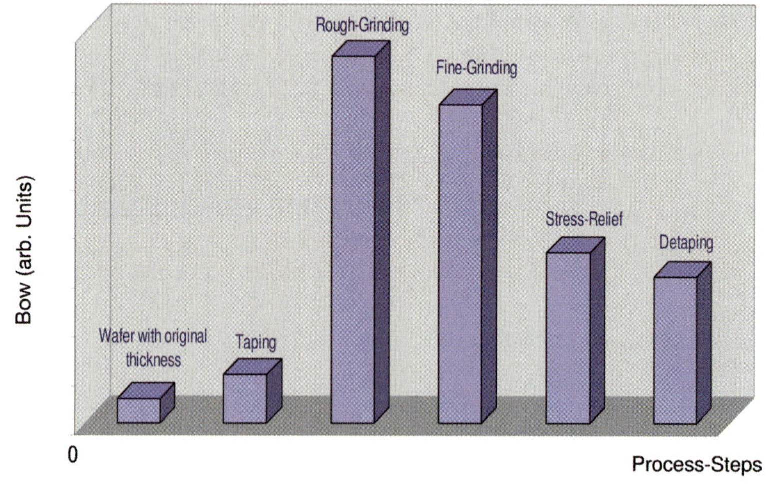

Figure 10.2 Wafer-bow: along the process flow of thinning. For a typical thickness >100 μm, for ≪100 μm the wafer could roll.

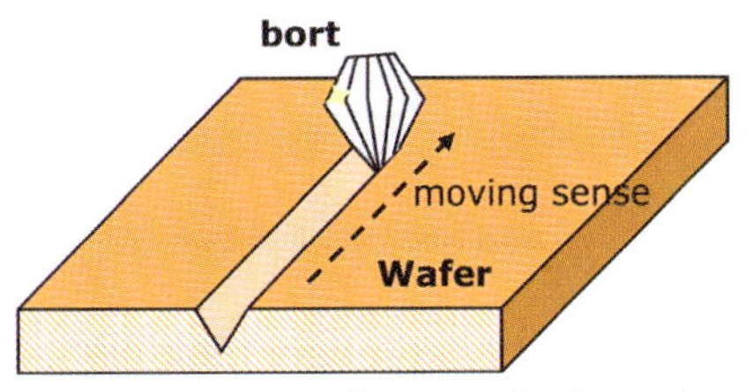

Figure 10.3 Plastic deformation ductile grinding.

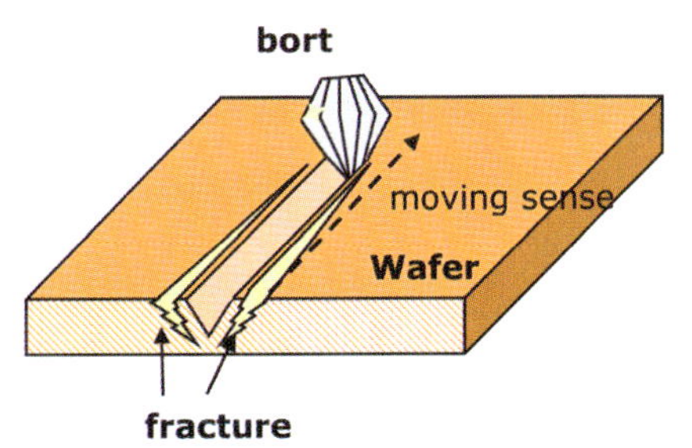

Figure 10.4 Brittle fracture cracks and defects, polycrystalline layer

The damage comes from the back-side grinding process. We would like to remove the silicon without damaging the rest of the wafer. The plastic deformation shown in Figure 10.3 is our goal. However, reality is closer to what is shown in Figure 10.4.

10.3 Grinding and Thinning

The process flow of thinning is shown in Scheme 10.1.

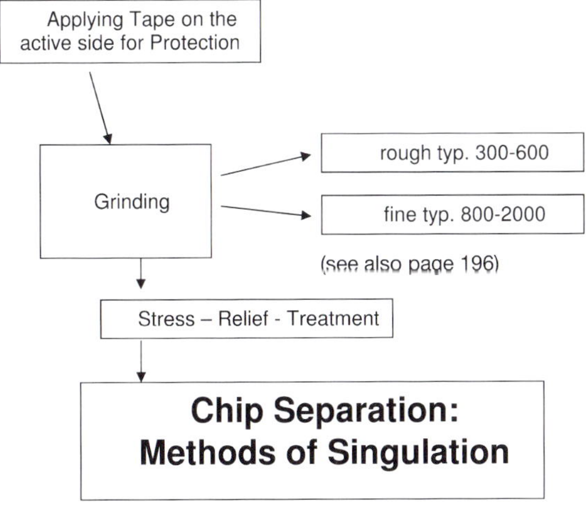

Scheme 10.1

We start by protecting the active side with a tape. Grinding is done in two steps, coarse grinding using diamonds around 20–80 μm in diameter and fine grinding using diamonds with around 1–8 μm in diameter. One of the key factors in a grinding wheel is the control of size distribution. It is essential to keep the grains in a certain range, as indicated by the mesh size. Mesh #600, for example, corresponds to a grain size of 20–30 μm.

10.3.1 Grinding Parameters

The central unit in this complex process of grinding crystal material is the abrasive tools (e.g. the grinding wheels). The essentials of a grinding wheel are grain, bond, pores and the sintering process combining these three (Figure 10.5).

The main parameters of grinding are the feed rate (material removal speed), wheel-roughness (grain-size, bond) and process cooling. If the process temperature is too high, the wafers could be destroyed or damaged; thus cooling is an important topic in wafer thinning. Table 10.1 shows some parameter ranges for both coarse and fine grinding.

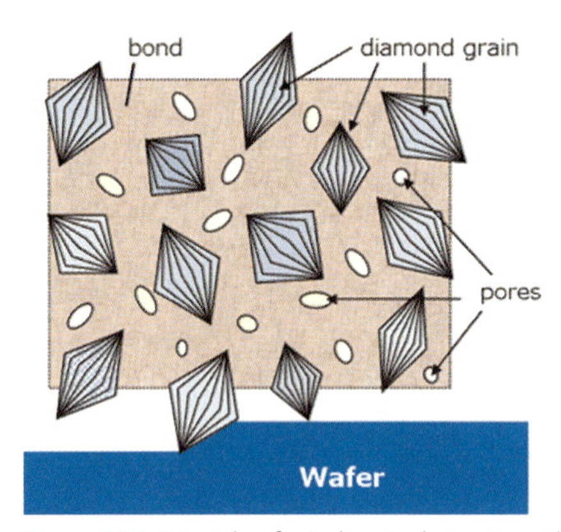

Figure 10.5 Principle of grinding tool structure: diamond grain, bond and pores.

Table 10.1 Ranges for some typical coarse and fine grinding parameters.

Parameter	Coarse grinding	Fine grinding
Wheel-roughness (mesh)	300–600	1000–3000
Diamond-diameters (μm)	10–80	1–8
Feed-speed ($\mu m\ s^{-1}$)	2–12	0.2–1.5
Bond/composite structure	Ceramic	Resin, some also ceramic
Silicon features	Coarse	Fine
Surface-roughness ($ra\ \mu m^{-1}$)	<0.2	<0.05
Breaking strength (N) (300 μm)	8	20

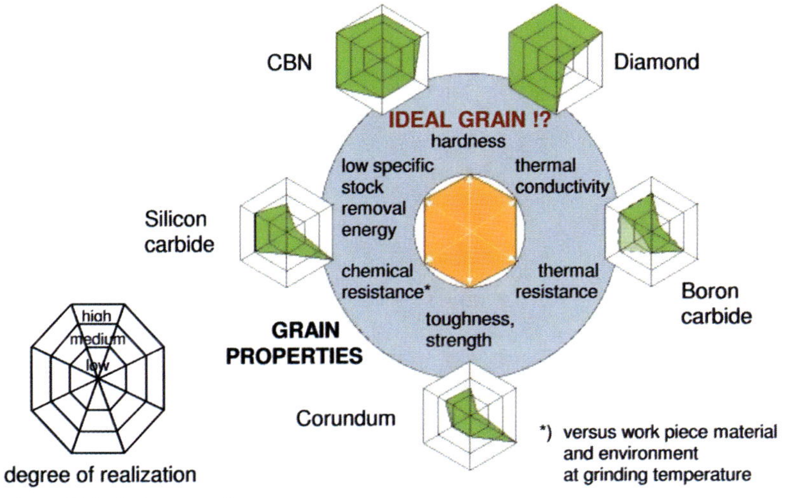

Figure 10.6 Grain properties.

Let us look at the different grinding materials. The grains have to meet several criteria (Figure 10.6). One alternative to the use of diamonds would be CBN (cubic bore nitride). To date, diamonds have proved to be most suitable for grinding silicon (and also other brittle materials like GaAs, SiC). Regarding the bond, mostly resin or ceramic (vitrified) are chosen, depending on the requirements (Figure 10.7).

10.3.2
Vice Versa Influences of Parameters

The main parameters of grinding are interdependent. There are vice versa influences of the breaking strength the chips show, wafer bow, coarse grinding feed speed, wheel hardness, wheel wear and fine grinding removal amount.

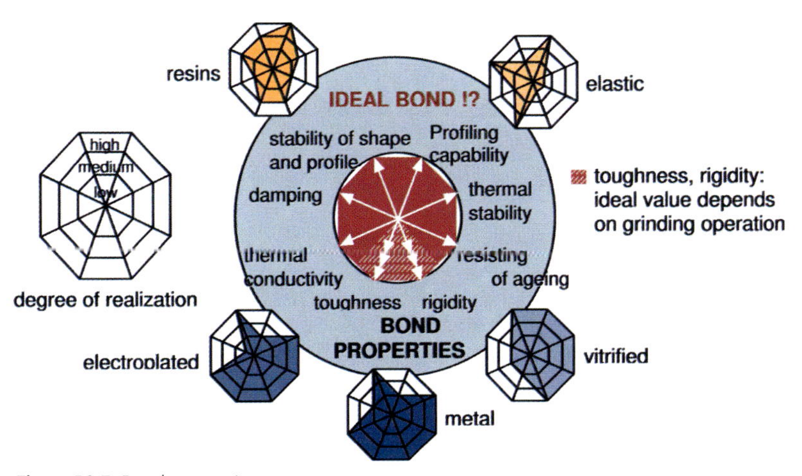

Figure 10.7 Bond properties.

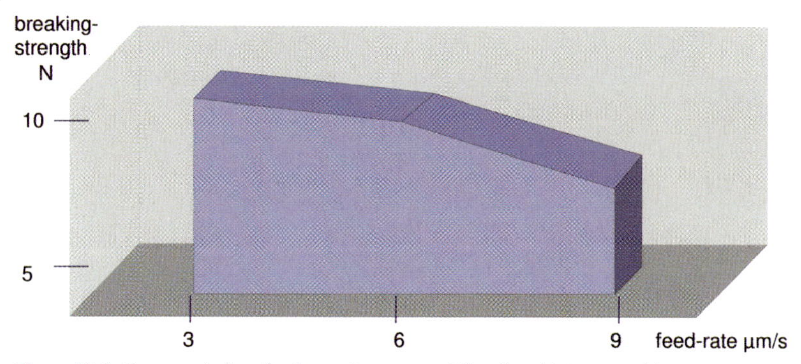

Figure 10.8 Coarse grinding feed-speed versus stability (breaking-strength).

If the hardness is too low, we get very high wheel consumption and the removal of material is not efficient. If the hardness is too high, however, the wafer reaches too high temperatures, leading to wafer breakage or damaging of the product. Coarse grinding has the main influence on bow/warpage and stability (breaking strength). There is a quite broad process window, depending on wafer size and product; for 8-inch wafers approx. 3–6 $\mu m\,s^{-1}$ (for fine grinding one tenth of the rate). The higher the feed speed the higher the wheel wear.

At faster grinding speeds we get fewer wafers out of a wheel. If we incline towards higher feed speeds, we risk lower stability (Figure 10.8), increased wafer bow (Figure 10.9), burning (you can see colourful oxide layers) and thus destroying the wafers. Grinding too slow can also lead to trouble: clogging of the wheel can be a consequence (the grains are embedded by the material ground, the removal rate decreases and the grinding pressure goes up drastically). There is also a vice versa influence between coarse grinding and fine grinding: the surface conditions left by the coarse grinding wheel have to be suitable for the fine grinding wheel. What we need is not "a good grinding wheel" but a good pair of grinding wheels. Of course, several wheels can be matched, but we have to check and make sure. Another dependency: the rougher we grind the lower the stability.

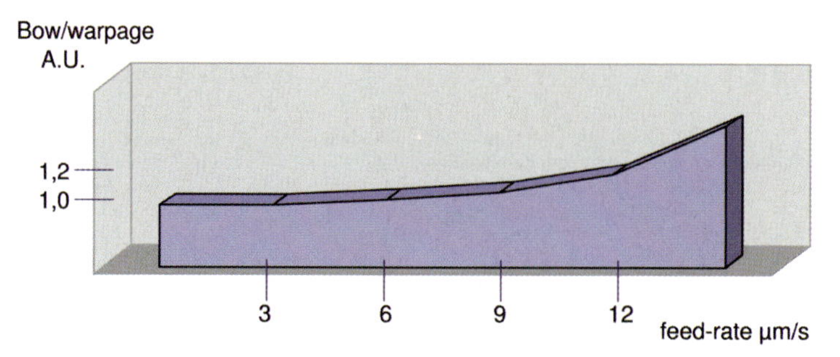

Figure 10.9 Coarse grinding feed speed versus bow/warpage.

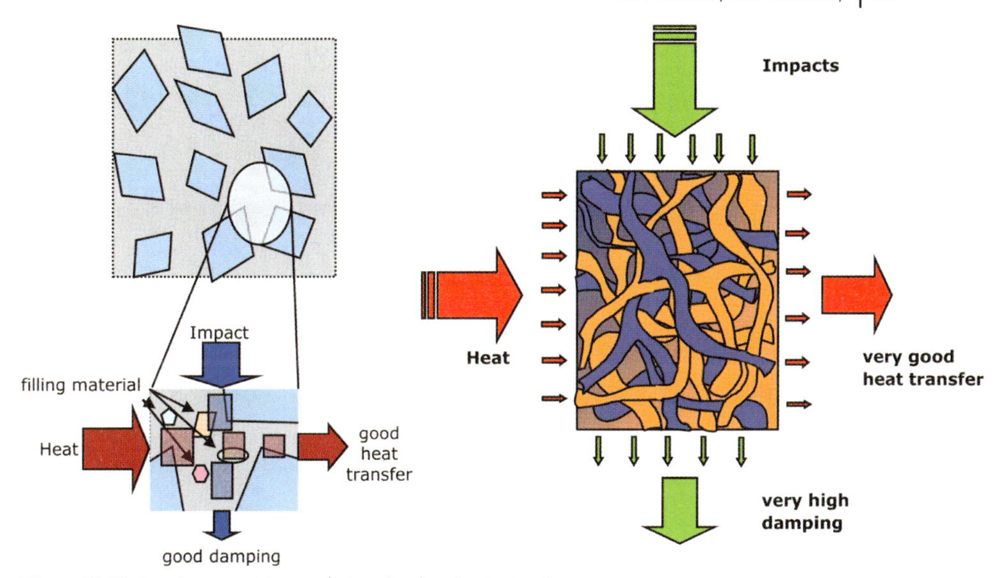

Figure 10.10 Bond composition and sintering is a key factor for high-performance grinding wheels.

Why are we coarse grinding at all – could we skip it? The answer is twofold: First, we cannot skip it as it is difficult to penetrate silicon with fine grind wheels. Second, if we manage to enter with a fine wheel, it is difficult to grind for hundreds of μms with most fine grinding wheels, and the process is quite slow. More generally, why are we grinding at all? Could we use different methods? We could, and in this chapter we discuss other methods, but for now the answer is: grinding is cheap, it is fast and it is an industrial standard.

Fine grinding compensates for part of the damage induced by coarse grinding. Nevertheless, after fine grinding, we have a certain level of damage left, which causes wafer bow (Figure 10.2).

Grinding wheels are under constant development. One approach for improvement is a bond system with very high damping and very high thermal conductivity [1], thus reducing vibrations and improving the cooling during grinding (Figure 10.10).

10.4 Stability and Flexibility

Why do we need a certain chip-strength? We need the stability for further processing the chips, for handling and for standing the stress that occurs in the different processes. During die bonding, for example, chips are grabbed by needles or nozzles or both. The chips have to survive this process step safely. Field applications also add further demands. For example, a chip in a security chip card has to withstand the

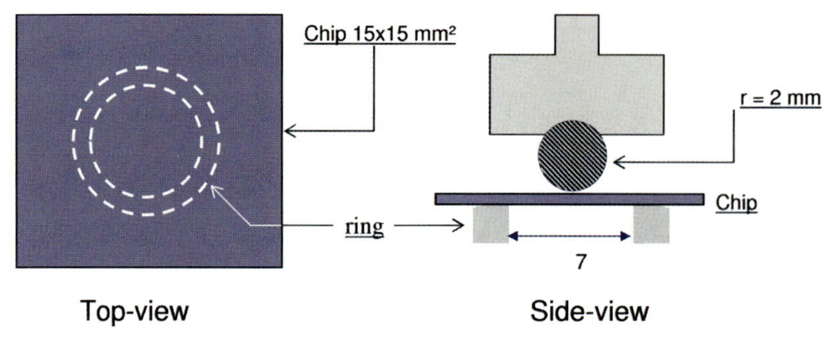

Figure 10.11 Breaking chips by the ball-ring method.

stress that comes from carrying and using it in daily life. This demands both stability and flexibility.

10.4.1 Measuring Breaking-Strength and Flexibility

10.4.1.1 Breaking Strength

There are several possibilities to investigate stability. Our approach is to measure breaking strength by the ball-ring method (Figure 10.11).

The ball is pressed by a force-giving system, moving forward slowly at 1 mm min^{-1}. The force and movement are measured continuously until the chip cracks.

10.4.1.2 Three- and Four-Points Bending Methods

Figure 10.12 illustrates the three- and four-points bending methods. A third rod (or double rod for four points) is pressed against the chip by a force-giving system (Figure 10.13). The measurement is similar to ball-ring. The system is controlled by a PC that continuously stores data for every chip of the lot. The dimensions shown in the Figure 10.12 are not fixed and can be adapted to the chips in test and the problem under investigation.

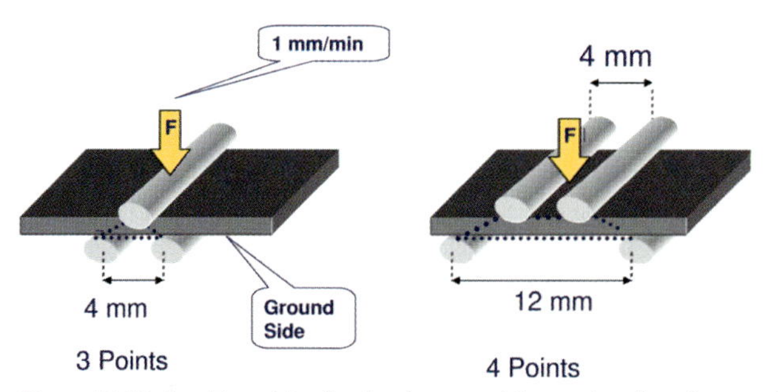

Figure 10.12 Breaking chips by the three- and four-points bending methods.

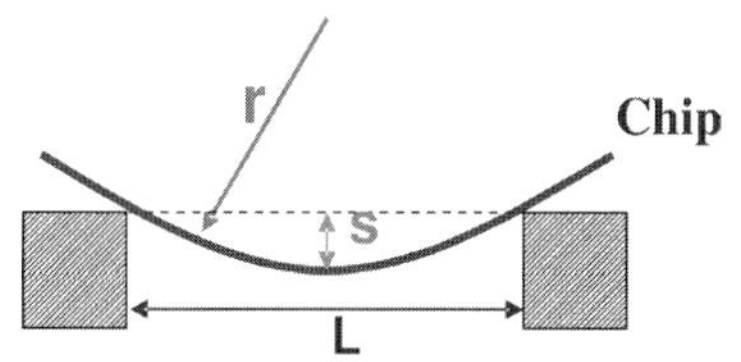

Figure 10.13 Measuring flexibility: Three-points bending **r** is the radius a chip can stand before it breaks. We track **s** during the breaking test. Given **s** when the chip breaks we calculate **r**. The radius a chip can stand before it breaks is proportional to the chip thickness.

10.4.2
Statistics and Evaluation

Following the process flow there are several steps where a certain mechanical stability is essential. Flexibility is also a key feature for die-bonding and the field-requirement simulated in several stress- and temperature cycle tests. Consequently, we need both characteristics and we need to perform statistical evaluations. First let us have a short look at statistical issues.

For brittle materials the distribution for breaking strength is unsymmetrical relative to the average force ($F_{average}$). It is common to use Weibull distribution (Figure 10.14). We use $F_{average}$ (arithmetic average) and sometimes F_{median} (force where 63.2% of the chips have cracked) to give a little more information about the distribution. Thus, for brittle fractures there is a range, which always has a broader shoulder towards smaller breaking strengths [7, 11].

To obtain reliable results we take 25 + chips per split. The security of our results we get is proportional to the square root of the number of chips we use per split in an experiment.

The classical way of presenting results is to produce a Weibull plot. This has a double logarithmic scale, showing breaking probability versus breaking strength (Figure 10.15).

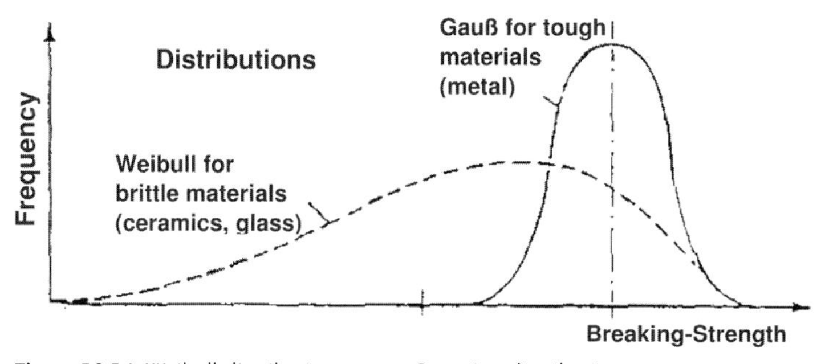

Figure 10.14 Weibull distribution versus Gaussian distribution.

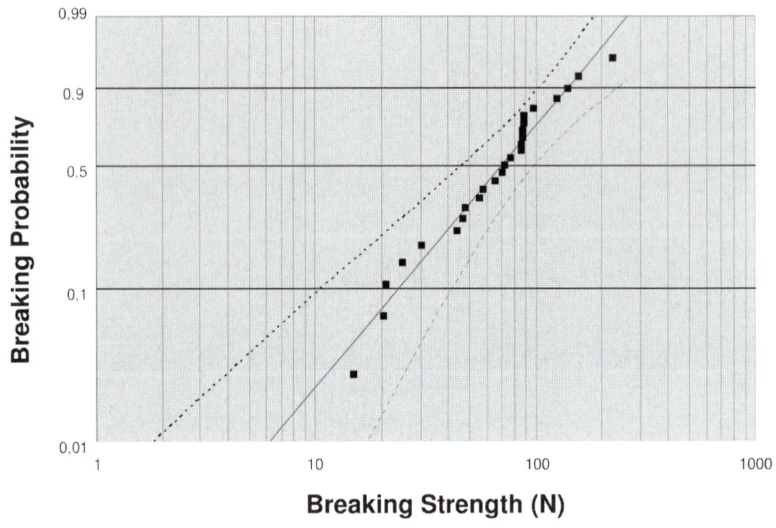

Figure 10.15 Weibull plot: breaking probability versus breaking strength.

10.5 Chip Thickness, Theoretical Model, Macroscopic Features

10.5.1 Chip Thickness

What happens to stability when we thin wafers by grinding? It decreases of course – but in what manner and to what extent? Figure 10.16 gives an overview of the relationship between thickness and stability.

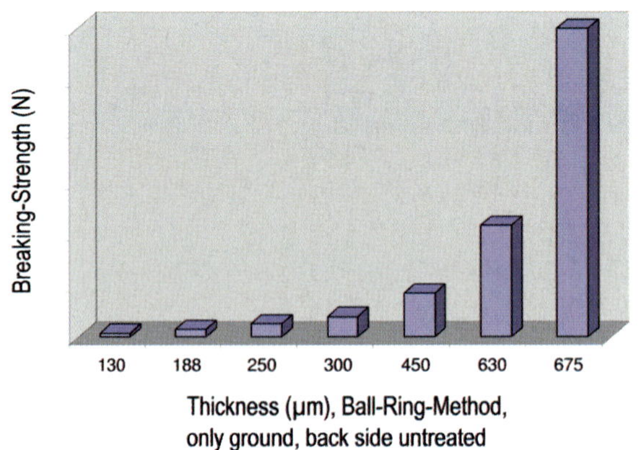

Figure 10.16 Stability (breaking strength) versus chip thickness for ground chips.

Two main aspects can be emphasized for wafers that are thinned by grinding only:

1. By taking off 7% of the original thickness we lose over 60% in breaking strength. So it is not only the thickness that decides about breaking strength; the most decisive influence comes from the back-side treatment.

2. Given a chip with ground back-side the breaking strength decreases rapidly with reducing thickness: if we reduce thickness to 30% (188 μm) the strength is reduced to 4–7%. To improve this situation we need to remove the damaged layer (stress release).

10.5.2 Theoretical Model

These facts are the consequence of the structure inside silicon after grinding. Therefore, we want to take a closer look at this structure (Figure 10.17).

Here we see the different regions, starting with the ground surface (on top) until we reach the active layer of the wafer, the chips, only some μm in thickness. A rough scale is given on the right-hand side of Figure 10.17, while the methods that can be used to investigate these layers is indicated on the left-hand side. Which zones of damage are relevant for macroscopic features like stability, flexibility, roughness and for the performance of the electronic device on the active side?

Several investigations have shown that thin chips down to at least 50 μm retain their performance, although there may be slight shifts, but within specification [2]. The electrical characteristics are influenced by bending stress. Most of these changes are reversible. However, we should keep the design robust for such applications. Consequently, in most cases, these stress and transition zones do not make any relevant contribution regarding device performance. Thus, these zones are not the target of our investigations, and for the moment we neglect them.

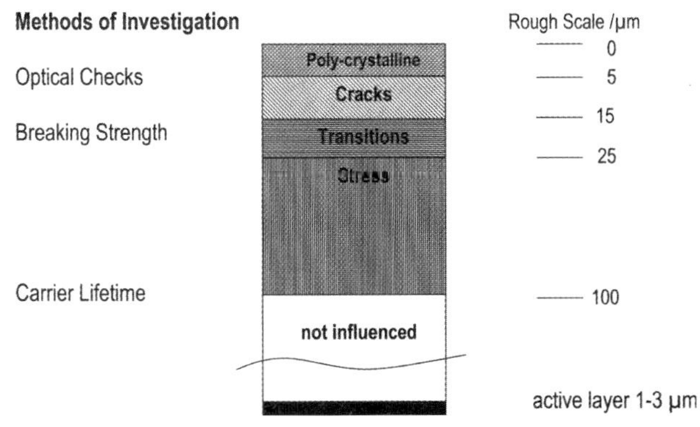

Figure 10.17 Classical structure inside silicon: zones of damage after mechanical thinning.

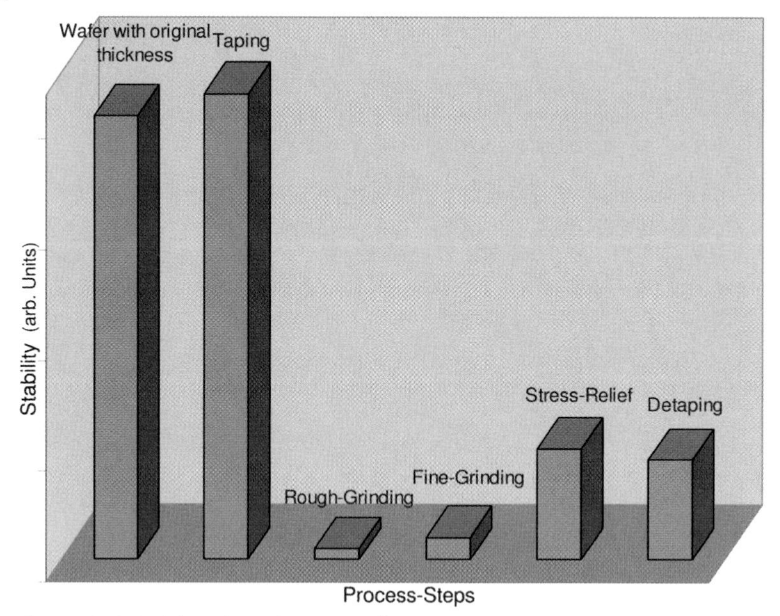

Figure 10.18 Stability: Along the process flow.

10.5.3 Macroscopic Features: Chip Strength, Flexibility, Roughness and Hardness

10.5.3.1 Chip Strength

Stability is one of the most important features of a chip. It influences both the yield during production and the reliability of the device after the chip has been placed into the package. Figure 10.18 shows the relative stability of a chip along the process flow of production.

To get rid of this damaged layer, created by grinding, different processes are in use. In Figure 10.19 we compare some of them in terms of the resulting breaking-

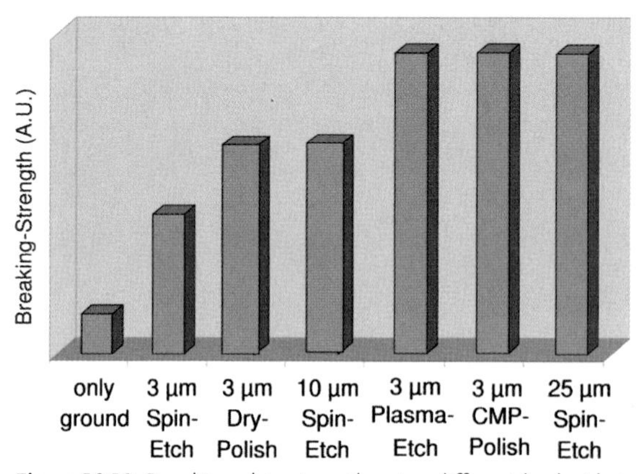

Figure 10.19 Resulting chip strength using different back-side treatments.

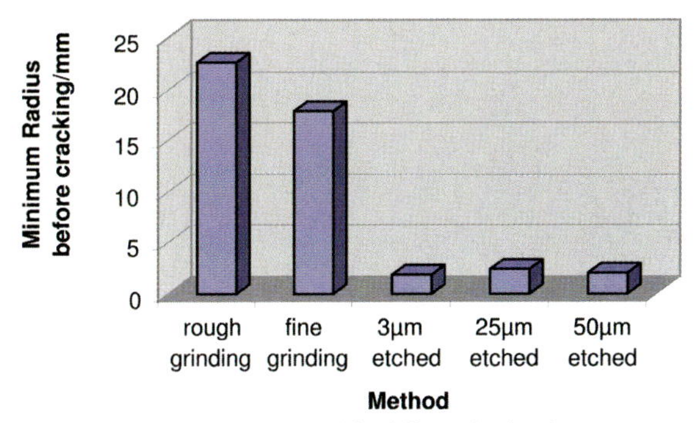

Figure 10.20 Flexibility achieved for different back sides (185 µm).

strength: saturation is reached by 20–25 µm, Spin-Etch; 3 µm, CMP-Polish; and 3 µm, Plasma-Etch.

The amount removed for saturation in stability depends on the method applied. The saturation limit for stability regarding a wet etch is 20–25 µm. The etch attacks the cracks, smoothening them, but also deepening them. Thus, a higher removal amount is necessary. This disadvantage can be turned to advantage. A wet etch can remove high amounts, even 100 µm, while with polishing it would be hard to remove 10s of µm.

10.5.3.2 **Flexibility**

For different Etch removal amounts during the back-side spin-etch we find that the Chips gain in strength upon increasing the Etch-removal amount from 3 to 25 µm. However, they do not gain much flexibility (Figure 10.20). Flexibility increases from coarse ground to fine ground to spin-etch, but we see no tremendous improvement for spin-etch removal amounts above 3 µm. A thinned wafer or a thinned chip consists of bulk material (assumed to be perfect silicon), some µms of active layers on the top and a damage layer on the back, as we saw above. Thus, we cannot describe the mechanical features of the chip by the modulus of elasticity for silicon. Looking at the experimental results we could say that due to damage relief treatment we gained stability (it takes more force to bend the chip) and due to the damage left the chip still cracks at a similar radius.

10.5.3.3 **Roughness**

The third important macroscopic feature we want to look at is roughness. The roughness of the back-side surface shows a certain correlation with breaking strength. Regarding die-bonding, back-side roughness also influences the distribution of glue and the resulting adhesion.

As a measure for roughness we use Ra, which can be seen as an average roughness within a measuring distance of some mm. The roughness of the chip back-side influences the adhesion for die-bonding and packaging. In addition, parameters of high-frequency devices can be influenced.

Roughness is reduced from coarse grinding to fine grinding to stress-release treatments (Figure 10.21). If we look through a microscope we can see a slight

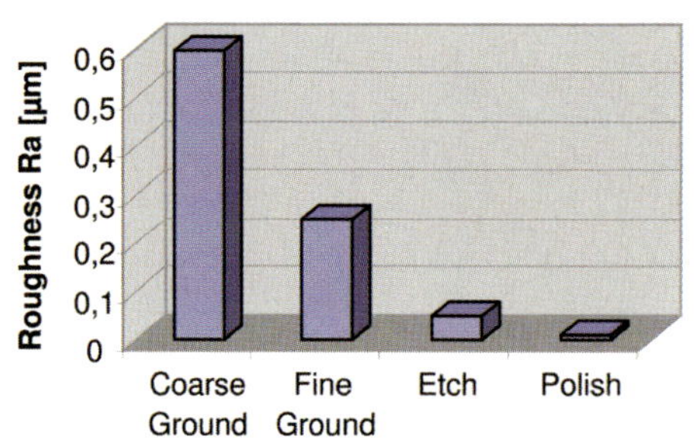

Figure 10.21 Roughness versus back-side treatment. (Ball-Park-Figures).

difference in appearance between only ground and ground plus plasma. A spin-etch results in a smooth mirror like surface whereas a plasma-etch gives only slight reductions in the roughness of the ground surface.

Coarse and Fine Grinding One of the most striking improvements in breaking strength is between rough- and fine-grinding. Taking off 20 µm by fine grinding results in a decisive increase of breaking strength. This is a decisive improvement in chip stability along the process flow of thinning. Fine grinding doubles the chip stability (Figure 10.22).

10.5.3.4 Hardness of Chip Surface

One of the parameters characterizing a surface is its hardness. The condition of a surface is also of interest for laminating issues like molding. We investigated back-side surface hardness to answer the following question: Does the hardness of a surface correlate with the extent of damage left by stress-release-treatments?

We carried out these investigations using a system that measures the material surface hardness while penetrating the material with a diamond pyramid. To get reliable results the depth we push into the material has to be at least one magnitude

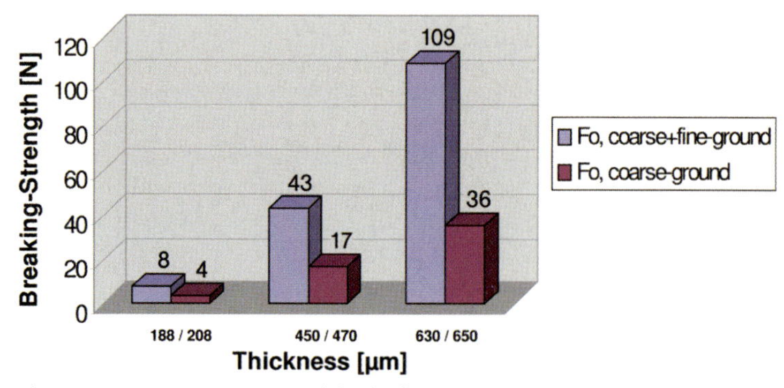

Figure 10.22 Improvement in stability by fine grinding. Ball-Ring-Method.

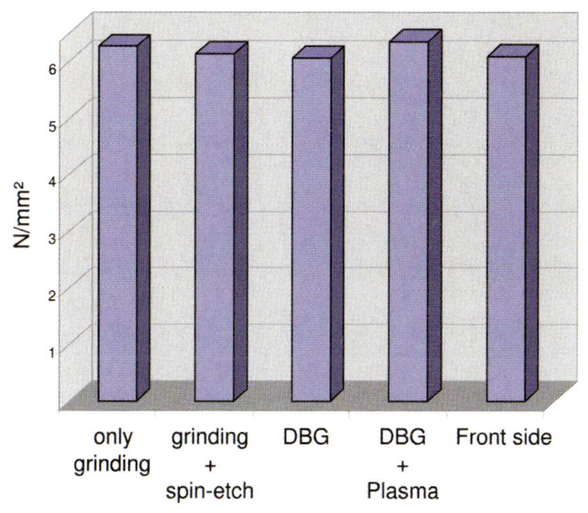

Figure 10.23 Hardness for different surface treatments.

above the measured roughness. We measure universal hardness according to DIN 50359.

Tests on blank silicon 8-inch-wafers show that back-side hardness seems to be independent of back-side treatment. The modulus of elasticity calculated from these hardness data is the same value found in the literature of 130 GPa for silicon(100) (Figure 10.23).

10.5.4
From Blank to Processed Chips: Changes?

For both stability (Figure 10.24) and flexibility (Figure 10.25) we see that processed chips show a reduced performance compared to blank unprocessed silicon. To give a ball-park figure, in general we find losses of 20–30%.

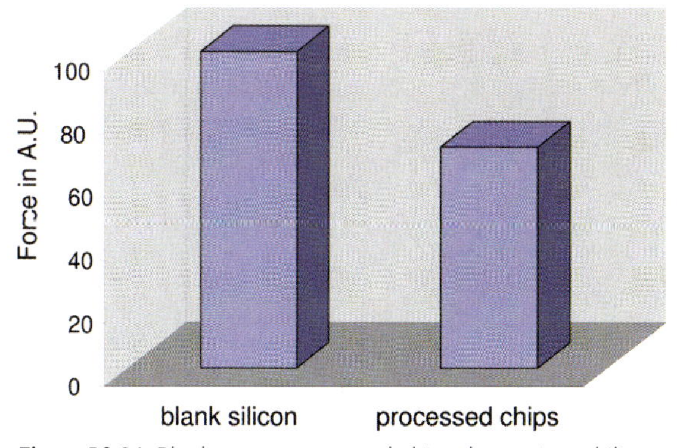

Figure 10.24 Blank versus processed chips: losses in stability.

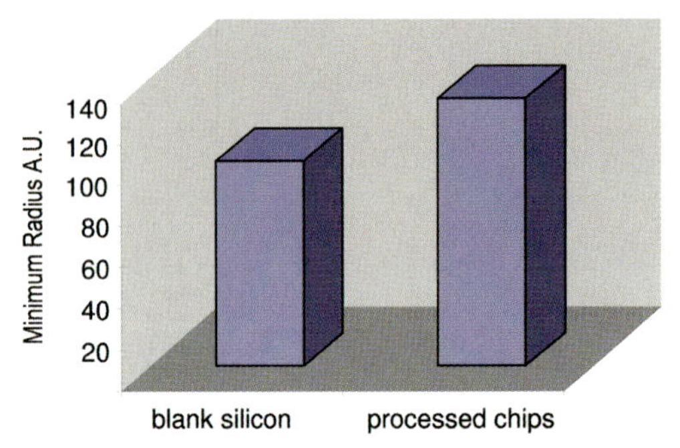

Figure 10.25 Blank versus processed chips: losses in flexibility.

This reduction is mainly due to the stress the product wafers see along the process flow. From thermal budget to chemical attacks by plasma, from wafer-handling in hundreds of process-steps to chip handling there are lots of impacts on chip stability. Thus, processed chips never reach the stability predicted by experiments performed with blank silicon wafers.

10.6
Stabilizing the Thin Wafer: Tapes and Carrier Systems

As a first rule we state: avoid handling whenever you can. If you cannot avoid handling, use supporting systems. Regarding thinning, the product range of semiconductor devices is divided into two areas.

First, thinning as last process-step, for example: memory, smart-cards, controllers, mobile communication; these products have no additional layers on the back. To avoid handling we recommend a cluster tool. For most tasks of this kind Separating by Thinning (SbT) technologies are suitable. For the handling inside the cluster tool, special damped cassettes, full surface contact handler or Bernoulli-handlers are in use. Result: we put in wafers and what we get out is separated chips on tape. The second area is processes on wafer back (to apply after thinning). Here the main applications are power devices processes: cleaning, metallization, ion-implantation. As handling after thinning is unavoidable we need a supporting system. Several quite different approaches have been presented:

1. Standard-Tapes: pressure sensitive and UV-tapes give a certain support.
2. Stiff tapes: good support but hard to detape.
3. Stiff tapes that become flexible after UV-irradiation chuck heating or through washing with hot water.
4. Supporting plates made of glass, ceramics, sapphire (Al_2O_3) or silicon. The wafers are fixed to these plates.

Several of these solutions are commercially available. If you cannot use SbT-technologies, one of these methods should be applied.

10.6.1
Special Tapes for Handling Wafers and Dies

- UV-Grinding-Tape: the tape applied on the active wafer side for device protection during thinning loses 90% of its adhesive strength through UV-irradiation, thus the tape is easy to remove afterwards.
- UV-Dicing-Tape: here the chips are easy to pick by the die-bonder as the dicing-tape loses its adhesive strength through UV irradiation.
- Detaping after mounting: the grinding tape is detaped after the wafer has been mounted to the dicing-frame; thus the wafer is always connected to a supporting tape.
- Special dicing-tape providing glue for die-bonding: the adhesive layer of the dicing tape goes with the die when picked. This same adhesive is then used for fixing the die in the die-bond-process.
- Special Grinding-Tape providing glue for Flip-chip Die-bonding: here the adhesive layer of the grinding tape remains at the wafer after removing the grinding tape. This adhesive layer on the active side of the chip can now be used in the flip-chip bonding process.
- Die-attach-film: the adhesive film is put on a wafer, diced with the chips and then used as adhesive layer during the stacking of these chips.

10.6.2
Carrier Systems

What kind of carrier should we use? Quite a view have been proposed: glass-plates, ceramics, compound materials, rigid tapes. It depends on the application, but you should differentiate the following.

- Lost carrier concept: the carrier is diced with the chips and built into the package.
- Recovered carrier concept: the carrier is recovered, wafer and carrier are separated again after this process sequence. Often the most suitable carrier is another blank wafer. The main advantages are that this carrier has the same material constants, is not expensive and can be used in every environment the active wafer survives.

The purpose of a supporting carrier is to enable the thin wafer to be handled as any ordinary wafer of original thickness. For example, you might have a thin wafer and you want to drill vias (Figure 10.26), or you need a firm support for applying processes on the back side of the wafer (Figure 10.27).

Figure 10.26 Support wafer for through silicon via.

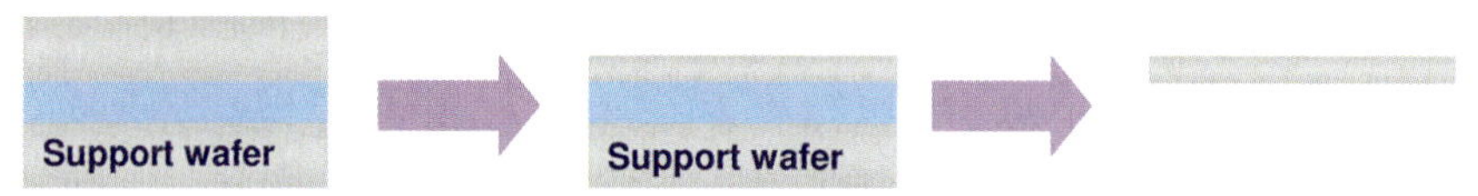

Figure 10.27 Carrier as handling support for thinning processes and processes applied to the back side.

After choosing the kind of carrier to use, thinning should no longer be a problem. But before you can do that we have to solve the second major topic: How to fix the wafer to the carrier?

Several adhesive systems are in use. Wax, for example, is complicated to apply, gives an optimum of fixture (Figure 10.28), but cannot withstand higher temperatures. Glue is hard to resolve, but simple to apply. Double sided tape is also not temperature stable, but can be a good solution if no high temperatures are involved. Electrostatic means need eventually recharging during process flow, but is simple and fast in fixing and separation.

Now that we are using a carrier, the focus changes: the new challenge is applying the carrier to the wafer and separating wafer and carrier after thinning and the processes on the wafer back side are done, without leaving any residues on the active devices. Processes on the back of the wafer can impose several constraints: the carrier system has to be suitable for the subsequent processes, for example cleaning, ion-implantation and metallization. This means that it has to withstand a vacuum, certain process temperatures and certain chemical media.

One way to stabilize a wafer is to add a ring on its edge. For some applications such a ring is tolerable. Recently, this idea was modified. Using a special grinding tool only the inner part of the wafer is thinned, leaving the outer rim as a stabilizing ring (Figure 10.29).

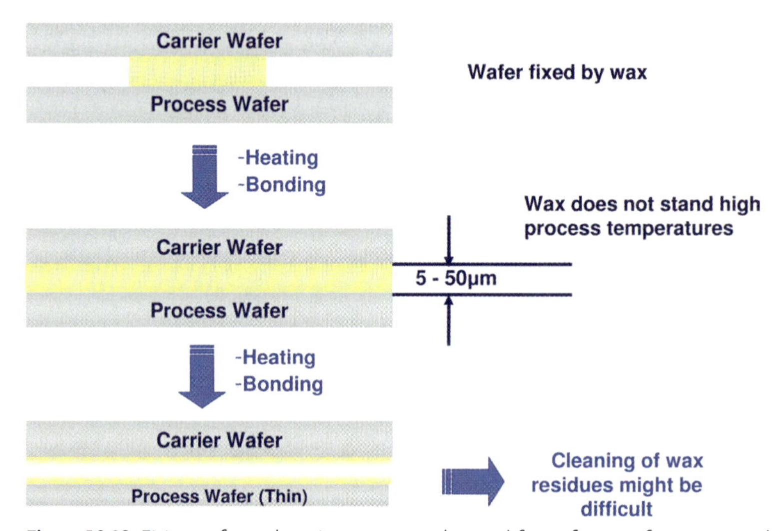

Figure 10.28 Fixing wafer and carrier: waxes can be used for wafer-to-wafer temporary bonding.

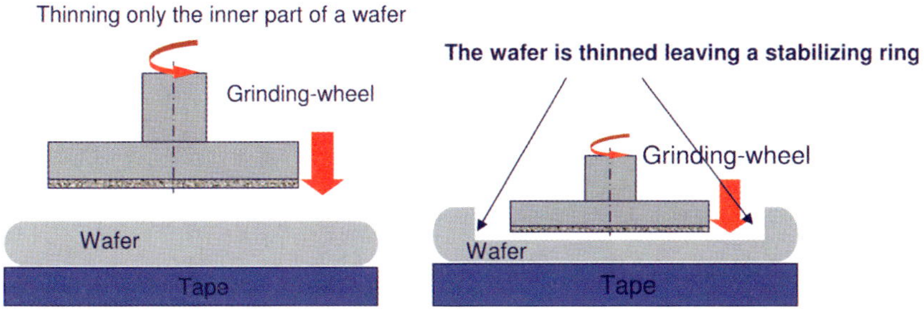

Figure 10.29 Stabilizing the wafer by a ring.

In this case the carrier takes shape during thinning. This method may not be suitable for each and every challenge, but it is a very good example of amending established methods of production [3].

10.7 Separating the Chips: Dicing Influencing the Stability

Separating the dice is the link between front-end and back-end. Most ground materials for IC-manufacturing in semiconductor industry are very brittle. This is the main reason for rough cutting edges. The cut itself, the rough edges and a certain security distance to the active chips result in a minimum "dicing street:" the necessary space left between the chips to separate them (Scheme 10.2).

10.7.1 Classical Mechanical Dicing

The state-of-the-art process is pure mechanics, but is nevertheless highly sophisticated (Figure 10.30).

Principal overview: the process flow of thinning

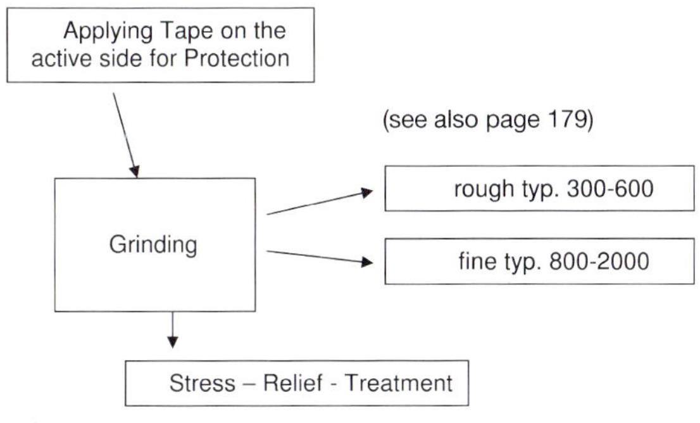

Scheme 10.2

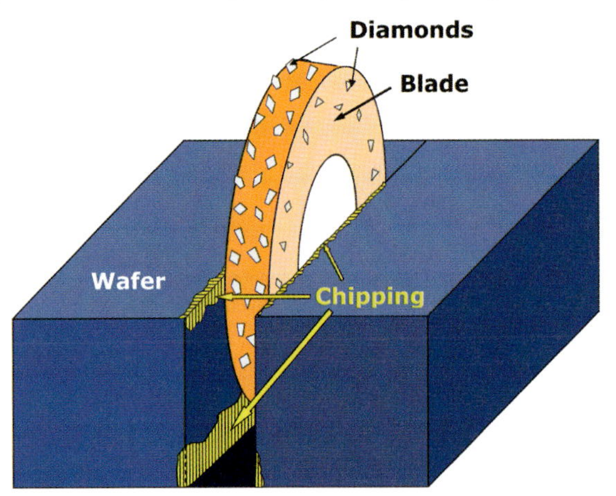

Figure 10.30 Illustration of the dicing process.

Minimizing the kerf-width can improve productivity because for small chips the number of systems per wafer can be increased. However, a certain die pitch can be needed for proper handling of the chips during die-bonding. If the dicing tape used is expandable the distance from chip to chip can be increased before die-picking.

Let us assume a small chip with an area of 1.5 mm^2 on a 6-inch wafer. Reducing the dicing street from 100 to 50 μm can result in a gain of nearly 15% (Figure 10.31). Thus, the goal for every chip manufacturer is clearly stated: a process with high yield and narrow streets.

Several processes have been developed in recent years:

- dicing through in one go
- various step-cut systems
- multiple blade dicers.

The state-of-the-art process is step cut (Figure 10.32). Here two dicing blades separate the chips. The first is normally a little broader and cuts into the silicon. The second

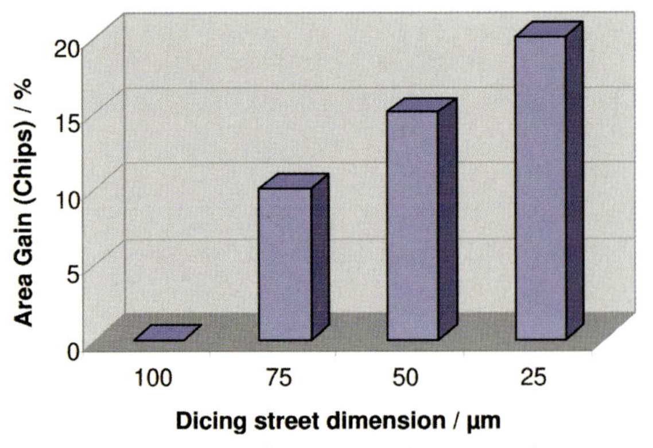

Figure 10.31 Gain in % of chips versus dicing street dimension.

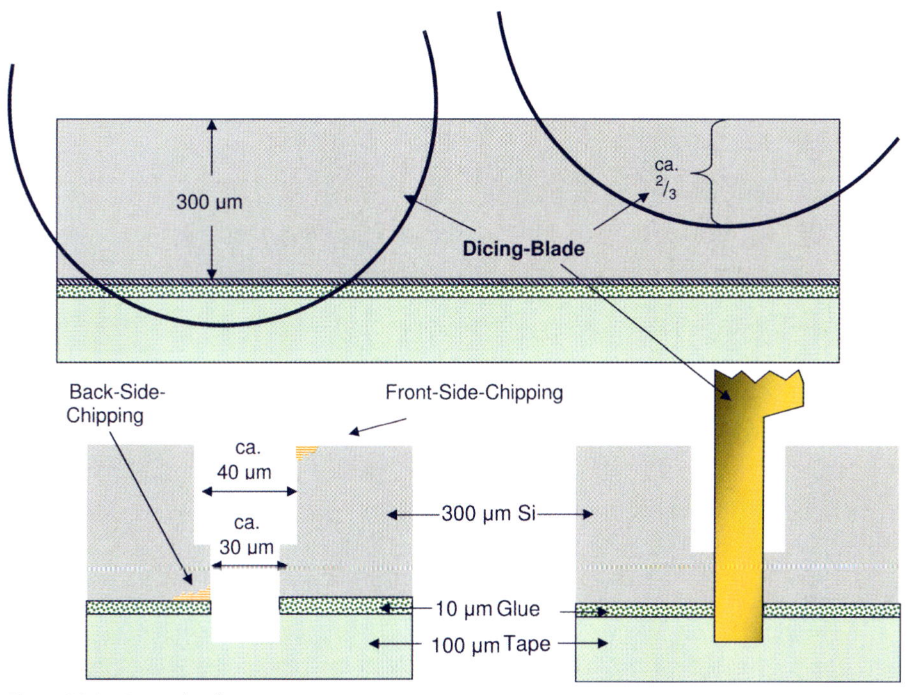

Figure 10.32 Principle of step cut.

cuts through the silicon, and also through the glue layer and into the tape. The dicing blade rotates at some 10 000 rpm and has a typical diameter of 2 inches.

If we put the dimensions to scale, we see that the scratches caused by the dicing blade are nearly parallel to the chip surface (Figure 10.33).

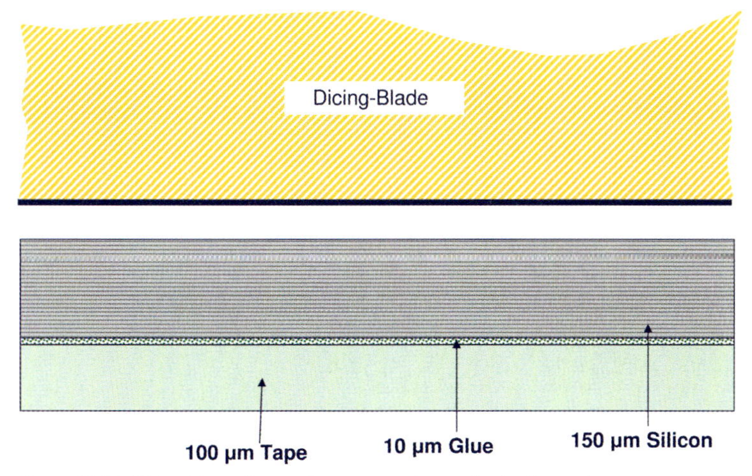

Figure 10.33 Chip side wall scratches are nearly parallel to the surface.

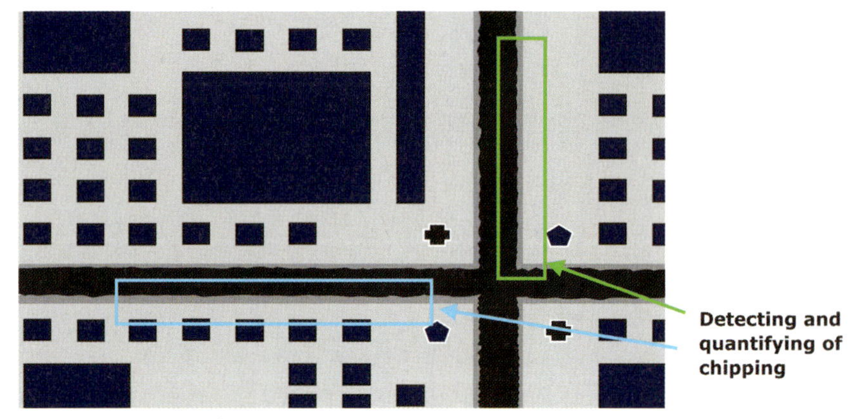

Figure 10.34 Control of dicing quality by visual systems.

The main problem of the dicing process can be summarized as *chipping* (Figure 10.32). Chipping means defects on the sides of the cut. In fact the process is more similar to grinding than to dicing with a knife. The cut creates a trench called a kerf.

How to control this issue? Quality can be controlled by visual inspection (Figure 10.34). The main advantage is objectivity. With the same system in different locations it is, for example, possible to compare results for outgoing and incoming inspections.

10.7.1.1 Damaged Sides of a Chip

Dicing is separation by grinding. A standard grinding-wheel (2000 mesh) and dicing-blade show the same diamond grid of 2–6 μm. The principle of material removal is comparable. The damage induced by dicing also lowers chip stability, as is known from the damage caused by the grinding process (Figure 10.35).

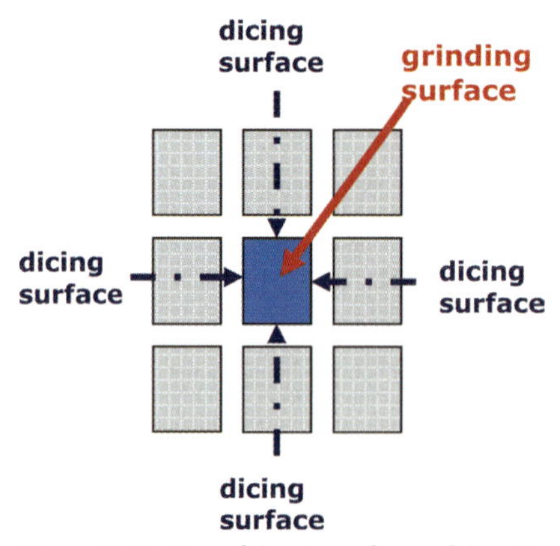

Figure 10.35 Five of the six surfaces of the chip are treated by either dicing or grinding.

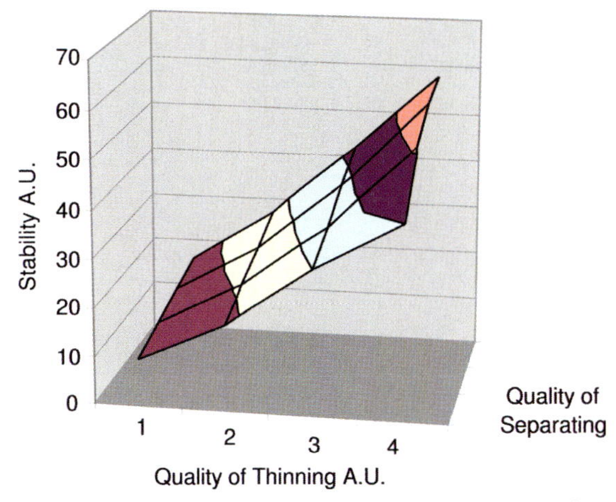

Figure 10.36 Quality of separation modulating the quality of thinning.

This side-wall surface damage consists of front side chipping, back-side chipping and side wall damage.

The separation of the chip creates additional surfaces. The chip can be seen as a cuboid. Five of the six sides of the cuboid are ground. Every scratch, chipping, and crack in a chip is a predetermined breaking point. The biggest contribution by far creating "rated breaking points" is the Thinning Process, which influences the whole surface area. The next decisive influence comes from separation. We often see the quality of separating as a factor modulating the quality of thinning (Figure 10.36).

10.7.2
Laser Dicing

In recent years laser dicing has been seen as the future technology for several niche, and some mass, applications, most importantly for thin chips (Figure 10.37). Two main aspects are essential in laser dicing processes:

- heat impact
- debris created.

The heat impact caused by the laser energy creates a heat affected zone of some microns. This zone creates a lot of tension, reducing chip stability. The laser evaporates the material in the dicing kerf. Regarding a dry-laser system we have to avoid condensation of debris to the active wafer-surface. This is normally achieved by applying a protection layer before laser dicing process starts. The debris can be taken off together with the protection layer. Such layers are also attached to the die side walls. The result is a rough surface covered with molten material.

A very interesting approach is the water-guided laser. Here the laser is coupled in a tiny thin water beam. The water beam helps in focussing the laser, cooling the process and cleaning the dicing kerf.

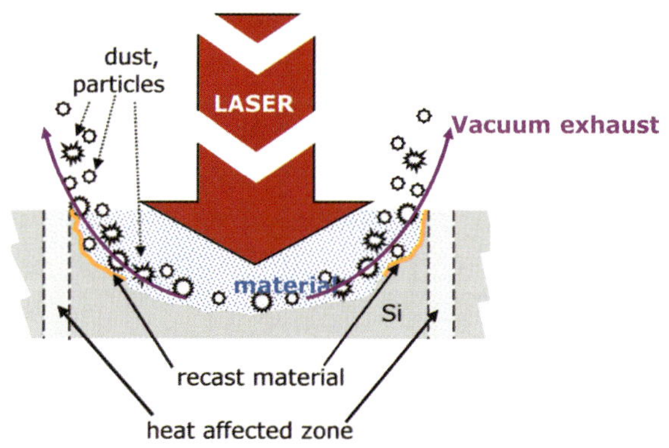

Figure 10.37 Principle of laser dicing process.

Laser dicing is also performed on dicing tape. Some applications use the standard tapes employed for classical dicing, some use special versions. Here we come to one of the most striking facts in introducing new technologies: we have to be compatible with the rest of the process flow. In this case this means the dicing tape must be transparent for the laser beam, otherwise the laser would also cut the tape and we could not carry out die-bonding. The tape has also to be suitable for the die-bonding process – in the end we need a complete production flow. New technologies are much more readily accepted if they do not cause numerous additional changes. Implementation in this case is also faster and less difficult.

10.7.2.1 **Scribing and Breaking Technologies**

Silicon is mechanically scratched and then separated by a breaking tool – this can also be carried out by support of laser. The laser creates a zone of damage (perforation line); singulation [2] is done by tape expansion or by a breaking tool (Figure 10.38).

10.7.2.2 **Separating the Chips in Parallel**

For some years technologies have been presented that separate all chips of a wafer in parallel. The basic idea is to apply a structured protection layer to the active chip, leaving the dicing streets open. Now it is possible to attack all dicing streets at the same time by chemistry, for example by plasma etch. Pros and Cons? Normally the removal rate is quite slow and the protection coating is a cost adder, as you have also to remove it afterwards. The smaller and thinner the chips, the more advantage these methods gain. For separating in parallel, the separation time does not depend on the number of chips per wafer (if enough reactant can be provided). For a real thin wafer (<100 μm), the removal rate will not be that decisive. In addition, for such application these methods can be very compatible. They cause only low heat impact and hardly any mechanical damage. Several methods have been presented (Figure 10.39) [5].

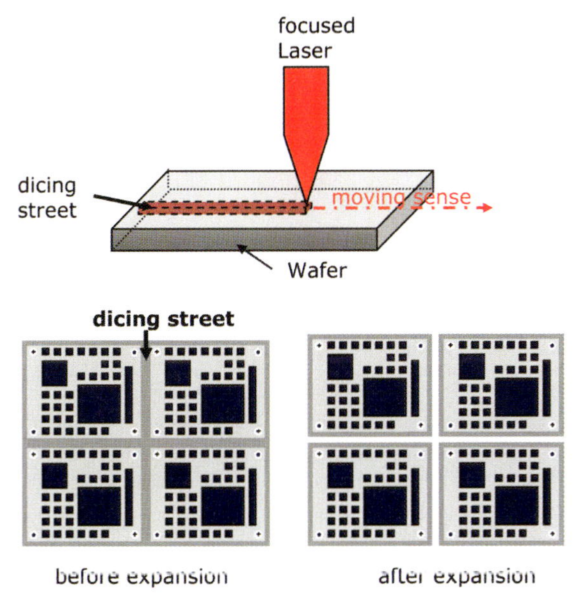

Figure 10.38 The laser creates a damage layer; chips are separated by expansion.

10.7.3 Comparing Methods of Separation

Regarding the different process flows for the singulation of chips, we compared several methods to gain an overview. We use separation by thinning flow. To

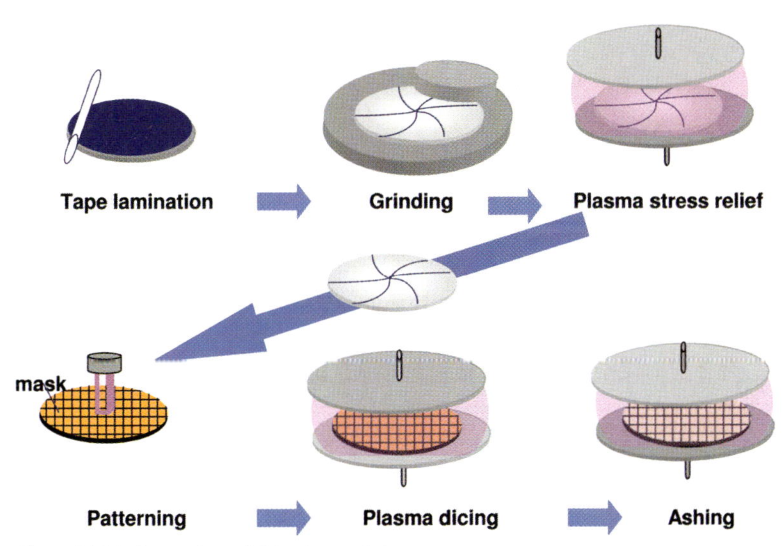

Figure 10.39 Separation of chips in parallel.

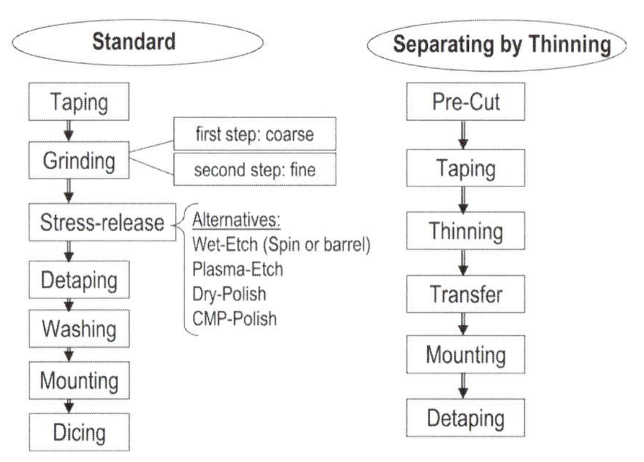

Figure 10.40 Comparison of two processes flows for thinning and singulating the wafers.

make it clear how this works, we can compare it with the standard method (Figure 10.40).

The stability of the device depends on the packaging surrounding the chip, the stability of the chip itself and the interaction between package and chip. Regarding chip stability, the final treatment (stress-release) is a key parameter. As we have seen, several stress relief treatments for increasing the stability of chips thinned by grinding already exist.

We chose the Dicing Before Grinding [6] process-flow as a vehicle to compare different separation technologies (Figure 10.41). We start with a pre-cut from the

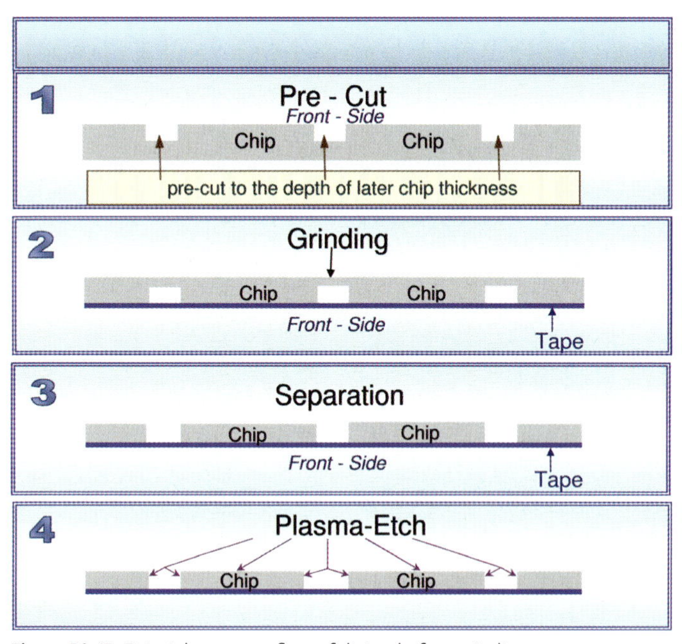

Figure 10.41 Principle process-flow of dicing before grinding.

front-side down to the chip thickness intended (1). Then the wafer is ground (2). This grinding-process is continued until it reaches the pre-cut (3), thus separating the chips. A plasma-etch is applied for stress-relief to back side and chip edges (4).

10.7.3.1 Breaking-Strength and Flexibility for Different Process-Flows

We compare the four following flows in back-side surface treatment for different wafer thicknesses (Figure 10.42) with regard to breaking strength and flexibility:

- only ground without additional back-side treatment
- ground and 10 μm Spin-Etch
- Dicing Before Grinding (DBG)
- DBG plus 3 μm Plasma-Etch.

Figure 10.42 shows that, through back-side treatments, the stability of ground chips can be improved greatly. Indeed, we can improve die-strength roughly by one order of magnitude through such treatments.

The same is true for flexibility. The chips reach a much smaller radius before they crack (Figure 10.43). In fact, the minimum radius a chip can stand is reduced by a factor of two and more by back-side treatments. This is due to the reduction of damage achieved. Especially, Dicing Before Grinding shows a tremendous improvement due to the reduction in chipping by smoothening the edges through a plasma.

What stability do we need for safe processing in back-end? Chips that are at least 300 μm can be processed without problems by nearly every back-end treatment. Thus, it should be sufficient to reach a stability equal to a 300 μm chip. Therefore, we look at this threshold in our stability plot of ground-only wafers (Figure 10.44).

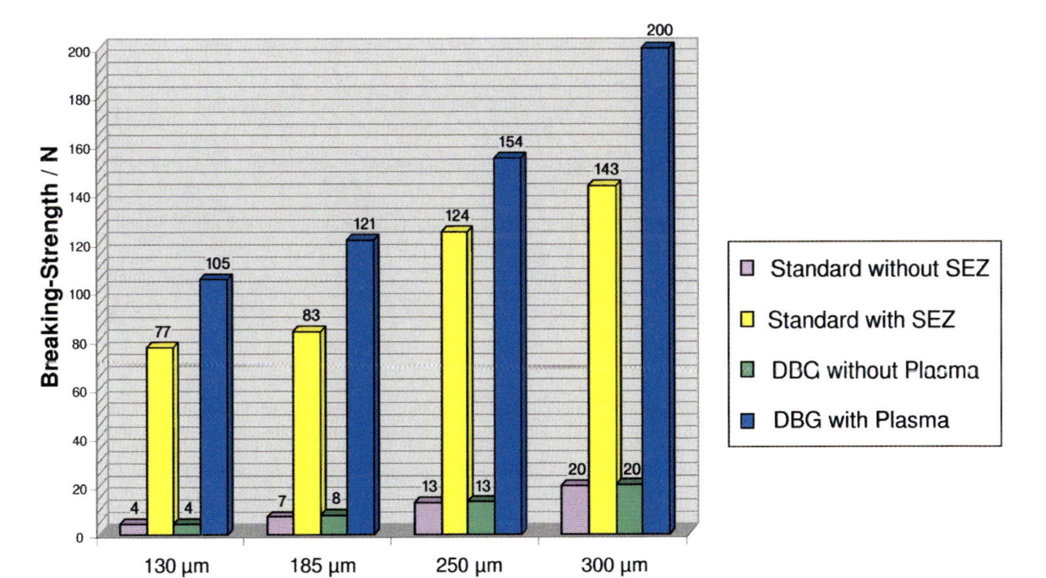

Figure 10.42 Chip-stability for different process-flows.

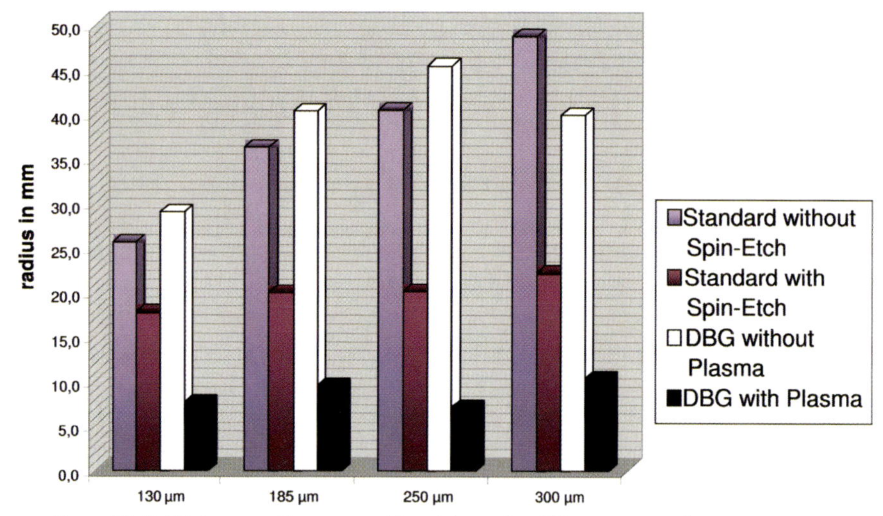

Figure 10.43 Minimum radius versus chip-thickness for different process-flows.

From Figure 10.44, chips reaching a stability of >20 N should be processable in most back-ends with sufficient yield. Putting this figure into the stability plot shown in Figure 10.45, it is obvious that down to about 100 µm chip-thickness the stability limit can be met by applying back-side treatments without any problem.

Some restrictions apply as the stability and flexibility may be sufficient for back-end processing but field-applications may impose additional constraints. Handling processes between thinning and back-end might call for higher stability. This limit can depend on chip-size and the back-end processes used.

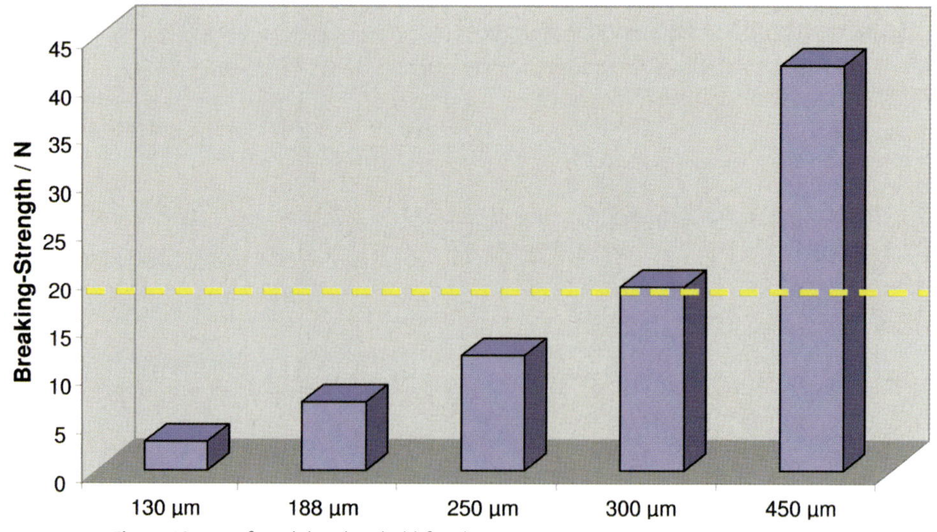

Figure 10.44 Safe stability threshold for chip processing.

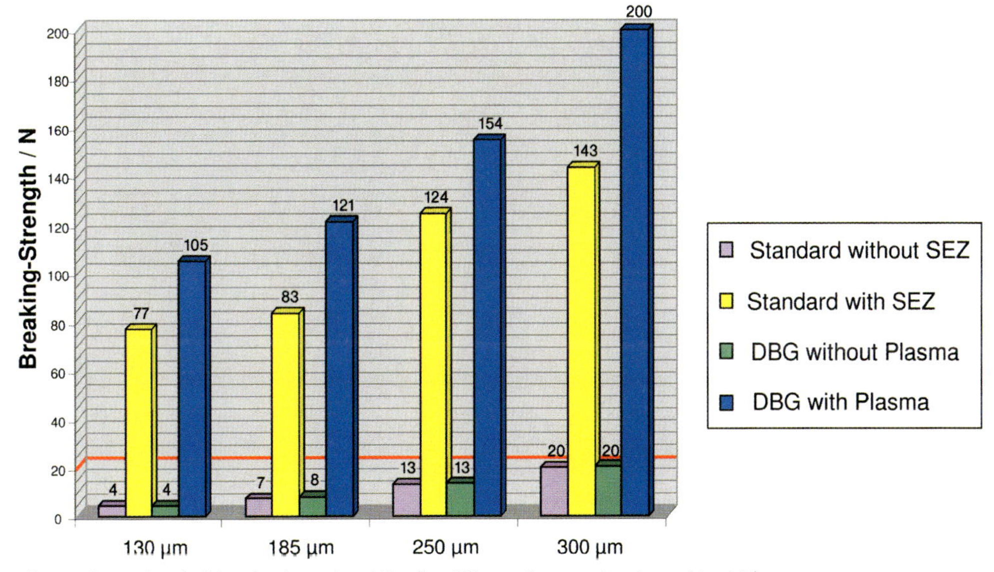

Figure 10.45 Threshold and achieved stability for different flows and various chip thicknesses.

The classical state-of-the-art process to singulate chips is mechanical dicing. The production of mainstream silicon devices involves singulation by this method. However, laser dicing is catching on – its main applications are for thin chips (beyond 150 µm), chips that are not rectangular and other semiconductor substrates like GaAs. Our investigations concentrate on silicon-based devices. We compared the damage induced by three singulation methods: dry laser dicing (UV-laser, 7 W), wet laser-dicing (IR-laser, 52 W, water-jet-guided) and mechanical dicing (dicing-blade S1440, 50 k rpm).

Figure 10.46 shows the die side walls created by the different dicing methods. The roughness seen on the side walls cut by laser comes partly from the heat impact and partly from the attachment of debris. To see what side wall damage is induced by the different separation methods, we compared the chip-stability reached. We used the DBG-process-

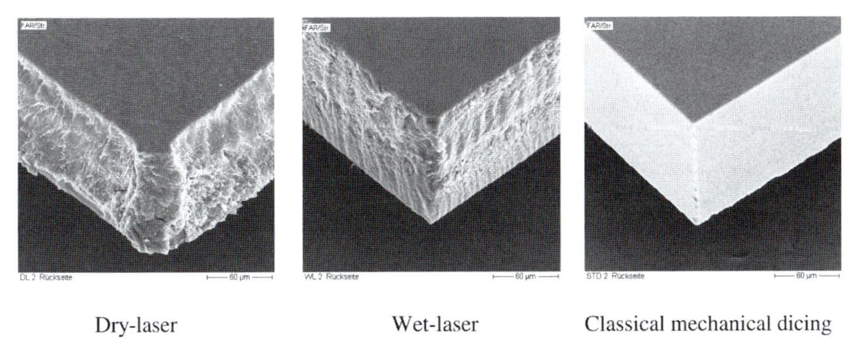

Figure 10.46 Die side wall surface resulting from different singulation methods.

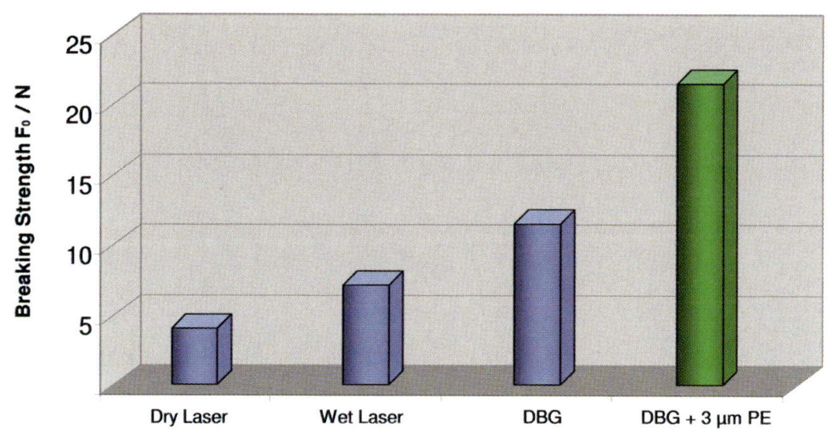

Figure 10.47 Stability achieved by different methods used for pre-cut. 125 μm Bare Silicon 15 × 15 mm 4 Point (4/12) Bending.

flow. The pre-cut was carried out by dry-laser (DL), wet-laser (WL) or mechanical dicing. The chips are 125 μm thick and we used the four-points bending method.

Laser technologies show less stability than mechanical dicing (Figure 10.47), but they also show less mechanical impact in separation. This is an advantage for thin and small chips. Lasers can reach kerf widths down to 20 μm, at least for wafer thicknesses beyond 100 μm. For mechanical dicing it would be difficult and economically not feasible to do 20 μm kerfs.

10.8 Conclusions

Regarding chip stability there is a second major player besides thinning technologies: chip separation. By thinning we create the back-side surface and by separating the chips we create the side wall surfaces. These processes induce damage to the back side and to the side walls. The more the damage in the back side, created by thinning, is reduced through stress-relief treatments the greater the influence of side wall damage can be seen. For mechanical dicing the main damage is chipping. Laser-dicing shows nearly no chipping, but it does show heat impact. These damages lead to reduced chip stability. Our investigations show clearly that the chip side wall condition has to be considered when it comes to stability.

10.9 Summary

No solution exists that can be applied to each and every challenge in thinning wafers. It is not feasibility, it is economic efficiency. Compatibility with existing production lines is important and often underestimated. The form of delivery has to be accepted by the subsequent production units, for example by the back-end or by your customers. This puts a lot of constraints on every solution. Stability is reduced tremendously with the reduction of thickness. Flexibility is as important as stability for several applications. Back-side treatments are the decisive processes in gaining stability and flexibility (up to one order of magnitude). A summary is presented below:

- Above 250 μm final thickness grinding in two steps (coarse and fine) is sufficient.
- For 250–100 μm without active back-side layers grinding plus stress release is adequate.
- Down to 50 μm, separating by thinning technologies can be applied.
- For thin wafers with active back-side layers carrier solutions are used.

For dicing and chip stability, the thinner the chips the more breaking-strength is influenced by chipping. This is the main area of advantage for new singulation technologies like laser dicing.

Most important for nearly any application of thin dies are the stability and flexibility of the chips and the bow of the thin wafers. Given a certain product these features are mainly influenced by silicon thickness, grinding parameters and back-side treatment. What method to use for thinning depends mainly on two facts: the final thickness of the chip and whether processes on the back side have to be applied. This is also product dependent and thus there exists quite a variety of solutions.

References

1. Yamagishi, K. (2006) Thin wafer Handling by DBG and Taiko Process, Disco, Annual Fraunhofer Forum, be-flexible, IZM Munich, www.be-flexible.com (accessed on March 2008).
2. Workshop on Ultrathin Silicon Packaging. (September 2002) Fraunhofer ISIT, Itzehohe, Germany.
3. Böge, A. (1990) *Mechanik und Festigkeitslehre, 21. Aufl*, Vieweg Verlag, Braunschweig.
4. Holz, B. (2004) Advanced production technologies for thinning and laser dicing of ultra thin wafers, Accretech, Annual Fraunhofer Forum, be-flexible, IZM Munich, www.be-flexible.com (accessed on March 2008).
5. Blumenauer, H. and Pusch, G. (1993) Bruchmechanik, Dt. Verlag für Grundstoffindustrie, Leipzig, 3. Aufl.
6. Hadamovsky, H.F. *et al.* (1990) Werkstoffe der Halbleiterindustrie, Dt. Verlag für Grundstoffindustrie, Leipzig, 2. Aufl.
7. Wilker, H. (2004) *Weibull-Statistik in der Praxis. Leitfaden zur Zuverlässigkeitsermittlung technischer Produkte, Norderstedt,* Books on Demand GmbH.
8. John, J.P. and McDonald, J. (1993) Spray etching of silicon in the HNO3/HF/H20 system. *Journal of the Electrochemical Society,* **140** (9), 2622–2625.
9. Arita, K. *et al.* (2006) Plasma etching technology for wafer thinning process, Panasonic. Known Good Die Workshop, Napa, California.
10. Priewasser, K. (2005) Thin chips manufacturing methods, Disco, Annual Fraunhofer Forum, be-flexible, IZM Munich, www.be-flexible.com (accessed on March 2008).
11. Sach, L. (1983) *Angewandte Statistik,* Springer Verlag.
12. Lehnicke, S. Rotationsschleifen von Si-Wafern, Promotion von 1999, Institut für Fertigungstechnik Uni Hannover, Forts chrittsberichte VDI, Reihe 2, Nr. 534.

13 Kröninger, W. and Mariani, F. (2006) Thinning and singulation: root-causes of the damage in thin chips. Proceedings 56th ECTC, San Diego, CA.

14 Kröninger, W., Wittenzellner, E. *et al.* (2002) Thinning silicon optimizing the grinding process regarding performance and economics, Infineon and Tyrolit, Annual Fraunhofer Forum, be-flexible, IZM Munich, www.be-flexible.com, (accessed on March 2008).

15 Kröninger, W., Perrottet, D., Buchilly, J.-M. and Richerzhagen, B. (Jan. 2005) Water Jet Guided Laser Achieves Highest Die Fracture Strength. *Future Fab International*, London, **18**, 157–159.

11
Overview of Bonding Technologies for 3D Integration

Jean-Pierre Joly

11.1
Introduction

Bonding is a key step for mainstream 3D Integration technology schemes developed so far that are based on Wafer to Wafer or Chip to Wafer stacking. It is generally performed before thinning down and eventual inter-strata interconnection processing and should therefore provide enough thermo-mechanical stability without delamination at the bonding interface during or after those post-process steps.

Different bonding approaches have been studied. The present chapter reviews those principles and the results obtained so far. We will not describe the alignment techniques that are used prior to bonding. This will be done in chapter 12.

Bonding techniques adapted to wafer or die bonding can be categorized in two different ways (Figure 11.1):

- Whether a layer is used specifically as an intermediate material between the two substrates to bond them together.
- Whether the bonding interface allows electrons to go through, which means whether dielectrics or conducting layers (mainly metals) are used at this interface.

The use of metallic layers is particularly interesting in 3D Integration to create direct area connections between the 3D strata. Meanwhile, as those connections must be isolated from each other the bonding interface between the substrates is not uniform in that particular case.

The choice of any bonding technique is strongly dependent on the 3D device requirements, such as the inter-strata connection density and the possibility or not to change the initial device lay-out. It is also linked to the whole integration flow scheme that is chosen.

Let us first describe in more detail the bonding principles listed in Figure 11.1.

Handbook of 3D Integration: Technology and Applications of 3D Integrated Circuits.
Edited by Philip Garrou, Christopher Bower and Peter Ramm

ISBN: 978-3-527-32034-9

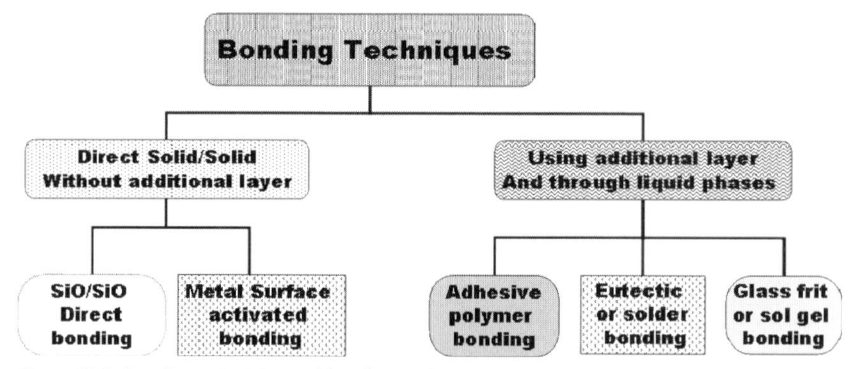

Figure 11.1 Bonding principles and bonding techniques.

11.2 Direct Bonding

11.2.1 Direct Bonding Principles

Bonding requires the creation of a given density of chemical bonds between the two pieces of material that should be linked. Let us take the example of two surfaces of the same material that are put into contact (Figure 11.2). If s is the local distance between the two surfaces, the local bonding pressure $q(s)$ is given by [1]:

$$q(s) = \frac{8W}{3\varepsilon}\left[\left(\frac{\varepsilon}{s}\right)^3 - \left(\frac{\varepsilon}{s}\right)^9\right]$$

where:

$$W = \gamma_1 + \gamma_2 - \gamma_{12}$$

ε is the inter-atomic distance in the material and γ_1, γ_2 and γ_{12} are, respectively, the surface energies of the two surfaces and the joint interface energy.

The last term varies with s to the power 9 and is related to the chemical bonding between atoms belonging to the two different pieces. This term indeed conditions the

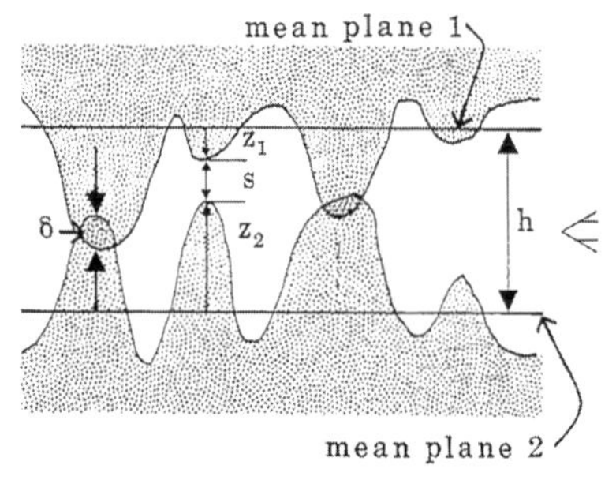

Figure 11.2 Schematic of two surfaces put into contact.

attraction (molecular adherence) between the two pieces, and this attraction is very dependent on the few contact points where the two surfaces are in contact at the atomic scale.

For any actual solid surface morphology without liquids, only a very low ratio of the surfaces is in intimate contact at the atomic scale. This means that the molecular adherence is often very low. It is, however, possible to increase the surfaces in contact at the atomic scale and, therefore, to obtain significant adherences in two different ways:

1. Surface polishing to reduce the surface roughness up to values close to the atomic scale (this is the principle used for Surface Direct Bonding).
2. Surface deformation; this is only possible for ductile materials through pressure and temperature assistance (this is called Surface Assisted Bonding: SAB).

11.2.2
Surface Direct SiO/SiO Bonding

Surface Direct Bonding has been mainly studied and implemented by using silicon dioxide layers. It is now presently used in industry for SOI (Silicon On Insulator) wafer production. It relies on the conventional wafer CMP (Chemical Mechanical Polishing) process.

For 3D Integration implementation, silicon dioxide, or eventually a silicate, is deposited and polished prior to bonding. This bonding system has been described from the chemical point of view by Stengl [2]. It has been demonstrated that the initial bonding energy is due to a couple of monolayers of water molecules that are spontaneously adsorbed on any silica materials and to hydrogen bonds between those water molecules and the oxygen atoms of the silicon dioxide (Figure 11.3a).

After annealing at medium-range temperatures (up to 800 °C) the bonding energy increases (Figure 11.3c). This corresponds to progressive gap closing. Water molecules diffuse progressively through the silica layers. At high temperature more and more direct Si–O bonds prevail and a full reconstruction of the silica frame is progressively achieved (Figure 11.3b). For 3D Integration, annealing temperatures cannot exceed 300–400 °C but it can be seen in Figure 11.3c that the bonding energy increase is already significant for those temperatures.

The bonding energy at room temperature is low but high enough to allow spontaneous bonding through the propagation of a bonding wave when dealing with normal wafers. A threshold of bonding energy, W, is indeed necessary to allow such a spontaneous bonding (Figure 11.4a). W should be larger than the energy expense (dU_E) per unit of surface (dA) related to the wafer elastic deformation [3].

$$W > \frac{dU_E}{dA}$$

As can be seen in Figure 11.4b, whatever the position on the wafer, for a typical wafer thickness of 0.75 mm this threshold value does not exceed 15 mJ cm^{-2},

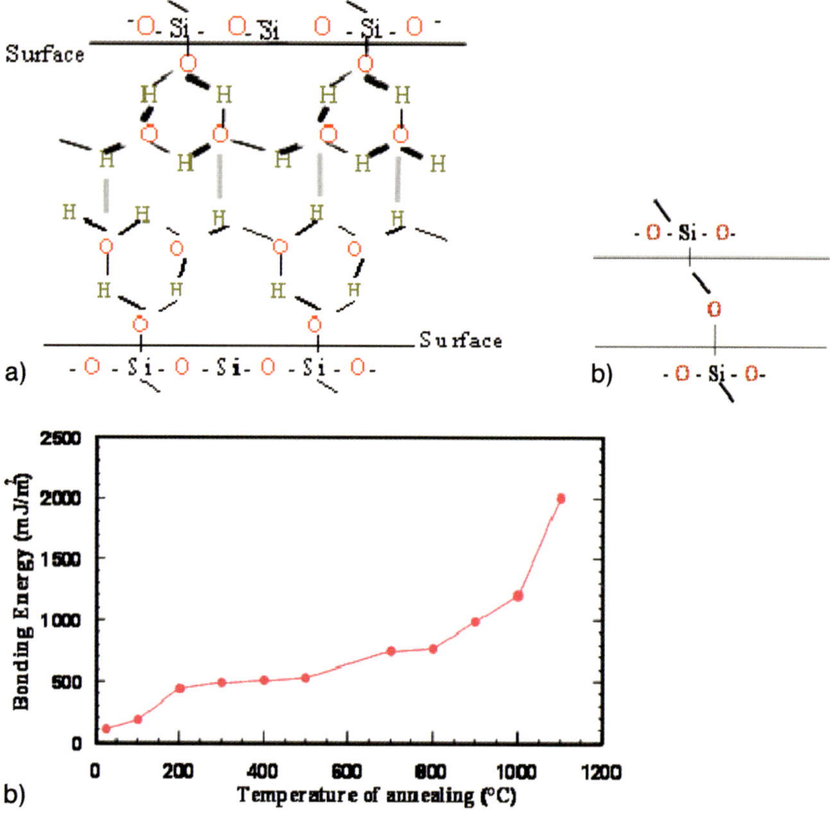

Figure 11.3 SiO/SiO direct bonding: the chemical bonds after annealing up to 800 °C (a) and after high temperature annealing >1000 °C (b) and the related evolution of bonding energy with the annealing temperature.

a value that can be easily obtained on polished SiO_2 layers. It is very sensitive to wafer thickness and can be even further reduced by thinning down one or two of the wafers.

The bonding energy is very dependent on the surface roughness and also on the chemical activity of the oxide layer towards water molecules. The influence of the initial roughness has been studied by Moriceau [4] (Figure 11.5). The authors have shown that if a certain amount of roughness is applied purposely through adequate processing it is possible to use Direct Bonding to bond and de-bond from a temporary template (actually a silicon wafer), even after high temperature processing while the wafer is bonded on a temporary silicon wafer template. This opens new possibilities in terms of 3D Integration.

As PECVD oxide is less dense than thermal oxide, higher water adsorption and therefore higher bonding energies can be obtained at room temperature after

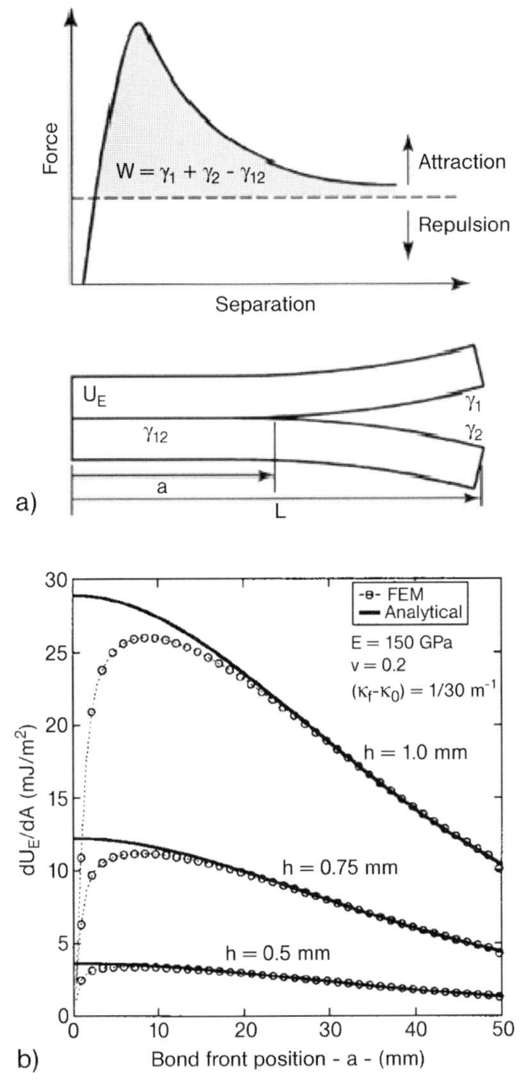

Figure 11.4 Conditions for spontaneous bonding of two wafers. The bonding energy should compensate for the elastic deformation energy (a). The variation of the bonding energy threshold against bond front position is shown in (b).

appropriate CMP [5] (Figure 11.6). Annealing at a very moderate temperature further increases the bonding strength. Direct Bonding is therefore fully compatible with any existing metallization and therefore fully compatible with 3D Integration.

If still higher bonding energy is needed the surface can be activated using plasma processing (Figure 11.7).

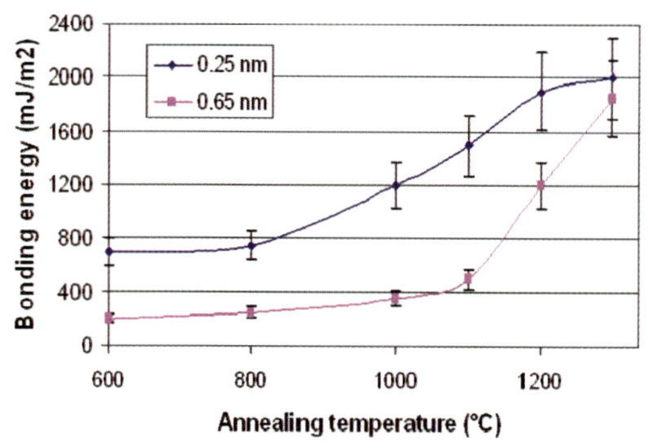

Figure 11.5 Variation of the SiO/SiO direct bonding energy with temperature and with the initial RMS surface roughness.

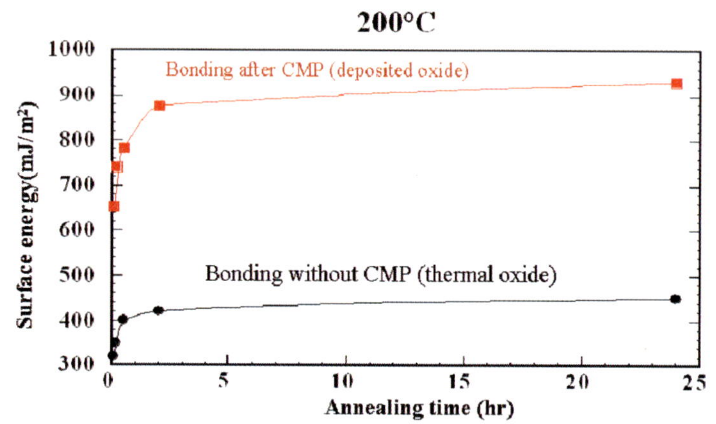

Figure 11.6 Bonding energy after low temperature annealing for thermally grown and vapor deposited (PECVD) oxides.

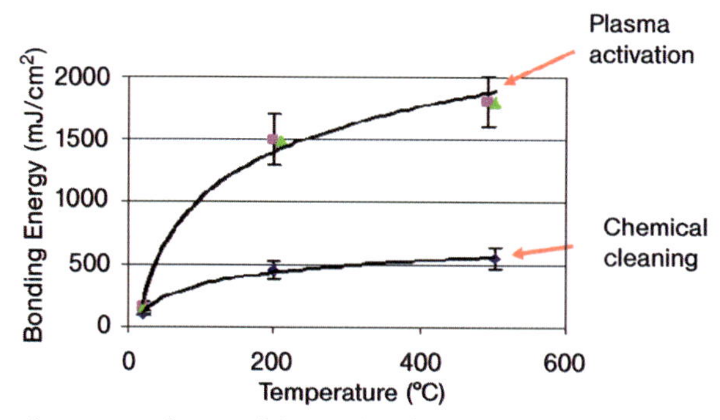

Figure 11.7 Influence of the initial surface treatment on SiO/SiO direct bonding energy.

For defect-free Direct Bonding the cleanliness before bonding should be maintained at a very high level. Any particle of sufficient size generates a bonding defect, which corresponds to a large area without bonding, where the substrates are elastically deformed around the particle. Modern cleaning technologies and modern clean room environment are generally good enough to keep the defect density under control.

In summary, Wafer to Wafer direct bonding based on SiO_2 layers deposited at low temperature can be used for 3D Integration providing there is:

- proper CMP polishing
- efficient particle removal through state-of-the-art cleaning.

Plasma activation can be used to enhance the bonding energy.

Oxide/oxide direct bonding is very stable regardless of the temperature and has been implemented on a large scale for SOI wafer fabrication.

11.2.3
Metal Surface Activated Bonding

Bonding metallic surfaces without melting and soldering is also possible but requires:

1. Careful removal of surface impurities or surface compounds and, especially, native oxides.
2. A way to increase the area of intimate contact between the two surfaces at the atomic scale.

This last requirement can be obtained in two different ways:

1. Through polishing for refractory metals in a similar way as described before for dielectrics.
2. Through plastic deformation and high pressure for ductile non-refractory metals; this is known as metal surface activated bonding or SAB.

Twordzylo [1] has established models of the variation of contact area with respect to the pressure applied to the specimen. Figure 11.8 gives an example for a highly deformable metal (aluminium). The importance of the plastic deformation is clearly highlighted.

The best way to get both clean surfaces and high contact areas for typical metals of interest such as copper is to implement *in situ* cleaning under a controlled atmosphere (e.g. under high vacuum) and to transfer the wafers under the same atmosphere in a bonding chamber under pressure.

Kim *et al.* [6] have used cluster equipment (Figure 11.9a) to achieve Cu/Cu bonding. Surface cleaning is performed in a first chamber using Ar plasma and oxide sputtering (Figure 11.9b). Alignment and pre-bonding without pressure is then performed in a second chamber. Finally, wafers are bonded under pressure in a third chamber. Figure 11.9c shows that the bonding is very efficient. It is sometimes difficult to distinguish the location of the bonding interface after bonding.

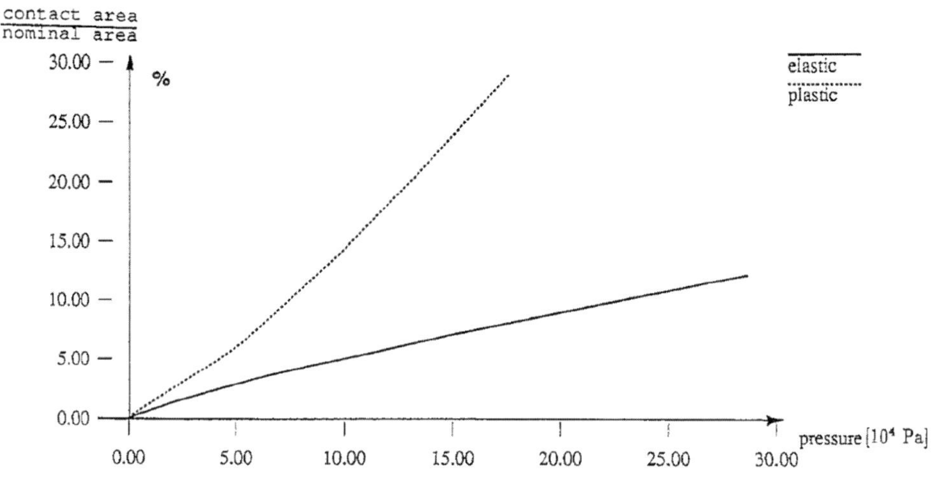

Figure 11.8 Contact area ratio versus pressure for two aluminium surfaces.

A bonding strength >6.47 MPa on $10 \times 10\,\text{mm}^2$ pieces has been measured in this particular case.

Wang *et al.* [7] have demonstrated that two different metals can be bonded together in a similar way. They have indeed bonded Au with Cu.

Noble metals such as gold have very stable surfaces that are not prone to chemical reactions with reactive gases. In this particular case it is not mandatory to clean the surface before bonding. The bonding under pressure can be performed under a standard atmosphere. This is the well-known "stud bumps" or Au-Au thermo-compression technology. A recent description has been given by Takahashi [8].

In summary, while metal surface activated bonding is a well-known technique for noble metals such as gold, it has been extended recently to other metals such as copper by using cluster tools and *in situ* dry cleaning.

11.3
Adhesive and Solder Bonding

A second range of bonding principles is based on adhesives or solders. A liquid phase (solution, gel or suspension) is placed or created between the two surfaces to be bonded and then solidified through polymerization, crystallization or firing. Use of a liquid alleviates the difficulty of having a large enough density of chemical bonds at the interface. It can indeed establish an intimate atomic contact with solid surfaces, providing they have large enough wetting properties. Careful surface preparation, which consists of removing any foreign particles and strongly reducing the surface roughness, is therefore less important when using liquid adhesives.

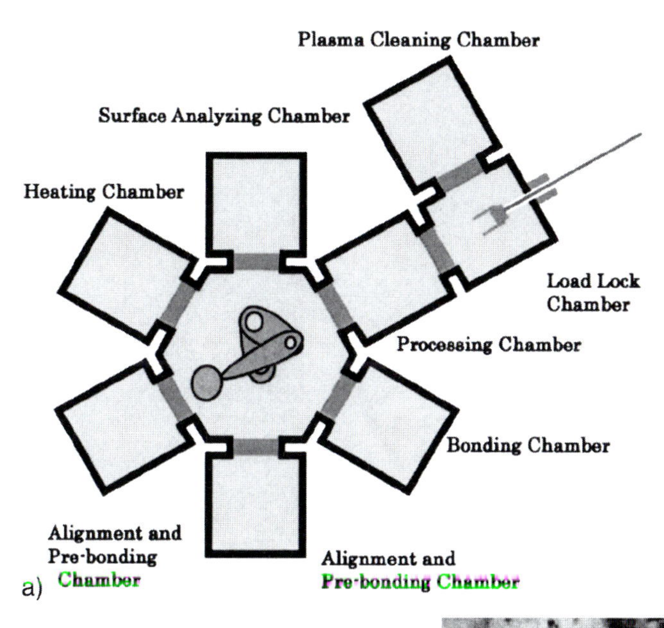

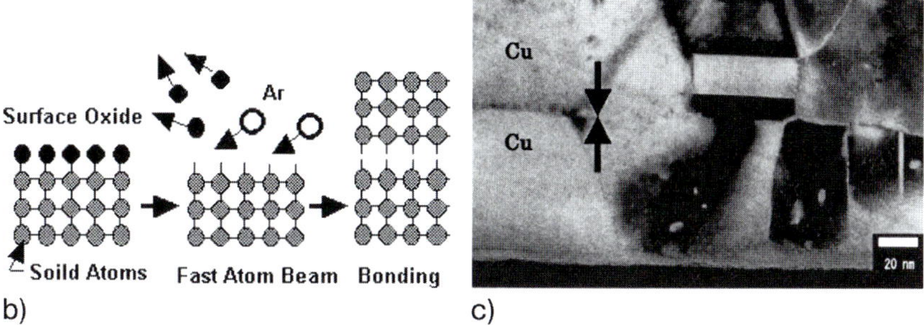

Figure 11.9 Cu/Cu activated bonding. (a) Cluster tool; (b) sputtering used for surface cleaning and (c) TEM cross section after bonding.

Polymer and solder bonding techniques are well known and currently used in traditional industries.

11.3.1 Polymer Bonding

The principle of polymer bonding is simple. A liquid organic solution is spun onto the surface of one substrate. The two substrates are put into contact. The solution is then polymerized at the reticulation temperature of the chosen polymer. A pressure can be applied between the two wafers in the meantime to prevent void formation.

Numerous polymers can be used, such as polyimide or resists, but the most popular is BCB (BenzoCycloButene) [9]. This polymer is well known in the Microelectronics industry and is already used for chip packaging. It is mechanically and

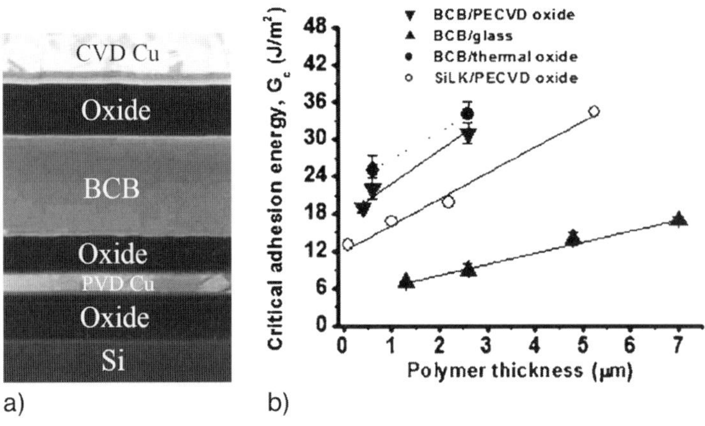

Figure 11.10 (a) SEM view of BCB bonding [10] and (b) critical adhesion energy of BCB bonding measured using the four-point bending method [9] .

thermally stable and can support any packaging process. An adhesion promoter is usually deposited before BCB spin coating. Layers for bonding are usually a few μm thick. The bonding process is performed at temperatures ranging typically from 200 to 250 °C. A SEM view of BCB bonding is shown in Figure 11.10a.

Kwon *et al.* [9] have shown by using the four-point bending method that high adhesion energy can be obtained with BCB bonding regardless of the material at the surface of the substrate (Figure 11.10b).

Chapter 13 of this book is devoted to BCB bonding for 3D.

Sol gels instead of polymers can be also used in a similar way for bonding purposes [10]. Hydrolysis at temperatures in the range 200–400 °C is used to create Si−O tetrahedra linking and solidification.

11.3.2 Metal Soldering or Eutectic Bonding

Soldering is also a very well-known technology that has been in use for many years for flip chip connections. It can also be used for high density 3D Integration, providing the alloys and the deposition techniques are modified.

The most well-known development of such a principle for 3D Integration is the so-called SLID (Solid-Liquid-Interdiffusion) concept [11]. It makes use of thin (in the few μm thickness range) metal layer melting and Sn/Cu intermetallic compound formation. Robust and low resistance area connections have been obtained using this approach. The technique is described in more detail in Chapter 14 of this book.

Copper posts are present on one side and (Cu + Sn) on the other side. The two substrates are brought into contact at temperatures higher than the Sn melting point, typically at 260 °C. At this temperature the metals can diffuse easily and can react to form an intermetallic compound (Cu_3Sn) according to the Cu−Sn phase diagram

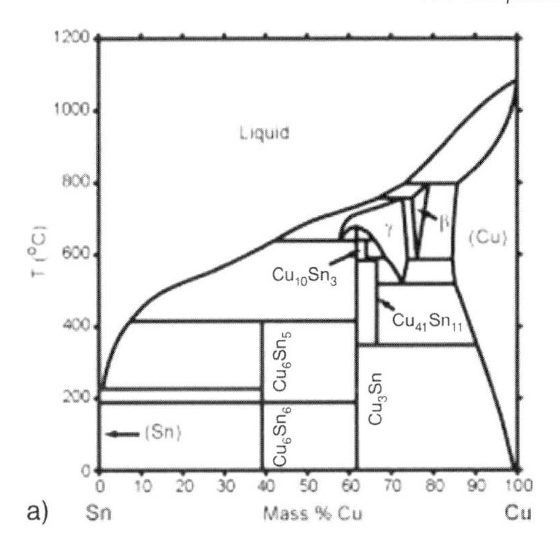

FIB of a 3 layer stack

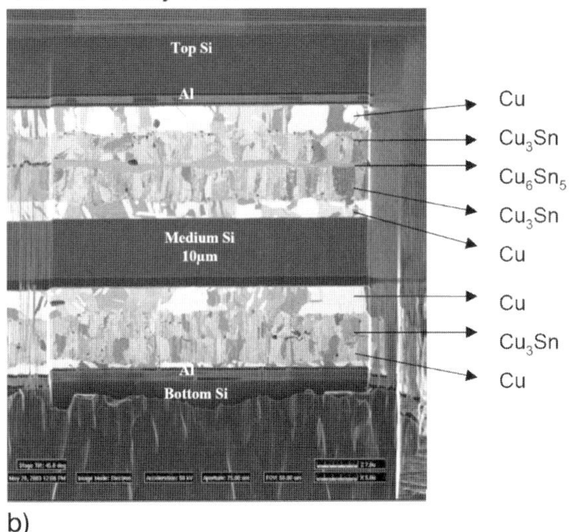

b)

Figure 11.11 Sn/Cu intermetallic compound formation: (a) Cu–Sn phase diagram and (b) image of the compound formed.

(Figure 11.11a and b). This compound is very stable and melts at temperatures higher than 600 °C.

11.4 Comparison of the Different Bonding Technologies

In summary, Table 11.1 lists bonding techniques that have been studied for 3D Integration. Indications are provided of the mechanical behavior (Bonding Energy),

Table 11.1 Comparison of bonding techniques that can be used for 3D wafer to wafer or chip to wafer stacking.

Bonding technology	Temperature (°C)	Bonding energy ($J\ m^{-2}$)	Direct area connection	Critical points	Advantage	In use in industry
Direct bonding (SiO_2/SiO_2)	200–1100	2		Surface cleaning and polishing	No added materials	√(SOI)
Metal surface activated bonding	Room temp.	20	√	*In situ* surface cleaning	No added materials	
Au–Au thermo compression	400	Mech. strength validated	√	Au contamin.	No added materials	√(Stud bumps)
Polymer	150–400	20		Thick glue layer	Simple	√(MEMS)
Sol gel	400	2				
Soldering	200–300		√	Unusual material		√(Packaging)

critical points, possibility of using bonding and area connection, process temperature and existing implementation at the industrial level.

References

1 Twordzylo, W., Cecot, W., Oden, J.T. and Yew, C.H. (1998) New asperity-based models of contact and friction. *Wear*, **220**, 113–140.

2 Stengl, R., Tan, T. and Gösele, U. (1989) A model for the silicon wafer bonding process. *Japanese Journal of Applied Physics*, **28**, 1735–1741.

3 Turner, K.T. and Spearing, S.M. (2002) Modeling of direct wafer bonding: Effect of wafer bow and etch patterns. *Journal of Applied Physics*, **92**, 7658–7666.

4 Moriceau, H., Rayssac, O., Aspar, B. and Ghyselen, B. (2003) The bonding energy control, an original way to debondable substrates, 7th International Symposium on Semiconductor Wafer Bonding, ECS Proceedings, PV 2003-19, pp. 49–56.

5 Di Cioccio, L., Biasse, B., Kostrzeva, M. *et al.* (2005) Recent results on advanced molecular wafer bonding technology for 3D integration on silicon, 9th International Symposium on Semiconductor Wafer Bonding, ECS Proceedings, PV 2005-6, pp. 280–287.

6 Kim, T.H., Howlander, M.M.R., Itoh, T. and Suga, T. (2003) Room temperature Cu–Cu direct bonding using surface activated bonding method. *Journal of Vacuum Science & Technology A-Vacuum Surfaces and Films*, **21**, 449–453.

7 Wang, Q., Hosoda, N., Itoh, T. and Suga, T. (2003) Reliability of Au bump–Cu direct interconnections fabricated by means of surface activated bonding method. *Microelectronics and Reliability*, **43**, 751–756.

8 Takahashi, K., Umemoto, M., Tanaka, N. *et al.* (2003) Ultra-high-density interconnection technology of three-dimensional packaging. *Microelectronics and Reliability*, **43**, 1267–1279.

9 Kwon, Y., Seok, J., Lu, J.-Q. *et al.* (2006) Critical adhesion energy of benzocyclobutene-bonded wafers. *Journal of the Electrochemical Society*, **153**, G347–G352.

10 Kwon, Y., Jindal, A., McMahon, J.J. *et al.* (2003) Dielectric glue wafer bonding for 3D ICs. *Materials Research Society Symposium Proceedings*, E5.8.1, 766.

11 Klumpp, A., Merkel, R., Ramm, P. *et al.* (2004) Vertical system integration by using inter-chip vias and solid-liquid-interdiffusion bonding. *Japanese Journal of Applied Physics*, **43**, L829–L830.

12
Chip-to-Wafer and Wafer-to-Wafer Integration Schemes

Thorsten Matthias, Stefan Pargfrieder, Markus Wimplinger, and Paul Lindner

12.1
Decision Criteria for 3D Integration

There is a huge variety of integration schemes for 3D interconnect based on Through-Silicon-Vias (TSVs). 3D interconnects enable the improvement of existing devices and device architectures in many different areas, like performance, functionality and feasibility. In some cases TSVs compete with existing architectures and in other cases they are the enabling technology for new devices. This results in various drivers for the implementation on TSVs and, as a further result, in various integration schemes. Standardization of manufacturing processes has not been started in this new field.

The biggest and most fundamental differentiator between the different manufacturing schemes is whether the integration is carried out by chip-to-chip (C2C), chip-to-wafer (C2W) or wafer-to-wafer (W2W) approaches. Each approach has been analyzed in great detail for several integration schemes. From a manufacturing point of view it seems that all three approaches are feasible and, depending on the application, reasonable. However, the specifics of the product will determine which approach has to be taken.

12.1.1
Different Wafer Sizes

One of the underlying principles of 3D integration is that different functional entities can be produced on different wafers. This allows a reduction in the process complexity of each individual wafer, thereby improving the individual yields and reducing the manufacturing costs. In turn, this permits a higher degree of utilization of existing equipment and reduces capital expenditures for implementing a manufacturing line for 3D devices. For C2C and C2W it is not necessary that the wafers have the same size, whereas for W2W this is a necessity.

Handbook of 3D Integration: Technology and Applications of 3D Integrated Circuits.
Edited by Philip Garrou, Christopher Bower and Peter Ramm

ISBN: 978-3-527-32034-9

12.1.2 Different Fabs

Splitting the individual functional entities on different layers enables the integration of chips or wafers out of different fabs and even from different suppliers. However, as many suppliers do not want to disclose the yield they achieve in their wafer fabs, it seems to be easier to buy chips than wafers.

12.1.3 Different Base Substrates

The beauty of wafer bonding is that it enables the integration of devices processed on different base materials. Logic and memory on silicon can be easily combined with RF devices on compound semiconductors.

12.1.4 Different Chip Size

This is a very important differentiator. For a wafer-to-wafer integration approach all the dies must have the same size in order not to waste silicon real estate. In reality this translates into two cases. Either the individual layers have the same functionality and therefore the same die layout like, for example stacking of memory layers. Or it is necessary to "design for stacking," meaning that the given size of the die determines the number of functional entities that have to be packed into one layer.

For chip-to-chip and chip-to-wafer integration schemes there is no such limitation in chip size. It is possible to stack several small dies on one large base die, taking full advantage of the modular approach of 3D integration (Figure 12.1).

12.1.5 Number of Stacked Layers

The W2W integration schemes allow stacking of many layers without any technical limits. After bonding, back-thinning and backside processing the resulting

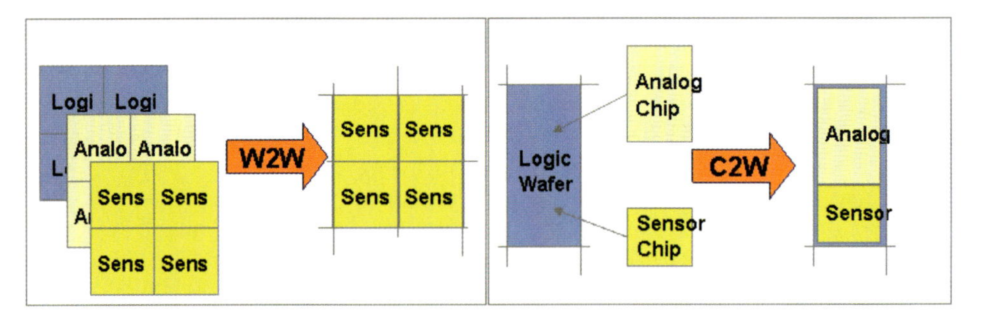

Figure 12.1 W2W integration requires identical chip size for all layers, whereas C2W allows staking of several small dies on the base chip.

stack is literally a wafer again and the complete integration scheme can be repeated multiple times. If required it is possible to thin the base wafer accordingly so that the resulting stack always has the geometric dimensions of a standard wafer.

For C2C and C2W, multiple repetitions are also possible, either requiring the same chip size or by using a pyramidal stacking approach. However, in these cases the geometry of the stack changes from layer to layer, which requires very flexible manufacturing equipment.

12.1.6
Modular Design

2D device architecture requires that all on-chip functionalities are integrated into the design of a specific chip. 3D device architecture enables splitting the individual functional entities into subgroups on different chips or wafers. All three main integration approaches are suited to support a modular design. The ability to integrate standard components with specialized ASICs reduces the design and testing costs significantly and in addition reduces the development time for new devices.

12.1.7
Yield Issue

The yield of 3D integration is the most controversial discussion within the industry. Stacking of dies has the inherent risk that if one layer is defective the whole device is defective. The cumulative yield of 3D devices is calculated as the multiplication of the individual layer yields (neglecting potential yield hits due to processes for stacking) and is therefore always worse than the individual wafer yield. However, a detailed analyses shows that it is, by far, not so obvious. 3D integration reduces the process complexity and the number of manufacturing steps, for example on-chip memory has, presently, to go through all the mask levels required for the logic processor; memory stacked on logic has significantly less processing steps. Therefore the individual wafer yields will be higher than for conventional 2D devices. Another advantage of 3D devices is that the die size is smaller than for 2D devices. During CMOS manufacturing one particle typically only impacts one die. Given a specific number of particles the yield of the wafer increases if the die size shrinks and the number of dies is higher. In addition, due to the reduced number of processing steps it can be expected that wafers for 3D devices have slightly reduced particle contamination. An interesting aspect of TSVs is that the high bandwidth given by the high density of vias allows failure tolerant device architectures, which increases the cumulative yield.

Testing of the dies prior to stacking, the *known good die* approach, significantly reduces the risk for 3D integration. However, depending on the integration scheme, it is necessary to be cautious. Probing systems typically scratch the surface 202 of the contact point to achieve good ohmic contact. However, as a result the surface of the contact points is damaged on the microscopic level. Such a roughened point can

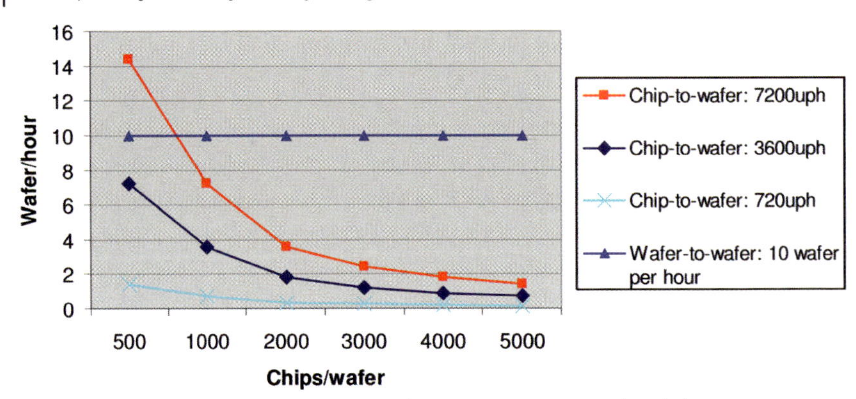

Figure 12.2 Throughput comparison: W2W alignment versus C2W pick-and-place.

cause problems during die or wafer bonding methods like Cu–Cu and silicon direct bonding, where a very flat surface is required.

12.1.8 Throughput

Throughput is a key criterion for manufacturability of any integration scheme. W2W integration schemes have the huge advantage that they apply parallel processing of all dies, whereas C2W as well as C2C apply sequential processing. The throughput difference becomes more significant with shrinking die size or larger wafer size (Figure 12.2).

12.1.9 Alignment

The achievable alignment accuracy of the layers to each other determines the possible density of interconnects.

In a C2W or C2C approach, an alignment accuracy of 5 ~ 15 μm is achieved today for a high throughput system with >1000 chips per hour. W2W has demonstrated alignment accuracies in the 1 μm range. Certain direct align processes (with live image of both wafers) achieve submicron alignment accuracy.

12.1.10 Cost

As for any new technology, TSVs and stacked dies will only be accepted by the market if the better performance can outweigh the price markup. The main cost contributors are:

1. Modeling, designing and simulating new architectures
2. Processing infrastructure: new equipment, process development, materials

3. Yield cost
4. Testing cost.

Within the scope of this book we will not discuss chip-to-chip integration schemes and technologies but, instead, focus on chip-to-wafer and wafer-to wafer approaches.

12.2 Enabling Technologies

12.2.1 Aligned Wafer Bonding

All envisioned process flows for 3D integration on wafer level contain wafer bonding as a key process step. Instead of wafer bonding there are competing technologies like growing an epitaxial layer of silicon on top of a fully processed device layer. These are very interesting methods, but they are not close to implementation in manufacturing lines yet.

During the 1990s the drivers for the development of wafer bonding methods were automotive sensor applications (MEMS, microelectromechanical systems) and silicon-on-insulator (SOI) wafer production. Over the last 20 years many wafer bonding methods have been developed and the feasibility for high volume manufacturing has been proven many times. Figure 12.3 shows an overview of the classes of wafer bonding methods. Some of the methods can be used for a wide range of applications, whereas others are very application specific.

Wafer bonding consists of two process steps. The first is the preparation of the wafer surface in terms of cleanliness, surface chemistry and/or intermediate layer properties. The second is the actual bonding. Depending on the wafer bonding method, specific ambient conditions or vacuum are applied, the wafers are bonded by mechanical pressure at elevated temperature. With silicon direct bonding a final annealing step is required, which can be performed as a batch process without applied mechanical pressure.

A particle between the wafer surfaces prevents intimate contact and bonding cannot occur. The resulting void is usually 2–3 orders of magnitudes larger than the particle itself. In the case of reflowing adhesives or eutectic metals going through the liquid phase during bonding, the particles can be embedded in the intermediate layer, while in all other cases the highest cleanliness is of utmost importance. Molecular bonding is especially sensitive to particle contamination. The molecular bonding methods are also very sensitive on the surface properties, specifically organic contamination and surface chemistry. These properties can be tailored by wet chemical or dry plasma activation. For manufacturing purposes it is necessary to control these properties tightly. Both surface chemistry and contamination occur over time after activation. To have the same time between activation and bonding for every wafer it is necessary to integrate the pre-processing modules into the production wafer bonding platform (Figure 12.4).

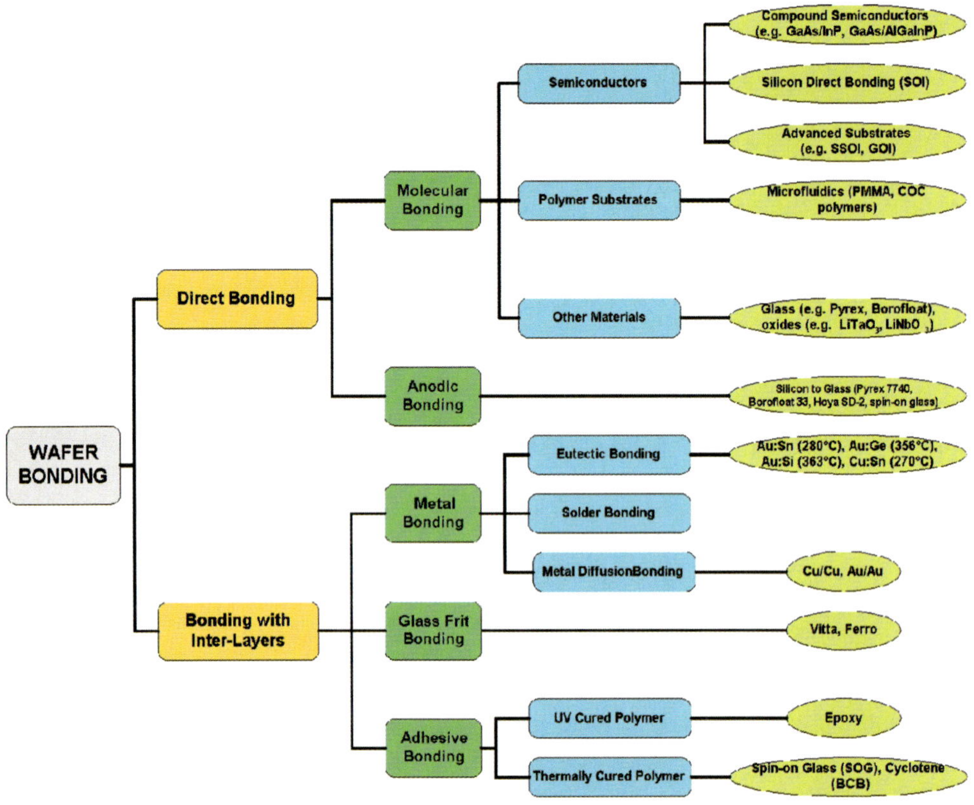

Figure 12.3 Overview of different wafer bonding methods; the main differentiator is whether an intermediate layer is used.

The key parameter for wafer bonding is the bond quality. Good bond quality means that high bond strength is achieved over the whole contact area. High bond strength guarantees that debonding or delamination does not occur either during the subsequent process steps and packaging or during the complete product life cycle. The key equipment parameters for high bond strength are pressure and temperature uniformity. The bond quality can be analyzed by several methods [1]. For manufacturing purposes, typically IR microscopy and scanning acoustic microscopy (SAM) are used.

Even though there are a large number of wafer bonding methods, only a few can be used for 3D integration. Many of the established methods are not compatible with CMOS:

1. For example, anodic, glass frit, Au-Au thermo-compression and the some eutectic bonding methods are not considered due to risk of metal ion contamination.
2. The bonding temperature has to be compatible with the thermal budget of the devices.
3. The TSV diameter is limited due to space constraints. This requires the intermediate layer to be very thin to minimize the aspect ratio of the TSV.
4. The bond layer needs very good thickness uniformity.

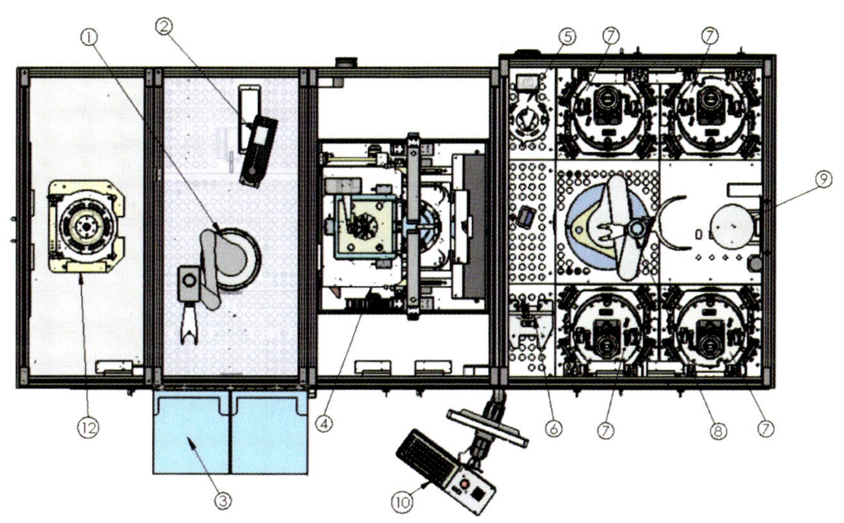

Figure 12.4 Footprint of the EVG Gemini wafer bonding platform with one or more pre-processing modules (left; parts 1–3, 12), the wafer-to-wafer alignment station and up to four bond chambers (part 7). ① Robot ② Pre Aligner ③ SMIF/FOUP loadport ④ SmartView Aligner ⑤ Unclamp station ⑥ Output cassette ⑦ Bond chamber ⑧ Door ⑨ External cooling station ⑩ User interface ⑫ Plasma activation chamber.

For production purposes the individual processes of wafer alignment and wafer bonding are separated into different process modules (Figure 12.5). First and foremost the technical requirements of rapid and uniform heating, high and uniform pressure and full flexibility regarding process gases or vacuum are not compatible with high precision alignment stages. The economical reasoning is that one alignment station with cycle times of 2–6 min can easily support several bond chambers with cycle times of 10–40 min. This process separation brings the inherent risk that during heat up of the wafer stack the wafer alignment can deteriorate. But the results in high volume manufacturing over the last 15 years have shown that high-performance wafer bonding systems can maintain the alignment accuracy safely.

12.2.1.1 **Alignment**

Wafer-to wafer alignment is a very important step for 3D integration – for many integration schemes high alignment accuracy is a necessary requirement. After bonding and subsequent processing the TSVs have to be in contact with the corresponding contact pads. The diameter of the vias and the size of the contact pads determine the alignment tolerance. The diameter of the vias is mainly determined by the via density and by the feasible aspect ratio for a given wafer thickness. To minimize the area of the chip consumed for 3D interconnects, applications with high via density require very small via sizes (<1 micron), whereas for devices with moderate via density larger via diameters are acceptable.

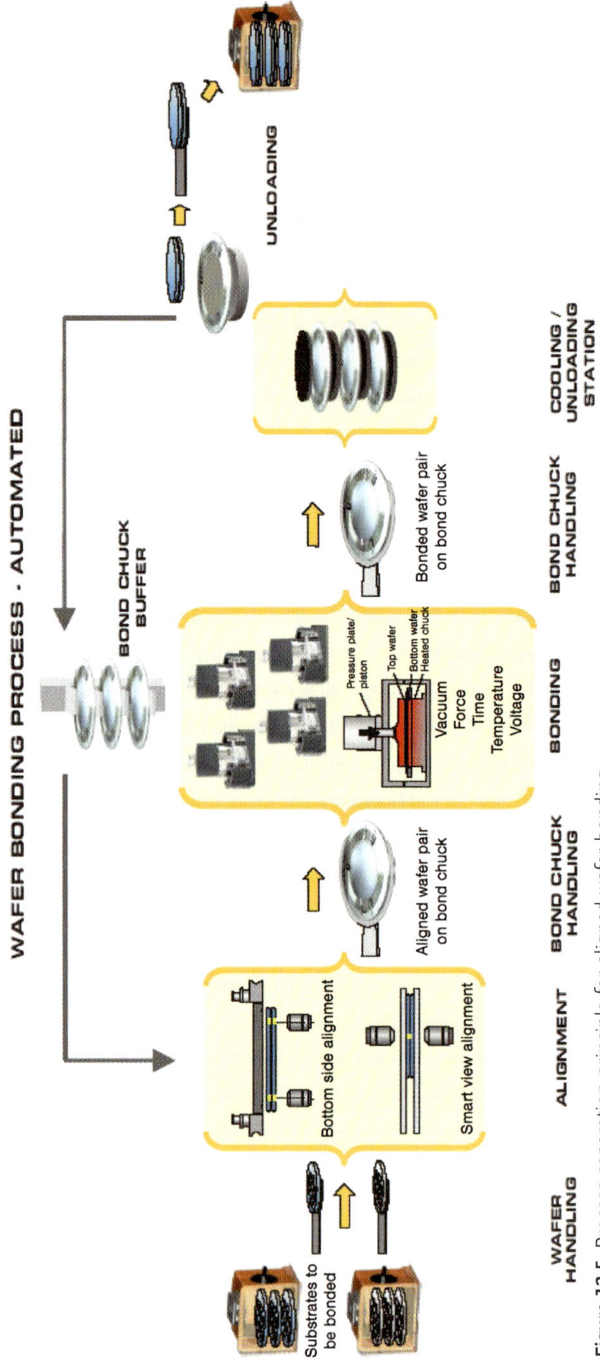

Figure 12.5 Process separation principle for aligned wafer bonding.

Wafer-to-wafer alignment is based on the alignment of two specific points, the alignment keys, on each wafer to each other. The alignment keys can be either on the front side of the wafers, within the bond interface or on the backside. The underlying geometrical principle is that the position of the wafer (position of the patterns on the wafer) is unambiguously determined by two points of the wafer. If the keys of both wafers are within the bond interface and at least one of the wafers is transparent for visible or infrared light, a *direct alignment* of the two wafers can be performed. Direct alignment means that a live image of both corresponding alignment keys is used for alignment. The feedback based on the live image enables a closed control loop and results in highest alignment accuracy.

Usually, the substrates used for 3D stacking are not transparent to visible light. However, if the substrates are very thin, for example for silicon in micron range, the light absorption is so low that the substrates become literally transparent. The alternative approach is the use of infrared light, which penetrate thick substrates (Figure 12.6). However, metal layers or high dopant concentrations are not possible for IR alignment and the wafers need to be backside polished. The best application for IR alignment is a two-layer stack. For multiple layer stacks the image quality degrades due to multiple diffraction, reflection and interference.

An *indirect alignment* method has to be used when either the wafers are not transparent or one set of alignment keys is on the wafer backside. An example would be a face-to-back integration approach, which is discussed below. As no live image is available, each wafer has to be aligned individually to two external reference points. Usually, the positions of the microscopes, specifically the center of the respective field of views, are taken as the reference points. Sequentially, each wafer is aligned to the reference points and finally the aligned wafers are brought into contact. It is very important that the microscopes are not refocused during the alignment procedure.

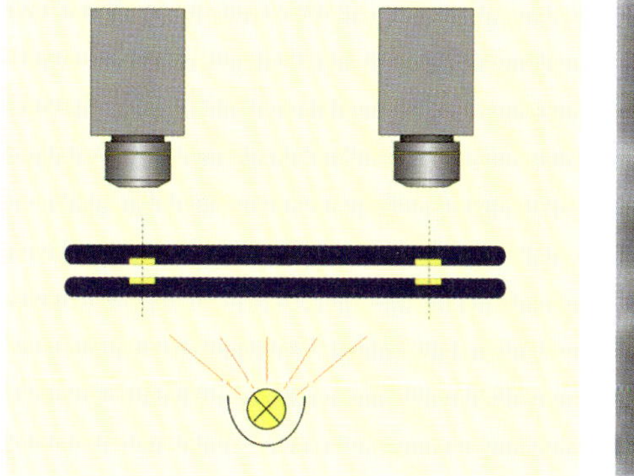

Figure 12.6 Transmission IR alignment principle (left): The bottom wafer is at a fixed position, whereas the top wafer is mounted on a x-y-θ stage. The alignment keys of both wafers are brought into registry. The live image (right) of both keys enables a closed control loop. An alignment accuracy of 0.5 μm (3σ) can be achieved with this method.

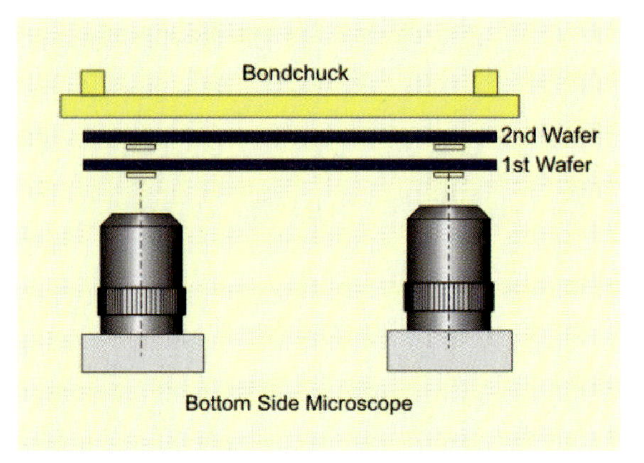

Figure 12.7 Backside alignment method: The two microscopes are the external reference points. Both wafers are sequentially aligned to these reference points and are then brought into contact and clamped on the bond chuck.

The optical axis of the microscopes is never perfectly parallel to the z-direction of the focus movement. Therefore any refocusing would shift the reference points for alignment.

With alignment keys on the wafer backside a *backside alignment* method has to be performed. The first wafer with the alignment keys in the bond interface is aligned to the two reference points. Then this wafer is raised along the z-axis and clears space for the second wafer. The second wafer has the alignment keys on the backside and is being aligned to the reference points (Figure 12.7). Finally the wafers a brought into contact, clamped on the bond chuck and are ready for transfer to the bond chamber.

The critical point is the wafer movement along the z-axis as it is a blind movement. The *SmartView alignment* method overcomes this limitation by the use of dual microscopes. Both microscopes have a common focal plane. The position of the respective optical axes to each other is determined by an *in situ* calibration that is performed at the beginning of each alignment. This offset is then taken into account by the alignment algorithm. The detailed process flow is illustrated in Figure 12.8. Both wafers are fixed on vacuum chucks. In the alignment position both wafer interfaces are in the same focal plane, which is guaranteed by a wedge compensation mechanism. The positions of the microscopes are the reference points for the alignment. First the bottom wafer is moved into a predefined alignment position and the microscopes are positioned according to the alignment keys. The position of the microscopes is locked and the bottom wafer chuck is moved away. Then the top wafer is moved into alignment position and aligned to the microscopes. The top wafer stage can be moved precisely in x, y, and θ. Due to the vector based pattern recognition the alignment procedure is very robust and can handle contrast and even slight shape variations from wafer-to-wafer.

The main advantage of the SmartView alignment is its flexibility and versatility. It is suitable for all types of wafers independent of the surface or bulk properties,

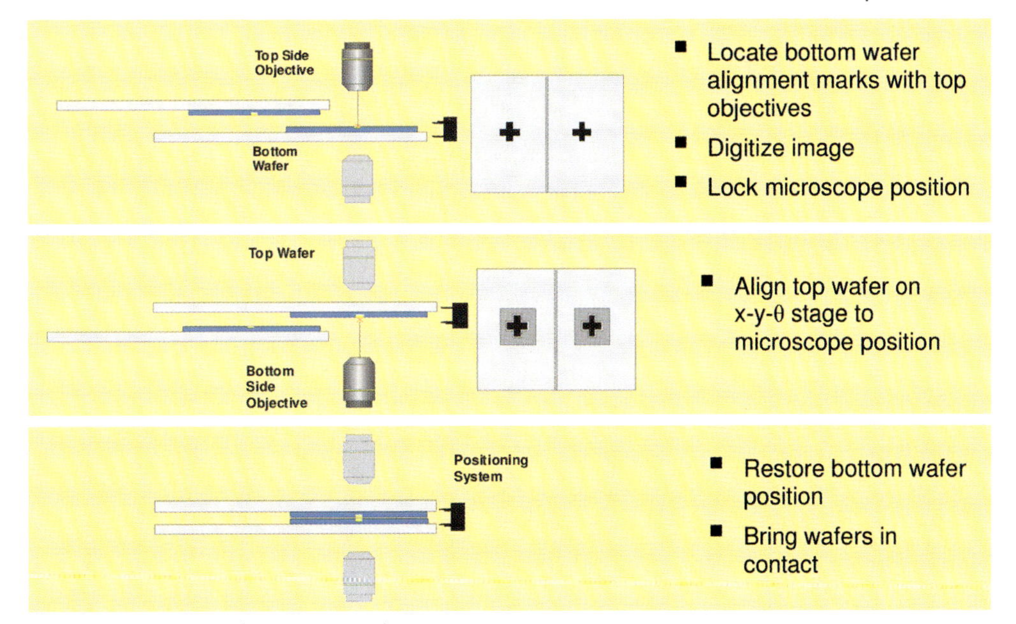

Figure 12.8 SmartView alignment principle.

independent of the thickness and it can be repeated multiple times for multilayer stacking without loss of quality.

Post bond alignment accuracy is based on the initial alignment accuracy and the impact of the bonding process. Table 12.1 shows that for many bonding methods the bonding process itself is the main contributor to the post-bond potential alignment error. Table 12.2 shows the current alignment capabilities and the expected improvements for the next two generations of production equipment.

There are three principle alignment errors: the translational misalignment, the rotational misalignment and the run-out error (Figure 12.9). Whereas the translational misalignment in x and y is a constant error for all the dies of the wafer, rotation and run-out correlate with the distance from the center and are therefore critical for 300 mm applications. Alignment accuracy is defined by the one die with the worst misalignment. An easy way to measure the alignment accuracy is by using Vernier structures (Figure 12.10). For opaque wafers, electrical test structures can be used for process control.

12.2.2 Bonding Methods

12.2.2.1 Cu–Cu

Metal–metal wafer bonding has the inherent advantage that the bond pads serve simultaneously as electrical and thermal interconnects. This allows integration schemes where the TSVs are processed prior to stacking. Cu–Cu thermo-compression

Table 12.1 Post-bond alignment accuracy is based on two factors: the initial alignment prior to bonding and the impact of the bonding process.

	Alignment system capability	
Alignment method	**Alignment accuracy at 3 sigma (μm)**	
Transparent wafer	±0.5	
SmartView face-to-face alignment	±1.3	
	Post bond alignment accuracy-add to alignment system capability	
Bonding technology	**Remark**	**Alignment accuracy at 3 sigma (μm)**
Anodic bonding	Optimum contrast metallized alignment key, 4″ wafer, CTE matched bond glass	±1
Glass frit bonding[a]	10 μm screen printed glass frit, Ferro 11–036, compressed to 4–6 μm during bonding process, 150 or 200 mm wafer	±5
Polymer thermo-compression bond	Thin spin coated adhesive (<1 μm), 150 or 200 mm Si-wafer	±0.6
Fusion bonds (Si + Si, SiO_2 + SiO_2)		±0.4 μm
Metal intermediate layer thermo-compression bond	Thin (<2 μm) metal layer to form an eutectic bond or metal to metal fusion bond (such as: Au–Si, Cu–Cu...)	±0.6

[a] Also applicable to other thick (>5 m) and reflowing intermediate layers.

bonding is the most prominent metal–metal bonding method for 3D interconnects. The main mechanisms for metal–metal thermo-compression bonding are metal ion diffusion and diffusional creep. The diffusion rate depends on the temperature, the pressure and the time. The maximal bonding temperature is limited by thermal budget of the devices. In addition, a higher bonding temperature increases the cycle time due to the longer heating and cooling ramps. In principle the applied pressure only depends on the equipment capabilities. Today's wafer bonder can apply up to 100 kN. However, due to the brittle nature of some new dielectrics it is not possible to go to ever higher pressure. Copper is a well understood material in modern CMOS manufacturing lines. As the metal layers of high-performance chips are, anyway, made of Cu it is a logical step to use the top Cu layer for wafer bonding. The Cu layers are created by the damascene process, which requires chemical-mechanical polishing (CMP) as the final process step. To obtain surface properties (flatness, roughness) required for wafer bonding, CMP requires that the Cu patterns are equally distributed across the wafer (no density changes) and that the Cu

Table 12.2 Roadmap for post-bond alignment accuracy; capabilities of current state-of-the-art equipment and targeted performance for next two generations of aligned wafer bonding production systems.

Roadmap for post-bond alignment accuracy	2007*	2008	2009
SmartView alignment accuracy (face-to face) (μm at 3σ)			
Room temperature bond	1.3	0.5	0.3
Heated bond (Cu–Cu)-200 mm wafer[a]	1.7	0.9	0.6
Heated bond (Cu–Cu)-300 mm wafer[a]	1.9	1.2	0.9
Through wafer alignment accuracy (visible light) (μm T 3 sigma)			
Room temperature bond	0.5	0.3	0.2
Heated bond (Cu–Cu)-200 mm wafer[a]	0.9	0.5	0.5
Heated bond (Cu–Cu)-300 mm wafer[a]	1.1	0.9	0.8

[a] Run out error correlates with wafer size.

patterns are very similar in size and shape. Therefore the density of Cu patterns required to achieve a reasonable flat surface is significantly higher than the density of bond pads required to obtain a strong wafer bond and also higher than the density of interconnects.

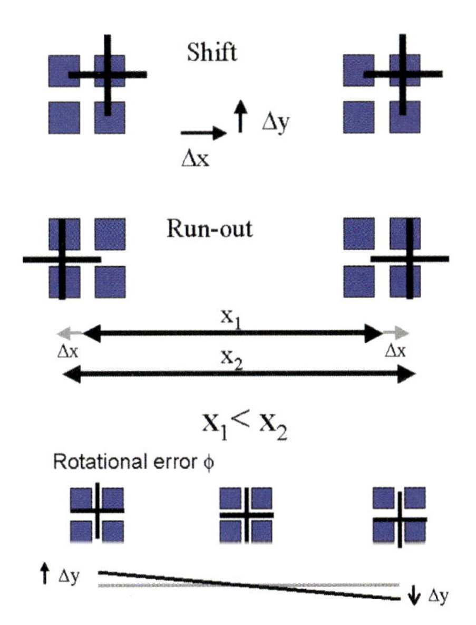

Figure 12.9 Potential alignment errors: Shift and rotational misalignment and run-out; the post-bond alignment is often an overlay of all three types. As rotational error and run-out scale with the position of the die, the alignment can be different for each die on the wafer. The alignment accuracy is defined by the die with the worst misalignment.

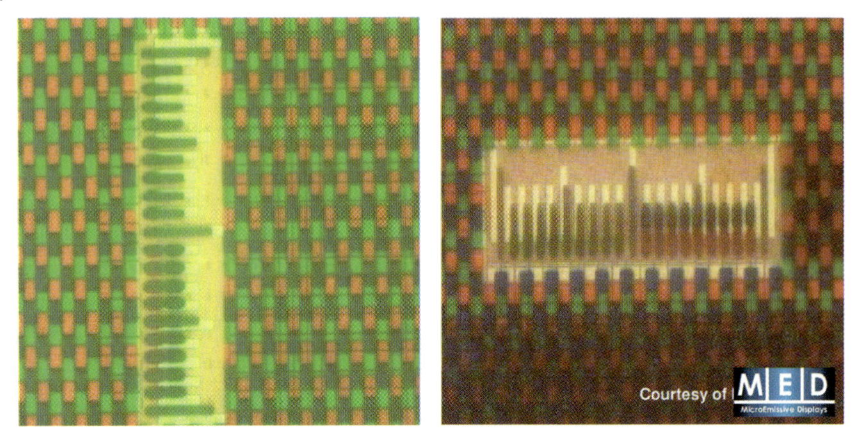

Figure 12.10 Vernier structures enable an easy optical check of alignment accuracy. This example shows filter alignment of a micro emissive display with <500 nm alignment.

Metal–metal thermo-compression wafer bonding initially was developed for MEMS applications such as hermetic capping of cavities. In these cases the wafer area consumed by the metal structures, for example gaskets, is very small, typically in the range 0.5–2% of the wafer area. In addition, the gaskets are usually significantly raised compared to the wafer surface. Both effects, small area ratio and raised structures, enable bonding schemes where real deformation of the metal structures occurs (= volume flow mainly based on creep). For 3D interconnects the area ratio of interconnects to whole wafer surface is significantly higher. The area for the Cu bond pads is typically in the range of 35–45% of the wafer surface. Due to the brittle nature of the dielectrics it is not possible to scale up the pressure to copy the bonding schemes established in MEMS. The reduced pressure shifts the bonding mechanism towards diffusion. There is no, or very limited, volume flow. This has very important consequences for the wafer bonding requirements. As the main mechanism is diffusion the contact pads have to be in contact right from the beginning. Therefore the specifications for surface roughness and thickness variation are very tight.

The key process parameters for Cu–Cu wafer bonding are temperature uniformity and pressure uniformity. The precise pre-bond alignment of the wafers has to be maintained during heat up. The better the temperature uniformity the faster heating ramps are possible. In addition, in-wafer-plane temperature non-uniformities would create local distortions and induce stress into the bond layer. Very good pressure uniformity is necessary to compensate for wafer bow and warp and also for thickness variations. It is important that all the bond pads start bonding at the same time. The effect of pressure uniformity on the bond quality for 300 mm wafer has been analyzed by Morrow *et al.* [2].

It is important to prevent oxidation of the Cu surfaces both prior to and during wafer bonding. The standard approach to avoid oxidation during heat up is by the use of reducing gas during bonding, usually a mixture of nitrogen with 4–10% hydrogen ("forming gas") [3].

Benzocyclobutene (BCB) An important process for 3D integration is BCB wafer bonding. BCB is available in different formulations; for 3D integration the non-photo-imageable formulation is usually employed. There are two main advantages to using an intermediate layer wafer bonding method. First the surface roughness requirements are significantly lower. BCB has very good surface wetting properties. In comparison to metal thermo-compression bonding a rough surface does not reduce the contact area for intermediate layer bonding as the two wafer surfaces do not get in intimate contact. This allows surface micro-roughness of up to 20 nm. In addition to this BCB reflows during heat up. Particles on the wafer surface are embedded in BCB and do not pose a problem for the bond quality.

However, the material reflow imposes some challenges for maintaining the alignment accuracy. Depending on the degree of pre-bond crosslinking the material goes through a liquid or sol–gel rubber type phase during heat up. A liquid layer between the wafers is always a problem as any shear force on one wafer would immediately produce a shift of one wafer, resulting in translational misalignment. However, studies by Niklaus *et al.* [4] showed that by tailoring the pre-bond cross-linking status it is possible to maintain the alignment accuracy during bonding. Another important aspect is the simultaneous heating of both wafers. BCB has a low thermal conductivity. During fast heating ramps each wafer gets more or less entirely heated by the respective heater as the BCB layer prevents thermal conductance between the wafers. Any temperature mismatch between top and bottom heater would result in asynchronous thermal expansion of the wafers, which would come at the cost of alignment accuracy. Therefore, simultaneous heating of top and bottom wafer, controlled by a closed-loop feedback system, is of highest importance for BCB aligned wafer bonding.

An interesting approach is the simultaneous bonding of Cu–Cu thermo-compression and BCB intermediate layer wafer bonding, which has been proposed and characterized by RPI [5]. A Cu damascene process with photo-imageable BCB is being used to manufacture a Cu/BCB surface patterning. During the wafer bonding process three different bonding methods are used: (i) Cu–Cu, (ii) BCB–BCB and (iii) Cu–BCB. All three methods belong to the group of thermo-compression bonding. Figure 12.11 shows a SAM image of Si wafers bonded with BCB.

Silicon Direct Bonding Silicon direct bonding (SDB) is a two-step process: aligned pre-bonding at room temperature and a high temperature annealing step. The wafer bond quality can be analyzed prior to annealing. If necessary the wafers can be debonded and bonded again. This is a main advantage over other wafer bonding methods for 3D as, usually, a void prevents subsequent wafer thinning. Another advantage is that the pre-bonding occurs at room temperature. Therefore run-out error cannot be induced during bonding and the achievable post-bond alignment accuracy with silicon direct bonding is better than with wafer bonding methods at elevated temperatures.

As the annealing can be done as a batch process, the cycle time is much shorter than for wafer bonding methods, which have to go through the complete bond cycle of heating–bonding–cooling. For some applications pre-bonding can be done in

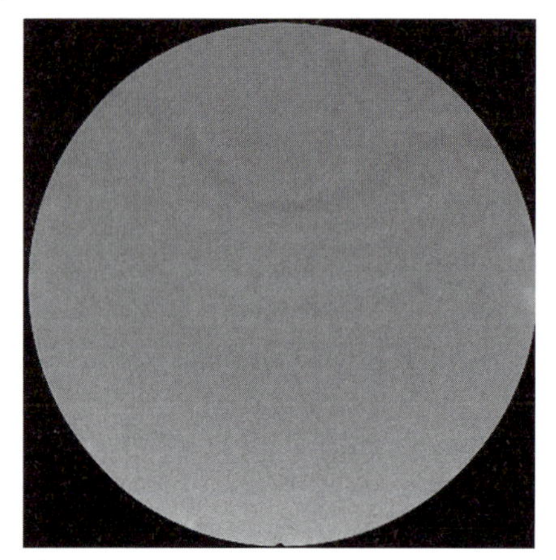

Figure 12.11 Scanning acoustic microscopy (Sonix AW Vision 3000) of 200 mm Si wafers bonded with BCB.

ambient conditions. This allows an *in situ* aligned bonding process, where the wafers bond directly within the alignment station (Figure 12.12).

The classical SDB methods require annealing schemes with peak temperatures of up to 1100 °C, which is of course not feasible for CMOS devices. Low temperature plasma activation modifies the wafer surface chemistry in such a way that the annealing temperature can be reduced to 200–400 °C. Due to this surface pre-treatment SDB methods are an option for 3D chip stacking. The surface activation lasts for tens of minutes and allows wafer-to-wafer alignment in ambient after activation. As for direct wafer bonding, the surface quality is of very high importance. Dedicated plasma activation chambers for wafer bonding are designed to prevent any potential surface damage (etching) effect during the plasma exposure. Figure 12.13 shows the process flows for aligned and non-aligned silicon direct bonding.

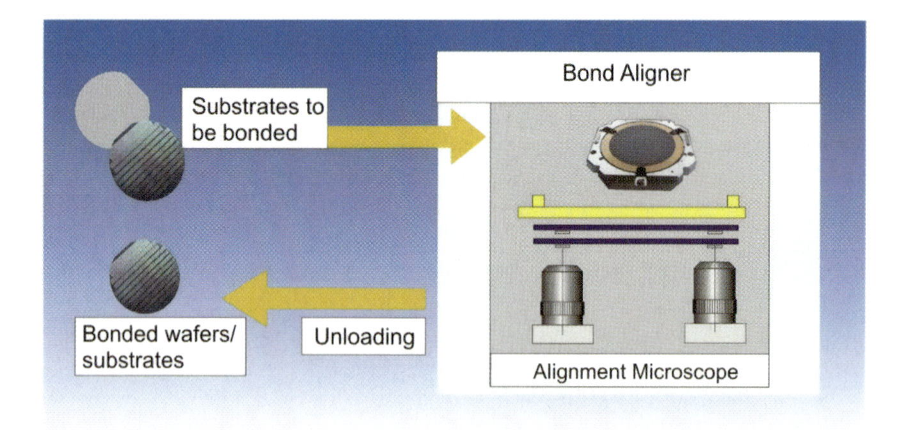

Figure 12.12 *In situ* bonding process for silicon direct bond applications.

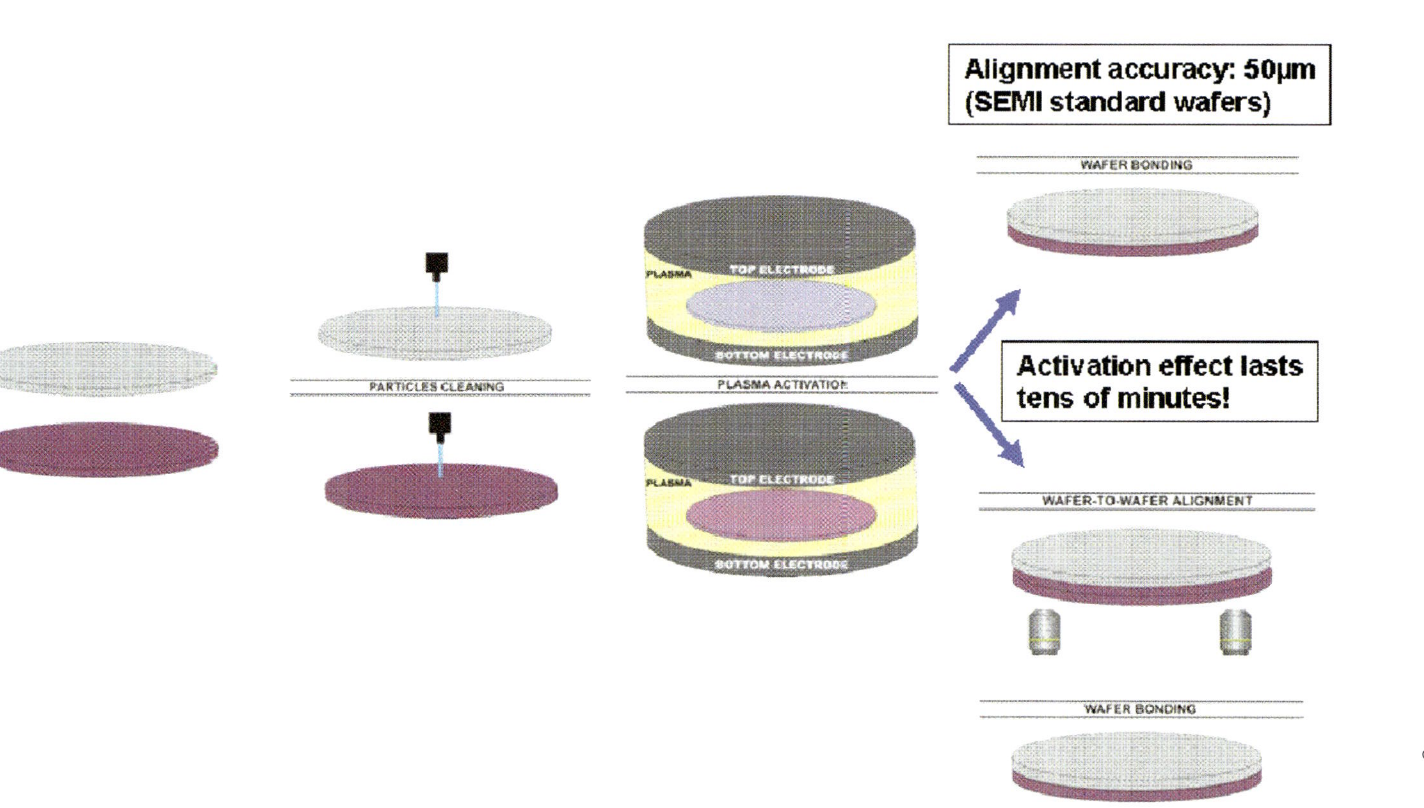

Figure 12.13 Process flow for silicon direct wafer bonding with LowTemp plasma activation. The activation effect lasts for tens of minutes.

The TSVs can be created either before or after wafer bonding. Creating the vias after wafer bonding has the advantage that the bonding process itself is very simple. The main requirement is a very smooth surface. Polishing of oxide layers is a very well understood and established process.

TSVs before bonding makes polishing the metal/oxide surface significantly more challenging. The advantage is that this process allows via first integration schemes. This method has been promoted by Ziptronix for their Direct Bond Interconnect (DBI) process.

12.2.3
Temporary Bonding/Debonding

Temporary bonding and debonding is a technology for the handling and processing of thin wafers. As such it is an enabling technology for backside processing for TSV devices. The basic concept of temporary bonding is that the device wafer is temporarily bonded to a carrier wafer prior to back thinning. The carrier wafer gives the device wafer mechanical stability and protects the wafer edge. In addition the carrier wafer prevents bending or warping of the thin device wafer. The geometry of the wafer stack can be tailored in such a way that the stack mimics a standard wafer. All these effects enable the producer to use standard fab equipment for further processing. The cost for carrier wafer based thin wafer handling is usually lower than the cost for dedicated thin wafer chucks, end-effectors and wafer cassettes on each individual tool. After backside processing the adhesive force between device wafer and carrier wafer is released and the thin wafer is either sent to assembly or is stacked on another wafer.

Temporary wafer bonding and debonding has been initially developed for compound semiconductor processing, as these materials are much more brittle than silicon, as well as for silicon based power devices. Figure 12.14 shows a typical process flow for temporary bonding.

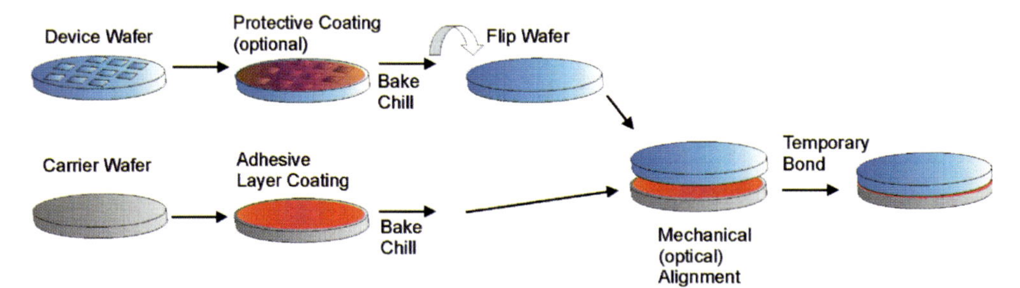

Figure 12.14 Standard process flow for temporary bonding; the device wafer is fully processed on the front side. The front side of the device wafer can be optionally coated with a protective layer. The carrier wafer gets coated with the spin-on adhesive and then both wafers are transferred to the bond chamber. Both wafers are aligned to each other and finally bonded at elevated temperature and usually under vacuum. This process flow can be used for any kind of carrier wafer properties. It is independent of wafer size, thickness and material.

There are various different intermediate materials for temporary bonding. The two main classes are spin-on adhesives and tape based materials. As temporary bonding is an auxiliary technology for subsequent processing, the choice of the correct intermediate material depends mainly on the following process requirements.

- *Chemical resistance:* The first question is the exposure to chemicals during the downstream processing. The intermediate material has to be either resistant against the chemicals or the exposure time to these chemicals has to be short enough, so that the bond strength and the edge protection are not negatively impacted. An important point is that the debond method is not impacted. An additional point is that the ability to clean the device wafer after debonding is not affected.
- *Maximal operating temperature and thermal budget:* These are two very important decision criteria. Of course, the properties of the intermediate layer must not degrade due to thermal exposure. It is sometimes possible to reduce the operating temperature of the downstream process; however, this usually comes at the cost of reduced throughput.
- *Edge protection:* The edge is the most vulnerable part of the thin wafer. The thinner the wafer the more important is edge protection. During subsequent processing, edge chipping could occur or a crack could be induced due to mechanical contact with the edge. For etch processes it is desirable to protect the edge to avoid further thinning of the edge ("razor blade edge"). Figure 12.15 shows the main difference in edge protection between tape materials and spin-on materials.
- *Debonding method:* For commercially available adhesives there are three classes of debonding mechanisms: thermal release, UV release and chemical release.

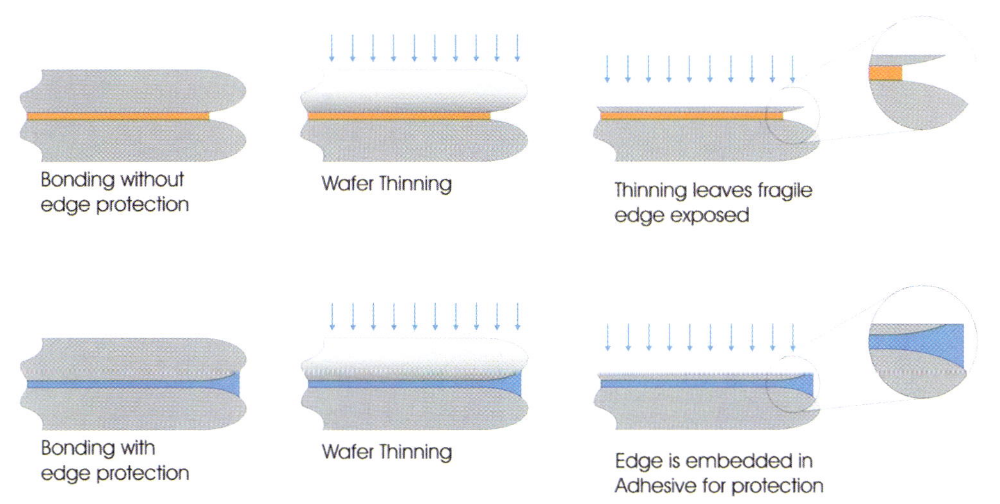

Figure 12.15 Tape adhesives usually do not provide edge protection (top row). For spin-on adhesives the optimal edge protection can be tailored by using the edge bead and controlled squeeze out during bonding (bottom row).

Chemical release, typically dissolution in solvent, has the main disadvantage that the thin wafer floats uncontrolled in the solvent bath after debonding, which is typically not compatible with the wafer thicknesses needed for TSVs. The release temperature of thermal release materials is higher than the maximum operating temperature, which is sometimes not compatible with the thermal budget of the devices. UV release materials require a transparent carrier wafer, which increases costs and has the disadvantage that the thermal expansion properties of device and carrier wafer are different. This may result in bow or warp of the stack. In addition, the thick carrier wafer can dominate the thermal expansion behavior of the whole stack.

Tape materials are usually debonded by a lift off process, whereas spin-on materials are typically debonded by a slide-off process (Figure 12.16).

12.2.3.1 Permanent Bonding of Thin Wafers

After thinning and backside processing the dies of the thin wafer are either singulated and then stacked in a C2C or C2W integration scheme or the thin wafer is permanently bonded to another device wafer. Permanent bonding can be done in both ways, either before or after debonding. The process flow for debonding after permanent wafer bonding is significantly easier as standard equipment can be used for alignment and bonding. The key criterion is whether the temporary adhesive is compatible with the permanent bonding temperatures. If this is not the case then the thin wafer has to be debonded from the carrier wafer prior to permanent bonding. This requires that the thin wafer is handled without support to alignment and bonding station. A hybrid solution is to transfer the thin wafer from one carrier to another carrier, to which the thin wafer is bonded with an adhesive, which is compatible with the following permanent bonding step.

12.2.4 Chip to Wafer Bonding

Chip to wafer 3D stacking is the principle alternative to wafer-to-wafer integration schemes. There is a fundamental difference between C2W integration based on wire

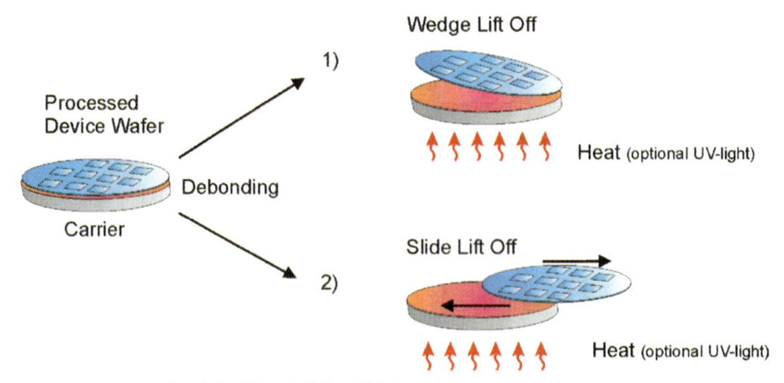

Figure 12.16 Wedge lift off and slide off debonding.

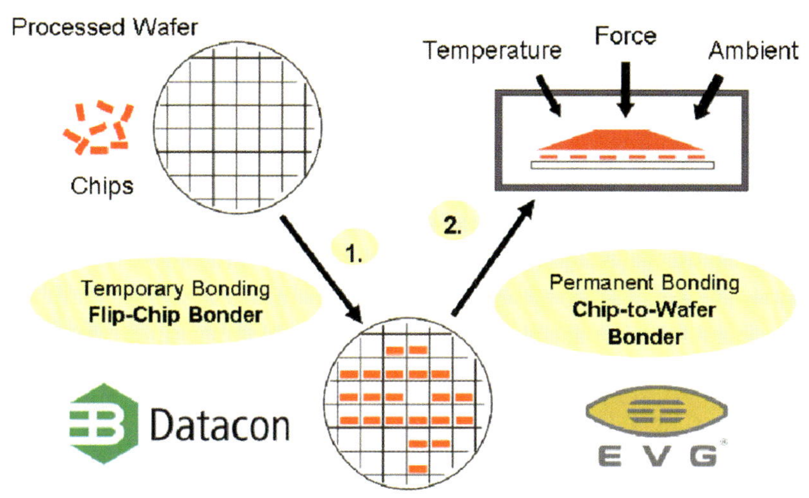

Figure 12.17 Process flow for a two-step chip-to-wafer integration.

bonding versus C2W based on TSV. This section reviews the technology for C2W based on TSVs. Approaches with via processing after stacking do not seem feasible for economic reasons. The processing of the vias (lithography, etching, plating, CMP,) requires parallel processing of all the chips, which is only possible with wafer-level processes. Via first integration requires a metal bond between the vias and the contact pads. However, metal bonding is a rather lengthy process as it involves either diffusional processes or in the case of eutectic bonding it requires heating and cooling cycles. One approach to achieve a reasonable throughput is to split the two main processes, chip-to-wafer placement and metal bonding.

The typical process flow is shown in Figure 12.17. The starting conditions are a fully processed device wafer and already singulated fully processed chips. The first step is the placement of the chips on the wafer. The chips are temporarily tacked to the base wafer by a high performance pick-and-place device bonder and then transferred to the chamber for permanent bonding.

Due to throughput requirements (see Chapter 1) it is important that only known good dies are used for further processing. As a consequence the base wafers are asymmetrically populated. In the EVG540C2W bonder the force application can be tailored in such a way that each chip only sees perpendicular forces. This guarantees that the chips stay in the alignment position during interconnect formation and bonding (Figure 12.18).

A necessary requirement for production purposes is that the singulated dies can come from different wafers. When several different chips are placed on a base die, these chips can even be out of different production lines or fabs. As a consequence it can not be assumed that the thickness of the chips is within tight limits. The bonding system has to be tolerant regarding chip thickness. Even though the chips might have different thicknesses, the force application has to be purely perpendicular. This is achieved by the use of a flexible compliant layer.

Figure 12.18 Taking full throughput advantage of the "Known Good Die" (KGD) approach requires that only good base dies are populated with dies. As a consequence the force application has to be controlled in such a way that only perpendicular forces are applied on each die. Any shear forces would deteriorate the alignment accuracy.

The bonding can occur in vacuum or in any required process gas. For the Cu-Sn SOLID (Solid–Liquid-Interdiffusion) process the use of a HCOOH enriched N_2 atmosphere has been recommended [6].

12.3
Integration Schemes for 3D Interconnect

Here we review the different integration schemes, with emphasis on the specific equipment and process requirements. For ease of description the following definitions are made:

- Wafer #1 is a fully processed device wafer with full thickness. In the context of the described integration scheme wafer #1 can also be already a stack of several layers.
- Wafer #2 is the second wafer either with full thickness or already thinned.
- In the context of integration schemes "face" refers to the front side of the device wafer. (Not to be confused with "face" in the context of wafer-to-wafer alignment, where it refers to alignment keys in the bond interface.)

In the literature the terms "via first" and "via last" are frequently used to distinguish between process flows. However, notably, these terms are not clearly defined and are used by different authors for different process flows. The two most common usages are (i) for via processing prior or after wafer thinning and (ii) for via processing prior or after chip stacking.

12.3.1
Face-to-Face Chip Stacking

Face-to-face chip stacking is a basic process flow for 3D integration. Substrates #1 and #2 are fully processed on the front side. After the necessary surface preparation they are bonded face-to-face. Partial or complete back-thinning of device wafers #2 is performed after wafer bonding. This brings the benefit that it is not necessary to use a carrier wafer for thin wafer handling. Device wafer #1 gives mechanical

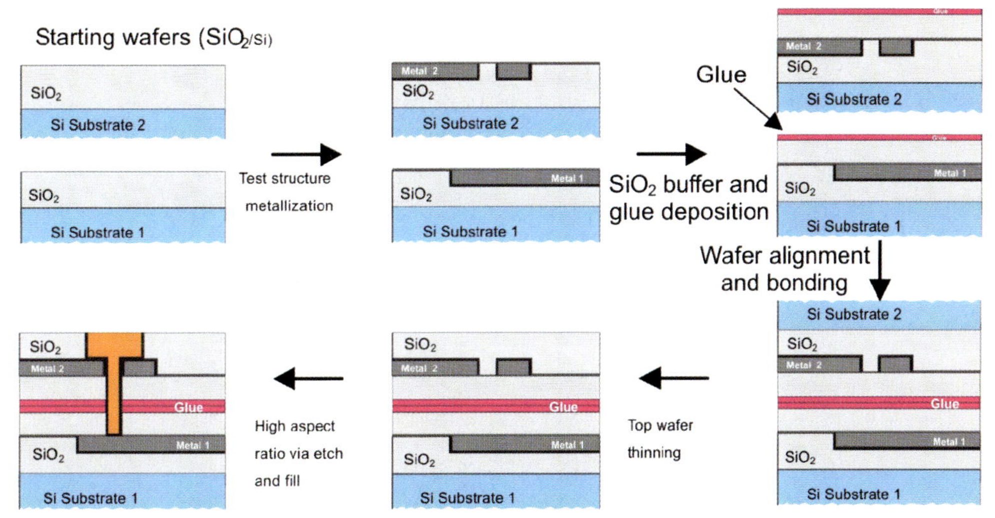

Figure 12.19 Principle of two-layer stack with thick wafers. In this case an adhesive wafer bonding process is illustrated with via processing after bonding. (Courtesy James Liu, Rensselaer Polytechnic Institute.)

stability for the whole stack during back grinding and subsequent backside processing.

The requirements for wafer bonding are very tight. As the back thinning is performed after the wafer bonding it is necessary that the bond interface is absolutely void free. Any void could result in wafer break-outs during back grinding. Therefore, a void is usually a "show stopper" for further processing. However, it has been verified that Cu–Cu and BCB bonding provide the necessary yields required for high volume manufacturing [2, 7].

The TSVs can be created either before or after wafer bonding.

A slight disadvantage of the face-to-face stacking approach is that the process flow cannot be repeated when going to multiple layers (Figure 12.19). Stacking a third layer on a two-layer face-to-face stack necessarily requires a *face-to-back* stacking approach. For multiple layers of the same functional units, such as stacked memory, it is necessary to design different chips and develop different process flows for every other layer. This is some disadvantage compared to a pure face-to-back stacking approach, where the same chip designs and process flows can be used for two-layer stacks as well as for eight-layer stacks (Figure 12.20).

12.3.2 Face-to-Back Chip Stacking

Integration schemes where the top metal layer of wafer #1 is connected to the TSVs on the backside of the thin wafer are described as face-to-back (F2B) chip stacking. For *two-layer stacks* face-to-back stacking always requires backside processing of wafer #2

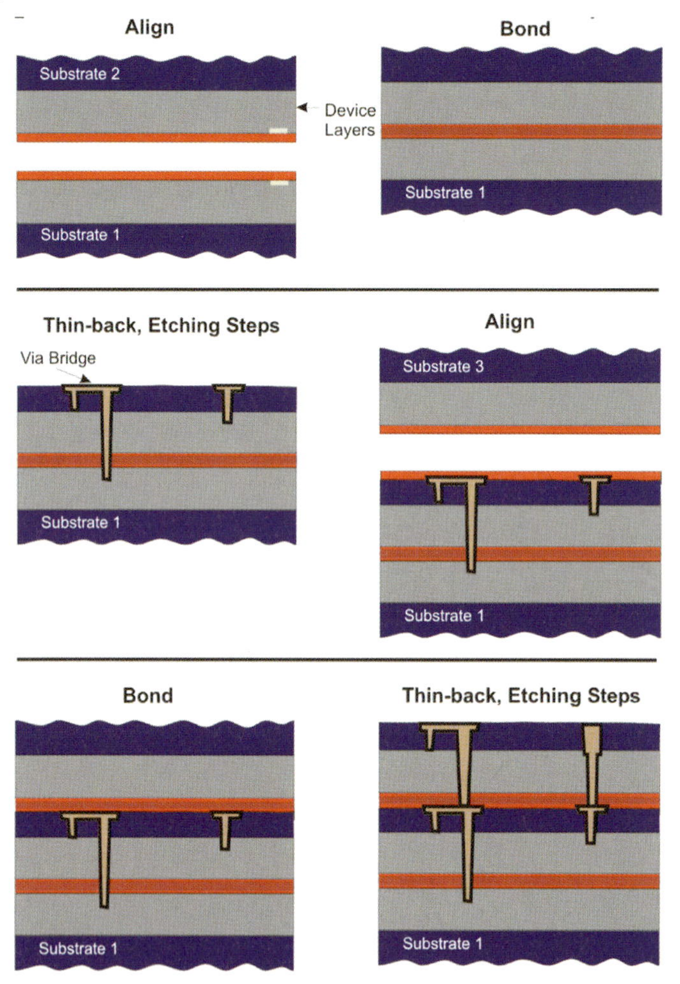

Figure 12.20 Principle of multilayer stacking with full thickness wafers. Substrates 1 and 2 are bonded face-to-face. After back thinning and backside processing of substrate 2, substrate 3 is bonded with the face-side to the backside of substrate 2 – a face-to-back integration. This switch from face-to-face to face-to-back is always necessary when multilayer stacking is carried out with thick wafers. This cartoon illustrates a bonding method with an adhesive intermediate layer like BCB. The vias are created after bonding.

prior to bonding. This results in the need to use a carrier wafer for the processing of the thin wafer (Figure 12.21).

For *multilayer stacks* there are two different integration schemes:

Integration scheme A: F2B-F2B-F2B-F2B
Integration scheme B: F2F-F2B-F2B-F2B

Integration scheme A is a repetitive manufacturing approach that allows taking the same processes and tools for Bond #1 to #2 as for #1 + 2 to #3 and so on.

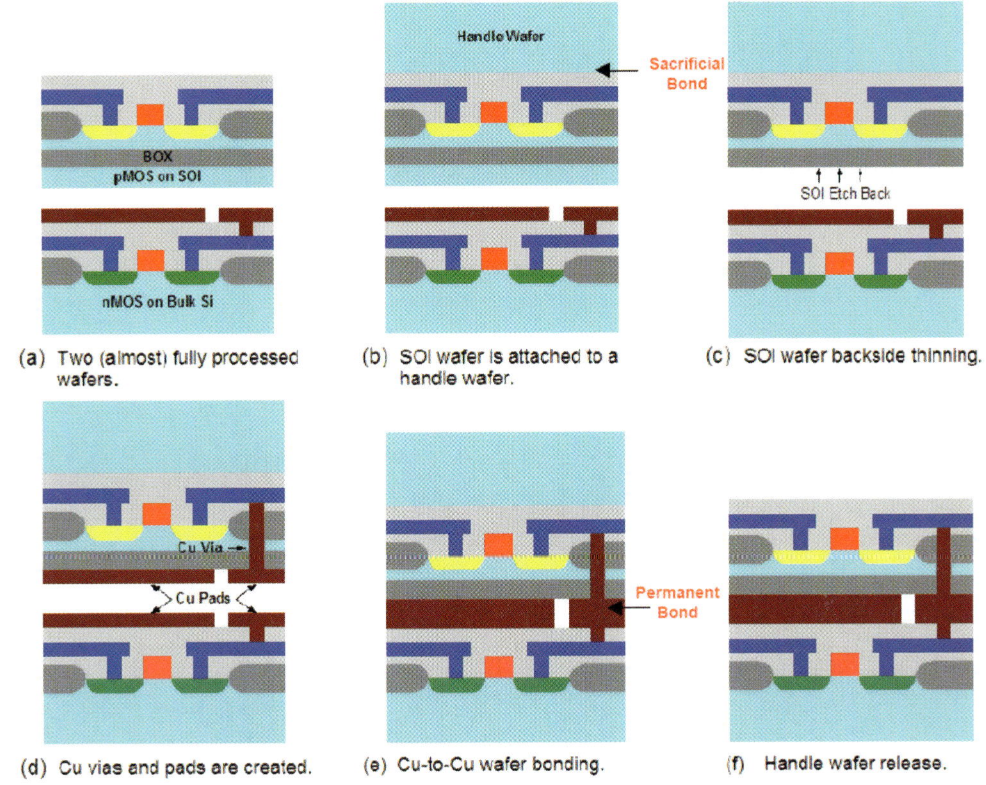

Figure 12.21 Two-layer chip stack integration scheme with Cu–Cu bonding and face-to-back stacking. This process sequence (a–f) can be repeated several times without modifications for multilayer devices. (Courtesy C.S. Tan, MIT, 3D conference, Tempe, June 14th/15th 2005.)

Manufacturing scheme A is feasible for W2W and C2W, even though the C2W approach is very challenging.

Integration scheme B starts with a face-to-face stacking and switches then to face-to-back stacking. This approach, quite cleverly, avoids the use of a carrier wafer. The entire backside processing for all layers is performed after bonding. Integration scheme B can only be implemented for W2W as thinning of stacked dies on a wafer does not seem feasible.

This manufacturing scheme tightens the requirements for wafer bonding, as a void would potentially cause wafer breakage during back-thinning. Wafer-to-wafer alignment has to be done by SmartView alignment or if possible by IR alignment. Owing to the use of full thickness wafers the processing yields should generally be higher than for approaches with thin wafers.

12.4 Conclusion

Through silicon vias and 3D chip stacking offer a promising path for the architecture of future devices. The key enabling technologies of wafer bonding, wafer alignment and thin wafer handling with a carrier wafer are well established in other silicon-based industries like MEMS. Depending on the application, the device and also price/performance structure of the product group, either chip-to-wafer or wafer-to-wafer integration schemes, are advantageous. At the moment there are a huge variety of integration schemes and processes. Once these processes are fully characterized for 3D applications regarding risk, yield, and performance it can be assumed that standardization of processes and process flows will start. However, all of the described processes have been shown to provide a viable solution for high volume manufacturing.

References

1 Tong, Q.-Y. and Gösele, U. (1998) *Semiconductor Wafer Bonding: Science and Technology*, John Wiley & Sons, Inc.

2 Morrow, P., Kobrinsky, M.J., Ramanathan, S., Partk, C.-M., Harmes, M., Ramachandrarao, V., Park, H.-M., Kloster, G., List, S. and Kim, S. (2005) Wafer-Level 3D Interconnects Via Cu Bonding. *Proceedings of the UC Berkeley Extension Advanced Metallization Conference (AMC 2004)*, Mater. Res. Soc. Symp. Proc, **20**, p.125–130.

3 Tadepalli, Rajappa. and Thompson, Carl, V. (2003) Quantitative characterization and process optimization of low-temperature bonded copper interconnects for 3-D integrated circuits. Proceedings of the IEEE 2003 International Interconnect Technology Conference, IEEE Catalog # 03TH8695, p. 36.

4 Niklaus, F., Kumar, R.J., McMahon, J.J. *et al.* (2005) Effects of bonding process parameters on wafer-to-wafer alignment accuracy in benzocyclobutene (BCB) dielectric wafer bonding. *Materials Research Society Symposium Proceedings*, **863**.

5 McMahon, J.J., Lu, J.-Q. and Gutmann, R.J. (2005) Wafer bonding of damascene-patterned metal/adhesive redistribution layers for via-first three-dimensional (3D) interconnect. Proceedings of IEEE 55th Electronic Components and Technology Conference (ECTC 2005), May 31–June 3, Florida. pp. 331–336.

6 Scheiring, C., Kostner, H., Lindner, P. and Pargfrieder, S. (2004) Advanced chip-to-wafer technology: enabling technology for volume production of 3D system integration on wafer level. (eds. Frank Niklaus, Ravi Kumar *et al.*), Proceedings IMAPS 2004, p. B10.8.1–B10.8.6.

7 Morrow, P., Park, C.-M., Ramanathan, S., *et al.* (May 2006) Three-dimensional wafer stacking via Cu–Cu bonding integrated with 65-nm strained-Si/Low-*k* CMOS technology. *IEEE Electron Device Letters*, **27** (5), 335–337.

13
Polymer Adhesive Bonding Technology

James Jian-Qiang Lu, Tim S. Cale, and Ronald J. Gutmann

Wafer bonding with intermediate polymer adhesives is an important fabrication technique for advanced microelectronics and microelectromechanical systems (MEMS) systems, such as three-dimensional integrated circuits (3D ICs), advanced packaging and microfluidics. In adhesive wafer bonding, the polymer adhesive bears the forces involved to hold the surfaces together. Compared to alternative wafer bonding approaches, the main advantages of adhesive wafer bonding include insensitivity to surface topography, low bonding temperatures, compatibility with standard integrated circuit wafer processing and the ability to join different types of wafers. In addition, adhesive wafer bonding is simple, robust and low-cost. This section reviews the state of the art of polymer adhesive wafer bonding technologies for 3D integration.

13.1
Polymer Adhesive Bonding Principle

Bonding two substrates together has long been an important process in the fabrication of both microelectronics systems and microelectromechanical systems (MEMS). The wide variety of wafer bonding techniques include direct bonding, anodic bonding, solder bonding, eutectic bonding, ultrasonic bonding, metal fusion bonding, thermocompression bonding, low-temperature melting glass bonding and adhesive bonding. In adhesive bonding, an intermediate adhesive layer is used to create a bond between two surfaces to hold them together. Although successfully used in many industries, including airplane, aerospace and car manufacturing, to join various similar and dissimilar materials, adhesive bonding did not have a significant role in early semiconductor processes. Recent research and development of adhesive wafer bonding includes bonding of large substrates using well defined and defect-free intermediate adhesive layers. Some applications require precise wafer-to-wafer alignment of the bonded wafer pairs. Recently developed reliable

Handbook of 3D Integration: Technology and Applications of 3D Integrated Circuits.
Edited by Philip Garrou, Christopher Bower and Peter Ramm

ISBN: 978-3-527-32034-9

and high yield adhesive bonding processes have made adhesive wafer bonding a generic, and in some cases enabling, wafer bonding technique for various applications [1].

In most commonly used adhesive wafer bonding processes, a polymer adhesive is applied to one or both of the wafer surfaces to be bonded (Figure 13.1). After joining the wafer surfaces that are covered with the polymer adhesive, pressure is applied to force the wafer surfaces into intimate contact. The polymer adhesive is then converted from a liquid or visco-elastic state into a solid state, which is typically done by exposing the polymer adhesive to heat or UV light.

The main advantages of adhesive wafer bonding include:

- Relatively low bonding temperatures (between room temperature and 450 °C, depending on the polymer material)
- Insensitivity to the topology of the wafer surfaces
- Compatibility with standard CMOS wafers
- Ability to join practically any wafer materials.

Adhesive wafer bonding does not require special wafer surface treatments such as planarization and extensive cleaning. Structures and particles at the wafer surfaces can be tolerated and accommodated to some extent by the polymer adhesive. While adhesive wafer bonding is a comparably simple, robust and low-cost process, concerns such as desirable properties of adhesive materials, limited temperature stability and limited data about long-term stability of many polymer adhesives in demanding environments need to be considered for use in 3D integration.

13.2 Polymer Adhesive Bonding Requirements and Materials

The polymeric adhesive layer for wafer bonding in 3D integration must provide a seamless interface and strong adhesion to prevent delamination, be sufficiently thin to minimize the via aspect-ratio, be processed at modest temperatures to avoid any effects on the device wafer performance and reliability, and be thermally and

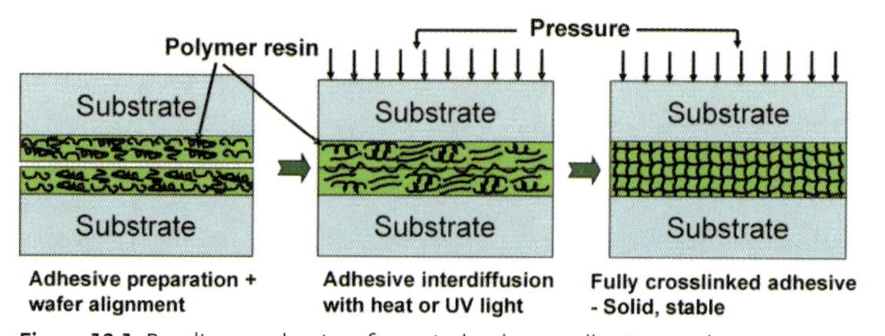

Figure 13.1 Bonding mechanism for typical polymer adhesive used in 3D integration.

mechanically stable after bonding. Desirable properties of the bonding adhesive include:

- Good adhesion and cohesion – adhesive to wafer (adhesion), and adhesive to adhesive (cohesion), to prevent delamination.
- No outgassing during bonding, to avoid void formation.
- High thermal and mechanical stability over the ranges of BEOL and packaging processing conditions; that is a high glass transition temperature and rigid structure after bonding.
- Low stress relaxation and creep.
- Low moisture uptake.
- Ability to form uniform and micron thick films over entire wafers.

Various polymers have been explored for wafer bonding, Reference [1] gives an intensive review. Based on the desired properties listed above, we selected and evaluated several polymers, such as Flare (Poly aryl ether), methylsilsesquioxane (MSSQ), benzocyclobutene (BCB), hydrogensilsesquioxane (HSQ) and vapor deposited polymer Parylene-N. These polymers can be deposited as thin films, have relatively well-known chemical and physical properties, and, importantly, are compatible with IC processing to some extent. Figure 13.2 shows some challenges of adhesive wafer bonding: voids due to outgassing from MSSQ, low cohesive bond strength with Parylene, and adhesive films that are too thin (<0.3 μm) to accommodate large particles with Flare, though the bond strength with Flare is very high.

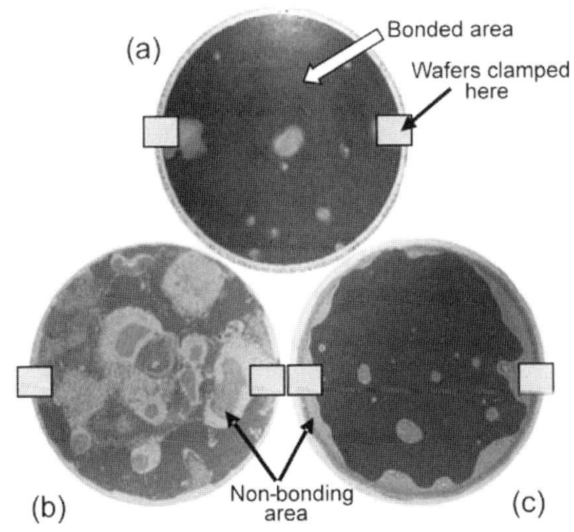

Figure 13.2 Photos of bonded wafers (200-mm Corning 7740 glass wafer to silicon wafer) using Flare (a), MSSQ (b) and Parylene-N (c). Highly contrasted areas show voids or non-bonding areas [2].

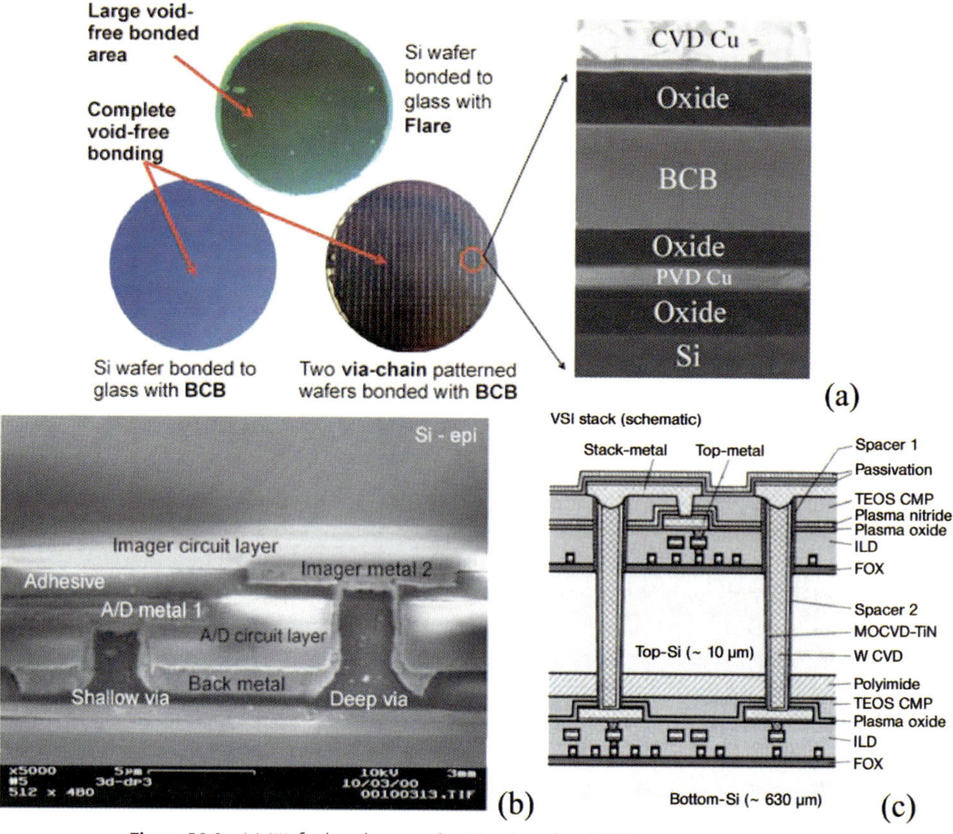

Figure 13.3 (a) Wafer bonding results: Corning glass 7740 bonded to prime Si with Flare (top), PG&G glass 1737 bonded to prime Si with BCB (left) and two via-chain patterned wafers bonded with BCB (right), with an SEM image showing a seamless bonding interface [3]. (b) An SEM of a 3D ring oscillator section, showing the adhesive bond between the imager and A/D wafers [4]. (c) Polyimide as a bonding adhesive [5].

Figure 13.3 shows several other examples of adhesive bonding used for 3D integration [3–5].

13.3 Wafer Bonding Technology Using Polymer Adhesives

Most 3D IC applications of adhesive wafer bonding require well defined and high yield bond interfaces, and often precise alignment of the bonded wafers. To repeatably achieve high quality bonding results, the bonding process and parameters, as well as wafer-to-wafer alignment where required, must be precisely controlled.

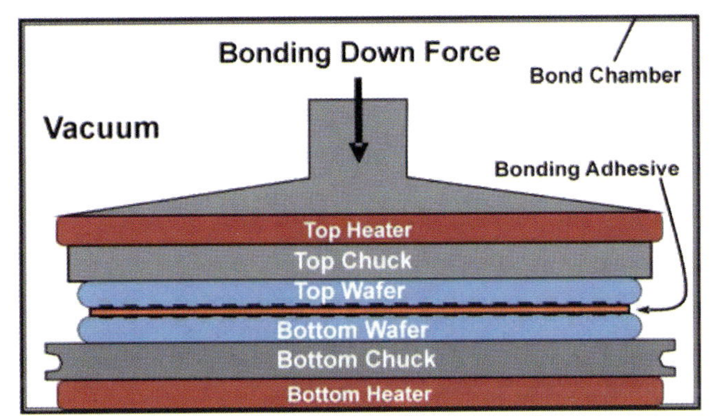

Figure 13.4 Schematic drawing of a typical commercial wafer bonder.

This is often accomplished with a cluster tool consisting of a wafer aligner and bonder. A wafer bonder typically consists of a vacuum chamber, a mechanism for joining the wafers inside the vacuum chamber, a wafer chuck and a bond tool (Figure 13.4). Bonding parameters such as bonding pressure, bonding temperature, chamber pressure and temperature ramping profile can significantly impact the resulting bonding quality and defect density. The wafer stack is placed between the bottom wafer chuck and the bond tool, or top wafer chuck. Thus, the wafer stack can be pressed together with the bond tool using a controlled pressure (force per wafer or bond area). The wafer stack can be heated through the bottom wafer chuck and the top chuck. The top chuck can be a stiff flat plate or stiff flat plate with a soft plate or sheet in between the top chuck and the wafer stack. Soft plates or sheets typically adapt better to non-uniformities of the wafer stack and thus distribute the pressure more evenly over the wafer stack.

Table 13.1 describes a typical process scheme for wafer bonding with an intermediate polymer adhesive. The process flow involves the use of bonding tool, and applies generally to bonding with thermoplastic polymer adhesives and with uncured (soft-baked) or to partially cured thermosetting polymer adhesives.

After wafer alignment and bonding, 3D integration is completed by wafer thinning and through silicon via (TSV) formation (also called, inter-wafer interconnect formation). Details of wafer thinning and TSV formation are described in Chapter 23.

13.4 Bonding Characterizations

The bond strength and the amount of void or defect formation at the bond interface in adhesive wafer bonding is influenced by the polymer adhesive, wafer materials, size and amount of particles at the wafer surfaces, wafer surface topography, polymer thickness, bonding pressure, degree of curing (level of polymerization) of the polymer adhesive, wafer thickness, polymer curing conditions and atmospheric

Table 13.1 Typical process steps for adhesive wafer bonding [1–3].

No.	Process step	Purpose of the process step
1	Cleaning and drying the wafers	Remove particles, contaminations and moisture from the wafer surfaces
2	Treating the wafer surfaces with an adhesion promoter (optional)	Adhesion promoters can enhance the adhesion between the wafer surfaces and the polymer adhesive
3	Applying the polymer adhesive to the surface of one or both wafers; patterning the polymer adhesive (optional)	The most commonly used application method is spin coating, with spray coating and vapor deposition as alternatives
4	Soft-baking or partially curing of the polymer	Solvents and volatile substances are removed from the polymer coating. Thermosetting adhesives may only be partially polymerized. Thermoplastic adhesives may be completely polymerized
5	Aligning wafers (optional)[a]	Two wafers are aligned using alignment marks on both wafers in a wafer aligner, and then clamped on the bottom bond chuck
6	Placing the aligned wafers (on the bottom chuck) in the bond chamber, establishing a vacuum atmosphere	If wafers are not aligned, the wafers can be placed on a bottom chuck in the bonder. Caution: to prevent voids, any trapped gasses at the bond interface should be pumped away before the bond is initiated
7	Applying pressure to the wafer stack with the bond tool	The wafer and polymer adhesive surfaces are forced into intimate contact over the entire wafer. The bonding pressure may be applied before or after the bonding temperature is reached
8	Ramping up to bonding temperature to remelt or cure the polymer adhesive while applying pressure	The hardening procedure depends on the curing mechanism of the used polymer adhesive and may take a few minutes to hours. The re-flow of the polymer adhesive is typically triggered through elevated temperature
9	Chamber cool-down, purge, and bond pressure release	To solidify the polymer adhesive, the bond pressure is not released until cooling down to certain temperature

[a] Wafer-to-wafer alignment prior to wafer bonding is a critical step in most of the 3D integration process flow. The alignment marks are usually metal patterns produced along with the top metal interconnects of the processed wafers that are to be aligned and bonded. The alignment marks can also be patterns of other kinds of materials (e.g. silicon oxide) or part of the patterned structures of the bonding polymer adhesive if the polymer adhesive is patterned prior to wafer bonding.

conditions in the bond chamber before wafer bonding. Various measures are taken to increase the bond strength and prevent voids and defects in the bonds. Examples of bonding characterizations are presented below, showing the bonding integrity and robustness.

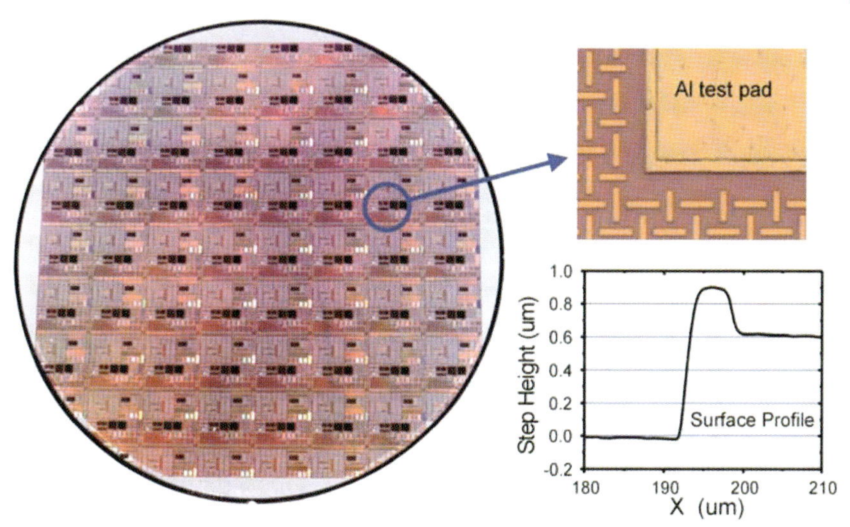

Figure 13.5 Semi-transparent Cu/oxide interconnect wafer bonded to a glass wafer using BCB after removal of Si substrate on a SEMATECH wafer (see text) [3, 6]. Though the surface profile (right) shows a step height of ~900 nm across the Al pads, void-free bonding is obtained and maintained after wafer thinning.

13.4.1 Optical Inspection Using Glass Wafer

Bonding defects and voids can be checked by optical inspection if wafers are bonded to a glass wafer with a coefficient of thermal expansion (CTE) that is well matched with that of silicon wafers (Figures 13.2 and 13.3a). The defect-free bonding of 200 mm damascene patterned wafers (Figure 13.3a) indicates that such wafer level non-planarity can be easily accommodated by a 2.6 μm thick BCB adhesive layer. In fact, we have regularly bonded wafers with relatively large topological features. Figure 13.5 shows a two-level copper interconnect test structure (provided by SEMATECH) bonded to a CTE-matched glass wafer. Though the surface profile shows a step height of ~900 nm across the Al pads on the processed wafers, void-free bonding is obtained and maintained after wafer thinning. The interconnect structures on such wafers are damage-free after the wafer bonding and thinning processes. In fact, the backside of the wafer with copper interconnect test structure has been ground and polished completely away (to the silicon-oxide interface) without adhesion failure of the BCB [3, 6].

13.4.2 Bonding Strength Characterization Using Four-Point Bending

In addition to having defect/void free bonding interfaces, it is critical to obtain high wafer bonding strength (critical adhesion energy). Four-point bending tests can be

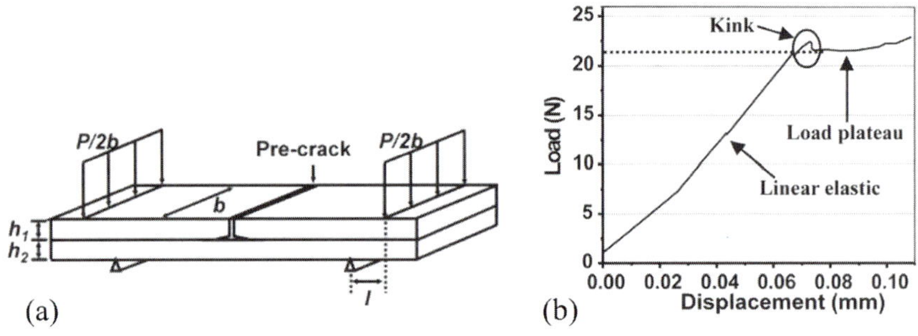

Figure 13.6 (a) Beam specimen geometry for four-point bending experiments. (b) Typical four-point bending result; a load–displacement curve [7, 8].

used to quantify bond strength and to identify a weak bond interface [7, 8]. This technique was developed to measure the critical adhesion energy at the interface in a beam specimen by analyzing a load–displacement curve. The beam specimen is mounted in a special fixture (Figure 13.6a) [7, 8]. In this fixture, a load cell measures applied load, and an actuator measures displacement of the "pins" that hold the specimen. The experiment essentially consists of measuring the load required to maintain a constant displacement rate. During displacement, the pre-crack propagates vertically to the weak interface, after which the crack proceeds along that interface. Figure 13.6b shows a typical load vs. displacement curve [7, 8].

Figure 13.7 shows the critical adhesion energy determined using four-point bending for different wafers bonded using BCB [6]. The critical adhesion energy of 32 J m^{-2} measured for BCB bonding of oxidized silicon wafers is well above the value of 22 J m^{-2} for BCB bonding of an oxidized silicon wafer to a wafer with Cu

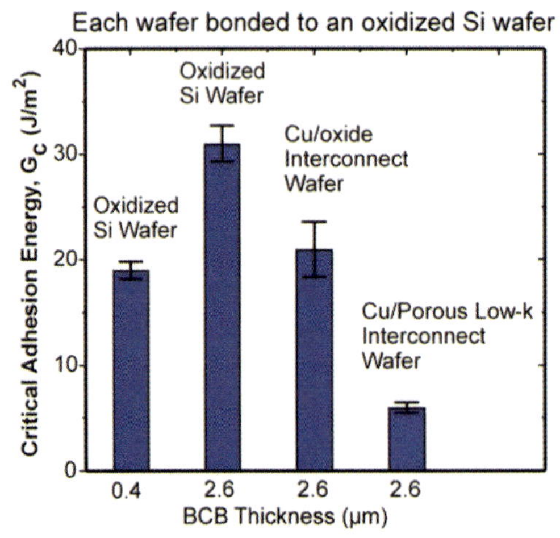

Figure 13.7 Bonding strength (critical adhesion energy) for oxide surface, Cu/oxide and Cu/low-k interconnect, showing the strength of BCB bonding [6].

interconnect structures with SiO_2 dielectric, and much higher than the value of $6\,J\,m^{-2}$ for BCB bonding of an oxidized silicon wafer to a wafer with Cu interconnect structures with JSR porous low-k dielectric [6]. Direct observations indicate that failure occurs within the interconnect test structure with either oxide or JSR porous low-k dielectrics, and not at either bonding interface. Even with a BCB bond layer as thin as 0.4 μm, the critical adhesion energy of $19\,J\,m^{-2}$ is sufficiently strong, that is much stronger than that of Cu/porous low-k interconnect structures.

13.4.3
Adhesive Wafer Bonding Integrity

The high critical adhesion energy of wafers bonded using BCB (Figure 13.7) is maintained after standard IC package reliability tests, such as die-level autoclave and liquid-to-liquid thermal shock (LLTS) [9]. The autoclave tests were conducted at conditions of 100% humidity, 2 atmospheres, and 120 °C for 48 or 144 hours. The LLTS tests were conducted between − 50 and 125 °C for 1000 cycles. In addition, the thermal cycling tests indicate that the wafer bonding using BCB is stable up to at least 400 °C [7].

While BCB as a dielectric bonding adhesive provides a defect-free bonding interface and sufficient bond strength with high temperature stability and packaging reliability as well as insensitivity to surface conditions (particles, roughness and planarity), it is also critical to evaluate bonding and thinning impacts on the electrical performance of processed wafers. The impact of these processes was evaluated with the double bonding and thinning procedure [6, 10]. Figure 13.8 shows a typical result of ring oscillator delay before and after double bonding/thinning and BCB ashing on wafers that have state-of-the-art 130 nm technology CMOS SOI test structures with four-level copper/low-k (organosilicate glass) interconnects and aluminium bond pads. Note that the silicon substrate of the CMOS SOI wafer was completely removed during the double bonding/thinning process; only the transistors, circuits and interconnects on the SOI layer with the buried oxide (BOX) layer are bonded on

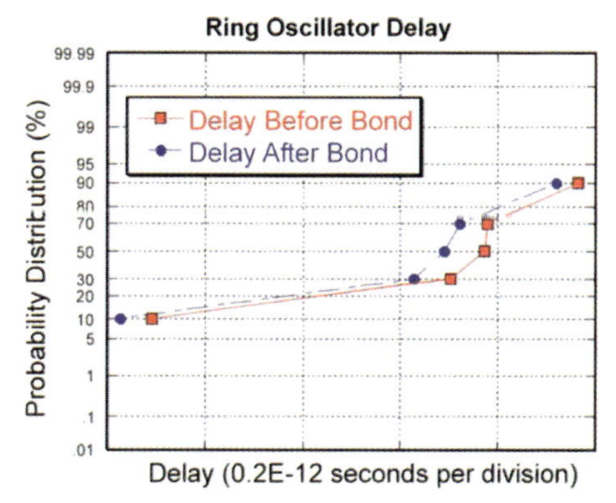

Figure 13.8 Ring oscillator delay before and after double bonding/thinning [10].

another silicon wafer by BCB (see Figure 23.8 in Chapter 23). All changes, including other tested parameters not shown in the figure, are less than one-third of the 10–90% spread in the distribution of original parameter values, indicating that the double bonding and thinning process flow has not significantly affected 130 nm technology CMOS device parameters [10].

13.5 Conclusions

Adhesive wafer bonding is a generic CMOS compatible technology that provides unique possibilities for fabrication and improvement of 3D integrated microsystems. Process schemes and parameters for adhesive wafer bonding with self-contained intermediate polymer films and with photolithographically patterned polymer adhesives (not covered here) are readily available in the literature. The main advantages of adhesive wafer bonding are the insensitivity to surface topography, the low bonding temperatures, the compatibility with standard integrated circuit wafer processing and the ability to join practically any kind of wafers. Adhesive wafer bonding requires no special wafer surface treatments such as planarization. Structures and particles at the wafer surfaces can be tolerated and accommodated to some extent by the polymer adhesive. In particular, BCB as a dielectric bonding adhesive provides a defect-free bonding interface, sufficient bond strength with high temperature stability and packaging reliability, and robust bonding and thinning integrity with minimum impacts on the electrical performance of processed wafers.

Acknowledgments

Rensselaer 3D integration research programs were supported by DARPA, MARCO, and NYSTAR through the Interconnect Focus Center (IFC). We gratefully acknowledge the contributions from many colleagues associated with the 3D group at Rensselaer, and the support of SEMATECH and Freescale Semiconductor for providing the processed wafers and electrical test results, as well as helpful discussions.

References

1 Niklaus, F., Stemme, G., Lu, J.-Q. and Gutmann, R. (2006) Adhesive wafer bonding. *Journal of Applied Physics (Applied Physics Review – Focused Review)*, **99** (3) 031101.

2 Lü, J.-Q., Kwon, Y., Kraft, R.P. *et al.* (2001) Stacked chip-to-chip interconnections using wafer bonding technology with dielectric bonding glues. 2001 IEEE International Interconnect Technology Conference (IITC), IEEE, pp. 219–221.

3 Lu, J.-Q., Lee, K.W., Kwon, Y. *et al.* (2003) Processing of inter-wafer vertical interconnects in 3D ICs, Advanced

Metallization Conference 2002 (AMC 2002), MRS Proceedings, (eds B.M. Melnick, T.S. Cale, S. Zaima and T. Ohta), Material Research Society Pittsburgh, 18, pp. 45–51.

4 Burns, J., McIlrath, L., Keast, C., *et al.* (5–7 Feb. 2001) Three-dimensional integrated circuits for low-power, high-bandwidth systems on a chip. 2001 IEEE International Solid-State Circuits Conference (ISSCC 2001), IEEE, pp. 268–269.

5 Ramm, P., Bonfert, D., Ecke, R. *et al.* (2002) Interchip viatechnology by using copper for vertical system integration. Advanced Metallization Conference 2001 (AMC 2001) (eds A.J. McKerrow, Y. Shacham-Diamand, S. Zaima and T. Ohba), Material Research Society Pittsburgh pp. 151–157.

6 Lu, J.-Q., Jindal, A., Kwon, Y. *et al.* (2003) Evaluation procedures for wafer bonding and thinning of interconnect test structures for 3D ICs. 2003 IEEE International Interconnect Technology Conference (IITC), June 2003, IEEE, pp. 74–76.

7 Kwon, Y., Seok, J., Lu, J.-Q. *et al.* (March 2005) Thermal cycling effects on critical adhesion energy and residual stress in benzocyclobutene (BCB)-bonded wafers. *Journal of The Electrochemical Society*, **152** (4), G286–G294.

8 Kwon, Y., Seok, J., Lu, J.-Q. *et al.* (Feb. 2006) Critical adhesion energy of benzocyclobutene (BCB)-bonded wafers. *Journal of The Electrochemical Society*, **153** (4), G347–G352.

9 Pozder, S., Lu, J.-Q., Kwon, Y., *et al.* (June 2004) Back-end compatibility of bonding and thinning processes for a wafer-level 3D interconnect technology platform. 2004 IEEE International Interconnect Technology Conference (IITC04), IEEE, pp. 102–104.

10 Gutmann, R.J., Lu, J.-Q., Pozder, S. *et al.* (2003) A wafer-level 3D IC technology platform. Advanced Metallization Conference in 2003 (AMC 2003) (eds G.W. Ray, T. Smy, T. Ohta and M. Tsujimura,), MRS Proceedings, Material Research Society, Pittsburgh, pp. 19–26,

14
Bonding with Intermetallic Compounds

Armin Klumpp

14.1
Introduction

To stack silicon layers in an economic way, the necessary process steps should be modular. This means that for every layer to be added the sequence of steps should be the same, without any influence from either the layer stack already built or the stack already formed. One possible influence while stacking could be the active shifting of underlying layers by re-melting the metal connections when placing the next layer on top. This shift can result in short cuts of electrical contacts at low pitch designs. Therefore these metal bonds should not re-melt after stacking. This can be realized by the use of high melting intermetallic phases. The process utilizes the diffusion of solid metal into the liquid phase of a lower melting metal and was already reported in 1966 by Bernstein *et al.* for the applications for bonding processes in integrated-circuit fabrication [1, 2]. His assigned acronym SLID (solid-liquid-interdiffusion) stands for this type of metallic bonding.

It is important not to confuse SLID with SOLID. Infineon and Fraunhofer IZM have developed within a common project the so-called SOLID technology for a face-to-face approach of die to wafer bonding using very thin Cu/Sn-Cu SLID layers [3]. The SLID process was applied for mechanical and electrical interconnects in combination with a through-silicon-via technology at the Fraunhofer IZM and introduced as ICV (Inter-Chip-Via) technology as a fully modular concept for vertical system integration optimized for chip-to-wafer stacking [4, 5].

14.2
Technological Concepts

Owing to the different thermal expansion coefficients of the devices and the metallic bond system, thermally induced stress is created, depending on the formation temperature. To lower the influence of this possible cause for reduced reliability,

Handbook of 3D Integration: Technology and Applications of 3D Integrated Circuits.
Edited by Philip Garrou, Christopher Bower and Peter Ramm

ISBN: 978-3-527-32034-9

the melting point of the lower melting metal should be as low as possible. In contrast, the reliability of the metallic bond is increased by high re-melting temperatures, as mechanical creeping is related to this re-melting point during temperature cycling. As a consequence the ideal metallic bonding system consists of a very low melting metal that vanishes completely by dissolving the high melting partner, resulting in a high melting alloy, the intermetallic phase.

14.2.1 Basic Material Selection

Some metal fulfill the characteristic "low melting." Usually, low melting is oriented on the stability of aluminium within a device, meaning a maximum temperature of 400 °C. Lead (Pb), bismuth (Bi), tin (Sn) and indium (In) are candidates, but lead is no longer accepted due to its toxicity when incorporated as metal or compound. Table 14.1 shows the melting points of these elements in descending order.

Often used refractory metals in the backend of device technology and their melting points are listed in Table 14.2. Nickel and gold are used, for example, for wire bond pads and copper as multilayer metallization in the device itself. Silver as pure metal sometimes is used as a thin capping layer to prevent the oxidation of copper.

Table 14.3 gives a selection of eutectic mixtures and the resulting melting temperatures. These material compositions can be purchased from different suppliers and show a wide selection of melting points.

In contrast to the available eutectic alloys, intermetallic compounds are formed by at least one low melting metal in contact with a high melting metal at elevated temperatures. The temperature preferably exceeds the melting point of the lower

Table 14.1 Melting points of low melting metals.

Metal	Melting point (°C)
Pb	327.5
Bi	271.3
Sn	231.9
In	156.6
Ga	29.8

Table 14.2 Melting points of refractive metals.

Metal	Melting point (°C)
Ni	1453.0
Au	1064.4
Cu	1083.0
Ag	961.9

Table 14.3 Melting points of eutectic alloys and their composition (wt%).

Eutectic alloy	Melting point (°C)
Au/Sn (80/20)	278
Ag/Sn	225
Ag/In (80/20)	206 [6]
Ag/In/Sn (3/20/77)	190
Bi/Sn (58/42)	139

Table 14.4 Melting points of intermetallic compounds.

Intermetallic compounds	Melting point (°C)
$AgIn_2$, Ag_2In	765–780 [6]
Ni_3Sn_2	1264 [7]
Ni_3Sn	1174 [7]
Cu_3Sn	676 [7]
Cu_6Sn_5	415 [7]
$AuIn_2$	454 [8]

melting metal to increase diffusion speed of the elements. Intermetallic compounds from the most promising combination of silver, copper, indium and tin are given in Table 14.4.

14.2.2 Principal Processing Scheme

To avoid polymeric underfillers, electrical contacts and mechanical joints can be formed at the same process steps. A temperature increase above the melting point of the lower melting metal and contacting opposite sides leads to symmetric interdiffusion and intermetallic compound formation. This is depicted schematically in Figure 14.1 for a binary metal system. The higher melting part of the metal layers is consumed by interdiffusion into the liquid. The ratio of this metal and the low melting metal should be chosen high enough to leave a remaining amount above the adhesion layers. Consuming all of the higher melting metal could result in adherence problems on the adhesion layer.

Figure 14.2 shows a top view of a daisy chain layout, illustrating electrical contacts (squares) next to mechanical joint areas (rectangles and frame), patterned at the same process step. The surface is smooth, for these pattern consist of copper only.

This idealized view of bonding results varies in appearance depending on the selected metal combinations and their reaction kinetics. Depending on the bonding temperature the intermetallic compounds can form needles or boulders, which are either crystalline or amorphous in appearance. The interface between remaining refractory metal and intermetallic compound does not remain smooth in general.

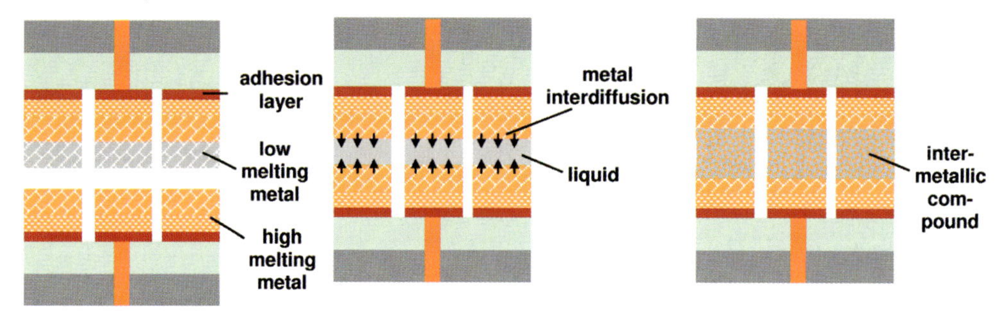

Figure 14.1 Schematics of a binary metal system. Electrical contacts with TSV and mechanical joints are formed at the same process steps. A temperature increase above the melting point of a lower melting metal and contact of opposite sides leads to symmetric interdiffusion and intermetallic compound formation. (Source: Fraunhofer IZMM.)

The formation of intermetallic compounds is, roughly, divided into two periods of time and reaction speeds. In the first period, liquid metal is in contact with the refractory metal. Dissolution in the liquid and diffusion is very fast. At seed points on the surface of the refractory metal the intermetallic compound with the highest formation energy is deposited. The consumed metal is further supplied by dissolution in the liquid. The deposited areas grow until the liquid phase is cut off from the refractory metal surface. From this point in time the second period of the bonding process begins. The reaction speed is dominated by diffusion of refractory metal through the intermetal compound of period one. This, of course, is more time consuming and takes longer until the lower melting metal is completely consumed. The liquid phase solidifies when the concentration of refractory metal shifts the melting point above the actual processing temperature. On further heating, this solidified phase transforms into the energetically next favored intermetallic compound. An example is shown in Figure 14.3. Here we have a cross section of a three-

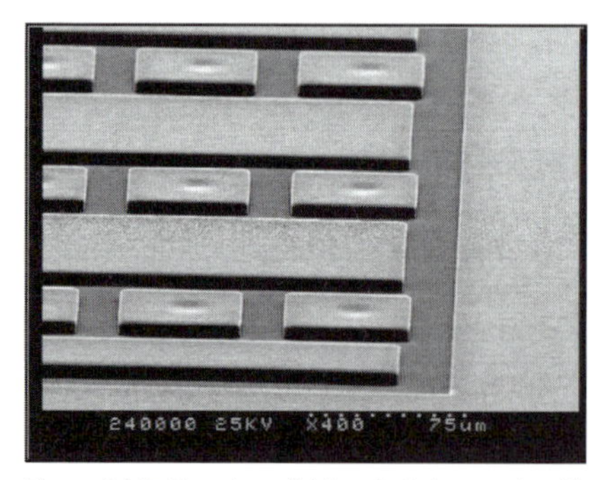

Figure 14.2 Top view of daisy chain layout that illustrates electrical contacts (squares) neighbored to mechanical joint areas (rectangles and frame), patterned at the same process step. (Source Fraunhofer IZMM.)

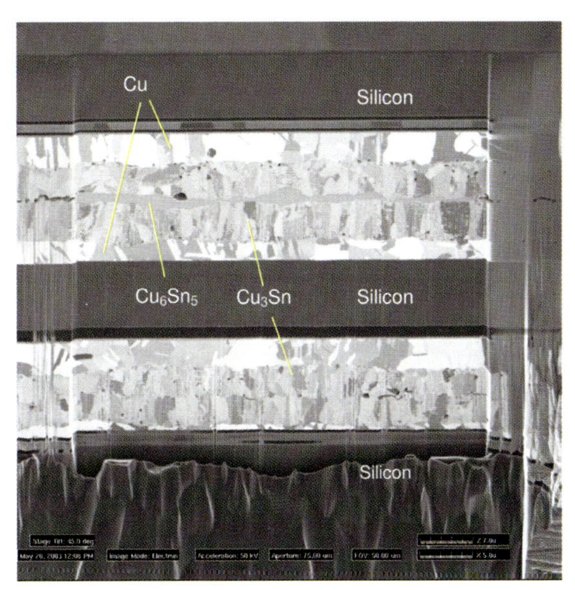

Figure 14.3 FIB-cross-section of a three-level silicon stack with two bond layers from Cu/Sn/Cu. The first layer (lower part) was temperature cycled twice, while the second layer (upper part) was heated only once. The first layer shows only the final state of intermetallic Cu_3Sn in contact with copper interfaces. The second layer has not completed its transformation; Cu_6Sn_5 is still sandwiched between Cu_3Sn. (Source Fraunhofer IZMM.)

level silicon stack with two bond layers from Cu/Sn/Cu. The first layer had undergone the process cycle of its own and the cycle of the second layer, while the second layer was heated only once. Therefore, in the first layer only the final state of intermetallic compound formation is visible. For Cu/Sn this means Cu_3Sn in contact with copper interfaces. In the second layer this transformation is not completed. Here, Cu_6Sn_5 is sandwiched between Cu_3Sn and only further heating will complete the formation.

Process temperature influences the appearance of the intermetallic compounds. In the presented example the temperature was 70 °C above the melting point of tin, which means a 300 °C peak value of a ramped process cycle. Cu_3Sn in this case is crystalline and shows grain growth perpendicular to the refractory metal copper.

14.2.3 Limiting Conditions for Applications

To reduce the process time (formation of a stable compound) the layer of the low melting metal should be as thin as possible, but topographic variations of the device can be compensated only within the range of this layer thickness. Therefore a suitable planarization of the device is mandatory.

To build electrical contacts and areas for mechanical joints, the selected metal combinations need to be patternable. Electroplating is an easy means to build pattern consisting of thin layer sequences in the cases of copper, nickel and tin. Indium is also reported with electroplating deposition but is mainly deposited with high vacuum evaporation. An example of preparing surfaces for low temperature soldering with indium is reported elsewhere [8]. Here multiple layers of Au and In are deposited on semiconductor wafers in one vacuum cycle to reduce In oxidation. The semiconductor dice are then bonded to substrates coated with Au. Above 157 °C, the indium layer melts and dissolves the Au layers to form a liquid–solid mixture. The diffusion process continues until the bond solidifies. Upon solidification, the bond has a melting temperature of 454 °C. However, the means for patterning are not described. A lift-off technique with resist patterns could be one choice.

Oxides on metal surfaces can prevent electrical contact between the opposite parts of a bondpad. The liquid metal pastes perfectly to the oxide surface, so good adhesion of the chips is observed, but as mentioned there is no electrical conducting path. This is clearly depicted in Figure 14.4, where the thin oxide layer on copper separates the lower part of the bond (converted into Cu_3Sn) from the untouched copper layer above. The oxide layer prevents copper from symmetrical diffusion from both sides and so nearly all the copper from the lower part is consumed in the intermetallic compound.

Oxidation of surfaces is sometimes prevented by deposition of noble metal layers on top of base metals. This leads to ternary metal systems with more complex phase diagrams.

The molecular volume of the intermetallic compounds is smaller than the sum of the volume of the metals before compound formation. This results in shrinkage of

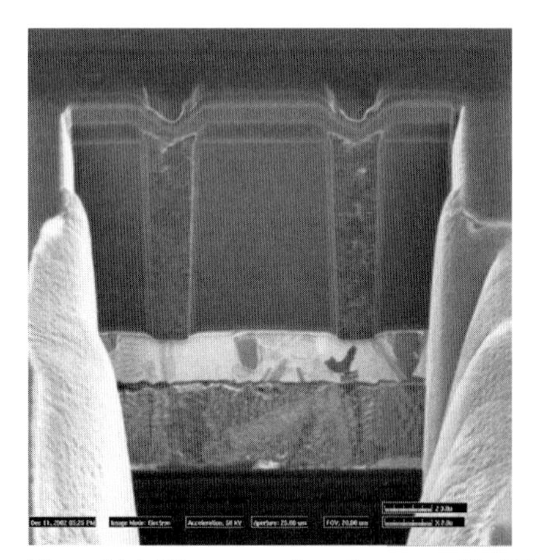

Figure 14.4 FIB-cross-section of tungsten-filled Through Silicon Vias (TSV) and SLID bonding metal combination of Cu/Sn/Cu. The upper part of the metal system is separated from the lower part by copper oxide, preventing symmetrical copper diffusion from both sides. (Source Fraunhofer IZMM.)

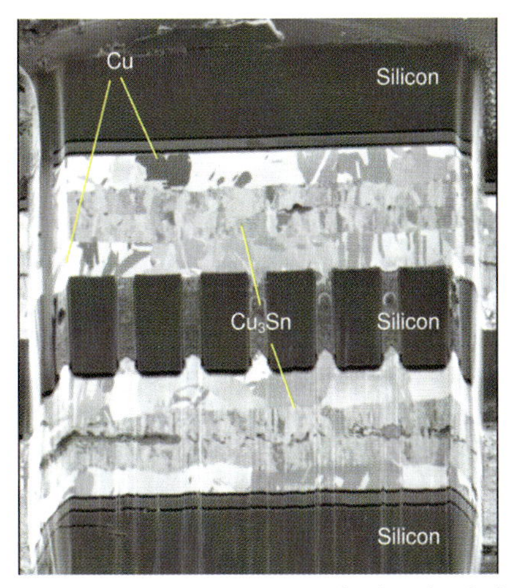

Figure 14.5 FIB-cross-section of a three-level silicon stack with two bond layers from Cu/Sn/Cu including TSV. The first layer (lower part of the picture) is placed on a topography level 300 nm lower than the surrounding chip level; the second layer (upper part) levels up with the chip. Voids and crevice are formed due to uncompensated shrinkage of the intermetallic compound Cu_3Sn. (Source Fraunhofer IZMM.)

the total bond layer thickness. Applying pressure is recommended to make sure the interfaces do not separate during the bonding process. But this is only valid if the shrinkage is equal all over the surface (that can mean the total wafer area). As soon as there are problems with topography variations, pressure does not help. Pressure mainly overcomes the bow of two substrates when bonding large areas.

Figure 14.5 gives an example, where the topography level of the bond area within the chip is about 300 nm lower than shown in Figure 14.3. Again this is the Cu/Sn/Cu metal system. Upon complete transformation of tin into the intermetallic compound the volume of the metal has shrunken. Although the layer of tin was initially 3 μm thick, it is not possible to compensate the 0.3 μm deep indentation of the chip layout at this location. Liquid tin has wetted both surfaces, as can be seen from the symmetrical formation of Cu_3Sn. Pressure does not improve the situation, because the chip has already mechanically contacted higher parts within the layout and can not be compressed further. As a consequence of the volume shrinkage of the compound, voids and crevices are formed.

The roughness of deposited layers influences the quality of bonds and process control when joining flat surfaces. To avoid inclusions the atmosphere must be controlled. Vacuum is the best choice followed by inert gases. If roughness comes from the low melting partner of the metal system, melting this layer before contacting with the opposite surface eases this problem. Figure 14.6 gives an example of a tin surface, deposited by patterned electroplating. The grains are very large, and so is the roughness.

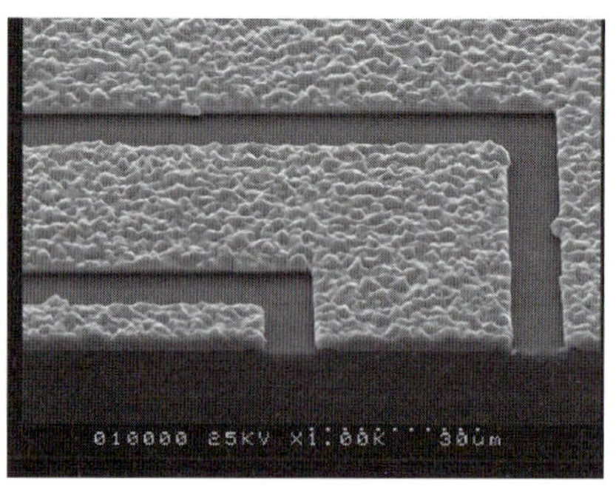

Figure 14.6 SEM of tin surface, deposited by patterned electroplating. The very large roughness is due to grain growth. (Source Fraunhofer IZMM.)

Correlated with the quality of the bond is, of course, its reliability. Although intermetallic compounds are more brittle, less ductile than the original metals, the failure of a bond when mechanically stressed lies not within the intermetallic compound. Instead, cracks propagate along the interfaces between either refractory metal or compound, between different compound phases or at the interfaces to the chip. The most critical case is the formation of voids or even cracks during bond formation due to volume loss as described above. These interfaces are primarily due to failures.

Figure 14.7 depicts details of bond pads with a good and a bad interface. Without voids due to proper intermetallic compound formation the chip tears off along the adhesion layer or even within the silicon of the device itself. In the presence of voids, due to poor contact during bonding (e.g. different topographic level) cracks propagate preferentially within these interfaces. Table 14.5 summarizes limiting conditions and measures to overcome them.

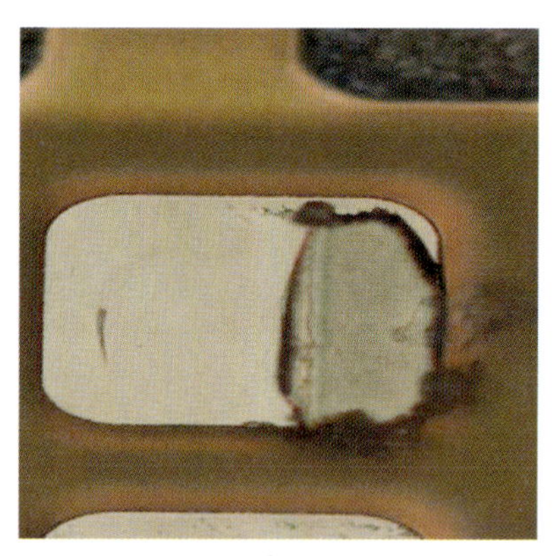

Figure 14.7 Details of bond pads with a good and a bad interface after shear tests. Without voids the chip tears off along the adhesion layer, or even within the silicon of the device itself (left-hand side). In the presence of voids, the chip separates at this interface (right-hand side). Due to roughness the surface appears black.

Table 14.5 Limiting conditions for applications and measures to overcome them.

Condition	Measure
Process time	Thin layer of liquid phase
Topography of chip or wafer	Planarization, pressure, chip layout
Patterning of selected metal combinations	Fit of design rules for pattern size
Oxidation of base metal surface	Pre-cleaning or noble metal deposition
Trapping of gases	Controlled atmosphere

14.3 Conclusion

Bonding with intermetallic compounds is a wide but promising field for successful stacking of multilayer devices. The metal system of copper and tin has already been under investigation for some time. Process flow and limitations are known, reliability values are under investigation. If process temperatures of 230–300 °C are still too high for some applications, alternative systems utilizing indium would be possible, giving access to temperature ranges starting from 158 °C.

References

1 Bernstein, L. and Bartolomew, H. (1966) Application of solid-liquid-interdiffusion (SLID) bonding in integrated circuit fabrication. *Transactions of the Metallurgical Society of AIME*, **236**, 405.

2 Bernstein, L. (December 1966) Semiconductor joining by SLID process, I. The systems Ag-In, Au-In and Cu-In. *Journal of the Electrochemical Society*, **113** (12), 1282–1288.

3 Huebner, H., Ehrmann, O., Eigner, M. *et al.* (2002) Face-to-face chip integration with full metal interface. Proceedings Advanced Metallization Conference (AMC 2002), Material Research Society Proceedings (eds B.M. Melnick *et al.*) Material Research Society, Pittsburgh, 18, p. 53.

4 Ramm, P., Klumpp, A. and Wieland, R. (2002) 3D-integration of integrated circuits by interchip vias and Cu/Sn solid-liquid-interdiffusion. Proceedings 6th VLSI Packaging Workshop of Japan, Kyoto, pp. 445–454.

5 Ramm, P., Klumpp, A., Merkel, R. *et al.* (2003) 3D system integration technologies. Proceedings Materials Research Society Spring Meeting, San Francisco (eds A.J.M. McKerrow, J. Leu, O. Kraft and T. Kikkawa), Material Research Society Proceedings, **766**, Warrendale, pp. 3–14.

6 Chuang, R.W. and Lee, C.C. (September 2002) Silver-indium joints produced at low temperature for high temperature devices. *IEEE Transactions on Components and Packaging Technologies*, **25** (3), 453.

7 Massalski, T.B. *et al.* (1990) *Binary Alloy Phase Diagrams*, ASM International, Materials Park, Ohio.

8 Lee, C.C., Wang, C.Y. and Matijasevic, G. (May 1993) Au-In bonding below the eutectic temperature. *IEEE Transactions on Components, Hybrids, and Manufacturing Technology*, **16** (3), 311–316, 29 references.

Handbook of 3D Integration

Edited by
Philip Garrou, Christopher Bower
and Peter Ramm

Further Titles of Interest

Ban P. Wong

Nano-CMOS Design for Manufacturability: Robust Circuit and Physical Design for Technology Nodes

2008

ISBN: 978-0-470-11280-9

Volker Saile, Ulrike Wallrabe, Osamu Tabata, Jan G. Korvink (Eds.)

LIGA and its Applications

Advanced Micro & Nanosystems Volume 7

2008

ISBN: 978-3-527-31698-4

Osamu Tabata, Toshiyuki Tsuchiya (Eds.)

Reliability of MEMS

Advanced Micro & Nanosystems Volume 6

2008

ISBN: 978-3-527-31494-2

Georges Hadziioannou, George G. Malliaras (Eds.)

Semiconducting Polymers

Chemistry, Physics and Engineering (2 Volumes)

2nd edition, 2007

ISBN: 978-3-527-31271-9

Volkan Kursun, Eby G. Friedman

Multi-voltage CMOS Circuit Design

2006

ISBN: 978-0-470-01023-5

Hagen Klauk (Ed.)

Organic Electronics

Materials, Manufacturing and Applications

2006

ISBN: 978-3-527-31264-1

Handbook of 3D Integration

Technology and Applications of 3D Integrated Circuits

Volume 2

Edited by
Philip Garrou, Christopher Bower and Peter Ramm

WILEY-VCH Verlag GmbH & Co. KGaA

The Editors

Dr. Philip Garrou
Microelectronic Consultants of North Carolina
3021 Cornwallis Road
Research Triangle Park, NC 27709-2889
USA

Dr. Christopher Bower
Semprius, Inc.
2530 Meridian Parkway
Durham, NC 27713
USA

Dr. Peter Ramm
Fraunhofer IZM
Hansastraße 27d
80686 München
Germany

Cover Description:
The artistic illustration explains the principle of 3D integration: Stacking and vertical interconnecting of device layers. The magnified portion depicts a cross section of a 3D integrated test chip which was fabricated at Fraunhofer Munich. The wafer-level 3D integration technology is based on bonding of thinned device substrates and a metallization process providing a high vertical wiring density. This technology presents a solution to performance and reliability limitations due to the wiring of microelectronic devices and enables completely new ultra-miniaturized products.

Picture: © Fraunhofer IZM, Munich

Library of Congress Card No.: applied for

British Library Cataloguing-in-Publication Data
A catalogue record for this book is available from the British Library.

Bibliographic information published by the Deutsche Nationalbibliothek
Die Deutsche Nationalbibliothek lists this publication in the Deutsche Nationalbibliografie; detailed bibliographic data are available in the Internet at http://dnb.d-nb.de.

Typesetting Thomson Digital, Noida, India
Printing Strauss GmbH, Mörlenbach
Binding Litges & Dopf Buchbinderei GmbH, Heppenheim

Printed in the Federal Republic of Germany
Printed on acid-free paper

ISBN: 978-3-527-32034-9

Contents

Volume 1

Handbook of 3D Integration: Technology and Applications of 3D Integrated Circuits.
Edited by Philip Garrou, Christopher Bower and Peter Ramm

ISBN: 978-3-527-32034-9

Preface

Many of readers of this book will doubtless see the title "Handbook of 3D Integration" and expect the two volumes to cover all manner of devices and packages that can be argued to be 3D. The editors are aware that stacked packages and stacked dies without through semiconductor vias (TSVs) are 3D structures, but we contend that these structures can ultimately be catalogued under "3D Packaging". The focus of this book is the technology and applications of "3D Integration," which we classify as the vertical integration of thinned and bonded silicon integrated circuits with vertical electrical interconnects between the IC layers. Most of the vertical interconnects discussed here will be through silicon vias (TSVs).

The editors are aware that the terminology "3D Integration" lacks specificity. Why not a handbook of "3D ICs" or "3D Silicon Integration" or "Vertical System Integration"? While each of these has pros and cons, we feel that the simple term "3D Integration" has already been accepted by many researchers, and it is very unlikely that new terminology will take hold.

These volumes intend to provide engineers and scientists with a timely and fairly comprehensive overview of the field. Although our goal was to be as complete as possible, there will clearly be some technical areas that are not covered. We do not cover monolithic growth approaches to 3D integration and we do not cover 3D integration of heterogeneous materials. Perhaps, as they mature, these technologies will be included in a future edition. The book is organized into five parts:

- Part I covers the processing technology for TSVs. This section includes chapters on deep reactive ion etching of TSVs, laser-drilled TSVs, TSV sidewall insulation, Cu electroplating and chemical vapor deposition of Cu and W.
- Part II covers wafer thinning and bonding technology. Included are chapters on thinning and singulation of silicon wafers, techniques and tools for wafer alignment and bonding, polymer bonding and intermetallic bonding.
- Part III covers the various integration processes being pursued across the globe. This section begins with a chapter that surveys commercial activity in 3D integration. Next, there is a grouping of chapters from universities and institutes that have distinct approaches to 3D integration. This group includes chapters from

Handbook of 3D Integration: Technology and Applications of 3D Integrated Circuits.
Edited by Philip Garrou, Christopher Bower and Peter Ramm

ISBN: 978-3-527-32034-9

Fraunhofer IZM, University of Arkansas, Japan's ASET consortium, CEA-LETI, MIT Lincoln Labs, IMEC, MIT and RPI. This part ends with a grouping of chapters from 3D integration start-up companies, including chapters from Tezzaron, Ziptronix and Zycube.

- Part IV covers design, performance and thermal management for 3D integration. Design for 3D is covered in chapters from NC State University, Fraunhofer IIS, Lincoln Labs and University of Minnesota. This part includes a chapter from Intel on testing of 3D circuits and a chapter from IBM on thermal management in 3D ICs.
- Part V contains chapters on specific applications of 3D integration. Individual chapter topics are on 3D microprocessors, 3D memory, sensor arrays, power devices and wireless sensor systems.

We would like to acknowledge all the authors of each chapter. It is the individual author contributions that made this work possible. We would also like to thank the authors for providing reviews of chapters. The editors are deeply grateful for the time and effort you each put into your chapters. We also wish to express our gratitude to those at Wiley-VCH who have been a great help in keeping us on schedule.

We hope that this book will serve as a valuable resource for practitioners of 3D integration. Based on the progress made over the last few years, we anticipate that the next decade will be a very exciting time to be working in this area.

November 2007

Phil Garrou, RTP, North Carolina
Chris Bower, RTP, North Carolina
Peter Ramm, Munich, Germany

List of Contributors

Sitaram Arkalgud
SEMATECH
2706 Montopolis Boulevard
Austin, TX 78741
USA

Brian Aull
Massachusetts Institute of Technology
Lincoln Laboratory
244 Wood Street
Lexington, MA 02420-9108
USA

Rozalia Beica
Semitool, Inc.
655 West Reserve Drive
Kalispell, MT 59901
USA

Robert Berger
Massachusetts Institute of Technology
Lincoln Laboratory
244 Wood Street
Lexington, MA 02420-9108
USA

Eric Beyne
IMEC
Kapeldreef 75
3001 Leuven
Belgium

Michiel A. Blauw
Eindhoven University of Technology
PO Box 513
5600 MB Eindhoven
The Netherlands

Christopher Bower
Semprius, Inc.
2530 Meridian Parkway
Durham, NC 27713
USA

Thomas Brunschwiler
IBM Zurich Research Laboratory
Advanced Thermal Packaging
Säumerstrasse 4
8803 Rüschlikon
Switzerland

Susan Burkett
University of Arkansas
Department of Electrical Engineering
3217 Bell Engineering Center
Fayetteville, AR 72701
USA

James Burns
Massachusetts Institute of Technology
Lincoln Laboratory
244 Wood Street
Lexington, MA 02420-9108
USA

Handbook of 3D Integration: Technology and Applications of 3D Integrated Circuits.
Edited by Philip Garrou, Christopher Bower and Peter Ramm

ISBN: 978-3-527-32034-9

Tim S. Cale
Rensselaer Polytechnic Institute
Mailstop CII-6015/CIE
110 8th Street
Troy, NY 12180-3590
USA

S. M. Chang
Industrial Technology Research
Institute of Taiwan
195 Chung Hsing Road
Chutung, Hsinchu
Taiwan 310, ROC

Barbara Charlet
CEA-LETI, MINATEC
Département Integration Hétérogene
Silicium
17, rue des Martyrs
38054 Grenoble Cedex 9
France

Nisha Checka
Massachusetts Institute of Technology
Lincoln Laboratory
244 Wood Street
Lexington, MA 02420-9108
USA

Chang-Lee Chen
Massachusetts Institute of Technology
Lincoln Laboratory
244 Wood Street
Lexington, MA 02420-9108
USA

Chenson Chen
Massachusetts Institute of Technology
Lincoln Laboratory
244 Wood Street
Lexington, MA 02420-9108
USA

Wouter Dekkers
NXP-TSMC Research Center
High Tech Campus 4
Mailbox WAG02
5656 AE Eindhoven
The Netherlands

Marc de Samber
Philips Applied Technologies
High Tech Campus 7
5656 AE Eindhoven
The Netherlands

Léa Di Cioccio
CEA-LETI, MINATEC
Département Integration Hétérogene
Silicium
17, rue des Martyrs
38054 Grenoble Cedex 9
France

R. Ecke
TU Chemnitz
Zentrum für Mikrotechnologien
Reichenhainer Straße 70
09126 Chemnitz
Germany

Günter Elst
Fraunhofer IIS
Design Automation Division
Zeunerstraße 38
01069 Dresden
Germany

Paul Enquist
Ziptronix
800 Perimeter Park, Suite B
Morrisville, NC 27560
USA

Andy Fan
Massachusetts Institute of Technology
Department of Electrical Engineering
77 Massachusetts Avenue
Cambridge, MA 02139
USA

Paul D. Franzon
North Carolina State University
Monteith GRC 443
ECE, Box 7914
Raleigh, NC 27695
USA

Philip Garrou
Microelectronic Consultants of North Carolina
3021 Cornwallis Road
Research Triangle Park, NC 27709-2889
USA

Pascale Gouker
Massachusetts Institute of Technology
Lincoln Laboratory
244 Wood Street
Lexington, MA 02420-9108
USA

Ronald J. Gutmann
Rensselaer Polytechnic Institute
Mailstop CII-6015/CIE
110 8th Street
Troy, NY 12180-3590
USA

David Henry
CEA-LETI, MINATEC
Département Intégration Hétérogene Silicium
17, rue des Martyrs
38054 Grenoble Cedex 9
France

Arne Heittmann
Qimonda AG
Gustav-Heinemann-Ring 212
81739 Munich
Germany

Thomas Herndl
Infineon Technologies
Operngasse 20b/32
1010 Vienna
Austria

David Heyes
NXP Semiconductors
Bramhall Moove Lane
Stockpat, Cheshire SK7 5B
UK

Thierry HILT
CEA-LETI
17, avenue des Martyrs
38054 Grenoble Cedex
France

Adrian Ionescu
Ecole Polytechnique Fédérale de Lausanne
Institute of Microelectronics and Microsystems
Electronics Laboratory
1015 Lausanne
Switzerland

Jean-Pierre Joly
CEA-LITEN, INES
Département des Technologies Solaires
50, avenue du Lac Léman
73377 Le Bourget du Lac
France

Craig Keast
Massachusetts Institute of Technology
Lincoln Laboratory
244 Wood Street
Lexington, MA 02420-9108
USA

Ervin (W. M. M.) Kessels
Eindhoven University of Technology
PO Box 513
5600 MB Eindhoven
The Netherlands

Armin Klumpp
Fraunhofer IZM
Hansastraße 27d
80686 Munich
Germany

Jeffrey Knecht
Massachusetts Institute of Technology
Lincoln Laboratory
244 Wood Street
Lexington, MA 02420-9108
USA

Werner Kröninger
Infineon Technologies AG
Postfach 10 09 44
93009 Regensburg
Germany

Yann Lamy
NXP-TSMC Research Center
High Tech Campus 4
Mailbox WAG02
5656 AE Eindhoven
The Netherlands

Patrick Leduc
CEA-LETI, MINATEC
Département Integration Hétérogene
Silicium
17, rue des Martyrs
38054 Grenoble Cedex 9
France

Paul Lindner
EV Group
Erich Thallner GmbH
DI Erich Thallner Straße 1
4782 St.Florian/Inn
Austria

W. C. Lo
Industrial Technology Research
Institute of Taiwan
195 Chung Hsing Road
Chutung, Hsinchu
Taiwan 310, ROC

James Jian-Qiang Lu
Rensselaer Polytechnic Institute
Mailstop CII-6015/CIE
110 8th Street
Troy, NY 12180-3590
USA

T. M. Mak
Intel Corporation
2200 Mission College Blvd., SC 12-604
Sauta Clara, CA 95052-8119
USA

Thorsten Matthias
EV Group
7700 South River Parkway
Tempe, AZ 85284
USA

Bruno Michel
IBM Zurich Research Laboratory
Advanced Thermal Packaging
Säumerstrasse 4
8803 Rüschlikon
Switzerland

Patrick Morrow
Intel Corporation
Mail Stop: RA3-252
5200 N.E. Elam Young Parkway
Hillsboro, OR 97124-6467
USA

Makoto Motoyoshi
ZyCube Co. Ltd.
ZyCube Sendai Lab.
519-1176 Aoba Aramaki, Aoba-ku,
Sendai-shi, Miyagi
985-0845 Japan

Sriram Muthukumar
Intel Corporation
Mail Stop: CH4-109
5000 W Chandler Blvd
Chandler, AZ 85226
USA

Pierre Nicole
THALES systèmes aéroportés
2 Avenue Gay Lussac
78851 Elancourt Cedex
France

Stefan Pargfrieder
EV Group
Erich Thallner GmbH
DI Erich Thallner Straße 1
4782 St. Florian/Inn
Austria

Robert Patti
Tezzaron Semiconductor Corp.
1415 Bond Street
Naperville, IL 60563
USA

Ulrich Ramacher
Infineon Technologies AG
Am Campeon 1-12
85579 Neubiberg
Germany

Peter Ramm
Fraunhofer IZM
Hansastraße 27d
80686 Munich
Germany

Rafael Reif
Massachusetts Institute of Technology
Department of Electrical Engineering
77 Massachusetts Avenue
Cambridge, MA 02139
USA

Thomas L. Ritzdorf
Semitool, Inc.
655 West Reserve Drive
Kalispell, MT 59901
USA

Fred Roozeboom
NXP-TSMC Research Center
High Tech Campus 4
Mailbox WAG02
5656 AE Eindhoven
The Netherlands

Mihai Sanduleanu
Philips Applied Technologies
High Technology Campus 7
5656 AE Eindhoven
The Netherlands

Sachin S. Sapatnekar
University of Minnesota
Department of Electrical
and Computer Engineering
200 Union Street
Minneapolis, MN 55455
USA

Anton Sauer
Fraunhofer IZM
Hansastraße 27d
80686 Munich
Germany

Leonard Schaper
University of Arkansas
Department of Electrical Engineering
3217 Bell Engineering Center
Fayetteville, AR 72701
USA

Peter Schneider
Fraunhofer IIS
Design Automation Division
Zeunerstraße 38
01069 Dresden
Germany

Stefan E. Schulz
TU Chemnitz
Zentrum für Mikrotechnologien
Reichenhainer Straße 70
09126 Chemnitz
Germany

Charles Sharbano
Semitool, Inc.
655 West Reserve Drive
Kalispell, MT 59901
USA

Herbert Shea
Ecole Polytechnique Fédérale de
Lausanne
Institute of Microelectronics and
Microsystems
Electronics Laboratory
1015 Lausanne
Switzerland

Antonio Soares
Massachusetts Institute of Technology
Lincoln Laboratory
244 Wood Street
Lexington, MA 02420-9108
USA

Vyshnavi Suntharalingam
Massachusetts Institute of Technology
Lincoln Laboratory
244 Wood Street
Lexington, MA 02420-9108
USA

Kenji Takahashi
Toshiba Corp.
1 Komukai Toshiba-cho, Saiwai-ku,
Kawasaki-shi, Kanagawa
212-8583 Japan

Maaike Taklo
SINTEF ICT
Microsystems and Nanotechnology
Gaustadalléen 23
0373 Oslo
Norway

Chuan Seng Tan
Nanyang Technological University
School of Electrical and Electronic
Engineering
50 Nanyang Avenue
Singapore 639798
Singapore

Kazumasa Tanida
Toshiba Corp.
1 Komu Kai Toshibacho, Saiwai-Ku
Kawasaki-shi, Kanagawa
212-8583 Japan

Mark E. Tuttle
Micron Technology, Inc.
Mail Stop 1-717
8000 S. Federal Way
Boise, ID 83707-0006
USA

Brian Tyrrell
Massachusetts Institute of Technology
Lincoln Laboratory
244 Wood Street
Lexington, MA 02420-9108
USA

Eric (F.) van den Heuvel
Philips Applied Technologies
High Technology Campus 7
5656 AE Eindhoven
The Netherlands

Emile van der Drift
Delft University of Technology
PO Box 5053
2600 GB Delft
The Netherlands

Richard (M. C. M.) van de Sanden
Eindhoven University of Technology
PO Box 513
5600 MB Eindhoven
The Netherlands

Eric van Grunsven
Philips Applied Technologies
High Tech Campus 7
5656 AE Eindhoven
The Netherlands

Co Van Veen
Philips Applied Technologies
High Technology Campus 7
5656 AE Eindhoven
The Netherlands

Jan F. Verhoeven
Philips Applied Technologies
High Technology Campus 7
5656 AE Eindhoven
The Netherlands

Susan Vitkavage
Lockheed Martin
5600 Sand Lake Road
Orlando, FL 32819
USA

Keith Warner
Massachusetts Institute of Technology
Lincoln Laboratory
244 Wood Street
Lexington, MA 02420-9108
USA

Josef Weber
Fraunhofer IZM
Hansastraße 27d
80686 Munich
Germany

Werner Weber
Infineon Technologies AG
Am Campeon 1–12
85579 Neubiberg
Germany

Bruce Wheeler
Massachusetts Institute of Technology
Lincoln Laboratory
244 Wood Street
Lexington, MA 02420-9108
USA

Robert Wieland
Fraunhofer IZM
Hansastraße 27d
80686 Munich
Germany

Markus Wimplinger
EV Group
7700 South River Parkway
Tempe, AZ 85284
USA

Jürgen M. Wolf
Fraunhofer IZM
Gustav-Meyer-Allee 25
13355 Berlin
Germany

Bernhard Wunderle
Fraunhofer IZM
Gustav-Meyer-Allee 25
13355 Berlin
Germany

Peter Wyatt
Massachusetts Institute of Technology
Lincoln Laboratory
244 Wood Street
Lexington, MA 02420-9108
USA

Donna Yost
Massachusetts Institute of Technology
Lincoln Laboratory
244 Wood Street
Lexington, MA 02420-9108
USA

III
Integration Processes

Handbook of 3D Integration: Technology and Applications of 3D Integrated Circuits.
Edited by Philip Garrou, Christopher Bower and Peter Ramm

ISBN: 978-3-527-32034-9

15
Commercial Activity

Philip Garrou

15.1
Introduction

Recently, there have been a significant number of 3D integration product development announcements and 3D integration has appeared on numerous corporate technology roadmaps [1].

It is increasingly clear that early adopters for 3D integration technologies will be so-called "chip-on-chip" technology that at first will use TSV for simple face-to-face two-chip stacks and imaging devices, which in their simplest form will use backside TSV for CMOS imaging chips.

The combination of stacking and TSV will likely appear first in memory stacks, FPGAs and memory on logic applications followed by memory on multicore microprocessors and then complete chip repartitioning and stacking.

Current DRAMs and NAND flash parts are built around 50- and 40-nm processes, respectively, but present DRAM and flash technologies face issues. There has been considerable concern that high speed memory chips such as DDR3 would suffer from performance limitations when connected in a stacked package using wire bonding technology.

Industry experts are pessimistic about shrinking design rules for NAND flash beyond the 32 nm generation because "... memory cells will be so small that operation will be unstable" and "... the big problem is not the transistors but the increase in delay" [2].

3D developments by Samsung, Micron, NEC and Tezzaron are all aimed towards cell phones and other portable devices having enough RAM to run high-definition video and other 3D graphics applications in the near future.

15.2
Chip-on-Chip Activity

For some, the first step in moving to 3D appears to be chip-on-chip (CoC) technology where chips are thinned and face-to-face bonded (Figure 15.1). This was first

Handbook of 3D Integration: Technology and Applications of 3D Integrated Circuits.
Edited by Philip Garrou, Christopher Bower and Peter Ramm

ISBN: 978-3-527-32034-9

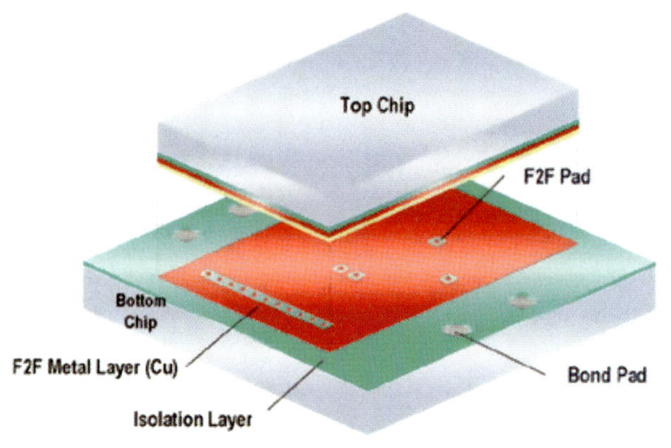

Figure 15.1 Infineon's SOLID face-to-face bonding technology [5].

proposed and built by MCM (multichip module) researchers at AT&T Bell Labs in 1998 [3]. When only two chips are to be bonded, this face-to-face solution requires no through silicon via as described in Chapter 3.

15.2.1 Sony

The microcontroller for the Sony Playstation has been fabricated using 90 nm merged DRAM process technology. At the end of 2005 Sony switched to CoC, bonding the DRAM directly to the logic. Reportedly, Sony was unable to anticipate lower cost by scaling from 90 to 65 nm merged DRAM process technology. "A massive capital investment would have been needed to drop the design rule to 65 nm. And even if we had single-chipped with DRAM and the logic, each demanding a different manufacturing process, it would have taken a long time to get the yield up" [4].

High-speed data transfer with high bandwidth is achieved when using CoC technology because of the direct memory to logic interconnect with microbumps. The microbumps provide more interconnects than wire bonding, and because they are only several dozen um in diameter they offer low parasitic capacitance, resistance and inductance, making it easier to raise the operating frequency. Since individual memory chips are used, the memory capacity limitations of merged DRAM are eliminated. The capacity of memory integrated into merged DRAM SoCs, for example, is no more than 128-Mbit, and will probably not exceed 256-Mbit even if the design rule is reduced to 65 nm [4].

Customers are demanding ever-larger amounts of memory for applications like HDTV, which used to get by with 64-Mbit or less of memory, but systems supporting 1080i imagery now come with 128- to 256-Mbit. As long as performance and function continue to improve in equipment, it will mean steadily increasing memory capacity, and eventually merged DRAM will be unable to handle the requirements [4].

Renesas and NEC both planned to offer CoC technology in packages in 2007. Renesas applications include communications equipment like servers and routers [4].

15.2.2
Infineon

Infineon 3D technology was initially developed in 2006 for Infineon's chip card IC products division. Key applications for the Infineon face-to-face die bonding technique (Figure 15.1) included mobile communication, credit/debit cards, prepaid telephone cards, and health insurance cards. The commercial introduction of these 3D products was delayed.

Infineon's prototype chip was a smartcard controller that combined 160 kbyte of non-volatile memory with a logic chip. In the SOLID process (jointly developed with Fraunhofer Munich – see Chapter 16) Infineon first reroutes the bond pads and defines copper/tin, contact pads, which allows a higher interconnect density than is possible with traditional bumping. A Datacon flip-chip bonder places known good die on the good die on the 300 mm base wafer, with an alignment accuracy of 10 μm (3 sigma), then attaches them with a temporary polymer adhesive at an effective rate of about 4000 units h^{-1}. An EVG wafer bonder then permanently bonds the chips at 270 °C and pressure for 1–2 hours. With Infineon's solid-liquid-interdiffusion technology (Chapter 14), the copper-tin forms a eutectic alloy that can then resist heat up to 600 °C, so that other chips could be added on top of the first bonded pair with the same process.

Infineon claims such technology will "help reduce the price of current chip solutions by up to 30%" and that it is suitable for chips with clock rates up to 200 GHz.

15.3
Imaging Chips with TSV

Image sensor device production has skyrocketed in recent years, fueled by the growth of cell phones with cameras. CMOS imaging sensors were not quick to take advantage of the form factor of WLP technology since WLP fabricate the interconnect structure (solder balls) on the face of the silicon devices and mount face down whereas CMOS sensing chips must mount face up. The obvious solution of having the active area of the sensor face up and the interconnect on the backside of the device is easily obtained by backside formation of TSV. Shellcase (Israel), recently purchased by Tessera, was the first to propose such technology with their opto-CSP technology [6]. Similar structures were recently commercialized by Schott glass [7].

Fujikura has announced that they are incorporating TSV to make image sensors for digital cameras. They claim this technique allows them to make devices $^1/_2$ as thin and $^2/_3$ the x–y dimensions of conventional devices [8].

Sanyo has also described image sensor modules developed for automotive control systems (Figure 15.2) [9].

Toshiba has described a process using laser ablation and dielectric film lamination to fabricate vertical TSV for low cost CMOS image sensor applications [10]. Patterned adhesive was used to attach a glass wafer to the CMOS image sensor wafer. The glass plate becomes part of the CMOS image sensor package, serving as a wafer handling

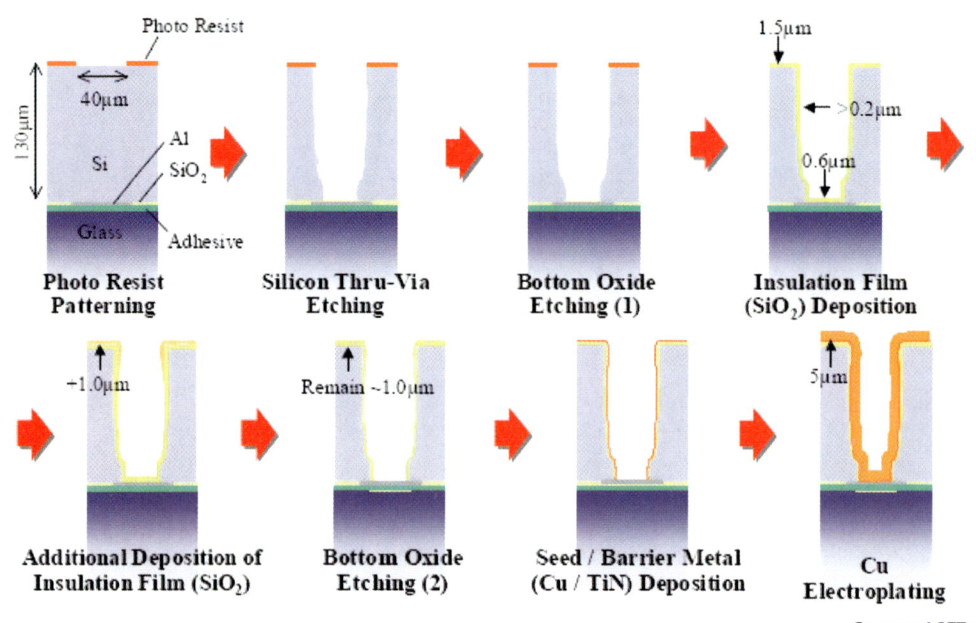

Figure 15.2 Sanyo process flow for generation of backside vias [9].

substrate and protection for the pixels from dust during assembly. The wafer is thinned by subsequent back-grinding and polishing.

Vias are drilled using a YAG laser from the back of the wafer with alignment to the back of the Ni capped Al pads. The laser ablates the silicon, the dielectric layer of silicon dioxide, and the Al stopping on the Ni. An epoxy based film resin is laminated under vacuum to achieve complete filling of the vias and void-free lamination on the back of the wafer.

A second set of vias are drilled, aligned with the center of the first vias using the same YAG laser equipment. The epoxy resin thickness must be maintained at 15 μm or more to maintain the needed insulating properties. The vias are slightly tapered to facilitate electroplating, which is employed to deposit 10 μm of Cu on the back of the wafer and inside the vias. This process is shown in Figure 15.3.

Zycube activity is covered in detail in Chapter 26.

15.4 Memory

15.4.1 Samsung

Samsung Electronics CEO Chang-Gyu Hwang reported that "... the integration of memory, logic, sensors, processors and software will be based on die stacking 3D

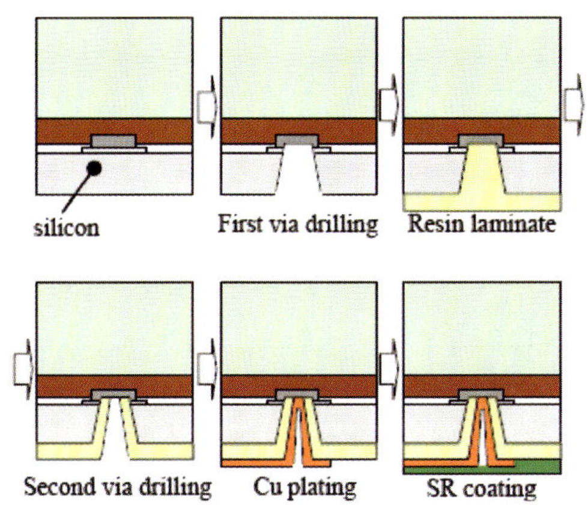

Figure 15.3 Toshiba image sensor package processing with TSV [10].

technology." He continued "... In the not-so-distant future we will probably have no choice but to make the transition in Si technology development from shrinking geometry to technologies capable of forming circuits in three dimensions. The semiconductor industry is on the verge of being reborn [11]."

In 2006 Samsung announced the company's first 3D prototypes based on its WSP (wafer-level stack process). A 16-Gbit memory device composed of eight stacked, 50-micron thick, 2-Gbit NAND flash die that are a combined 0.56 mm in height was fabricated. The stacked device showed a 15% smaller footprint and is 30% thinner than an equivalent wire-bonded solution. WSP also reduced the length of the interconnects, resulting in an approximately 30% increase in performance due to reduced electrical resistance. Samsung's WSP technology uses lasers to form the TSV, which reportedly reduces production cost significantly as it eliminates the typical photolithography-related processes required for mask-layer patterning.

The stack is shown in Figure 15.4 and the interconnect is shown in detail in Figures 15.5 and 15.6.

Samsung reported that it would apply its WSP technology to the production of NAND-based memory cards for mobile applications and other consumer electronics.

In 2007 Samsung also fabricated an all-DRAM stacked memory package using TSV connections between chips. The device consists of four 512-megabit DDR2 DRAMs that total 2 gigabits. Samsung said it can make up to 4 gigabyte DIMM. The packaged stack is 1.4 mm thick. Samsung claims the DRAM stacking was more difficult than their initial stacking of NAND flash die [13].

The TSV connected memory die reportedly eliminates the performance degradation seen in multi-chip packages with high-speed memory chips operating at speeds of 1.6 Gbits s^{-1} or more connected by current WB technology. Samsung says such technology has potential for cell phones (because space is so limited) but is also developing the process for "next-generation computing systems in 2010 and beyond" [14].

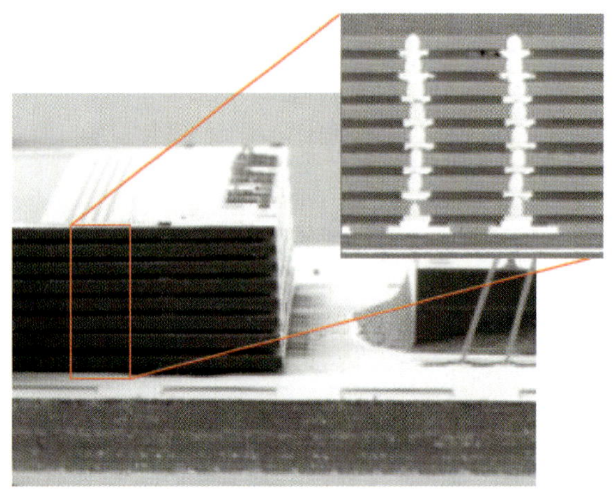

Figure 15.4 Samsung 16 GBIT NAND flash, eight-chip stack with TSV [12].

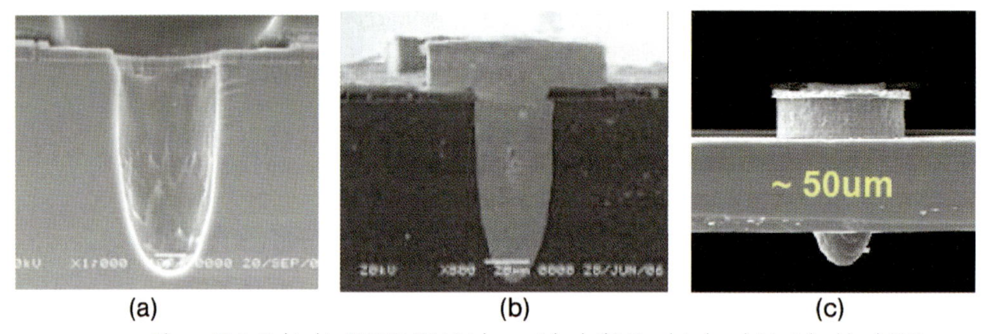

Figure 15.5 Eight-chip NAND TSV (a) laser etched, (b) Cu plated and (c) etched back [12].

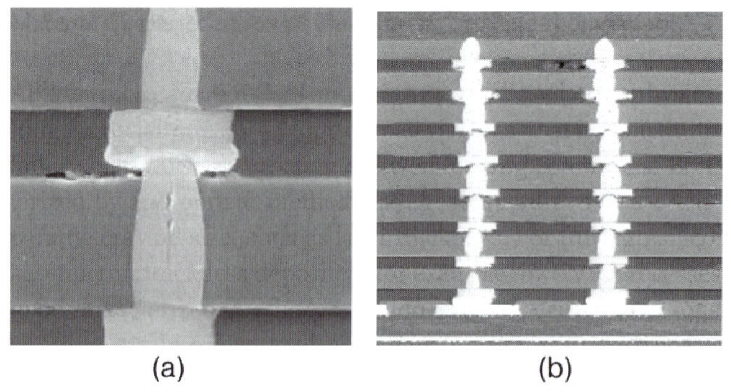

Figure 15.6 Eight-chip NAND TSV (a) bonding and (b) full eight-layer stack [12].

Samsung proposes the following applications:

- small form factor NAND memory cards
- small form factor high speed DRAM modules
- memory merged SiP (memory + ASIC).

15.4.2 Elpida

Elpida has developed stacking technology with poly silicon FEOL technology in their joint NEDO study with NEC and Oki. A complete description of this technology is given in the NEC discussion below. Elpida predicts that by 2010, at the 32 nm node, they can achieve 3× the memory density that would be obtained at this node by using non TSV technology (Figure 15.7).

15.4.3 Tezzaron & Chartered

Chartered and Tezzaron Semiconductor have announced an agreement to manufacture Tezzaron 3D devices in volume [16]. Chartered connects stacked wafers with hundreds of thousands of TSV which Tezzaron calls Super-Contacts. The wafers are then aligned with a precision of 0.5-micron and bonded using Cu–Cu bonding. Tezzaron activity, and a complete description of their technology, is covered in further detail in Chapter 24.

15.4.4 NEC

In 2006 NEC Electronics, Elpida Memory and Oki Electric published the results of their joint 3D integration program "Stacked Memory Chip Technology Development Project" that was supported by NEDO (New Energy and Industrial Technology Development Org) in Japan. Elpida supplied die design and DRAM fabrication,

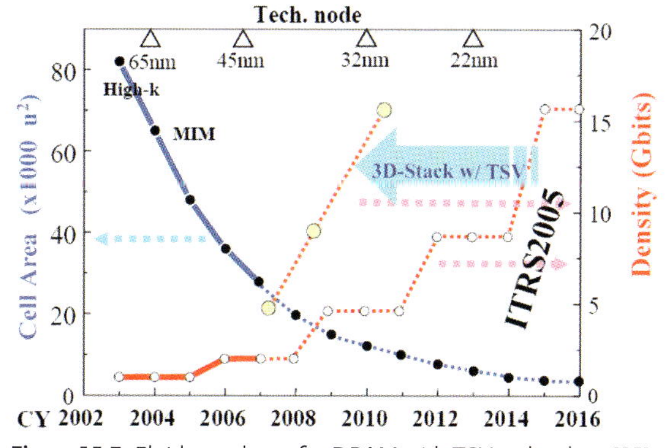

Figure 15.7 Elpida roadmap for DRAM with TSV technology [15].

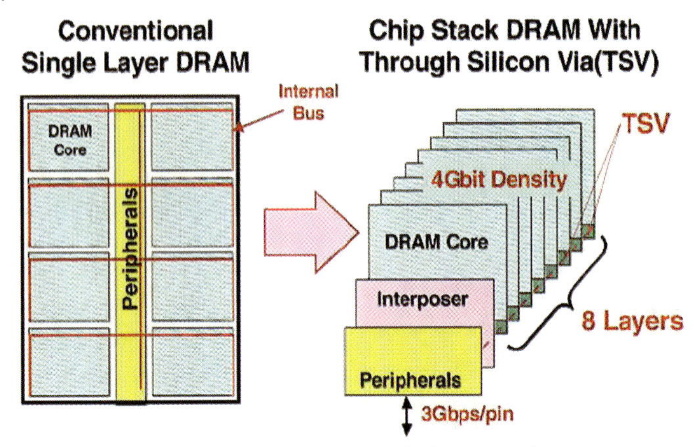

Figure 15.8 Conventional 4G DRAM vs stacked DRAM with TSV [17].

NEC supplied ASIC fabrication and interposer design and Oki supplied die stacking and package over molding.

By using 3D stacking technology, the memory capacity of a single DRAM package was increased without waiting for the next silicon generation [18]. For example, 4 G of DRAM capacity was achieved by stacking eight, 50 micron thick, 512 M DRAM chips. This will allow handheld devices to carry as much memory as a computer and should help meet the increasing demands of high-definition video and graphics. A schematic of the structure is shown in Figure 15.8.

As shown in Figures 15.8–15.10, it is a FEOL, vias first process with poly-silicon conductor that is joined by copper–tin eutectic bonding.

Figure 15.9 shows the stacked DRAM process flow. Since Elpida carries out the DRAM fabrication after the TSVs are formed, doped poly-Si is employed as the

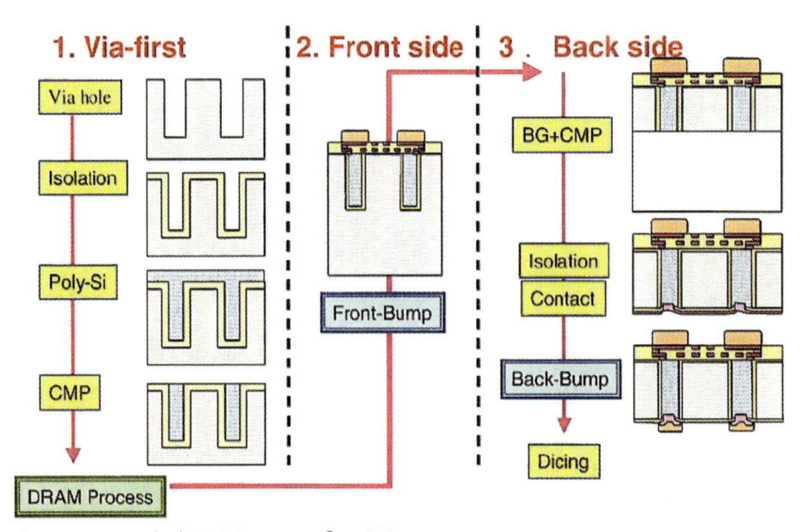

Figure 15.9 Stacked DRAM process flow [17].

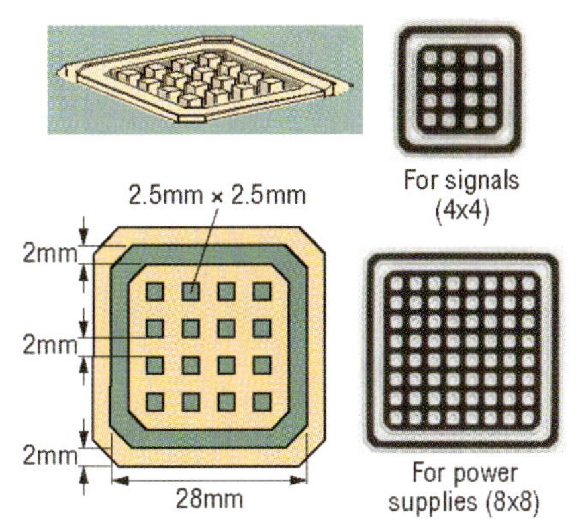

Figure 15.10 Annular vias for signal and power distribution [17].

conductor. They use annular via structures; 4×4 array vias for signal and 6×6 or 8×8 vias for power-surrounded by a trench ring to reduce parasitic capacitances (Figure 15.10). Each via is 2 μm in diameter, so that they can be uniformly filled with poly-Si in a relatively short deposition step. After via etch, isolation, and poly-Si deposition, CMP provides the smooth top surface for subsequent DRAM processing.

When DRAM fabrication is completed, copper micro-bumps are formed. The wafer is temporarily bonded to a support and the backside of the wafer is ground down to 50 μm to expose the back of the TSV. Si_3N_4 insulation is deposited by CVD, the insulation is patterned and the copper microbumps are plated [20 μm (height) ×30 μm (diameter)]. Figure 15.11 details the micro bump bonding technology. The wafer is subsequently removed from the support material, moved to dicing tape and is diced.

The DRAM stack can be packaged (Figures 15.12 and 15.13). This packaging, for which they have coined the acronym SMAFTI (smart feed through interposer), can also be used for bonding of memory to logic (Figure 15.14).

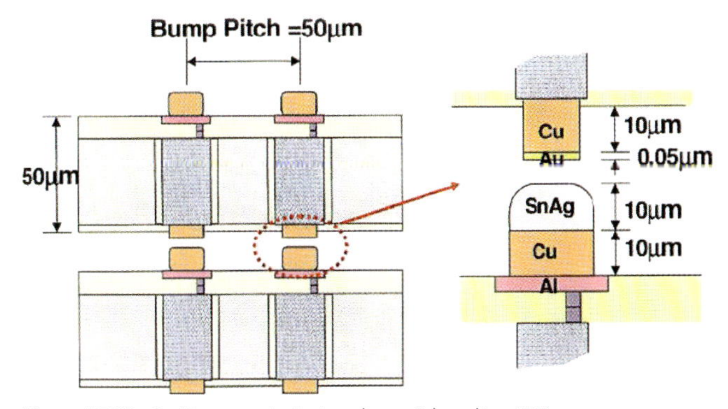

Figure 15.11 Cu/Sn eutectic "micro-bump" bonding [17].

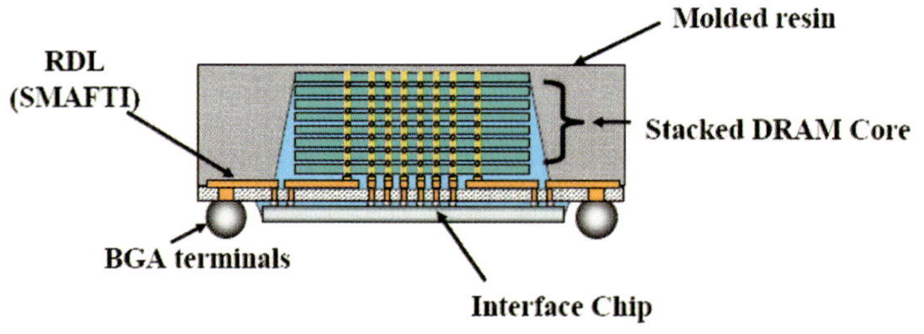

Figure 15.12 Proposed packaging of DRAM stack [19].

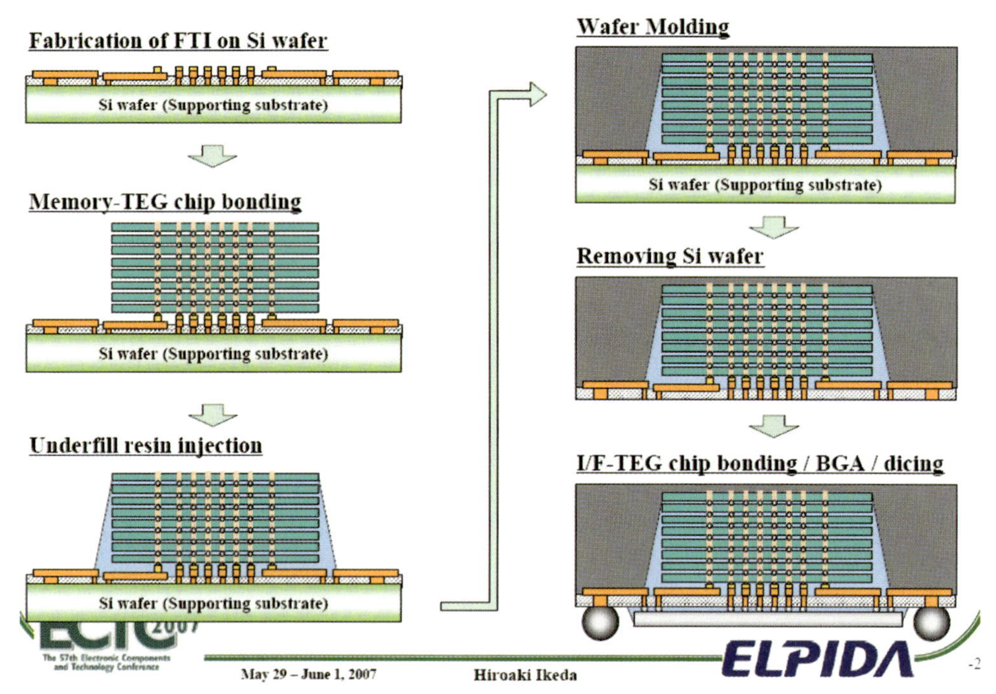

Figure 15.13 DRAM stack package process sequence [20].

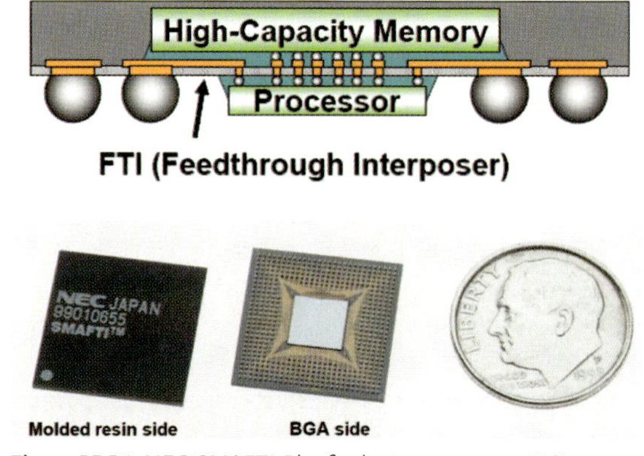

Figure 15.14 NEC SMAFTI Pkg for logic + memory [20].

15.4.5
Micron

In 2006 Micron described their "osmium" wafer-level packaging technology, which they indicated would be applied to both their semiconductor devices and CMOS image sensor lines. Though little detail was given, the technology was reported to include TSV [they call through-wafer interconnects (TSV)], redistribution layer technology, and wafer-level encapsulation. The use of 40 μm TSVs, to vertically connect peripheral bond pads on the thinned die stacks, reportedly improves the electrical R & C parasitics. In addition, the thinned silicon substrate substantially reduces the final package height. Micron indicated that memory devices and image sensors could be fabricated without leadframes and substrates, thus lowering Micron's packaging cost, now estimated at 15–25% of the finished product cost [21]. Their image sensor technology using TSV is known internally as iWLP (image wafer level package).

Marc Tuttle has indicated that:

> ... from a DRAM memory perspective, performance is the dominant driver for 3-D structures .... As DRAM data rates climb above 1 Gbit/s, this becomes much more important due to the electrical parasitics, which tend to be substantial for conventional stacking structures using wire bonding. For NAND memory, however, the main market pressure for 3D comes from size rather than data rate performance (which is about an order of magnitude lower than DRAMs). To fit eight or more NAND ICs in one package for the latest mobile products, it is critical to minimize the stacking height, as well as the overall length and width [22].

15.5
Microprocessors & Misc. Applications

15.5.1
Intel

In late 2006 Intel displayed a 300 mm wafer of 80 core microprocessors. Intel noted that TSV was the technology that allowed them to overcome latency issues and achieve transfer rates between the processor and memory of up to a terabyte per second (Figure 15.15). CEO Paul Otellini indicated that within five years such chips would be in production. Using TSV, Intel mates processor cores directly to 256 kB of SRAM. Otellini said that "... TSV will make for an enormous increase in overall system performance, likely even greater than the inclusion of 80 cores on a single die". They indicated that TSV could be used in various Intel chips, not just the "terascale" chip.

Intel is looking at both "logic + memory" stacking, which includes stacking cache or main memory onto a high-performance logic device and "logic + logic" stacking

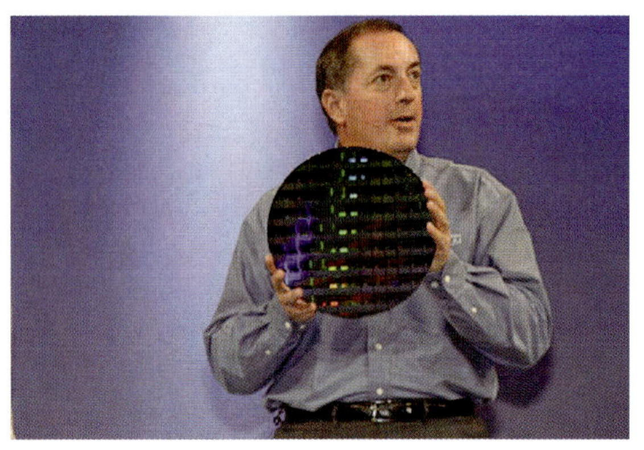

Figure 15.15 Intel's 80 core multiprocessor with TSV [23].

that involves splitting a logic area between two or more layers and requires much tighter pitches than "logic + memory" stacking [24–27]. Chapter 34 gives more detail on the use of 3D Integration in the next generation of microprocessors.

Technical concerns include 3D stacking of strain-enhanced Si devices and low-K dielectrics, both of which are sensitive to stresses. 3D stacking typically involves thinning device layers to less than 100 μm, which makes such layers more prone to stress effects [24]. Another challenge involves thermal management issues caused by 3D stacking since microprocessors have a high power density than other applications and heat dissipating paths are limited. Although thermals are *potentially* worse in a stacked microprocessor configuration, they are manageable through "intelligent thermal design" [25].

Intel's main goal is reportedly to reduce interconnect length since a significant amount of the power in a microprocessor can be consumed in backend interconnect wire. 3D stacked layers exhibit high bandwidth, low latency, and low power interfaces. The wire reduction by 3D integration provides opportunities to trade off performance, power, and area [25].

In published work Intel show Cu–Cu bonding for mechanical and electrical interconnection. Figure 15.16 shows a cross section of an SRAM memory chip

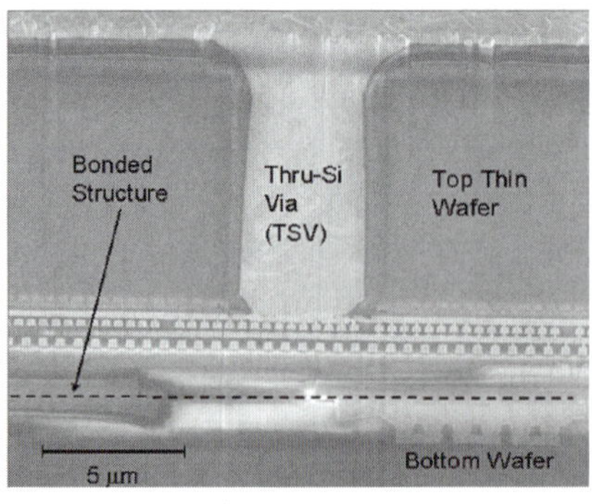

Figure 15.16 Intel TSV [26].

Cu–Cu bonded to a processor and connected by a 5 μm via formed vias last on the stack [26].

15.5.2
IBM

IBM recently released their TSV technology to their manufacturing group. IBM is running chips using the through-silicon via technology in its manufacturing line and will begin making sample chips using this method available to customers in the second half of 2007, with production in 2008. The first application of this through-silicon via technology will be in wireless communication chips that will go into power amplifiers for wireless local area network (LAN) and cellular applications. In 2008, IBM will sell production quantities of a power amplifier that sports as many as 100 direct metal links to a power ground plane. This reportedly could lower power consumption by as much as 40% (thus increasing battery life of portable products) for a device that is a key component in cell phones and Wi-Fi adapters.

IBM said it also plans to apply 3D technology to a range of chips, including those running in business, government and scientific servers and supercomputers. IBM plans to apply the TSV technique in wireless communications chips, Power processors, Blue Gene supercomputer chips and high-bandwidth memory applications [27].

More specifically, IBM plans to use the technology to link a microprocessor to its ground plane to stabilize power distribution across the chip. That will require more than 100 vias to voltage regulators and other passives. They estimate that it could cut a CPUs power consumption by 20% [28].

IBM is also reportedly going to use this technology in its 65 nm fab to bond SRAM to a processor for its power based server line and then as a high-bandwidth link between CPUs and memories. IBM is already converting the custom processors used in its Blue Gene supercomputers into through-silicon via packaging. The new chips will mate directly with cache memory chips. A prototype SRAM using the technology is being fabricated in IBMs 300-mm production line using 65-nm process technology [29].

As with Intel, IBM will also need this technology in multicore processors due to bandwidth limitations between the microprocessor and the memory and power distribution issues. As more and more cores are added to chips it becomes increasingly difficult to deliver uniform power to each one. By stacking vertically and reducing the interconnect length, IBM feels it can overcome these problems.

IBM has been heavily supported by DARPA for the development of tools and technologies through the 3D IC program in the Microsystems Technology Office (MTO).

IBM has not indicated what process flow it plans to employ in production. Much of the early information it has published on 3D technology involved vias last processing on SOI wafers/devices (Figure 15.17).

In more recent publications IBM has also detailed Cu–Cu bonding studies similar to MIT and Intel [32]. Copper interconnects were fabricated with a standard BEOL damascene process followed by oxide CMP processes to recess the oxide level

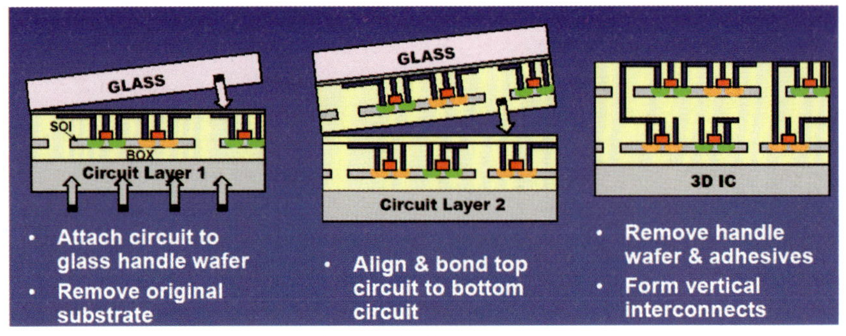

Figure 15.17 IBM SOI processing scheme [30, 31].

to 40 nm lower than the Cu surface. They concluded that Cu–Cu bonded wafers with a slow temperature ramp rate (6 °C min^{-1}) have better bonding quality than those with a fast rate (32 °C min^{-1}); application of a small force prior to temperature ramping and high bonding down-force during bonding improves bond quality and the quality of the bonded interface improves with increasing interconnect pattern density.

References

1 Garrou, P. (April 2007) Posturing and positioning in 3D ICs. *Semiconductor International*, 88.

2 Ooishi, M. (April 2007) *Vertical stacking to redefine chip design*, Nikkei Electronics Asia.

3 Low, Y., Frye, R. and O'Connor, K. (1998) Design methodology for chip-on-chip applications. *IEEE Transactions CPMT, Part B*, **21**, 298.

4 Uno, M. (February 2007) *Chip-on-chip offers higher memory capacity speed*, Nikkei Electronics Asia.

5 Gruber, W. (Feb. 2004) Turning chip design on its head, www.synopsys.com/news/compiler/art1lead_infineou-feb04.

6 Garrou, P. (2000) Wafer level chip scale packaging an overview. *IEEE Transactions Advances Packaging*, **23**, 1521.

7 Garrou, P. (December 2006) Opto-wafer level packaging, (o-WLP) for CMOS imaging sensors, *Semiconductor International*, p. sp10.

8 Japan firm pushing image sensor shrinking process, Solid State Technology. (online, accessed November 2007)

9 Umemoto, M., Kameyama, K., Suzuki, A. *et al.* Novel through Si process for chip level 3D integration, 1st Int Workshop on SoP, SiP Sept. 2005 Ga Tech, Atlanta Ga.

10 Sekiguchi, M. *et al.* (2006) Novel low cost integration of through chip interconnection and application to CMOS image sensor. Proceedings ECTC, p. 1367.

11 Hwang, C.G. (2006) New paradigms in the silicon industry. IEEE IEDM.

12 Lee, K. (2006) Next generation package technology for higher performance and smaller systems. 3D Architectures for Semiconductor Integration and Packaging Conference, Burlingame CA.

13 (4/22/2007) Samsung develops new, highly efficient stacking process for DRAM. *Semiconductor International.* (online, accessed November 2007)

14 Clendennin, M. (April 23rd 2007) Samsung uses direct metal links in DRAM stacks, EE Times.

15 Ikeda, H. (2007) 3D Stacked DRAM using TSV. ECTC Plenary Session.

16 LaPedus, M. (June 12th 2007) Chartered and Tezzaron partner on 3D devices. EE Times (on line).

17 Mitsuhashi, T. *et al.* (2007) Development of 3D processing process technology for stacked memory. Enabling Technologies for 3D Integration, MRS Symposium Proceedings 970 (eds C. Bower, P. Garrou, P. Ramm and K. Takahashi), Material Research Society, Pittsburgh, pp. 155.

18 Kawano, M. *et al.* (2006) A 3D packaging technology for 4 Gbit stacked DRAM with 3 Gbps data transfer. IEEE IEDM.

19 Kurita, Y. *et al.* (2007) A 3D stacked memory integrated on a logic device using SMAFTI technology. ECTC, pp. 821.

20 Ikeda, H. (2007) 3D stacked DRAM using through silicon via, ECTC plenary session.

21 Davis, J. (August 18 2006) Micron takes wraps off packaging innovation, *Semiconductor International* (online).

22 Tuttle, M. Micron, Personal communication.

23 www.intel.com. (accessed on November 2007)

24 Morrow, P., Park, C., Ramanathan, S. *et al.* (2006) Three-dimensional wafer stacking via Cu–Cu bonding integrated with 65-nm strained-Si/low-*k* CMOS technology. *IEEE Electron Device Letters*, **27**, 335.

25 Black, B. *et al.* (February 2007) 3D design challenges. IEEE-Proceedings of the International Solid State Circuits Conference, p. 410.

26 Morrow, P., Black, B., Kobrinsky, M. *et al.* (2007) Design and fabrication of 3D microprocessors, Enabling Technologies for 3D Integration, MRS Proceedings 970 (eds C. Bower, P. Garrou, P. Ramm and K. Takahashi), Material Research Society, Pittsburgh, pp. 91–130.

27 Merritt, R. (April 12th 2007) IBM readies direct chip-to-chip links, EE Times (on line).

28 LaPedus, M. (April 16th 2007) IBM preps 3D stacks for the market, EE Times (on line).

29 Stokes, J. (April 12 2007) IBM goes vertical with chip interconnects, www.ArsTechnica.com.

30 Guarini, K.W. *et al.* (2002) Electrical integrity of state-of-the-art 0.13 lm SOI CMOS devices and circuits transferred for three-dimensional (3D) integrated circuit (IC) fabrication. *IEE IEDM Tech. Digest*, 943.

31 Topol, A.W. *et al.* (2006) Three dimensional integrated circuits. *IBM Journal of Research and Development*, **50**, 491.

32 Chen, K. *et al.* (2006) Structure, design and process control for Cu bonded interconnects in 3D integrated circuits. IEEE IEDM.

16
Wafer-Level 3D System Integration

Peter Ramm, M. Jürgen Wolf, and Bernhard Wunderle

16.1
Introduction

16.1.1
Drivers for the Introduction of 3D System Integration

In general, the introduction of 3D integration technologies into production of microelectronic systems will be driven by:

1. Form factor: reduction of system volume, weight and footprint.
2. Performance: improvement of integration density and reduction of interconnect length, resulting in improved transmission speed and reduced power consumption.
3. Low-cost fabrication: reduction of processing costs for, for example, mixed technologies products.
4. New applications: for example, ultra-small wireless sensor systems.

The introduction of very advanced microelectronic systems, such as 3D microprocessors [1] and 3D integrated image processors [2, 3], will be mainly driven by performance enhancement as an emerging solution to the on-chip "wiring crisis" caused by signal propagation delay. Suitable 3D integration technologies will be used in IC production to overcome the performance bottleneck caused by the predicted fundamental obstacles in backend-of-line [4].

Figure 16.1 shows a roadmap [5] for 3D integration concepts and applications (Fraunhofer IZM). Through silicon via technology is a key element for performance improvement of memory/processor stacks, image sensors, wireless sensor systems and advanced processors, in long term even enabling 3D CPUs with fully 3D architectures.

The potential for low-cost fabrication is the key for future applications of 3D integration. Today, fabrication of Systems-on-a-Chip (SoC) is based on embedding multiple technologies by monolithic integration into one silicon substrate. But

Handbook of 3D Integration: Technology and Applications of 3D Integrated Circuits.
Edited by Philip Garrou, Christopher Bower and Peter Ramm

ISBN: 978-3-527-32034-9

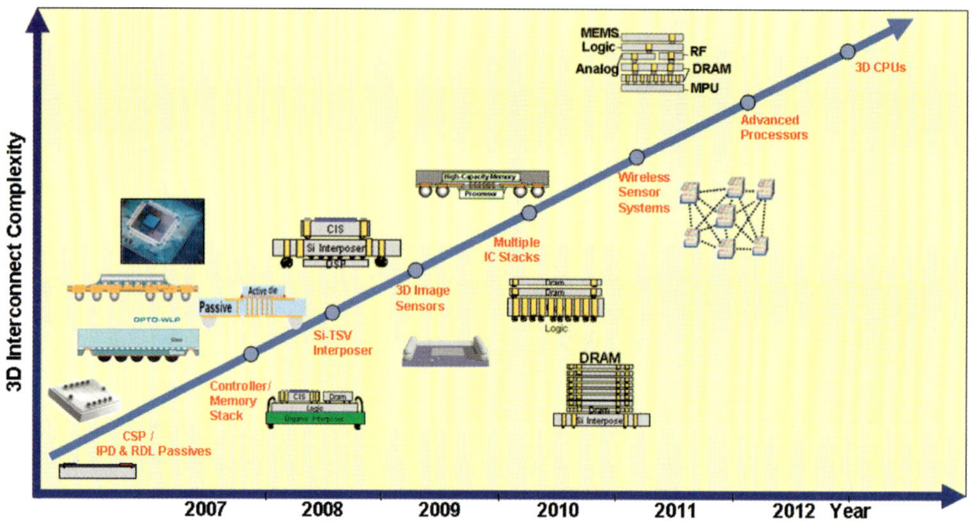

Figure 16.1 Roadmap of 3D integrated systems (evolution of 3D interconnect performance in diverse applications. Source: Fraunhofer IZM).

there are serious disadvantages. The chip partition with the highest complexity drives the process technology, leading to a "cost explosion" of the overall system. In contrast, suitable 3D integration technologies enable the integration of different optimized base technologies with the potential of low-cost fabrication through high yield and smaller IC footprints: Device stacks (e.g. controller and memory) fabricated with optimized 3D integration technologies will show reduced production costs in competition to monolithic integrated SoCs. Even so-called heterogeneous integration, the combination of devices fabricated on different substrate materials (e.g. Si, SiGe and GaAs) can be performed by 3D system integration in a cost-effective way.

Furthermore, new multifunctional micro-electronic systems are realized by 3D system integration: Ultra-small smart systems for applications such as distributed wireless sensor networks (e-Grain™, e-CUBES®) [5, 6]. For future applications, such systems for ambient intelligence will be highly miniaturized. 3D integration technologies have to be applied because of their relevant benefits: extreme system volume reduction, reduction of power consumption (for lifetime enhancement), reliability improvement and low-cost fabrication for meeting high volume market requirements.

16.1.2 Technological Concepts

Several technology concepts for 3D integration have been introduced since the 1980s – some of them came already to production. In the last few years die stacking technologies have come into large volume fabrication. The large spectrum of 3D integration technologies can be classified in different ways. Regarding the performance of the enabling technologies three main categories are reasonably chosen:

1. Stacking of packages or substrates
2. Die stacking (without through semiconductor vias – TSVs)
3. Vertical system integration (with through semiconductor vias – TSVs).

Package-on-Package (PoP), and Package-in-Package (PiP) technologies as well as die stacking with wire bonding for inter-chip interconnects are in production by several companies (e.g. Intel, Hitachi, Sharp, Amkor, ASE, Philips, NXP, ST Microelectronics, Tessera, Infineon) today.

There is long-standing experience and a large spectrum of 3D packaging technologies at Fraunhofer IZM and the Technical University of Berlin in Germany [7]. As of this writing, these technologies (stacked packages, stacked dies) are commonly categorized as 3D packaging, in order to distinguish them from 3D integration with through semiconductor vias (TSVs) which Fraunhofer introduced as "Vertical System Integration" – VSI®.

The so-called "Chip in Polymer" approach is based on embedding thinned chips in a printed circuit board [8].

In order to improve performance and enable low-cost fabrication, wafer-level die stacking technologies are in advantage. In corresponding approaches as e.g. "Thin Chip Integration" (TCI) technology, thinned dies are embedded and interconnected in a polymer layer with a modified multilayer thin film wiring on wafer-level.

Fraunhofer in Munich has been active in the area of 3D integration since the mid-1980s. First 3D integrated CMOS circuits were fabricated using re-crystallization of poly-silicon layers [9]. Since the early 1990s Fraunhofer Munich has been focusing 1998 their R&D on wafer stacking technologies using vertical inter-chip interconnects through the silicon substrates (inter-chip vias or through silicon vias). The so-called Vertical System Integration (VSI®) is characterized by very high density vertical inter-chip wiring with freely positioned through silicon vias. The corresponding fabrication of 3D ICs is mainly based on thinning, adjusted bonding and vertical metallization of device substrates.

The following sections will focus on technologies which take advantage of wafer level processing to achieve the highest miniaturization degree, excellent electrical performance and high volume fabrication.

16.2 Wafer-Level 3D System Integration Technologies

Wafer level packaging technologies, e.g. Chip Size Packages (CSP) with redistribution layer (RDL), are already used in production lines for a number of applications. The advantageous ability to realize further components on large-area wafers through combination of additional technological steps is a decisive criterion for further development of this technology. 3D integration technologies at wafer-level are of advantage to satisfy the need to increase performance and functionality while reducing size, power consumption and cost of the system [7, 10]. In consequence, Fraunhofer IZM's R&D focusses on wafer-level 3D integration technologies for both die-to-wafer stacking (without TSVs) and vertical system integration with TSVs).

16.2.1 Die-to-Wafer Stacking

16.2.1.1 Flip Chip and Face to Face

A so-called face-down wafer level circuit integration technique is based on an extension of a standard redistribution technology for wafer-level-CSPs. In this approach a base chip at wafer level will be used as an active substrate for smaller and thinner ICs. These thinned active components are assembled in flip chip fashion on the base IC wafer using a flip chip technique with solder bumps. This technology uses thin film techniques to reroute the peripheral I/Os of the base wafer according to the electrical requirements to an array of under bump metallization (UBM) pads. A low electrical resistivity of the rewiring metallization is achieved by electroplated copper. A low-k photo-definable polymer layer (e.g. BCB) is used as the dielectric layer. The redistribution layer can also be used for the integration of passive devices (e.g. resistors, inductors, capacitors, filters, etc., Figure 16.2).

The leadfree solder bumps for the flip chip assembly are realized by electroplating. After flip chip assembly the die will be protected by encapsulation e.g. glop top or over molding. Large solder balls (>300 μm) on the base chip can be utilized for the interconnection to the next package level. Figure 16.2 shows the principle of the integration concept.

An alternative low cost process for wafer-level die stacking without through silicon vias was reported by Infineon Technologies and Fraunhofer IZM [11]. The

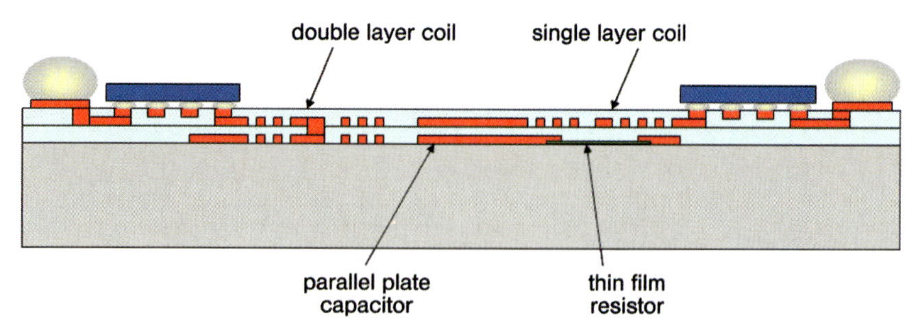

Figure 16.2 Principle of chip on chip integration technology (flip chip technique) with integrated passives in the redistribution layer on base circuit.

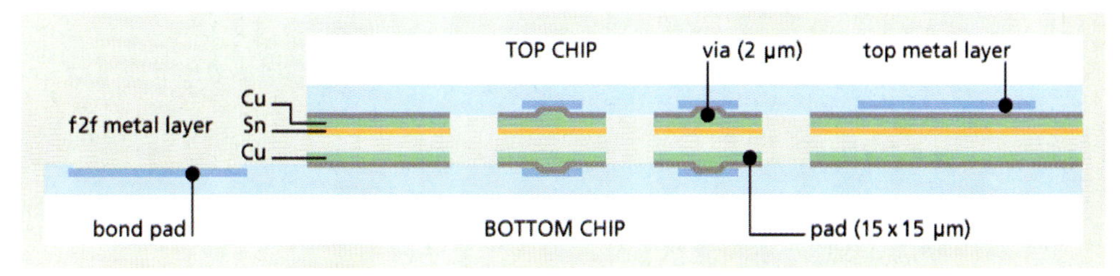

Figure 16.3 Schematic of the face-to-face die stacking technology SOLID. (Source: Infineon [11]).

"Face-to-Face" die stacking technology "SOLID" is well suited for fabrication of stacks with two vertically integrated device layers. Both, the mechanical bond and the electrical connections between the bottom and the top chip are realized by solid-liquid-interdiffusion – SOLID – soldering of thin electroplated and structured Cu/Sn layers. Figure 16.3 shows a schematic cross section of the chip stack with typical via and pad dimensions and Cu/Sn as face-to-face (f2f) metal system.

16.2.1.2 **Wafer Level Thin Chip Integration (TCI)**

A different approach, the thin chip integration concept (TCI), avoids the flip chip assembly of devices with solder bump interconnections on the silicon substrate. Key elements of this technology are extremely thin, completely processed and tested wafers and chips (KGD). In contrast to other packaging techniques the TCI concept uses thinned chips (thickness $< 20\,\mu m$), mounted "face-up"on a base chip by adhesive bonding e.g. with a thin polymer layer [12]. This technique offers excellent electrical properties of the wiring system and the interconnects for the active and passive devices. Besides form factor improvement, the signal transmission time, e.g., for high speed memory modules, will be reduced compared to the flip chip or single chip packages.

The process for TCI modules starts up with one type of bottom wafer carrying larger base chips. The completely processed device wafers for the top IC have to be mounted on a carrier substrate by a reversible adhesive bond and undergo a backside thinning process until the thinned wafers show a remaining thickness of approximately $20\,\mu m$. The thinned chips are mounted on the active base wafer (bottom wafer) and covered by a photosensitive dielectric layer. BCB as dielectric layer offers excellent electrical properties, high temperature stability, and very low water up-take with a medium curing temperature. A thin film redistribution layer (Cu) is used to interconnect the top and carrier circuits (Figure 16.4) [13]. The wiring layer is covered by a passivation layer where the same dielectric material is used. The similar process sequence can be used to integrate a second active device on top of the base wafer. Finally, the deposition of an under bump metallization and a solder bump deposition complete the manufacturing of the TCI module. Figure 16.5a/b shows a schematic construction of the TCI-Module.

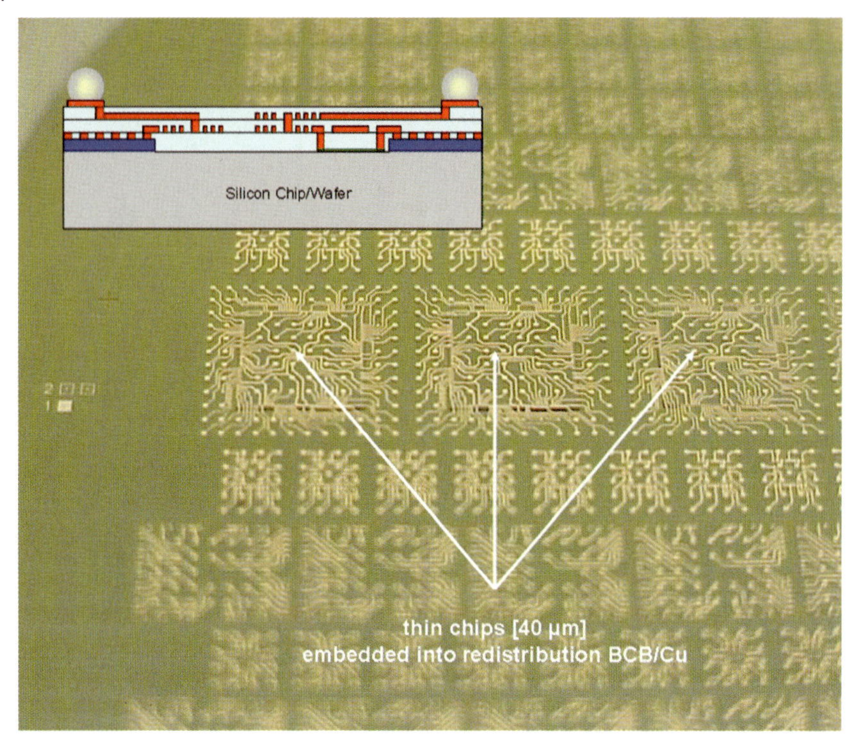

Figure 16.4 Thin silicon devices (40 μm) embedded in wafer level redistribution layer (RDL) [13].

Using a silicon interposer with metallized through silicon vias (TSV) as a carrier substrate allows stacking of multiple TCI modules (Figure 16.5 c/d). For the realization of the through silicon vias (TSV) in a silicon interposer or silicon device wafers different approaches for the via etching (e.g. wet etching, DRIE or laser drilling) and via metallization (e.g. CVD-W, CVD-Cu, ECD-Cu, doped silicon or metal paste) can be performed. The selection of the technology is determined by system requirements (e.g. via density, electrical resistivity, etc.).

Metal filling of TSV using electroplating is especially useful for via sizes between 5 μm and 20 μm, which is in a special focus for silicon interposer with TSVs as a carrier substrate. After via etching using e.g. DRIE and sidewall isolation, the seedlayer can be applied by CVD (e.g. Cu or W) or an adequate sputtering process, e.g. Ti/W:Cu.

Figures 16.6 a/b show a through silicon via (diameter 15 μm filled with Cu by electrodeposition using a thin CVD tungsten/sputtered Cu seedlayer) [14, 15]. The copper will also be deposited on the wafer front side during the via plating, which will be removed by a later etching step. Depending on the via sizes and depth (aspect ratio-ASR) a wafer thinning (grinding, CMP, etching) from the backside is required to get access to the metallized vias. The IO terminals on the backside are realized by standard thin film processing (polymer – Cu metal) followed by solder ball placement. A mechanical support of the interposer during backside processing can be provided by a temporary bonding on a carrier substrate (e.g. silicon or glass).

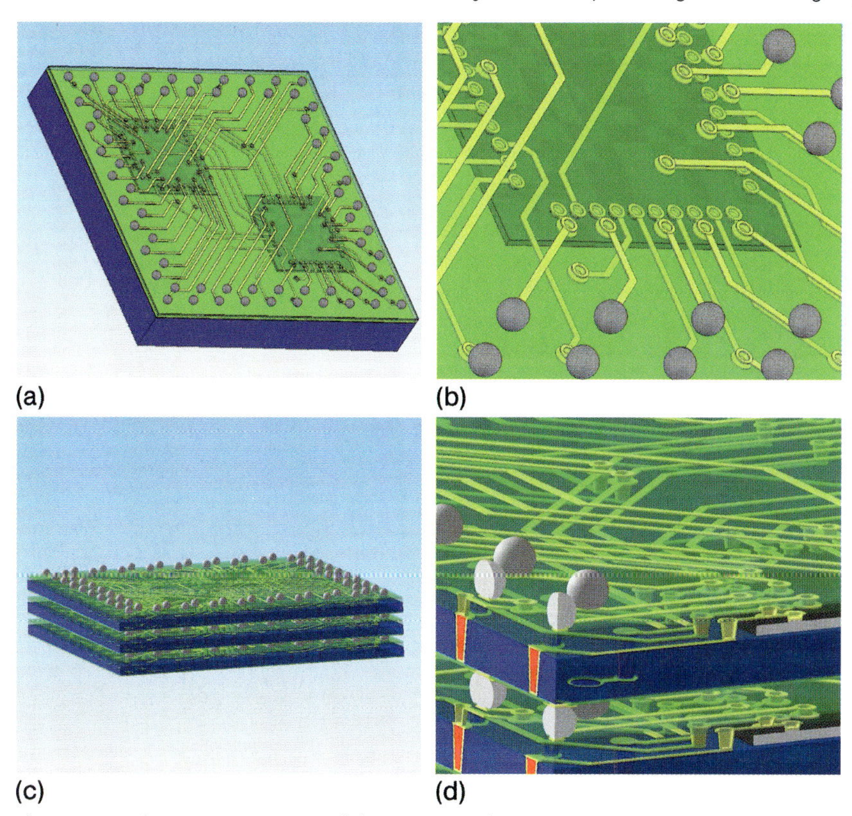

Figure 16.5 Schematic construction of the TCI approach, (a) example with two embedded die; (b) detailed view. (c) Schematic of a stackable TCI module through silicon vias and embedded active devices; (d) detailed view.

16.2.1.3 **Passive Device Integration**

One of the key advantages of the TCI approach is the integration of embedded passive devices into the redistribution layer very close to the active devices, which results in minimal parasitics [15] and overall size reduction of the module. In [16] a TSV silicon interposer technology for RF application is described which takes advantage from passive device integration into the redistribution layer (RDL) and flip chip assembly of active devices e.g. transceiver. The values of the integrated passive devices into the RDL layer depends on the chosen materials, e.g. dielectric, resistor material, and design rules. Typical values are in the range of 1–80 nH for inductors, <10 pF/mm^2 for polymer dielectrics, <10 nF/mm^2 for Ta_2O_5 for MIM capacitors and 100 Ω/sq. for resistors (NiCr). In [17, 18] a process for flexible polymer layers with integrated passive devices is described where individual layers are finally stacked (Figure 16.7).

The schematic construction of the thin film build-up substrate is shown in Figure 16.8. On the top and bottom sides 5 μm thick nickel serves as pad terminal metallization. Two copper layers are used to form the internal routing of the

(a)

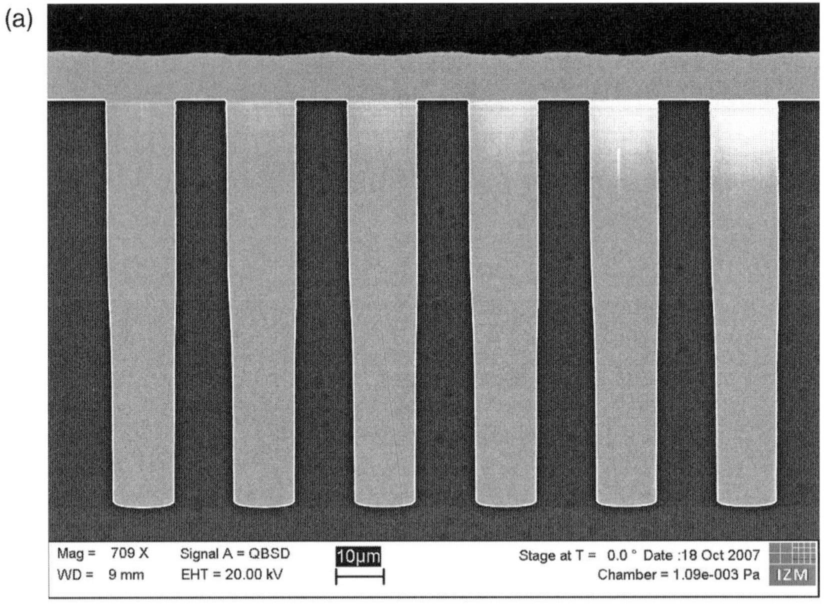

(b)

Figure 16.6 (a) Copper filled TSV by electroplating, size: 15 μm diameter, 65 μm depth. (b) Detail of copper filled TSV with CVD silicon oxide as passivation and a sputtered TiW seed/barrier layer.

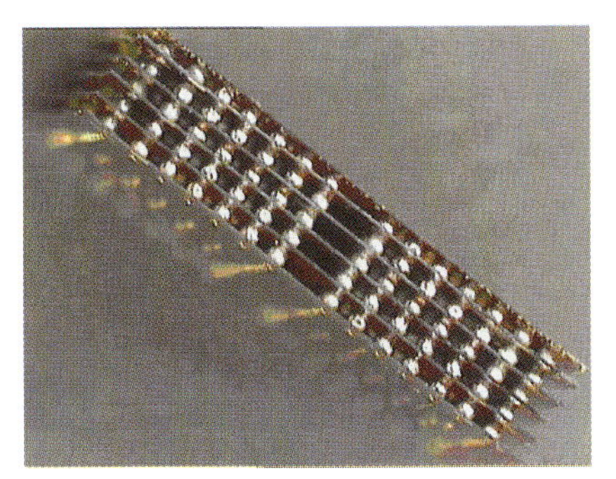

Figure 16.7 Stack of flexible polymer layers with integrated passive devices structures [17].

substrates. Nickel chromium was used as resistive material for the integrated resistors. Three thin film polyimide layers provide the mechanical stability for assembly and stacking. The middle layer serves as interlayer dielectric, the two outside dielectric layers provide for device passivation and solder stop. The total thickness of the build-up is in the range of 50 μm.

The polymer layer substrates were realized using thin film technology on six inch silicon support wafers. The separation of the substrates from the carrier wafer, which is the second-last step in the process sequence, is based on tackling a release layer by a solvent.

Durimide® 7320, a photo-imageable polyimide precursor from Fuji-Film Electronic Materials was used as the dielectric layer. The material shows excellent flex properties as a freestanding film after release from the temporary carrier substrate. The spin-on applied Durimide precursor has a cinematic viscosity of 5800–6400 $mm^2 s^{-1}$ and solids content of 41 wt%. The final layer thickness can be set

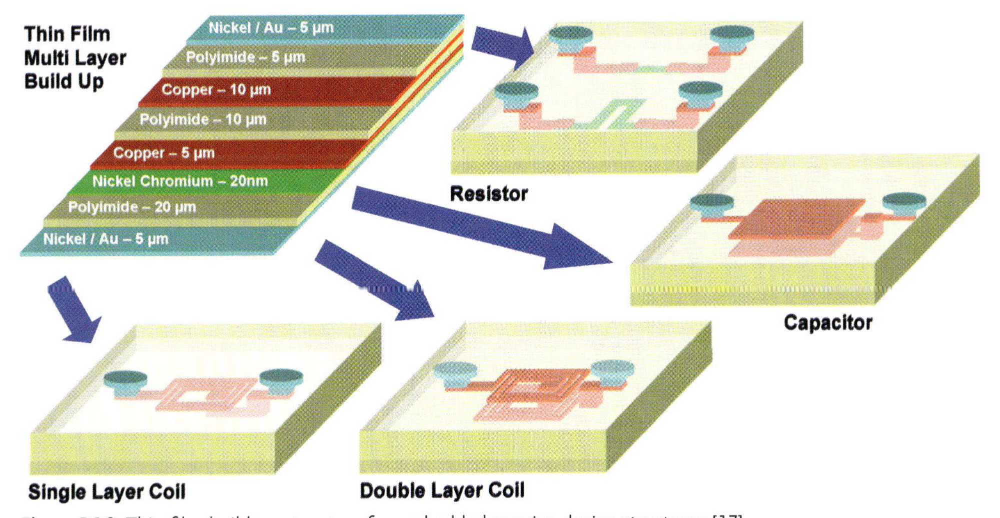

Figure 16.8 Thin film build-up structure for embedded passive device structures [17].

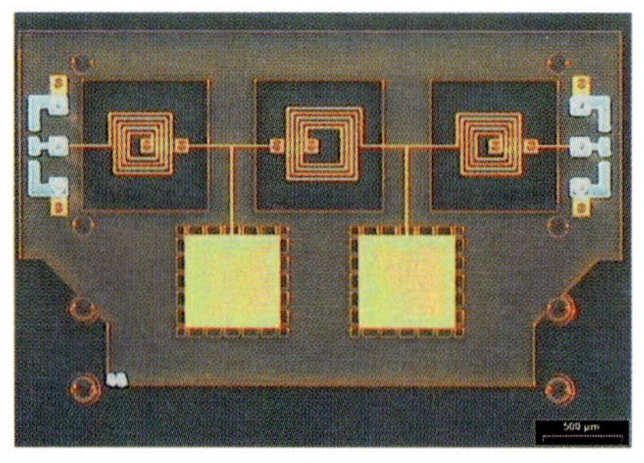

Figure 16.9 Low pass filter (Integrated Passive Device, IPD) with three inductors and two capacitors.

by spin-on parameters in a wide range of values from 4 μm to 30 μm according to coverage requirements. A hot-plate prebake at 100 °C, which removes part of the solvents, prepares the layer tack free for further handling and contact exposure. Broadband UV-exposure at mask aligner crosslinks the exposed precursor areas, while unexposed areas can be removed in a tank development step to deliver via openings in the dielectric. To convert the photo-defined precursor into the final polyimide, a high-temperature curing step is required. 60 minutes at 350 °C are sufficient to ensure full imidization. In a sequential multilayer build-up, ramping to a lower level (280–300 °C) is recommended for every additional layer to enable interlayer crosslinking (adhesion) and to avoid turning the flex brittle.

Figure 16.9 shows a fabricated filter [17] before release from the temporarys substrate as well as their simulated and measured attenuation (Figure 16.10) behavior after substrate release. Each filter includes three coils and two capacitors, which are connected via micro strip lines to a Tschebyscheff low pass. The six coils have nominal inductance values of 2 × 2.7 nH, 1 × 3.8 nH, 2 × 6.1 nH and 1 × 8.6 nH. The four capacitors have nominal values 2 × 0.34 pF and 2 × 0.96 pF. The electric

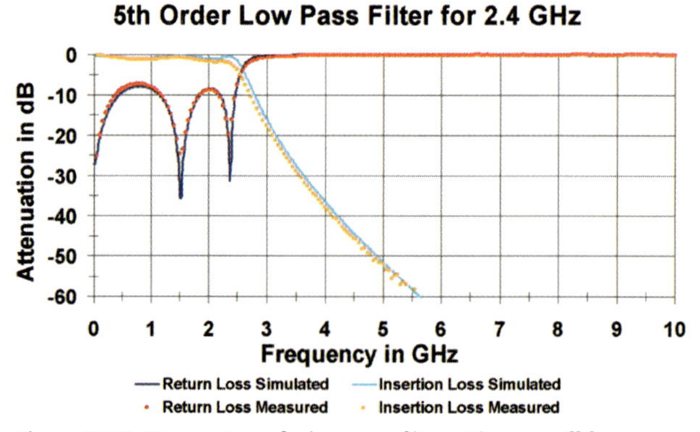

Figure 16.10 Attenuation of a low pass filter with a cut-off frequency of 2.4 GHz.

measurements values of the realized filter match very well the simulation for these high-frequency structures, indicating that the capacitors were realized with the correct interlayer polyimide thickness of 10 μm between the copper electrodes.

16.2.2 Vertical System Integration

3D integration technologies based on bonding and vertical inter-chip wiring of stacked thinned device substrates using free positioned (area) through semiconductor vias (TSVs) can be called vertical system integration. VSI® is characterized by the potential of very high density vertical interconnects with the use of standard silicon wafer processes (mainly backend-of-line). It is commonly used to distinguish between "via first" and "via last" concepts: the formation of through semiconductor vias before stacking of the device substrates shall be defined as "via first", after stacking as "via last". It is also reasonable to distinguish between concepts allowing vertical system integration of ready processed device substrates and concepts with the need to modify the basic device fabrication technology. Depending on the availability of devices (e.g. on wafer-level), the production issues (e.g. modifications of the base technologies) and the specific application, either the one or the other can be of advantage. Moreover, the choice of the basic substrate stacking principle is the key for the introduction of 3D integration for most applications.

VSI concepts are in principle suitable for both wafer-stacking and chip-stacking. In general, technologies largely relying on standard wafer fabrication show a comparatively favorable cost structure. Consequently, wafer-level VSI technologies are most promising. Figure 16.11 shows the principle of the two corresponding stacking concepts: wafer-to-wafer stacking and chip-to-wafer stacking.

For wafer stacking approaches, the step raster on the device wafers must be chosen identically. This is easily fulfilled for 3D integration of devices of the same kind (e.g. stacked memories) but in the general case of non-identical device areas, the handicap of processing with identical step raster would result in active silicon loss and, in consequence, increase the fabrication cost per die. The use of chip-to-wafer stacking approaches in principle allows for the vertical integration of known good dies to known good dies (Figure 16.11, right-hand side).

The common key element of all technologies for VSI is the formation of TSVs. Fraunhofer Munich's key competence is the realization of a reliable vertical inter-chip via metallization using area through silicon vias with high aspect ratios, providing high 3D interconnect densities. Furthermore, their R&D focuses on the other major requirements for vertical system integration: a precise thinning technology including a cost-efficient handling concept and a suitable bonding process. In general, there are, mainly, three bonding schemes used for wafer-level 3D system integration:

1. Fusion bonding
2. Metal bonding
3. Adhesive bonding

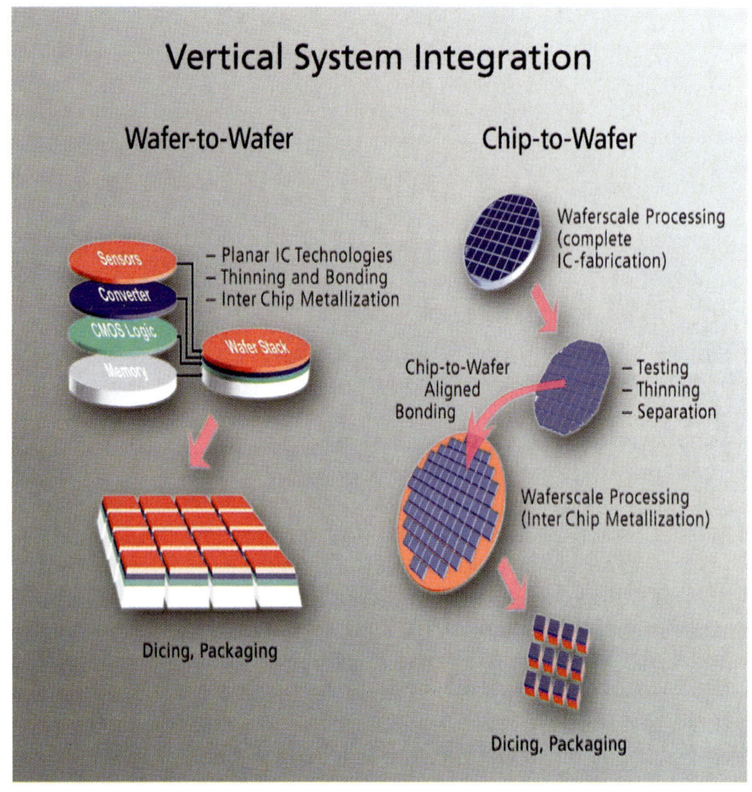

Figure 16.11 Vertical system integration: wafer-to-wafer and chip-to-wafer stacking concepts.

Fraunhofer has developed several concepts for vertical system integration, mainly in the area of adhesive bonding and metal bonding and with both "via first" and "via last" approaches. The following sections describe two main stream technologies. The so-called ICV-SLID technology, based on copper/tin bonding, is a pure "via first" process. The so-called ICV technology, based on polyimide bonding, shows a mixture of both characteristics, "via first" and "via last".

16.2.2.1 **Vertical System Integration Using Adhesive Bonding – The ICV Technology**

Fraunhofer Munich, in cooperation with Infineon Technologies, have developed a wafer-level vertical system integration technology based on low temperature bonding with polyimide as intermediate layer and a 3D metallization process that provides a very high density vertical wiring between the thinned device wafers by the use of W- or Cu-filled through silicon vias. The so-called Inter-Chip-Via (ICV) technology is described in detail in the literature [19–22]. Figure 16.12 shows the corresponding schematic of a vertically integrated device stack with polyimide (PI) as intermediate layer and inter-chip vias through all layers of the thinned top Si

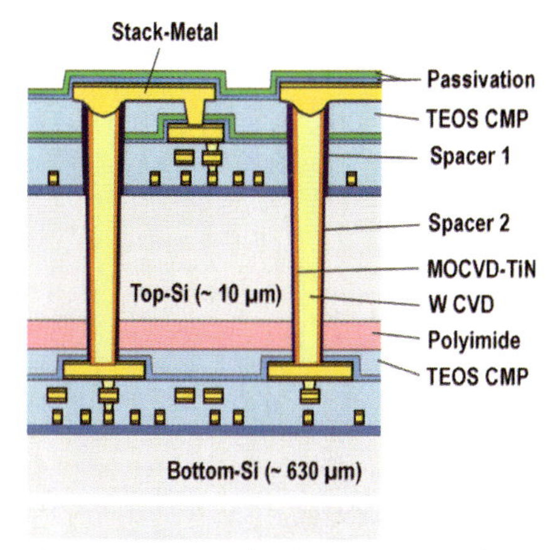

Figure 16.12 ICV technology: schematic of a vertically integrated device stack.

substrate (through silicon via), providing a vertical interconnect between metallization levels of both devices.

The ICV concept in principle is suitable for both wafer stacking and chip to wafer stacking for vertical system integration [23]. In fact, the ICV technology is characterized by a very high vertical interconnect density but not optimized for chip-to-wafer stacking. The process flow shows both aspects, "via first" and "via last" concept. Although the through vias are pre-processed before thinning and stacking, there is a need for additional process steps after stacking, giving reason for the use of more complex techniques on stacked chip-on-wafer level (e.g. resist technology for very large substrate topographies). But certainly, the wafer-stacking approach is well suited for 3D integration of devices with identical foot print.

Prior to wafer stacking, through silicon vias (TSVs) with typically 1–3 μm diameters are prepared on the top wafer. The high aspect ratio TSVs are etched through all dielectric layers and deep into the silicon. The wafer is then temporarily bonded onto a handling substrate by using a glue polymer and thinned with high uniformity until the TSVs are opened from the rear. After optical alignment of the stabilized top wafer versus a polyimide coated bottom wafer, the durable polyimide bond is established at approximately 400 °C and the handling substrate is removed. Now, the inter-chip vias are opened to the bottom wafer's metallization, laterally isolated with highly conformal O_3/TEOS oxide (see Chapter 6) and finally metallized. The metallization can be performed using electroplating or MOCVD. For void-less filling of high aspect ratio vias, highly conformal MOCVD processes are of advantage. Excellent results were achieved using both CVD tungsten and CVD copper (in both cases with CVD-TiN as seed layer) [24]. Subsequently, for so-called metal plug formation a suitable metal etch back process is applied. The lateral electrical connection of the metal-filled through silicon via with a metallization level of the top wafer is performed by opening contact windows on the top wafer followed by a standard metallization and passivation. Finally, the bond pads are opened and the 3D integrated device stacks can be tested, diced and packaged by use of standard procedures.

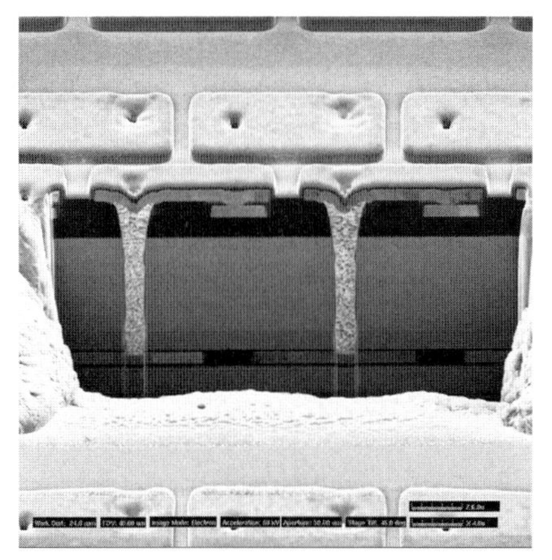

Figure 16.13 FIB of a vertically integrated test chip structure, showing $2.5 \times 2.5\,\mu m^2$ inter-chip vias (contact chain with 10 000 TSVs).

Figure 16.13 shows a FIB cross section of a vertically integrated test structure (contact chain with 10 000 TSVs) and Figure 16.14 a CVD-W filled through silicon via in detail ($16\,\mu m$ deep, $2.5 \times 2.5\,\mu m^2$). Typical interconnect resistances are in the range of $1\,\Omega$.

Figure 16.14 ICV technology – FIB of a 3D test structure with CVD-W through silicon vias ($2.5 \times 2.5\,\mu m^2$).

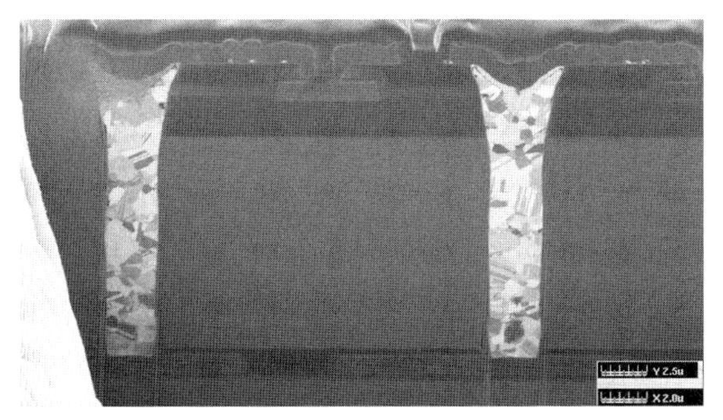

Figure 16.15 ICV technology – FIB of a 3D integrated test structure with CVD-Cu through silicon vias.

The corresponding result with CVD-Cu filled through silicon vias is shown in Figure 16.15.

16.2.2.2 **Vertical System Integration Using Solid-Liquid-Interdiffusion Bonding – The ICV-SLID Technology**

In general, technologies largely relying on standard wafer fabrication processes show a comparatively favorable cost structure. However, regarding the aligned bonding process, chip area issues may speak against stacking of device wafers: the loss of active silicon area caused by processing of wafers with identical step sizes for devices with different chip sizes increases the total cost per stacked die. For chip-to-wafer stacking approaches the starting materials can be completely processed wafers, too. Known good dies of the top wafer are aligned bonded to the known good dies of a bottom wafer after wafer-level testing, thinning and separation. In favorable approaches, this process step represents the only one on chip-level within the total vertical system integration sequence. The subsequent processing for vertical metallization is on the wafer-scale again.

In consequence, Fraunhofer Munich IZM has focused on the development of a VSI technology with no need for additional process steps on the stacked chip level. Well suited for chip-to-wafer stacking concepts are VSI technologies based on metal–metal bonding, such as Cu–Cu bonding [25] or intermetallic compound bonding (see Chapter 14). The so-called ICV-SLID concept [26] is based on the bonding of top chips to a bottom wafer by very thin soldering pads (e.g. Cu/Sn) which provide both the electrical and the mechanical interconnect by solid-liquid-interdiffusion (SLID) soldering [27]. The ICV-SLID concept is a non-flip concept ("back-to-face", b2f). The through silicon vias are fully processed – etched, isolated and metallized – prior to the thinning sequence, with the advantage that the later stacking of the separated known good dies to the bottom device wafer is the final step of the 3D integration process flow. Using fully processed devices, ICV-SLID can be categorized as post backend-of-line "via-first" 3D integration process. As a fully modular concept, it allows the formation of multiple device stacks. Figure 16.16 shows a schematic cross section of a vertically integrated circuit in accordance with the modular back-to-face concept, also indicating the stacking of the next level chip.

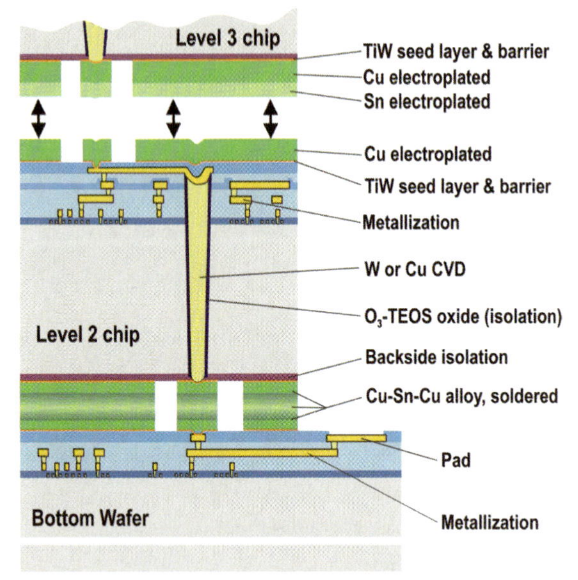

Figure 16.16 Schematic cross section of a 3D integrated circuit according to the modular ICV-SLID technology.

Detailed description of the technology development and results have been presented in several publications [7, 28–30]. A typical process sequence for the fabrication of 3D integrated circuits according to the ICV-SLID chip-to-wafer stacking concept is described briefly. The first essential step of the ICV-SLID process flow is the formation of through silicon vias. The etching, lateral isolation and metal filling of the vias are performed on wafers with standard thickness, thus basically resulting in high-yield fabrication of through silicon vias. The TSVs are connected to the contact wiring of the devices by standard metallization (aluminium or copper, depending on the technology). The process sequence for the formation of the metallized inter-chip vias is described briefly: The TSVs with typically 1–3 μm feature size are prepared on a fully processed and tested device wafer by etching through all passivation and multi-level dielectrics layers, followed by a deep silicon trench etch. For lateral via isolation, a highly conformal CVD of O_3/TEOS-oxide is applied (see Chapter 6) and the through silicon vias are metallized by using e.g. MOCVD of tungsten (MOCVD-TiN as barrier layer) and etch back for metal plug formation. Figure 16.17 shows an array of W-filled through silicon vias with 12 μm depth and $2 \times 2\,\mu m^2$.

The lateral electrical connection of the W-filled TSVs with the uppermost metal level of the device is performed by standard metallization. Figure 16.18 shows the schematic cross section of the device with metallized vias. At this stage all processes for the formation of the vias are completed on wafer-level.

The devices are now ready for wafer-level test and selection. The top wafer is then temporarily bonded to a handling wafer and thinned with very high uniformity using precision grinding, wet chemical spin etching and a final CMP step until the W-filled vias are exposed from the rear (Figure 16.19).

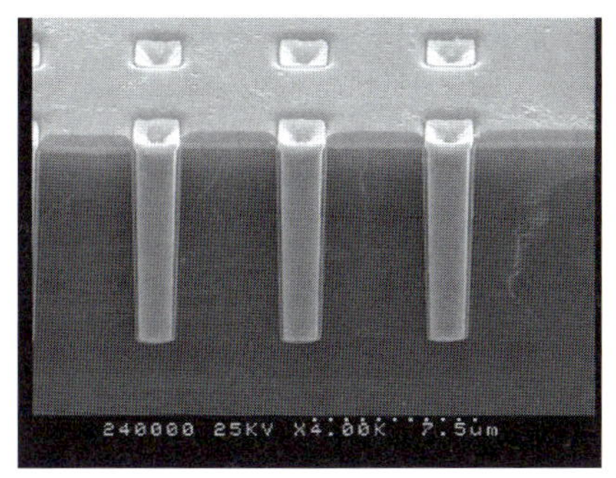

Figure 16.17 Top device with W-filled through silicon vias (12 μm depth, $2 \times 2\,\mu m^2$).

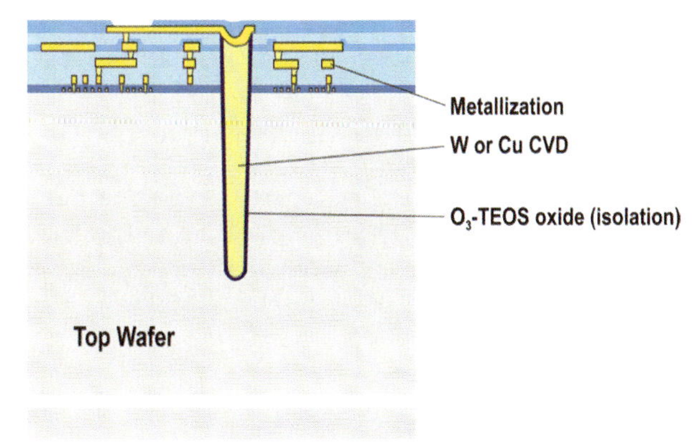

Figure 16.18 ICV-SLID technology – Schematic cross section of the device with metallized vias, pre-processed on wafer-level.

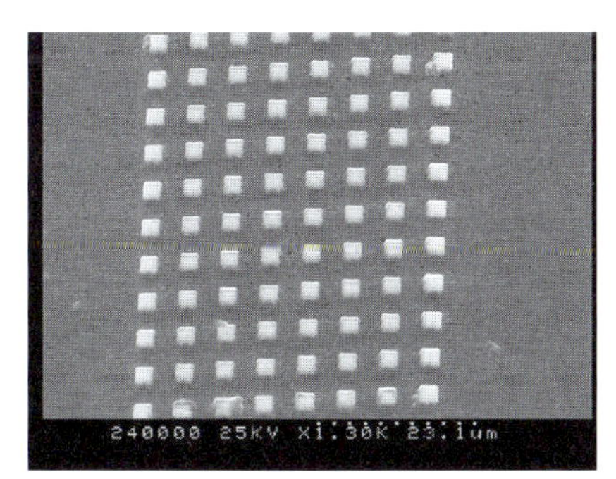

Figure 16.19 Bottom side of a thinned top device stabilized by the handling substrate, showing CVD-tungsten plugs, opened from the rear.

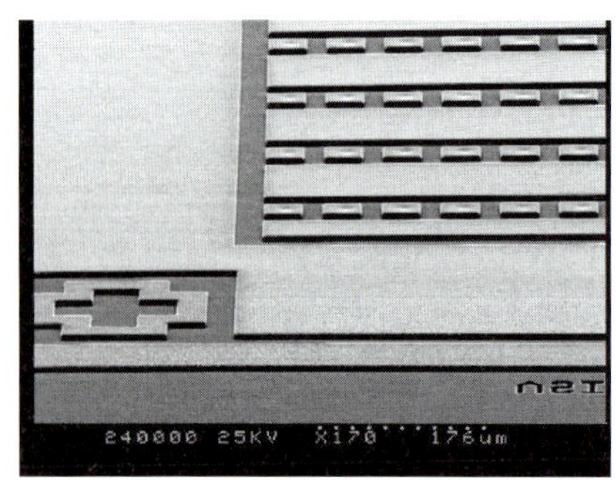

Figure 16.20 Top side of the bottom wafer with through-mask electroplated copper.

After deposition of dielectric layers for electrical isolation and opening etch to the W-filled TSVs, through-mask electroplating of a typically 8 μm thin copper/tin bilayer is applied on the rear side of the thinned wafer. The surface is completely covered with the soldering metal; electrical contacts are formed by isolation trenches in the Cu/Sn layer and the remaining areas that are not used for electrical means serve as dummy areas for mechanical stabilization of the device stack. The bottom wafer is through-mask electroplated with Cu as the counterpart metal of the soldering metal system. Figure 16.20 shows the surface of the bottom wafer, including alignment structures.

After dicing, the selected known good dies – stabilized with the handling substrates – are picked and placed to the bottom wafer by use of a chip-to-wafer bonder (preferably equipment providing high throughput at a high alignment accuracy). Figure 16.21

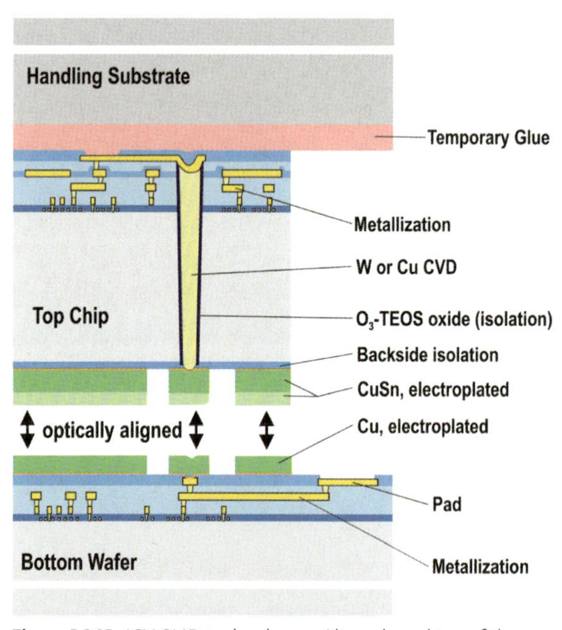

Figure 16.21 ICV-SLID technology – Aligned stacking of the thinned and stabilized top device to the bottom device wafer.

shows the aligned stacking of the stabilized thinned chip to the bottom device wafer, both prepared for the SLID bonding by structured metal layers.

The mechanical bond and the electrical contact of the transferred chips are performed in one step by a soldering technology called Solid-Liquid-Interdiffusion (SLID) which is reported in detail in Chapter 14. For the Cu/Sn SLID metal system the process is outlined here: During the soldering step at a temperature of approximately 300 °C, the liquid Sn is interdiffused by solid copper, finally forming the intermetallic phase Cu_3Sn. This so-called ε-phase is thermodynamically stable with a melting point above 600 °C. Using appropriate film thicknesses, tin is consumed and the solidification is completed within a few minutes, leaving unconsumed copper on both sides. Figure 16.22 shows a FIB of a 3D integrated test structure processed according to the ICV-SLID technology after soldering and removal of the handling substrate. The W-filled TSVs are interconnected by Al wiring to the metallization of the top device and by the above-described soldering metal system to the metallization of the bottom device. The schematic in Figure 16.21 indicates that the ICV-SLID technology is a pure "via first" concept: the through silicon vias are ready processed – etched, metallized and connected – prior to the chip-to-wafer stacking, with the key advantage of no need for further 3D integration processes after SLID bonding.

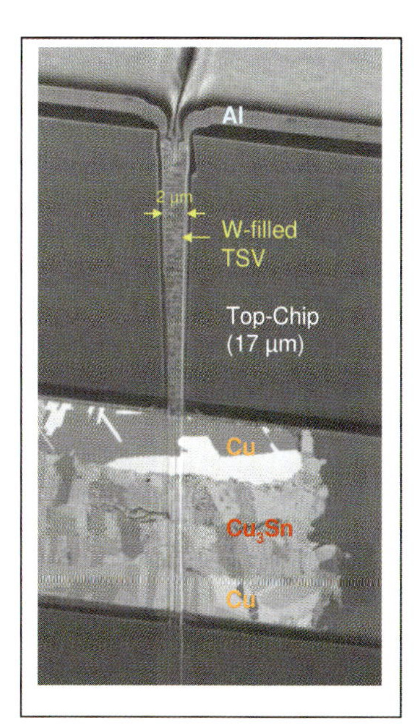

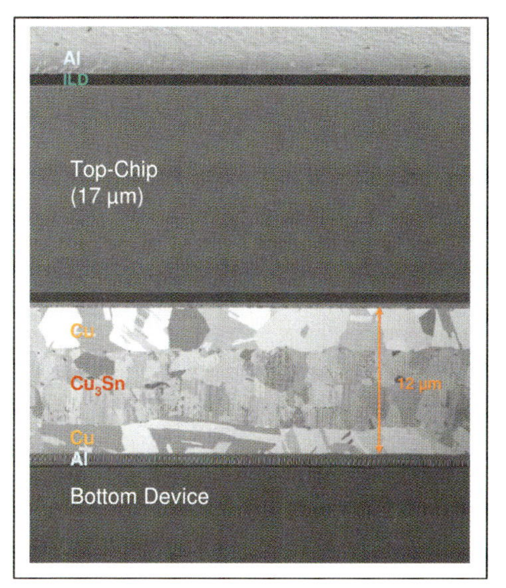

Figure 16.22 ICV-SLID technology – FIB of a 3D integrated test structure, showing a cross section of a stacked chip with CVD-W filled through silicon vias connected to the bottom wafer by Cu/Sn SLID metallization (Cu/Cu_3Sn/Cu) Left: Region with TSV; Right: Cu/Sn SLID bond in detail.

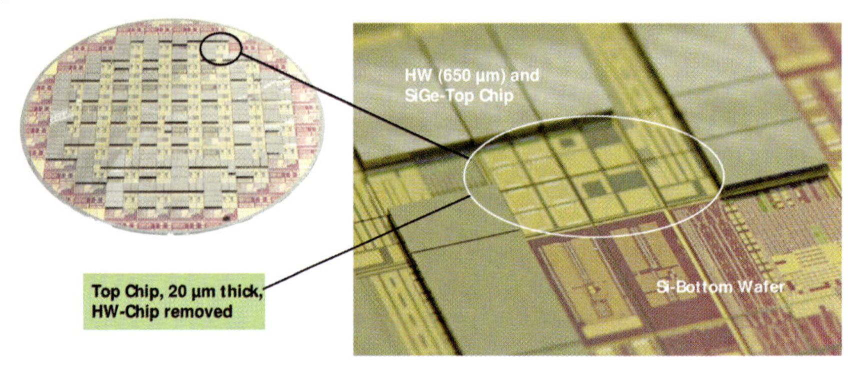

Figure 16.23 Vertical system integration – Chip-to-wafer stacking of strained Si CMOS devices (SiGe chips) and standard Si CMOS devices (Si wafer) by ICV-SLID technology.

As a representative result for the application of heterogeneous integration, Figure 16.23 shows stacked devices of different fabrication technologies on a 200 mm wafer. Strained Si CMOS top chips, fabricated on SiGe substrates, are 3D integrated by use of the ICV-SLID technology with standard CMOS devices on the bottom Si wafer. The stacked 20 µm thin SiGe chips can be seen on the top of the Si device wafer with partly removed handling substrates (HW-chips, thickness: 650 µm).

Void-less metallization of the high-ratio aspect through silicon vias can can be achieved either by CVD of tungsten or copper. Fraunhofer IZM is pursuing both ICV-SLID technology based on W- and Cu-filled through silicon vias, depending on the applications (e.g. RF requirements). Besides the electromagnetic performance, reliability issues for the different applications also have to be considered for the selection of the applied metal system.

16.3 Reliability Issues

Besides the electromagnetic performance, thermo-mechanical reliability issues for the different applications also have to be considered for the selection of the applied metal system and its geometry. Mechanical stresses and strains, thermally induced by the mismatch in the coefficients of thermal expansion (CTE) of the individual materials, are liable to damage materials, interfaces and interconnects during processing or operation [31]. They are a lifetime-limiting factor and have to be considered in the design of 3D-integrated microelectronic systems. By means of a "physics-of-failure"-based approach, using experimental methods for material characterization and test as well as Finite Element (FE) simulations, it is possible to generate a lifetime model for a specific failure mechanism and use it for the prediction of reliability as a function of given design variables and loading conditions [32].

16.3.1
Failure of 3D Integrated Systems

Some typical interconnect failures concerning 3D integrated systems, which can be traced back to thermo-mechanical loading during either packaging, assembly or operation, can be seen in Figure 16.24.

To understand and then predict possible failures, one has to obtain a lifetime model for the respective failure mechanism. Such a model, which is at the heart of every successful lifetime prediction, is a theoretical framework linking failure analytical data of strong experimental support with simulation results including material properties as a function of loading conditions and technological parameters. These are usually empirical or semi-empirical relationships between a measured (often statistical) quantity, for example, a mean number of cycles to failure (N_f) or a mean crack length and a corresponding calculated failure criterion. A good example for a lifetime model is the rather simple and often used Coffin–Manson relation (Equation 16.1) for low cycle fatigue failure of solder joints [33], which uses an accumulated (equivalent) creep strain ε_{cr} (or also a dissipated energy) per thermal cycle as failure parameter.

$$N_f = c_1(\varepsilon_{cr})^{-c_2} \tag{16.1}$$

where c_1 and c_2 are two empirical material constants. This relationship, once calibrated, can then be used for numerical lifetime prediction. The Coffin–Manson model can also be employed using periodic stress or plastic strain as failure

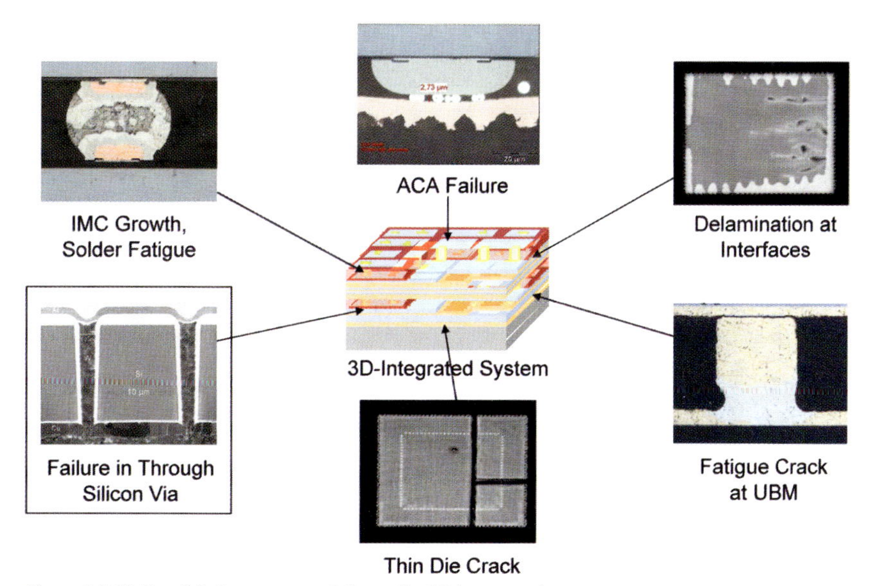

Figure 16.24 Possible interconnect failures for 3D-integrated systems. Here, the focus is put on through silicon vias.

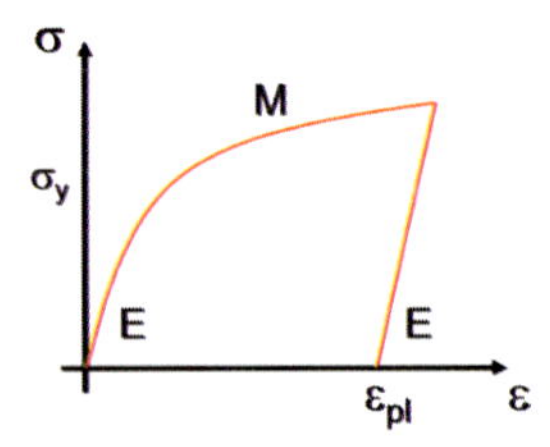

Figure 16.25 Elastic–plastic material behavior.

parameter (for a review on solder joint reliability, see, for example, Reference [34]). For such calculations, the simulation requires knowledge of material data. This is exemplified for an elastic-plastic material behavior in Figure 16.25.

Tensile loading beyond the yield stress σ_y induces plastic deformation and further hardening, characterized by a tangent modulus M. Upon unloading, the material responds again elastically (characterized by Young's modulus E), leaving an irreversible plastic strain ε_{pl}. During this inelastic process, energy is dissipated within the deformed material, which accumulates damage. As a stress state within a real package is seldom uni-axial, it is often advantageous to use the equivalent quantities in Equation 16.1 due to their scalar connection with shear-dominated contributions in the deviatoric tensor-invariants of the stress (or strain) tensor. Hence these quantities can serve as failure criterion for mainly shear stress induced failure mechanisms involving, for example, dislocation motion or grain boundary sliding.

Other prominent failure mechanisms depicted in Figure 16.24 that involve crack propagation, such as interface delamination, may not be treated by such an accumulative damage approach but require an explicit treatment of cracks by the theoretical framework of fracture mechanics instead. The remainder of this section is dedicated to a screening for critical points and qualitative assessment of through silicon via reliability by calculation of stresses and plastic strain for later linking to a lifetime model.

16.3.2
Material Characterization of Thin Layers by Nano-Indentation

All lifetime prediction by simulation that is based on a physics-of-failure approach hinges crucially on knowledge of material data, that is, constitutive material laws and failure behavior as a function of given loading conditions. Both need to be characterized by appropriate methods, as these quantities also often depend on process parameters and size, such as layer thickness. For through silicon vias thin layers of sub-micron thickness deposited, for example, by CVD processes or sputtering need to be characterized. As the fabrication of test-specimens for tensile testing seems to be impractical here, it is advantageous to opt for the method of nano-indentation, which offers a convenient means to obtain material data like Young's modulus (E) and hardness [35]. However, for implementation into an FE-code it is necessary to

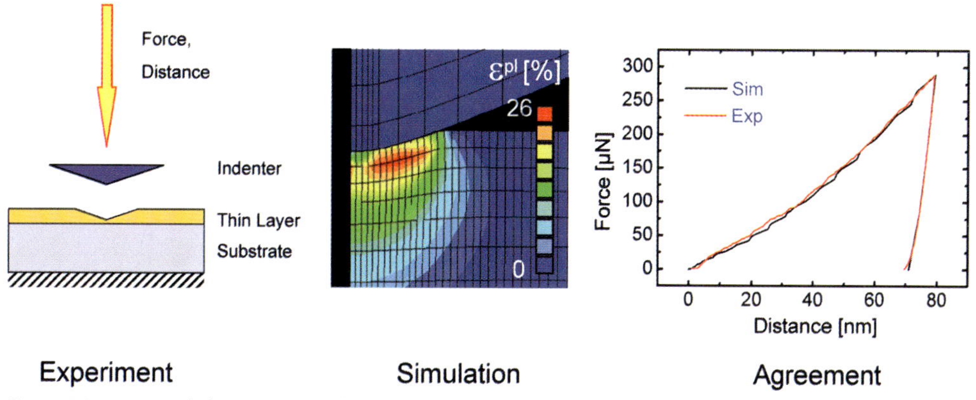

Figure 16.26 Material characterization by nano-indentation: Extraction of elastic-plastic data by means of Finite-Element simulation.

also provide the yield stress σ_y as well as tangent modulus M to describe an elastic–plastic material as depicted in Figure 16.25. To obtain the desired quantities, it is possible to use a combined approach based on nano-indentation and simulation, where the three material parameters are systematically varied by an optimization routine until best agreement with the experimental force–distance curve is reached. This procedure, depicted in Figure 16.26, is necessary as the nano-indenter does not produce a simple but rather a complex stress and strain state. As can be inferred from the good agreement of simulation and experiment, the elastic–plastic material data should be valid for the simulation of a through silicon via.

In the given case, the 800 nm thin AlSiCu layer was sputtered onto a monitor wafer under real process conditions to then undergo the nano-indentation measurement. The whole process, including calibration [36], is outlined in Reference [37]. By this method, values of $E = 93$ GPa, $\sigma_y = 190$ MPa and $M = 1400$ MPa were obtained for the AlSiCu layer. For the remaining material data please refer to Table 16.1 or Reference [37].

Table 16.1 Material properties; all material properties are given at room temperature.

Material	*E* (MPa)	Poisson ratio	CTE (ppm K^{-1})	σ_y (MPa)	*M* (MPa)
Cu	90.000	0.35	16.5	180	6.600
Cu_3Sn	100.000	0.3	18.0	*480*	*5.800*
AlSiCu	93.000	0.3	24.0	*190*	*1.400*
W	145.000	0.28	4.4	*1600*	*10.000*
Si	168.000	0.22	2.8	—	—
SiO_2	75.000	0.28	0.55	—	—

16.3.3
Thermo-mechanical Simulation of Through Silicon Vias

After implementation of material laws and data into the FE-code, the model can be generated. Such a model requires two features for the envisaged study: first, a parametrical layout to allow rapid variations with respect to geometry and material and, second, a successive build-up option to permit a realistic simulation of the process flow. The latter is necessary to capture the correct stresses evolving at bi-material interfaces due to thermal mismatch on cooling from the respective process temperature. Figure 16.27 shows the thermal load history according to the different thermal process steps like, for example, layer deposition, etching or structuring [37], starting from the thermal oxide formation on the wafer.

Figure 16.28 shows the FE-model and simulation results for the through silicon via using rotational symmetry. It was modeled after Figures 16.16 and 16.22, leaving out the TiW seed layers due to their thinness. Then, four different configurations of interest were simulated (where the abbreviations refer to Figure 16.28):

1. The standard layout, using 20 μm thick silicon (STD)
2. A variant using low temperature processes for oxide formation and sputtering (Low T)
3. A variant using 50 μm thick silicon, that is, longer vias (Tck Si)
4. A variant using copper instead of tungsten as filler material (Cu).

Here, the most important results of the FE-simulations [37] are depicted for equivalent stress and equivalent plastic strain, where always the maximum value at the most critical location is given after all process and thermal cycling steps. As the magnitude of stress or plastic strain can be taken as a failure criterion, design optimization involves their possible reduction. In the first place this can be achieved

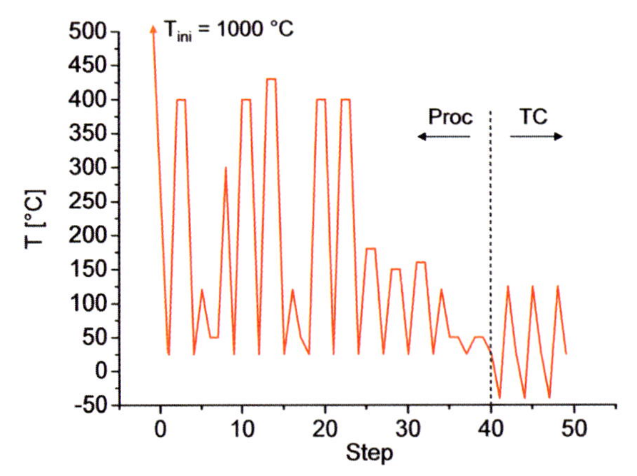

Figure 16.27 Thermal load history as used for simulation input. For the evaluation of thermo-mechanical stress, the process steps are followed by three thermal cycles ($T = -40$ to 125 °C).

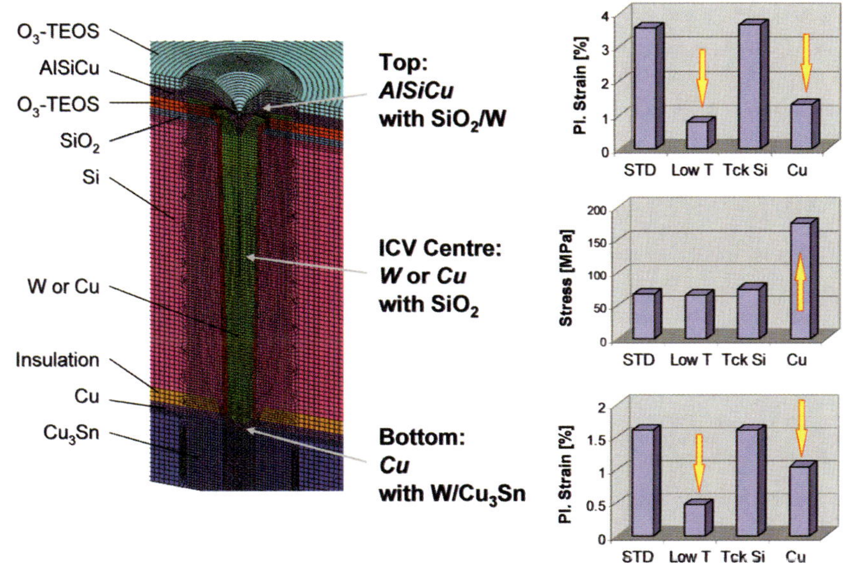

Figure 16.28 Results of Finite-Element simulation: Identification of critical points with calculated maximum equivalent stress and maximum accumulated equivalent plastic strain, respectively. Arrows indicate favorable/unfavorable changes as design parameters are varied.

by a low temperature process as this minimizes thermal mismatch, which is expectedly dominant for the AlSiCu/SiO_2/W interface at the top of the via, causing the AlSiCu to deform plastically, that is, stresses at low temperature are of the order of the yield stress. Using copper as filler also reduces this thermal mismatch significantly. At the centre of the via copper has the inverse effect due to a larger local mismatch to the adjacent SiO_2 than tungsten, thus giving rise to a larger stress. Copper as a filler may hence represent a higher reliability risk, although it has to be verified experimentally if the calculated stress is already critical. At the bottom end of the via the copper layer accumulates plastic strain during processing. However, during thermal cycling, no periodic increase in accumulated plasticity occurs, in that there is only a small reliability risk to be expected there. The same is true for the configuration using thicker silicon: the stresses depend on local CTE-mismatch.

As no experimental lifetime data are available, the presented numerical results can only be interpreted qualitatively. In comparison it would be possible to determine if a calculated stress or strain is critical or not, calibrating the lifetime model. As no periodic plastic strain per cycle occurs, only a stress delta over one periodic thermal cycle may be used as failure parameter in the sense of a Coffin–Manson based approach. For verification on the micrometre scale, the simulated stresses can be experimentally determined by local residual stress measurement, see for example, Reference [38].

16.4
Conclusions

There are several reasons for the introduction of 3D system integration technologies. Besides the advantage of enabling microelectronic systems with smaller volumes, footprints and weights, there are mainly three major drivers for advanced 3D integrated products:

1. Performance improvement
2. Cost reduction for fabrication of multifunctional systems
3. Enabling of completely new ultra-miniaturized products.

Consequently, no single 3D integration technology is suitable for the fabrication of the large variety of 3D integrated systems. Moreover, even one single product may need several different technologies for cost-effective fabrication. For example a wireless sensor system consisting of MEMS, ASICs, memories, antennas, and power modules can be fabricated in a cost-efficient way by application of specific optimized 3D technologies for the integration of the different sub-modules.

Fraunhofer contributes on all levels of 3D system integration. Successful market entry of 3D system integration will be determined by the achieved performance improvement and the profitability in relation to the total system cost. Manufacturing based on wafer level fabrication processes shows in general a comparatively favorable cost structure and performance. Consequently, Fraunhofer has focused R&D on wafer-level processes: wafer-level die stacking (without through silicon vias) and vertical system integration – VSI (without through silicon vias). In the first category, chip-embedding concepts, such as the so-called thin chip integration technology (TCI), are well-established at Fraunhofer IZM and the Technical University of Berlin. These techniques provide vertical integration densities of typically 10^2–10^4 cm^{-2} and are well-suited for many applications, such as, for example, integration of active and passive components, resulting in performance and reliability enhancement. Several wafer-level 3D integration technologies based on through silicon vias have been developed at Fraunhofer Munich since the early 1990s. The focus of the R&D was on "via first" concepts based on adhesive bonding or metal bonding schemes. The so-called Inter-Chip Via (ICV) technology is based on bonding with polyimide as intermediate layer. Optimized for wafer-to-wafer stacking, it provides a very high vertical interconnect density of some 10^6 cm^{-2}. In most applications, wafer yield and chip area issues speak against wafer stacking. Therefore, chip-to-wafer stacking concepts utilizing known dies only are of advantage for 3D system integration of devices with non-identical chip sizes or/and low wafer yield. Fraunhofer IZM Munich has developed a post backend-of-line "via first" 3D integration process, the so-called ICV-SLID technology based on metal bonding using solid-liquid-interdiffusion (SLID) soldering. The SLID metal system provides both the mechanical and electrical connection in a single step. The ICV-SLID fabrication process is well suited for cost-effective production of both high-performance applications (e.g. 3D microprocessor) and highly miniaturized multifunctional systems; the latter preferably in combination with wafer-level die stacking, as, for example, TCI technology. The fabrication of distributed wireless sensor systems (e.g. e-CUBES®) [39] is a typical example of the need for such mixed approaches.

Besides electromagnetic performance, reliability issues for the different applications also have to be considered for the selection of the applied metal system. Suitable thermo-mechanical simulations are a crucial ingredient in the process of technology development – for example to identify the locations of highest loading during processing and operation by monitoring stress and plastic strain building up due to thermal mismatch. It can thus be shown that with W-filled through silicon vias the maximum stresses and strains are observed not in the bulk region of the through silicon via but in the upper part of the via between IC metal layer and the tungsten filler, while the use of copper-filler induces more stress in the through silicon via itself. The consequence, process steps are adjusted to minimize stress to assure higher reliability.

Short Summary

Successful market entry of 3D system integration will be determined by the achieved performance improvement and the profitability in relation to the total system cost. Manufacturing based on wafer level fabrication processes shows in general a comparatively favorable cost structure and performance. In consequence, Fraunhofer focused the R&D on wafer-level processes: Wafer-level die stacking (without through-Si vias) and vertical system integration - VSI® (with through-Si vias). Chip-to-wafer stacking concepts utilizing known dies only, are of advantage for 3D system integration of devices with non-identical chip sizes or/and low wafer yield. Fraunhofer Munich developed a “via first” 3D integration process, the so-called ICV-SLID technology based on metal bonding using solid-liquid-interdiffusion (SLID) soldering. The SLID metal system provides the mechanical and the electrical connection, both in one single step. The ICV-SLID fabrication process is well suited for the cost-effective production of both, high-performance applications (e.g. 3D micro-processor) and highly miniaturized multi-functional systems (e. g. e-CUBES®).

Besides the performance, also reliability issues have to be considered for the different applications. Thermo-mechanical simulations, e. g. for vertical interconnects based on tungsten and copper filled through-Si vias, respectively, represent a crucial ingredient in the technology development for 3D integration.

Acknowledgments

The wafer-level 3D system integration technologies and simulations described in this chapter were partly developed in projects supported by the German Government and the European Commission.

References

1 Morrow, P. *et al.* (2007) Design and fabrication of 3-D microprocessors. Material Research Society Symposium Proceedings 970, MRS 2006 Fall Meeting, Boston (eds C.A. Bower, P.E. Garrou, P. Ramm and K. Takahashi), Materials Research Society, Warrendale, Pennsylvania, pp. 91–102.

2 Koyanagi, M. *et al.* (2001) Neuromorphic vision chip fabricated using three-dimensional integration technology. IEEE International Solid-State Circuits Conference ISSCC.

3 Koyanagi, M. (2003) Three dimensional integration technology by wafer-to-wafer stacking. Proceedings International Workshop on 3D System Integration, Munich.

4 International Technology Roadmap for Semiconductors (ITRS), http://public.itrs.net, 2007 release.

5 Wolf, M.J., Schacht, R. and Reichl, H. (2004) The “e-grain“ concept building blocks for self-sufficient distributed microsystems. *Frequenz*, **58** (3–4), 51–53.

6 http://www.ecubes.org.

7 Ramm, P., Klumpp, A., Merkel, R., Weber, J., Wieland, R., Ostmann, A., Wolf, J.M. and Reichl, H. (2003) 3D system integration technologies. Material Research Society Proceedings 766, MRS 2003 Spring Meeting, San Francisco (eds A.J. McKerrow, J. Leu, O. Kraft and T. Kikkawa), Materials Research Society, Warrendale, Pennsylvania, pp. 3–14.

8 Löher, T., Neumann, A., Pahl, B., Patzelt, R., Ostmann, A. and Reichl, H. (2006) Laminate concepts for chip embedding: process technologies and reliability results. EMPS, 4th European Microelectronics and Packaging Symposium, Terme Catez, Slovenia, pp. 21–24.

9 Seegebrecht, P., Bollmann, D., Buchner, R., Csepregi, L., Haberger, K., Klinger, S., Klumpp, A., Ramm, P., Schreil, M., Seidl, A., Seitz, S., Sigmund, H. and Weber, J. (1990) Dreidimensionale Integrierte Schaltungen, Contract NT-2731, Fraunhofer-Institut für Festkörpertechnologie (IFT), München; Bundesmisisterium für Forschung und Technologie, Bonn, http://opc4.tib.uni-hannover.de:8080, “NT-2731-A3”, published final report (01.03.1987–31.12.1989), München.

10 Reichl, H. and Wolf, M.J. (6–7 November 2006) Hetero system integration – challenges and requirements for packaging MHSI 2006, Sendai, Japan.

11 Hubner, H., Aigner, M., Gruber, W., Klumpp, A., Merkel, R., Ramm, P., Roth, M., Weber, J. and Wieland, R. (2003) Face-to-face chip Integration with Full Metal Interface.Proc. Advanced Metallization Conference AMC 2002, San Diego(eds B.M. Melnick, T.S. Cale, S. Zaima and T. Ohta), Materials Research Society, Warrendale, Pennsylvania, pp. 53–58.

12 Landesberger, C., Reichl, H., Ansorge, F., Ramm, P. and Ehrmann, O. (2004) Multichip Module and Method for Producing a Multichip Module. EP 1 192659 B1, DE 100 11 005 B4.

13 Toepper, M., Baumgartner, T., Jordan, R. and Reichl, H. (2006) The Role of Thin Film Polymers in SiP Applications. Symposium on Polymers for Microelectronics. May 3–5, 2006. Wilmington, Delaware, USA.

14 Wolf, M.J., Ramm, P. and Klumpp, A. (2007) 3D-Integration TSV Technology. EMC 3D Technical Symposium, Eindhoven, Netherlands.

15 Wolf, M.J., Ramm, P. and Reichl, H. (2007) 3D-System Integration on Wafer Level. SEMI Technology Symposium 2007, International Packaging Strategy Symposium (IPSS), Tokyo, pp. 9-3–9-9.

16 Binder, F. (2007) Low Cost Si Carrier – 3D for High Density Modules. 3D Architectures for Semiconductor Integration and Packaging, San Francisco, CA.

17 Zoschke, K., Buschick, K., Scherpinski, K., Fischer, Th., Wolf, J., Ehrmann, O., Jordan,

R., Reichl, H. and Schmueckle, F.J. (2006) Stackable thin film multi layer substrates with integrated passive components. 56th Electronic Components and Technology Conference, San Diego, Kalifornien USA, pp. 806–813.

18 BMBF-Verbundprojekt (2005) Autarke verteilte Mikrosysteme. FKZ: 16SV1656.

19 Ramm, P. and Buchner, R. (Sep. 22 1994) Method of making a vertically integrated circuit, US Patent 5,766,984, priority, [DE].

20 Kühn, S., Kleiner, M., Ramm, P. and Weber, W. (1995) Interconnect capacitances, crosstalk and signal delay in vertically integrated circuits. International Electron Device Meeting IEDM 1995, Washington, IEDM Tech. Digest, pp. 249–252.

21 Ramm, P. *et al.* (1997) Three Dimensional Metallization For Vertically Integrated Circuits, in Microelectronic Engineering 37/38 (eds S. Namba, J. Kelly and M. van Rossum), Elsevier Science, pp. 39–47.

22 Ramm, P., Bonfert, D., Gieser, H., Haufe, J., Iberl, F., Klumpp, A., Kux, A. and Wieland, R. (2001) Interchip via technology for vertical system integration. Proc. Int. Interconnect Technology Conf. IITC 2001, San Francisco, pp. 160–162.

23 Ramm, P. and Buchner, R. (Sep. 22 1994) Method of making a three-dimensional integrated circuit, US 5,563,084, priority, [DE].

24 Ramm, P., Bonfert, D., Ecke, R., Iberl, F., Klumpp, A., Riedel, S., Schulz, S.E., Wieland, R., Zacher, M. and Gessner, T. (2002) Interchip via technology using copper for vertical system integration. Proceedings Advanced Metallization Conference AMC 2001, Montreal (eds A.J. McKerrow, Y. Shacham-Diamand, S. Zaima and T. Ohba), Materials Research Society, Warrendale, Pennsylvania, pp. 159–165.

25 Morrow, P. *et al.* (2005) Wafer-level 3D interconnects via Cu bonding. Proc. Advanced Metallization Conference AMC 2004, San Diego (eds D. Erb, P. Ramm, K. Masu and A. Osaki), Materials Research Society, Warrendale, Pennsylvania, pp. 125–130.

26 Ramm, P. and Klumpp, A. (May 27 1999) Method of vertically integrating electronic components by means of back contacting, US Patent 6,548,391 priority, [DE].

27 Bernstein, L. and Bartolomew, H. (1966) Application of solid-liquid-interdiffusion (SLID) bonding in integrated-circuit fabrication. *Transactions Met Society AIME*, **236**, 405.

28 Klumpp, A., Merkel, R., Wieland, R. and Ramm, P. (2003) Chip-to-wafer stacking technology for 3D system integration. Proceedings Electronic Components and Technology Conference ECTC 2003, New Orleans, pp. 1080–1083.

29 Ramm, P., Klumpp, A., Merkel, R., Weber, J. and Wieland, R. (2004) Vertical system integration by using inter-chip vias and solid-liquid-interdiffusion bonding. *Japanese Journal of Applied Physics*, **43** (7A), 829–830.

30 Ramm, P. (2006) 3D system integration: enabling technologies and applications. Extended Abstracts of the International Conference on Solid State Devices and Materials SSDM 2006, Yokohama, pp. 318–319.

31 Lau, J.H. (1993) *Thermal Stress and Strain in Microelectronic Packaging*, Van Nostrand Reinhold, New York.

32 Wunderle, B. and Michel, B. (2006) Progress in reliability research in the micro and nano region. *Journal of Microelectronics Reliability*, 1685–1694.

33 Manson, S.S. (1966) *Thermal Stress and Low Cycle Fatigue*, McGraw-Hill, New York, USA.

34 Dudek, R. (2006) Characterisation and modelling of solder joint reliability, in Mechanics of Microelectronics (eds G.Q. Zhang, W.D. van Driel and X.J. Fan), Springer, pp. 377–468.

35 Cheng, Y.-T. and Cheng, C.-M. (1999) Can stress–strain relationships be obtained from indentation curves using conical and pyramidal indenters? *Journal of Materials Research*, **14**, 3493–3496.

36 Oliver, W.C. and Pharr, G.M. (1992) An improved technique for determining hardness and elastic modulus using load and displacement sensing indentation experiments. *Journal of Materials Research*, **7** (6), 1564–1583.

37 Wunderle, B., Mrossko, R., Wittler, O., Kaulfersch, E., Ramm, P., Michel, B. and Reichl, H. (2007) Thermo-mechanical reliability of 3D-integrated microstructures in stacked silicon Material Research Society Symposium Proceedings, 970, MRS 2006 Fall Meeting, Boston (eds C.A. Bower, P.E. Garrou, P. Ramm and K. Takahashi), Materials Research Society, Warrendale, Pennsylvania, pp. 67–78.

38 Vogel, D., Sabate, N., Wunderle, B., Keller, J., Michel, B. and Reichl, H.(Nov 1–4 2005) Nanoreliability for mechanically loaded devices. International Congress of Nanotechnology 2005, San Francisco, USA.

39 Ramm, P. and Sauer, A. (2007) 3D integration technologies for ultrasmall wireless sensor systems – the e-CUBES project. *Future Fab International*, **23**, 80–82.

17
Interconnect Process at the University of Arkansas

Susan Burkett and Leonard Schaper

17.1
Introduction

Three-dimensional (3D) integration is one of many solutions being explored to solve the current interconnect problems in the semiconductor industry. In recent years, with feature size reduction and increased transistor performance, on-chip interconnect technology has remained relatively the same, except for an increase in the number of interconnect layers. Interconnect latency has become the primary limit on both the performance and the energy dissipation of gigascale integration (GSI) [1]. IBM announced on April 12, 2007 their use of through silicon via (TSV) technology as a way to extend Moore's Law by stacking chip components that would ordinarily be placed next to one other. This news is the culmination of intensive 3D integration research over the past few years. Numerous efforts by universities, research institutes, and companies have taken place and a summary of their approaches is provided in Reference [2]. Although the methods may differ, the primary concept remains the same, which is to create communication paths among relatively distant points with low latency while merging multiple technologies. 3D integration offers the opportunity to significantly reduce the length of the longest global interconnects that are currently present in 2D interconnections.

Several approaches have been taken to achieve interconnects along the z, or vertical, axis. Formation of through-wafer interconnects may involve etching through ~300–500 m of silicon [3–5]. The resulting vias will typically have a diameter in the range 25–50 μm. These wafers do not require special handling due to the wafer thickness; however, highly conformal coatings are needed for via lining materials because of the high aspect ratio. Another approach is to form a blind via without etching all the way through the wafer and then remove the remaining silicon from the back side. This can be done with smaller diameter vias [6] or vias in the range listed above [7]. This process can create very thin wafers, depending on the via etch depth. If a wafer is <~50 μm, a handling wafer is required. An important processing consideration in this connection scheme is the order in which the via is created in terms of bonding the chips. One approach, to create the array of vias before bonding chips, is commonly used in industry and by many research

Handbook of 3D Integration: Technology and Applications of 3D Integrated Circuits.
Edited by Philip Garrou, Christopher Bower and Peter Ramm

ISBN: 978-3-527-32034-9

groups [8–12]. The vias are coated with insulation and barrier materials typically before filling with copper. The wafer is thinned from the back until the metallized vias are exposed. The vias can also be etched after wafers are thinned and polished, in SOI substrates, using the oxide as an etch stop [13]. Another technique involves via formation after chips are individually stacked [14]. When creating via interconnections, copper is the preferred metallization since it provides lower resistance, higher allowed current density and increased scalability.

At the University of Arkansas, through silicon vias (TSVs) are a critical piece in the test vehicle for a project that involves 3D integration of a silicon control IC and a GaAs transmit/receive (T/R) MMIC. TSVs are currently formed on dummy silicon wafers during the development phase. Once this phase is complete, vias will be formed after the fabrication of Si-based electronic devices is complete. This approach requires that processing temperatures be kept relatively low (<350 °C) and that processing be compatible with existing devices. TSVs on dummy wafers are currently fabricated by etching blind vias in the range of 10–25 μm diameter and to depths of 75–125 μm. The vias are lined with insulating, barrier and seed films and filled by copper electroplating. The process wafer is bonded to a carrier wafer using an adhesive layer. Mechanical grinding removes the bulk of the silicon from the wafer back side and is followed by a gentler mechanical polishing step. The vias are exposed from the back side through a reactive ion etch step and contact pads are formed over the vias.

Our specific application requires stacking Si and GaAs chips using an appropriate thermal management scheme to allow cooling between chips. An array of copper posts provides compliance and space between the dissimilar substrates for coolant fluid flow. A copper-tin intermetallic compound is used for the joining process. Figure 17.1 shows a schematic illustration of the integrated GaAs-Si-based electronic system [15]. The following sections detail the process flow for TSV fabrication, the chip assembly process and system integration.

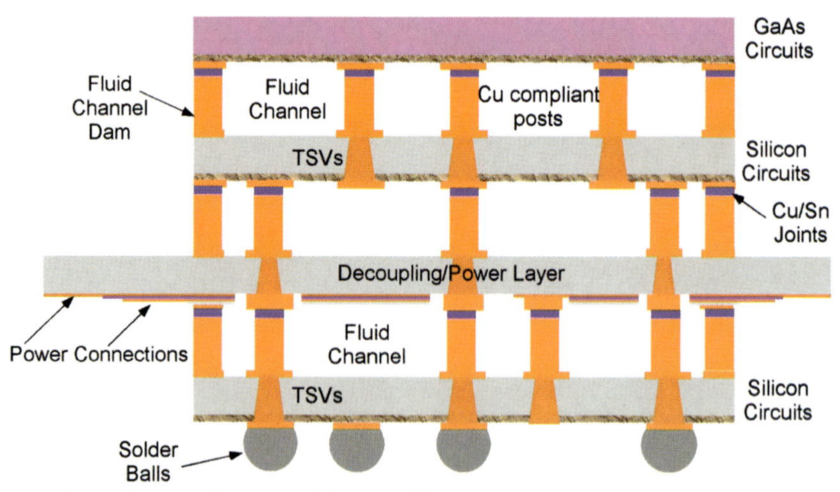

Figure 17.1 Schematic illustration of the assembled 3D Si-GaAs system using the TSV process developed at the University of Arkansas.

17.2 TSV Process Flow

The goal is to develop a robust process for the creation of vertical interconnects in a manner compatible with current IC fabrication technology. The formation of a TSV involves several processing steps with the major areas including via formation, via lining, via filling and back side processing.

17.2.1 Via Formation

The creation of vertical interconnects requires formation of vias deep into the silicon substrate using deep reactive ion etching (DRIE) techniques. All DRIE processing is conducted in an STS Multiplex Advanced Silicon Etcher (ASE) with an inlet for helium cooling of the wafer from the back side. The specifications for the silicon substrates used in these studies are: crystal orientation of ⟨100⟩, resistivity values in the range 0.5–1.0 Ω cm, and an average wafer thickness of 375 μm. The Bosch process [16] alternates many short etch and passivation cycles to allow preferential etching in a vertical direction while the via sidewalls are protected with a fluoropolymer thin film that is deposited during the passivation cycle. In successive etch steps, the polymer coating deposited on the via base will be removed before the coating on the sidewalls. There is a high selectivity of silicon to the photoresist masking material (8 μm) that results in deep anisotropic etching of the substrate. Etch and passivation gases are SF_6 and C_4F_8, respectively. Since the process is cyclic and alternates between etch and passivation steps, scalloping of the sidewalls occurs due to the isotropic nature of the sulfur hexafluoride based etch cycle. Scalloping on the via sidewall has been observed to decrease in size as the coil power decreases [17]. The silicon etch rate depends on feature size and the amount of silicon area that is exposed although it is generally in the range 1–5 $\mu m\,min^{-1}$. DRIE processing is controlled by several parameters that include etch/passivation gas flow rate, etch/passivation cycle time, coil/platen powers and chamber pressure by means of an automatic pressure control (APC) angle setting.

The anisotropic nature of DRIE processing provides a vast number of features and devices that can be fabricated, allowing the microelectromechanical systems (MEMS) area to explode. This process is designed for high aspect ratio etching of silicon, producing via sidewall angles that measure 90°. Conventional RIE techniques generally produce a via with a slightly tapered profile and this process has been used in our laboratory [18, 19] for via formation when the via diameter was much smaller, allowing vias with similar aspect ratios although handling issues associated with thin wafers have to be considered. Mask selectivity is considerably lower for RIE than for DRIE and is therefore not used to etch deep vias. The formation of a slightly tapered, or sloped, via sidewall is beneficial in the TSV process developed at the University of Arkansas due to the subsequent deposition of lining materials. Lining materials are deposited by plasma enhanced chemical vapor deposition (PECVD) and sputtering techniques. Sputter deposition is the limiting process in terms of conformal coverage of materials. Other CVD techniques would allow better coverage of the via although, often, the processing temperature may not be compatible with TSV processing. In other approaches to vertical interconnect formation, MOCVD has been used for conformal seed layer deposition [5].

To achieve a tapered via, there should be excess etching at the via opening compared to the via base. A modified Bosch process using a seven-module approach with varying etch and passivation cycles assists in achieving this profile. The modified Bosch process recipe has seven different modules with constant etch cycle times and variable passivation cycle times and is given in Table 17.1. Each module is expected to achieve a different etch profile. These profiles are formed by controlling the passivation layer deposited on the via sidewall, a critical protective film for preventing or allowing etching to occur at the proper angle. Process parameters include: SF_6 gas flow = 112 sccm; C_4F_8 gas flow = 85 sccm; argon gas flow = 18 sccm; APC angle = 60°; platen power is 12 W for the etch cycle and 0 W for passivation; coil power is 200 W for both etch and passivation cycles. As stated previously, low coil power reduces scallop size and is therefore set to 200 W. The platen power elevates plasma ion flux and energy in addition to improving the directionality of the plasma reactive particles [20]. A reduced chamber pressure also improves the directionality of energetic ions [21]. At an APC angle of 60°, the pressures recorded are 18 and 13 mTorr for etch and passivation cycles, respectively. Notably, when the pressure is lower (APC $\leq$ 50°), the sidewall angle of the via approaches 90° and when the pressure is higher (APC > 65°) sidewall bowing is observed [22]. Figure 17.2 shows a fully etched via using the seven-module recipe [23]. The profile shows that some lateral etching occurs at the via opening. This via has a sidewall angle of 86°. The vias are easily etched to greater depths by increasing the number of cycles.

17.2.2 Via Lining

Once vias are formed by DRIE processing, a 1 μm thick layer of silicon dioxide is deposited at 250 °C inside to provide insulation between the metal filled vias and the

Table 17.1 Modified Bosch recipe using a seven-module approach developed at the University of Arkansas, showing the etch and passivation times.

Modules	No. of cycles	Etching time (s)	Passivation time (s)
1	5	18	5
2	5	18	7
3	5	18	9
4	6	18	11
5	6	18	13
6	6	18	15
7	6	18	17

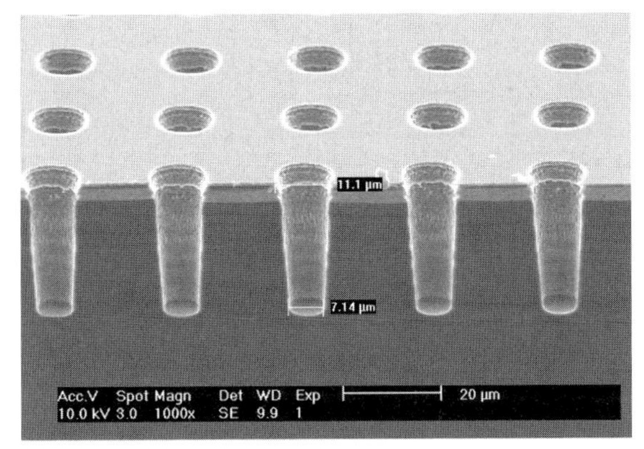

Figure 17.2 SEM cross sectional image of vias etched by the seven-module Bosch recipe, showing the tapered profile and lateral etching at the via opening.

silicon substrate. Plasma enhanced chemical vapor deposition (PECVD) is used to deposit the insulation layer at a deposition rate of $0.1\,\mu m\,min^{-1}$. After oxidation, tantalum nitride (TaN) barrier (0.5 µm) and copper seed layers (1.0 µm) are deposited by DC sputtering. TaN is reactively sputtered at a pressure of 5 mTorr, power of 2500 W, using a mixture of high purity Ar and N_2 gases (3 : 1). Rutherford backscattering spectroscopy (RBS) confirms a N : Ta ratio of 1.08 for the films [23, 24]. TaN prevents diffusion of copper into silicon and improves the adhesion between materials [25, 26]. Comparison studies of Ti, TiN, and TaN barrier metal materials have indicated the superior performance of TaN as well as good electromigration resistance if the interface between the barrier material and Cu is carefully controlled [27]. The copper seed layer plays a critical role in the subsequent electroplating process.

Obtaining a conformal film inside the via by sputter deposition is challenging. The ability to do this successfully depends primarily on the via profile. Sputtering requires the deposition surface to have some incident angle with respect to the trajectory of the atoms. Non-uniform or non-continuous coverage of the copper seed layer inside the via will pose a challenge for proper via filling in the electroplating step. We have observed that vias may be filled with copper even when incomplete seed layer coverage exists. The copper will fill the space inside the via although the adhesion of the copper to the areas that do not contain adequate seed layer coverage is poor. To illustrate this aspect, a series of experiments to study the coverage of the copper seed layer involved terminating the electroplating process at different times after the copper seed layer deposition in vias with nearly straight sidewalls. When the copper seed layer is present inside the via, copper will plate and forms a layer thick enough to be easily observed under an optical microscope. After this step, a clear epoxy is applied on the top of the sample and is allowed to cure in a vacuum oven at room temperature for 9 hours. This step allows the epoxy to fill the vias and holds the copper in place when dicing. Figure 17.3a–c [28, 29] shows a series of via cross sections with

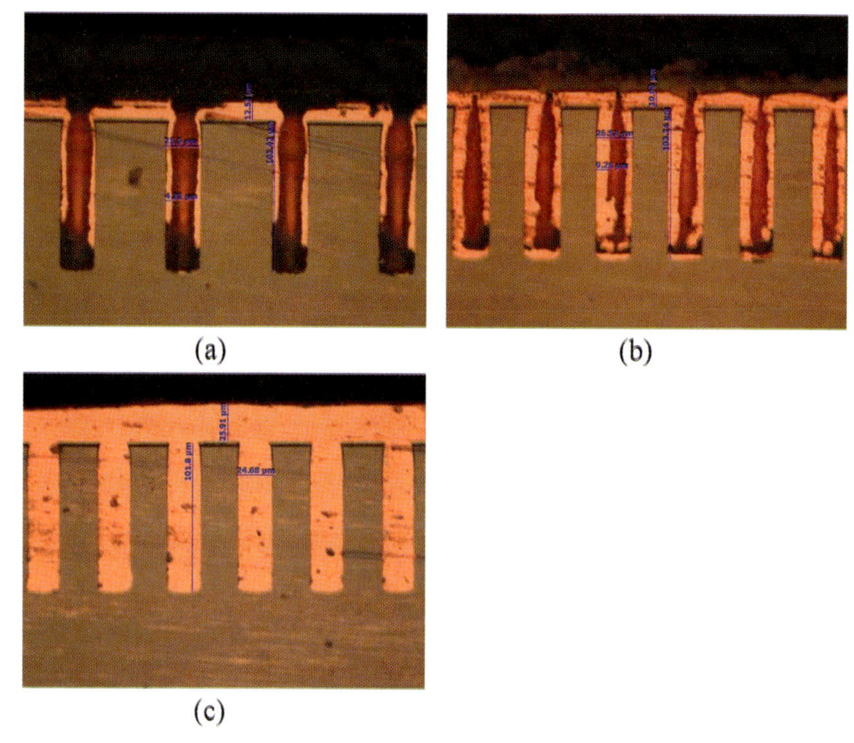

Figure 17.3 Optical microscope cross sectional images of vias lined and partially filled by electroplating for variable times: (a) 30 min; (b) 2 hours; and (c) 4 hours.

copper plating times of 30 min, 2 hours, and 4 hours, respectively. Early plating stages show non-continuous seed layer along the via sidewall and copper barely touching portions of the via base. However, if plating is carried out long enough, the vias will become filled with copper. The resulting adhesion of copper to the via is poor, however, and the copper is easily removed from the via base while dicing. Therefore, the formation of a tapered, or sloped, via is essential for conformal seed layer coverage, which is a prerequisite for good adhesion and void-free via filling.

17.2.3 Via Filling

Once a conformal seed layer is achieved, electroplating processes using a sulfate-based electrolyte and additives provided by Enthone Inc. are used to fill the vias. The aspect ratio of the features results in a non-uniform distribution of current density, one that is much higher at the corner openings than inside the vias. This leads to a higher deposition rate at the via corners, causing the vias to close without complete filling if DC plating is used. This results in voids inside the via, a phenomenon commonly observed in DC plating of features with relatively high aspect ratios. To overcome this, periodic pulse reverse (PPR) plating combined with organic additives is used. In PPR plating, copper at high current density areas is removed during a short reverse pulse to balance the deposition rate over the entire wafer. A period of off-time with no positive or reverse pulse allows for a restabilization of the plating bath. Compared to DC plating, PPR plating at the same average current density gives a uniform plating thickness and high throwing power. To further refine the copper deposit to obtain a bright surface as well as a smooth film with improved physical properties, organic additives are necessary. There are three basic types of additives used in acid copper plating: carriers, brighteners, and levelers. Brighteners, small molecular weight sulfur-containing compounds, can easily reach inside the vias, accelerate the plating reaction in this area, and promote the filling of copper in small feature sizes [28]. Brighteners play a very important role in void-free via filling. Both carriers and levelers are large molecular weight polymers that moderate the activity of brighteners on the wafer surface [28]. Because of their size, they do not play a role in via filling for small feature sizes. Levelers displace brightener species in high current density sites (protrusion areas) to suppress the plating rate at these regions and reduce excess growth of copper. The organic additives both accelerate the deposition rate inside the vias and suppress the rate at via openings on the surface.

Filling vias to steer clear of voids requires a bottom-up superconformal filling process in which the via has a higher deposition rate at the base than the sidewall. IBM uses the term superfilling to describe this process [30]. This filling procedure relies heavily on the role of additives, and the mechanism to predict superfilling of small features is called the curvature enhanced accelerator coverage (CEAC) mechanism [31–33]. This mechanism has been well studied for filling vias and trenches. Bath composition, pulse waveform and bath temperature influence significantly the filling of high aspect ratio features and the resulting film quality [34–36]. Agitation also influences the filling process. In our TSV process, a fountainhead plating fixture ensures that the plating solution reaches the inside of the vias [23]. A pulse profile consisting of 100 ms forward : 10 ms reverse : 100 ms off-time and current densities of 10 mA cm^{-2} forward, 15 mA cm^{-2} reverse have been experimentally determined to allow superfilling vias (Figure 17.4) [37]. The excess copper on the wafer surface can be removed by chemical mechanical polishing, or, in our case, a leveling waveform.

17.2.4 Back Side Processing

After electroplating to fill the vias, the process wafer is attached face down to a carrier wafer for bulk and fine removal of silicon from the wafer back side. Mechanical grinding, polishing and RIE processes allow exposure of the vias from the back side of the process wafer. The carrier wafer is necessary after wafer thinning processes for support and handling purposes. During development work on TSV processing, liquid crystal polymer (LCP) film was the bonding agent used to laminate process and carrier wafers. LCP is sandwiched between process and carrier wafers while heating under a pressure of 12.5 Tons. At $\sim$300 °C, the polymeric material softens and acts as a good adhesive layer between the wafers. This material forms an excellent bond that stands up to the aggressive grinding processes. However, our process requires

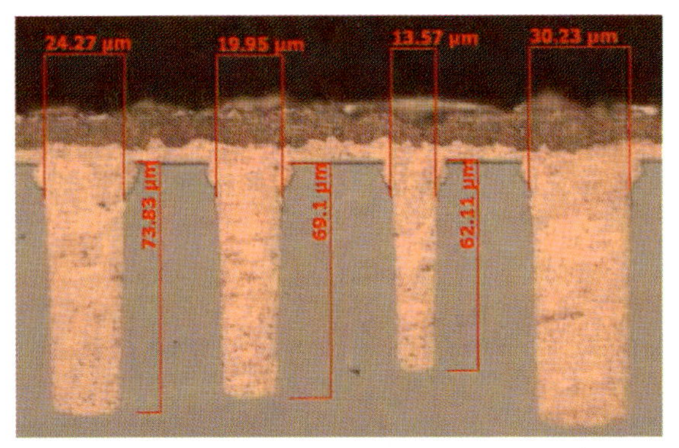

Figure 17.4 Optical microscope cross sectional image of variable diameter vias filled with copper by electroplating using a PPR waveform in combination with organic additives to the copper electroplating bath.

debonding at a later stage for chip stacking. An adhesive manufactured by Brewer Scientific is now used for bonding due to the inability to debond these wafers when using LCP [38]. The thermoset adhesive has operating temperatures in the range of room temperature to 400 °C. The material is applied by spin coating at 2000 rpm (4 μm), soft baked at 180 °C for 3 min and hard baked at 200°C for 3 min. Bonding takes place at 160 °C for 2 min under vacuum. Debonding occurs at 400 °C for 40 min in a nitrogen atmosphere.

After bonding, the process wafer is thinned from the back side using a Logitech PM5 polishing tool. Mechanical grinding performs the bulk removal of silicon until wafer alignment marks become visible. The process of grinding until wafer alignment marks are visible takes ~2 hours. Grinding with a 9 μm abrasive slurry provides an etch rate of approximately 2.5 $\mu m\ min^{-1}$. The alignment marks used for photolithographic processes are significantly larger than the via feature size. Therefore, they are etched deeper than the array of vias, forming a natural process stop. A slurry grit size of 0.3 μm is used to mechanically polish the wafer surface to a smooth finish. This process removes material at a much slower rate (0.25 $\mu m\ min^{-1}$). To successfully smooth and planarize the wafer surface, a 1 hour polish is performed. Wafer polishing is performed after grinding to assist in wafer planarization and improve the surface finish from damage created in the mechanical grinding step.

After mechanical grinding and polishing, approximately 5 μm of silicon remains until the vias are exposed from the back side. Vias are exposed using RIE (SF_6/O_2-based gas chemistry) and a photoresist mask to selectively etch silicon in the area directly above the vias. RIE is used in this step rather than DRIE because the bonding adhesive is a possible contaminant in the DRIE tool. This etch step creates additional "vias" in the area directly above the previously formed DRIE via that eventually connect once the silicon and lining materials are removed (Figure 17.5) [7]. The masking of silicon material except in the area of the vias eliminates the silicon loading effect observed when blanket etching of the entire wafer is used to expose the via [7]. Before RIE processing, the wafer is placed in a copper etch solution (10% H_2SO_4, 5%

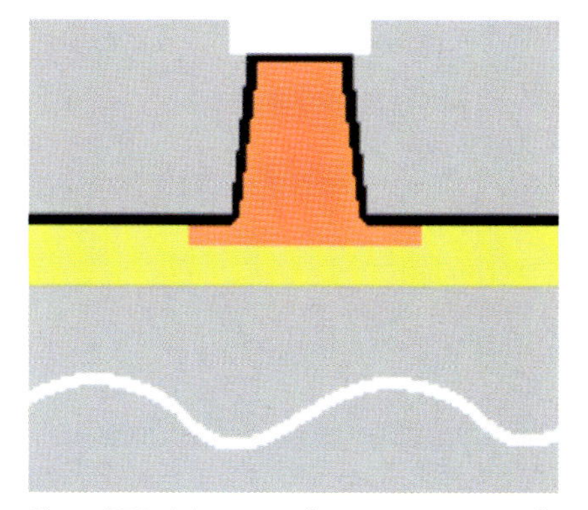

Figure 17.5 Schematic of a via cross section from the process wafer back side. Insulation, barrier, and seed films will be etched where the vias are exposed and Cu electroplated to create a metal filled TSV.

H_2O_2) for 20 s to remove any copper that may have smeared from the alignment marks. After rinsing with deionized (DI) water, hexamethyldisilizane (HMDS) and 2 µm of AZ4110 photoresist is applied to the wafer by spin coating. The photoresist mask is then exposed and developed in AZ4000 developer solution (3 parts water to 1 part developer) for 90 s only in the regions over the vias. To expose the vias, RIE processing with the following parameters is used: SF_6 flow of 40 sccm, O_2 flow of 35 sccm, power of 180 W, pressure of 180 mTorr and etching of 5–10 min or until the smallest diameter vias become visible under an optical microscope.

Once the vias are exposed, the photoresist is removed using a hot vapox solution. Silicon dioxide is deposited by PECVD and functions as the insulation layer between silicon and metallization as described previously. HMDS is applied and photolithography defines the top of the copper pads. Buffered oxide wet etching is used to etch oxide for 2 min over the copper, leaving a connection path for contact pads. The photoresist is removed and sputter deposition of Ti followed by Cu is performed. Pads providing connection to vias are created by a patterning step. After developing photoresist, contact pads are created by copper electroplating. The copper plating is terminated when photoresist and the copper pads are planar. A profilometer is used for this inspection. Since copper oxidizes easily, nickel and gold are electroplated (1 µm each) on this surface. The nickel serves as an intermediate layer to improve the resistivity and adhesion between copper and gold. Copper and Ti are removed to create the test structures. Gold on the contact pads acts as a mask during the copper and Ti etch. Finally, the wafer is cleaned in DI water and is ready for electrical testing.

17.2.5
Electrical Testing

For testing electrical integrity, a test mask is used with three configurations designed for determination of via chain continuity, single via resistance, and via isolation. Figure 17.6 shows a schematic illustrating the test structures [7]. Test structures for

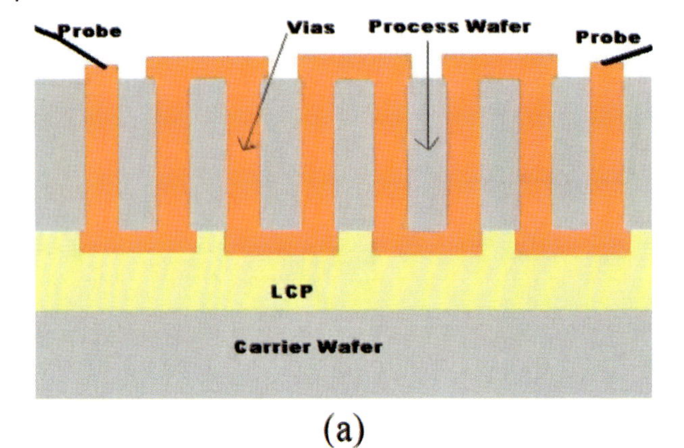

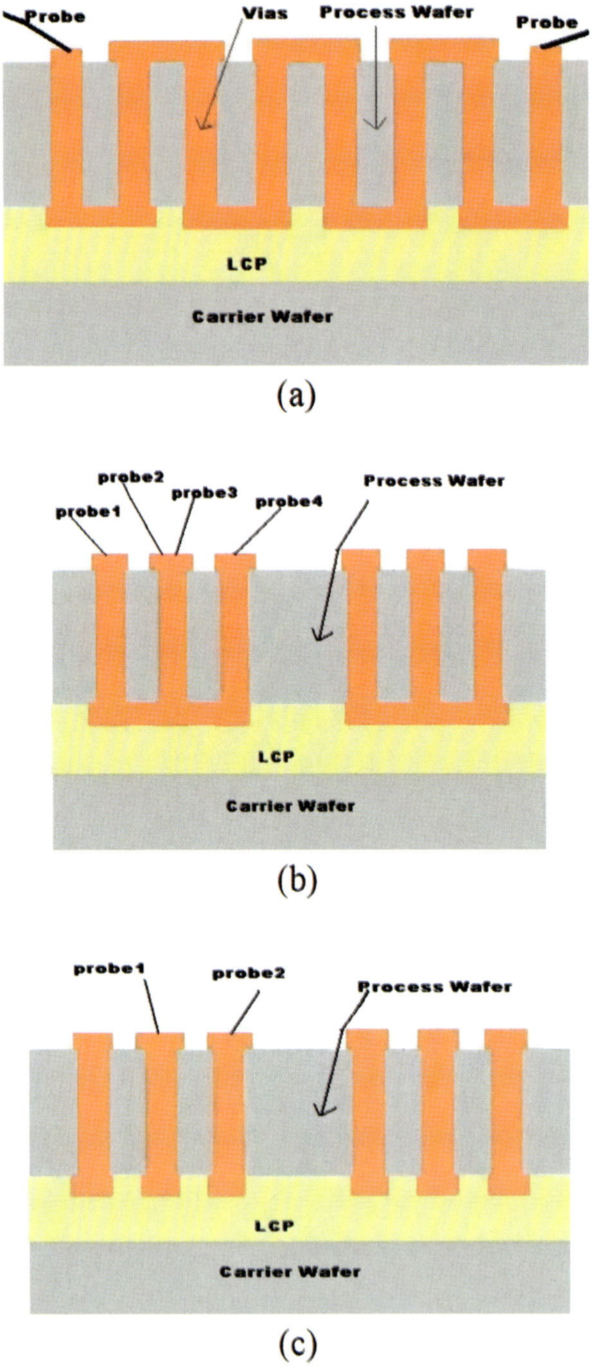

(a)

probe2
probe3
probe1
probe4
Process Wafer
LCP
Carrier Wafer

(b)

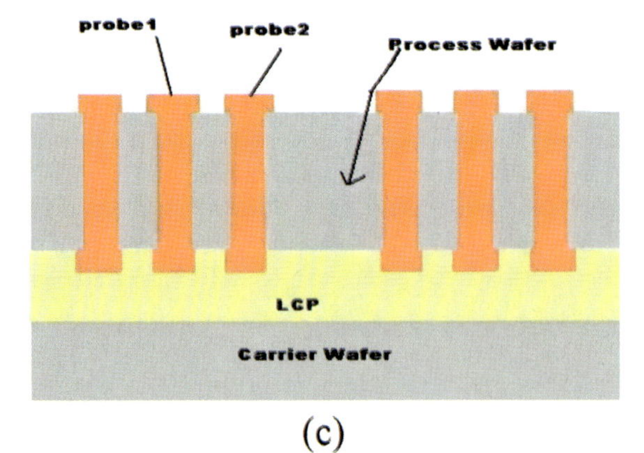

(c)

Figure 17.6 Schematic showing cross sections of the electrical test structures: (a) via chain quality test, (b) single via resistance measurement and (c) isolation testing between two vias.

Table 17.2 Resistance (R) measurements for three TSV test configurations.

Via diameter (μm)	Single via R (measured) (Ω)	Single via R (theoretical) (Ω)	Short via chain R (40 vias) (Ω)	Isolation R (MΩ)
10	0.017	0.016	1.35	270
15	0.007	0.007	0.73	260
20	0.005	0.004	0.7	225
25	0.004	0.003	0.58	200

vias of 10, 15, 20 and 25 μm diameters are tested. The tests allow confirmation that the vias are completely interconnected and that the copper is isolated from the silicon wafer by the silicon dioxide layer. Testing is performed while the process wafer is still attached to the carrier wafer.

Table 17.2 shows the average value of several measurements taken across the test wafer. Results show that the single via resistance of the smaller diameter via is greater than that of the larger diameter via, as expected since resistance is inversely proportional to area. Theoretical values for single via resistance also agree well with experimentally measured values. Chain continuity is very good although via resistance for chains of 40 vias is not exactly 40× the single via resistance. This is because vias are interconnected through connection pads which contribute some small amount of resistance. A cross section of a via chain is shown in Figure 17.7 [7]. The measurement of extremely high resistance values between two neighboring vias of any diameter size proves that the vias are isolated from each other and from the silicon.

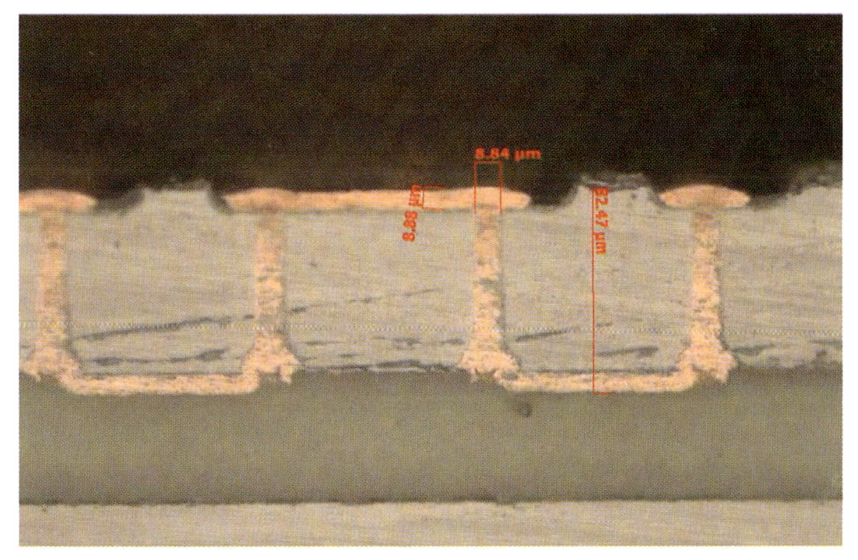

Figure 17.7 Cross section of a TSV chain created for electrical testing purposes.

17.3
Chip Assembly

Chip stacking approaches vary considerably at different organizations and include bonding at the wafer level as well as the chip level. Copper pillar bumps are increasingly used in high-end packaging instead of solder balls because of their compliance [39] and the fact that they require less space, resulting in decreased pitch sizes. In our system, Si and GaAs layers will be connected by TSVs and an array of copper posts that provide compliance between dissimilar materials and allow coolant to flow between the layers. The area between chips for fluid flow is important for heat removal and to provide integrated decoupling and power distribution (see Figure 17.1). Associated yield is also improved by this approach because individual die can be tested before assembly.

Copper posts are fabricated by DC electroplating into a deep mold formed by resist. At the same time a copper dam is formed around most of the circuit layer. This is shown schematically in Figure 17.8 [15]. The copper posts are 35 μm in diameter and 100 μm high. Preliminary calculations indicate that it is possible to remove 20 W per layer for a 1 cm^2 chip stack with 1 g s^{-1} fluid flow (3M PF-5070) at a velocity of 0.6 m s^{-1} with a pressure drop of $\sim$1 psi through the 100 μm high channel [15]. The posts provide space for coolant to flow and the dam contains the fluid. Numerical simulations have been performed to investigate the impact of post height and diameter on the resulting stress. Figure 17.9 shows that as post height increases from 5 to 40 μm a 20% reduction in stress is observed [40]. When the height exceeds 40 μm, there is only an additional 10% benefit. The diameter of the post exhibits less impact on the computed stress values. Thus the post/dam height has been chosen to be 100 μm to provide coolant flow with low pressure drop.

Tin plating (3 μm) is used to cap the posts, forming a Cu_3Sn intermetallic compound that behaves in a similar way to soldering by way of a solid-liquid-interdiffusion process [41, 42]. The compound forms at 300 °C but will not remelt below 600 °C. Development of the copper posts and copper dam is almost complete [43] while chip assembly and system integration are well under way.

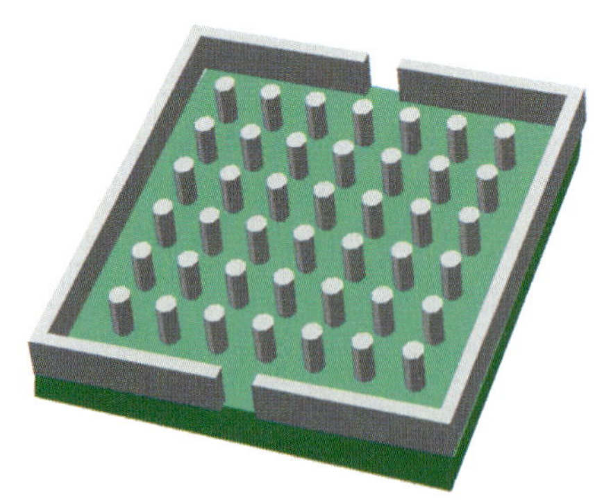

Figure 17.8 Schematic showing the copper posts that provide compliance and allow for coolant to flow through and be contained by the dam structure.

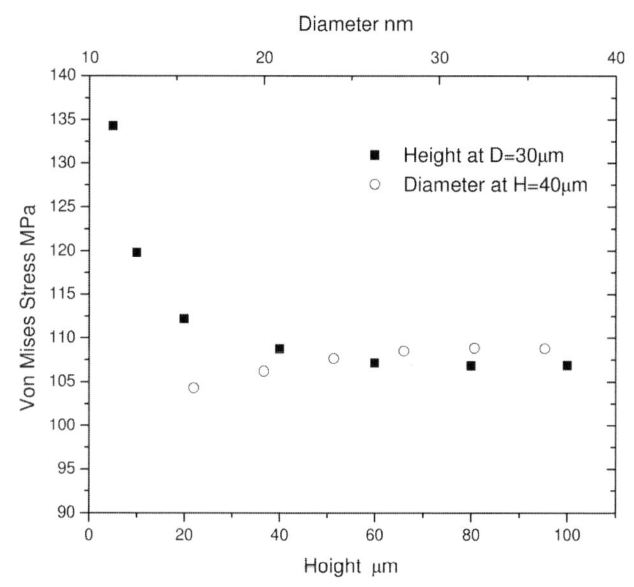

Figure 17.9 Von Mises Stress versus copper post height and copper post diameter.

To fabricate copper posts, a resist mold placed on a sputtered Ti/copper layer is required. SU-8 photoresist was initially used to create the mold although the resist remained on the copper posts after the resist removal process. This created challenges in removing the remaining residue. KMPR 1000 series resists supplied by MicroChem Corporation are now used to create the mold for copper posts. This is a negative resist with similar characteristics to the SU-8 resist. Although the KMPR 1000 series photoresist is easier to remove, it also becomes crosslinked at the base of the via. This prevents copper from plating in that area. A possible reason for the crosslinking in this area is light scattering from the metallic surface. To address this problem, an optical "cut-off" filter was used. The filter eliminates UV wavelengths below 350 nm. The KMPR resist requires this filter because the compound is sensitive to very short wavelengths. Using the optical filter reduces crosslinking significantly. During Cu post processing, KMPR resist is applied by a dual spin coating at 2800 rpm to achieve a 100 μm thick resist film. Between each coating, the wafer is baked at 100 °C and cooled to room temperature before application of another layer of resist. The initial soft bake time is 5 min while the second soft bake time is 18 min. The exposure dose is 1000 mJ cm^{-2}. The post-exposure bake at 105 °C is for 4 min.

Copper plating at a current density of 20 mA cm^{-2} for 5.5 hours allows plating of copper to fill the resist mold. Figure 17.10 shows SEM images of (a) a single copper post and (b) an array of copper posts created using this resist. The posts are robust; initial testing indicates that they adhere very well to the surface and can withstand a large manual force (see Figure 17.11). Daisy chain test vehicles of copper posts with connecting links have been assembled to examine yield and connectivity. Initially, the links had the posts capped with electroplated tin (Figure 17.12). However, the assembled test structures showed very poor yield. Failure analysis revealed many

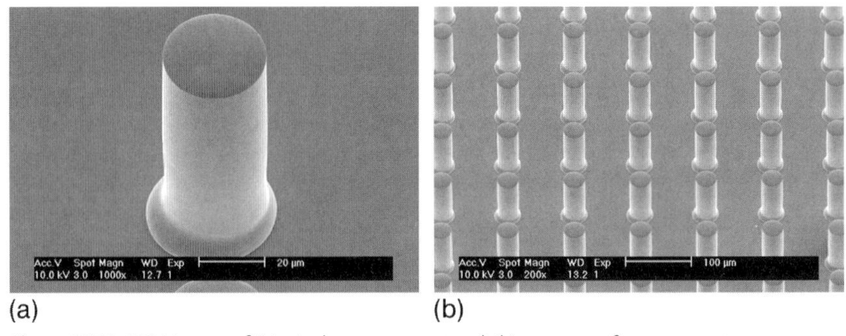

Figure 17.10 SEM image of (a) single copper post and (b) an array of copper posts.

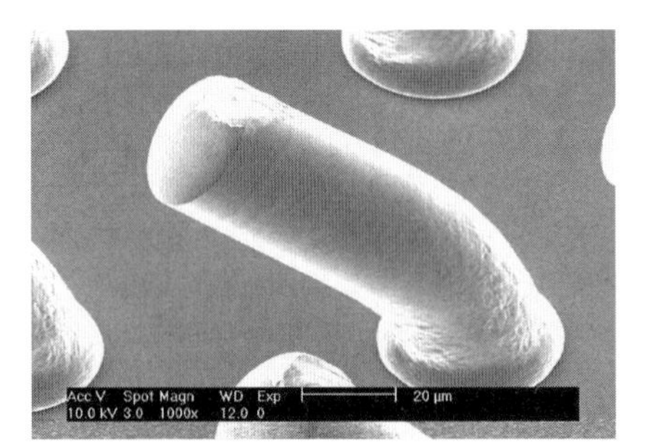

Figure 17.11 SEM image of a copper post bent under manual force using tweezers.

unbonded posts, caused by height variations in copper plating. Thus a polishing step was introduced, in which the copper posts are polished while the KMPR photoresist is still in place. This step is carried out at the wafer level and results in very uniform posts. Since the posts are the same height as the KMPR, to plate tin it is necessary to apply thin (4 µm thick) photoresist and pattern with the post/dam mask before the tin plating step.

Current test vehicles have the tin plated on the connecting links that are bonded to the posts. Before bonding on the flip chip bonder, the posts are dipped in an 8 µm high reservoir of flux. The flux reservoir is made of silicon, with the shallow flux well slightly larger than the test vehicle chip etched by DRIE. The flux is doctor bladed into

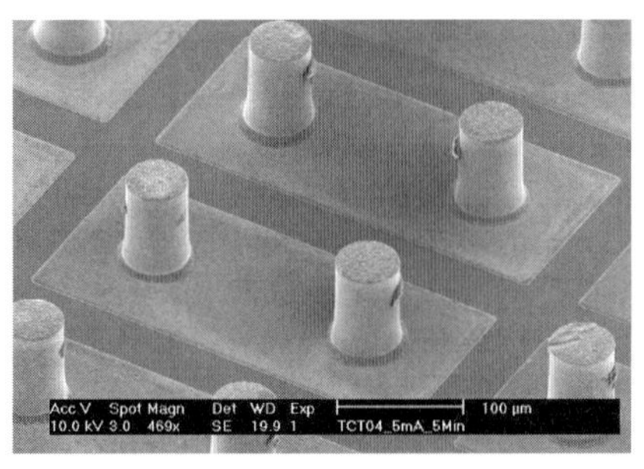

Figure 17.12 SEM image of tin-capped copper posts on metal traces.

Figure 17.13 Image of a cross section of an assembled post structure.

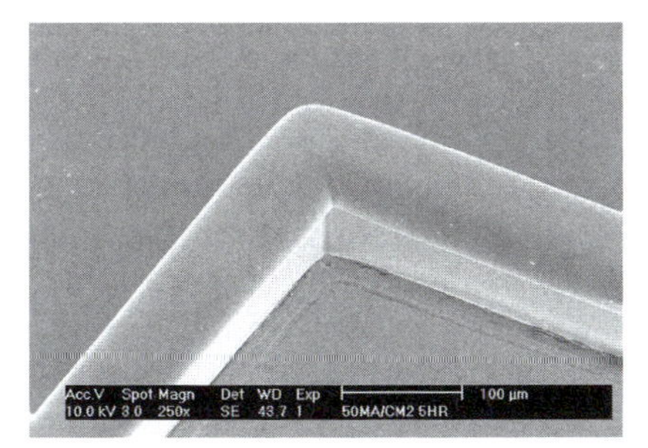

Figure 17.14 SEM image of the fluid dam.

this well to ensure uniform depth, and thus uniform flux deposition on the posts. Assembly is performed at 300 °C with ~1 kg of pressure. Figure 17.13 shows a cross section of an assembled post structure. Test vehicles of post chains include multiple chains of 40, 80 and 160 posts, with over 2000 total posts tested on the 30 μm diameter test vehicle. Several assemblies have been tested with only one open post, which is a very high yield.

Figure 17.14 shows a photograph of the fluid dam. Formation of the Cu/Sn intermetallic must be continuous for the length of the dam, to maintain fluid channel integrity. Pull tests were done and parts were examined to ensure continuous joining, and an assembled test vehicle pair was filled with fluid under pressure with no leaks occurring. At the time of writing, work is proceeding on a full four-layer test vehicle that will be used to demonstrate full electrical and thermal 3D capability.

17.4 System Integration

The system to be integrated in this project consists of a silicon-based control IC and a GaAs transmit/receive (T/R) MMIC. The T/R MMIC is fabricated using a 0.5 μm gate length E/D pHEMT process on 100 μm thick GaAS wafers. This process offers high capacitance/area MIM capacitors, small via grounds, and controlled, repeatable gate length transistors. Enhancement mode pHEMTS are used in amplifier stages, and depletion mode pHEMTS in switches. The custom silicon ASIC provides DACs for bias control of the amplifiers, a serial-to-parallel data interface/shift registers/latch circuits for controlling the phase shifters and T/R switches, and a limited amount of on-board memory. The clock speed is 10 MHz and the silicon IC provides power on/off reset control and bias sequencing and also provides ESD protection for bias and control line ports. TSVs provide the necessary interconnections through and routing across the control IC and the 100 μm high copper posts will connect the silicon and GaAs chips together. A thick BCB coating on top of the MMIC helps provide handling and environmental protection.

17.5
Summary

A through silicon via (TSV) process has been developed at the University of Arkansas to enable 3D integration of a system consisting of silicon and GaAs-based electronic circuitry. In the silicon-based electronic wafer, vias are formed by deep reactive ion etching, lined by PECVD and sputter deposition processes for insulation and barrier/seed layers, and copper filled by electroplating. This process wafer is passivated and then attached to a carrier wafer by an adhesive that will allow debonding at an elevated temperature at a later stage. Bulk and fine removal of silicon from the back side of the process wafer is performed by mechanical grinding and polishing. The blind vias are finally exposed from the back side by a reactive ion etch step. Insulation and barrier layers are removed to allow the via to be copper plated, giving the TSV. Contact pad connections are formed. Once TSVs are constructed, copper posts are electroplated into a deep resist mold. At the same time, a copper dam for coolant containment is formed. The process wafer is debonded, scribed, and chips are assembled using a flip chip bonder. The system to be integrated in this project is a silicon IC control chip and a GaAs T/R MMIC. This integration is underway and accomplishments will be reported in the future.

Acknowledgments

This work was supported by AFRL under Grant No. FA8650-04-2-1619 and the authors acknowledge the graduate students and technical staff at the University of Arkansas High Density Electronics Center (HiDEC) and the Engineers at REMEC Defense & Space.

References

1 Meindl, J.D., Davis, J.A., Zarkesh-Ha, P. *et al.* (2002) Interconnect opportunities for gigascale integration. *IBM Journal of Research and Development*, **46**, 245–263.

2 Garrou, P. (2005) Future IC's Go Vertical, *Semiconductor International*, Semiconductor Packaging Edition.

3 Soh, H.T., Yue, C.P., McCarthy, A. *et al.* (1999) Ultra-low resistance, through-wafer

via (TWV) technology and its applications in three dimensional structures on silicon. *Japanese Journal of Applied Physics*, **38**, 2393–2396.

4 Chow, E.M., Chandrasekaran, V., Partridge, A. *et al.* (2002) Process compatible polysilicon-based electrical through-wafer interconnects in silicon substrates. *Journal of Microelectromechanical Systems*, **11**, 631–640.

5 Burkett, S.L., Qiao, X., Temple, D. *et al.* (2004) Advanced processing techniques for through-wafer interconnects. *Journal of Vacuum Science & Technology B*, **22**, 248–256.

6 Schaper, L., Burkett, S., Spiesshoefer, S. *et al.* (2005) Architectural implications and process development of 3-D VLSI Z-axis interconnects using through silicon vias. *IEEE Transactions on Advanced Packaging*, **28**, 356–366.

7 Rowbotham, T., Patel, J., Lam, T. *et al.* (2006) Back side exposure of variable size through-silicon vias. *Journal of Vacuum Science & Technology B*, **24**, 2460–2466.

8 Ozguz, V., Marchand, P. and Liu, Y. (2000) 3D stacking and optoelectronic packaging for high performance systems. International Conference High Density Interconnect Syst. Packaging, **4217**, pp. 1–3.

9 Engelhardt, M. (2002) Vertically integrated circuits (VIC): a 3D technology for advanced SOCs? Proceedings AVS 3rd International Conference Microelectronic Interfaces, pp. 19–22.

10 Klumpp, A., Merkel, R., Wieland, R. and Ramm, P. (2003) Chip-to-wafer stacking technology for 3D system integration. Proceedings Electronic Components Tech. Conference, pp. 1080–1083.

11 Tanaka, N., Yamaji, Y., Sato, T. and Takahashi, K. (2003) Guidelines for structural and material-system design of a highly reliable 3D die-stacked module with copper through-vias. Proceedings Electronic Components Tech. Conference, pp. 597–602.

12 Gutmann, R.J., Lu, J.-Q., Pozder, S. *et al.* (2004) A wafer-level 3D IC technology platform. Advanced Metallization Conference, pp. 19–26.

13 Guarini, K.W., Topol, A.W., Ieong, M. *et al.* (2002) Electrical integrity of state-of-the-art 0.13 μm SOI CMOS devices and circuits transferred for three-dimensional (3D) integrated circuit (IC) fabrication. Technical Digest – Int. Electron Devices Meeting, pp. 943–945.

14 Markunas, B. (2002) Mixing signals with 3-D integration. *Semiconductor International.*

15 Schaper, L., Burkett, S., Gordon, M. *et al.* (2006) Systems in miniature: meeting the challenges of 3-D VSLI, 8th VLSI Packaging Workshop in Japan.

16 Läermer, F. and Schilp, A. (1994) German Patent No. DE-4241045C1; U.S. Patent No. 5,501,893 (1996).

17 Polamreddy, S., Figueroa, R., Burkett, S.L. *et al.* (2005) Sloped sidewall DRIE process development for through silicon vias. IMAPS International Conference Dev. Packaging.

18 Spiesshoefer, S., Rahman, Z., Vangara, G. *et al.* (2005) Process integration for through-silicon vias. *Journal of Vacuum Science & Technology A*, **23**, 824–829.

19 Figueroa, R.F., Spiesshoefer, S., Burkett, S. and Schaper, L. (2005) Control of sidewall slope in Silicon vias using SF_6/O_2 plasma etching in a conventional RIE tool. *Journal of Vacuum Science & Technology B*, **23**, 2226–2231.

20 Gomez, S., Jun Belen, R., Kiehlbauch, M. and Aydil, E.S. (2004) Etching of high aspect ratio structures in Si using SF_6/O_2 plasma. *Journal of Vacuum Science & Technology A*, **22**, 606–615.

21 Abdolvand, R. and Ayazi, F. (2005) Single-mask Reduced-gap Capacitive Micromachined Devices. 18th IEEE International Conference MEMS, pp. 151–154.

22 Abhulimen, I.U., Polamreddy, S., Burkett, S., Cai, L. and Scharper, L. (2007) Effect on process parameters on via formation in Si

using Deep Reactive Ion Etching (DRIE), *J. Vac. Sci. Technol B*, **25**, 1762–1770

23 Spiesshoefer, S., Patel, J., Lam, T. *et al.* (2006) Copper electroplating to fill blind vias for three-dimensional integration. *Journal of Vacuum Science & Technology A*, **24**, 1277–1282.

24 Patel, J. (2006) Through Silicon Vias Process Integration with concentration on a Diffusion Barrier, Backside Processing, and Electrical Characteristics, The University of Arkansas, M.S. Thesis, Chapter 2.

25 Rossnagel, S.M. (2002) Characteristics of ultrathin Ta and TaN films. *Journal of Vacuum Science & Technology B*, **20**, 2328–2336.

26 Tsai, M.H., Sun, S.C., Tsai, C.E. *et al.* (1996) Comparison of the diffusion barrier properties of chemical-vapor-deposited TaN and sputtered TaN between Cu and Si. *Journal of Applied Physics*, **79**, 6932–6938.

27 Hayashi, M., Nakano, S. and Wada, T. (2003) Dependence of copper interconnect electromigration phenomenon on barrier metal materials. *Microelectronics Reliability*, **43**, 1545–1550.

28 Lam, T. (2006) Electroplating And Reactive Sputtering For Through-Silicon Via (TSV) Applications, The University of Arkansas, M.S. Thesis, Chapter 2.

29 Abhulimen, I.U., Lam, T., Kamto, A. *et al.* (2007) Effect of via profile on seed layer deposition for Cu electroplating. IEEE Region 5 Conference, pp. 102–104.

30 Andricacos, P.C., Uzoh, C., Dukovic, J.O. *et al.* (1998) Damascene copper electroplating for chip interconnections. *IBM Journal of Research and Development*, **42**, 567–574.

31 Moffat, T.P., Wheeler, D., Huber, W.H. and Josell, D. (2001) Superconformal electrodeposition of copper. *Electrochemical and Solid State Letters*, **4**, C26–C29.

32 Josell, D., Wheeler, D., Huber, W.H. and Moffat, T.P. (2001) Superconformal electrodeposition in submicron features. *Physical Review Letters*, **87**, 016102-1–016102-4.

33 Moffat, T.P., Wheeler, D. and Josell, D. (2004) Superfilling and the curvature enhanced accelerator coverage mechanism. *Electrochemical Society Interface*, **13**, 46–52.

34 Kim, B., Sharbono, C., Ritzdorf, T. and Schmauch, D. (2006) Factors affecting copper filling process within high aspect ratio deep vias for 3D chip stacking. Proceedings Electronic Components Tech. Conference, pp. 838–843.

35 Lee, H.-J. and Lee, D.N. (2002) Effects of current waveform and bath temperature on surface morphology and texture of copper electrodeposits for ULSI. *Materials Science Forum*, **408–412**, 1657–1662.

36 Sun, J.-J., Kondo, K., Koamura, T. *et al.* (2003) High aspect ratio copper via filling used for three-dimensional chip stacking. *Journal of the Electrochemical Society*, **150**, G355–G358.

37 Burkett, S., Schaper, L., Rowbotham, T. *et al.* (2006) Materials aspects to consider in the fabrication of through-silicon vias. *Materials Research Society Symposium Proceedings*, **970**, Y06-01.

38 Puligadda, R., Pillalamarri, S., Hong, W. *et al.* (2007) High-performance temporary adhesives for wafer bonding applications. *Materials Research Society Symposium Proceedings*, **970**, Y04-09.

39 He, A., Bakir, M.S., Allen, S.A.B. and Kohl, P.A. (2006) Fabrication of compliant, copper-based chip-to-substrate connections. Proceedings Electronic Components Tech. Conference, pp. 29–34.

40 Boyt, D., Abhulimen, I.U., Gordon, M.H. *et al.* (2007) Finite element analysis of power dissipation and stress in 3-D stack-up geometries. IEEE Region 5 Conference, pp. 194–198.

41 Huebner, H., Ehrmann, O., Eigner, M. *et al.* (2002) Face-to-face chip integration with full metal interface. Proceedings Advanced Metallization Conference, pp. 53–58.

42 Benkart, P., Kaiser, A., Munding, A. *et al.* (2005) 3D chip stack technology using through-chip interconnects. *IEEE Design & Test of Computers*, **22**, 512–518.

43 Schaper, L.W., Liu, Y., Burkett, S. and Cai, L. (2007) Assembly and cooling technology for 3-D VLSI chip stacks. IMAPS Device Packaging Workshop, Scottsdale, Arizona.

18
Vertical Interconnection by ASET

Kenji Takahashi and KazumasaTanida

18.1
Introduction

Ultrahigh density packaging is an essential technology for ubiquitous computing in the future. Three-dimensional chip stacking with vertical interconnections through Si chips is potentially the best technique of semiconductor system integration. The technology is attracting the interest of package engineers due to its high capability of signal transmission, and it is expected to ultimately become a way of integrating various devices [1–6]. Figure 18.1 shows the process flow and Figure 18.2 shows a sample made by the 3D chip stacking technology. However, there has been some criticism that the through-via fabrication processes are too expensive, or that dissipation of heat dissipation by the chip stacking structure is poor.

During the five-year national project of "Electronic System Integration," ASET (Association of Super-Advanced Electronics Technologies) has almost overcome most of the challenging issues of the high-productivity process of Cu electrodeposition and thin-wafer-handling issues.

Section 18.2 gives an overview of the fabrication process, while Section 18.3 discusses high-speed Cu electrodeposition. The basic Cu plating bath composition allowed a very low current density so that the vias were filled without any voids. The optimization of the concentration of the electrodeposition solution and the use of a new current application pattern and an aeration technique decreased the electrodeposition time to as short as one hour.

Section 18.4 deals with thin-wafer-handling issues. The wafer was bonded to a SEMI standard size glass substrate with UV curable adhesive tape prior to back grinding so that the glass-bonded wafer could be processed safely without any modification of the conventional equipment. The debonding equipment employed new mechanisms of initial delamination and gentle release of the thin wafer from the glass. The results of the debonding quality are presented.

Handbook of 3D Integration: Technology and Applications of 3D Integrated Circuits.
Edited by Philip Garrou, Christopher Bower and Peter Ramm

ISBN: 978-3-527-32034-9

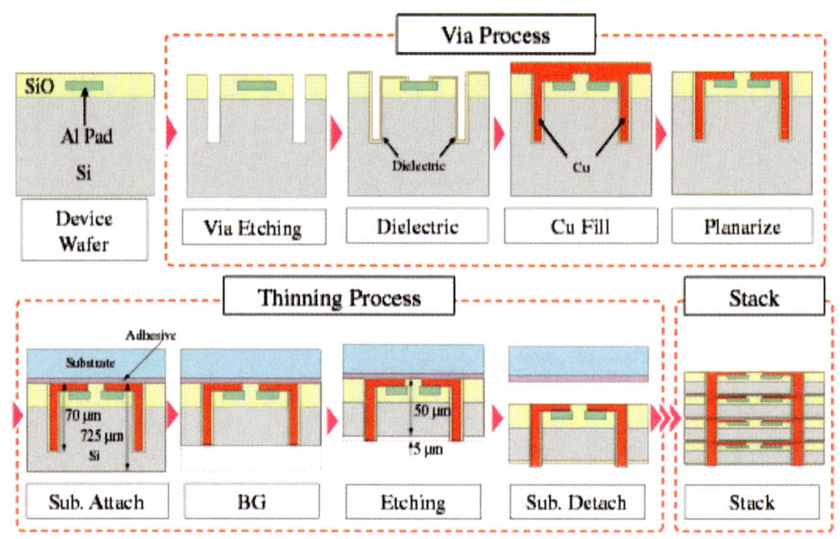

Figure 18.1 Process flow of 3D chip stacking.

Section 18.5 covers 3D chip stacking and Section 18.6 concerns thermal issues. The thermal management of the 3D chip stacking structure is an essential issue. Thermal analysis accompanied with experiments on the structure is presented. Various case studies on factors that affect heat dissipation, such as interchip encapsulant, dielectric layer and thermal vias, are summarized. Furthermore, a novel cooling interface for high-power application is examined.

In Section 18.7 we examine the electrical performance of the vertical interconnections. The final section (Section 18.8) introduces an application of Cu through-via technology to a commercially available image sensor. The successful results proved that the through-via structure and process are promising future system integration technologies.

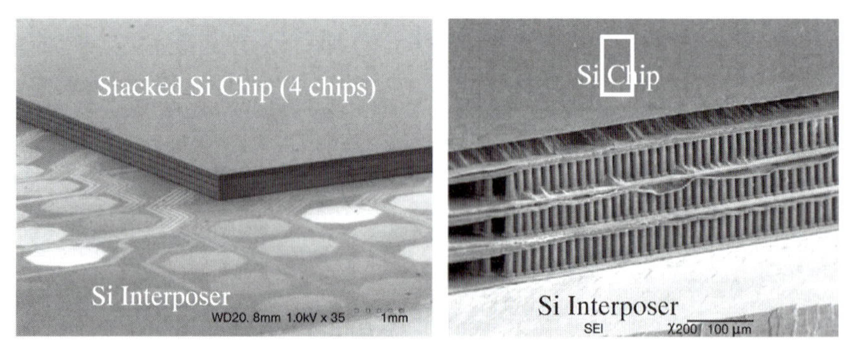

Figure 18.2 Stacked chips by the 3D chip stacking technology.

18.2 Fabrication Process Overview

As shown in Figure 18.1, the fabrication process involves through silicon via fabrication, wafer thinning, wafer back processing and chip stacking.

The through silicon via fabrication process was constructed so that commercially available equipment could be used without major modification. Almost all of the equipment is used in the LSI interconnection process. The main processes are:

1. Si etching by reactive ion etching (RIE)
2. Plasma chemical vapor deposition (CVD) of SiO_2 to cover inside the via
3. Metal-organic CVD (MO-CVD) of TiN and Cu as barrier and seed metal layers
4. Electrodeposition of Cu to fill the via
5. Chemical mechanical planarization (CMP) of Cu
6. Wafer thinning and bumping
7. Chip stacking.

Through vias 10 μm-square and 70 μm deep were etched by RIE. The vias showed good surface morphology. Sidewalls, covered by plasma CVD-SiO_2, showed acceptable coverage. The SiO_2 was 0.2 and 0.5 μm thick at the sidewall and bottom, respectively, when the thickness of that on the surface was 1.6 μm.

MO-CVD was performed for TiN (10 nm) and Cu (150 nm) deposition instead of the PVD method. Overhang prohibited TiN/Cu deposition by PVD. MO-CVD showed excellent coverage at the sidewall and the bottom. The actual thickness was from 0.12 μm to 0.16 nm. Through via filling was attempted by $CuSO_4$-based copper electroplating. Extra Cu on the wafer surface was removed by Cu-CMP. The Cu damascene process formed the interconnection between through electrode and the device pad at the same time. A high removal rate CMP slurry was developed so that the thick Cu film could be removed effectively.

The wafer was thinned down to 50 μm by backgrinding and plasma etching to expose the bottom of the Cu through via. The thinned wafer was attached on a glass wafer by double-sided UV film. Low-temperature deposition of SiN was applied to the back of the wafer to avoid thermal damage of the UV film. After removing SiN on the top of Cu extrusion by CMP, the wafer was gently detached from the glass wafer and transferred to dicing tape.

The wafer was diced to chips and the chips were picked up from the tape. A high precision flip-chip bonder was used to stack the chips with through vias. A Cu-Sn intermetallic bonding mechanism was used for the interconnection. A very small amount of encapsulant resin was applied by jet dispenser.

18.3 Via Filling by Cu Electrodeposition

In previous work [7–10], 10 μm-square, 70-μm deep vias were filled completely by Cu electrodeposition under the conditions of high Janus Green B (JGB) concentration and a two-step current application pattern. The electrodeposition time was as short as 3.5 hours, which is considerably shorter than the previous 12 hours taken for complete filling. However, the electrodeposition time should be less than 1 hour for the cost of the process to be comparable to those of other processes. The present study investigated a specially designed bath, pulse reverse current, two-step current application, and O_2 bubbling prior to electrodeposition.

18.3.1 Experimental

18.3.1.1 Sample and Plating Apparatus

Si chips with 10 µm-square/70 µm-deep vias were prepared for the experiments. The surface of the chip was conformably covered with chemical vapor deposition (CVD) SiO_2, metal organic-chemical vapor deposition (MO-CVD) TiN, and MO-CVD Cu. The chip size of the sample was about 2 cm square. The Si chip was set on the rotating disk electrode (RDE) of the electrodeposition apparatus (Figure 18.3), and appropriate current was applied between the sample and the anode. The rotation rate of the RDE was 1000 rpm.

The basic bath composition was $CuSO_4$-$5H_2O$ and H_2SO_4. The specially designed bath included bis(3-sulfopropyl) disulfide (SPS), suppressor B (SPR B), leveler A (LEV A) and Cl^- as additives.

Pulse reverse (PR) current was applied in all experiments. The PR cycle was on (200 ms)/reverse (10 ms)/off (200 ms). Two-step current application was also examined, in which a relatively low current density from the beginning and a high current density during the last 10 min of plating were applied.

18.3.1.2 Experiments

1. The concentrations of SPR B and LEV A were investigated. Combinations of SPR B (0.5, 5, 10 ppm) and LEV A (0.2, 0.5, 1.0 ppm) were used. The current density was consistent at 5 mA cm^{-2} throughout the plating process. The electrodeposition time was 90 min.

2. Two-step plating was applied to the bath with optimum concentrations of SPR B and LEV A. The concentrations of H_2SO_4 and SPS were also optimized separately. The bath includes 25 g per litre H_2SO_4, 2 ppm SPS, 70 ppm Cl^-, 5 ppm SPR B, and 0.2 ppm LEV A. The current density in the first step was 6 mA cm^{-2} and that in

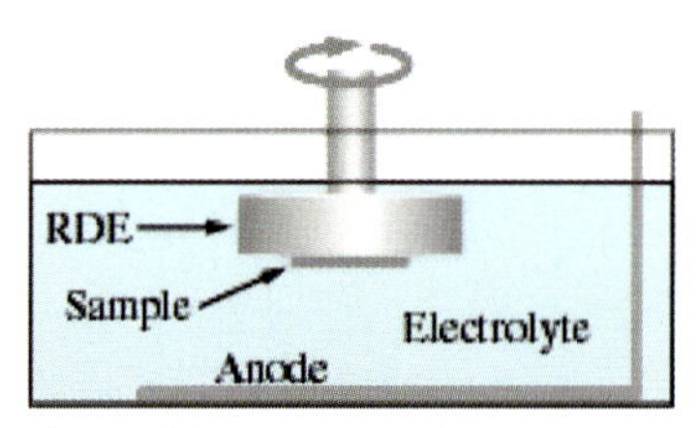

Figure 18.3 Experimental apparatus.

the second step was 15 mA cm^{-2}. The electrodeposition times of the first step and the second step were 50 and 10 min, respectively.

3. The bath was bubbled with O_2 for 1 hour prior to electrodeposition to reduce the concentration of Cu^+, which acts as an inhibitor of the effective electrodeposition of Cu^{2+}, in the bath. Bubbling of N_2 and incorporation of H_2O_2 were also tested. The bath was the same as that used in Experiment 2. The current was applied at 6 mA cm^{-2} for 75 min. Finally, O_2 bubbling and two-step plating was combined to realize perfect filling in 1 hour. The current density in the first step was 6 mA cm^{-2} and that in the second step was 15 mA cm^{-2}.

18.3.2 Results and Discussion

1. The results showed that the concentration of LEV A was the dominant factor. Conditions with more than 0.5 ppm of LEV A yielded seams or voids. With 0.2 ppm of LEV A, all conditions of SPR B yielded no seams or voids in the vias. Thus, the optimum concentrations of SPR B and LEV A were 5 and 0.2 ppm, respectively.

2. Figure 18.4 shows the result of the experiment. No significant void is observed in the vias. However, narrow seams can be found in the lower portion of the vias. The current density applied in this experiment was the highest value ever adopted. According to the experimental experience, the voids or seams remaining at the middle or bottom of the vias should be corrected by improving the "bottom-up" characteristics of the bath. The bath composition is the best one for deep and high-aspect-ratio via filling.

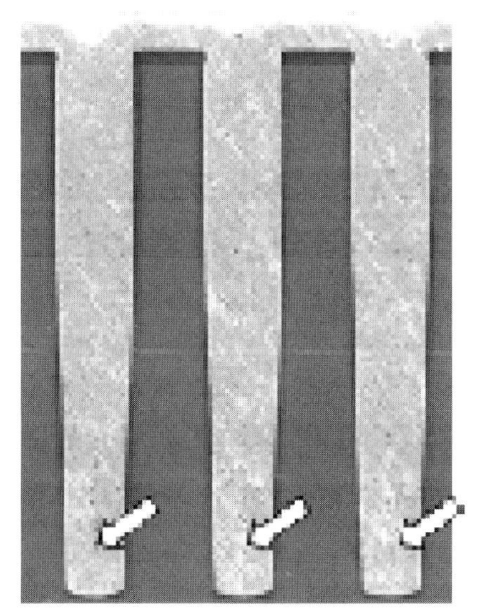

Figure 18.4 Cross-sectional SEM micrograph obtained by Experiment 2.

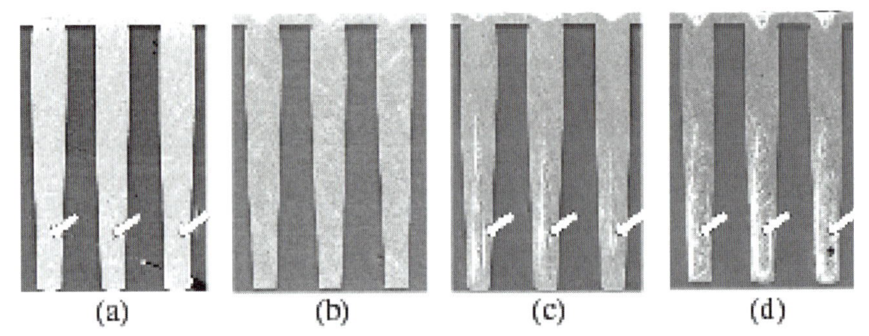

Figure 18.5 Cross-sectional SEM micrographs of the vias obtained by Experiment 3: (a) no pretreatment, (b) O_2 bubbling, (c) N_2 bubbling and (d) H_2O_2 added to the bath.

3. Figure 18.5 shows electrodeposition results of O_2 bubbling, N_2 bubbling and H_2O_2 at 6 mA cm^{-2}. O_2 bubbling showed no void or seam in the vias as expected. An additional experiment showed that the O_2 bubbled bath yielded a larger current value than did the N_2 bubbled bath, indicating that the use of an O_2 bubbling accelerated the electrodeposition rate of Cu.

Figure 18.6 shows the kinetic and transport models of Cu and complexes [11, 12]. Reaction of 3-mercapto-1-propane sulfonic acid [MPS, $HS(CH_2)_3SO_3H$], a monomer of SPS, with Cu^{2+} generates the Cu^+ and Cu(I)-thiolate complex [$Cu(I)S(CH)(CH_2)_2SO_3$] in the via and surface of cathode as described by:

$$2Cu^{2+} + MPS \rightarrow 2Cu^{+} + Cu(I)\text{-thiolate} + 3H^{+}$$

Cu^+ is transported to the outside of the via, where the oxygen concentration is higher than that in the via, and consumed with oxygen in a boundary layer between the surface of the cathode and the bulk solution, as shown by:

$$2Cu^{+} + O_2 + 2H^{+} \rightarrow 2Cu^{2+} + H_2O_2$$

Cu(I)-thiolate complex remains in the via and acts as an effective accelerant for bottom-up filling. Adding H_2O_2 to the bath degraded the quality of filling. This indicates that the above reaction is reversible.

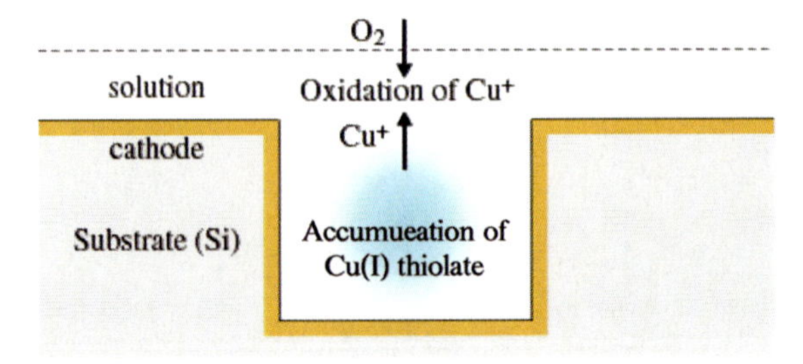

Figure 18.6 Kinetic and transport models.

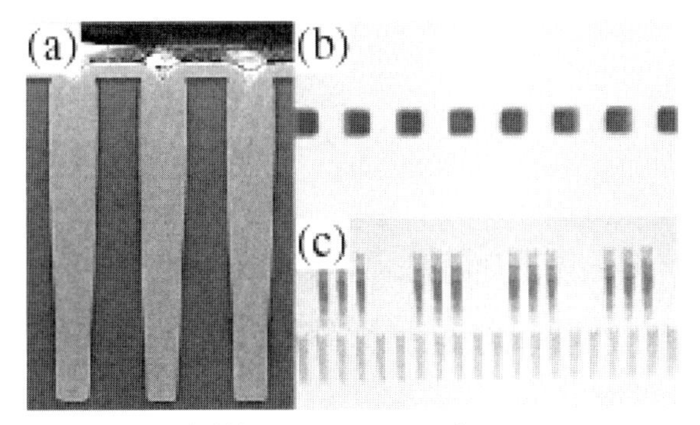

Figure 18.7 O_2 bubbling pretreatment and two-step electrodeposition result: (a) cross-section SEM micrograph, (b) vertical view by X-ray inspection and (c) slanted view by X-ray inspection.

Figure 18.7 shows a scanning electron microscope (SEM) micrograph and X-ray photographs of O_2 bubbling pretreatment and two-step electrodeposition at 6 mA cm^{-2} for 50 min, as well as at 15 mA cm^{-2} for 10 min. Complete filling was confirmed by both cross-sectional and vertical inspection.

Finally, Cu electrodeposition into deep and high-aspect-ratio vias successfully filled them completely within as little as 60 min.

18.4 Handling of Thin Wafer

The handling method of a thinned wafer is one of the most critical technologies in realizing 3D chip stack with through-vias. Glass substrate was used as wafer carrier. This requires a thermoresistive and removable adhesive, low-temperature processing, wafer bonding equipment and wafer debonding equipment. Double-sided ultraviolet (UV) curable adhesive was used to bond the wafer to the glass. The decomposition temperature was about 120 °C, which is sufficiently higher than the highest temperature during the wafer-back process. The glass was a commercially available one and was 725 μm thick [13].

18.4.1 Wafer Debonding Method

The adhesion strength of UV tape can easily be degraded by UV irradiation. However, with a hard substrate and fragile wafer debonding, the wafer cannot be safely debonded from the substrate unless initial delamination at the interface is appropriately induced [14]. A thin blade could be used to prompt such initial delamination. Figure 18.8 outlines the method. The delamination front is easily propagated by lifting the ring frame. However, the wafer is discontinuously bent at the delamination front as the delamination progresses across the wafer. Thus, the possibility of wafer

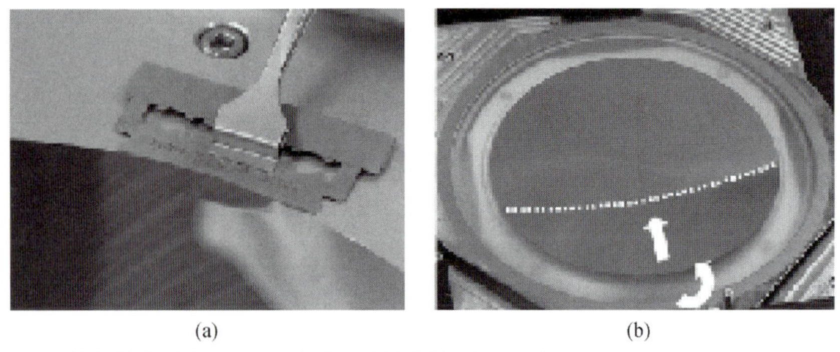

(a) (b)

Figure 18.8 Wafer debonding method: (a) initial delamination is induced by a blade; (b) the ring frame is lifted to propagate delamination.

breakage at the delamination front by *in situ* measurement of the wafer height, calculation of the maximum stress and measurement of wafer strength were estimated.

18.4.2 Estimation of Tensile Stress

Figure 18.9 shows models of wafer debonding.

The tensile stress can be roughly estimated by regarding the wafer as a circular cantilever where the delamination front is the fixed end. Figure 18.9b–d show simplified models for the calculation.

Two laser displacement sensors are used to extrapolate the height at the wafer end and the delamination distance. 50- to 200-μm-thick wafers were used to investigate

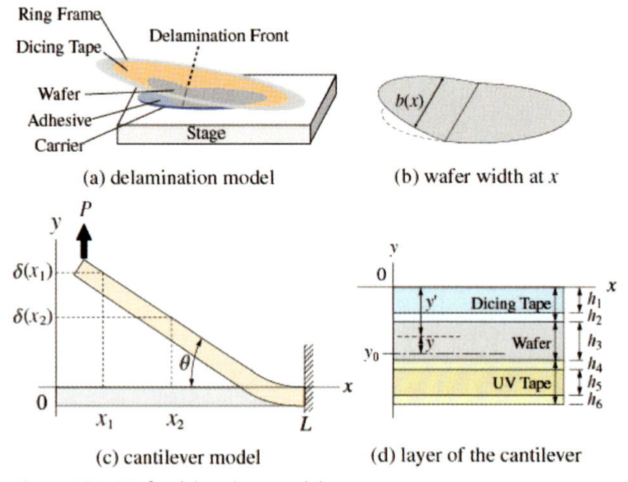

Figure 18.9 Wafer debonding models.

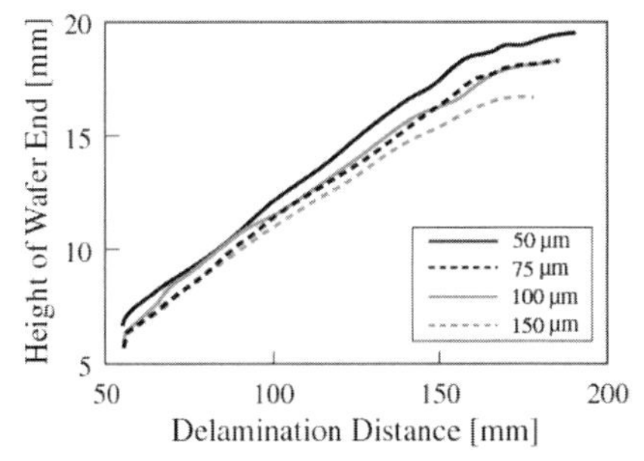

Figure 18.10 Relationship between delamination length and height of the wafer end.

the effect of wafer thickness. Figure 18.10 shows the relationship between the extrapolated delamination front and the extrapolated height of the wafer end. A thinner wafer shows a slightly higher wafer edge, but the differences from those of the thicker wafers are not large. The 200-μm-thick wafer could not be measured because it could not be fixed onto the stage because of the large lifting load.

Figure 18.11a shows the calculated results of stress at the delamination front against delamination distance. The maximum stress is observed in the early stage of delamination, and it gradually decreases as delamination propagates. Figure 18.11b is a plot of the maximum stress at the delamination front against wafer thickness. The maximum stress at the delamination front was 26 MPa for 50-μm-thick wafer.

18.4.3
Strength of Thinned Chip

There are many reports on Si fracture strength; however, through-vias may act as crack initiators when the wafer is bent. Thus, the strength of Si with and without through-vias as reference was measured.

Sample chips were 10.4-mm-square, with 50 μm thick Si covered with 1.4-μm-thick Si_3N_4 film as back passivation. Through-via samples had 10-μm-square through-vias, filled with electrodeposited Cu, located periphery of the chip with a

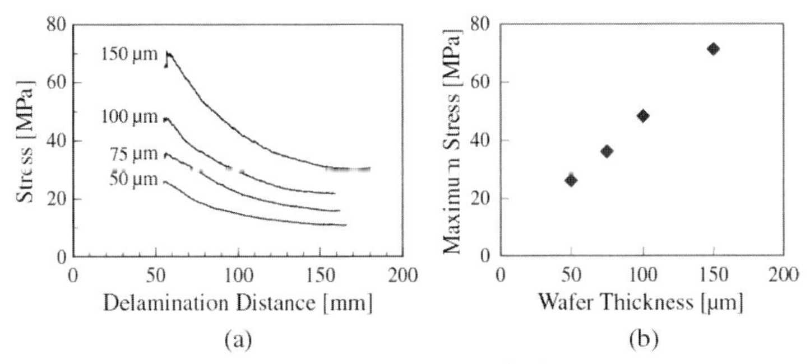

Figure 18.11 Stress calculation results: (a) relationship between delamination distance and stress for 50- to 150-μm-thick wafers; (b) wafer thickness and maximum stress.

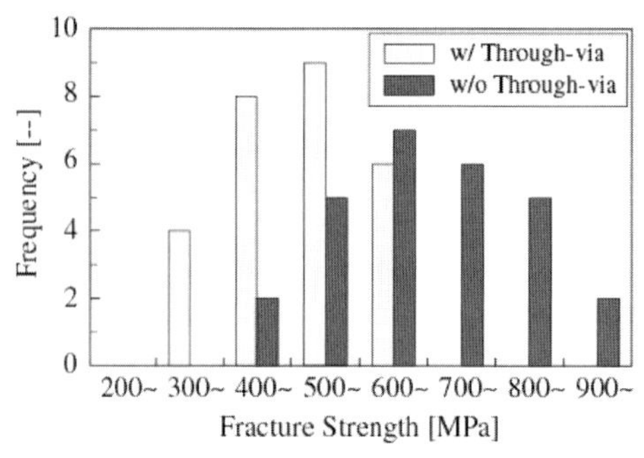

Figure 18.12 Fracture strength of Si with (w/) and without (w/o) through-vias.

20 μm pitch. The SEMI Standard G86-0303 compatible three-point bending test using an universal testing machine (EZ-Graph, Shimadzu Corp.) was adopted to measure the strength. Figure 18.12 shows the results, revealing that the strength of Si with through-vias (413 MPa) is lower than that without through-vias (600 MPa). Furthermore, the deviation of the strength of Si with through-vias is relatively small. The tested pieces showed obvious differences. Samples without through-vias shattered into small fragments, but those with through-vias cracked only into two pieces. These results indicate that the through-vias act as crack initiators.

18.4.4 Discussion

The fracture strength of Si with through-via is ten times higher than the maximum stress (Figure 18.11b). This suggests that the wafer is free from cracking during delamination. However, the wafer may crack if the delamination front propagates inconsistently for any reason, such as uneven load, voids or defects at the interface. Furthermore, the relationship between delamination distance and height of the wafer edge showed no significant differences against wafer thickness. This implies that the wafer lifting speed is not synchronized to the delamination propagation speed. In the measurement system, the restriction of the laser sensors positions limits the minimum delamination distance, but the stress becomes higher when the delamination distance is small. It is necessary to improve the sensor position to estimate the stress values for small delamination distances.

18.5 3D Chip Stacking

18.5.1 Technical Issues of 3D Chip Stacking

The vertical interconnection was fabricated by chip stacking process, which include Cu bump bonding (CBB) utilizing Cu-Sn diffusion for connecting Cu TVs. The

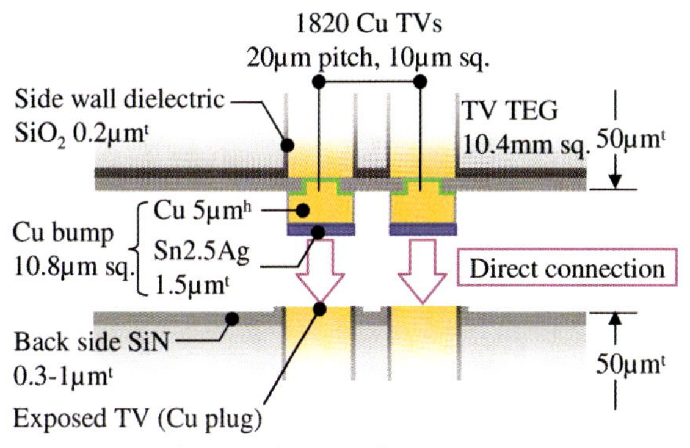

Figure 18.13 Schematic diagram of CBB structure.

CBB is expected to be a simple interconnection to directly connect Cu TVs without formation of bumps on the chip backside (Figure 18.13). However, there are two technical issues to realize the CBB process. One is to achieve complete diffusion and reduction of the Sn alloy to the Cu-Sn intermetallic compound (IMC) layer considered to be the optimum minute interface structure [15]. Another important issue is the influence of a Cu oxide layer, formed on exposed Cu from the chip back side.

However, for 3D chip stacking LSI with minute interconnection, the gaps between devices become >10 μm. It is difficult to encapsulate by conventional capillary flow process, because the temperature and injection balance of resin at each layer are difficult. Non-conductive particle paste (NCP) utilizing small shot perform (SSP) process is one solution [16]. The encapsulation conditions at each layer are the same, because the encapsulation and bonding bumps are performed at the same time, after dispensing NCP onto the interposer or the chip backside.

18.5.2
Bondability on 20-μm-Pitch Interconnection

First, to consider the controllability of generation of Cu_3Sn, Cu-Sn diffusion in a micro-region such as minute interconnection in 20-μm-pitch was assessed, utilizing a micro Cu/Sn2.5Ag bump over Sn melting point, which have not been studied before. To study quantitatively the growth kinetics of interfacial Cu_3Sn, the micro bumps were annealed at 250, 300, 350 °C for various times on heat stage of high-precision flip-chip bonder (FC-1000, Toray Engineering Co., Ltd.), which realize 20-μm-pitch high accuracy bonding. In addition, Cu_3Sn thickness was measured by cross-sectional scanning electron microscopy (SEM) analysis. Figure 18.14 shows a plot of Cu_3Sn thickness against the square root of annealing time. As shown in the figure, the Cu_3Sn thicknesses were fitted by linear regression, and showed a parabolic dependence at various temperatures. As a result, 1.5-μm-thickness Sn2.5Ag was completely diffused into Cu_3Sn in approximately 20 s at 350 °C between Cu bump and Cu plug in the case of a symmetrical structure between Cu bump and Cu plug.

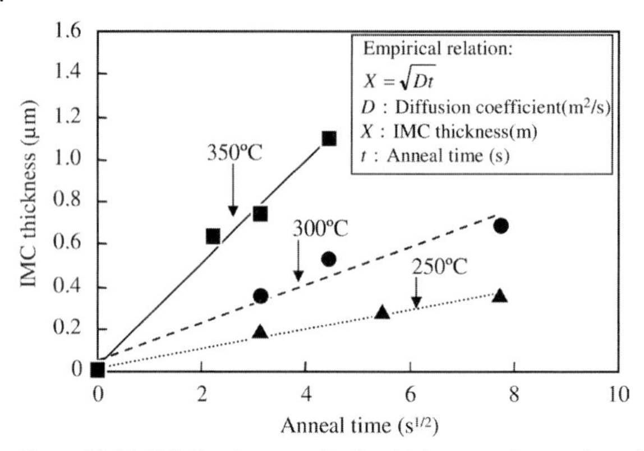

Figure 18.14 Relation between Cu_3Sn thickness and anneal conditions.

To study these phenomena further, the activation energy for the growth of Cu_3Sn was calculated by means of an Arrhenius relation:

$$D = D_0 e^{-Q/kT} \tag{18.1}$$

where D_0 is the diffusion constant, Q is the activation energy, k is the Boltzmann constant, T is absolute temperature, and D is the diffusion coefficient given by the slope of each line in Figure 18.15. The figure shows the Arrhenius plot (ln D against $10\,000/T$). The activation energy of IMC growth we observed was 0.88 eV, and the trend of IMC growth in this study is the same as that seen by Vollweiler [17], who evaluated total growth of Cu_3Sn (ε-phase) and Cu_6Sn_5 (η-phase) in 70Bi-30Sn alloy at the same temperature. Moreover, the result is fits well with other studies at temperatures lower than the Sn melting point [18–22]. Therefore, the Cu-Sn diffusion was well controlled and the optimum bonding interface structure in minute interconnection can form.

Next, to evaluate the influence of Cu oxide to bondability, the 20-μm-pitch Cu/CoC model was applied to measure the bonding strength of Cu bump interconnection (Figure 18.16). The Si chip was 10.4-mm-square and 50-μm-thick. The Si interposer

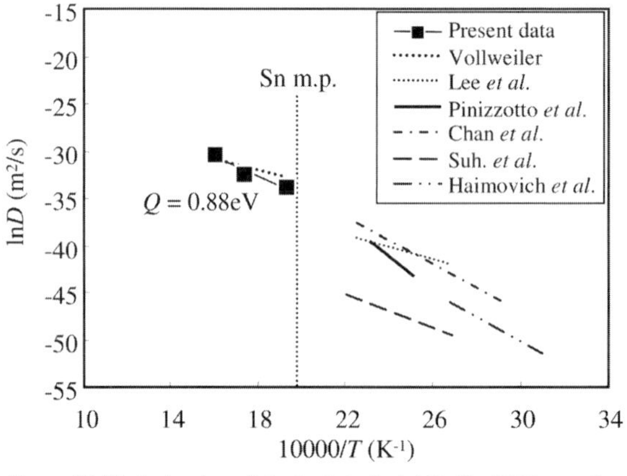

Figure 18.15 Arrhenius plots for interfacial Cu-Sn IMC growth.

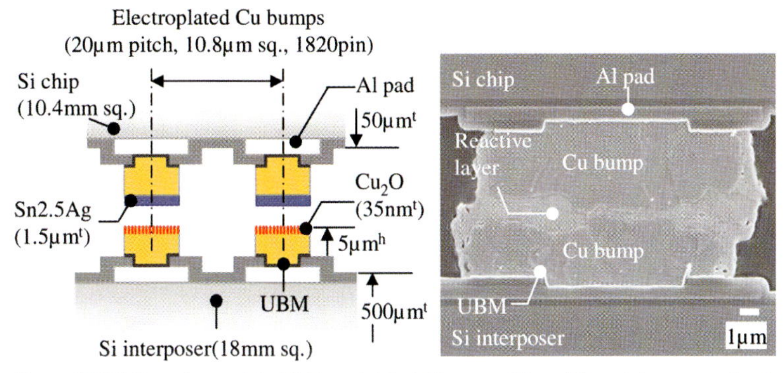

Figure 18.16 Experimental Cu/CoC model for influence of Cu oxide on interconnection bondability.

was 18-mm-square and 500-μm-thick. The 1820 electroplated Cu bumps, 10.4-μm-square and 5 μm high with 1.5-μm-thick electroplated Sn2.5Ag layer, were formed on the aluminium pads through under bump metallurgy (UBM) in the 20-μm-pitch, allocated on the periphery of the Si chip. The Cu bumps without Sn2.5Ag layer formed on Si interposer were supposed to be Cu plug exposed from the chip backside. In advance, the Cu oxide contamination on exposed Cu plug during flip-chip bonding was measured by Auger electron spectroscopy (AES). It was confirmed that approximately several tens of nanometres Cu oxide (Cu_2O) was formed on the Cu plug surface without holding the specimen in vacuum. Therefore, Cu bumps with an oxide layer corresponding to the Cu plug were prepared, and removed by application of a 5% HCl cleaning solution for a measured period of time. The bonding conditions were set as 350 °C and 60 s for complete diffusion and reduction of Sn alloy to Cu_3Sn; conditions that were defined as standard bonding conditions. The bonding force at 49.0 N per chip was constant. N_2 gas was flowed during the bonding process to prevent oxidation of molten Sn at the bonding interface.

The influence of Cu oxide on interconnection bondability as the bonding tensile strength was confirmed clearly (Figure 18.17). It is considered that Cu oxide formed

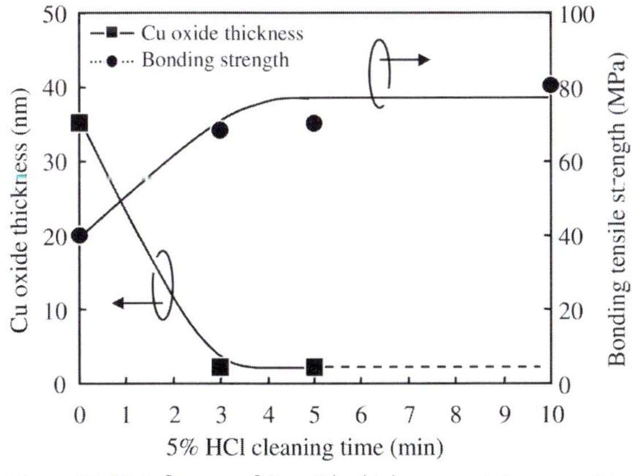

Figure 18.17 Influence of Cu oxide thickness on interconnection bondability.

Table 18.1 Bonding conditions for evaluation of bonding parameters.

Parameter	Bonding temperature	Bonding time
Bonding temperature (°C)	240, 270, 300, 350, 400, 450	350
Bonding time (s)	60	5, 10, 30, 60
Bonding force (N)	49	49
N_2 gas flow volume ($\mathrm{L\,min^{-1}}$)	10	10

on an exposed Cu plug during bonding must be reduced to several nanometers for next chip stacking. Furthermore, a dry etching process for Cu oxide was studied. It utilized the Ar RIE sputter unit (V-1000, Yamato Scientific Co., Ltd.) at sputter conditions of RF power at 1000 W, Ar flow at 15 sccm and driving pressure at 10 Pa. As the result, the Cu oxide layer was removed as well as for a 1-min acid etching treatment. In addition, the batch processing of multiple specimens was expected by utilizing this process. Therefore, an effective Cu oxide removal process for the sequential chip stacking was established.

Bonding temperature and time are considered to have a significant impact on bondability in terms of Cu-Sn diffusion. Therefore, these parameters were evaluated utilizing the Cu/CoC model shown in Figure 18.16. The Cu oxide layer on the surface of a Cu bump was removed by Ar sputtering in advance. The Si chip was mounted onto the Si interposer at each condition after fine-loop control of the recognition sequence at 80 °C by flip-chip bonder. Table 18.1 summarizes the bonding conditions for evaluation of bonding parameters. After the bonding under various bonding conditions, the bonding tensile stress strength was measured. In addition, the breaking point was determined to confirm the weak point of Cu bump interconnection. Next, to clarify the state of the IMC, the cross section of the bonding interface was analyzed by field emission scanning electron microscopy (FE-SEM) and field emission transmission electron microscopy (FE-TEM).

Figure 18.18 shows the dependence of bondability on bonding temperature, confirming that the bonding temperature was effective for increasing bonding

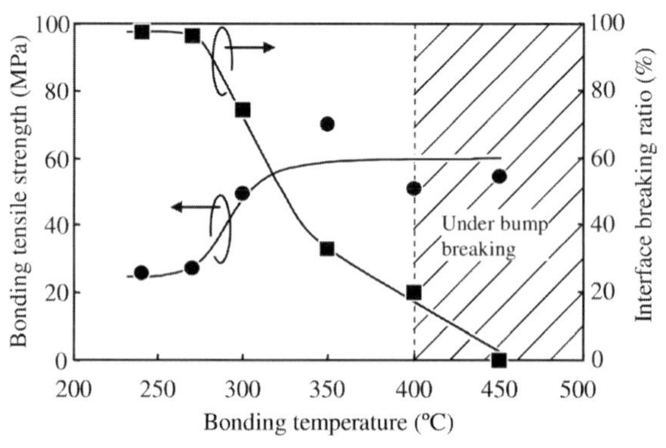

Figure 18.18 Influence of bonding temperature on interconnection bondability.

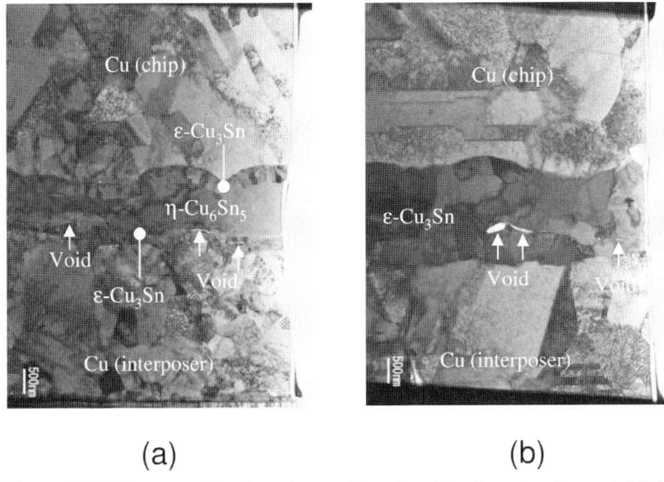

Figure 18.19 Cross-sectional analyses of bonding interface structure: at (a) 240 and (b) 350 °C.

strength, and that the bonding strength was saturated at 350 °C. In addition, the extent of breakage at the bonding interface, considered to be composed of IMC, also decreased with rising bonding temperature. In terms of breaking mode, breakage occurred in the bulk Cu bumps or under the bump structure that included that of the UBM and aluminium pad in addition to the breaking interface below 350 °C. However, delamination at the UBM boundary and the lower part of aluminium pad was observed frequently at 400 °C, and the delamination mode was completely dominant at 450 °C. It was assumed that the adhesive strength of the under bump structure decreased relative to the bonding interface composed of IMC. To clarify the state of the IMC, bonding interface at 240 and 350 °C were analyzed by TEM because of the significant difference in the bondabilities shown in Figure 18.8. Figure 18.19 shows the TEM image at the bonding interface. The Sn alloy was completely reduced and the single Cu_3Sn layer was confirmed at 350 °C. However, minute voids were observed in some portions of Cu_3Sn layer. With respect to the existence of void, the electroplated Cu bump surface was a patterned indented surface, and the voids were considered to remain in molten Sn after the upper and lower bumps were connected. Meanwhile, a multilayered structure consisting of $Cu_3Sn/Cu_6Sn_5/Cu_3Sn$ was found between both Cu bumps at 240 °C. The upper and lower Cu_3Sn layer thicknesses were approximately 0.33 μm, and this agrees well with the corresponding result of Figure 18.14. The results indicate that the IMC state at the bonding interface is the governing factor of bondabilities of Cu bump interconnection in the 20 μm pitch; 350 °C is the optimum bonding temperature.

Figure 18.20 shows dependence of bondability on bonding time, confirming that the bonding strength and interface braking ratio were saturated after 10 s. Figure 18.21 shows the SEM image of a magnified portion at the bump edge of the bonding area at 10 s. The alloy layer at bonding interface consisted of a single Cu_3Sn layer, and the bonding strength corresponded to the bonding temperature studied under the same conditions. However, according to Figure 18.14, 20 s at

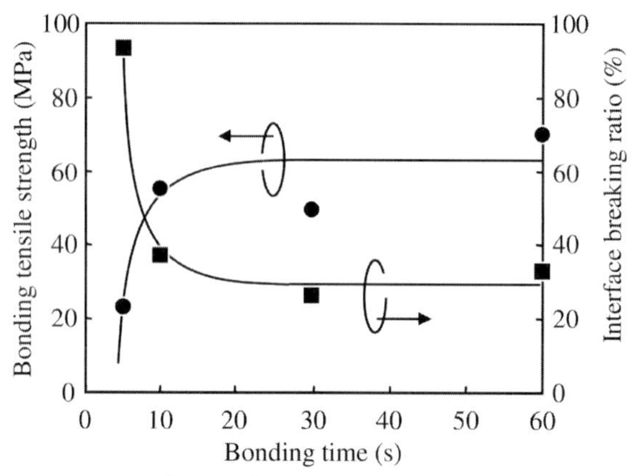

Figure 18.20 Influence of bonding time on interconnection bondability.

350 °C was necessary to diffuse a 1.5-μm-thickness of Sn2.5Ag to Cu_3Sn completely in the case of the symmetrical structure between Cu bumps. In fact, the molten Sn flowed to the bump edge and the diffusion area was increased. Therefore, early Sn reduction at the bonding interface was achieved.

18.5.3
NCP Preform Process for Layered Micro Thin Gaps

Figure 18.22 shows the structure of the experimental model for the fundamental evaluation of the NCP preform process. The Si chip was 10-mm-square and 50-μm-thick. The Si interposer was 18 mm-square and 500-μm-thick. The 1844 electroplated Au bumps, 12-μm-square, 7.5-μm high and 5-μm-height, were formed on the aluminium pads through under bump metallurgy (UBM) in the 20-μm-pitch, allocated on the periphery of the upper chip and on the interposer at the same location. NCP was preformed onto the interposer before bonding. Filler particles were not included so as to prevent bonding effects caused by intervening particles.

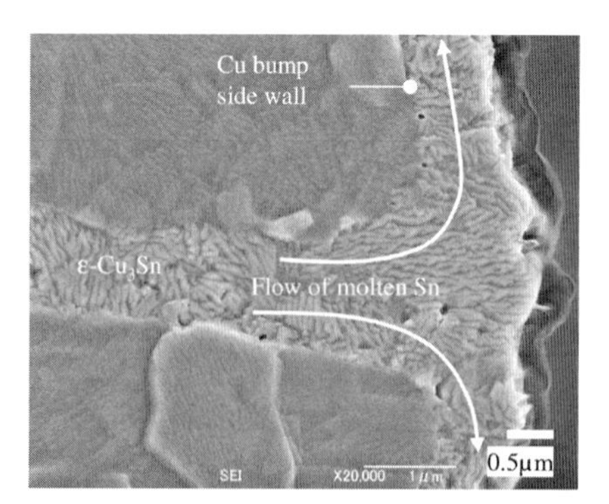

Figure 18.21 Cross-sectional analysis of bonding interface structure at 10 s.

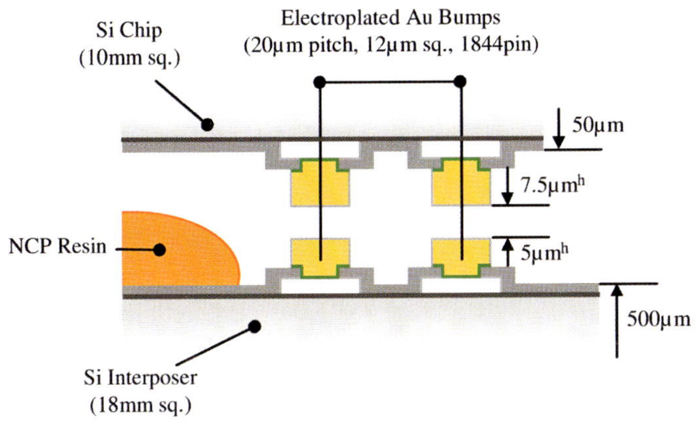

Figure 18.22 CoC structure for evaluation on NCP preform process.

However, it maintained a low coefficient of thermal expansion (CTE) and high Tg for thermal stress control. In addition, it had low viscosity to control bonding force easily. Table 18.2 summarizes the material properties of the NCP. Figure 18.23 shows the structure of the jet dispenser and SSP flow [23]. In the basic mechanism of jet generating, NCP is given kinetic moment by the air piston needle, which instigates coaxial head valve movement up and down. Excellent small shot dispense was realized, with high productivity for continuous shots. Moreover, no sensing of base height is a great advantage for 3D stacked devices, which have different base heights for each layer. The theoretical total amount of NCP was calculated as 0.87 ± 0.088 mg by geometrical simulation of this experimental model. In this case the sensetive weight control under 0.017 mg is demanded to dispense an optimal pattern and the mass of one dot was achieved as 0.010 mg by SSP. Moreover, this small shot dispense system gave a maximum displacement of less than 0.1 mm. After NCP was dispensed onto the interposer, the upper chip was mounted onto the interposer by the two-step thermo-compression bonding process. A NCP was flowed to the chip edge at 80 °C. Then, NCP was cured for 60 s at 240 °C. The bonding force was constant during the bonding process. Table 18.3 summarizes the bonding conditions. The dispense pattern of NCP was carefully evaluated by infrared (IR) microscopy to determine that

Table 18.2 Material properties of NCP.

Item	Condition	Measured value
Viscosity	25 °C	19.1 Pa s
Gel Time	150 °C	98 s
CTE (below Tg)	TMA	69×10^{-6}
CTE (over Tg)	TMA	176×10^{-6}
Tg	TMA	157 °C
Elastic modulas@25 °C	DMA	4.6 GPa
Elastic modulas@150 °C	DMA	1.9 GPa

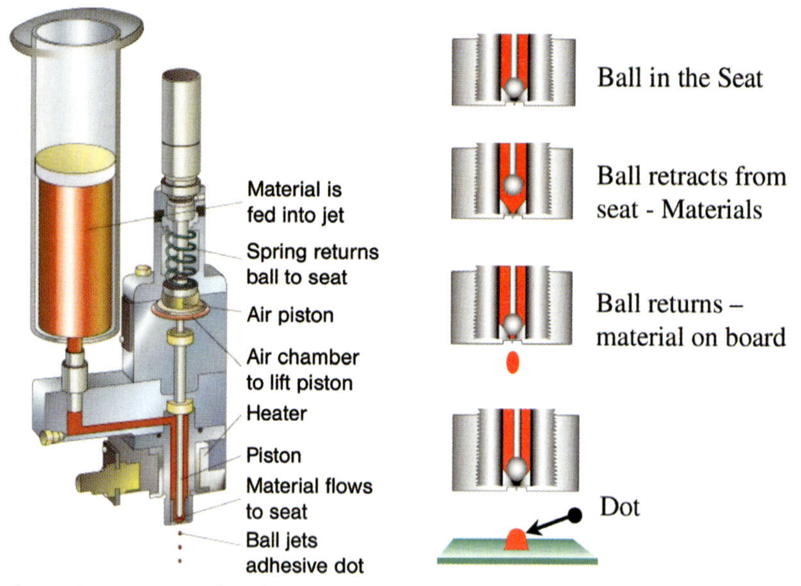

Figure 18.23 Structure of jet dispenser and SSP flow.

the NCP had wetted and spread through the whole gap without creeping up the bonding tool along the side of the thin Si chip. The cross section of the bonding interface was analyzed to confirm rejection of NCP between upper and lower bumps. The chip backside warpage after bonding was measured by laser displacement meter to confirm the possibility of continual chip stacking.

Figure 18.24 shows a test dispense pattern (diagonal cross type), which supplied the center with a large amount of NCP. The total amount of NCP, composed of 88 dots, was equivalent to 0.88 mg per chip, which agrees with the theoretical total amount. Figure 18.25 shows an IR microscopy image of the micro thin gap through the chip and interposer at chip edge after bonding. The micro thin gap was almost encapsulated by the peripheral interconnection area, and no void and cracking were found. Although, creeping up was found at the edge of CoC area, it was considered to be solved by further optimization of amount of NCP and the dispense pattern utilizing the small shot dispense technology. Figure 18.26 shows an SEM image of the cross section of the bonding area after bonding. As shown in the image, equable plastic modifications of the upper and lower Au bumps were confirmed. The gap between the Si chip and interposer was about 8 μm. Although some minute voids

Table 18.3 Bonding conditions for two-step thermo-compression bonding.

Parameter	Step 1	Step 2
Bonding force (N)	50	50
Bonding temperature (°C)	80	240
Bonding time (s)	10	60

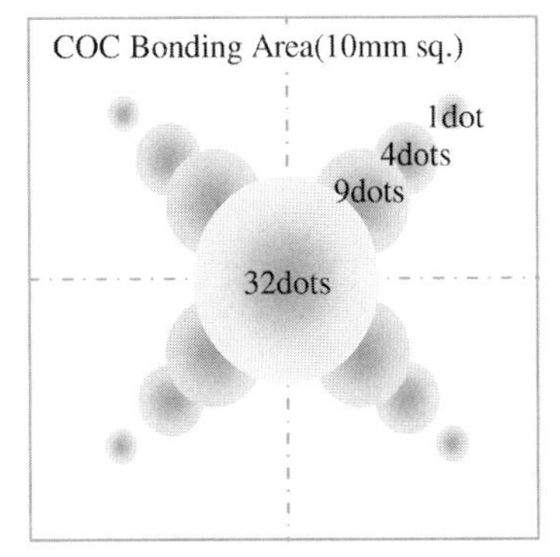

Figure 18.24 Optimal NCP dispense pattern.

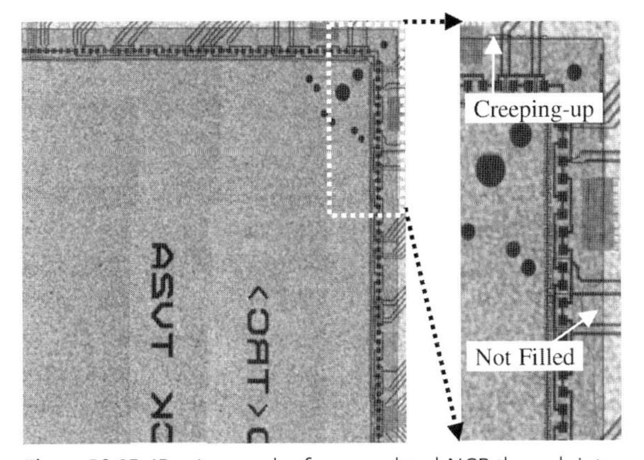

Figure 18.25 IR micrograph of encapsulated NCP through interposer and chip.

were confirmed at the bonding interface, no NCP resin was found by energy dispersive X-ray spectrometer (EDX) analysis. Figure 18.27 shows the chip backside warpage after bonding measured by laser displacement meter. As shown, the warpage was very small (less than 3 μm) because the bonding force was maintained during bonding and curing. According to analysis of the chip backside warpage, it

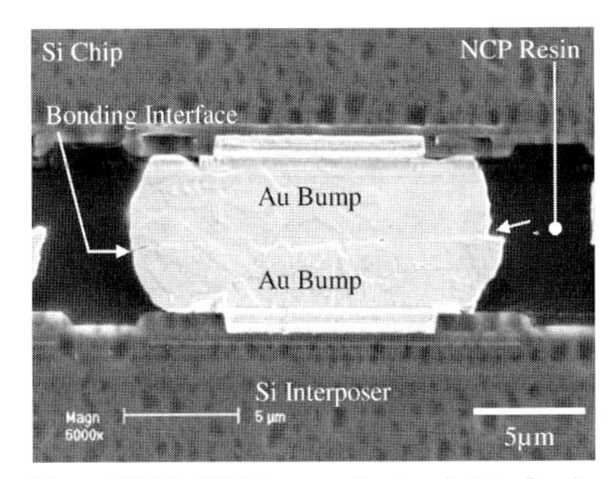

Figure 18.26 SEM image of micro joint after bonding.

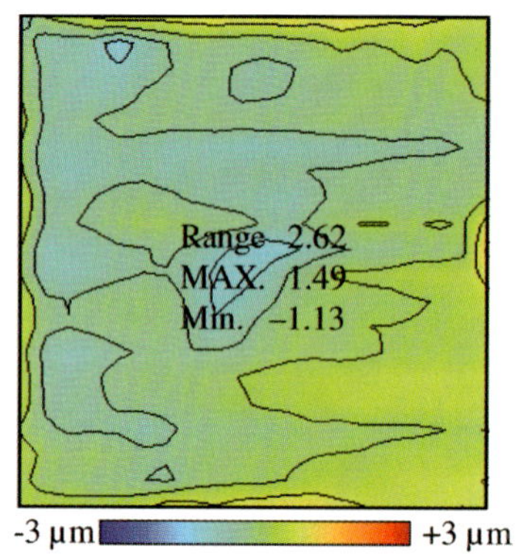

Figure 18.27 Chip warpage after bonding.

was considered that co-planarity of bumps could be controlled for each layer. Therefore, the NCP preform process enabled the realization of void-less encapsulation and stacking of the chip onto the stable bonding area. In addition, this process is a possible encapsulation process for 3D chip stacking LSI with the minute interconnections in 20 μm pitch.

18.5.4 Fabrication of Vertical Interconnection

Figure 18.28 shows the schematic model of Cu/3D interconnection and an SEM image of the Cu plug at the chip-backside. The TV-chip was 10.4-mm-square and 50-μm-thick and 10-μm-square Cu TVs were fabricated in a 20-μm-pitch, allocated on the periphery of the Si chip. The Si interposer was 18-mm-square and 500-μm-thick. The 1820 electroplated Cu bumps, 10.4-μm-square and 5 μm high with a 1.5-μm-thick electroplated Sn2.5Ag layer, were formed on the Cu TVs through UBM. The Cu plugs exposed from the chip backside were formed through wafer thinning, SiN-CVD and SiN-CMP processes. Table 18.4 shows the fabrication flow of vertical

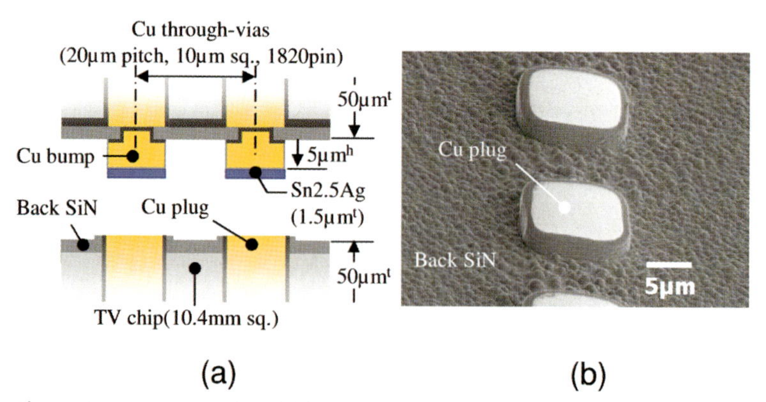

Figure 18.28 Experimental model for 3D chip stacking reliability: (a) schematic model of Cu-TV interconnection and (b) SEM image of Cu plug at chip-backside.

Table 18.4 Fabrication of 3D chip stacking structure.

Process	Target Spec	Methods	Conditions
TEG / TV chip / Interposer	No extrusion No creeping-up	Small shot perform (SSP)	Preform quantity: under 0.4 mg/ 10.4 mm sq.
	Integrity bonding with single IMC (Cu_3Sn)	Cu bump bonding (CBB)	Temp.: 350 °C
	Remove Cu oxide from exposed Cu	Ar sputtering (RIE)	Time: 60 s Force: 49.0 N N_2 flow: 10 L min^{-1} RF power: 1000 W
TEG / TV chip / TV chip / TV chip / Interposer	Void-free underfill	Vacuum encapsulation	Time: 10 min Ar flow: 15 sccm Drive pressure: 10 Pa Pressure: 5.33 kPa Temp.: 80 °C Post cure: 125, 150 °C/30 min

interconnection by the chip stacking process. The encapsulating resin was dispensed onto the Si interposer before bonding by SSP. Then, the first layer TV test element group (TEG) finished TV fabrication process was mounted onto the Si interposer. This method was expected to prevent the chip backside warping by means of flowing resin through almost micro thin gap and then curing the resin during pressuring by flat bonding tool. Bonding conditions were set to achieve complete diffusion and reduction of the Sn alloy to Cu_3Sn and curing resin. The Cu oxide layer was formed on exposed Cu plug during bonding process as previously noted, and removed utilizing Ar sputtering under the conditions used to etch the Cu surface at approximately 80 nm. Subsequent layers were then mounted sequentially under the same conditions. After bonding all layers, the remaining gaps between each layer were encapsulated utilizing the vacuum encapsulation unit (Century-VE, Asymtek), which was expected to prevent the formation of voids at the bonding interface due to air trap. Finally, the encapsulated resin was cured. Each set of conditions for the chip stacking process are summarized in Table 18.4.

Figure 18.29 shows an SEM image of the entire cross section of a 3D chip stacking structure with 20-μm-pitch multilayer vertical interconnection, fabricated using the chip stacking process. The three 50-μm-thick TV TEGs with 20-μm-pitch Cu TVs and the 50-μm-thick Bump TEG were successfully mounted onto the Si interposer with a

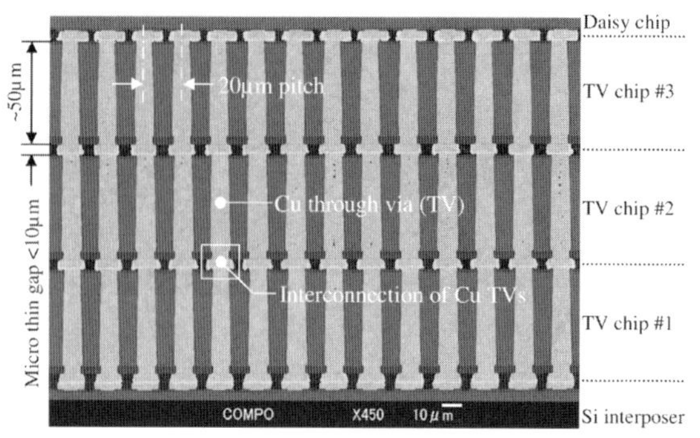

Figure 18.29 SEM image of entire cross-section of 3D chip stacking with 20 μm-pitch-multilayer vertical interconnection.

high accuracy. The four layered micro thin gaps (<10 μm) between Si chips were fully encapsulated with side-filled underfill resin; the two-step encapsulation method consisting of preform and post filling resin was considered to be most suitable for the encapsulation of 3D chip stacking LSI with micro thin gap. Figure 18.30 shows a magnified micrograph of the CBB structure connecting Cu TVs of first and second TV chip. As shown, the exposed Cu TV of first TV TEG was connected directly with a Cu bump of the second TV chip through the Cu_3Sn layer. The backside silicon nitride layer shielded Si substrate from the molten Sn flowed to the bump edge. In addition, it was confirmed, that molten Sn flowed to the side wall of Cu bump without horizontally extending, and it was expected to prevent the short between Cu bumps due to molten Sn. As a result, this interconnection was expected to achieve highly reliable bonding with TVs. The consistent fabrication of vertical interconnection utilizing chip stacking process was thus realized.

18.5.5
Reliability of Vertical Interconnection

The 20-μm-pitch minute interconnection of CoC structure utilizing electroplated Au bump connected metallurgically by thermo-compression bonding has been investigated [24]. It was found that a stress caused by thermal mismatch in the thickness

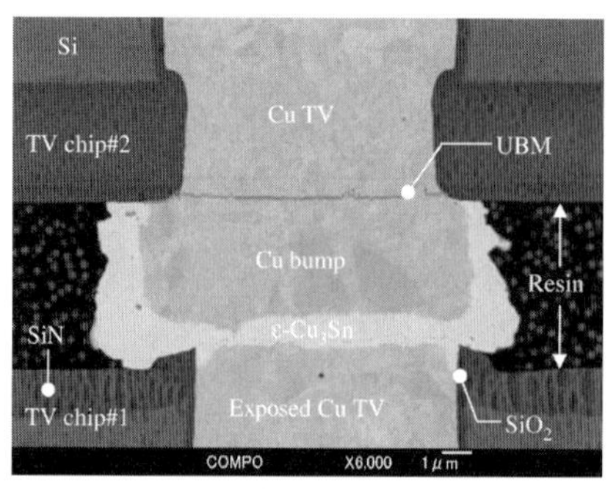

Figure 18.30 Magnified micrograph of CBB structure connecting Cu TVs of first and second TV chip.

direction between the interconnection and the resin cannot be neglected even for a Si on Si structure [25]. More than 1% maximum equivalent plastic strain caused at Au bump interconnection cannot be acceptable, and non-filler resin was certain to break the interconnection. The resin property was considered as one of the most important issues; it was difficult to encapsulate the low stress resin containing over 60 wt% filler particle into the micro thin gap without voids. In contrast, for the chip stacking process, the optimized CBB interconnection through Cu-Sn IMC was expected to achieve high reliability against stress arising from structure. In addition, it was possible to encapsulate the gap without void using a 55 wt% filler particle resin by means of a two-step encapsulation method as previously noted. Consequently, two structural models were applied to evaluate the vertical interconnection reliability systematically. One is the CoC structure, to evaluate the 20-μm-pitch CBB reliability under various encapsulation conditions. The high rate filler content resin with the filler at 55 wt% (Resin A), confirmed as having acceptable reliability on Au bump interconnection, was applied as reference; the silica-filler particles were on average 0.3 μm in diameter, with a sharp size distribution to make it relatively easy to encapsulate the micro thin gap. Moreover, non-filler particle resin (Resin B), which affects the interconnection seriously, was applied to assess CBB capability. In addition, the 3D chip stacking structure was used to evaluate vertical interconnection reliability. This was a four-layered structure composed of Cu TVs connection utilizing the CBB and the reliable Resin A (Figure 18.31). Finally, the possibility of practical application of vertical interconnection was verified in terms of reliability. The temperature cycle test (TCT) was employed for effective evaluation of each structure, especially for the CTE. The TCT was applied from −40 to 125 °C. The daisy chain circuit through the interconnection was formed with 64 separable chains at each sample. The samples were electrically monitored by measurement of the interconnection resistance through the daisy chain patterns during the test and a 10% resistivity change was set as a criterion.

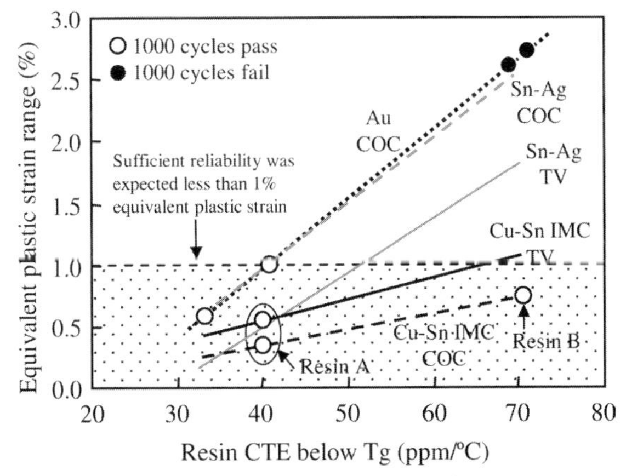

Figure 18.31 Simulation results of equivalent plastic strain range of various interconnection structures and reliability test results.

Table 18.5 Reliability test results (sample count).

Temperature cycle test (TCT) cycles	COC w/ resin A	COC w/ resin B	3D w/ resin A
100	0/5	0/5	0/5
300	0/5	0/5	0/5
600	0/5	0/5	0/5
1000	0/5	0/5	0/5
1500	0/5	0/5	0/5

Table 18.5 shows the reliability test results. Acceptable reliability for TCT at 1500 cycles was confirmed with all samples. These results accorded with the forecast of the mechanical effect of the interface material on the interconnection reliability utilizing finite element method (FEM) analysis for temperature cycling. Figure 18.32 shows the equivalent plastic strain range (ε_{eq}) on interconnection for each structure, particularly focusing on differences in CTE of the underfill resin. As seen in the results, ε_{eq} depends on the CTE of the resin under each condition. This dependence is reduced when the bonding interface is composed of Cu bump with IMC. It is considered that the stress caused by the thermal mismatch in thickness direction between the minute interconnection and the resin cannot be neglected even for a Si-on-Si structure, as previously noted. Although the stress causes plastic deformation of the Au bump easily, the rigid Cu and IMC layer are negligibly deformed. The ε_{eq} of the Cu bump and IMC layer is increased in the case of 3D structure with Cu TVs. Because the thermal deformation of Cu TV is much larger than that of a Si chip, the deformation of Cu TV pulls the Cu bump and IMC layer, which are negligibly deformed by stress of resin. The evaluated structures in this study are plotted in the figure, and these structures are considered as acceptable in terms of thermal stress of resin. Figure 18.32 shows a cross sectional analysis of a CBB structure connecting Cu TVs of first and second TV TEG after TCT for 1500 cycles. As shown, no defect was observed at the bonding interface. The Kirkendall voids due to Cu-Sn diffusion was prevented at test condition because of bonding interface composed of Cu_3Sn single

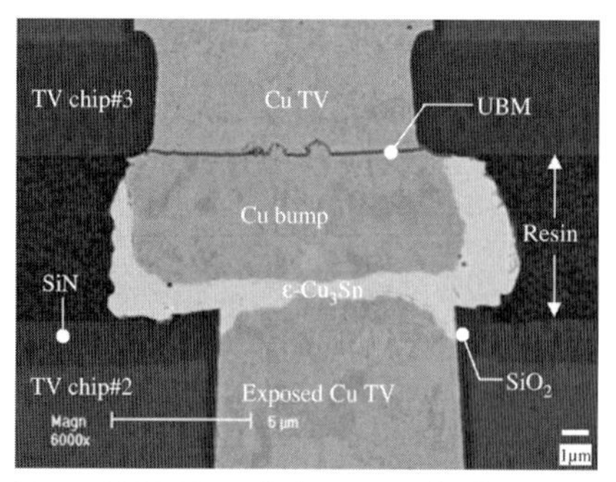

Figure 18.32 Magnified micrograph of CBB structure connecting Cu TVs of first and second TV chip after TCT at 1500 cycles.

IMC layer without Sn and Cu_6Sn_5. It was considered that complete diffusion and reduction of Sn alloy to a Cu_3Sn layer during flip chip bonding was preferable for minute interconnection reliability. Therefore, it was confirmed that CBB has a substantial capability for 20-μm-pitch minute interconnection of Si-on-Si structures, and the practical application for vertical interconnection fabrication by chip stacking process is possible.

18.6 Thermal Performance of Chip Stack Module

The thermal design of the 3D chip stack is one of the most important issues in the integration of high-performance LSIs [26–28]. The temperature rise of the chip stack is not very serious if heat dissipation from one chip is several watts. However, the many unclear problems concerning future high-performance system LSIs include tens of watts of power consumption, increasing number of chips to stack, and low thermal conductivity materials such as low-k dielectric.

The following subsections describe measurement of the thermal resistance of actual stacked chips, estimation of the effect of the passivation layer, and a proposed novel cooling interface for high-power applications.

18.6.1 Measurement of Thermal Resistance

To examine the fundamental thermal performance of a chip stack module, Θ_{jc} of the four-chip-stack module was measured using the test chips, including circuit elements for heat generation and temperature sensing. The chip was 10 mm square and 50 μm thick, and 1820 Cu bumps and Cu through vias were fabricated on its periphery with a 20 μm pitch. Every Cu bump is directly connected to the Cu via of the next chip, except for the top-layer chip. The lowest chip is bonded to the Si interposer. The gap between the chips, about 5 μm, was filled with underfill resin. The Θ_{jc} is defined here as the thermal resistance from the surface of the top-layer chip through the back of the interposer (Figure 18.33a). To avoid the thermal influence from the surroundings and to measure the simple Θ_{jc} of the four-chip stack, the specimens were set up in a thermal insulator and cooled by a commercially available fan heatsink attached to the back of the interposer. The heating area was 5×5 or 6×6 mm to investigate the effect of heat flux density.

Figure 18.33b shows the result of Θ_{jc} measurement for the two sizes of heat-dissipated areas as a function of the number of stacked chips. First, one can see that Θ_{jc} increases by about 0.3 K W^{-1} per layer with increasing number of stacked chips in the case of "6×6 mm-heated." The Θ_{jc} increment for a one-chip stack is rather small compared with those of "package-stack" modules. This is because Θ_{jc} of the chip-stack module is mainly due to the thin underfill layers [26]. Secondly, the increase in Θ_{jc} strongly depends on the heated area of the chip surface. In Figure 18.33b, the difference between the two sizes of heating conditions amounts

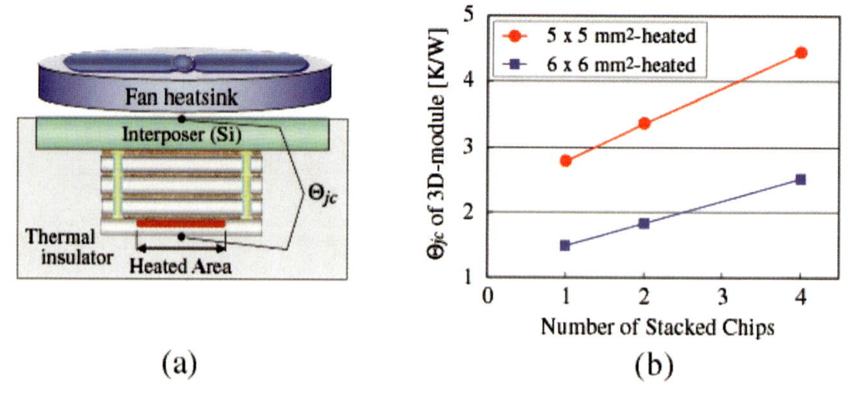

Figure 18.33 (a) Method and (b) results of thermal resistance measurement.

to over 1 K W^{-1}. The strong dependence on the heated area is attributed to the poor in-plane spreading resistance caused by the large aspect ratio of the outline of the ultrathin chip. This thermal uniqueness due to the large aspect ratio is noteworthy because Si is commonly regarded as a high thermal conductivity material.

18.6.2
Effect of Passivation Layer

To break down the internal thermal resistance of the 3D-stack module, the finite volume method (FVM) was used in carrying out thermal conduction analyses. The structural models calculated were (i) no bump, no through-via; (ii) bumps, no through-via; and (iii) bumps, through-vias. The conditions with and without a 10-μm-thick passivation layer were applied for the structural model.

Figure 18.34 shows the details of the internal thermal resistance (Θ_{int}) of the one-chip stack without passivation, where the passivation (PV) layer means surface device circuit layers composed of, for example, circuit-element, wiring and dielectric layers.

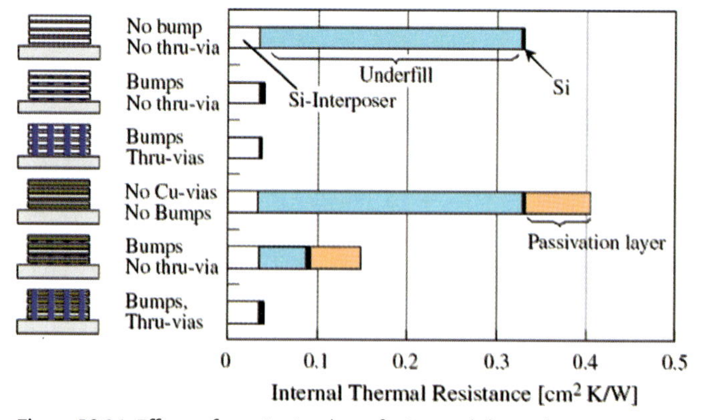

Figure 18.34 Effects of passivation layer for internal thermal resistance.

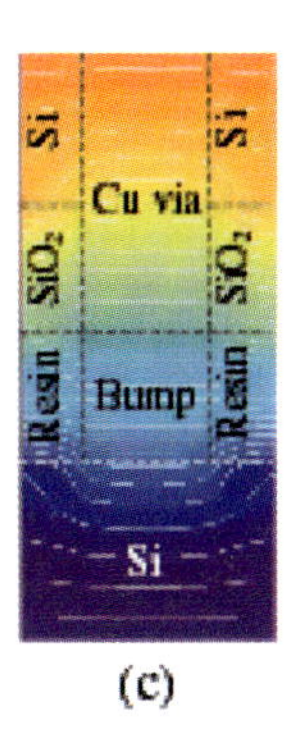

Figure 18.35 Temperature contours around a bump: (a) w/o passivation, w/o Cu via; (b) w/passivation, w/o Cu via; (c) w/ passivation, w/Cu via.

In both cases, Θ_{int} for the underfill layer accounts for more than 70% of the total Θ_{int}. The second major contribution is the Θ_{int} of the PV layer. The thermal effect of Cu vias varies depending on the thickness of the PV layer. For 0 μm-passivation, bumps act as thermal paths connecting the Si chip with the Si interposer so effectively that their combination with through-vias exerts no additional effect on Θ_{int} reduction. In contrast, for 10 μm-passivation, the combination of through-vias and bumps effectively produces a synergistic effect. This is because Cu vias function as thermal vias so effectively against the PV layer whose thermal conductivity is about one-hundredth that of Si. Figure 18.35 shows some of the calculated temperature contours around a bump. Figure 18.35a and b indicate that the PV layer seriously suppresses heat dissipation. Figure 18.35c shows that a through-via acts as an effective thermal path even with 10 μm-passivation. Accordingly, thermal bumps should be applied in combination with through-vias for high-power LSI devices with considerable multilayered wiring.

18.6.3
Investigation of a Novel Cooling Interface

These results confirm that, by applying the through-via/bump combination, Θ_{int} can be reduced to around 0.1 $cm^2\,K\,W^{-1}$ per layer in a 3D-stack module. Therefore, for lower-power applications where the power per chip is less than 10 W and the maximum stack number is 5, a conventional cooling method is applicable, such as the attachment of a heatsink or a heat pipe.

However, if high-power chips (>10 W) are mounted in a 3D-module with a high stack number, the increase of thermal resistance due to the accumulation of Θ_{int} is considerable. In such a case, inner layer chips may overheat if direct cooling is not available. Another cooling interface is necessary by which individual chips are equally cooled in parallel.

Unfortunately, an ultrathin (50 μm) chip has extremely anisotropic thermal characteristics because of the large aspect ratio. Therefore, the well-known "edge

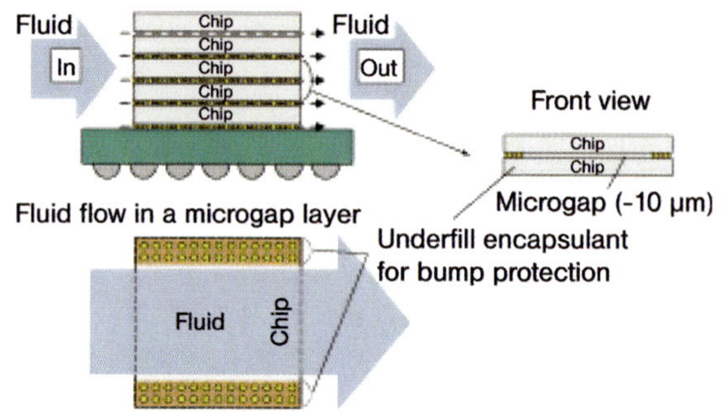

Figure 18.36 Schematic model of microgap cooling for chip-stacking module.

cooling effect" using the side surface of chips is not very efficient as a cooling interface. Consequently, the only available option that realizes "parallel cooling" is convection cooling, utilizing the microgap between stacked chips (Figure 18.36). The 3D-stack modules have "built-in" flow channels between every pair of adjacent chips. If a fluid flow can be injected into every chip-to-chip gap, effective "parallel cooling" can be realized more naturally.

To roughly estimate the thermal effect of parallel cooling utilizing the microgaps, the temperature contours for a four-chip-stacked module were computed by FVM to obtain a preliminary estimation. In this calculation, each chip was assumed to consume 25 W (100 W for four chips).

Figure 18.37 shows the computed temperature contours for two different cooling conditions: (i) conventional cooling using top, side and bottom surfaces; (ii) novel cooling using microgap surfaces. A uniform heat transfer coefficient (5000 W m^{-2} K^{-1}), which corresponds to that of water flowing in a 2-mm-dia. tube at 1 cm^3 s^{-1}, was applied to the surfaces of both models as the boundary conditions. For the microgap cooling, the heat transfer coefficient was applied only to the inner surface of the four gaps.

Accelerated heat current flowing effectively out through the gap surfaces can be seen in the case of microgap cooling, while heat concentration occurs in the region of innerlayer chips with conventional cooling. This result is quite expected, but is very

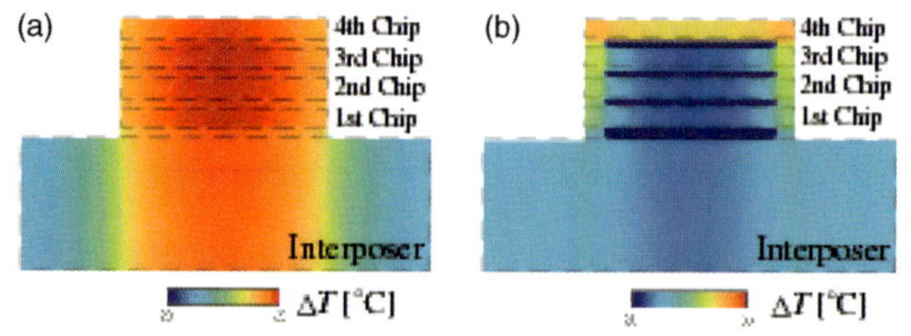

Figure 18.37 Simulation results of high-power chips: (a) conventional cooling; (b) novel microgap cooling.

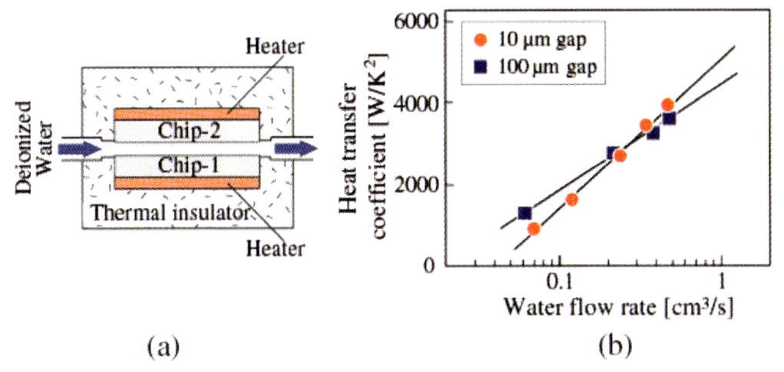

Figure 18.38 Evaluation of thermal characteristics of microgap cooling: (a) experimental scheme; (b) results of heat transfer coefficient.

useful for understanding the effect of the parallel cooling interface when series-connected high-power chips must be cooled in a chip-stacked module.

These "built-in" flow channels in the chip-stacked module are very narrow (about 10 μm) and the size is similar to microchannel heatsinks. For microchannel cooling, many researchers, including Tuckerman and Pease [29], confirmed that fluid flowing into a narrow microchannel shows an eccentric and excellent heat transfer characteristic.

However, there are many unknown factors to be clarified regarding microgap cooling (about 10 μm) because the gap is extremely thin. To quantify the heat transfer characteristic for the microgap between chips, a simple experiment was carried out. Figure 18.38a shows the experimental scheme for determining the heat transfer coefficient. Two Si chips were bonded by the flip-chip bonding method. Two kinds of samples were prepared, one with a 10-μm-gap and the other with a 100-μm-gap. Deionized water was injected into the gap using a peristaltic pump. Two chips were heated by the heaters on the back. The results shown in Figure 18.38b indicate that a water flow rate of 1 $cm^3\ s^{-1}$ realizes a heat transfer coefficient of 5000 $W\ m^{-2}$, the value used in the simulation above. Notably, a small gap shows a high heat transfer coefficient at higher flow rates. The inversion point of the two kinds of gaps is concerned with the difference in transient heat transfer. Although the experiment has not covered a water flow rate higher than 1 $cm^3\ s^{-1}$, the unique thermal characteristic is positive for microgap cooling in the 3D-stack modules.

18.7
Electric Performance of Vertical Interconnection

Two important evaluations were used to investigate the performance of vertical interconnections. One is of the DC performance of resistance through multilayered daisy chain circuits, and the other is AC performance of high-speed transmission performance utilizing self-oscillating circuits by test element group (TEG) chip. The results are described below.

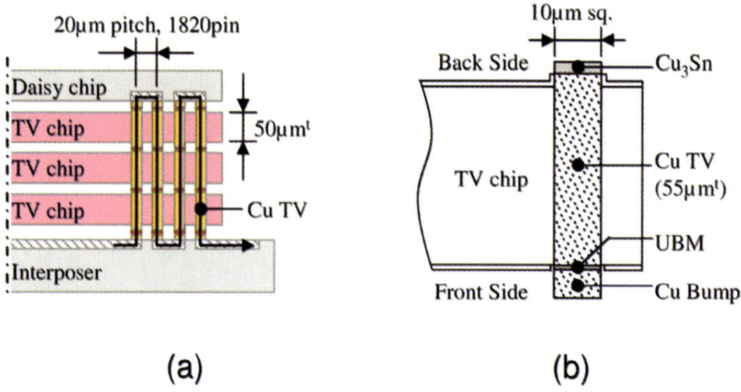

Figure 18.39 Experimental model for DC performance of 3D chip stacking: (a) daisy chain circuit and (b) Cu-TV interconnection.

18.7.1
DC Performance through Multilayered TVs

The DC performance of a multilayered daisy chain circuit was measured. Figure 18.39 shows daisy chain circuit, which consists of an interposer, TVs chip and daisy chip on the top layer. The daisy chains of CoC structure and stacked structure with TVs chip from single to three layers were measured by the Kelvin method.

Figure 18.40 shows a plot of the resistance of the daisy chain circuit against the number of TVs. The change in resistance was proportional to the number of TV chip layers. As a result of the proportional coefficient of resistance increase, the resistance of the vertical interconnection was only 15.4 mΩ per layer, which is very close to the theoretical value (12.4 mΩ). This means that the chip stacking achieved ideal electric contact with no failure, with an interface that is easy to control, as the Cu bump and Cu plugs project from Si substrate. Therefore, the measurements showed that very

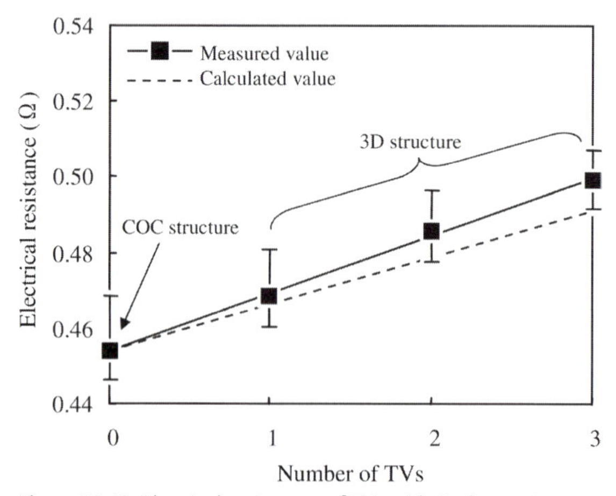

Figure 18.40 Electrical resistance of TVs with Cu bump interconnection.

high-speed data transmission inhibits CR delay and also could improve voltage degradation through long power line and ground bounds in large-scale high performance LSI, thereby contributing to making possible very large-scale, high-performance system integration.

18.7.2
AC Performance through Multilayered TVs

To determine AC electric performance, the signal delay of the 3D structure through the TV was measured using built-in circuits of the TEG, TV chip and interposer. The circuits fabricated on the TEG were inverters and connecting traces. The inverters had internal feedback traces or were connected to the pads for external feedback lines. The interconnection size and process were the same as in the previous subsection. Table 18.6 shows the structures and equivalent circuits. Structure 1 was composed of only the single TV chip, the TEG chip, and the interposer. Structure 2 consisted of the three TV chips. The internal circuit was to characterize the inverter, which had an on-chip trace that connected by Al trace directly between the input and the output of the inverter. The external circuit was designed to feedback the signal through TVs and Cu traces. The output signal of inverter was connected directly without buffer. The input buffer was through a charge up protection diode on the tri-state buffer output. The supplied voltage for TEG chip was 3.3 V. The oscillating frequency was measured by a digital storage oscilloscope (LC574AL, LeCroy Corp.).

Table 18.7 shows the results of self-oscillating frequency. The signal delays through the TVs were calculated from the frequency. The frequency data were normalized to acquire the signal delay through the TV by substituting the data into

Table 18.6 Structure and equivalent circuit of sample.

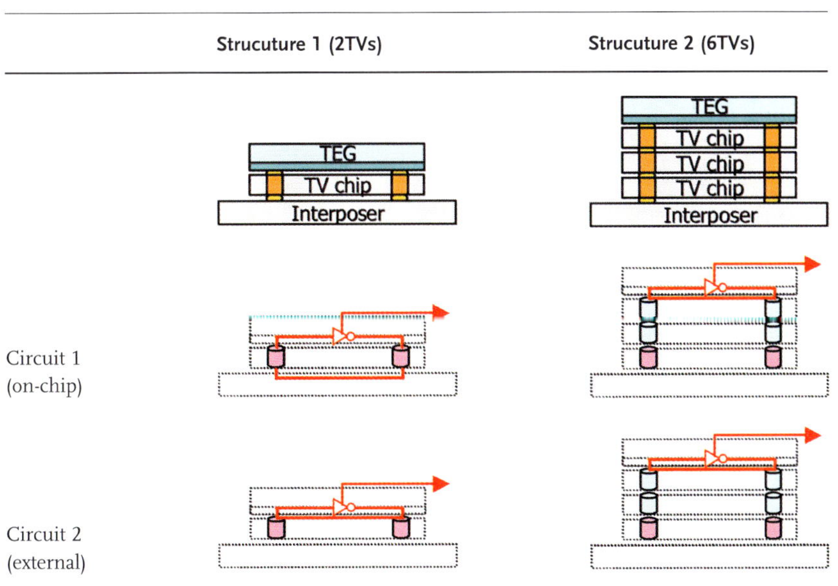

Table 18.7 Self-oscillation frequency (MHz) with several TVs structure with Cu bump interconnection.

Circuit models	1	2	3	Average
On-chip w/2 TVs	618.8	627.2	—	623.0
External w/2 TVs	590.3	598.8	—	594.6
On-chip w/6 TVs	644.6	637.4	613.3	631.8
External w/6 TVs	612.4	603.8	583.4	599.9

Equation 18.2.

$$\tau = \frac{1}{2f^{\text{external}}_{\text{6TVs}}}\left(1 - \frac{f^{\text{external}}_{\text{6TVs}}}{f^{\text{on-chip}}_{\text{6TVs}}}\right) - \frac{1}{2f^{\text{external}}_{\text{2TVs}}}\left(1 - \frac{f^{\text{external}}_{\text{2TVs}}}{f^{\text{on-chip}}_{\text{2TVs}}}\right) \quad (18.2)$$

where τ is the signal delay through the TVs, f is the observed frequency, the subscript denotes the structure type of the samples (Single and Triple) and the superscript expresses the circuit type of the sample (Internal and External Circuit). The measured delay of a TV was 0.9 ps, which is comparable to the value from simulation assuming the Si around the TV was a conductor [30]. Therefore, as the delay was very small, at least under few GHz, a device will be able to connect any devices on another layer, as if they are fabricated on a chip. The result shows that the through-via will demonstrate excellent capability for high-speed data transmission.

18.8
Practical Application of Through-vias

The through-via structure has long attracted the interest of the semiconductor community. Unfortunately, application of the technology has been very limited. Some advanced research reported that the through-via structure could be applied for image sensors and signal processors [31, 32]. However, very few studies on commercially available chips have been presented.

The through-via technology was applied to charge coupled device (CCD) wafers that were assembled into commercially available CCD modules for built-in cameras of cellular phones. Major technical challenges were via etching control, low temperature SiO_2-CVD, low temperature TiN/Cu-CVD. The via etching is substantially different from the development described before in terms of etch stopper. In this case, an Al pad worked as an etch stopper at via bottom. Low temperature processes were required to protect organic materials on the CCD wafers from thermal degradation. With sophisticated process control of RIE, SiO_2-CVD and TiN/Cu-CVD, through-vias were successfully fabricated. Electrodeposited Cu film covered the wafer backside and inside of the vias conformally (Figure 18.41a).

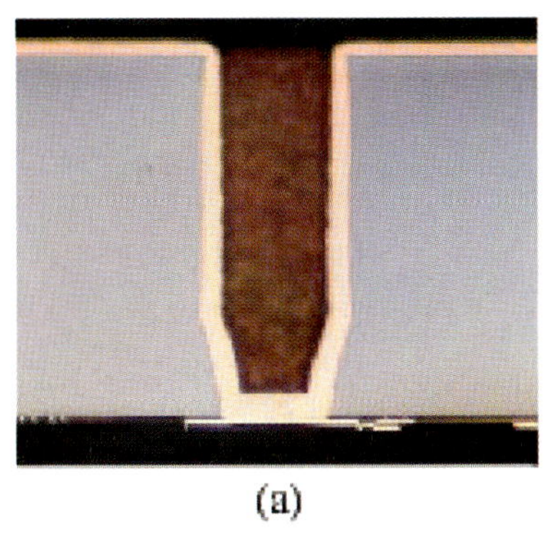

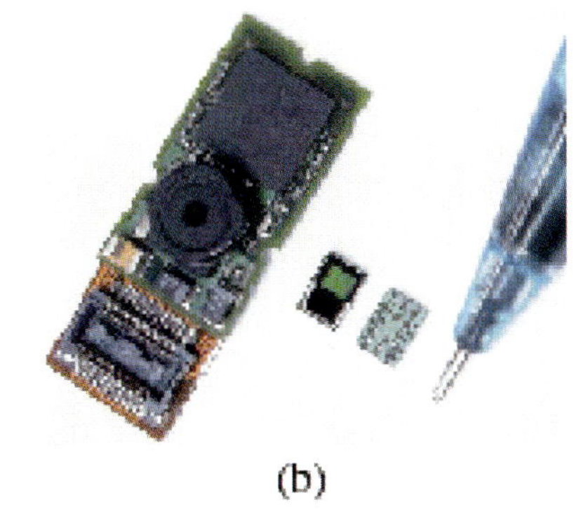

(a) (b)

Figure 18.41 Through-via technology applied to CCD: (a) Cross-sectional micrograph of a through-via fabricated in a CCD chip. (b) WLP processed CCD chips with through-vias and a CCD module for cellular phones in which the CCD chip was assembled.

The wafer backside could be easily processed in wafer level package (WLP) fabrication lines. Figure 18.41b shows the WLP processed CCD chips, and CCD camera modules for cellular phones in which the CCD chips were assembled. The camera modules demonstrated identical imaging characteristics to those in volume production, with extremely high yield and reliability.

These results show that the through-via structure and the fabrication process achieved a practical-use grade and proved the usefulness of this technology.

18.9 Conclusion

Intensive studies of three-dimensional chip stacking and ultra-fine-pitch interconnection have established that these technologies are very effective for future system integration technologies. The high-productive processes, novel wafer handling technique, and thermal design guidelines will accelerate extensive development of 3D chip stacking. Moreover, the success of the imaging chip will encourage packaging engineers to expand the application areas of through-via technology. Semiconductor chips with through-vias might become ubiquitous in the near future.

Acknowledgment

This study was performed under the basic plan of "Research and Development of Ultra High-Density Electronics System Integration" supported by NEDO (The New Energy and Industrial Technology Development Organization). This chapter includes all the researcher's activity of the Tsukuba Research Center of Electronic System Integration Technology Research Dept., ASET.

References

1 Takahashi, K., Terao, H., Tomita, Y., Yamaji, Y., Hoshino, M., Sato, T., Morifuji, T., Sunohara, M. and Bonkohara, M. (2001) Current status of research and development for three-dimensional chip stack technology. *Japanese Journal of Applied Physics*, **40**, 3032–3037.

2 Matsumoto, T., Kudoh, Y., Tahara, M. *et al.* (1995) Three-dimensional integration technology based on wafer bonding technique using micro-bumps. Extended Abstracts 1995 International Conference Solid State Devices and Materials, pp. 1073–1074.

3 Ramm, P., Bollmann, D., Braun, R. *et al.* (1997) Three dimensional metallization for vertically integrated circuits. *Microelectronic Engineering*, **37/38**, 39–47.

4 Lu, J.-Q., Kumar, A., Kwon, Y. *et al.* (2001) 3-D integration using wafer bonding. Conference Proceedings Adv. Metallization Conf. 2000 (eds: Edelstein, D., Dixit, G., Yasuda, Y., Ohba, T.), Materials Research Society, Warrendale, PA, USA, pp. 515–521.

5 Sasaki, K., Matsuo, M., Hayasaka, N. and Okumura, K. (2001) 128Mbit NAND flash memory by chip-on-chip technology with Cu through plug. 2001 International Conference Electron. Packaging Proc., Japan Institute of Electronics Packing, Tokyo, Japan, pp. 39–43.

6 Spiesshoefer, S. and Schaper, L. (2003) IC Stacking Technology using Fine Pitch, Nanoscale through Silicon Vias. Proc. 53rd Electron. Components and Technol. Conference, IEE, Piscataway, N. Y., USA, pp. 631–633.

7 Kondo, K., Okamura, T., Oh, S.-J. *et al.* (2003) Copper via filling electrodeposition of high aspect ratio through chip electrodes used for the three dimensional packaging. *Journal of Japan Institute of Electronics Packaging*, **6**, 596–601 [in Japanese].

8 Sun, J.-J., Kondo, K., Okamura, T. *et al.* (2003) High-aspect-ratio copper via filling used for three-dimensional chip stacking. *Journal of the Electrochemical Society*, **150**, G355–G358.

9 Kondo, K., Yonezawa, T., Tomisaka, M. *et al.* (2003) Copper electrodeposition of high-aspect-ratio vias for three dimensional packaging. Extended Abstracts 2003 International Conference Solid State Devices Mater., pp. 380–381.

10 Kondo, K., Yonezawa, T., Taguchi, Y. *et al.* (2003) Time shortening of through electrode Electrodeposition for three dimensional packaging. Proceedings 13th Microelectron. Symposium (MES 2003), Japan Institute of Electronics Packing, Tokyo, Japan, pp. 256–259 [in Japanese].

11 Barkey, D., Kondo, K., Matsumoto, T. and Wu, A. (2003) Effects of aeration on additive interaction in copper deposition. Symp. Metallization Processes in Semicond. Device Fabrication at the National AIChE Meeting (ed. Landau, U.), American Institute of Chemical Engineering, Cleveland, OH 44106.

12 Kondo, K., Matsumoto, T. and Watanabe, K. (2004) Role of additives for copper damascene electrodeposition-experimental study on inhibition and acceleration effect. *Journal of the Electrochemical Society*, **151**, C250–C255.

13 Ueno, M., Marusaki, K., Taguchi, Y. *et al.* (2003) Proceedings 17th Jpn. Inst. Electron. Packaging Annual Meeting, pp. 231–232 [in Japanese].

14 Ueno, M., Egawa, Y., Fujii, T. *et al.* (2004) Proceedings 18th Jpn. Inst. Electron. Packaging Annual Meeting, pp. 71–72 [in Japanese].

15 Tanida, K., Umemoto, M., Tomita, Y. *et al.* (2003) Micro Cu Bump Interconnection on 3D Chip Stacking Technology. Extended Abstracts 2003 International Conference Solid State Devices Materials, pp. 378–379.

16 Umemoto, M., Tanida, K., Tomita, Y. *et al.* (2002) Non-metallurgical bonding technology with super-narrow gap for 3D stacked LSI. Proceedings of The 4th Electron. Packaging Technol. Conference, IEE, Piscataway, N. Y., USA, pp. 285–288.

17 Vollweiler, F.O.P. (1993) MA thesis, Naval Postgraduate School, Monterey.

18 Lee, Y.G. and Duh, J.G. (1999) Interfacial morphology and concentration profile in the unleaded solder/Cu joint assembly. *Journal of Materials Science*, **10**, 33–43.

19 Pinizzotto, R.F., Jacobs, E.G., Wu, Y. *et al.* (1993) The dependence of the activation energies of intermetallic formation on the composition of composite Sn/Pb solders. Annual Proceedings International Reliab. Phys. Symposium, pp. 209–216.

20 Chan, Y.C., Alex, C.K. So and Lai, J.K.L. (1998) Growth kinetic studies of Cu–Sn intermetallic compound and its effect on shear strength of LCCC SMT solder joints. *Materials Science and Engineering*, **55**, 5–13.

21 Suh, M.-S. and Kwon, H.-S. (2000) Growth kinetics of Cu–Sn intermetallic compounds at interface of 80Sn–20Pb electrodeposits and Cu based leadframe alloy, and its influence on the fracture resistance to 90°-bending. *Japanese Journal of Applied Physics*, **39**, 6067–6073.

22 Haimovich, J. (1993) Cu–Sn intermetallic compound growth in hot-air-leveled tin at and below 100 °C. *AMP Journal of Technology*, **3**, 46–54.

23 Babiarz, A. J. (2006) Jetting small dots of high viscosity fluids for packaging applications. Semiconductor International, August 2006, SP-2–SP-8.

24 Tanida, K., Umemoto, M., Morifuji, T. *et al.* (2003) Au Bump interconnection in 20 μm pitch on 3D chip stacking technology. *Japanese Journal of Applied Physics*, **42**, 6390–6395.

25 Umemoto, M., Tomita, Y., Morifuji, T. *et al.* (2002) Superfune flip-chip interconnection in 20 μm-pitch utilizing reliable microthin underfill technology for 3D stacked LSI. Proceedings 52nd Electron. Comp. Technol. Conference, IEE, Piscataway, N. Y., USA, pp. 1454–1459.

26 Yamaji, Y., Ando, T., Morifuji, T. *et al.* (2001) Thermal characterization of bare-die stacked modules with Cu through-vias. Proceedings 51st Electron. Components and Technol. Conference, IEE, Piscataway, N. Y., USA, pp. 730–737.

27 Nakamura, T., Yamada, Y., Morooka, T. *et al.* (2002) Thermal analysis of self-heating effect in three dimensional LSI. Extended Abstracts 2003 International Conference Solid State Devices and Mater., pp. 316–317.

28 Kalyanasundharam, J. and Iverson, R.B. (2002) Application of a global-local random-walk algorithm for thermal analysis of 3D integrated circuits. Conf. Proceedings, Adv. Metallization Conference 2002 (eds : Melnick, B.M., Cale, T.S., Zaima, S., Ohta, T.), Materials Research Society, Warrendale, PA, USA, pp. 59–65.

29 Tuckerman, D.B. and Pease, R.F.W. (1981) High-performance heat sinking for VLSI. *IEEE Electron Device Letters*, **EDL-2**, 126–129.

30 Sato, T. (2002) Integrated System in Low Power Drive Report II, 78 [in Japanese].

31 Lee, K.W., Nakamura, T., Sakuma, K. *et al.* (2000) Development of three-dimensional integration technology for highly parallel image-processing chip. *Japanese Journal of Applied Physics*, **39**, 2473–2477.

32 McIlrath, L.G. (2002) High performance, low power three dimensional integrated circuits for next generation technologies. Extended Abstracts 2002 International Conference Solid State Devices and Mater., pp. 310–311.

19
3D Integration at CEA-LETI

Barbara Charlet, Lèa Di Cioccio, Patrick Leduc, and David Henry

19.1
Introduction

Leti research in 3D integration started in the early 1990s as an approach to packaging with vertical memories stacking and flip-chip interconnections [1]. As a result, in the last 15 years, several approaches with inter-chip bump connections, including the various later approaches of micro-bumps, have been developed and combined with through wafer vias or deep vias. Furthermore, different techniques for heat management, advanced materials and chip and MEMS stacking have been developed for the packaging approach. At the same time, in the 1990s, wafer level integration with direct wafer bonding was studied and improved to develop thin layer and circuit transfers [2]. During the same period the smart-cut process was developed in parallel and technologically validated for successful industrial transfer to the well-known SOITEC company. Leti continues its research involvement with SOITEC and is also involved in research programs with ST Microelectronics, Freescale, Tracit, Atmel and other industrial partners and research institutes, collaborating on heterogeneous thin layer transfer by smart-cut or direct wafer bonding and bonded interface mastering. These activities were largely developed for various materials and substrate architectures and continue to contribute to 3D integration development. The advanced front- and back-end development platforms existing in the Leti clean room facilities enable a synergizing of expertise to further 3D integration in different field applications such as key processes development for circuits transfer and interconnection, front-end integration, heterogeneous integration for microelectronic, optoelectronic and MEMS applications.

19.2
Circuit Transfer for Efficient Stacking in 3D Integration

The choice of processes implemented for 3D ICs integration has a very important role in stacked 3D system performance [3, 4]. One of the key processes in 3D integration is

Handbook of 3D Integration: Technology and Applications of 3D Integrated Circuits.
Edited by Philip Garrou, Christopher Bower and Peter Ramm

ISBN: 978-3-527-32034-9

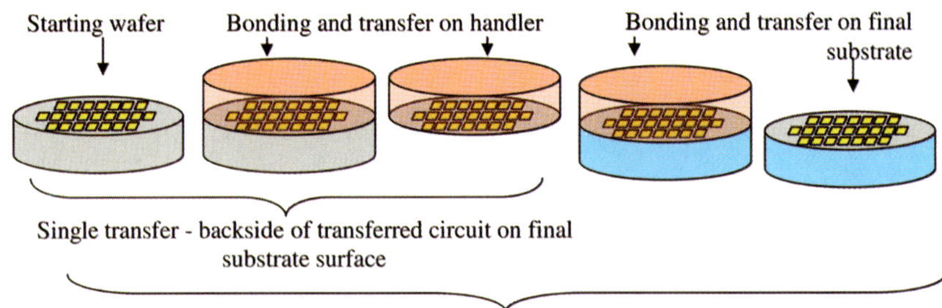

Figure 19.1 Schematic diagram of a vertical integration using a single or double circuit transfer onto a receiver substrate [9].

strata stacking, which can be realized by wafer-to-wafer, die-to-wafer or die-to-die bonding. SiO_2/SiO_2 covalent bonding of substrates has become a well-controlled process since the introduction of SOI substrate in volume production [5]. Good yield and circuit compatibility is obtained after a low temperature interface stabilization process [6]. This generic layers transfer process can be applied to the various cases of 3D integration schemes, with the objective to also stack dies or wafers. Figure 19.1 shows a schematic of a patterned layer transfer using either non-aligned or aligned bonding. The same transfer scheme will be applied to die stacking [7, 8]. The challenge is to obtain a perfect bonding quality since wafers have a relative high incoming surface topography. Depending on the chosen strategy for the circuit transfer one or both of above transfer stages will be repeated for multilayer stack integration. The complexity of patterned wafers or dies used for circuits transfer also contributes in the choice of integration scheme. For this reason, before defining a 3D integration flow chart, each individual process will be carefully examined regarding its technological compatibility and integrated layer properties.

We have demonstrated various 3D integration schemes related to specific applications and implementing different key processes (alignment, bonding, thinning and inter-chip interconnection) at wafer or die level. Developments are detailed below of the key processes that allow the 3D realization of different front- and back-end applications.

19.3 Non-Destructive Characterization of Stacked Layers

Processes development for 3D integration also presents challenging issues in characterization setups and methods. More often, the characterization techniques are adapted from 2D integration. However, in many cases specific development and methodology considerations are needed. Considering the complexity of stacks and the cost of components, non-destructive methods are essential. We discuss hereafter several non-destructive methods and set-up possibilities applied to stacked ICs.

19.3.1 Stacked Interface Checking

Two main cases can be considered for bonded checking interface quality:

- After patterned chips alignment, but before the bonding interface final stabilization (e.g. by annealing) – non-destructive characterization allows the possibility to select the defectively bonded wafers or die, having no required interfacial specifications (e.g. in alignment) or defects (bubble due to particles). These dies or wafers can be debonded and then reprocessed for correct interface achievement. By one adapted characterization the processed dies or wafers will be spared.
- After bonded interface annealing – non-destructive characterization allows the choice of correctly stacked dies or wafers and permits selection for subsequent steps of post processing.

Due to the stacked substrate thickness, and for most cases the non-transparent nature of substrates, embedded interface characterization is not possible (or not precise enough) with standard characterization tools. For several characterization techniques, like stacked interface defect evaluation or patterns alignment accuracy measurement, an infrared (IR) camera or microscope is required to evaluate the checked characteristics, even if this method has many limitations for metal patterned areas and rough surface wafers. Another possibility of non-destructive interface characterization is offered by scanning acoustic microscopy, which can identify the lack of adherence at a stack interface [10].

Figure 19.2 illustrates the interface and edge defects observed between two patterned wafers after direct bonding via SiO_2 layer.

19.3.2 Alignment Accuracy Measurement

Non-destructive characterization and control is also crucial for precise alignment accuracy evaluation during the integration of stacked circuits or micro structured

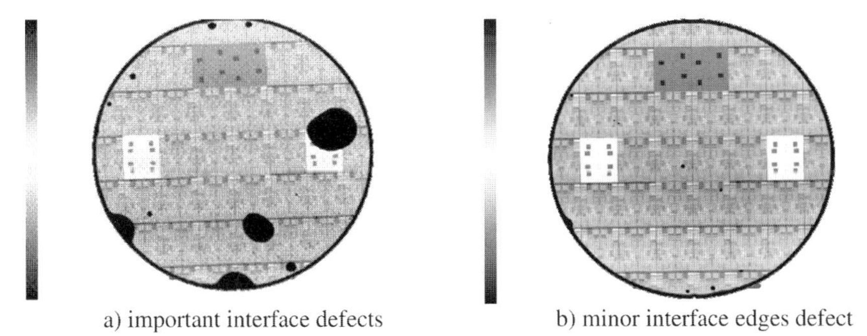

a) important interface defects
b) minor interface edges defect

Figure 19.2 Interfacial defects inspection by scanning acoustic microscopy on the 8-inch stacked patterned wafers. Inspection was performed after the wafer bonding with alignment.

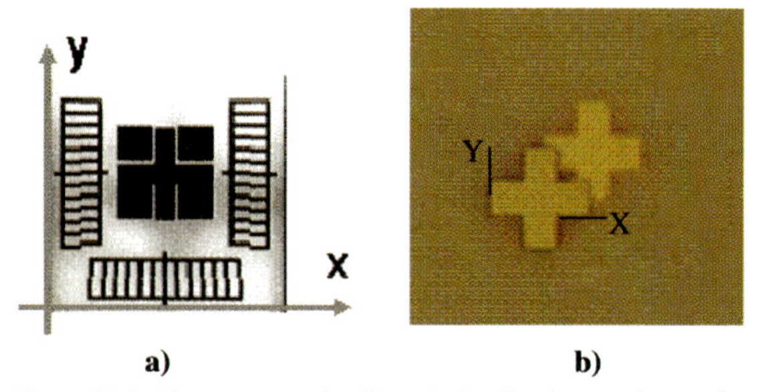

Figure 19.3 Alignment marks of bonded wafers having the x and y axis misalignment. (a) IR microscopy picture, showing the alignment cross and verniers through the non-transparent substrate. (b) Optical microscopy picture of two strata alignment key viewed through the thinned transparent substrate.

components. For a non-transparent stack, IR microscopy is, presently, a unique non-destructive possibility to estimate alignment accuracy, because acoustic wave microscopy is not precise enough (due to the measurement wavelength). Misalignment values can be measured with a vernier-type structure (Figure 19.3). Figure 19.4 shows an example of the accuracy of patterned wafer alignment measurement, obtained on 200 mm patterned wafers aligned by two different alignment tools.

The latest alignment tool improvements, such as the EVG – "Smart View" machine [11], indicate a good opportunity to achieve submicron accuracy in the near future.

19.3.3 Thinned Stack Characterization

Backside wafer thinning process development needs strong interaction with non-destructive characterization. It is particularly appreciated for *in situ* thickness monitoring during the grinding, etching or polishing process. The main stages of

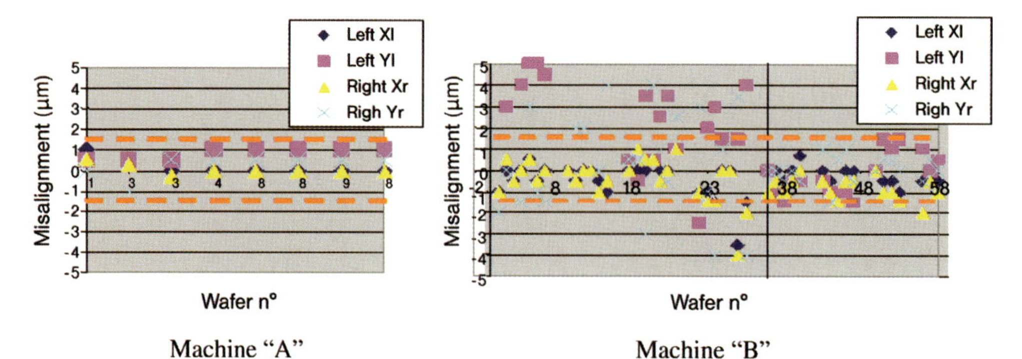

Figure 19.4 Patterned wafers alignment accuracy measured with two different alignment tools. The 200 mm wafers interface is fixed by SiO_2 direct bonding.

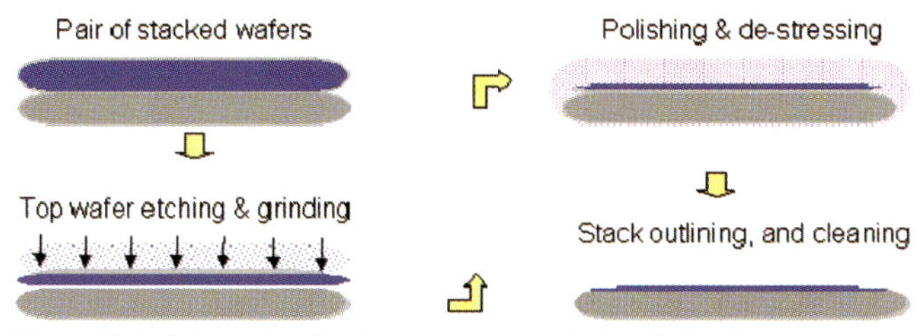

Figure 19.5 Schematic of main process steps in bonded wafers backside thinning down.

one side bonded wafers thinning process are schematically represented in Figure 19.5.

At all stages of this process the upper thinned layer thickness and quality must be controlled, particularly after the fine thickness achievement. Furthermore, the thinned stack post-processing requires the compatible stacked layers quality. Several non-destructive methods are used in ICs technology, such as AFM, SEM, TEM, and so on, but most of them examine the wafer surface locally. As an example of whole wafer surface inspection we report an optical method based on laser light scattering. This method offers the possibility to gather information about surface roughness, particle contamination and subsurface layer local defects. Figure 19.6 shows the surface quality measured on a Surfscan SP2uv from KLA Tencor equipment, based on laser scattering using haze measurements [12]. However defects classification

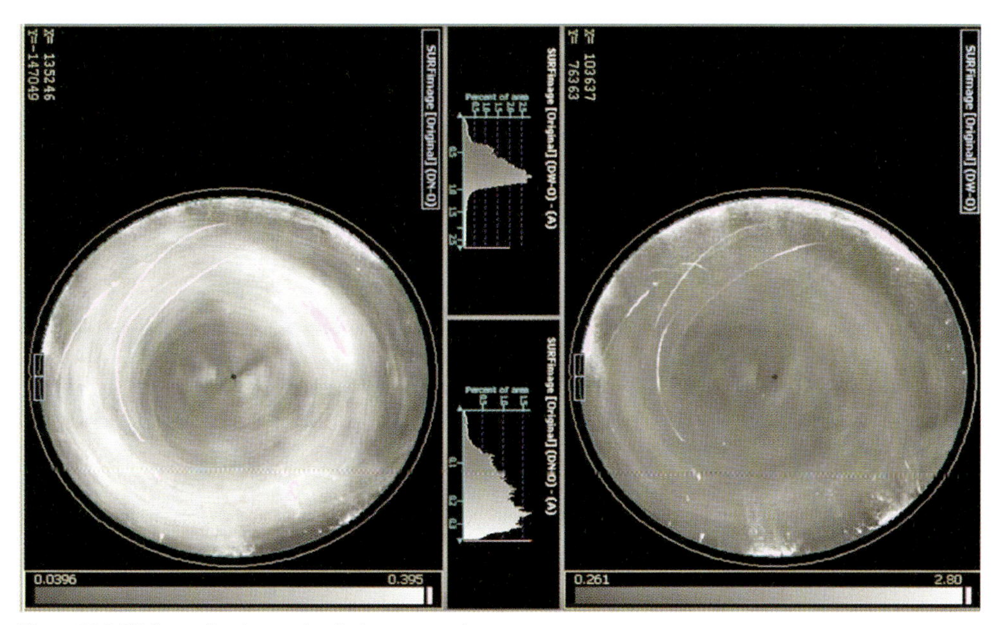

Figure 19.6 Wafer surface inspection by laser scattering set-up – two different sensitivity scans provided on the backside thinned wafer of a double strata stack. The surface roughness variations, scratch areas and local defects can be identified.

acquires complementary characterization, this surface characterization method can offer a good basis for comparative study of thinned wafers surface quality.

This chapter is not an exhaustive report on the study of non-destructive methods used for characterization but, instead, highlights the fact that 3D integrated progress needs the adaptation of existing methods as well as the development of new non-destructive characterization methods.

19.4 Example of 3D Integration Application Developments

Implementation of advanced ICs to 3D integration is an example of technology where multiple process schemes can be developed using various innovative materials, processes and tools. New equipment opportunities, like precise alignment, efficient bonding or stacking, vias opening and filling, make new technological approaches for 3D integration possible [13]. First, electrical data and simulation models have demonstrated very promising performance of these new engineered substrates and circuits, such as increased speed and frequency and reduction of power consumption and dissipation [14]. Furthermore, heterogeneous substrate stacking enables new device architectures and system functionalities. We review some approaches to 3D integration, developed in collaboration with research institutes and industrial partners. A few examples of 3D circuit integration and performance are discussed.

19.4.1 Through Silicon Via Filled with Doped Polysilicon – Advanced Packaging [15]

One of the key technologies for 3D integration is Through Silicon Via (TSV). It is a very promising technology in advanced packaging, for the replacement of wire bonding. This approach is very relevant, especially in terms of size reduction, circuit performance enhancement and cost reduction. For fully integrated products, such as SiP, SoP, 3D component integration (e.g. memory stacking) or MEMS structure packaging, this technology is mandatory [16]. A key point for TSV technology development is its capability to be integrated in a classic CMOS process flow, without any influence on such technological steps as lithography or annealing. This TSV technology, called pre-process via or via-first [17], has to be completely compatible with subsequent standard CMOS processes [18]. This new via-first technology requires several specific developments for design and each step of via manufacturing. First, a specific design based on annular via has been developed and the via resistance has been optimized by calculation [15]. Then, a new process has been developed and improved after the first morphological characterizations, especially for deep etching and void-free polysilicon filling [19]. Finally, morphological characterization and electrical tests have been performed. Figure 19.7 presents the general process flow of the polysilicon filled TSV technology.

Silicon deep etching was performed in a STS HRM tool with an adapted Bosch process. A 5 μm thick positive resist (JSR 335) or 1.4 μm thick TEOS oxide was used

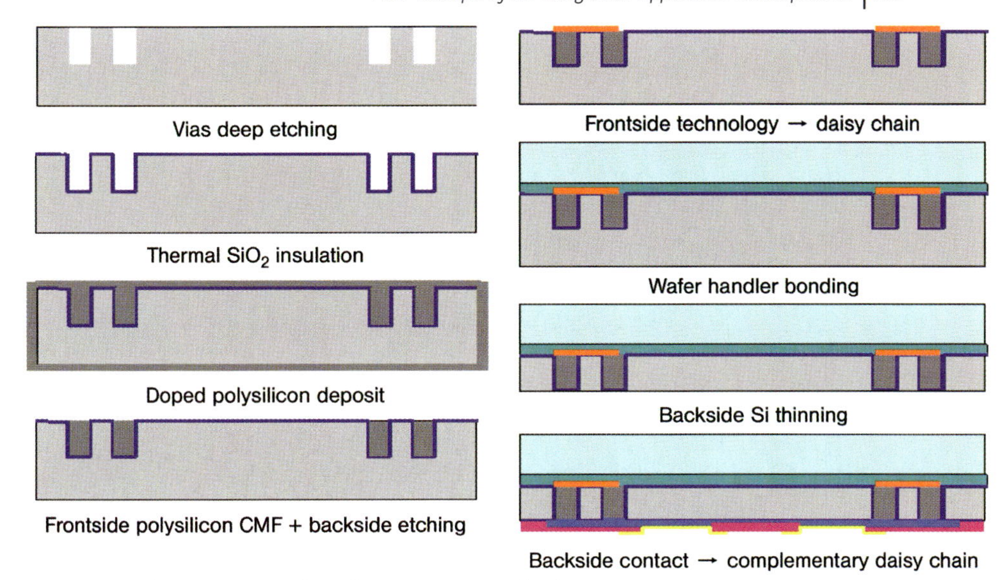

Figure 19.7 Schematic process flow of pre-process via technology.

for the mask. The current via aspect ratio (between 15 and 35), need a particular process to prevent premature etch stopping. This process development has been reported previously [15].

A cross-sectional view of a 5 μm width via is presented in Figure 19.8a, and two cross sections corresponding to two processes conditions are also presented in Figure 19.8b and c.

After deep etching of the trenches and stripping of the mask, an insulation layer was formed by using a classical silicon thermal oxidization. The process temperature was 1000 °C in a steam ambient. The targeted SiO_2 thickness was 0.5 μm.

Subsequently, polysilicon via filling was performed. The major objective of the vias filling was to obtain void-free trenches, to assure a good contact resistance and to prevent reliability issues. Two different doped polysilicons have been used to reach

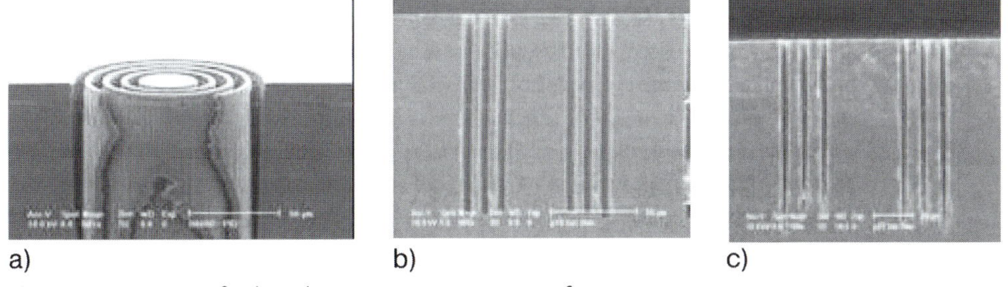

Figure 19.8 SEM view of a cleaved 5 μm via (a), cross section of trenches etched by anisotropic via process (b) and tapered via process (c).

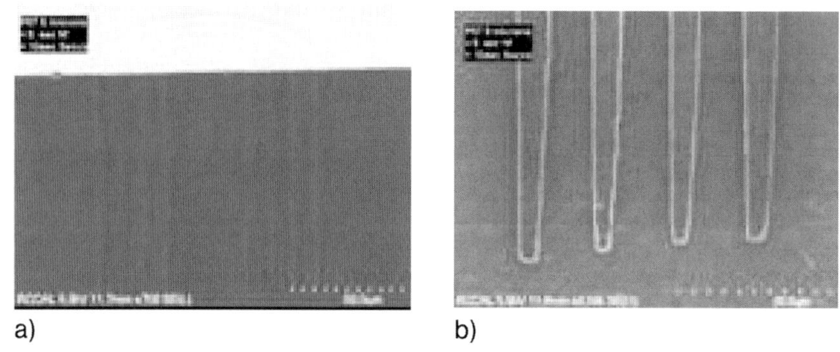

Figure 19.9 SEM cross section after void-free filling process.

these objectives. Both materials are n-type phosphorous-doped polysilicon and the deposits were performed using a LPCVD technique in a TEL furnace. All deposit parameters are given in Reference [15].

Using the optimized process, we obtained void-free trenches for 5 μm wide vias. Figure 19.9a presents a SEM cross section of a 5 μm wide via. A view of the bottom of the trenches is also presented in Figure 19.9. There are no voids and no defects in this area and the SiO_2 insulation layer is perfectly conformal.

Electrical tests of completely patterned vias have been performed on the wafers having the via filled with two types of polysilicon.

The conclusions of this study are:

- For all geometries, the via resistance is less than the initial requirement (1 Ω).
- For the optimized geometry (5 μm trenches – 1300 μm^2 surface), the mean via resistance value was 0.227 Ω ± 13% at 3σ. This value has been measured on 130 via.
- The final resistances of via with both polysilicon were quite close but the difference between the two was due to differences in material resistivity.
- The difference between measurement and calculation was due to contact resistance between vias and pad, which cannot be calculated easily.

In this work, specific designs for via-first achievement have been defined and a complete technology has been developed to create this type of via structure. Many technological issues, such as voids or stress aspects, have been solved during this work and a daisy-chain test vehicle has been demonstrated. The electrical results obtained with this technology were extremely encouraging: the resistance of the optimized geometry was close to 0.250 Ω compared to an initial requirement of 1 Ω. Moreover, the uniformity of resistances was 13% at 3σ.

19.4.1.1 Neo-Wafers Concept – Advanced Packaging [20]

The neo-wafers approach is used when chips need to be processed again at the wafer level after electrical testing, wafer backside thinning and dicing. Selection of the tested good dies and integration to the common support by an adhesive layer were carried out to supply the chips for interconnection rerouting on the consolidated wafer structure. Figure 19.10. depicts the main stages of the neo-wafer rebuilding

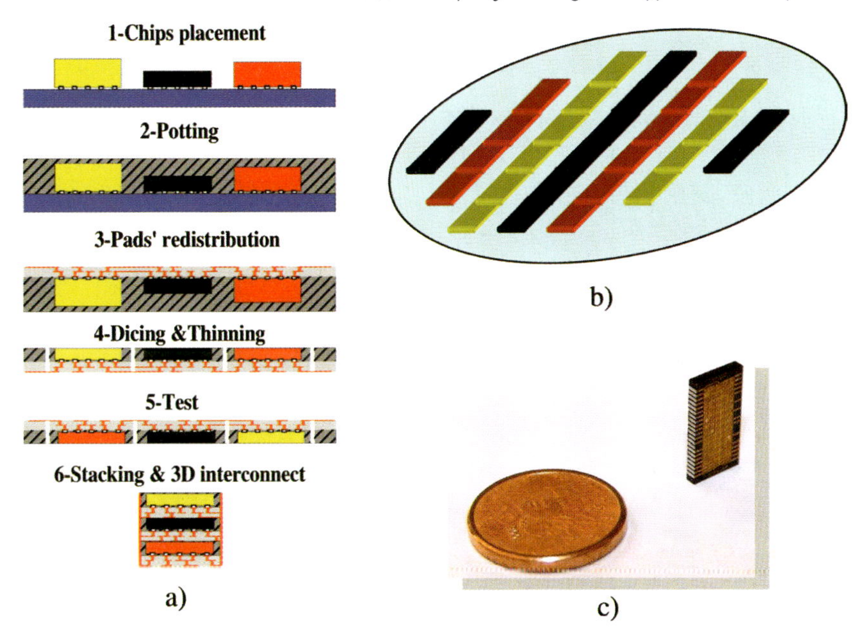

Figure 19.10 Schematic diagram of fabrication flow for rebuilding a wafer (a); representation of a rebuilt wafer (b); and 3D stacked memories – compared to one cent of € money (c).

process for 3D structure formation with selected chips. The 3D structure can be built using the wafer level approach for pads redistribution, chips thinning, dicing and then the known good dies (KGD) were stacked and interconnected by vertical copper connections, forming the final 3D structure. The high performance of the stacked structure was confirmed by a 1 Gbit memory demonstrator.

19.4.2 Die-to-Wafer Integration for Opto-Electronics Application [21, 22]

The maturity of direct wafer bonding processes, largely used on various materials and geometries, gives an opportunity for die-level heterogeneous material integration. Hetero-integration of InP, 50 mm wafers, on silicon has already been developed [8]. For this a silicon dioxide layer is deposited and then processed on InP(1 0 0) with an adequate epitaxy layer. Thanks to this preparation, the bonding of InP/SiO_2 on Si/SiO_2 wafers is similar to Si/SiO_2 on Si/SiO_2 bonding [7, 21]. The same process has been used for die bonding. Figure 19.11 shows the die-to-wafer approach, where stacking is achieved by direct die-to-wafer bonding. The dies are obtained by a mechanical dicing of 360 μm thick InP substrate containing an epitaxial heterostructure. The minimal bonded die size is 1 mm^2. The epitaxial stack contains the etch-stop layers helpful in post-bonding substrate removal. The molecular bonding assures a good adhesion even at room temperature [22]. At room temperature, a bonding energy around 200–250 mJ m^{-2} has been measured using the Maszara blade

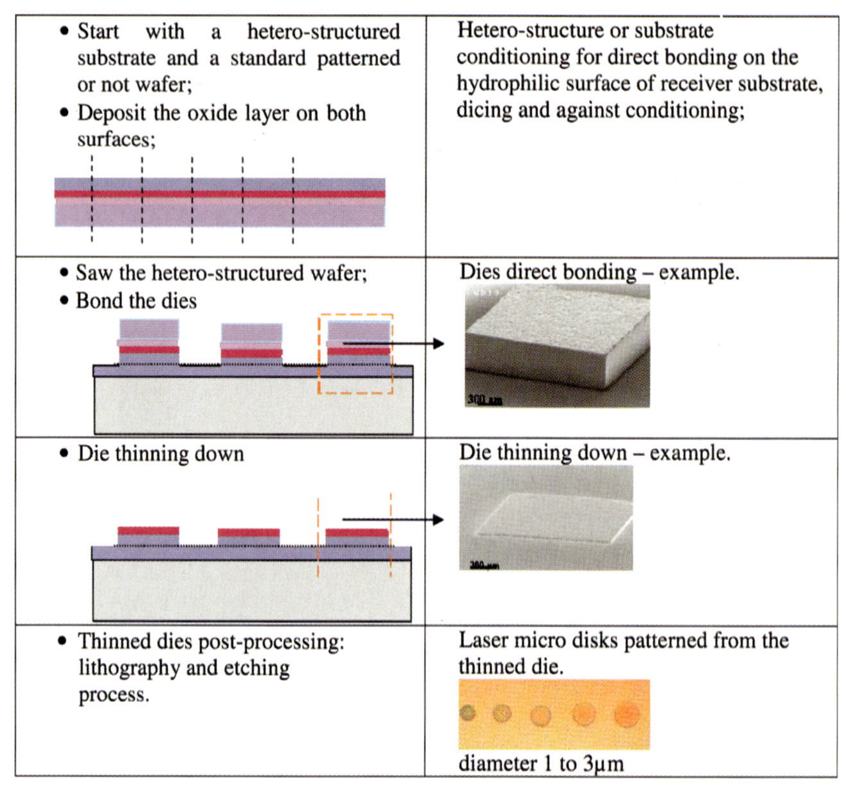

Figure 19.11 Schematic description of die-to-wafer integration and illustrated examples of achieved structures.

opening method [23]. Thermal annealing at 200 °C improves the interface links and provides a high bonding quality (bonding energy >700 mJ m^{-2}).

When the bonded die is an epitaxial stack including an etch-stop layer, the initial InP substrate can be mechanically thinned down to several micrometres and the remaining InP substrate and the sacrificial InGaAs layer can be chemically and selectively back-etched. Using this approach we bonded a die containing $InAs_{0.65}P_{0.35}$ having a 6 nm thick Single Quantum Well (SQW) confined between 120 nm thick InP barriers. In this case, the final thickness of the reported die with a SQW is thinned to 256 nm.

19.4.3 Examples of Wafer-to-Wafer 3D Integration

19.4.3.1 Double Gate MOS Transistor [24, 25] – FEOL Application

3D integration has been evaluated and considered as a solution for submicron technology IC, where the RC delay becomes one of dominant factors [26]. The front end of line (FEOL) integration benefits of FE compatible processes like layer transfer include the new nanoscale architecture. Multi-gate devices are the most promising architectures to fulfil the roadmap targets for sub-32 nm nodes. Among them, planar

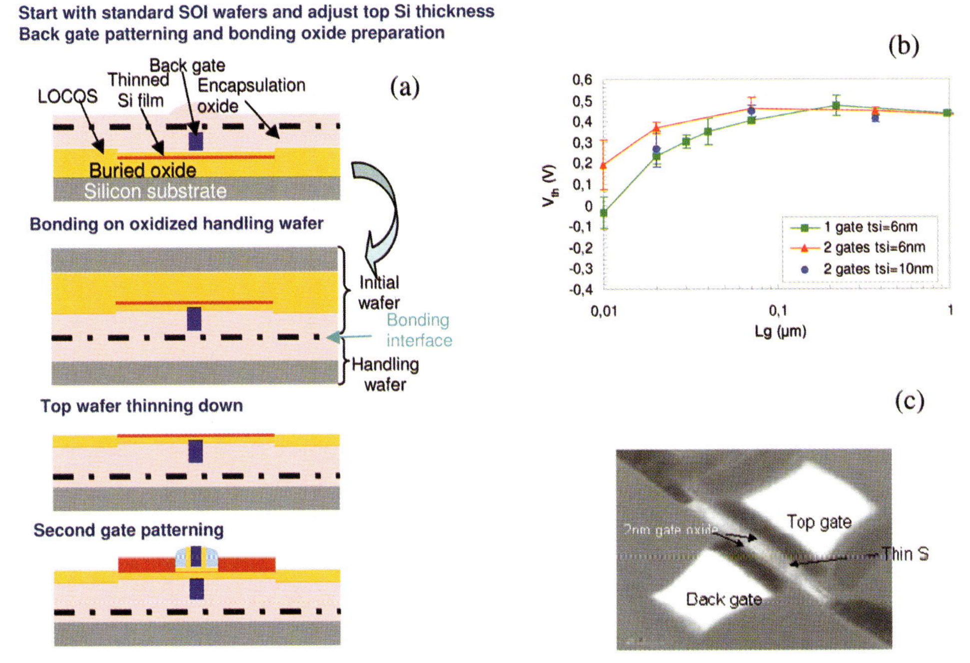

Figure 19.12 Schematic of the fabrication flow for the non-self-aligned planar Double Gate MOS transistor (a), V_{th} performances comparison (b) and TEM picture of integrated DGMOS structure (c).

double gate MOS transistors offer an ability to naturally integrate strained Si required to enhance the transport properties of ultra-scaled devices. A planar double gate CMOS transistor with a 40 nm metal gate was achieved using direct molecular wafer bonding. Figure 19.12 shows the non-self-aligned process performed for demonstrator achievement.

19.4.3.2 High Density Inter-chip Connections [27, 28] – BEOL Application

To reduce global interconnect lengths, chip form factor and to increase bandwidth a high density interconnection technology was developed using circuits stacking by direct bonding. Several potential technical concerns have emerged from the first attempts to realize this integration and from modeling. Wafer-to-wafer bonding quality, that is alignment precision, interface quality, and vias patterning, has a large impact on final stack yield and performance.

The studied integration is a two-strata face-to-face stacking using wafer-to-wafer SiO_2/SiO_2 molecular bonding and copper "via last" process to connect both circuits (Figure 19.13).

Process development was performed using 90 nm nodes patterned wafers with STI (Shallow Trenches Isolation), PMD (planarized middle dielectric) and an especially designed last metal level. The top wafer was realized on an SOI substrate

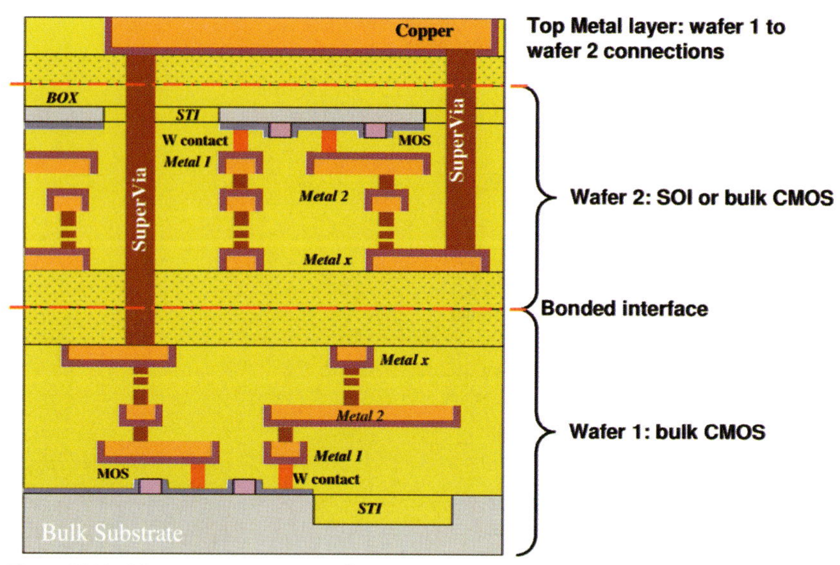

Figure 19.13 Schematic cross-section of 3D integration in the case of SOI substrate as top wafer [27].

and the bottom wafer was bulk silicon. On both wafer surfaces, 800 nm PECVD SiO_2 was deposited. After this surface preparation, with chemical-mechanical planarization and cleaning, the wafers are bonded face-to-face at room temperature and then stabilized by annealing at $T < 400\,°C$. The wafer bonding technique and low temperature interface stabilization have been reported in different studies [7, 9, 14]. The resulting bonding interface is also strongly dependant on the stack deformation and bonding environment. The bonding interface was characterized using a non-destructive method such as IR microscopy and acoustic scanning microscopy (before the top SOI wafer thinning) and by optical microscopy and SEM (after backside silicon removal of the top SOI substrate). Figure 19.14 shows images of the bonded

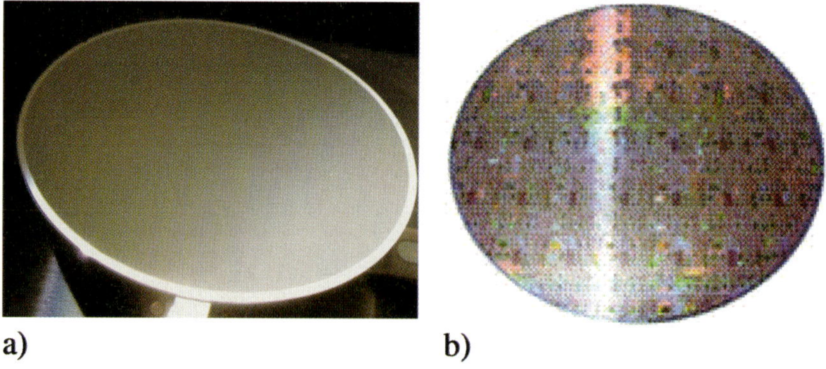

Figure 19.14 Two strata stacked face-to-face circuits' after (a) blanket wafer thinning and edge outlining and (b) top SOI wafer silicon removal.

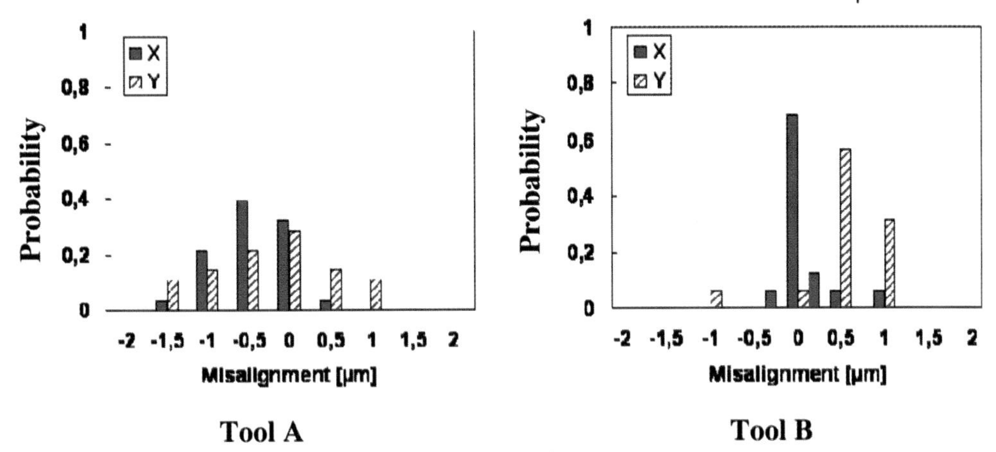

Figure 19.15 Misalignment evaluation of the two alignment tools used for wafer-to-wafer aligned bonding of patterned interconnections.

and thinned wafer. The process consists of several thinning steps followed by an edge outlining to completely remove backside silicon from the SOI wafer.

Alignment accuracy of the staked wafers was measured in two dies within a wafer diameter close to the edges. The same misalignment values are obtained after the wafer bonding at room temperature and after the stack annealing. The performances of two alignment set-ups, from two different companies, are compared in Figure 19.15. The graphs show the XY misalignment values dispersion, where alignment accuracy is $\pm 1.5\,\mu m$ for tool A and $\pm 1.0\,\mu m$ for tool B. As we can see the alignment accuracy is strongly dependant on the tool performance. It depends also on the wafer internal stress and deformation. Actually, these parameters are the same as those that contribute to the bonded interface quality. For this study, we controlled the wafers flatness and used substrates with very small deformation ($<20\,\mu m$) in warp and bow.

After alignment and bonding, the backside SOI wafers were thinned down. Silicon was removed by mechanical grinding, chemical-mechanical polishing and chemical etching using tetramethylammonium hydroxide (TMAH). The buried oxide layer of SOI was a good stop-etch to effectively preserve the transferred top layer. Inter-strata interconnections were realized by plasma etching process through the stacked strata (Figure 19.16). The vias were then filled with a TaN barrier and copper [28].

Figure 19.17 shows the cross-section of inter-strata interconnects. It can be seen that the supervias make good contact to both the top and bottom metal layers and are void free. The supervia shown has an aspect ratio (AR) of 3.1 : 1. In same process assembly, supervias with an AR of up to 4.5 : 1 have been fabricated. The supervias were patterned at Freescale (Austin, TX), etched at CEA-LETI (Grenoble, FRANCE) and then filled at Freescale.

The via chain conductivity was measured by electrical probe. All supervia dimensions except the sub-micronic resulted in good yield [28]. Those inter-strata interconnects demonstrate the feasibility of high density interconnections with a supervia

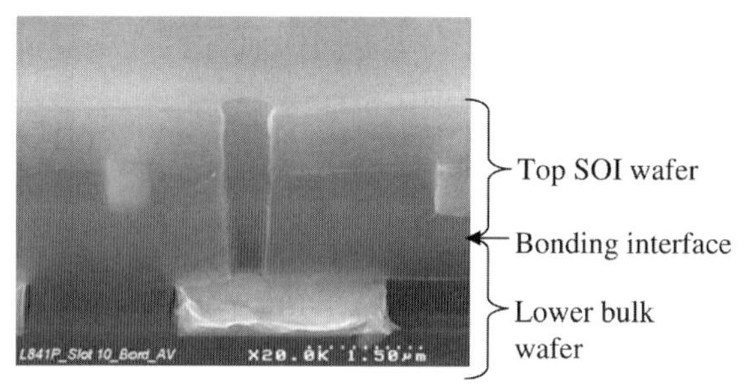

Figure 19.16 Deep via etched through two stacked face-to-face wafers after top SOI substrate removal.

pitch of ~5 μm. The pitch value is mainly due to the bonded wafers alignment and post-processing tolerance.

This first step of complete circuit transfer and post-processing opens the way to the next circuit strata integration, to achieve high-performance architecture.

19.4.3.3 **Chip-to-Chip Capacitive Interconnections at Wafer Level Integration [29, 30] – BEOL Example**

Chip-to-chip capacitive interconnections were achieved using the wafer-level integration approach. The assembly consists of post-processing standard technology MOS manufactured wafers that have a symmetrical last integration level lay-out. The sequential process flow required for this implementation is given in Figure 19.18a. The main process steps of this integration are: a low temperature PECVD oxide deposition on the top of devices of each wafer to be planarized; then, the resulting insulation layer is adjusted to the required thickness for capacitances. To create the capacitance stack, the wafers are precisely aligned face-to-face with micron precision (Figure 19.18c) and bonded by molecular direct bonding; the SEM photograph in Figure 19.18b. highlights the bonding interface with the 400-nm inter-electrode oxide. The backside of the top wafer is then thinned down, to prepare the structure for I/O via by lithographical definition of alignment with the buried chips architecture of I/O pads. Finally, the vias are opened through the bulk silicon, Front-end and

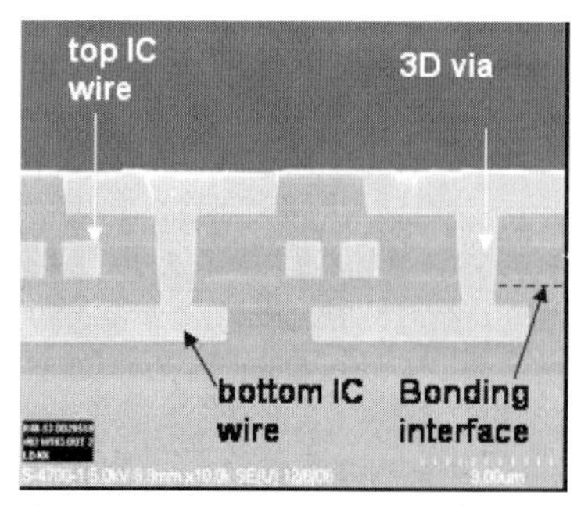

Figure 19.17 SEM cross section of copper supervias that connect the two stacked face-to-face circuits.

a) Main stages of capacitive interconnect integration
Start with FE & BE integrated up to last metal CMOS wafers;
Dielectric deposition, planarization &
surface thickness adjustment;

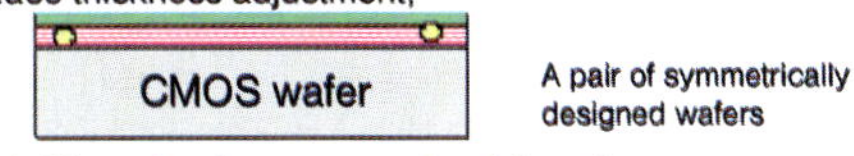

CMOS wafer
A pair of symmetrically designed wafers
Backside wafer alignment marks elaboration;
Aligned direct bonding;

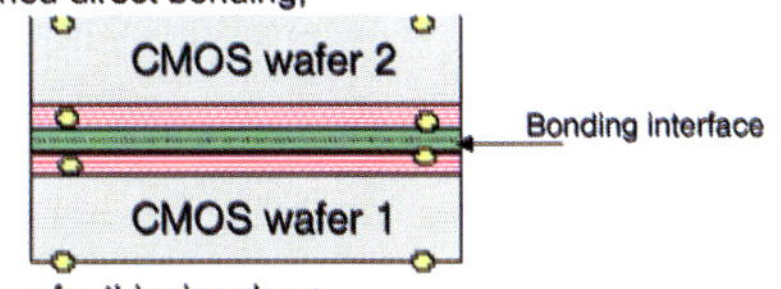

CMOS wafer 2
Bonding interface
CMOS wafer 1
Top wafer thinning down,
Aligned I/O via lithography on the thinned surface;

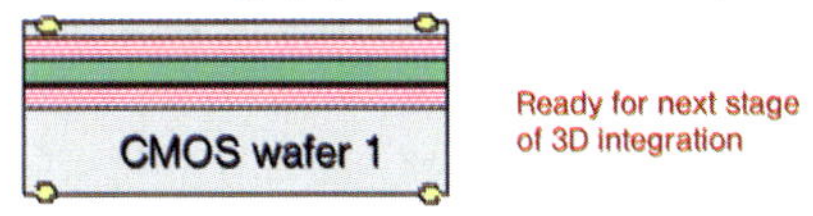

CMOS wafer 1
Ready for next stage of 3D integration
I/O vias opening through the Silicon and stacked layers
Stacked structure dicing for packaging.

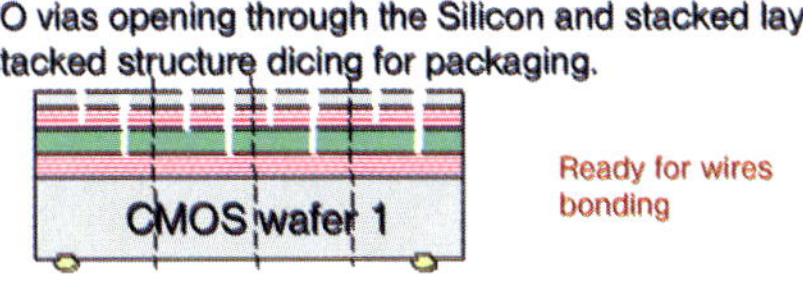

CMOS wafer 1
Ready for wires bonding
b) Bonded and thinned circuits SEM picture

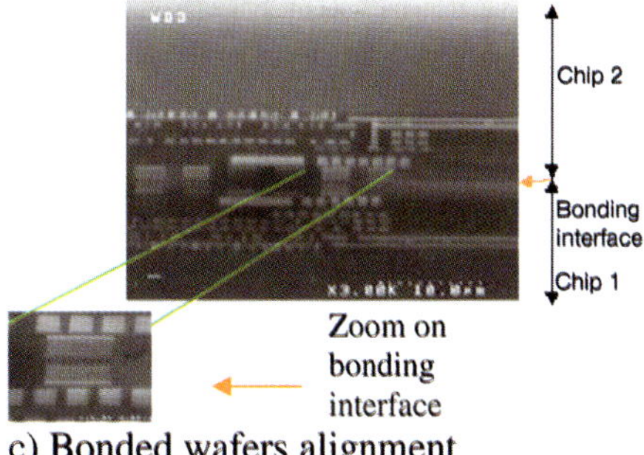

Chip 2
Bonding interface
Chip 1
Zoom on bonding interface
c) Bonded wafers alignment accuracy checking

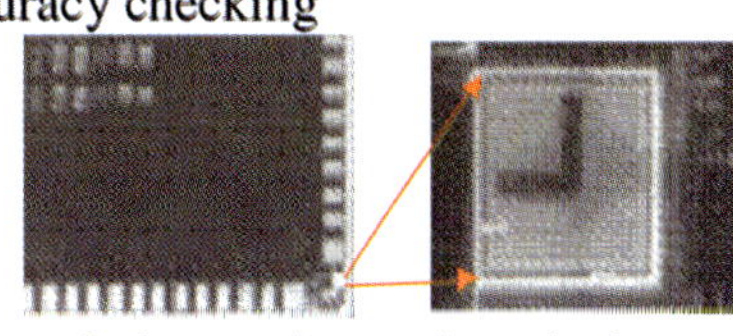

d) Aligned via opening and stacked chips packaging on the test ceramic.

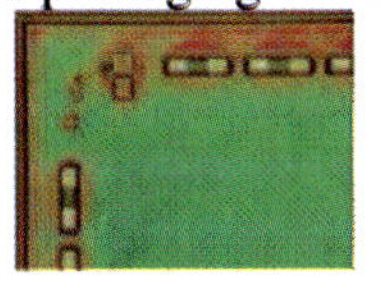

Figure 19.18 Schematic of fabrication flow for attaining inter-chip capacitive interconnects, and a few illustrated examples of wafer level process integration (see text for details).

Back-end integrated stack and bonding interface, making the via for buried I/O pads on both bonded wafers devices. The opened pad areas are shown in the optical image in Figure 19.18d. The stacked wafers are diced and packaged in standard test ceramics by gold wires for capacitive interconnections tests.

For the first time, the capacitively interconnected chips demonstrator was successful, thanks to aligned direct bonding with adapted dielectric permittivity and the small inter-electrode gap (~400 nm).

The packaged dies were tested with a dedicated test-setup to fully characterize at-speed the communication structures [28]. The inter-chips communication performances are summarized in Table 19.1. The communication tests of three different

Table 19.1 Communication bandwidth of different size capacitors as inter-chips capacitive interconnections.

Channel size (μm)	25 × 25	15 × 15	8 × 8
Max rate/pin (Gbps)	1	1.2	1.23

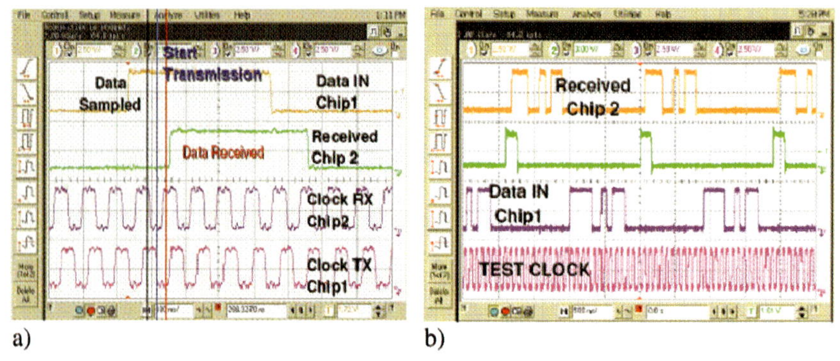

Figure 19.19 Measurement waveforms related to the test interface on inter-chips communication capacitors.

capacitive structures are provided in the frequency range 10–25 MHz and with a 2.5 W power supply. The results confirm the good functionality of all capacitor dimensions, as shown in Figure 19.19a and b.

The results presented in this study demonstrate the first reliable approach for wafer level integrated stack having the capacitive interconnected chips. These capacitive interconnections were implemented from standard 0.13 μm CMOS technology and open the way for a new interconnect approach in 3D integration technology.

19.5 Summary

Enabled processes for 3D integration are open now for a wide range of substrates, circuits and stacking configurations. The heterogeneous substrates and circuits will be stacked using mature processes such as direct wafers or dies bonding, wafers or dies alignment, backside thinning, and deep via elaboration. This background of 3D integration processing allows new smart structures integration and opens the way for future advanced devices. The LETI improvements in 3D integration reported in this chapter show the many possibilities in technological approach and device integration. A new approach of TSV for advanced packaging, compatible with CMOS high temperature process, has been developed. Chip stacking was achieved as an advanced packaging by neo-wafers rebuilding. The heterogeneous die-to-wafer integration opens the way to a new generation of optoelectronic devices. At wafer level integration, Front-end new architecture devices, Bonded Double Gate MOS, were realized and Back-end patterned wafers were interconnected with high density interconnections. For the first time, capacitive inter-chip connections were successfully integrated for wireless connections using standard CMOS processed wafers. The 3D integration approach is now ready for different applications and stacking configurations: chip-to-chip, chip-to-wafer or wafer-to-wafer. Integration availability has been proven and can be used for new challenging circuits and systems.

Acknowledgments

We gratefully acknowledge the efforts of all CEA –LETI colleagues involved in 3D integration theme developments and especially N. Kernevez, M. Fayolle, M. Zussy, T. Enot, B. Biasse, M. Kostrzewa, M. Vinet, M. Heitzmann, H. Moriceau, G. Poupon, N. Sillon, and from the Crolles 2 Alliance consortium G. Passemard, R. Jones, S. Pozder, R. Chaterjee, D. Thomas, A. Martin. Especial thanks are for T. Matthias and O. Bobenstetter from EVGroup for a fruitful demo collaboration.

We thank the CEE support and all the collaborators of following projects: PICMOS-project FP6-2002-IST-1-002131; High Tree – project IST2001-38931-; NESTOR – project (IST-2001-37114); WALORI – project IST- 2001- 35366; WALPACK – project PIDEA 01-131; NALIM – project and EPIX – NET European Network of excellence.

References

1 Massit, C.G. and Nicolas, G.C. (1995) High performance 3D MCM using silicon microtechnologies. Proceeding ECTC 21–24 May, Las Vegas, Nevada.

2 Biasse, B., Zussy, M., Giffard, B. and Aspar, B. (1999) SOI circuit transfer on transparent substrate by molecular adherence. Journées nationales de microélectronique et optoelectronique, France, 1 June.

3 Garrou, P. (2005) 3D integration: a status report. Proceeding of 3D Architecture for Semiconductors Integration and Packaging, June 13–16, RTI International, Burlingam, Tampe, Arizona.

4 Rhett Davis, W., Wilson, J., Mick, S. *et al.* (November–December 2005) Demystifying 3D Ics: The pro and cons of going vertical. *IEEE Design & Test of Computers*, **22**, 498–510.

5 Yoshimi, M. and Mazure, C. (2004) Solid-State and Integrated Circuits Technology Conference Proceedings, Vol 1, 18–21 Oct. pp. 258–261.

6 Guarini, K. *et al.* (2003) The impact of wafer-level layer transfer on high performance devices and circuits for 3-D IC fabrication. International Symposium on Thin Film Materials, Process and Reliability, ECS, PV 2003-13, p. 3790.

7 Moriceau, H. *et al.* (2003) The International Symposium on Semiconductor Wafer Bonding, ECS Proceedings PV 2003, pp. 19–49 and p. 101.

8 Di Cioccio, L., Migette, M., Zussy, M. *et al.* (2006) Proceedings of the 2nd Workshop on Wafer Bonding for MEMS Technology, Halle Germany, p. 13.

9 Aspar, B., Lagahe-Blanchard, C., Sousbie, N. *et al.* (2006) New generation of structures obtained by direct wafer bonding of processed wafers. in *Semiconductor Wafer Bonding 9; Science, Technology and Application.* ECS Transactions, 3 (6), 79–90.

10 Fournel, F., Moriceau, H. and Beneton, R. (2006) Low temperature void free hydrophilic or hydrophobic silicon direct bonding. in *Semiconductor Wafer Bonding 9; Science, Technology and Application*, ECS Transactions, 3 (6), 139–146.

11 Matthias, T., Linder, P., Pelzer, R. and Wimplinger, M. (2004) Trend in aligned wafer bonding for MEMS and IC wafer-level packaging an 3D interconnect technologies. IWLPC 2004, San Jose, October 10–12.

12 Holsteyns, F. *et al.* (2003) Monotoring and qualification using comprehensive surface haze information. IEEE International

Symposium on Semiconductor Manufacturing, pp. 378–381.

13 Bonkohara, M., Motoyoshi, M., Kamibayashi, K. and Koyanagi, M. Current and future three-dimensional LSI integration technology by chip on chip, chip on wafer and wafer on wafer. *Material Research Society Symposium Proceedings*, **970**, p. Y03–03.

14 Burns, J.A. and Chen, C.K. *et al.* (October 2006) A wafer-scale 3D circuit integration technology. *IEEE Transaction on Electron Devices*, **53** (10), 2507–2516.

15 Henry, D., Baillin, X., Lapras, V. *et al.* (2007) Via first technology development based on high aspect ratio trenches filled with doped polysilicon. Proceedings of the 57th Electronic Components and Technology Conference, Reno, Nevada, May 27–June 01.

16 Umemoto, M. *et al.* (May 2004) High-performance vertical interconnection for high density 3D Chip Stacking Package. Proceedings of the 54th Electronic Components and Technology Conference, Las Vegas, Nevada, pp. 616–623.

17 Ok, S.J. *et al.* (August 2003) High density, high aspect ratio through-wafer electrical interconnect vias for MEMS packaging. *IEEE Transactions on Advanced Packaging*, **26** (3).

18 Andry, P.S. *et al.* (2006) A CMOS compatible process for fabricating electrical through vias in silicon. ECTC 2006, San Diego 30-05/02-06.

19 Lietaer, N. *et al.* (2006) Development of cost effective high density through wafer interconnect for 3D Microsystems. *Journal of Micromechanics and Microengineering*, **16**, 29–34.

20 Souriau, J-Ch., Lignier, O., Charrier, M. and Poupon, G. (2005) Wafer level processing of 3D system in package for RF and data applications ECTCE – 2005.

21 Di Cioccio, L., Migette, M. Zussy, M. *et al.* (2006) Proceedings of the 2nd Workshop on Wafer Bonding for MEMS Technology, Halle Germany 13.

22 Kostrzewa, M., Di Cioccio, L., Zussy, M. *et al.* (2005) InP dies transferred onto silicon substrate for optical interconnects application. *Sensors and Actuators, A*, **125**, 411–414.

23 Di Cioccio, L., Jalaguier, E. and Letertre, F. (2004) Compound semiconductor heterostructures by smart-CutTM: SiC on insulator, QUASICTM substrates, InP and GaAs, in *Heterostructures on Silicon, Wafer Bonding – Applications and Technology*, Springer Series in Materials Science, Springer, **75**, 263–314.

24 Widiez, J., Daugé, F., Vinet, M. *et al.* (2004) Proceedings of IEEE International SOI Conference, p. 185.

25 Vinet, M. *et al.* (2004) Planar double gate CMOS transistors with 40 nm metal gate for multipurpose applications. Proc. IEDM.

26 Fitzgerald, E.A. *et al.* (2005) Engineered substrates and their future role in microelectronics. *Material Science & Engineering B*, 124–125.

27 Leduc, P., de Crécy, F., Fayolle, M. *et al.* (2007) Challenge for 3D IC integration: bonding quality and thermal management – IITC. Proc. IITC.

28 Chatterjee, R., Fayolle, M., Leduc, P. *et al.* (2007) Three dimensional chip stacking using a wafer-to-wafer integration. Proc. IITC.

29 Charlet, B., di Cioccio, L., Dechamp, J. *et al.* (2006) Chip-to-chip interconnections based on the wireless capacitive coupling for 3D integration. *Microelectronic Engineering*, **83**, 2195–2199.

30 Fazzi, A., Mangani, L., Mirandola, M. *et al.* (2007) 3D capacitive interconnections for wafer-level and die-level assembly. *IEEE Journal of Solid-State Circuits*, **42**, p. 2270–2282.

20
Lincoln Laboratory's 3D Circuit Integration Technology

James Burns, Brian Aull, Robert Berger, Nisha Checka, Chang-Lee Chen, Chenson Chen, Pascale Gouker, Craig Keast, Jeffrey Knecht, Antonio Soares, Vyshnavi Suntharalingam, Brian Tyrrell, Keith Warner, Bruce Wheeler, Peter Wyatt, and Donna Yost

20.1
Introduction

Lincoln Laboratory has developed a wafer-scale three-dimensional (3D) integrated circuit technology whereby 3D chips are constructed by transferring, bonding together, and electrically connecting the active sections of integrated circuits (ICs) that were fabricated on silicon-on-insulator (SOI) substrates [1]. Layer-transfer techniques were originally developed at Lincoln Laboratory to transfer thin GaAs strips fabricated on a GaAs template for the fabrication of solar cell devices [2]. This concept was later applied to the fabrication of displays by Kopin Corp [3] who transferred ICs fabricated on SOI substrates; the transferred layer consisted of the buried oxide (BOX), the thin SOI film, and the multilevel interconnect. Further work at Kopin Corp., Northeastern University, and Lincoln Laboratory led to an enhanced wafer-scale 3D integrated circuit technology. The building blocks of that 3D technology were SOI circuit fabrication, low-temperature wafer–wafer adhesive bonding, transfer of SOI analog and digital circuits to photodiode imaging circuits, and electrical connection of the circuit structures with through-silicon vias (TSVs) [4]. This technology led to the first demonstration of a "true" 3D circuit with the successful operation of a 3D 64 × 64 visible imager [5]. More recently, numerous institutions have also used various layer-transfer techniques to develop 3D integration technologies [6–11].

The technology used to construct the first 3D imager had serious limitations. The adhesive bond limited subsequent processing temperatures to less than 200 °C. This prevented using chemical vapor deposition (CVD) tungsten as an electrical connection between layers since it is deposited at 475 °C, and also eliminated the post-processing sinter near 400 °C that was required to reduce multilevel via

Handbook of 3D Integration: Technology and Applications of 3D Integrated Circuits.
Edited by Philip Garrou, Christopher Bower and Peter Ramm

ISBN: 978-3-527-32034-9

resistances and to repair oxide damage from plasma processes. The adhesive bond also limited a 3D-IC to two functional layers since an additional bonding and transfer process would disturb the initial adhesive bond. The TSV design required deep silicon etch and oxide refill processes, and outgassing from the adhesive bond prevented scaling the TSV size to less than 6 μm. Finally, the wafer–wafer alignment equipment had an overlay error of 2 μm, which set the pitch of the TSVs to 10 μm minimum. As a result, a new 3D-IC technology was developed at Lincoln Laboratory to overcome these limitations.

20.2 Lincoln Laboratory's Wafer-Scale 3D Circuit Integration Technology

20.2.1 3D Fabrication Process

3D circuits are fabricated on 150-mm SOI substrates using a 180-nm fully depleted SOI process that includes mesa isolation of transistors and three levels of metal interconnect. A new term, "tier", was adopted to distinguish among design layers, physical layers, and transferred layers of a 3D-IC and is the functional section of a wafer that consists of the active silicon, the interconnect, and for an SOI wafer, the BOX. A tier is approximately 8 μm thick. The 3D assembly process and a 3D chip consisting of three tiers are illustrated in Figure 20.1. The process begins by transferring tier 2 to the base tier, tier 1, after face-to-face infrared alignment, oxide-oxide bonding at 275 °C, and removal of the handle silicon to expose the BOX of tier 2. The BOX is used as an etch stop for the silicon etch to produce a uniformly thin active layer and is an essential step in the 3D assembly technology. For this reason all circuits to be transferred must be fabricated with SOI substrates. The handle silicon of a transferred tier is removed by grinding the silicon to a thickness of about 70 μm followed by a silicon etch in a 10% TMAH solution at 90 °C. Since the ratio of silicon to BOX etch rates in TMAH is 1000 : 1, the handle silicon is removed without attacking the BOX and without introducing a thickness variation in the transferred tier, a factor that is essential when forming the vertical connections between tiers, the 3D vias. In both etches the edge is protected to ensure that the wafer can be handled by cassette-to-cassette equipment and that the silicon removal process does not attack the oxide-oxide bond. 3D vias are designed to lie in the mesa isolated regions of the tiers so that lining the vias with a deposited dielectric is not required to achieve insulation between the vertical connections.

The 3D vias are patterned and etched through the BOX and deposited oxides to expose metal contacts in both tiers. The 3D vias are then filled with tungsten that is planarized by chemical mechanical polishing (CMP) to electrically connect the two tiers. The metal contact in the upper tier is an annulus with a 1.5-μm opening that also functions as a self-aligned hard mask during the plasma etch of the oxide beneath it to reach the metal land in the lower tier. To fully land the 3D via, the size of the metal pad, and thus the pitch of the vertical interconnect, is proportional to twice the wafer–wafer misalignment.

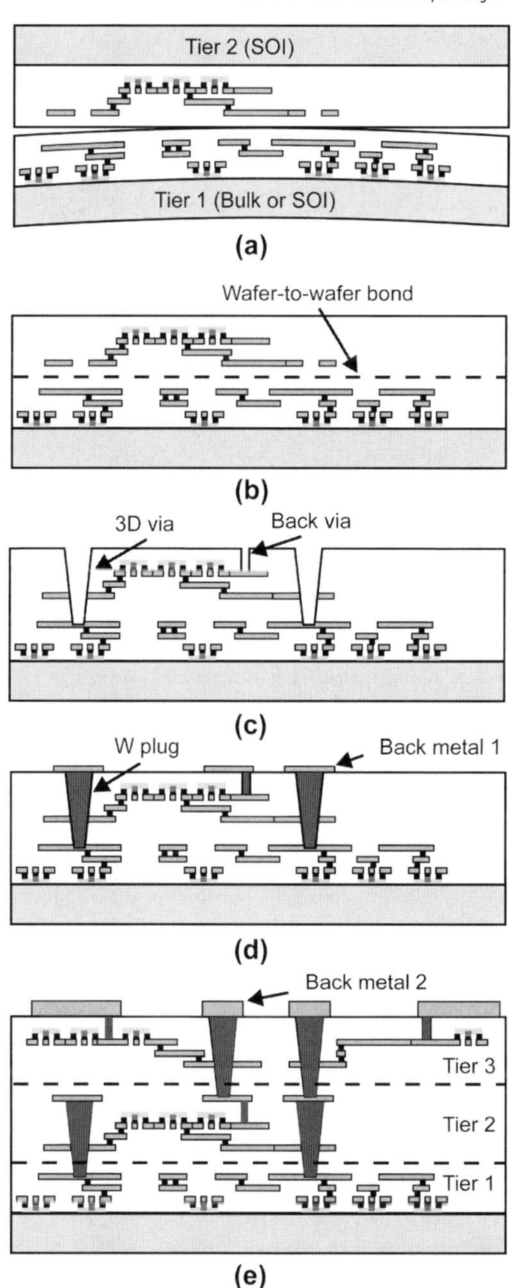

Figure 20.1 Assembly process for a 3D chip: (a) Two completed circuit wafers are planarized, aligned and bonded face to face; (b) the handle silicon is removed; (c) 3D vias are etched through the deposited BOX and the field oxides; (d) tungsten plugs are formed to connect circuits in both tiers; and (e) after tier 3 is transferred, bond pads are etched through the BOX for testing and packaging.

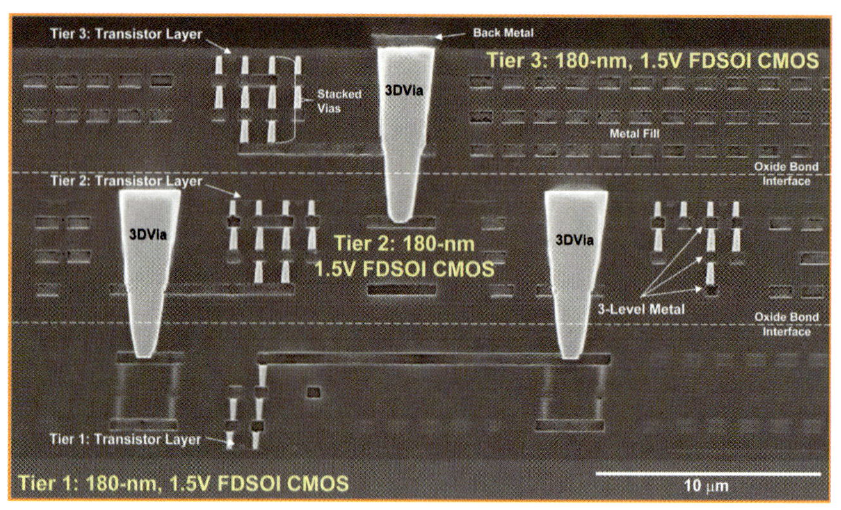

Figure 20.2 Cross-sectional SEM of a three-tier ring oscillator in which inverters in each FDSOI tier are electrically connected with 3D vias. Note that the 3D vias are located in the isolation (field) region between transistors.

A third tier, tier 3, can then be added to the tier 1–2 assembly using the same processes, except that the front side of tier 3 is bonded to the BOX of tier 2, and 3D vias connect the top-level metal of tier 3 to metal pads on the BOX of tier 2. The bond pads are etched to expose the back of the first-level metal for probing and wire bonding. If the 3D chip is a digital circuit, the bond pads are etched through the BOX and deposited oxides of tier 3. If it is a back-side-illuminated imager, tier 1 is a bulk silicon detector wafer in which photodiodes were fabricated. An additional transfer to a carrier wafer is then required, and bond pads are etched after thinning the back side of the detector wafer to tune the silicon to the required optical absorption. The cross-sectional scanning electron micrograph (SEM) of a three-tier ring oscillator shown in Figure 20.2 illustrates interconnections between tiers and the compactness possible with the 3D technology.

20.2.2
3D Enabling Technologies

The building blocks of Lincoln Laboratory's 3D circuit integration technology are SOI circuit fabrication, precision wafer–wafer alignment, low-temperature wafer–wafer oxide bonding, and electrical connection of the circuit structures with dense vertical interconnections.

An SOI technology is required for all tiers that are to be transferred, but a fully depleted SOI (FDSOI) technology [12] is preferred since FDSOI is a low-power technology and is particularly advantageous for 3D circuits to minimize heat generation within the transferred tiers. The principal technology features are a 400-nm BOX thickness, 40-nm SOI channel thickness, mesa isolation, elimination of sidewall leakage with sidewall implants, 4-nm gate oxide, dual-doped polysilicon gates, oxide-nitride spacers, cobalt salicide, three levels of aluminium-based interconnect with planarized dielectrics, and tungsten plug contacts and via connections. All features are defined using a 248-nm CANON stepper with a 22 × 22-mm field of view. Mesa isolation simplifies the placement and fabrication of vertical connections between tiers since the connections are formed in the field-isolation regions. This 180-nm technology has an unloaded inverter delay of 30 ps per stage at 1.5 V; f_T for the n- and p-FETs are 65 and 40 GHz, respectively.

The 3D via pitch is a critical factor in the viability of the 3D technology since for optimal circuit density the minimum pitch of the 3D vias should be comparable to that of the 2D vias that connect multilevel metal layers. The 3D via pitch is determined by the 3D via size, which is dependent upon the achievable oxide-etch aspect ratio, by the design rule that the 3D vias be fully landed, and by the wafer–wafer alignment tolerance, which is primarily constrained by the overlay capability of the wafer–wafer alignment system. The pitch, P, can be written as shown in Equation 20.1:

$$P = 2 \times WA + 3DV + MS \tag{20.1}$$

WA is the wafer–wafer alignment capability, 3DV is the diameter of the 3D via at the landing pad, and MS is the minimum metal-metal spacing. Initially, a modified mask aligner with an overlay tolerance of ±2 μm was used, which, combined with a 1.5-μm opening in the annulus, led to a 5.5-μm-square landing pad and a 6-μm pitch. To scale the inter-tier connections, modern wafer stepper technologies were used to design and fabricate a precision wafer–wafer alignment system whose overlay goal was ±0.25 μm [13]. The basic components of the precision alignment system are two InGaAs cameras with subpixel edge detection for infrared alignment; a six-axis piezoelectric stage with nanometer-scale resolution to provide X, Y, Z, Θ, tilt, and tip motions while tier 2 is aligned to a fixed tier 1; and a precision XY air-bearing stage controlled by an environmentally compensated laser interferometer to map wafers before alignment. The latter is intended to measure the grid distortion of each wafer and select the best match between wafers and, eventually, to compensate for distortion between wafers by substrate heating. The system is shown in Figure 20.3a, and the alignment repeatability data of Figure 20.3b show that an overlay of ±0.35 μm has been achieved. Alignment is measured using box-in-box alignment marks fabricated in the same metal layers that contain the 2D alignment marks, and the features are measured after substrate removal using a commercial overlay tool [14]. The overlay goal of ±0.25 μm will require wafer mapping before 3D integration in order to bond wafers with equivalent grid distortion.

The wafer–wafer bonding process in 3D integration has three requirements. The first is that the room-temperature bond be sufficiently strong to prevent wafer slippage between the wafer alignment and wafer-bonding processes, since the alignment and a 275 °C heat treatment take place in two separate instruments. Second, the bond temperatures must not exceed 500 °C, the upper limit of the aluminium-based interconnect. Finally, the bond must be sufficiently strong to withstand the 3D fabrication process. CMOS wafers to be bonded are coated with 1500 nm of a low-temperature oxide (LTO) deposited by low-pressure chemical vapor

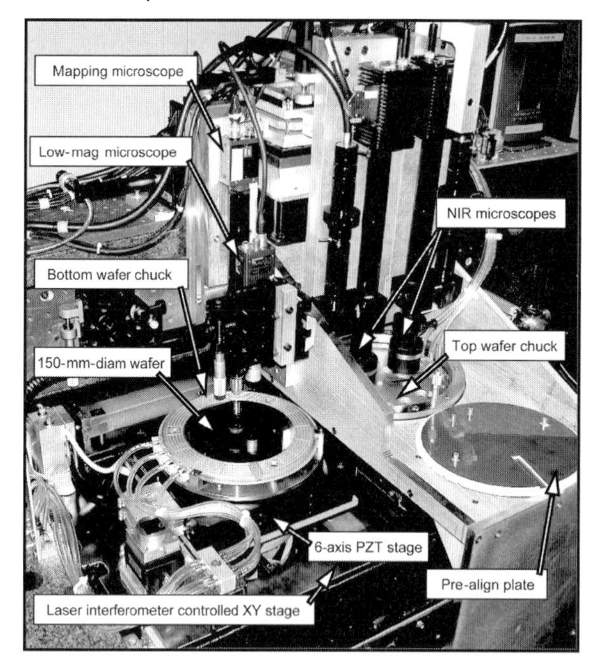

(a)

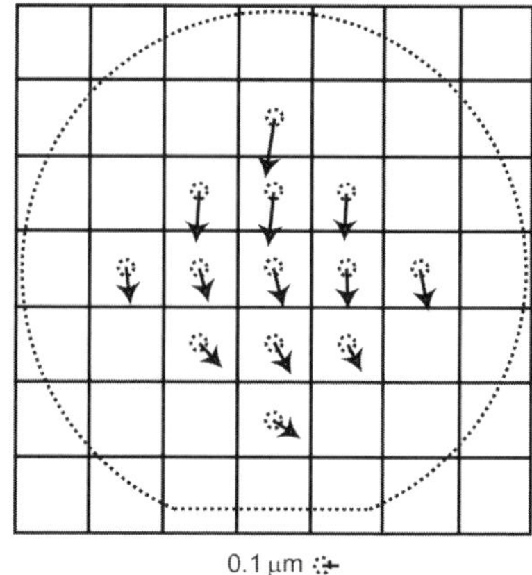

	X	Y	
Mean:	0.042	-0.24	μm
3-Sigma:	0.187	0.272	μm
Min:	-0.062	-0.422	μm
Max:	0.136	-0.093	μm
Orthog:		-3.13	ppm
Scaling:	0.05	-3.73	ppm

(b)

Figure 20.3 (a) Precision wafer–wafer alignment system and (b) vector plot of wafer–wafer alignment using the precision aligner for two 150-mm-diameter bonded wafers with a 22-mm die size. The alignment error was modeled in terms of displacement, orthogonality and scaling. Successive alignment measurements show a 3σ repeatability of <0.35 μm.

deposition (LPCVD) at 430 °C. 1000 nm of the oxide is removed by CMP to planarize and smooth the surfaces to a roughness of <0.4 nm RMS (as measured by atomic force microscopy using a 10-μm square scan). The wafers are immersed in H_2O_2 at 80 °C for 10 min to remove any organic contaminants and to activate the surfaces with a high density of hydroxyl groups [15], after which the wafers are rinsed and spun dry in nitrogen in a standard rinse/dryer. The wafers are aligned and bonded by initiating contact at the center of the top wafer. When the surfaces are brought into contact, weak (~0.45 eV) hydrogen bonds are created at the interface (Si–OH : HO–Si), the bond interface propagates radially within 2–5 s to the edge of the wafer pair, and after 30 s the wafer pair can be removed from the aligner without disturbing the bond and wafer alignment. The bond strength is increased by a thermal cycle that creates covalent bonds at the interface from the reaction:

$$\mathrm{Si{-}OH : HO{-}Si \rightarrow Si{-}O{-}Si + H_2O}$$

with the Si–O bond having a bond energy of 4.5 eV.

Optimal thermal cycle parameters for this particular bonding technique were determined by measuring bond strengths in the temperature range 150–500 °C.

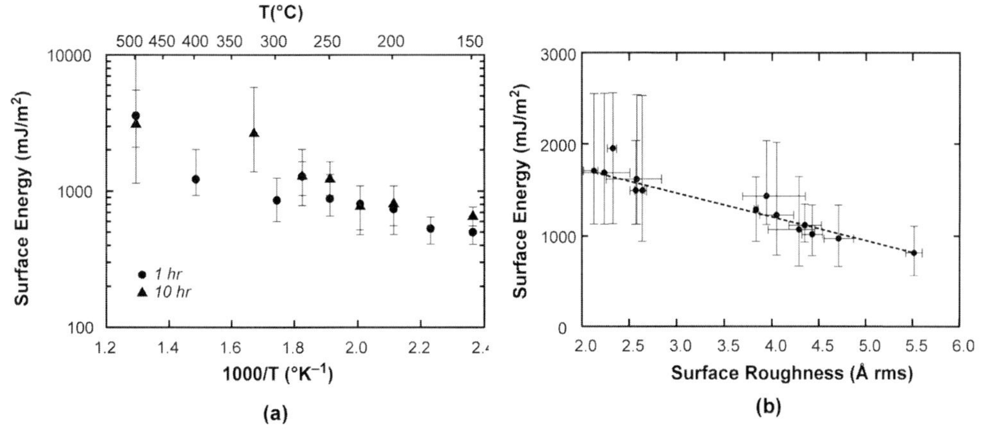

Figure 20.4 (a) A thermal cycle is required to increase the wafer–wafer bond strength of the low-temperature bond process to at least 1000 mJ m^{-2}. Heat treatments greater than 300 °C may cause bubbles in the bond interface; the measured roughness in (a) was 0.4–0.5 nm, but improved CMP processes have led to a reduction in surface roughness (b) and increased bond strengths.

Figure 20.4a is an Arrhenius plot of the measured surface energy for 1- and 10-h cycles. Surface energies were measured using the blade insertion technique [16], and crack lengths were measured using infrared inspection at four points along the edge of the wafer pair. A 1-mm measurement error was assumed, and the maximum and minimum bond energies were determined after subtracting the 1-mm error from the shortest and adding 1 mm to the longest cracks, respectively. A 275 °C, 10-h thermal cycle was chosen as optimum since a surface energy of about 1000 mJ m^{-2} is required to allow removal of the handle substrate and subsequent 3D via processing without disturbing the aligned pair. Infrared inspection determined that voids were formed for bonding temperatures above 300 °C as a consequence of water being formed and trapped within the bonding regions. Removal of the handle silicon of a transferred tier removes a barrier to water diffusion, and as a result voids are not formed during back-end-of-the-line (BEOL) processes that reach 475 °C. The effect of surface roughness on bond strength (Figure 20.4b) indicates that low-temperature bonding requires careful control of the CMP process [17]. An additional factor to be controlled in the bonding process is surface planarity, which restricts surface steps to less than 50 nm.

The unrestricted placement of 3D vias is an integral part of the 3D technology. A 3D via, shown in Figure 20.5, consists of a metal annulus in the upper tier, a metal land in the lower tier, and a tungsten plug that electrically connects the two features. The plug is formed in an oxide hole that is plasma etched with a resist mask that is aligned to the metal annulus with a mean and 3σ overlay error less than 100 and 300 nm, respectively. The annulus is a unique feature of the design since it forms the top electrical contact and masks the oxide etch to the metal land. After resist removal,

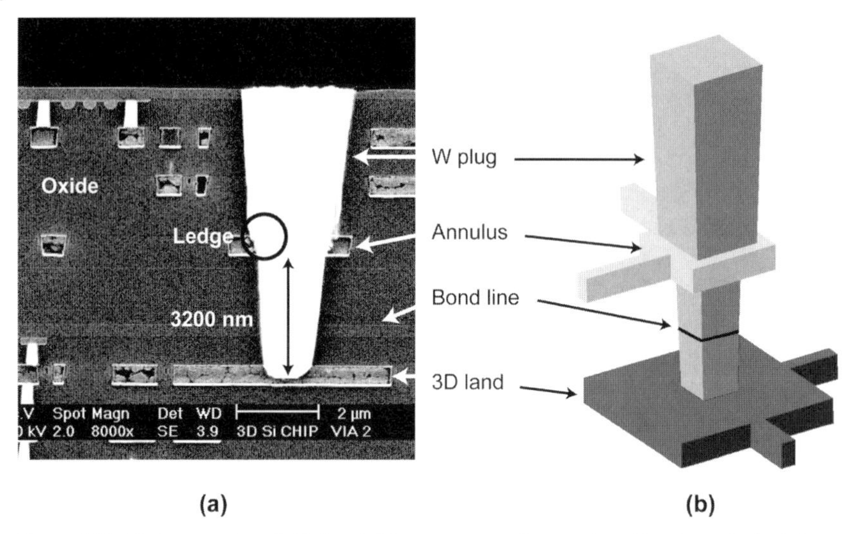

Figure 20.5 (a) Cross-sectional SEM and (b) isometric drawing of a 3D via. The essential features shown are the metal annulus in metal 3 of the upper tier, the 3D land in metal 1 of the lower tier, and the tungsten plug that electrically connects the two tiers. Not shown is the 2.5-µm resist pattern that masks the oxide etch and determines the size of the via to the plane of the annulus. In (a) the ledge, which is the overlap of the plug and the metal annulus, is circled.

tungsten is deposited and planarized by CMP, leaving approximately 8 µm of tungsten in the hole. The present 3D via design consists of a 3-µm square landing pad, a 1.5-µm square annulus opening, and a 1.75-µm square resist window called 3D cut. The initial 3D chip design that used these inter-tier connections achieved a 3D-via pitch of 6 µm as compared to the 26-µm pitch used in the 3D imager first reported [6].

The oxide via is etched in a Trikon Technologies low-pressure, high-density, helicon-based cluster tool to achieve a high aspect ratio etch [18]. The etch conditions are a balance between polymer deposition used to achieve a vertical sidewall and wafer bias to sputter the deposited polymer from the base of the via during etching. A section of the metal annulus, called a ledge, is exposed during the oxide etch, and the interaction of the plasma with the ledge affects the etch profile below the plane of the annulus. The metal interconnect is a deposited stack of titanium/aluminium-silicon/titanium/titanium nitride, and after tier transfer the metal stack is inverted and titanium is exposed to the plasma. 3D via chains were fabricated to optimize the process. 3D via resistance and yield were analyzed as a function of etch chemistry, system pressure, RF bias and 3D via design features. Initial 3D via chain tests had high yield and low resistance to the annulus but low yield to the landing pad on the lower tier. Cross-sectional SEM photos of 3D vias showed that the width of the metal ledge had been substantially reduced; the vias were pinched off below the annulus metal, but structures etched without the annulus displayed no via pinch-off. The metal ledge was eroded as a result of chemical and physical etching since titanium and titanium nitride are volatile in fluorine-based chemistries, but aluminium

Table 20.1 3D via etch chemistry.

	3D via etch	
	Oxide	**Metal**
Source power (W)	1750	2750
Bias power (W)	1300	250
Pressure (mTorr)	3.5	6
C_4F_8 (cc)	30	—
CH_2F_2 (cc)	40	—
CO (cc)	24	—
Ar (cc)	80	—
Cl_2 (cc)	—	75
BCl_3 (cc)	—	10
N_2 (cc)	—	15

fluoride, a plasma etch reactant, is not and so a high wafer bias was necessary to remove it. In addition, the etch chemistry produces a polymer when the oxide etch reaches the metal annulus and interacts with titanium, leaving a metal organic deposit on the via sidewalls, leading to via pinch-off. A yield model constructed from test devices with varying 3D cut and annulus sizes indicated that the 3D via chain yield was proportional to the annulus opening and inversely proportional to the ledge width, with the annulus opening being the major yield determinant factor. The yield of a 10 000 3D via chain increased from 35 to 100% by minimizing the exposed ledge, adding a Cl_2-based metal etch *in situ* to etch the ledge, and adding a small amount of O_2 to the oxide etch to remove the etch polymer. The current process results in a 3D via with a median resistance of 0.75 Ω. Table 20.1 summarizes the oxide etch chemistry.

20.2.3
3D Technology Scaling

The overlay of features to be vertically connected between tiers places constraints beyond wafer–wafer alignment. In 3D technology, variations between wafers in x-y die placement and rotation add to the overlay error of the 3D via, and the error is doubled if one tier is flipped with respect to the other. The first constraint requires that the first lithographic level stepped on each wafer to be 3D integrated must place the center of the exposed die pattern coincident to the wafer's center to within 100 μm. This requirement ensures a uniform bond at the tiers' edge and aids wafer handling. The second constraint is to place the die parallel to the wafer flat to within 5 ppm. To assist the alignment process of the first level, features are included in the reticle to measure pattern placement and field rotation. The exposure stepper's alignment algorithm ensures that subsequent lithographic levels satisfy the die placement criteria. The third constraint is tier distortion that occurs during 2D and 3D fabrication. Distortion degrades the registration between two tiers and, if not controlled, can result in 3D vias that are not fully landed since the placement of the metal annulus in the upper tier cannot compensate for different grid distortions in the two tiers. Note that in 2D lithography grid distortion can be accommodated during level-level registration by the alignment algorithm. Registration between two tiers is measured with the overlay tool, and the total overlay error is de-convolved into translation, rotation, expansion and orthogonality errors in the x and y axes. A viable 3D technology thus requires control of tier distortion throughout 2D fabrication as well as during 3D fabrication. Wafer bow caused tier distortion and was minimized by depositing stress-compensation oxides on the back sides of wafers before wafer–wafer alignment. Tier distortion during wafer–wafer alignment also occurred because of wafer deformation by a vacuum hold-down chuck and was minimized by redesigning the chuck. The 3D technology has been scaled as a result of the efforts to control overlay error. One example is the rule for spacing of 3D cuts in tier 2, ST2, to prevent shorting between tungsten plugs that connect separate 3D lands in tier 1 (Equation 20.2):

$$\mathrm{ST2} = \mathrm{MS} + \mathrm{WA} + \mathrm{VS} - \mathrm{3D\ Cut} \qquad (20.2)$$

MS is the minimum spacing between the 3D land metal features, 0.3 μm; WA is the wafer–wafer alignment error, 1 μm; VS is the required size of the tungsten plug in contact with the 3D land, 1 μm; and 3DCut is the resist window, 1.75 μm. The preceding work led to a reduction in spacing from 3.55 to 1.55 μm and a reduction in the size of the 3D land from 5.5 to 3 μm.

In 3D integration, tier 3 is bonded to tier 2 face-to-back (Figure 20.1), and the 3D vias connecting those tiers penetrate the SOI and polysilicon planes. Exclusion zones that reduce the density of SOI active and polysilicon interconnect are required to prevent tungsten to SOI and polysilicon shorts. The spacing rule, ST3, is dominated by the wafer–wafer alignment error (Equation 20.3):

$$\mathrm{ST3} = \mathrm{WA} + \mathrm{PS} + \mathrm{AE} \qquad (20.3)$$

WA is the wafer–wafer alignment error; PS is the spacing between the tungsten plug and SOI, 0.175 μm; and AE is the accumulated alignment and feature size budget, 0.185 μm. Combining PS and AE with WA leads to a spacing between the 3D cut on tier 3 and SOI on tier 2 of 1.35 μm. The effect of WA on the SOI and polysilicon density on tier 2 can be removed by the addition of a metal land on the BOX of tier 2 that is connected to metal 1 of tier 2 by the standard 300-nm via (Figure 20.6). An additional stepper alignment is required, but the spacing between the via, BV0, and SOI is reduced to 0.35 μm since the wafer–wafer alignment error has been removed from the overlay budget.

20.3 Transferred FDSOI Transistor and Device Properties

The 3D integration process creates FDSOI transistors in tiers 2 and 3 that differ from those in tier 1 as well as conventional SOI transistors since transistors in those tiers

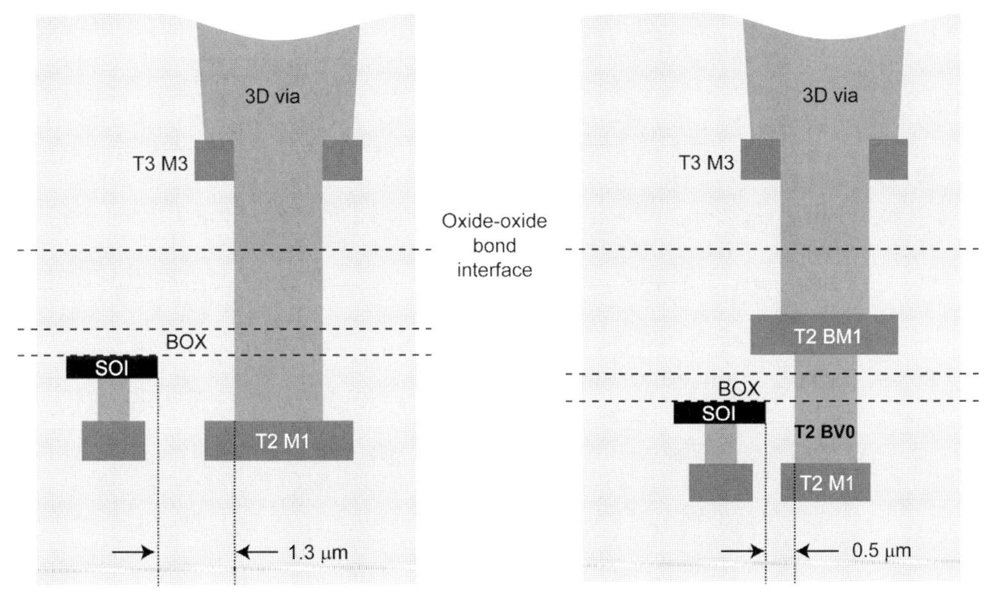

Figure 20.6 Inclusion of back metal (BM1) and a back via (BV0) on tier 2 reduces the 3D via–SOI spacing from 1.3 to 0.5 μm.

do not have a silicon substrate. As a result, the body potential of a transistor in a transferred tier is less well determined since the potential is a function of fringing electric fields from the source and drain, the field from the gate, the integrated charge in the body, the integrated electric field from oxide charge above the body, and the field from any electrodes above the transistor. Test transistors and circuits were designed for testing devices before and after 3D integration in both tiers 2 and 3 in order to determine the effect of 3D integration and the absence of a substrate electrode on their properties [19]. Test results showed that removal of the substrate had a minimal effect on device properties. Ring oscillators and 8×8 multipliers were similarly characterized, with the data indicating no change in delays per stage or operating frequency.

Of particular interest is the effect of radiation-induced charge trapped in isolation oxides, that is field, trench and buried oxides, on circuits from which the substrate was removed as part of a 3D imager for space applications. In FDSOI transistors the main effect of total dose irradiation is a threshold shift due to charge trapping in the BOX because the front and back oxides are capacitively coupled. Removing the substrate after bonding a FDSOI tier to a handle wafer reduces the radiation-induced threshold voltage shift and offstate leakage of FDSOI nFETs, and thinning the BOX further decreases the nFET sensitivity to irradiation [20]. The threshold shifts of transistors in the three tiers of a 3D chip were evaluated as a function of total integrated dose (TID) using 10-keV X-rays. As shown in Figure 20.7, for nFETs in tier 1 the decrease in threshold voltage was similar in magnitude to that of nFETs in a conventional FDSOI wafer, while for nFETs in tiers 2 and 3 the shifts were less.

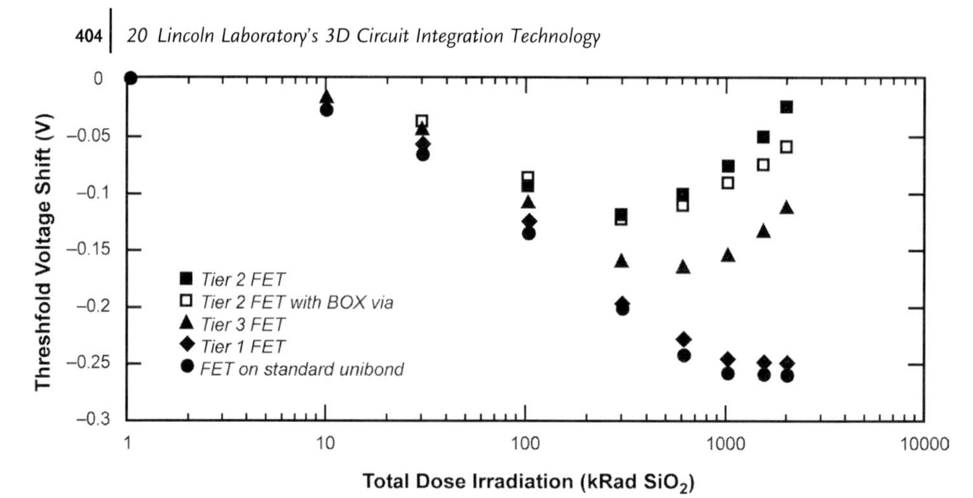

Figure 20.7 TID-induced threshold voltage shift for nFETs on tiers 1, 2 and 3 of a 3D chip with the transistors in the offgate bias conditions during irradiation. Each transistor has 20 fingers with L = 0.18 μm, W = 10 μm. The red data points are typical data for an nFET on a single standard SOI wafer.

Since both tiers 2 and 3 were flipped and bonded onto tier 1, these results are consistent with experiments of single-tier wafers bonded to handle wafers and the substrate removed (see Figure 18 of Reference [20]). These data indicate that the integration of multi-tier wafers reduces the radiation-induced threshold voltage shift for transistors in the upper two tiers and does not degrade the total dose tolerance of transistors in the first tier.

The effects of heat generation in tiers 2 and 3 of a 3D chip is an area of current research since devices in those tiers are embedded within dielectric layers and the heat generated is more difficult to remove than heat generated in tier 1. The effects of self-heating on transistors in a 3D circuit stack were compared by measuring the actual temperature of each tier as a function of power consumption using planar pn junction diodes as temperature sensors and SOI resistors as the heat source. The effectiveness of various heat-sinking techniques was also studied using vias that were etched through the BOX and filled with tungsten at the same time as the contacts were formed [21]. A diagram of the structure studied is shown in Figure 20.8a along with heat transfer paths. Two types of heat sinks were evaluated. One used metal pads at the top of the 3D stack to radiate the heat into air, and the other used tungsten-filled vias through the BOX to dissipate the heat into the substrate. Note that design rule constraints require the heat to be transferred through standard metal interconnects and vias, which have fairly high thermal resistance because of their small size, before reaching the heat sink. The measured temperatures of each tier (Figure 20.8b) indicate that the temperature of tier 3 is the highest and can exceed 250 °C at 300 mW, which is equivalent to approximately 125 W mm^{-2} of surface area. Although both BOX vias and the top

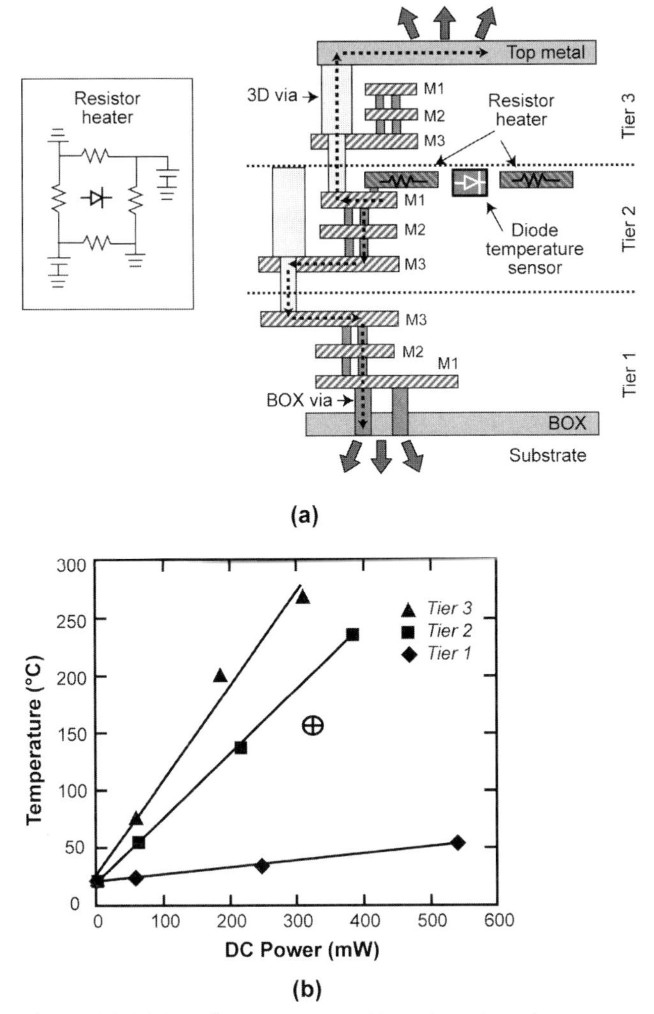

Figure 20.8 (a) Heat flow was generated by resistors in each tier and (b) caused a temperature increase as measured by pn diodes in each tier. The encircled "+" shows the temperature with metal pads on the 3D vias of tier 3.

metal pads contributed to temperature lowering, the top metal is more effective in removing heat. Similar measurements with transistors showed that adding top metal or vias through the BOX can help dissipate the heat, but the improvement in drive current and transconductance is less than 10%. Fortunately, thermal impacts on FET performance are small and SOI-3D integrated circuits are more forgiving with high-temperature operation than bulk silicon because of significantly reduced junction area. These results show that 3D-ICs should be designed with circuits that dissipate the greatest power placed in tier 1 and the least power in tier 3.

20.4
3D Circuit and Device Results

The "low-hanging fruit" of 3D technology is focal plane design and fabrication because the imaging tier has a 100% fill factor, and the analog and digital processing tiers are below the imaging tier. Using Lincoln Laboratory's 3D technology discussed above, we were successful in the design, fabrication and operation of avalanche photodiode (APD) imagers and 3D visible imagers.

20.4.1
3D-LADAR Chip

The 3D-LADAR chip [22] is a laser radar (LADAR) imager based on Geiger-mode APDs integrated with high-speed all-digital CMOS timing circuits. A target is illuminated with a laser, and the chip measures the arrival time of a single reflected photon that is incident on an APD such that the diode avalanches and produces a digital pulse that functions as a stop signal to a fast digital counter in the pixel circuit. This "photon-to-digital conversion" yields quantum-limited sensitivity and noiseless readout, enabling high-performance LADAR systems. Previous visible LADAR systems [23] used 32 × 32 silicon APD arrays epoxy bonded to bulk 0.35-μm CMOS foundry chips. The pixel spacing was 100 μm and was limited primarily by the area required for the pixel circuitry. The circuits achieved a 0.5-ns timing quantization. The 3D-LADAR chip is a 64 × 64 imager composed of three tiers with 50-μm pixel spacing. Tier 1 is the 30-V Geiger-mode APD array and is the imaging tier. It is fabricated in a bulk wafer with an epitaxial layer for improved quantum efficiency. Tier 2 is fabricated in a 3.3-V, 0.35-μm SOI process. It contains circuitry that arms and disarms an APD and generates a stop signal when the APD has avalanched. Tier 3 is fabricated in a 1.5-V, 0.18-μm SOI process. It includes a 9-bit pseudorandom counter that is connected to a 3-bit register in tier 2 to improve the timing resolution of the pixel. The isometric drawing of a single pixel in Figure 20.9 shows the five 3D via

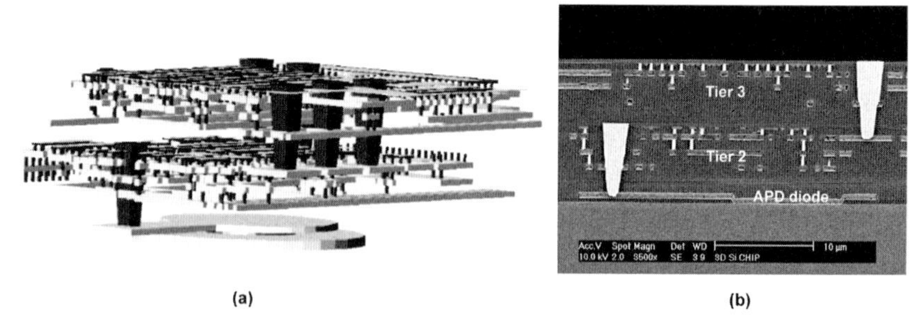

(a) (b)

Figure 20.9 (a) Isometric drawing of a 3D-LADAR pixel, showing a single 3D via connecting the APD diode to tier 2 and five 3D vias between tiers 2 and 3; there are 88 3.3-V FDSOI transistors in tier 2 and 138 1.5-V transistors in tier 3. (b) Cross-sectional SEM of a pixel intersecting two 3D vias.

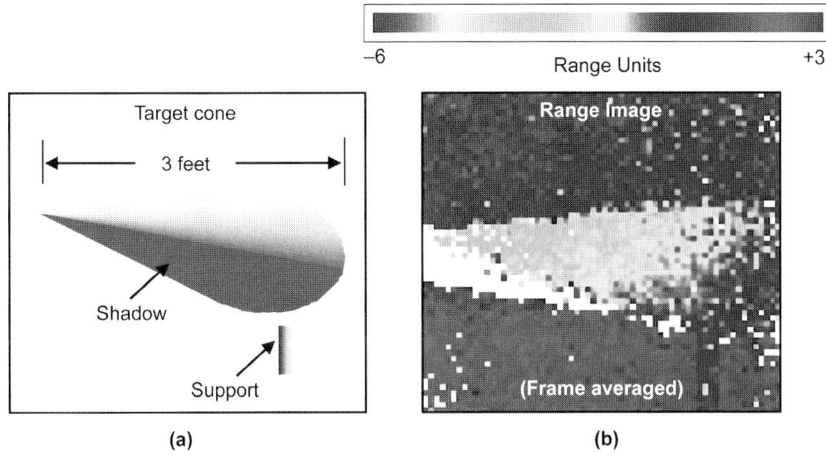

Figure 20.10 (a) A cone was illuminated by a 4-kHz pulsed laser, $\lambda = 532$ nm. The reflected pulses were incident on the 3D 64×64 LADAR imager located 5 m from the cone, where the pulses were detected, multiplexed and sent to an off chip processor to create the image shown in (b). The range units of the 9-bit counter of tier 3 correspond to ~0.1 m per count.

connections between tiers and the APD arming, pixel timing and control circuitry that are beneath the image plane. Tier 1 would be thinned after 3D integration in order that the LADAR chip performs as a back-side imager, similar to back-side-illuminated charge-coupled devices (CCDs). Figure 20.10 shows an imaging target, which is a cone, and a range image of the cone produced by the 64×64 imager after processing 210 frames to eliminate spurious counts. These results indicate that with a redesign to improve power distribution, the combined effect of the array size, the reduced pixel size and the faster timing circuitry will result in a ~80× reduction in voxel volume compared to previous [23] designs. This 3D-LADAR chip represents the first functional three-tier circuit with active circuits and devices on all tiers and is the first demonstration of three different process technologies integrated into one 3D chip.

20.4.2
1024 × 1024 Visible Imager

A two-tier 1024×1024 visible imager with 8-µm pixels has been demonstrated [24]. Tier 1 is a p^+n photodiode, and tier 2 is a FDSOI tier operated at 3.3 V. This is the most dense 3D imager circuit developed using this 3D circuit technology and illustrates the realization of a 100% imager fill factor by 3D technology. Each pixel of the 1024×1024 array includes a reverse-biased p^+n diode (in tier 1), a reset transistor, a source follower transistor, and a select transistor (in tier 2) (Figure 20.11b). Measured responsivity from the PMOS reset imager with added in-pixel capacitance was ~2.7 µV per e^- and from the NMOS reset imager was ~9.4 µV per e^-,

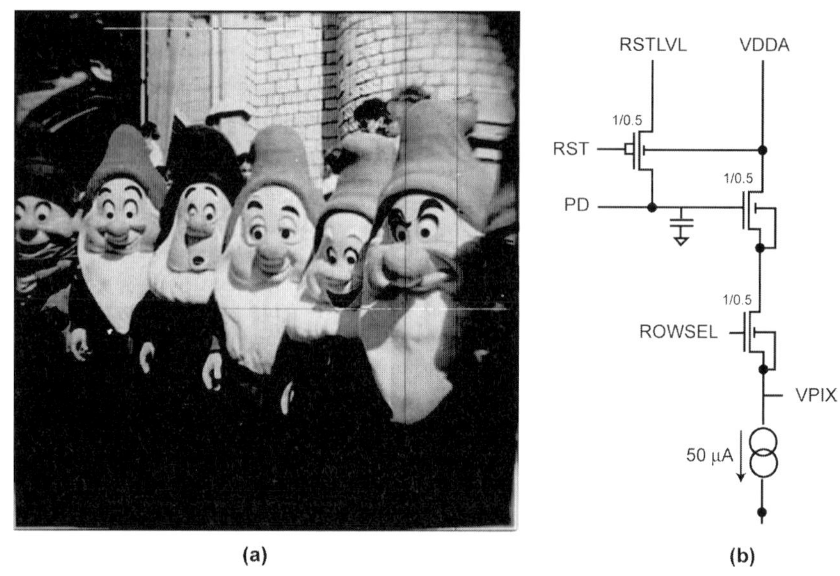

Figure 20.11 (a) Output from a 1024 × 1024 3D imager with 8-μm pixel. (b) The pixel circuit in tier 2, the 3.3-V FDSOI tier. The photodiode (PD) is in tier 1. The imager has a 100% fill factor.

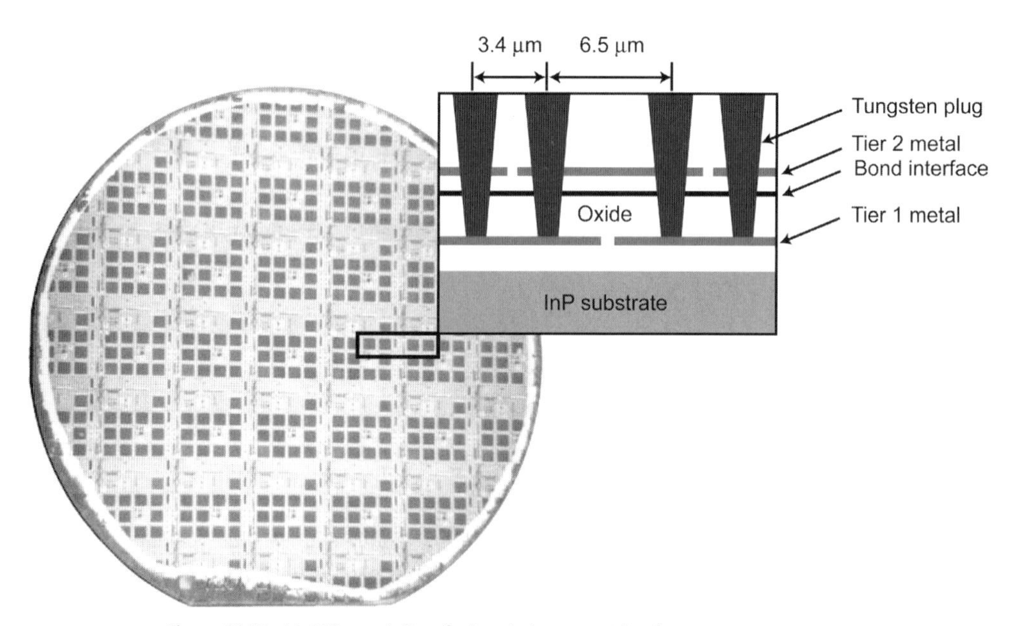

Figure 20.12 (a) 150-mm InP wafer bonded to an oxidized silicon wafer with (b) 3D via chains with 10 000 vias per die. The chain yield was 100%, and the average 3D via resistance was less than 1 Ω.

corresponding to a charge-handling capacity of 350 000 and 85 000 e^-, respectively. Measured pixel operability is in excess of 99.9%; the principal yield detractor was column or row dropouts and was not due to defective 3D vias. The high degree of pixel functionality is seen in Figure 20.11a, an image acquired by projecting a 35-mm slide onto the CMOS circuit side of the 3D integrated imager.

20.4.3 Heterogeneous Integration

The feasibility of heterogeneous integration of dissimilar materials while building 3D-IR focal planes has been demonstrated recently [25]. Six-inch indium-phosphide wafers were bonded to oxidized silicon substrates (Figure 20.12) and high-yield 3D via chains fabricated to show the applicability of the 3D technology to materials other than silicon. This is a major step towards the integration of mixed-materials in 3D-ICs for digital and imaging applications.

20.5 Summary

Functional 3D circuits of two and three tiers were fabricated with the wafer-scale 3D circuit integration technology. The technology permits the unrestricted placement of dense-vertical interconnections between tiers and the fabrication of 3D circuits with different technologies and eventually different materials. The advantages of the technology are obvious for imaging applications where a 100% fill factor can be obtained with a chip that includes complex pixel circuits. Current work is directed towards the application of 3D technology to mixed-signal applications as part of the DARPA-funded Multiproject program as well as the development of techniques to control and remove heat from the 3D stack. Results from the first DARPA Multi-project program suggest that: (i) 3D static random-access memories (SRAMs) will have higher bandwidth performance due to reduced interconnect delay, (ii) 3D field-programmable gate arrays (FPGAs) will offer greater design and programmability flexibility, (iii) 3D interconnect tiers will provide reduced signal delay and latency and better RF isolation between and within tiers and (iv) we can expect micro-electro-mechanical systems (MEMS) to be integrated into 3D structures.

Acknowledgments

The work was sponsored by the Defense Advanced Research Projects Agency under Air Force contract #FA8721-05-C-0002. Opinions, interpretations, conclusions, and recommendations are those of the authors and are not necessarily endorsed by the United States Government.

The authors wish to acknowledge the dedication and persistence of the Microelectronics Laboratory staff and the editorial assistance of Karen Challberg.

References

1 Burns, J.A., Aull, B.F., Chen, C.K. *et al.* (2006) A wafer-scale 3-D circuit integration technology. *IEEE Transactions Electron Devices*, **53** (10), 2507–2516.

2 McClelland, R.W., Bozler, C.O. and Fan, J.C.C. (1980) TP-A2 the cleft process: A technique for producing many epitaxial single-crystal GaAs films by employing one reusable substrate. *IEEE Transactions Electron Devices*, **27** (11), 2188.

3 Sailer, P.M., Singhal, P., Hopwood, J. *et al.* (1997) Creating 3D circuits using transferred films. *IEEE Circuits Devices Magazine*, **13** (6), 27–30.

4 Burns, J., McIlrath, L., Hopwood, J. *et al.* (2000) An SOI-based three-dimensional integrated circuit technology. IEEE International SOI Conference Proceeedings, pp. 20–21.

5 Burns, J., McIlrath, L., Keast, C. *et al.* (2001) Three-dimensional integrated circuits for low-power, high-bandwidth systems on a chip. Digest Tech. Papers IEEE International Solid-State Circuits Conference, 453, pp. 268–269.

6 Reif, R., Fan, A., Chen, K.-N. and Das, S. (2002) Fabrication technologies for three-dimensional integrated circuits. Proceedings IEEE International Symposium Quality Electronic Design, pp. 33–37.

7 Chan, V.W.C., Chan, P.C.H. and Chan, M. (2000) Three dimensional CMOS integrated circuits on large grain polysilicon films. Tech. Digest IEEE International Electron Devices Mtg., pp. 161–164.

8 Fukushima, T., Yamada, Y., Kikuchi, H. and Koyanagi, M. (2005) New three-dimensional integration technology using self-assembly technique. Tech. Digest IEEE International Electron Devices Mtg., pp. 359–362.

9 Lea, R., Jalowiecki, I., Boughton, D. *et al.* (1999) A 3-D stacked chip packaging solution for miniaturized massively parallel processing. *IEEE Transactions Advanced Packaging*, **22** (6), 424–432.

10 Fukushima, T., Yamada, Y., Kikuchi, H. and Koyanagi, M. (2005) New three-dimensional integration technology using self-assembly technique. Tech. Digest IEEE International Electron Devices Mtg., pp. 359–362.

11 Topol, A., Tulipe, D., Shi, S. *et al.* (2005) Enabling SOI-based assembly technology for three-dimensional (3D) integrated circuits (ICs). Tech. Digest IEEE International Electron Devices Mtg., pp. 363–366.

12 MITLL Low-Power FDSOI CMOS Process Design Guide, Revision 2006:7, (2006) Advanced Silicon Technology Group, MIT Lincoln Laboratory, 244 Wood St., Lexington, MA 02420.

13 Warner, K., Chen, C., D'Onofrio, R. *et al.* (2004) An investigation of wafer-to-wafer alignment tolerances for three-dimensional integrated circuit fabrication. IEEE International SOI Conference Proceedings, pp. 71–72.

14 Metra 2100 Process Engineer's Manual, Optical Specialties Inc., (1993).

15 Warner, K., Burns, J., Keast, C. *et al.* (2002) Low-temperature oxide-bonded three-dimensional integrated circuits. IEEE International SOI Conference Proceedings, pp. 123–124.

16 Maszara, W.P., Goetz, G., Caviglia, A. and McKitterick, J.B. (1988) Bonding of silicon wafers for silicon-on-insulator. *Journal of Applied Physics*, **64** (10), 4943–4950.

17 Chen, C.K., Warner, K., Yost, D.R.W. *et al.* (2007) Scaling three-dimensional SOI integrated-circuit technology. IEEE International SOI Conference Proceedings, p. 87–88.

18 Knecht, J., Yost, D., Burns, J. *et al.* (2005) 3D via etch development for 3D circuit integration in FDSOI. IEEE International

SOI Conference Proceedings, pp. 104–105.

19 Burns, J., Warner, K. and Gouker, P. (2001) Characterization of fully depleted SOI transistors after removal of the silicon substrate. IEEE International SOI Conference Proceedings, pp. 113–114.

20 Gouker, P., Burns, J., Wyatt, P. *et al.* (2003) Substrate removal and BOX thinning effects on total dose response of FDSOI NMOSFET. *IEEE Transactions Nuclear Science* **50** (6), 1776–1783.

21 Chen, C.L., Chen, C.K., Burns, J.A. *et al.* (2007) Thermal effects of three dimensional integrated circuits stacks. IEEE International SOI Conference Proceedings, p. 91–92.

22 Aull, B., Burns, J., Chen, C. *et al.* (2006) Laser radar imager based on three-dimensional integration of Geiger-mode avalanche photodiodes with two SOI timing-circuit layers. Digest Tech. Papers IEEE International Solid-State Circuits Conference, pp. 304–305.

23 Aull, B.F., Loomis, A.H., Gregory, J. and Young, D. (1998) Geiger-mode avalanche photodiode arrays integrated with CMOS timing circuits. IEEE Annual Device Research Conference Digest, pp. 58–59.

24 Suntharalingam, V., Berger, R., Burns, J.A. *et al.* (2005) Megapixel CMOS image sensor fabricated in three-dimensional integrated circuit technology. Digest Tech. Papers IEEE International Solid-State Circuits Conference, pp. 356–357.

25 Warner, K., Oakley, D.C., Donnelly, J.P. *et al.* (2006) Layer transfer of FDSOI CMOS to 150 mm InP substrates for mixed-material integration. International Conference Indium Phosphide Related Materials, pp. 226–228.

21
3D Integration Technologies at IMEC

Eric Beyne

21.1
Introduction

The mission of micro-electronic interconnect technology is to interconnect an ever increasing number of passive and active circuits. Over the years, interconnects have evolved into a hierarchical structure of interconnect levels, from the transistor to the system level. On the chip level, these hierarchal levels are referred to as local, intermediate and global interconnect layers. In the off-chip packaging community, levels are traditionally referred to as wafer (0), package (1st), printed circuit board (2nd), rack (3rd) and system (4th) levels. 3D Integration strategies may be defined at each of these layers of the wiring hierarchy, each resulting in a different complexity (density) and technology (cost) solution. Of particular interest are those 3D technologies that seek to realize 3D-connectivity at the package, bond pad, global, intermediate and local interconnect levels (see also Figure 21.1).

Clearly, 3D connectivity density will have to strongly (exponentially) increase as 3D interconnects move to the lower levels of the interconnect hierarchy. Therefore the optimum technology platform at each layer of the wiring hierarchy will differ. At IMEC, we therefore categorize the different 3D-technologies according to the related manufacturing technology infrastructure: [1–3]

- 3D-SIP: packaging infrastructure
- 3D-WLP: wafer-level packaging infrastructure
- 3D-SIC and 3D-IC: IC-foundry infrastructure.

The 3D-SIP technology includes the wire-bonded stacked die packages and the "package-on-package", PoP, 3D stacks. It is currently the most mature technology and in high volume production. A relatively low packaging density characterizes 3D-SIP.

The 3D-WLP technology is based on wafer-level packaging infrastructure, as used for flip chip bumping and redistribution. Using additional technology elements developed for MEMS-technology, such as deep anisotropic Si-etching, 3D electrical connections can be realized at the wafer level (0-level).

Handbook of 3D Integration: Technology and Applications of 3D Integrated Circuits.
Edited by Philip Garrou, Christopher Bower and Peter Ramm

ISBN: 978-3-527-32034-9

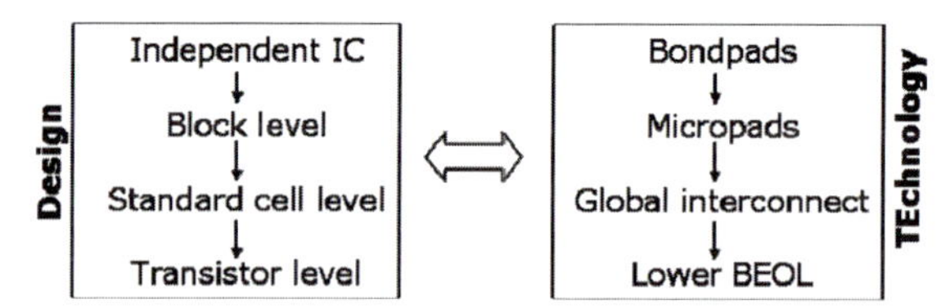

Figure 21.1 Relation between IC-design hierarchy levels and technology interconnect layers.

The 3D-SIC approach uses the Si-foundry technology to create very high density vertical interconnects. This type of 3D stacks can be divided in two main classes, one that addresses the global interconnect layers, 3D-SIC, and one that realizes 3D interconnects at the lowest intermediate and local interconnect levels, 3D-IC.

The 3D-SIC or "stacked-IC" class typically interconnects relatively large circuit blocks or "tiles" in the third dimension, similar to the 2D global interconnects on a typical SOC ("System-on-chip") die (Figure 21.2). The 3D-SIC can therefore also be considered to be a "3D-SOC" solution. It allows for re-use of IP-blocks and can be well integrated in the current state-of-the-art SOC-design methodologies. A possible application for this approach is the combination of logic and memory (Figure 21.3). Most applications require a combination of logic and memory. When large amounts of memory are needed, the memory is realized as a separate die, using a high density, optimized memory technology. Owing to the use of large busses on the logic and memory die and the use of off-chip interconnects, only a relatively slow and

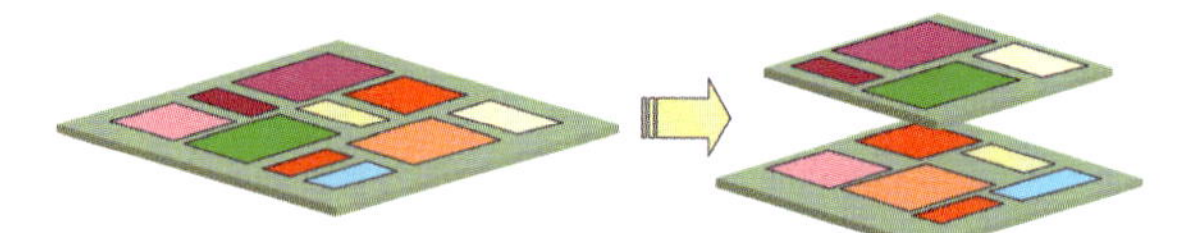

Figure 21.2 Conceptual view a 3D stacked SOC. Functional "tiles" on the die are rearranged in multiple die that are vertically interconnected, resulting in much shorter global interconnect lines.

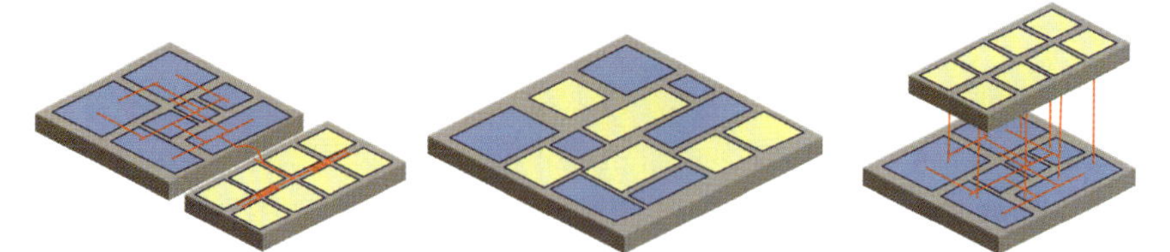

Figure 21.3 Different approaches for combining logic and memory circuits. Left: 2D interconnect between logic and memory die; center: (2D-SOC) combined logic and memory device; right: "heterogeneous 3D-SOC" stacking of a memory and logic device with 3D interconnects between individual logic tiles and memory banks.

power-hungry interconnect between memory and logic is possible. To overcome these limitations, for example, for real-time data processing applications, a SOC approach is typically used. Although not optimal for the integration of high-density memory, the IC logic technology is used for integrating large amounts of memory. This allows for allocating smaller pieces of memory (memory-banks) to specific logic blocks. Distance between logic and memory is short, resulting in the required performance. The integrated memory is, however, of the same performance as dedicated memory technologies would offer. In particular, a much larger die area is consumed by the memory cells, resulting in a die area that is significantly larger than the case with two-die solutions. 3D interconnect technology may solve this problem, by allowing for logic "tiles" on a first die to directly access memory banks on a memory chip. In this case the number of 3D connections required from the memory die to the logic die will increase by an order of magnitude compared to the I/O count of standard memory devices. Similarly, as for the example shown in Figure 21.2, this approach uses 3D interconnects as "global-on chip" interconnect layers to realize a "heterogeneous 3D-SOC" structure.

The 3D-IC class aims at interconnection in 3D of circuits at a lower level of granularity, even at the logic gate and transistors level. This requires 3D-interconnects at the local BEOL wiring hierarchy. According to Rent's rule, the required number of 3D interconnects will increase exponentially with reducing interconnect level. Therefore, an extremely large 3D wiring density must be achieved for 3D-IC technologies. This requires extremely small and narrow pitch 3D interconnects, similar to the transistor and even contact dimensions. Otherwise the area blocked by the 3D connections will be very large compared to the available die area, significantly reducing the active device density. The 3D-IC technology requires a wafer-to-wafer bonding approach and is sequential in nature. Wafers with FEOL layers will be stacked first with local 3D interconnects realized after bonding each individual wafer level. Only when all "local" layers are finalized will the BEOL intermediate and global interconnect layers be added. These layers are common for all layers. Considering the fact that the BEOL layers are one of the significant bottlenecks for the success of future die-shrinks, it can be anticipated that this will also be the case for 3D-ICs, therefore requiring a large numbers of interconnect planes. For the reasons outlined above, a "3D-SOC" 3D-SIC approach is more likely to be successful and economically viable than a pure 3D-IC approach.

21.2 Key Requirements for 3D-Interconnect Technologies

Clearly, the wire-bonded die-stack will not offer the generic 3D packaging solution for advanced systems and scaled semiconductor devices. Several requirements for 3D packaging and interconnection technologies can, however, be put forward. These are discussed below.

The technology should allow for high-density 3D interconnects. Peripheral interconnects, such as wire-bonds, are inherently limited in density as the pitch has to decrease linearly with the wiring demand, whereas in an area-array situation the pitch has only to decrease with the square root of the wiring demand.

High speed and low power applications both demand shorter interconnects with low parasitic capacitance. For high speed, low inductance is also of high importance.

Any 3D-technology that crosses through the Si-die area should have minimal impact on the FEOL (Front-end-of-line, the active die area with the transistors) and the BEOL (Back-end-of-line, the on-chip interconnect layers). Large numbers of large 3D-via connections may block large die-areas, where active circuits and interconnects must be excluded. This will cause the die to require a large Si-area on the wafer, increasing the die cost and defeating the purpose of 3D stacking, which is to miniaturize the system and shorten the circuit interconnect lengths. Actually, the loss in die area may be considerably larger than the actual via size, as the routing of the circuit cells becomes more complicated when large areas of the BEOL are blocked.

A 3D stacking technology should allow for different die sizes. In general, when assembling die made in different technologies, using combinations of existing and newly designed die, it is very unlikely that the die sizes will match. Furthermore, when using heterogeneous integration of various technologies, the wafer size of the different die may not match (300, 200, 150 mm and even smaller diameter sizes will continue to co-exist for different technologies such as memory, logic, analog, RF, high voltage and compound semiconductors). Wafer-to-wafer bonding for 3D stacks will, therefore, be limited to 3D IC-stacks where all layers are realized in the same or similar technology. The main applications for this are memory stacks with a single type of memory and high performance logic wafers, where advanced CMOS-wafers are vertically stacked to allow for packing more transistors in a synchronous region of space.

A significant threat to 3D packaging is the so-called "known-good-die" (KGD) problem. When combining n untested die from wafers with a die yield Y_i, the compound yield of the structure will be $Y_m = Y_s^{n-1} \times Y_i^n$ (with Y_s the yield of the stacking process). As an example, illustrated in Figure 21.4, combining three die with a device yield Y_i of 80% and a stacking yield Y_s of 95%, results in a module yield Y_m of only 46%. The cost of such a wafer-to-wafer stacking process is therefore dominated by the cost of the lost good die ("scrap"-cost). Technologies that allow for die-to-die or die-to-wafer bonding may introduce a component-screening test to significantly increase the confidence level in the die to a "Good-enough-die" level. This can be done using relatively simple test schemes, such as IDDQ testing and visual inspection of contacts. For the example above, raising the die yield level to 95% would result in a more acceptable 85% module yield.

3D stacking schemes should also consider the thermal management of the module. The key issue for thermal management in an electronic system is how to transfer the heat generated by a localized heat source (the active silicon die area) to the ambient environment, generally the air surrounding the devices. By stacking the die in a small volume the total heat dissipation of the system may be reduced, due to the shorter interconnect lengths. However, the local heat density in the system will dramatically increase. The thermal problem becomes twofold: getting the heat out of the stack to the "package" boundaries and getting the heat from the small package to

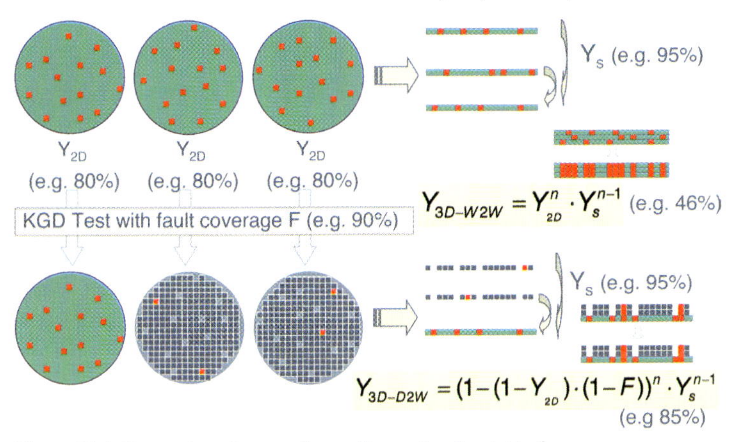

Figure 21.4 Comparison impact die quality on the die yield of wafer-to-wafer or die-to-wafer stacking.

the environment. The first problem requires the use of highly thermal conductive materials in the package and the use of thin layers in the package build-up, in particular for the electrically insulating and "gluing" layers, which generally posses a poor thermal conductivity compared to metals or silicon.

Finally, last but not least, the process for realizing a 3D stacked device should be cost-effective. The main implication of this factor is that no single generic 3D-packaging solution will be possible. The 3D technology should be chosen to fit the requirements set forward by the application. A technology with a very high 3D wiring density may be "overkill" for an application requiring only a moderate number of interconnects.

To achieve a cost-effective 3D process, several technology process requirements may be put forward:

- The technology should maximize collective processing. This favors a wafer-level approach. Although at some stage in the process individual die will need to be handled, because of compound yield issues (which also significantly impact cost).
- The process should maximize the amount of parallel processing:
  - Wafers (die) should be prepared separately for 3D stacking.
  - The process should allow for die screening to obtain "good-enough-die", such as using self-test and IDDQ testing methods.
  - Preferably a Die-to-Wafer placement is performed (with KGD on KGD), followed by a collective bonding step of the individual die at the wafer level.
  - The 3D stack is built by repeating this process with a minimum of sequential processing steps.

Processes that are fully sequential (e.g. wafer-to-wafer bonding, followed by a contact formation process, followed by additional sequential wafer-to-wafer bonding and contact formation processes) suffer from additional yield loss inherent to processes with a large number of sequential steps and also due to the large process strain put on the die that are placed first in the stack.

21.3
3D Technologies at IMEC

21.3.1
3D-SIP for System-Level Miniaturization

3D-SIP is of particular interest when it is used as a stacking technology of SIP packages. Consider a system composed of several clearly defined sub-systems. Each sub-system could be integrated in a system-in-a-package fashion, using the appropriate packaging technology for that particular subsystem. At the end, the SIP subsystems can be stacked in the 3rd dimension by a collective process, creating a 3D-SIP system solution.

As the layers of the 3D stack are actual sub-systems, only a modest 3D interconnect density is required to interconnect the different subsystems. Also testing of the different SIP layers is greatly simplified.

An example of such 3D-SIP integration scheme, realized at IMEC, is a fully integrated low power RF transceiver shown in Figure 21.5 [1]. This device measures only 7 × 7 mm. It consists of two CSP-type devices (CSP = Chip-Scale-Package). The top CSP is realized using IMECs RF-MCM-D technology with integrated passives [4, 5]. The bottom CSP is a double-sided high-density printed circuit board with a high density flip chip die on the bottom side and several discrete passive components mounted on the topside. The connection of this bottom part to the top part is obtained by using solder balls on the topside of the bottom laminate and encapsulation of the topside devices.

Figure 21.5 Fully integrated low power RF radio, measuring $7 \times 7 \times 2.5$ mm, realized by 3D stacking of CSP packages. 3D joining using micro-bumps.

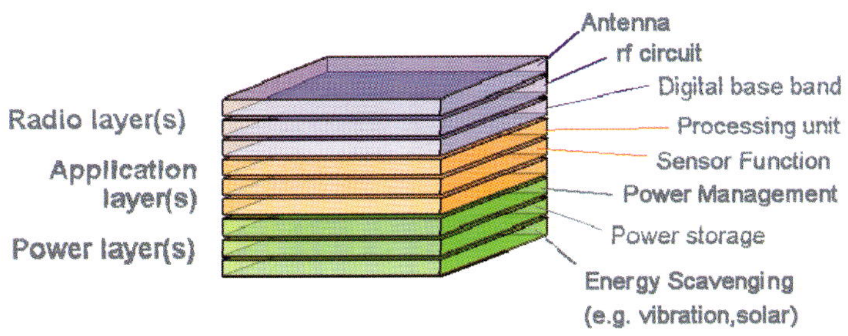

Figure 21.6 Schematic representation of a 3D-SIP concept "eCube", for the realization of distributed, fully autonomous "ambient intelligent" systems. Each layer in the stack is a fully integrated SIP sub-system.

A particularly interesting application area for 3D-SIP is the realization of distributed, fully autonomous systems for realizing so-called "ambient intelligence" systems [6]. These are sometimes referred to as smart-dust, e-grains or e-cubes. As shown in Figure 21.6, such systems can be divided into clear subsystems: the radio (antenna, RF-front-end base band), the main application (processor, sensors, actuators) and the power management (regulation, storage, generation). Each of these functions can be realized as a SIP-subsystem. These may be very small (a few mm to a few cm), enabling its realization using wafer level processing technologies. These 2D-subsystems can be stacked on top of each other, realizing a dense 3D-SiP system [7]. Figure 21.7 shows such an "e-cube" wireless autonomous module with a volume of only one 1 cm^3 realized at IMEC.

A further evolution of 3D-SIP technology is the embedding of Si-die and SMD components in the 3D-SIP interposer substrate using sequential build-up board technologies.

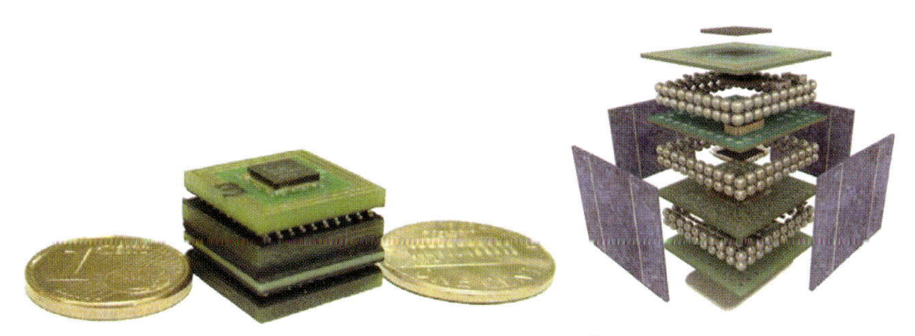

Figure 21.7 Photograph (left) and schematic (right) view of 1 cm^3 eCube (14 × 14 mm), developed at IMEC for a medical application. The 3D-SIP consists of a RF-SIP, with integrated antenna, a low-power DSP SIP, a 19 channel EEG/ECG sensor die, a power SIP. Small solar cells power scavengers, to provide for long time autonomy, may be added on the module sidewalls.

21.3.2
3D-WLP

Wafer level packaging technologies may be very cost-effective to realize 3D stacks. The main goal is to interconnect chips in the 3rd dimension at the bond pad-circuit level. The required connection density is, therefore, of the same order as the current chip-I/O interconnect densities.

The simplest implementation is the face-to-face bonding using flip-chip or micro-bump connections. Two different 3D-WLP options are considered:

- Realization of 3D-through silicon via connections, followed by stacking by a method similar to flip chip mounting.
- Stacking thin die on-top of other die or substrates and contacting both die using a multilayer thin film technology.

Both technologies are studied at IMEC and described in more detail below.

21.3.2.1 **3D-WLP Using Through-Si Vias**

The most common approach used for realizing through-Si vias consists of the following steps:

- Etching of a “blind” via hole in the Si-wafer using the Bosch RIE-ICP etching method.
- Dielectric isolation of the Si-hole: using of CVD oxide or nitride passivation.
- Metallization of Si-holes by realizing a solid metal via-“plug”. Typically Cu electroplating is used for the via filling, followed by a CMP polishing step to remove the excess Cu plated on the wafer.
- Back grinding of the wafer, exposing the Cu plug, finalizing the 3D via-process.

This process has been shown to be effective for realizing high-density 3D via connections. However, several important issues remain:

- Only a thin insulation layer is used between the Si substrate and the Cu-plug. This results in a rather high electrical capacitance of the through hole connection, exceeding the capacitance of standard wire-bond pads.
- A rather thick Cu plug is used in the Si-via hole. Due to the large CTE mismatch between Si and Cu this will cause significant thermo-mechanical stresses upon thermal cycling [8].
- Electroplating of Cu to fully fill the Si-via is a complex process that requires a long process time for each wafer. The use of a CMP polishing step further increases the cost of this technology.

As a solution to these shortcomings, we propose a modified 3D-via build-up (Figure 21.8) [9, 10]:

- The thin CVD insulating layer is replaced by a 2–5 μm thick polymer isolation layer, deposited by CVD, spin or spray coating.

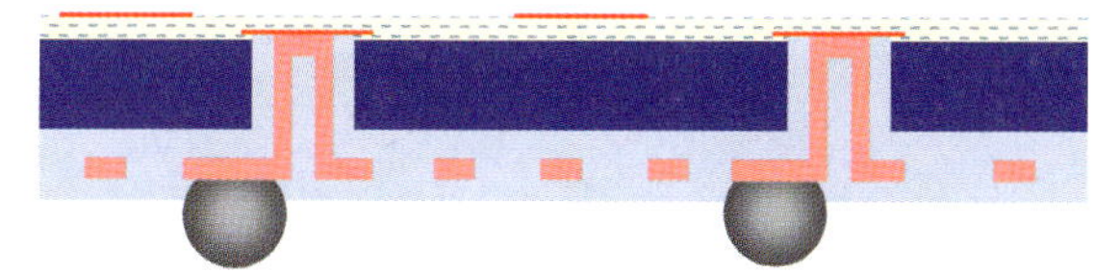

Figure 21.8 Schematic representation of IMECs 3D-WLP through-Si via approach.

- A conformal Cu-plating is used to realize the via connection, similar to blind vias in build-up PCB boards via, but with smaller dimensions.
- The remaining key-hole in the via is filled with a polymer coating.

In addition, the via is realized after thinning the Si wafer, therefore no etching through the BEOL of the Si-die is required. For many applications, such as advanced CMOS nodes with Cu/low-k back-end structures, etching through the BEOL after wafer fabrication is very difficult or even impossible. This proposed process flow is further illustrated in Figure 21.9. The method has some significant advantages:

- Lower cost by simplifying the processes, reducing the process time and reducing the required equipment capital investment.
- Strong parasitic capacitance reduction by using a thicker low-k isolation layer, allowing for high speed and RF 3D connections.
- Strongly reduced thermo-mechanical stresses by using "open" copper metallizations and the use of lower modulus dielectric materials: "compliant" through-hole structure [8].

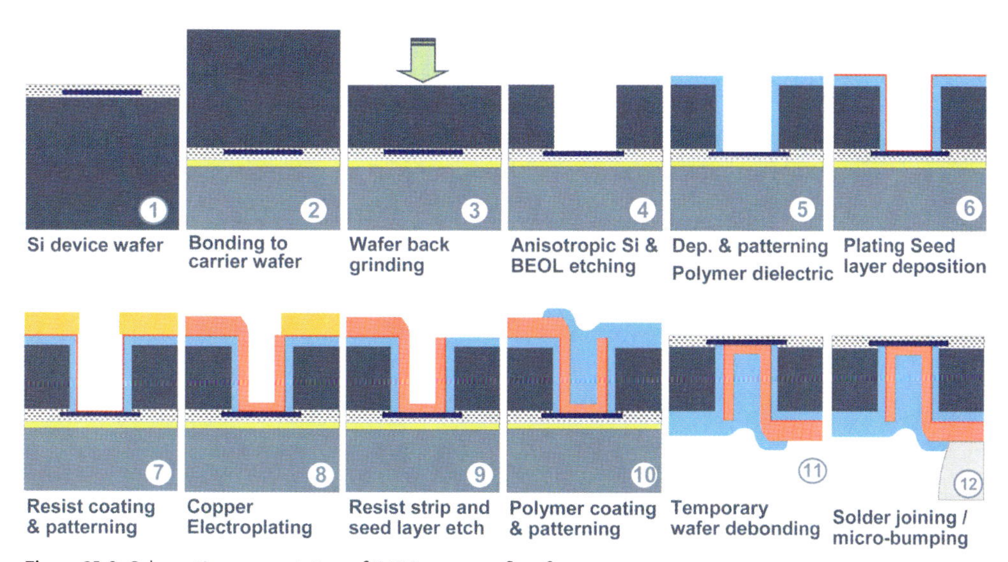

Figure 21.9 Schematic representation of IMECs process flow for realizing 3D-WLP through-Si via connections.

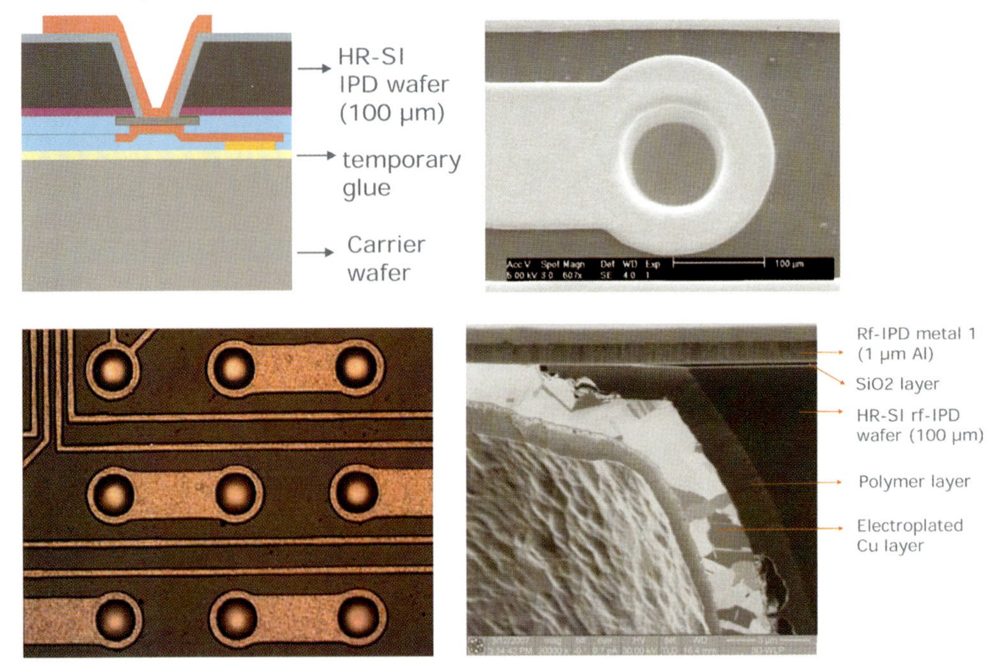

Figure 21.10 Application of IMECs 3D-WLP through-Si via connections to an RF-integrated passive device (IPD) application on high resistivity Si wafers.

- Compatible with common wafer-level packaging redistribution and bumping technologies.

An application example of 3D-WLP through-Si vias is given in Figure 21.10 [10].

21.3.2.2 3D-WLP By Ultrathin Chip Stacking (UTCS)

A different approach to 3D stacking consists in stacking thin die on active device wafers and using multilayer thin film technology to interconnect the thin die with the host wafer. IMECs ultrathin-chip-stacking (UTCS) approach [11, 12] uses very thin (10–20 μm) Si die, embedded in a redistribution technology. Such an approach allows for a high level of system level flexibility and integration density. This technique also allows for stacking of die with largely varying dimensions, as well as the integration of thin film passive components in a 3D interconnect stack.

To realize a multilayer die UTCS stack, four basic processes are used.

1. Ultra-thin chip-on-chip, UTCoC, process
2. Ultra-thin chip embedding, UTCE, process
3. Ultra-thin chip-in-flex, UTCF
4. UTCF-stacking process for more than two die layer UTCS structures.

The UTCoC process is shown schematically in Figure 21.11. A wafer with active, tested die is bonded to a temporary silicon carrier wafer. The scribe lanes of this wafer

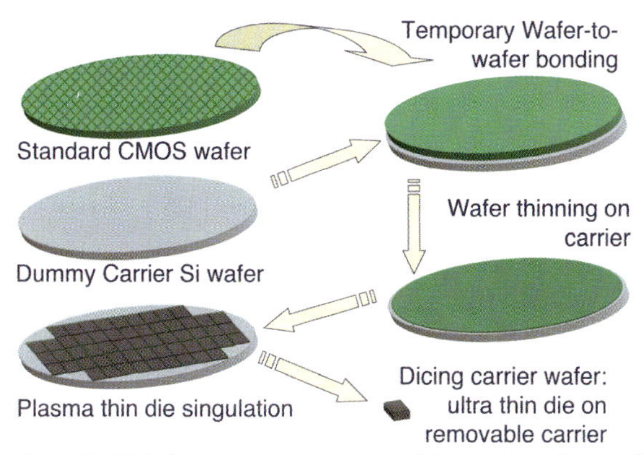

Figure 21.11 Schematic representation of the ultrathin chip-on-die, UTCoC, process.

have been pre-diced or etched to become shallow trenches. Using a combination of coarse and fine grinding, the active wafer is thinned to a thickness of 15–20 μm. Plasma etching is used to remove any remaining Si damage and to obtain the desired final thickness. As the wafer was pre-scribed, the thin die are separated from each other on the carrier wafer. The next step consists in dicing the carrier wafer to obtain the UTCoC chips for further processing.

For the UTCE process, shown schematically in Figure 21.12, either an active wafer or another dummy carrier wafer is used as host substrate. A polymer glue-layer, such as BCB, is spun on the wafer. A flip-chip die bonder is used to place with high alignment accuracy KGD-UTCoC chips on KGD host wafer die. The actual bonding of these die (polymer glue curing under pressure) is performed collectively at the wafer level. The next step is the collective removal of the sacrificial layer and the temporary carrier chips of the UTCoC die (e.g. thermal or chemical). Subsequent steps consist of depositing thin film dielectric layers for isolation and thin film, electroplated copper patterns for electrical connection. At the end of this process a two-layer UTCS stack is obtained (Figure 21.13). An example of such a two-die layer

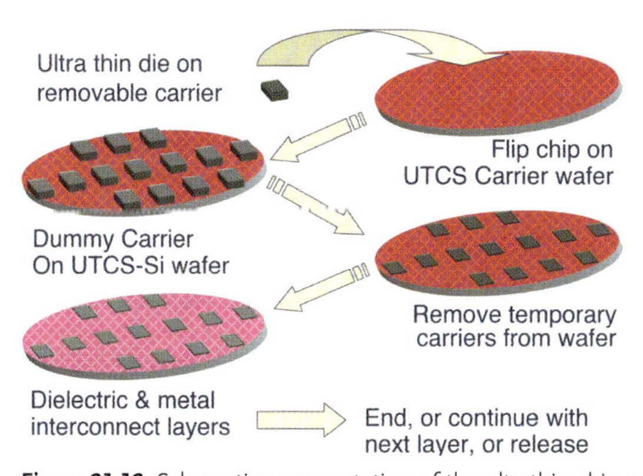

Figure 21.12 Schematic representation of the ultrathin chip embedding, UTCE, process.

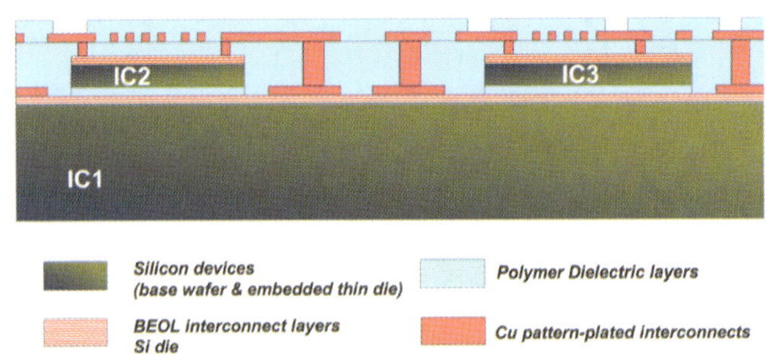

Figure 21.13 Schematic representation of an Ultra-Thin-Chip-Stack, UTCS, with two layers of die.

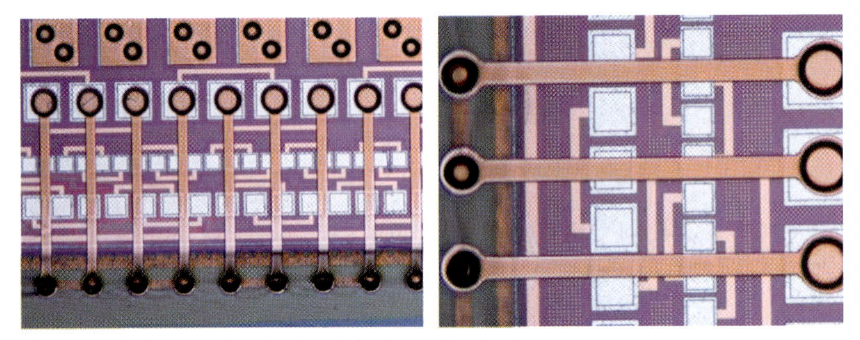

Figure 21.14 Photographs of embedded, 20 μm thick Si-die, transferred to host substrate and electrically connected to that substrate using the UTCE die embedding technique.

structure is shown in Figure 21.14. By using thin film lithography, a very high number of interconnects can be realized between the die. This allows for additional interconnects between sub-sections of the die. In the proposed UTCS structure, via connections in the 3rd dimension are realized in the area around the chips. As these thin film vias are realized with pad sizes smaller than 50 μm, a very high interconnect density in the third dimension is obtained.

If, for the UTCE process, a sacrificial substrate is used, this substrate may be removed, effectively resulting in a very thin flex foil (10–30 μm) with embedded active die – the UTCF (Ultra-Thin-Chip-in-Foil). For realizing an n-layer UTCS stack, it is possible to stack UTCF films on UTCE stacks, using micro-bump flip chip connections (Figure 21.15). This results in a more a very versatile 3D-interconnect stack with very high 3D interconnectivity. In contrast to the traditional die-stacking technologies, the dies that are stacked can have arbitrary sizes at any layer and do not need any particular process or design adaptation.

21.3.2.3 3D-SIC "Cu-Nail" Technology

3D-SIC technology uses IC-foundry infrastructure to realize through-Si via connections. Most approaches in this field realize such through-vias after finalizing the IC process. Our approach differs from this by introducing a new process step during IC

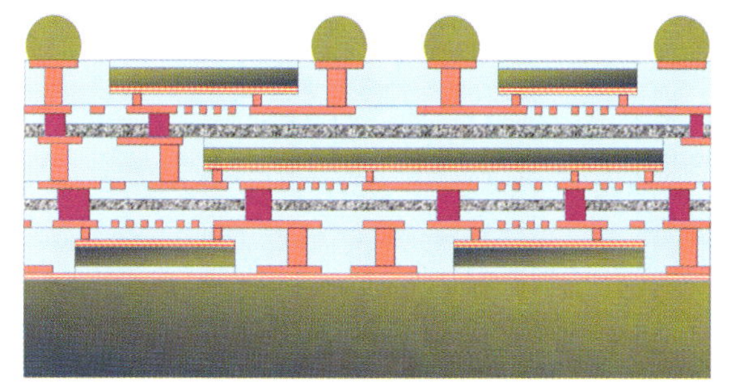

Figure 21.15 n-Layer UTCS stack, realized using parallel processing combining one UTCE and n-2 UTCF layers.

processing, the so-called "Cu-nail" [13]. This Cu-nail is processed after the FEOL process (transistors), but before the BEOL process (multilayer damascene interconnect layers) (Figure 21.16). The Cu-nail is realized by plasma etching a ±25 μm deep Si hole with a diameter of 3–5 μm. A modified Cu single damascene process is used to fill the hole. A CVD oxide layer is used as thin dielectric insulating layer and a CMP stop layer. A TaN barrier is then deposited. The via hole is then filled with electroplated copper. CMP is used to remove the Cu "overburden". After this process, standard BEOL is used to finalize the Si-die.

After finalizing the wafer process and wafer test, the wafer is mounted on a temporary carrier and thinned down to a Si-thickness of only 10 μm. During this process, the "Cu-nails" are exposed on the wafer backside (see Figures 21.17 and 21.18) [14].

The 3D stacking is performed as a die-to-wafer bonding process, similar to the 3D-WLP process mentioned above. However, in this case Cu–Cu direct bonding is used,

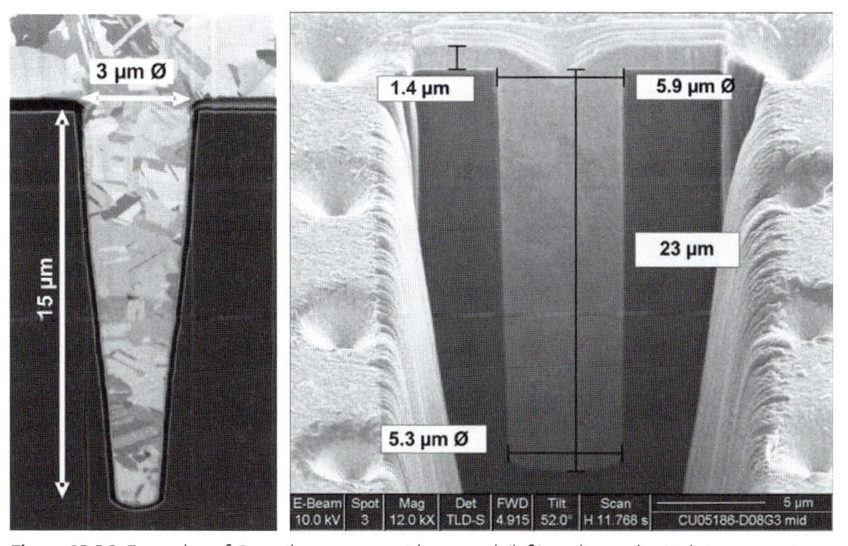

Figure 21.16 Examples of Cu-nail structures with tapered (left) and straight (right) geometries.

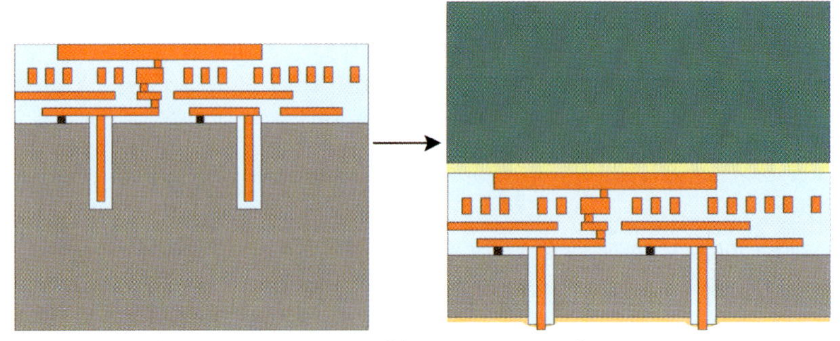

Figure 21.17 Schematic representation of the "3D-SIC" Cu-nail via process. Left: standard CMOS wafer with "Cu-nail" before the BEOL process. Right: thinned CMOS chip on carrier chip with exposed Cu-nails.

rather than a solder joint or micro bump connection [15]. The stacking process consists of a fast die-to-wafer alignment and placement, followed by a collective wafer-level Cu/Cu bonding process. It can easily be repeated to obtain multi-die stacks. (See Figure 21.19.)

The main advantages of this 3D-SIC approach are:

- Minimal impact on the CMOS wafer design and processing:
  - only a small exclusion area on the FEOL (small via holes),
  - no impact on the BEOL wiring,
  - small number of additional process steps needed, only one additional litho step, resulting in a low process cost.

- Parallel processing route: wafers are prepared for 3D stacking and only KGD die are stacked involving a minimum number of processing steps. This is required to achieve high 3D-module yields and low cost processing.
- A very high density 3D-interconnect is possible, as the Cu-nails are only a few micrometer in diameter. Densities up to, and exceeding, $104\,\text{mm}^{-2}$ are feasible.
- Very thin interface layers allow for a low thermal resistance [16].

Figure 21.20 shows a FIB-cross section of a Cu-nail connection. In Figure 21.21, an array bonded Cu-nails is shown (after wet etch removal of the Si from the bonded die).

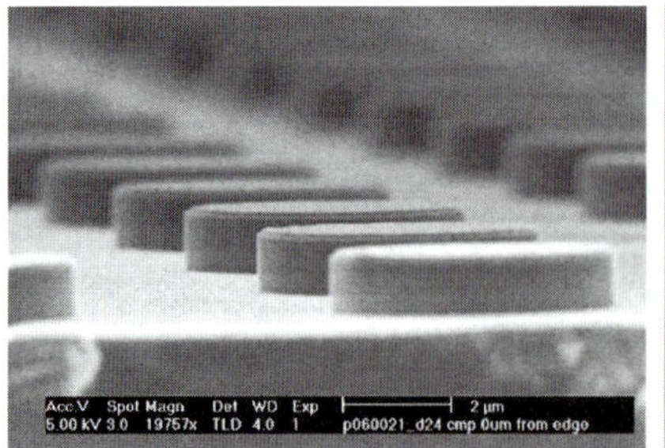

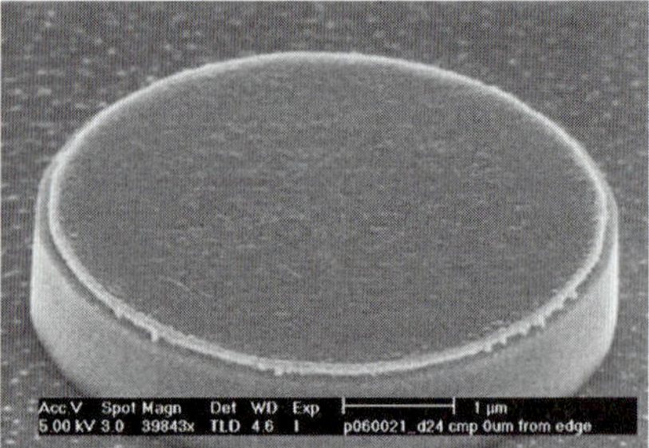

Figure 21.18 Pictures of exposed Cu-nails on the die back-side after thinning.

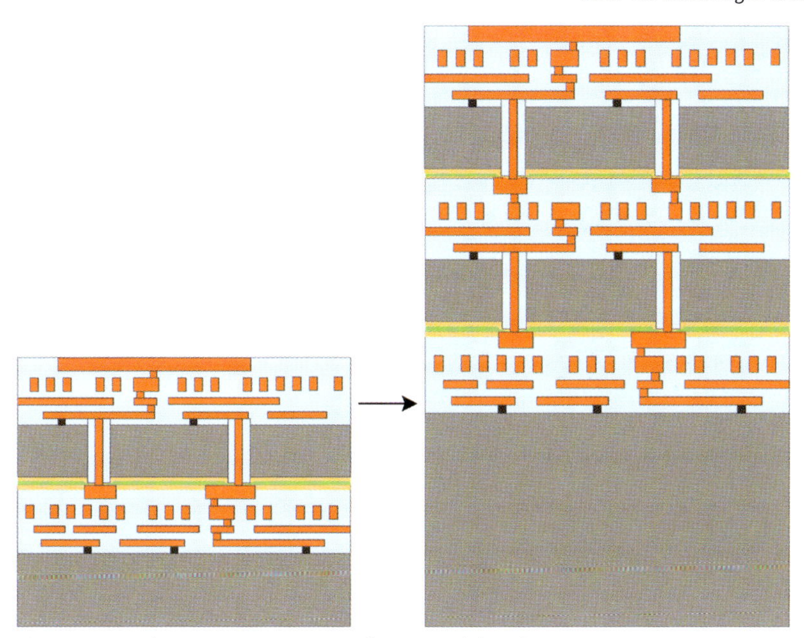

Figure 21.19 Schematic representation of a two- and three-layer 3D-SIC stack using Cu–Cu bonding.

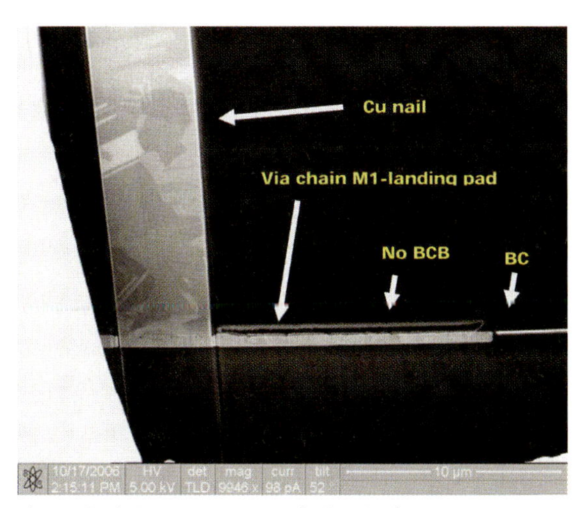

Figure 21.20 FIB cross-section of a Cu–Cu thermo-compression bonded Cu-nail. The BCB glue layer is also partially visible.

Figure 21.21 Bonded Cu-nails revealed after a selective wet-etch of the Si die bonded on a base metal wafer (daisy-chain test structure). The suspended metal bridges are Metal 1 interconnects of the etched-away top die.

21.3.2.4 **3D Research Roadmap of IMEC**

In the previous section, IMECs approaches to 3D-SIP, 3D-WLP and 3D-SIC technologies were discussed. Table 21.1 compares these different approaches. A common aspect of all technologies is the need for die thinning technologies down to 50 μm and less. For the 3D-SIC and UTCS technologies, die will be thinned down even further. Figure 21.22 gives a schematic representation of IMECs R&D roadmap.

Table 21.1 Classification and comparison of different 3D interconnect technologies at different levels of the interconnect hierarchy.

	3D-SIP	**3D-WLP**		**3D-SIC**
Technology	**Package interposer**	**WLP, Post-passivation**		**Si-foundry, post FEOL**
3D interconnect	**Package I/O**	**UTCS embedded die**	**Si-through cvias**	**Si-through “Cu nail” vias**
Interconnect density	“Package-to-package”	“Around” die	“Through” die	“Through” die
Peripheral	2–3 mm^{-1}	10–50 mm^{-1}	10–25 mm^{-1}	25–100 mm^{-1}
Area-array	4–11 mm^{-2}	100–2.5 k mm^{-2}	16–100 mm^{-2}	400–10 k mm^{-2}
3D Si Via pitch	—	—	40–100 μm	<10 μm
3D interconnect pitch	300–500 μm	20–100 μm	—	—
3D Si Via diameter	—	—	20–40 μm	1–5 μm
Die thickness	>50 μm	10–50 μm	40–100 μm	10–50 μm

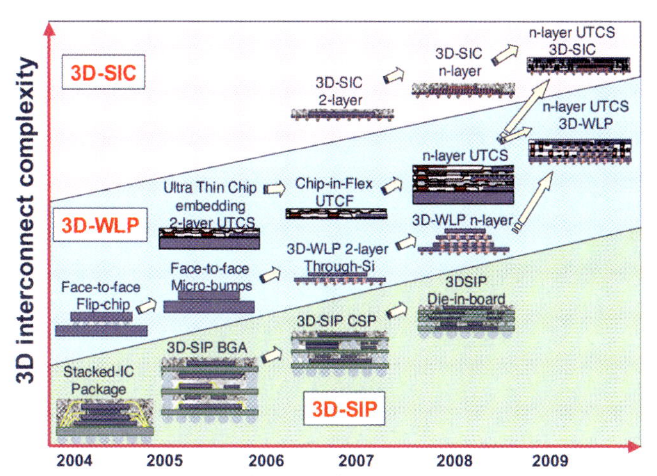

Figure 21.22 IMECs 3D packaging and interconnection roadmap for 3D-SIP, 3D-WLP and 3D-SIC technology families.

References

1. Beyne, E. (2004) 3D interconnection and packaging: Impending reality or still a dream? Proceedings of the IEEE International Solid-State Circuits Conference, ISSCC 004, 15–19 February 2004, San Francisco, CA, USA, pp. 138–145.
2. Beyne, E. (2006) The rise of the 3rd dimension for system integration. Proceedings of the International Interconnect Technology Conference – IITC, 5–7 July 2006, San Francisco, CA, USA, pp. 1–5.
3. Beyne, E. (2006) 3D system integration technologies. Proceedings of the IEEE Symposium on VLSI Technology, Systems, and Applications (VLSI-TSA), 24–26 April 2006, Hsinchu, Taiwan, pp. 19–25.
4. Beyne, E. (2006) Interconnect and packaging technologies for realizing miniaturized smart devices, in *AmIware. Hardware Technology Drivers of Ambient Intelligence* (ed. S. Mukherjee), Springer, Dordrecht, Chapter 3.1, pp. 107–123.
5. Pieters, P., Vaesen, K., Brebels, S. *et al.* (2001) Accurate modeling of high-Q spiral inductors in thin-film multilayer technology for wireless telecommunication applications. *IEEE transactions on Microwave Theory and Techniques, MTT,* **49**, pp. 589–599.
6. Baert, K., Gyselinckx, B., Torfs, T. *et al.* (2006) Technologies for highly miniaturized autonomous sensor networks. *Microelectronics Journal*, **37** (12), 1563–1568.
7. Stoukatch, S., Winters, C., Beyne, E. *et al.* (2006) 3D-SIP integration for autonomous sensor nodes. Proceedings of the 56th IEEE Electronic Components and Technology Conference, 30 May–2 June 2006, San Diego, CA, USA, pp. 404–408.
8. Gonzalez, M. *et al.* (2005) Influence of dielectric materials and via geometry on the thermomechanical behaviour of silicon through interconnects. Proceedings of the 10th Pan Pacific Microelectronics Symposium, SMTA, Hawaii, January 25–27.
9. Patent US 2004 0259292 A1-US 10817763.
10. Sabuncuoglu Tezcan, D., Pham, N., Majeed, B. *et al.* (2006) Sloped through wafer vias for 3D wafer level packaging. Proceedings of the 57th IEEE Electronic Components and Technology Conference,

ECTC 2007, May 29–June 1, Reno, NV, USA.

11 Patent 1999-010. US 673,099,7 B2 and EP 101,462,0 A2.

12 Beyne, E. (2001) Technologies for very high bandwidth electrical interconnects between next generation VLSI circuits. IEEE-IEDM 2001 Technical Digest, December 2–5, Washington, D.C., USA, S23-p3.

13 Swinnen, B., Ruythooren, W., De Moor, P. *et al.* (2006) 3D integration by Cu–Cu thermo-compression bonding of extremely thinned bulk-Si die containing 10 μm pitch through-Si vias. Technical digest IEEE International Electron Devices, Meeting, IEDM, 11–13 December 2006, San Francisco, CA, USA.

14 De Munck, K., Vaes, J., Bogaerts, L. *et al.* (2006) Grinding and Mixed Silicon Copper CMP of Stacked Patterned Wafers for 3D Integration. MRS Fall Meeting Symposium Y: Enabling Technologies for 3-D Integration, 26–29 November 2006, Boston, MA, USA.

15 Ruythooren, W., Stoukatch, S., Lambrinou, K. *et al.* (2006) Direct Cu–Cu thermo-compression bonding for 3D stacked IC integration. Proceedings of the IMAPS International Symposium on Microelectronics, 8–12 October 2006, San Diego, CA, USA.

16 Chen, C., Vandevelde, B., Swinnen, B. and Beyne, E. (2006) Enabling SPICE-type modeling of the thermal properties of 3D-stacked IC's. Proceedings of the IEEE Electronics Packaging Technology Conference, EPTC, 6–8 December 2006, Singapore, pp. 492–499.

22
Fabrication Using Copper Thermo-Compression Bonding at MIT

Chuan Seng Tan, Andy Fan, and Rafael Reif

22.1
Introduction

This chapter describes a fabrication method for multilayer 3D ICs initially proposed and developed at the Microsystems Technology Laboratories (MTL) at MIT. This approach to 3D integration is based on metallic wafer bonding, that is, low-temperature direct copper-to-copper (Cu-to-Cu) thermo-compression bonding [1]. A thin device layer (with interconnects) is attached on top of a substrate wafer that contains an underlying device layer (with interconnects) in a back-to-face (or face up) manner. An SOI wafer is used to achieve a thin device layer and the thin layer handling is accomplished with an additional handle wafer. Inter-layer connections are established using short vertical vias. In subsequent sections of this chapter, we begin by describing low-temperature wafer-to-wafer bonding using Cu as the bonding medium in Section 22.2. Building on work on Cu wafer bonding, a process flow of a back-to-face double-layer ICs stack is proposed and illustrated in Section 22.3. This scheme presents several technological challenges, which are discussed in Section 22.4. The salient features of this approach are also included wherever necessary.

22.2
Copper Thermo-Compression Bonding

This section studies low-temperature (400 °C and below) thermo-compression bonding of copper thin films coated on blanket wafers. As the name implies, thermo-compression bonding involves mechanical pressing and heating of the wafers. Two wafers can be held together when the Cu thin films bond together to form a uniform bonded layer. For this technique to be applicable to wafers that carry device and interconnect layers, an upper bound of temperature step is set at 400 °C to prevent undesired damages to the interconnects. The bonding procedures are

Handbook of 3D Integration: Technology and Applications of 3D Integrated Circuits.
Edited by Philip Garrou, Christopher Bower and Peter Ramm

ISBN: 978-3-527-32034-9

described and Cu grain microstructure evolution as a result of bonding and subsequent annealing is also monitored by cross-sectional study using SEM and TEM. Based on this study, the bonding mechanism is proposed.

The main objective of this Cu thermo-compression bonding study is to explore its suitability for utilization as a permanent bond that holds active device layers together in a multilayer ICs stack. Copper is chosen based on its promising properties and this is elaborated in Section 22.4. Note that using Cu as the bonding layer to form a multilayer IC stack serves two purposes: (i) a mechanical bond that joins device layers together and (ii) an electrical bond that establishes conductive paths between device layers. Therefore, the quality of the bonded Cu layer must be examined against the above requirements.

22.2.1 Bonding Procedures

Thermo-compression bonding of Cu-coated wafers has been demonstrated and characterized on blanket 150 mm silicon wafers. Thermal oxide (500 nm) is grown on all wafers prior to metallization. The next step is the deposition of tantalum (50 nm) and copper (300 nm) in an e-beam deposition system. Tantalum (Ta) is used as a diffusion barrier to prevent Cu diffusion into the device layers, which might degrade the electrical integrity of the device layers in an actual process [2]. A pair of as-prepared wafers is aligned face-to-face, clamped together on a bonding chuck and loaded into a wafer bonder. After N_2 purge, the chamber is evacuated and a down force is applied on the wafer pair. The temperatures of the chuck and top electrode are ramped up to 300 °C and maintained at this temperature. The contact force is 4000 N when the wafer pair is in full contact at 300 °C and the bonding step lasts for 1 hour. On 150 mm wafers, this contact force is equivalent to a contact pressure of about 226 kPa. After bonding, the bonded pair is removed from the bonding chamber and then annealed in atmospheric N_2 ambient at 400 °C for 1 hour to allow further Cu inter-diffusion and grain growth so that higher bond strength can be achieved.

22.2.2 Bonding Mechanism

Figure 22.1a is a scanning electron microscopy (SEM) image of an as-prepared Cu-coated wafer prior to bonding. The cross section of the bonded Cu layer sandwiched within two oxide layers is shown in Figure 22.1b. Evidently, the two Cu bonding layers merge and a homogeneous bonded layer is obtained. To understand the microstructure of the bonded Cu layer, transmission electron microscopy (TEM) analysis has been performed on this sample. Figure 22.2 shows the cross-sectional TEM image of the bonded layer. As can be seen, large Cu grains, which often extend beyond the original bonding interface, are obtained after bonding and annealing. Dislocation lines are also found in the Cu grains. A possible bonding mechanism that gives rise to the above grain structures has been proposed [3, 4]. From the TEM image, it is evident

(a)

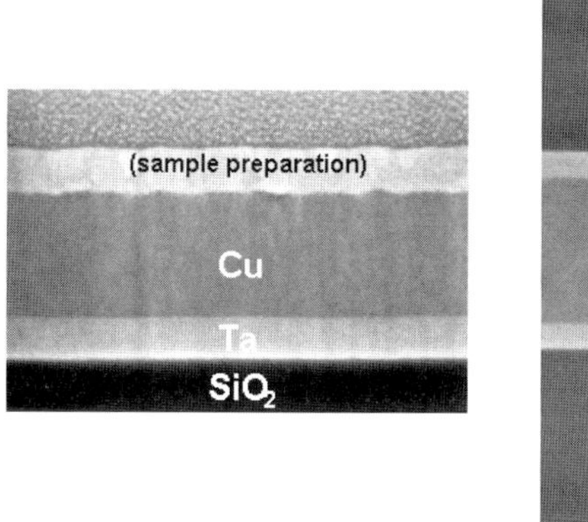

(b)

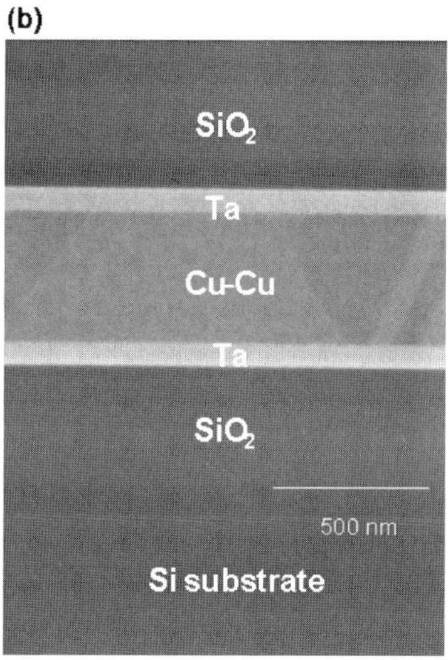

Figure 22.1 SEM images of bonded Cu layer sandwiched between oxide layers. Cu-coated wafers were bonded at 300 °C for 1 hour followed by an anneal at 400 °C for 1 hour.

that there is substantial grain growth during bonding and annealing. The jagged Cu–Cu interface suggests that inter-diffusion between two Cu layers has taken place. During bonding and subsequent annealing, Cu layers are in intimate contact under the applied pressure. At process temperatures between 300 and 400 °C, Cu atoms

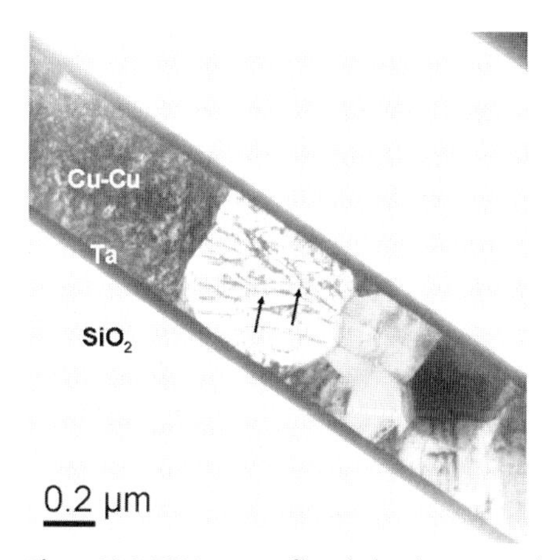

Figure 22.2 TEM image of bonded Cu layer. Note that the bonding Cu layers merge and a homogeneous Cu layer is obtained after bonding and anneal. Grain structures that extend across the original bonding interface are observed. Dislocation lines (marked with arrows) are clearly seen in the grains.

acquire sufficient energy to diffuse rapidly, and Cu grains begin to grow. At the bonding interface, diffusion can happen across the bonding interface and grain growth can progress across the interface. After sufficiently long duration, large Cu grains on the order of 300–500 nm are obtained, and a homogeneous bonded Cu layer is formed. While the above was performed on Cu thin films deposited in an e-beam system, similar bonding behavior is also observed on electroplated Cu thin films widely used in a manufacturing environment.

One of the key requirements, apart from particle-free surfaces, is the surface roughness of wafers prior to bonding. The RMS surface roughness of the Cu-coated wafers in the above experiment is about 2 nm. This measurement is performed using atomic force microscopy (AFM). If a poly-silicon layer is intentionally inserted between Cu and oxide layers, Cu thin films are found to bond and form a homogeneous layer despite an RMS roughness of 8–10 nm due to insertion of the poly-silicon layer [5]. For comparison, oxide surfaces must have RMS roughness below 1 nm for successful fusion bonding in the case of silicon dioxide wafers. The fact that Cu wafer bonding can be accomplished at a much higher surface roughness is due to the contact force and heating used in the above experiment.

22.3 Process Flow

Although there are various 3D process flows based on wafer bonding, most wafer bonding schemes share four common considerations:

1. The bonding medium ("glue layer") of choice, which is a permanent bond that holds active device layers together.
2. The method for Si substrate thinning.
3. The wafer-to-wafer alignment accuracy.
4. The interlayer electrical interconnection method.

For back-to-face bonding, one also has to consider the temporary bond between donor and handle wafers and its release. The remainder of this section focuses on these considerations as they apply to the 3D integration scheme highlighted in this chapter.

In the MIT's 3D integration scheme, multiple device wafers are sequentially bonded to each other in a back-to-face manner using low-temperature Cu thermo-compression bonding. Figure 22.3 depicts our definition of a 3D circuit, in which two device layers are both bonded and electrically interconnected using matching Cu pads (the bonding interface). The ideal case for such architecture is to have the smallest possible lateral dimensions for the Cu pads and interlayer vias to ensure high via density. In reality, wafer-to-wafer alignment tolerances during bonding and the maximum aspect ratio of the vias one can create will ultimately be the size-limiting determinants of these vias/Cu pads. As shown in Figure 22.3, when the top device layer is a thin SOI, the aspect ratio of the inter-wafer vias can be relaxed for ease in fabrication, while still maintaining a relatively high vertical density across the

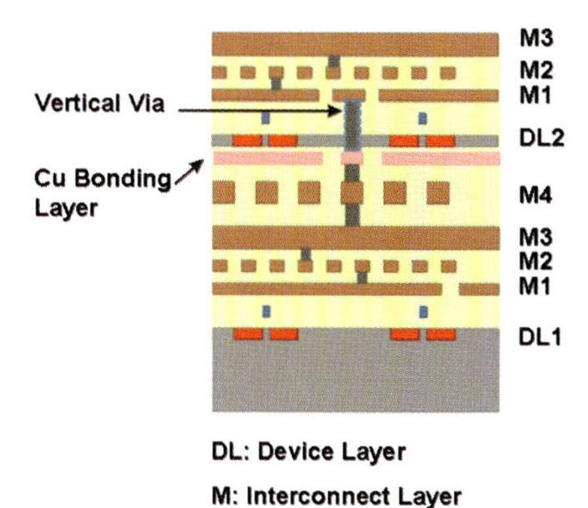

Figure 22.3 Schematic of a back-to-face double-layer 3D IC fabricated using low-temperature copper wafer bonding. From [9] (© 2002 IEEE).

wafer. To recapitulate, the goal of this process flow is to create a double-layer ICs stack by transferring a thin device layer (with interconnects) from a top SOI wafer on a bottom device substrate. In principle, this flow can be repeated to form a multilayer stack as needed.

Figure 22.4 is a flowchart of such a process. It begins with a top SOI device wafer (also called a donor wafer). To lay down the device layer on the bottom wafer (called a substrate wafer) so that it is facing up in the final 3D stack, the SOI substrate must be removed. For ease of thin film handling, the SOI wafer has to be attached to a handle wafer, which provides mechanical support during subsequent process steps. Note that the bond between the donor SOI wafer and the handle wafer is a temporary bond that has to be de-bonded at the end of the flow. Therefore, one has to choose a bond that is sufficiently strong to hold the SOI wafer during processing and, also, the same bond can be released readily at the end of the flow. Upon complete donor SOI substrate removal, vias and Cu pads are formed from the backside. The thin SOI layer, which is attached to the handle wafer, is then aligned and permanently bonded

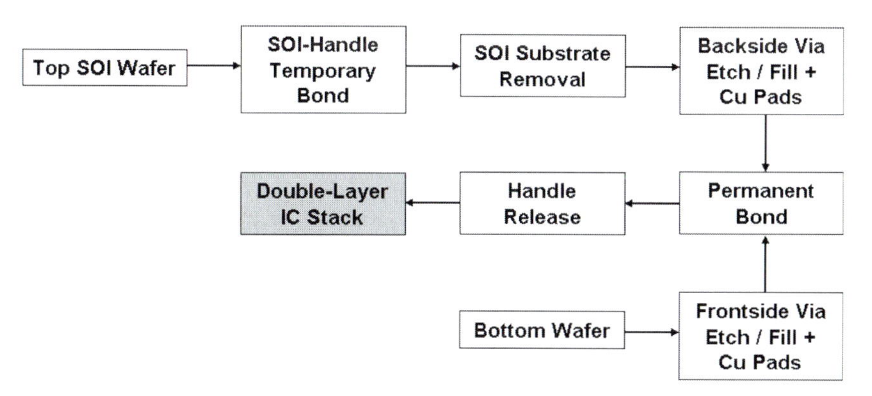

Figure 22.4 MIT process flow chart.

to a substrate device wafer that has matching Cu pads. Finally, the handle wafer is released to form a 3D stack. A progression to multilayer stack can be achieved by successive layer transfer. Process details of each of the above steps are discussed in the following.

22.3.1 Handle Wafer Attachment

This 3D scheme begins with a typical SOI substrate (e.g. 100 nm SOI/400 nm BOX) that contains both CMOS devices and its corresponding multi-level interconnects. Next, in order to stack this SOI device layer on top of another device layer on a base substrate, it first has to undergo backside substrate thinning up to the BOX layer. Since the device and interconnect layer is only a few μm thick, one has to be able to handle it properly. For mechanical support during grind-back, it is imperative to attach the SOI to a handle wafer with a special adhesive that is (i) strong enough to withstand vigorous shearing force from grinding; (ii) chemically inert to hot aqueous hydroxide solution that is used during wafer etch-back; and (iii) can easily be de-bonded by other means to release the handle from the 3D stack. Because hot chemical solutions will usually delaminate organic polymers from their substrates, it is very difficult for CMOS-compatible polymer adhesives to satisfy all three criteria. However, these requirements can be met with a careful choice of metallic bonding layers. Two options are studied in the following:

1. Zr/Cu-Cu/Zr: One can choose Zr/Cu–Cu/Zr layers as the temporary bond. The Cu–Cu bond acts as the adhesion layer that attaches the donor SOI wafer to the handle wafer. On the other hand, zirconium (Zr) layers can be used as the "release" medium. This is because Cu can withstand HF corrosion but Zr dissolves extremely rapidly in dilute HF (much faster than SiO_2).

2. Al/Ta/Cu-Cu/Ta/Al: A second option is to use Al as the release layer. This is because hot HCl has very different selectivities for etching Al, Ta, and Cu. HCl is the optimum choice as release agent because Cu is able to resist hot HCl corrosion if one minimizes the concentration of oxidizing species within the liquid solution. For extra protection, a Ta layer is added since it is also impervious to hot HCl. On the other hand, hot HCl can destroy the Al layer more readily. Again, the Cu–Cu bond is used to hold the SOI wafer to the handle wafer.

The ability of Ta, Cu, Zr and Al to withstand corrosion of different solutions is qualitatively summarized in Table 22.1. The handle wafer attachment scheme is shown in steps 1 and 2 in Figure 22.5.

The process begins with a back-end-of-line (BEOL) completed SOI donor wafer (also known as top wafer). This SOI wafer carries device and interconnect layers. The donor wafer is passivated by an overlayer of 500 nm PECVD oxide and followed by oxide CMP to achieve global planarization. Once passivated and planarized, the SOI substrate is ready for handle wafer attachment. Since the handle wafer, which is silicon, is exposed to hydroxide solution (e.g. tetramethylammonium hydroxide –

Table 22.1 Corrosion rate of Ta, Cu, Zr and Al in various chemical solutions.

	HF	Hot KOH/TMAH	Hot HCl
Ta	High	Low	Very low
Cu	Very low	Low	Low (non-aerated)
Zr	Very high	Very low	Very low
Al	Very high	Very high	Very high

TMAH) during SOI substrate thinning, it must be protected against corrosion. The protection can be achieved by thermally oxidizing the handle wafer to a thickness of about 500 nm since TMAH etches oxide extremely slowly. Note that oxidation of handle wafer is carried out before SOI wafer attachment. Then, both the SOI wafer and handle wafers are metallized according to the sequences in (i) or (ii) above. With no need for wafer-wafer alignment, the SOI and the oxide handler are bonded at 300 °C in N_2 ambient with persistent contact pressure of 4000 N for 30 min. The pair undergo a further anneal at 400 °C in N_2, in which the Cu–Cu interface forms a strong bond to ensure that the bonded pair withstand subsequent mechanical substrate grinding without separation.

22.3.2 Substrate Etch-Back and Backside Via Formation

The process steps discussed in this section are depicted in steps 3–6 in Figure 22.5. With the handle wafer in place, the next task involves the complete removal of the SOI substrate using a combination of mechanical grinding and aqueous chemical

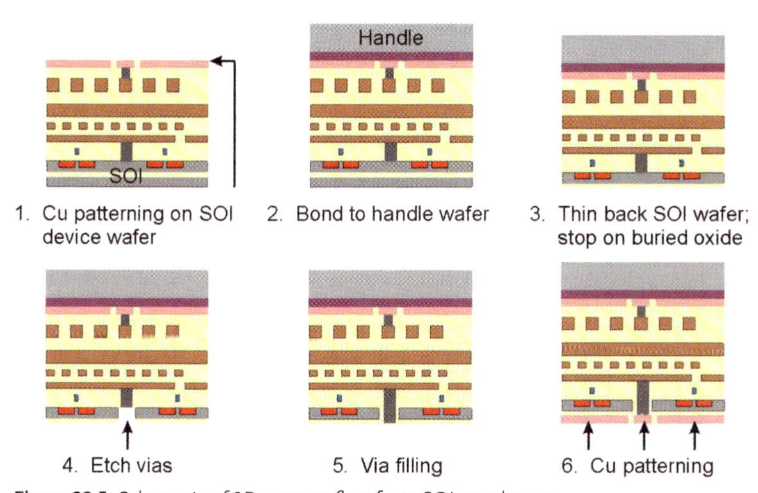
Handle
SOI
1. Cu patterning on SOI device wafer
2. Bond to handle wafer
3. Thin back SOI wafer; stop on buried oxide
4. Etch vias
5. Via filling
6. Cu patterning

Figure 22.5 Schematic of 3D process flow from SOI attachment to handle wafer, SOI substrate thin-back, back side via formation and fill and Cu bonding pads formation. From [9] (© 2002 IEEE).

etching, which will selectively stop on the BOX layer. Note that mechanical grinding is a rather abusive step that has no selectivity between oxide and silicon, and it leaves behind a rough surface. To use this combination effectively, 550 μm of the SOI substrate is removed using mechanical grinding and this step is stopped at about 100 μm from the BOX layer. The remaining 100 μm of Si is stripped easily using 12.5 wt% TMAH solution at 85 °C for about 100 min. Since TMAH has excellent selectivity towards oxide [6], the etching stopped on the BOX layer. Therefore, SOI wafers are attractive choice as the BOX layer serves as excellent etch stop layer. By using SOI wafers, one can achieve an ultra-thin layer to be used in layer transfer.

At the TMAH concentration mentioned above, the oxide layer provides a near-perfect protection of the handle wafer against TMAH attack. Zirconium is virtually impervious to hydroxide attack until the solution temperature reaches beyond 120 °C. For a bonded Cu-Cu layer, however, interface delamination at the wafer edges is observed. There are two ways to prevent such Cu corrosion. One can either reduce the exposure time to hydroxide attack or add a Cu corrosion inhibitor to the etch solution.

In contrast, the Al release layers undergo severe corrosion during TMAH etch-back if not protected properly. This would cause an unwanted premature separation of handle wafer from the SOI wafer. Therefore, a scheme is needed to circumvent this problem. Instead of depositing the layers inside the laminate structure in option 2 one after another, the Al layer deposition can be first made to have an exclusion ring around the wafer edges. Then, one can protect this Al layer from any form of corrosion by depositing the Ta/Cu over-layers without exclusion rings.

Upon completion of SOI substrate removal, interlayer vias are created by etching through the BOX, SOI, and ILD 1, finally stopping on metal layer M1. This is followed by PECVD oxide sidewall passivation and via filling using a damascene process. Finally, 50/300 nm thick Ta/Cu pads, with lateral dimensions on the order of 3 to 5 μm, are patterned right on top of the interlayer vias. Auxiliary Cu pads, which have no role in inter-wafer communication, can also be patterned to increase the total surface area for subsequent Cu–Cu wafer bonding. The size of the pads and their separation are limited by the alignment accuracy.

22.3.3 Wafer-Wafer Alignment and Bonding

The thinned SOI layer after the process steps in Figure 22.5, complete with backside interlayer vias and Cu bond pads, is now ready to be bonded to another CMOS device substrate, presumably also with its own set of multilevel interconnects and matching Cu/Ta bond pads that mirror those of the thinned SOI. The thinned SOI layer is mechanically supported by the handle wafer during subsequent steps.

Both wafer-to-wafer alignment and bonding are performed in a wafer aligner and bonder. Since the system has an inherent 1–3 μm of misalignment, any Cu/Ta bond pads less than or equal to 3 μm across are unacceptable. Thus, wafer-to-wafer alignment is the ultimate factor in determining the interlayer via density. With better optical alignment systems, it is possible to decrease the Cu/Ta pad sizes down

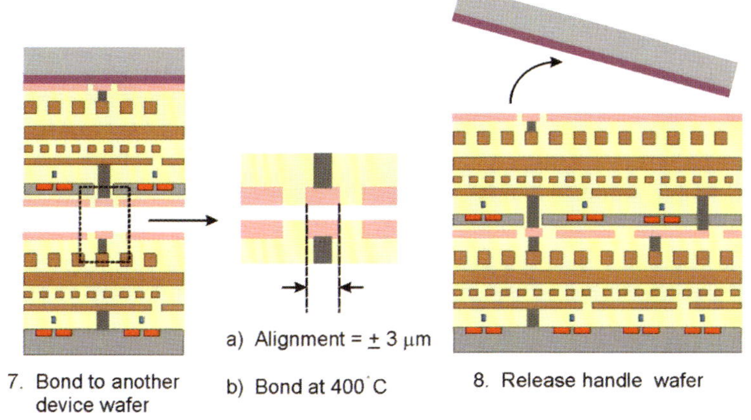

Figure 22.6 Schematic of 3D process flow showing thinned SOI device layer transfer on substrate wafer. From [9] (© 2002 IEEE).

to around 1 μm or so, which corresponds to a substantial increase in via density. When the wafers are properly aligned, the pair is clamped and transferred to the bonding chamber, where both substrates are heated to 300 °C in N_2 ambient and hot-pressed for 30 min under 4000 N of contact force. Further anneal at 400 °C for 30–60 min in N_2 completes the Cu–Cu bond. This process is illustrated in step 7 in Figure 22.6.

22.3.4 Handle Wafer Release

The final step of the 3D integration scheme is releasing the handle wafer from the top SOI layer (step 8 of Figure 22.6 and Figure 22.7).

Recalling from Section 22.3.1, the oxide handle wafer is attached to the SOI substrate with a Zr/Cu–Cu/Zr bond in option 1. To destroy this metal bond, one can immerse the wafer stack in dilute water/HF (10 : 1) solution. Zirconium, like Ti,

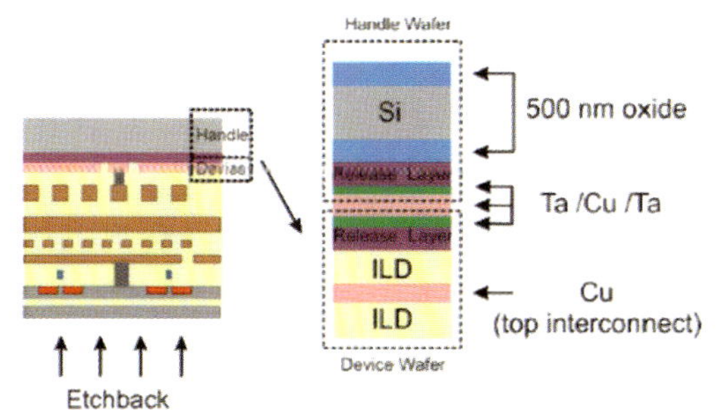

Figure 22.7 Release structure for handle wafer formed by a bonding medium sandwiched between two release layers. From [9] (© 2002 IEEE).

succumbs to HF attack at an extremely high rate, much higher than the rate of SiO_2 degradation. With vigorous agitation, Zr undercutting across the whole wafer can occur, thereby releasing the oxide handler from the finished 3D stack. This release method becomes challenging when the wafer size increases as one need to allow for sufficiently longer time for complete HF encroachment into the Zr layer. Longer soaking times in HF can subject the oxide layers in the 3D stack to undesired corrosion, hence destroying the structures.

In option 2 with Al as the release layer, a mixture of HCl solution at 100 °C is used to destroy the Al release layers. The process begins by removing the Cu/Ta bilayer within the exclusion ring. This can be accomplished by using a dilute HF + nitric acid solution. Once the protective Cu/Ta bilayer in the exclusion ring is removed, hot HCl is used to remove the Al release layer. By switching to hot HCl, one can ensure that the oxide layers on the 3D stack are not attacked by HF excessively.

However, this method proves challenging for thin Al layer on large wafers. Within the first hour of acid release, the initial transient acid encroachment rate (into the Al layer) is fast and furious. However the resulting acid encroachment rate drops and encroachment distance reaches an asymptotic value. Therefore, wafer level handle release with acid encroachment is extremely challenging due to surface tension effects. Although a hot HCl solution can be used to destroy the Al release layers and cause the collapse of the laminate structure, the acid encroachment distance within the release layer cavity is not deep enough to facilitate the separation of the bonded pair. Therefore, this method of wafer release can be improved by: (i) forming channels in the release layers between the dies to assist HCl encroachment, (ii) forming additional release cavities by deep-RIE etching from the top of the handle wafer or (iii) reducing the substrate size further, such as on a die-scale.

When the handle wafer is released from the 3D stack, further metallization can be performed and a progression to a multilayer stack can be accomplished by repeating the layer transfer steps discussed above.

22.3.5
Alternative of Temporary Bonding and Release

Alternatively, an oxide-oxide bond can be used to attach the SOI donor wafer to the handle wafer. This method has the advantage that the bond is materially distinct from the permanent Cu–Cu bond used for electrical and mechanical connectivity.

In this layer transfer method, the finished SOI wafer is coated with 1 μm of Plasma-Enhanced Chemical Vapor Deposition (PECVD) oxide. Since the PECVD oxide is relatively porous, a low-temperature densification is performed at about 400 °C for 12 hours in inert N_2 ambient. This step allows degassing from the porous SiO_2, which is detrimental to the bonding, to take place prior to bonding. Surface smoothness is a very critical factor that determines successful wafer bonding. It is well known that the PECVD oxide has very rough surface. Therefore, it is necessary to smooth the oxide surface by means of chemical mechanical polishing (CMP). It is also essential to ensure a particle-free post-CMP oxide surface originated from the CMP slurry. Since the SOI wafer-thinning step, which is essentially Si etching, comes partly in the form of TMAH chemical etch, the Si handle wafer needs to be protected against TMAH attack by growing a 500 nm of oxide layer on the handle wafer. A high dose and high energy of H_2^+ ions is implanted into the handle wafer with the expected range of about 0.5 μm. Hydrogen plays a role in wafer exfoliation during handle wafer release.

Prior to bonding, the SOI and the handle wafers are chemically cleaned, rinsed in DI water and dried. The two wafers are bonded face-to-face, that is thermal oxide (on handle wafer) and PECVD oxide (on SOI wafer) bonding with no requirement for precision alignment at room temperature. To enhance the bonding strength, the bonded pair is annealed below 300 °C. We have successfully demonstrated the ability to bond thermal oxide to PECVD oxide with careful surface preparation [7]. Note that the annealing temperature must be controlled carefully to prevent premature hydrogen-induced wafer splitting. The SOI wafer is now ready for thinning. Upon formation of vertical electrical vias and Cu pads on the BOX, the SOI wafer is aligned to the bottom wafer (a FEOL-finished wafer with matching Cu bonding pads) and bonded. This can be done at 300 °C and 4000 N for 30 min. Post bonding, the stack is annealed in N_2 ambient for 30 min at 400 °C to allow further Cu grain growth that will enhance the bond. This step also leads to handle wafer release as well. Under this condition, hydrogen will cluster together and cause lateral cracking, hence releasing the handle wafer from the actual 3D device stack. The remaining thin Si (~0.5 μm) can be stripped away easily by a quick (~5 s) dip in TMAH solutions. The concept behind this method is illustrated in the schematic in Figure 22.8. The above has been successfully demonstrated on blanket films. Figure 22.9 shows a cross-sectional SEM of the final 3D stack [8].

In the method described above, there are three temperature steps that must be controlled carefully to provide sufficient temperature window for each process step to take place. One has to ensure that the bonded oxide wafer pair is annealed below the temperature at which hydrogen-induced wafer exfoliation takes place. The annealing temperature of the oxide bonded wafer pair can be lowered substantially by careful surface preparation, such as using plasma activation. The wafer splitting temperature is directly related to the hydrogen implant dose. In the above demonstration, the wafer splitting occurs rapidly at 400 °C. Similarly, one must also ensure that the

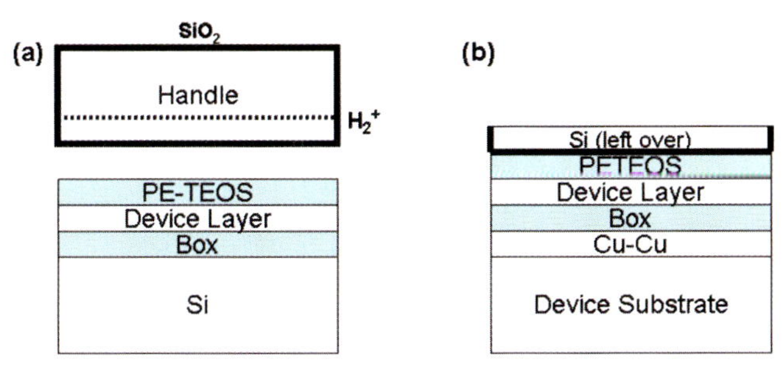

Figure 22.8 Schematics showing the concept of using oxide fusion bond as the temporary bond between SOI wafer and handle wafer, and hydrogen-induced wafer splitting for handle wafer release.

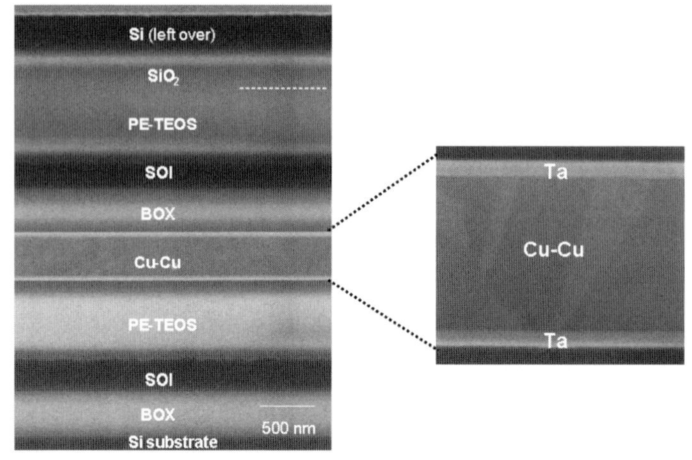

Figure 22.9 An SEM showing the silicon layers stack obtained after handle wafer release. Handle wafer is released at the peak of the implanted hydrogen profile. From [8] (© 2005 IEEE).

permanent bonding of Cu pads between the SOI wafer and substrate wafer is performed at less than 400 °C.

Further discussions related to this 3D flow can be found in References [9–11].

22.4 Discussion

In this section, several salient features of the 3D process flow presented above are discussed.

22.4.1 Metallic Bonding Medium

Using metal as the bonding medium is an attractive choice because the bonding medium acts as an electrical bond to establish a conductive path between active layers in a 3D IC, and as a mechanical bond to hold the active layers together reliably. In the 3D flow presented above, the bonds formed by Cu pads that connect the vertical vias act primarily as electrical bonds while the mechanical bonds are formed by the auxiliary Cu pads. Metallic bonding also inherently allows a "via-first" approach for vertical integration, and hence relaxes the aspect ratio requirement on interlayer vias. At the same time, a metal interface allows additional wiring and routing.

Using metal as bonding interface is also an attractive choice because metal is a good heat conductor and this will help circumvent heat dissipation problems that may be encountered in 3D ICs. Heat generated in the upper layers can be conducted through the metallic bonding layer to the substrate wafer and then to the heat sink. Another advantage offered by a metal bonding interface is that the metal layer can act as a ground shield if properly grounded, hence providing better noise isolation between device layers on the stack.

22.4.2
The Choice of Copper

Copper is a metal of choice because it is a mainstream CMOS material, and it has good electrical ($\rho_{Cu} = 1.7\,\mathrm{m\Omega\,cm}$ vs. $\rho_{Al} = 2.65\,\mathrm{m\Omega\,cm}$) and thermal ($K_{Cu} = 400\,\mathrm{W\,m^{-1}\,K^{-1}}$ vs. $K_{Al} = 235\,\mathrm{W\,m^{-1}\,K^{-1}}$) conductivities and longer electro-migration lifetimes than aluminum-based lines. Most importantly, Cu bonds to itself under the correct conditions.

As pointed out above, the Cu–Cu bonding interface serves both as a "glue" layer and electrical contact points between different device planes. Therefore, reliability of the Cu–Cu bond, in terms of both its mechanical and electrical properties, is of utmost importance. There are several methods to examine the bond strength of bonded Cu layers, including tensile pull test, four-point bending test [12], razor blade test [1], die sawing test [13] and ultimately the wafer thinning test [5]. Wafer pairs bonded at temperatures between 300 and 400 °C can withstand die sawing (1 cm × 1 cm size) and wafer thinning (a combination of grinding and TMAH etch). Bond uniformity across the wafer improves when bonding is performed at higher temperature. Several wafer conditions can adversely affect the bond, including wafer bow, surface particles and surface oxidation.

Initial results on determining the Cu–Cu interface specific contact resistance have shown great promise, but more work is needed to fully characterize the electrical properties of the bond [14].

Unlike oxide fusion bonding, which can occur at room temperature by initially forming hydrogen bonds, Cu thermo-compression bonding does not occur at room temperature under normal conditions. Copper layers bond under "harsh" conditions, by applying contact force and heating up to 300–400 °C. When the wafer pair is cooled to room temperature, large tensile stress is induced in the Cu layer due to thermal mismatch. Interfacial voids have been observed in bonded blanket Cu layers [15]. The interfacial voids are close to the original bonding interface, as shown in the SEM image in Figure 22.10. Thermal stress analysis predicted tensile stresses higher than the yield stress in the bonded Cu layer at room temperature. Under such stress conditions, the bonded Cu layer deforms and stress relaxation can result in interfacial voids. Formation of interfacial voids can seriously degrade electrical and mechanical properties of the bonded Cu layer. This points to the need for careful

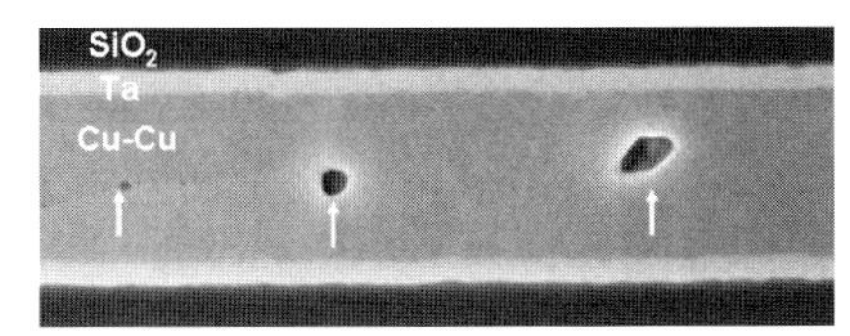

Figure 22.10 Observation of voids in the bonded Cu layer. This layer was bonded at 300 °C for 1 hour.

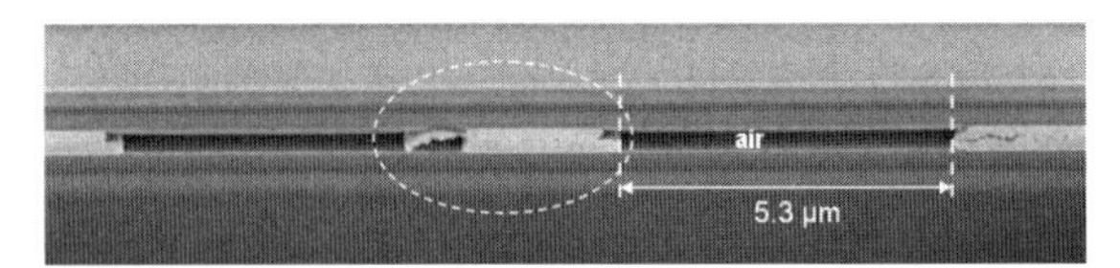

Figure 22.11 Cross sectional SEM image shows ~2.0 μm bonded Cu lines. This SEM shows that it is possible to achieve a misalignment of <1.0 μm with current set-up. From [16] (© 2007 Materials Research Society).

study of the origin of the void formation so that counter-measures can be implemented.

Since a Cu–Cu bond does not form instantaneously at room temperature like in the case of oxide fusion bonding, an aligned wafer pair must be mechanically clamped and transferred into the bonding chamber, depending on the set-up. This transfer can cause the alignment to change. During heating, top and bottom wafers, especially if they are not symmetrical in terms of materials and structural designs, can expand differently, hence introducing additional errors to the alignment accuracy.

Since Cu is a conductive medium, a continuous Cu bonding layer between active layers is of no practical application. In an actual multilayer 3D ICs implementation having Cu as the bonding medium, Cu bonding should be done in the form of pad-to-pad or line-to-line bonding with proper electrical isolation. Figure 22.11 shows a cross section of Cu lines (2.0 μm) that are successfully bonded [16]. The spacing between bonded lines is 5.3 μm and it is filled with air. Interfacial voids are observed in the bonded lines and they can lead to serious reliability concerns. The bonding process should be optimized to minimize the formation of void. Another reliability concern is the empty space between the bonded lines that might reduce mechanical support between the active layers. Moisture in the empty space can also potentially corrode the bonded Cu lines. One solution is to form damascene Cu lines and to perform hybrid bonding of Cu and dielectric such as oxide.

22.4.3 Back-to-Face Bonding Orientation

The MIT's 3D process is based on a modular flow where one can construct a multilayer stack by successive bonding of device layer on top of an existing vertical base stack. The process is also designed such that exposure of the continually-growing base stack to mechanical or chemical attack is minimized. In this back-to-face bonding fashion, the SOI wafer is thinned and bonded to the substrate wafer, hence eliminating the potential damage from the SOI thinning step to the whole 3D base stack. The process allows devices from the "top" substrate to be made from any given material or technology, including strained-Si, III–Vs, or nanostructures that can tolerate the required bonding temperatures.

Owing to the need to handle an ultra-thin layer before it is laid on the substrate, a temporary bond to a handle wafer and subsequent release is required. Two options are presented in Section 22.3. When using metallic release layer and chemical release agent, one needs to identify a suitable chemical solution with the required selectivity. Ideally, the release agent must be able to remove the release layer at a much faster rate than the bonding layer. This method is attractive because the release temperature is usually at most 100 °C and it is possible to recycle the handle wafer. However, this method has proven challenging when releasing larger wafer sizes, hence limiting its use to smaller sample size at the die level. In contrast, the release method using hydrogen-induced wafer splitting is an attractive method in that the release mechanism takes place across the entire wafer almost simultaneously and it is possible to recycle the handle wafer. However, one must keep the Cu permanent bonding temperature low enough so as not to cause premature wafer splitting.

22.5 Summary

This chapter describes a method of forming multilayer ICs stack. The method is based on low-temperature thermo-compression bonding of wafer coated with Cu thin films. To achieve thin device layer for transfer, SOI wafers are used as the donor wafers. Since the top device layer is transferred on the substrate facing up, a temporary bond to a handle wafer and subsequent release is needed. Several salient features of this 3D flow are highlighted and discussed.

References

1 Fan, A., Rahman, A. and Reif, R. (1999) Copper wafer bonding. *Electrochemical and Solid-State Letters*, **2** (10), 534.

2 Holloway, K. and Fryer, P.M. (1990) Tantalum as a diffusion barrier between copper and silicon. *Applied Physics Letters*, **57** (17), 1736.

3 Chen, K.N., Fan, A., Tan, C.S. *et al.* (2002) Microstructure evolution and abnormal grain growth during copper wafer bonding. *Applied Physics Letters*, **81** (20), 3774.

4 Chen, K.N., Fan, A. and Reif, R. (2001) Microstructure examination of copper wafer bonding. *Journal of Electronic Materials*, **30** (4), 331.

5 Tan, C.S. and Reif, R. (2005) Multi-layer silicon layer stacking based on copper wafer bonding. *Electrochemical and Solid-State Letters*, **8** (6), G147.

6 Chen, P.H., Peng, H.Y., Hsieh, C.M. and Chyu, M.K. (2001) The characteristic behavior of TMAH water solution for anisotropic etching on both silicon substrate and SiO_2 layer. *Sensors and Actuators, A* **93**, 132.

7 Tan, C.S., Fan, A., Chen, K.N. and Reif, R. (2003) Low-temperature thermal oxide to plasma-enhanced chemical vapor deposition oxide wafer bonding for thin-film transfer application. *Applied Physics Letters*, **82** (16), 2649.

8 Tan, C.S., Chen, K.N., Fan, A. and Reif, R. (2005) A back-to-face silicon layer stacking for three-dimensional integration. IEEE International SOI Conference, Honolulu, HI, October 3–6, pp. 87–89.

9 Reif, R., Fan, A., Chen, K.N. and Das, S. (2002) Fabrication technologies for three-dimensional integrated circuits. Proceedings of the International Symposium on Quality Electronic Design (ISQED), 33, San Jose, CA.

10 Reif, R., Tan, C.S., Fan, A. *et al.* (2003) 3-D interconnects using Cu wafer bonding:

technology and applications. Advanced Metallization Conference 2002, Materials Research Society, pp. 37.

11 Reif, R., Tan, C.S., Fan, A. *et al.* (2004) Technology and applications of three-dimensional integration, at 206th Electrochemical Society Fall Meeting, Honolulu, HI, 2004. Dielectrics for Nanosystems: Materials, Science, Processing, Reliability, and Manufacturing, (eds R. Singh, H. Iwai, R.R. Tummala and S. Sun), The Electrochemical Society Proceedings Series, PV 2004-04, The Electrochemical Society, Pennington, NJ.

12 Tadepalli, R. and Thompson, C.V. (2003) Quantitative characterization and process optimization of low-temperature bonded copper interconnects for 3-D integrated circuits. Proceedings of the IEEE International Interconnect Technology Conference, 36, San Francisco, CA. pp. 261–276.

13 Chen, K.N., Tan, C.S., Fan, A. and Reif, R. (2004) Morphology and bond strength of copper wafer bonding. *Electrochemical and Solid-State Letters*, **7** (1), G14.

14 Chen, K.N., Fan, A., Tan, C.S. and Reif, R. (2004) Contact resistance measurement of bonded copper interconnects for three-dimensional integration technology. *IEEE Electron Device Letters*, **25** (1), 10.

15 Tan, C.S., Reif, R., Theodore, D. and Pozder, S. (2005) Observation of interfacial voids formation in bonded copper layer. *Applied Physics Letters*, **87** (20), 201909.

16 Tan, C.S., Chen, K.N., Fan, A. *et al.* (2007) Silicon layer stacking enabled by wafer bonding. MRS Fall Meeting, Boston, MA, November 27–December 1 2006. Enabling Technologies for 3-D Integration. (eds C.A. Bower, P.E. Garrou, P. Ramm and K. Takahashi), Material Research Society Symposium Proceedings, Volume 970. Material Research Society. pp. 193–204.

23
Rensselaer 3D Integration Processes

James Jian-Qiang. Lu, Tim S. Cale, and Ronald J. Gutmann

23.1
Introduction

Rensselaer 3D integration research can be traced back to research on wafer-level integration systems in the 1980s [1], when the wafer size was 4 inches and millimeter-sized through silicon vias (TSVs) were proposed using laser drilling or vertical dopant diffusion. Rensselaer renewed its 3D integration research in 1999, as part of the Interconnect Focus Center (IFC) supported by MARCO, DARPA and NYSTAR. The main goal of this IFC research was to address global interconnect delay issues associated with 2D chips, using IC technologies developed through the 1990s, such as copper damascene patterning. An additional driver is the potential for 3D systems to facilitate the increasing complexity associated with new materials and processing techniques that are needed to increase the performance and functionality of integrated circuits. The RPI 3D group introduced a monolithic wafer-level 3D platform and its processing technology on 8″ wafers [2, 3]. Four key processes, namely, wafer alignment, wafer bonding, wafer thinning and inter-wafer via formation, and two major technology platforms were pursued, namely, "via-last" [2, 3] and "via-first" 3D platforms [4]. Many academic and industrial research groups are developing 3D technologies (see other chapters of this book); this chapter summarizes the research on major 3D platforms and technologies developed at Rensselaer and in collaborations with research partners.

23.2
Via-Last 3D Platform Using Adhesive Wafer Bonding and Cu Damascene Inter-Wafer Interconnect

Figure 23.1 shows a representation of a via-last platform for monolithic 3D hyper-integration,[1] which uses a wafer aligner for wafer-to-wafer alignment, a dielectric

1) The term "hyper-integration" means to integrate various materials, processing technologies, and functions beyond the ultra-large scale integration (ULSI) or gigascale integration.

Handbook of 3D Integration: Technology and Applications of 3D Integrated Circuits.
Edited by Philip Garrou, Christopher Bower and Peter Ramm

ISBN: 978-3-527-32034-9

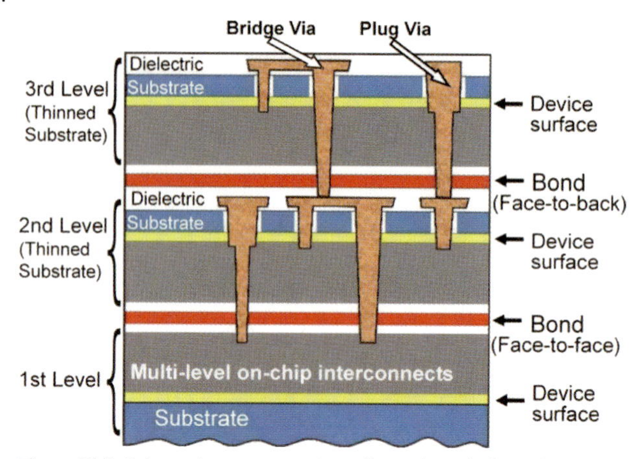

Figure 23.1 Schematic representation of a via-last platform for monolithic 3D hyper-integration using adhesive wafer bonding and Cu damascene inter-wafer interconnect, showing bonding interface, vertical inter-wafer vias (plug- and bridge-type), and “face-to-face” and “face-to-back” bonding [2].

bonding adhesive for wafer-to-wafer attachment, a three-step thinning process for thinning the “top” wafer, and copper damascene patterning for inter-wafer interconnection [2, 3]. To start a 3D IC using this platform, two processed wafers with active device layers and multilevel on-chip interconnects are aligned to tolerances within 1 μm and bonded using a dielectric adhesive under conditions compatible with CMOS processing and packaging. The top wafer in the two-wafer stack is thinned to ≤10 μm by a three-step thinning process, that is backside grinding, CMP and wet etching to an etch stop (e.g. an implanted layer, epi layer, oxide layer, or a buried oxide (BOX) layer in SOI technology). Subsequently, “bridge-type” and/or “plug-type” inter-wafer interconnects are formed using a Cu damascene patterning process. This inter-wafer damascene patterning process, developed jointly by Rensselaer and University at Albany [5], involves high-aspect-ratio (HAR) via etching, Cu/barrier deposition and CMP. Thus, a long distance interconnect that might run a centimeter across a conventional 2D chip may be replaced by a 2–10 μm (vertical length) HAR via between chips. Repeating this process flow, the third wafer (or more) can then be aligned, bonded, thinned, and interconnected. Three key advantages of this approach compared to other bonded-wafer 3D approaches are:

- the ability of the dielectric adhesive to accommodate wafer-level non-planarity (e.g. wafer bow) and particulates at the bonding interfaces;
- no handling wafers are required – thinned silicon is not transferred as in other wafer-level 3D approaches;
- stacks of three or more wafers can be fabricated without changing the processing approach.

Similar to other 3D platforms, there are four key processing challenges, namely wafer-scale alignment accuracy, bonding integrity, wafer thinning and leveling control, and inter-wafer interconnection. Each of these processes must be compatible with semiconductor processing constraints such as pressure and temperature. These processes are summarized in the following section.

23.3 Via-Last 3D Platform Feasibility Demonstration: Via-Chain Structure with Key Unit Processes of Alignment, Bonding, Thinning and Inter-wafer Interconnection

An inter-wafer interconnect via-chain structure was designed to develop 3D unit processes and demonstrate the feasibility of our via-last approach to 3D processing with 200 mm wafers [2, 3]. Figure 23.2 shows a process flow for the via-chain structure fabrication. Conventional Cu damascene BEOL processes are used for Cu metallization on the bottom wafer (M1) and the top wafer (M2), while bridge metal (M3) and inter-wafer vias are fabricated using an inter-wafer damascene process. Based on screening tests of various dielectric bonding adhesives, benzocyclobutene (BCB) was selected as the baseline wafer bonding. BCB is a low-k polymer, which has been used in various semiconductor applications and is compatible with most CMOS processes. Void-free bonding is routinely obtained with BCB on both blanket and patterned wafers, as described in Chapter 13. The three-step thinning process was used to thin the top Si wafer: (1) Si substrate mechanical grinding, (2) Si polishing to remove damage and stresses caused by grinding and reduce top wafer thickness to ~35 μm and (3) wet chemical etch to remove the remaining silicon. Tetramethylammonium hydroxide (TMAH) is used to etch Si due to its compatibility with CMOS processing and excellent etch selectivity for Si relative to SiO_2. The inter-wafer interconnects are formed by deep via etching and clean, chemical vapor deposition

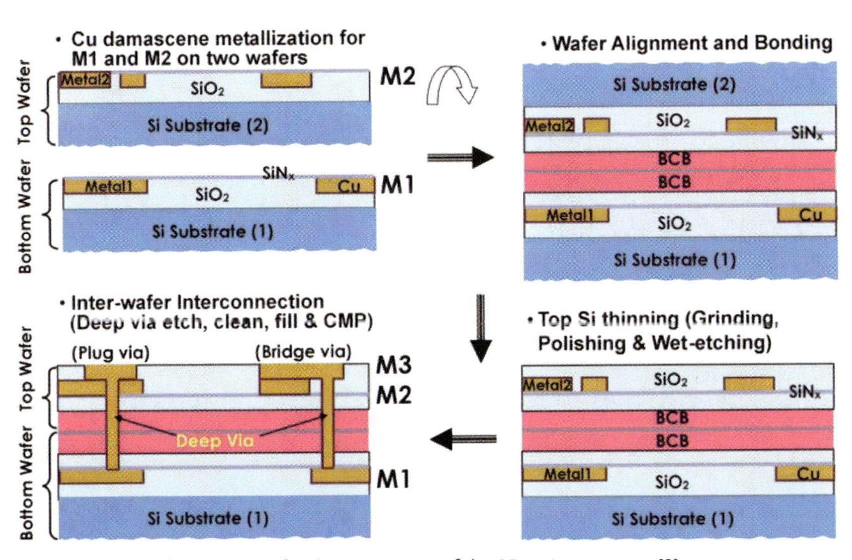

Figure 23.2 Via-chain process for demonstration of the 3D unit processes [2].

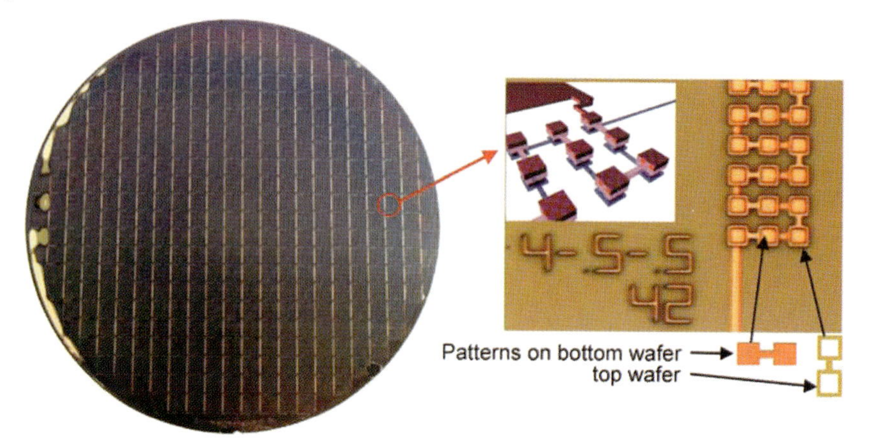

Figure 23.3 Image of 3D via-chain test structures on 200 mm wafer after Cu-damascene patterning on two wafers, wafer alignment and bonding using BCB, Si backside grinding/polishing and wet etching of the top wafer. Irregular patterns near the wafer edge were induced by wafer-scale nonuniformity of the processing tools for oxide etch and Cu sputter deposition [6].

(CVD) of TaN liner, Cu CVD fill and CMP. Additional process information on the 3D via-chain fabrication is presented elsewhere [5].

As shown in Figure 23.3, two Cu damascene patterned wafers were face-to-face aligned, bonded and thinned, revealing wafer-scale and die-scale alignment within 1 μm [6]. These promising results indicate that a very high density of inter-wafer interconnects (via diameter ~2 μm) is feasible with this via-last 3D platform. Thinning of the backside of the top silicon wafer in the bonded wafer stack to ~35 μm by mechanical grinding and CMP showed no difference in the patterned wafer surface quality before and after top silicon wafer thinning [7]. The bonded and

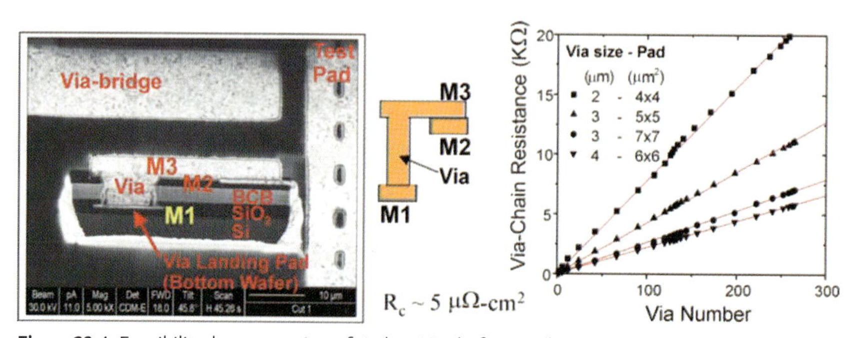

Figure 23.4 Feasibility demonstration of via-last 3D platform with a via-chain structure: (left) a FIB cross-section of the via-chain structure showing Cu metallization on bottom wafer (M1) and top wafer (M2), bridge metal (M3) and the inter-wafer vias; (right) via-chain resistance vs. via-chain length (via number) for nominal via size of 2, 3, 4 μm [5].

thinned wafer pair showed good silicon thickness uniformity without edge cracks, revealing good bonding and thinning integrity of the stacked wafers.

Figure 23.4 shows an FIB cross-section and the via-chain resistances from a functional wafer pair [5]. Though the via-chain specific contact resistance of $\sim 5\,\mu\Omega\text{-cm}^2$ is much larger than expected, the linear relationship between the chain resistance and chain length indicates that continuous and uniform 3D via-chains are demonstrated for nominal inter-wafer via sizes of 2, 3, 4 and 8 μm. The high contact resistance is attributed to inadequate cleaning of via-etch residue prior to metallization.

23.4 Via-First 3D Platform with Wafer-Bonding of Damascene-Patterned Metal/Adhesive Redistribution Layers

A via-first 3D technology platform [4, 8, 9] currently under investigation at Rensselaer employs wafer bonding of damascene-patterned metal/adhesive redistribution layers on two wafers. This approach provides inter-wafer electrical interconnects (via-first) and adhesive bonding of two wafers in one unit processing step. A conceptual schematic of the via-first approach is shown in Figure 23.5. Copper/tantalum (Cu/Ta) and BCB are selected as the metal and adhesive to demonstrate the feasibility of this via-first 3D approach. This diagram shows a damascene patterned Cu/BCB "redistribution layer" over the uppermost metal layer of a second wafer, which is then flipped, aligned and bonded to a patterned Cu/BCB layer on the first

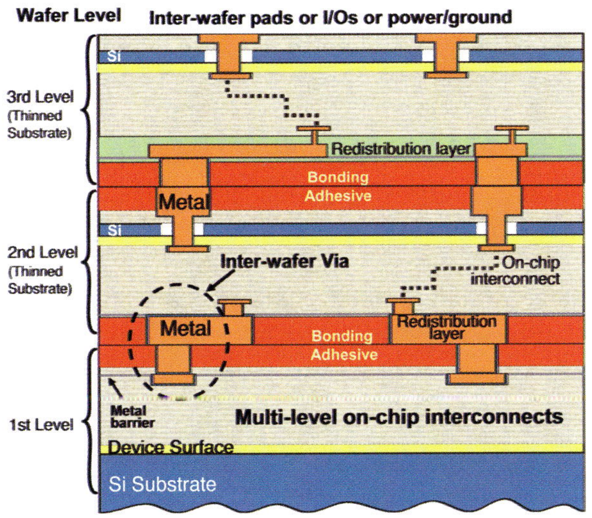

Figure 23.5 Schematic of Cu/BCB redistribution layer bonding for via-first 3D integration (Cu–Cu bonds provide inter-wafer interconnects and BCB–BCB adhesive bonds provide mechanical wafer attachment) with two options of the redistribution layers, as shown on the 2nd and 3rd wafers, respectively [4, 8, 9].

wafer. Note that the patterned Cu/BCB layer on the first wafer can also be a Cu/BCB redistribution layer if needed (not shown in Figure 23.5). The substrate of the face-down bonded second wafer is then thinned. The process can be extended to multiple wafer stacks by etching through the thinned second wafer of the bonded pair to create another damascene patterned layer, which mates with a third wafer. Note that the patterned Cu/BCB layer on the thinned second wafer substrate can also be a Cu/BCB redistribution layer if needed (not shown in Figure 23.5).

Moreover, an extra Cu/oxide (or Cu/BCB, or other metal/dielectric) redistribution layer, for example that over the uppermost metal layer of the third wafer as shown in Figure 23.5, can be added prior to patterning process of any Cu/BCB bonding layer. This simplifies the patterning of the Cu/BCB bonding layer because only Cu bonding posts (vias) are needed, and offers a simple bonding scheme. That is, with minimal misalignment, one always bonds Cu posts to Cu posts and BCB field to BCB field. This minimizes undesirable contact (i.e. bonding) of long Cu lines with BCB field – see the bonding layers between second and third wafers shown Figure 23.5. This extra redistribution layer provides additional redistribution capability compared with the approach of combining Cu bonding vias and the redistribution layer (e.g. the Cu/BCB redistribution layer on front side of the second wafer as shown in Figure 23.5).

This via-first approach using wafer bonding of damascene-patterned metal/adhesive redistribution layers provides:

- both electrical and mechanical inter-wafer connections/bonds (combining advantages of both BCB/BCB and Cu/Cu bonding);
- thermal management options: Cu/BCB redistribution layers can serve as a thermal conductor and/or spreader (with large percentage of Cu area), or as a thermal insulator (with large percentage of BCB area), or as a thermal conductor for some selected areas and a thermal insulator for other areas;
- high inter-wafer interconnectivity while allowing larger alignment tolerance by eliminating deep inter-wafer vias;
- redistribution layers for inter-wafer interconnect routing for wafers, on which the inter-wafer interconnect pads are not matched, which further reduces the process flow and is compatible with wafer-level packaging (WLP) technologies.

This approach is attractive for applications of monolithic wafer-level 3D integration (e.g. 3D interconnect, 3D ICs, wireless, and smart imagers, etc.) as well as wafer-level packaging, mechanical-electrical micro-systems (MEMS), optical MEMS, bio-MEMS and sensors.

The bonding process for such a technology platform is challenging, as various surfaces are exposed, including the dielectric adhesives, diffusion barriers and electrical conductors. Ideally, all should be capable of being bonded to one another without interfering with the electrical characteristics of the Cu-to-Cu interconnection. Surface preparation techniques for improving adhesion of BCB to Si, Si_3N_4 and Cu have been discussed in the literature [10], but not with respect to wafer bonding. Further, wafer bonding of soft-baked BCB is well documented for use in 3D applications (see Chapter 13), as well as damascene patterning of Cu in fully cured BCB [11]. Considering these factors, a partially-cured BCB layer offers the best compromise between patterning capability for Cu/BCB redistribution layer and bond quality of BCB-to-BCB bonding.

23.5 Via-First 3D Platform Feasibility Demonstration: Via-Chain Structure with Cu/BCB Redistribution Layers

Fabrication of resistive via chains using the via-first redistribution layer approach to 3D integration, which is reported in Reference [4], is summarized here. First, a layer of damascene patterned Cu in BCB was fabricated on each of two 200 mm Si wafers as follows. After 2 μm thick thermal oxide was grown, BCB was spun onto these substrates to a nominal thickness of 1.2 μm, and partially cured in a coat/bake track modified to include a nitrogen purge. The BCB partial cure temperature was 250 °C and hold time was 60 s, resulting in ~55% BCB crosslinking. The partially cured BCB was photolithographically patterned in an i-line stepper, and then etched in an inductively coupled plasma (ICP) etcher using C_4F_8 and oxygen as reactive species. A Ta liner and Cu were sputtered over the patterned BCB at a low power level suitable for deposition over a polymer dielectric. Chemical-mechanical planarization (CMP) was carried out on this film stack using commercially available slurries and pads until the pattern was well defined. Post CMP brush cleaning was done with deionized water and PVA brushes. These wafers, with single-level damascene patterned redistribution layers, were then aligned and subsequently bonded in a vacuum chamber. The bond process included a mechanical downforce of 10 000 N, and a temperature ramp to 250 °C and soak for 60 min, followed by a further ramp to 350 °C and soak for 60 min, and finally cooling to room temperature. One of the bonded wafers was thinned to a nominal thickness of 50 μm through grinding and polishing, followed by a wet chemical etch in TMAH, which has high Si-to-SiO_2 selectivity, to completely remove the remaining silicon.

After the above fabrication procedure, the bonded pair was sectioned, allowing a lot split for characterization of structural and electrical properties. The structures were observed using cross-sectional focused ion beam/scanning electron microscopy (FIB/SEM). One resulting cross-section of the bonded area near the wafer center is depicted in Figure 23.6 [4, 8]. This figure shows a bonded Cu-to-Cu interface, a bonded BCB-to-BCB interface and a Cu-to-BCB interface that appears to be in intimate contact, but may not be well-bonded across the wafer. For electrical characterization, the surface oxide (originally the isolation layer under the Cu/BCB redistribution layer) is removed to allow access to the via-chains. A two-point probing setup was used to measure electrical resistance on several inter-wafer vias, and optical microscopy was used to measure the overlap area. The contact resistance was on the order of $1 \times 10^{-7}\,\Omega\,cm^2$.

Mechanical characterization on bonded wafers using partially-cured BCB shows a bond strength in a range of 15–30 J m^{-2}, depending on the processing conditions. The

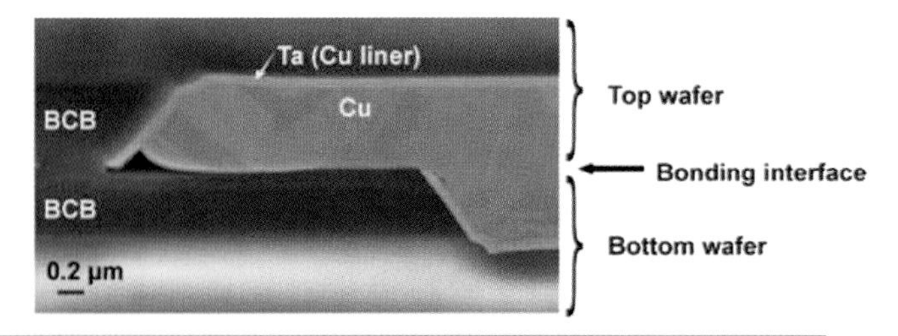

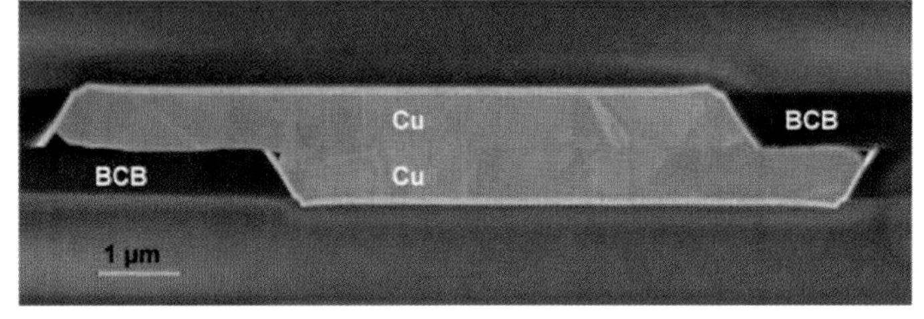

Figure 23.6 Cross-sectional FIB/SEM image of bonded damascene patterned Cu/BCB wafers showing well bonded Cu-to-Cu, BCB-to-BCB interfaces [4, 8].

bond strength in our via-last 3D baseline processing is close to 32 J m^{-2} as found when using soft-baked BCB (see Chapter 13). Even the lowest bond strength is about a factor of 3 greater than needed for damascene-patterned IC interconnects (5 J m^{-2}). These results are quite promising, and demonstrate the feasibility of this via-first 3D approach. However, key challenges need to be overcome, and are currently under investigation. These challenges include:

1. optimizing the BCB partial-curing process, to provide sufficient bond strength and serve as a damascene patterning-compatible dielectric material for Cu/BCB redistribution layer fabrication;
2. achieving wafer-level, feature-scale planarity of the Cu/BCB redistribution layer fabricated by single-level chemical-mechanical planarization (CMP);
3. developing post-CMP surface treatment and wafer bonding schemes to facilitate bonding of Cu-to-Cu, BCB-to-BCB and Cu-to-BCB over the entire wafer in one unit process step;
4. lowering the electrical resistivity of the inter-wafer Cu interconnection;
5. developing reliable redistribution layer design and fabrication protocols.

23.6 Unit Process Advancements

23.6.1 Wafer-to-Wafer Alignment

Precise alignment on the wafer level is one of the key challenges affecting the performance of 3D inter-wafer interconnects. Since 1999, we have closely collaborated

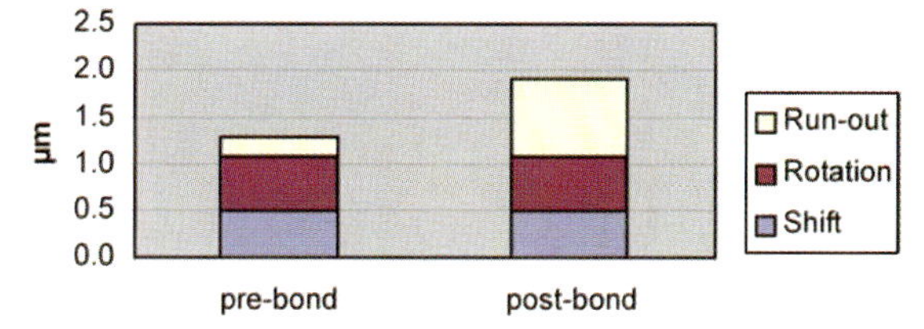

Figure 23.7 Pre-bond vs. post-bond alignment accuracy – 20 tests using 8'' Si wafers [12].

with EVGroup to develop and improve our wafer alignment accuracy. The world's-first SmartView aligner was manufactured and installed at Rensselaer in 2000. Various approaches have been investigated to improve the alignment accuracy [12–14], including alignment mechanisms, alignment key design and thermal mismatch control to improve the mechanical alignment accuracy, partially-cured BCB bonding to improve the bonding-induced misalignment [13], and keyed structures for fine alignment to reach sub-micron alignment accuracy [14].

Figure 23.7 shows a few factors that affect alignment accuracy: 20 pre-bond and post-bond alignments of Si-to-Si wafers (200 mm) were analyzed to determine the contributions of the shift, rotation and run-out between these two wafers on the alignment accuracy [12]. Since the post-bond run-out exhibits a large effect on the alignment accuracy, several measures have been taken to control the post-bond run-out. For example, fine control of the temperatures on both wafers and/or keyed structures help reduce the post-bond run-out [13, 14].

23.6.2
Adhesive Wafer Bonding

Adhesive wafer bonding has been intensively investigated as described in Chapter 13. Key collaborations with SEMATECH and Freescale demonstrated that our BCB bonding and wafer thinning processes do not affect Cu/low-k interconnect structures or 130 nm technology node CMOS devices and circuits [15–17]. Figure 23.8 shows a FIB image near the saw path of a CMOS SOI wafer after a double bonding and thinning process, BCB ashing and sawing [16]. While additional research and development is needed to demonstrate the robustness of inter-wafer interconnected wafers, these research collaborations with industrially supplied and characterized test wafers have demonstrated process potential.

23.6.3
Oxide-to-Oxide Bonding

Oxide-to-oxide bonding has been intensively studied for various purposes, such as for 3D integration, as described in Chapters 11 and 20. We have investigated oxide-to-oxide wafer bonding for 3D integration and for bonded silicon-based lasers. Compared to other bonding methods that we have investigated (e.g. adhesive, Cu, or Ti based bonding), preliminary results show that oxide-oxide bonding requires

Figure 23.8 FIB cross section of bonded CMOS SOI wafer using BCB; note that (1) the silicon substrate of the CMOS SOI was completely removed and (2) another silicon wafer bonded on top of the CMOS SOI wafer was also completely removed and bonding adhesive BCB was ashed, the CMOS devices and circuits were electrically tested [16].

extremely smooth and clean oxide surfaces and special surface treatments. Thermal oxide bonding requires thermal annealing at very high temperatures ($\sim$1000 °C). While PETEOS oxide with careful surface treatment (e.g. fine CMP, chemical or plasma surface activations) is favorable for low temperature bonding, outgassing from the PETEOS during bonding is a concern because it might be responsible for low bonding yield.

23.6.4
Copper-to-Copper Bonding

In addition to the Cu bonding research summarized in Figure 23.6, other Cu bonding approaches have been investigated at Rensselaer. The SEM image in Figure 23.9 shows excellent bonding between a 4 μm Cu interconnect post and Cu pad, obtained in collaboration with IBM [18]. When proper pattern design (e.g. size and density) and bonding conditions (e.g. temperature ramp-up control) are utilized, bonding failures are not observed after a dicing test [18].

In an attempt to lower the Cu bonding temperature, Cu nanorods were used as the Cu bonding interface [19]. This was motivated by the lower sintering temperature of Cu nanorods relative to Cu films. Figure 23.10 shows a SEM top view of Cu nanorod arrays and cross-sectional FIB/SEM image of the bond region after bonding wafers that are covered by Cu nanorods. The bonding was performed at 400 °C under a downforce of 10 000 N over the 200 mm wafer pair. We have performed Cu nanorod based bonding at temperatures ranging from 200 to 400 °C in combination with external pressure (i.e. a downfore of 10 000 N over the 200 mm

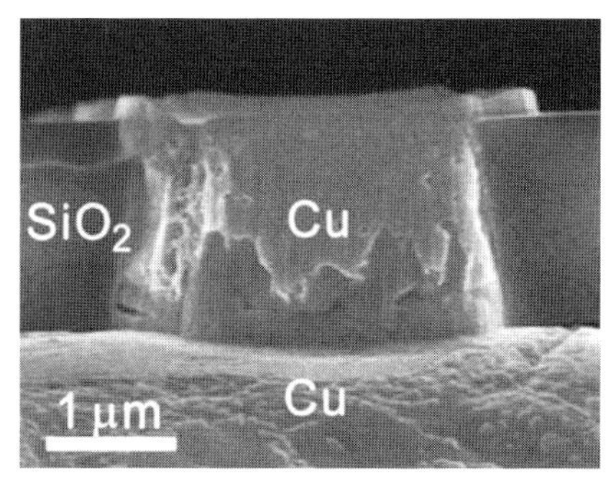

Figure 23.9 SEM image of a 4 μm Cu interconnect bonded to Cu pad, showing a high quality bonding interface [18].

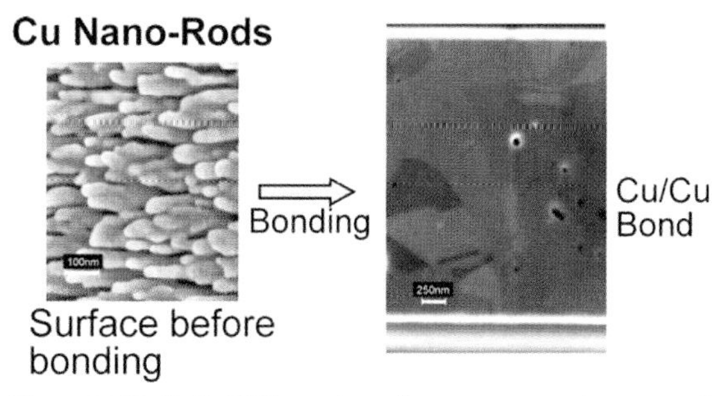

Figure 23.10 (Left) SEM top view of copper nanorod arrays and (right) cross-sectional FIB/SEM image of copper nanorods bonded at 400 °C [19].

wafer pair). The microstructure evolution shows that the bonding structure is approaching densification with increased bonding temperature, and the interface is eliminated at a bonding temperature of 400 °C. Enhancement of the sintering behavior of Cu nanorod arrays using an external pressure was demonstrated at 200 °C; the Cu nanorod arrays did not sinter at 200 °C without extermal pressure [19].

23.6.5
Titanium-Based Wafer Bonding

To further explore BEOL-compatible metal-based wafer bonding, and understand the fundamental bonding mechanisms, we investigated wafer bonding with various metals. Very promising wafer bonding results were achieved using titanium (Ti) as the bonding intermediate for Ti/Ti, Ti/Si, Ti/SiO_2 wafer bonding [20, 21]. As shown in Figure 23.11, strong and nearly void-free wafer bonds at Ti/Si interfaces are obtained at 300 to 450 °C. Titanium plays important roles in overcoming kinetic barriers associated with low-temperature wafer bonding. Strong Ti/Si-based wafer

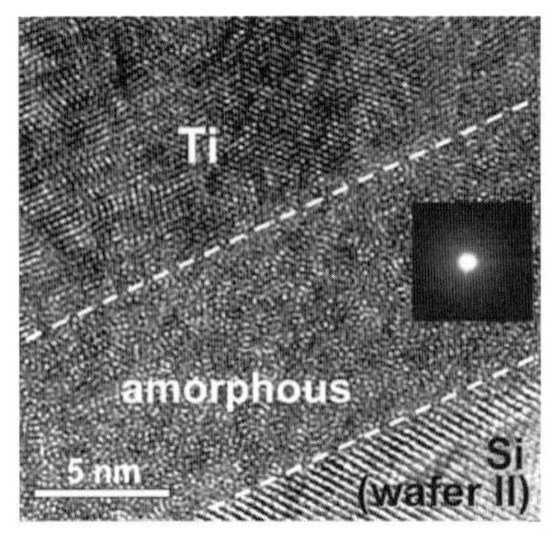

Figure 23.11 Cross-sectional HRTEM micrograph of a Ti/Si wafer pair bonded at 400 °C, showing an approximately 8 nm thick amorphous layer at the bonding interface. The inset is the halo pattern from the amorphous layer under diffraction mode [20].

bonding at $T \leq 450\,^{\circ}\mathrm{C}$ is attributed to a solid-state amorphization. The promising results offer new opportunities for applications in microelectromechanical systems (MEMS) and three-dimensional integrated circuits (3D ICs) because Ti is widely used in the IC manufacturing, and has excellent bio-compatibility with tissue, bone and blood. Thus, Ti-based bonding appears quite attractive for 3D integration of bio-MEMS devices.

23.7 Carbon Nanotube (CNT) Interconnect

Much attention has been focused recently on replacing copper with carbon nanotube (CNT) bundles for interconnect applications due to CNTs' excellent electrical properties. They are among the few materials that could exceed the $1 \times 10^{7}\,\mathrm{A\,cm^{-2}}$ current density that the ITRS predicts will be needed at the 32 nm generation; they appear to exhibit ballistic conduction, whereby electron scattering from defects or impurities is absent or negligible. A key requirement for use of CNT as an interconnect alternative is that CNTs must be closely-packed to reach a conductivity beyond that of copper [22, 23].

A technique to grow closely-packed CNT bundles has proven elusive, so we focused instead on densifying bundles that have already been grown. We demonstrated a post-growth processing method to significantly increase the site density, and thereby further reduce the resistance, of CNT bundles [23]. This is a critical step towards CNT interconnect for use in future ICs. The densified CNTs can be further processed to cut off the ends of the densified ones through mechanical polishing. These as-cut bundles can be used as basic building blocks for IC interconnects. As an example, Figure 23.12 shows a schematic process flow to make such densified bundles and

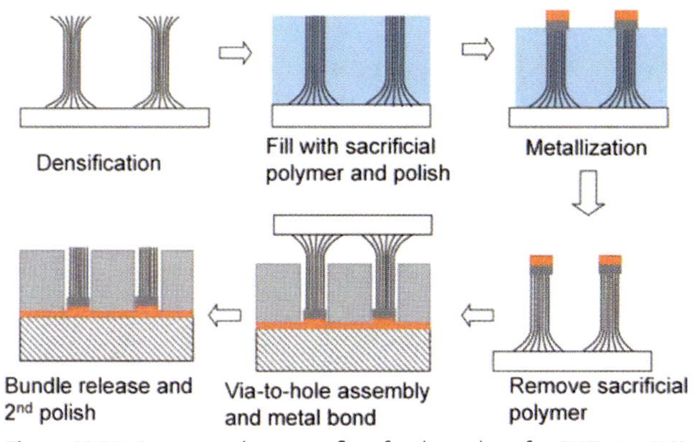

Figure 23.12 A proposed process flow for through-wafer CNT vias [23].

their implementation in 3D IC integration as through-wafer vias for electrical and/or thermal paths.

In the densification process, CVD-grown, vertically oriented CNTs are immersed in an organic solvent, and then withdrawn from the liquid bath by solvent evaporation. Consequently, the individual nanotubes are compacted by aggregating into higher-density bundles by capillary coalescence, and remain together by van der Waals forces. Figure 23.13 shows SEM images of CNT bundles before and after such a densification process. As reported in Reference [23] this technique can increase the density of the bundles by a factor of 5×–25×, depending on bundle height, diameter, pitch and specific CNT properties. Using CMP to remove the ends of the densified CNTs creates cylindrical or nail-shaped bundles that can be used as building blocks for CNT interconnects because denser CNTs means more conduction channels are available, and thus less resistance is expected.

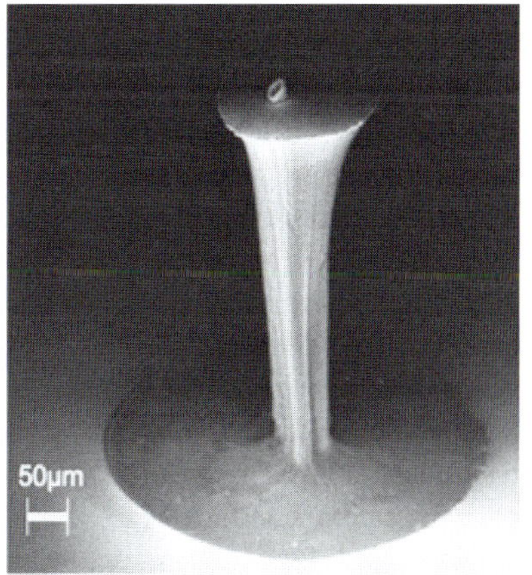

Figure 23.13 CNT bundles (left) before and (right) after densification [23].

23.8
Summary

Various unit processes developed and characterized at Rensselaer (and collaborating organizations) have been described. Our main focus has been BCB bonding with Cu inter-wafer interconnects (i.e. both TSVs and vias at the bonding interface), fabricated with both via-last and novel via-first process flows. While the via-last process flow has established BCB as a viable bonding adhesive and simplifies the bonding process, the via-first process flow is simpler while incorporating an inter-wafer redistribution layer. Alternative inter-wafer interconnect technologies described (including Cu nanorods, Ti, and CNTs) offer promise of improved electrical, thermal and mechanical properties.

Acknowledgments

Rensselaer 3D integration research programs were supported by DARPA, MARCO and NYSTAR through the Interconnect Focus Center, as well as by SRC, EVGroup, IBM, Freescale and SEMATECH. We gratefully acknowledge the contributions from many colleagues associated with the 3D group at Rensselaer and from our collaborators at the University at Albany, EVGroup, IBM, Freescale Semiconductor and SEMATECH.

References

1 Tewksbury, S.T. (1989) *Wafer-Level Integrated Systems: Implementation Issues*, Springer.
2 Lü, J.-Q., Kumar, A., Kwon, Y. *et al.* (2001) 3D Integration using wafer bonding. *Advanced Metallization Conference 2000 (AMC 2000)*, (eds D. Edelstein, G. Dixit, T. Yasuda and Y. Ohba), MRS Conference Proceedings Series, Volume 16, Material Research Society, pp. 515–521.
3 Lu, J.-Q., Cale, T.S. and Gutmann, R.J. (August 2005) Wafer-level three-dimensional hyper-integration technology using dielectric adhesive wafer bonding. in: Materials for Information Technology: Devices, Interconnects and Packaging, (eds E. Zschech, C. Whelan and T. Mikolajick), Springer-Verlag Ltd, London, pp. 386–397.
4 McMahon, J.J., Lu, J.-Q. and Gutmann, R.J. (May 31–June 3 2005) Wafer bonding of damascene-patterned metal/adhesive redistribution layers for via-first three-dimensional (3D) interconnect. IEEE 55th Electronic Components and Technology Conference (ECTC 2005), IEEE, pp. 331–336.
5 Lu, J.-Q., Jindal, A., Kwon, Y. *et al.* (2003) 3D system-on-a-chip using dielectric glue bonding and Cu damascene inter-wafer interconnects. International Symposium on Thin Film Materials, Processes, and Reliability, (eds G.S. Mathad, T.S. Cale, D. Collins, M. Engelhardt, F. Leverd and H.S. Rathore), The electro-chemical society, PV2003-13, pp. 381–389.
6 Lu, J.-Q., Rajagopalan, G., Gupta, M. *et al.* (2004) Planarization issues in wafer-level 3D integration. Advances in Chemical-Mechanical Polishing, MRS Symposium Proceedings, Volume 816 pp. 217–228.

7 Lu, J.-Q., Kwon, Y., Jindal, A. *et al.* (2003) Dielectric glue wafer bonding and bonded wafer thinning for wafer-level 3D integration. Semiconductor Wafer Bonding VII: Science, Technology, and Applications, (eds F.S. Bengtsson, H. Baumgart, C.E. Hunt and T. Suga), ECS, PV 2003-19, pp. 76–86.
8 Lu, J.-Q., McMahon, J.J. and Gutmann, R.J. (2006) Via-first inter-wafer vertical interconnects utilizing wafer-bonding of damascene-patterned metal/adhesive redistribution layers. Proceedings CD of 3D Packaging Workshop at IMAPS Device Packaging Conference, Scottsdale, AZ, IMAPS, pp. 148.
9 Gutmann, R.J., McMahon, J.J. and Lu, J.-Q. (2006) Damascene patterned metal/ adhesive redistribution layers. Enabling Technologies for 3-D Integration, (eds. C.A. Bouer, P.E. Garrou, P. Ramm, K. Takahashi, MRS Symposium Proceedings Volume 970, MRS, pp. 206–214.
10 Garrou, P., Scheck, D., Im, J.-H. *et al.* (August 2000) Underfill adhesion to BCB (Cyclotene™) bumping and redistribution dielectrics. *IEEE Transactions on Advanced Packaging*, **23** (3), 568–573.
11 Price, D.T., Gutmann, R.J. and Murarka, S.P. (1997) Damascene copper interconnects with polymer ILDs. *Thin Solid Films*, **308–309**, 523–528.
12 Wimplinger, M., Lu, J.-Q., Yu, J. *et al.* (2004) Fundamental limits for 3D wafer-to-wafer alignment accuracy, Materials, Technology, and Reliability for Advanced Interconnects and Low-k Dielectrics. (eds R.J. Carter, C.S. Hau-Riege, G.M. Kloster, T.-M. Lu and S.E. Schulz2004 MRS Proceedings Volume 812, pp. F6.10.1–F6. 10.6.
13 Niklaus, F., Kumar, R.J., McMahon, J.J. *et al.* (Feb 21 2006) Adhesive wafer bonding using partially cured benzocyclobutene (BCB) for three-dimensional integration. *Journal of The Electrochemical Society*, **153** (4), G291–G295.
14 Lee, S.H., Niklaus, F., McMahon, J.J. *et al.* (2006) Fine keyed alignment and bonding for wafer-level 3D ICs, Materials, Technology and Reliability of Low-k Dielectrics and Copper Interconnects. (eds T.Y. Tsui, Y.-C. Joo, A.A. Volinsky, M. Lane and L. MichaelsonMaterial Research Society Proceeding Volume 914, pp. 0914–F10-05.
15 Lu, J.-Q., Jindal, A., Kwon, Y. *et al.* (June 2003) Evaluation procedures for wafer bonding and thinning of interconnect test structures for 3D ICs. 2003 IEEE International Interconnect Technology Conference (IITC), IEEE, pp. 74–76.
16 Gutmann, R.J., Lu, J.-Q., Pozder, S. *et al.* (2003) A wafer-level 3D IC technology platform. Advanced Metallization Conference in 2003 (AMC 2003), (eds G.W. Ray, T. Smy, T. Ohta and M. Tsujimura), MRS Proceedings, pp. 19–26.
17 Pozder, S., Lu, J.-Q., Kwon, Y. *et al.* (June 2004) Back-end compatibility of bonding and thinning processes for a wafer-level 3D interconnect technology platform. 2004 IEEE International Interconnect Technology Conference (IITC04), IEEE, pp. 102–104.
18 Chen, K.-N., Lee, S.H., Andry, P.S. *et al.* (Dec 2006) Structure design and process control for Cu bonded interconnects in 3D integrated circuits. 2006 IEEE International Electron Devices Meeting (IEDM 2006), San Francisco, CA, IEEE (2006), pp. 367–370.
19 Wang, P.-I., Karabacak, T., Yu, J. *et al.* (2006) Low temperature copper-nanorod bonding for 3D integration. Enabling Technologies for 3-D Integration, (eds. C.A. Bouer, P.E. Garrou, P. Ramm, K. Takahashi, MRS Symposium Proceedings Volume 970, MRS, pp. 225–230.
20 Yu, J., Wang, Y., Lu, J.-Q. and Gutmann, R.J. (August 2006) Low-temperature silicon wafer bonding based on Ti/Si solid-state amorphization. *Applied Physics Letters*, **89**, 092104.
21 Yu, J., Wang, Y., Moore, R.L. *et al.* (2007) Low-temperature titanium-based wafer bonding: Ti/Si, Ti/SiO_2, and Ti/Ti. *Journal of The Electrochemical Society*, **154** (1), H20–H25.

22 Naeemi, A., Sarvari, R. and Meindl, J.D. (June 2006) On-chip interconnect networks at the end of the roadmap: limits and opportunities. IEEE 2006 International Interconnect Technology Conference (IITC 2006), Burlingame, CA, IEEE, p. 221–223.

23 Liu, Z., Bajwa, N., Ci, L. *et al.* (June 2007) Densification of carbon nanotube bundles for interconnect application. IEEE 2007 International Interconnect Technology Conference (IITC 2007), Burlingame, CA, IEEE, pp. 201–203.

24
3D Integration at Tezzaron Semiconductor Corporation

Robert Patti

24.1
Introduction

Many different bonding processes can be used for 3D integration, each of which carries its own problems and advantages. Processes generally fall into four basic types: (i) oxide or dielectric bonding, (ii) glues such as BCB, (iii) solder attach and (iv) metal bonding. Some processes employ combinations of these types, notably the DBI process from Ziptronix and the BCB/copper process developed at RPI. Tezzaron's process uses metal bonding – specifically, copper-to-copper.

Another 3D process distinction is chip-to-wafer vs. wafer-to-wafer. For all types of bonding, wafer-to-wafer processing provides a lower cost than chip-to-wafer. Chip-to-wafer does offer the benefits of a Known Good Die (KGD) approach, but KGD only works if the individual dies can be tested. Where there are several thousand or more vertical interconnects per die, completely testing each die is probably impossible – one may need to resort to "kinda good die." Also, the process of testing leaves scrapes, gouges, ridges on the pad surfaces. This damage is likely to cause issues in subsequent 3D integration – it may indeed cause more failures than would be encountered with untested and unknown bad dies. Each case must be looked at under its own merits and requirements. Although Tezzaron's process can accomplish both chip-to-wafer and wafer-to-wafer bonding, this chapter will focus on the wafer-to-wafer process flow.

24.2
Copper Bonding

Metal-to-metal bonding is a very common and well-known manufacturing process. It provides both mechanical attachment and electrical connectivity in one step. Another advantage is that metal bonding processes produce no out-gassing, unlike organic glues. Of all the metals, gold forms bonds most readily, requiring very little pressure or heat. Gold bonding is commonly used for providing seals in semiconductor

Handbook of 3D Integration: Technology and Applications of 3D Integrated Circuits.
Edited by Philip Garrou, Christopher Bower and Peter Ramm

ISBN: 978-3-527-32034-9

packaging and its use for 3D integration dates back more than 20 years (e.g. US patent #4612083, July 17, 1985). The major disadvantage of using gold in ICs is that the metal migrates aggressively. Copper is second only to gold for metal bonding. Copper bonding requires higher pressure and temperature than gold, but it costs less, exhibits a far greater bond strength and is much more easily contained within a semiconductor device. Copper thermal diffusion bonding is used for everything from refrigerator coils to US quarters. The process obviously works well, as one never hears of US quarters falling apart.

24.2.1 Advantages of Copper Bonding

Copper is already a standard part of normal CMOS processing. It provides excellent electrical and thermal dissipation characteristics and it is easily planarized using existing CMP technologies. Copper-to-copper bonding provides excellent bond strength. The bond can be produced at temperatures as low as 280 °C in a very reproducible process, which has been easily replicated at multiple foundries. Copper bonding produces good yields at a very low cost.

24.2.2 Disadvantages of Copper Bonding

Typically, copper thermal diffusion bonding is done at 375 °C. Thermal expansion of the wafers is a significant alignment issue during bonding. For 200 mm wafers, a two-degree temperature differential between the wafers can result in ~1-μm alignment error.

Copper is not generally available as a full wafer face process; therefore, there are issues with edge defects and possible peeling. Tezzaron uses edge grinding to control and eliminate this issue, but additional edge dies may be sacrificed as a result of the grinding.

24.3 Yield Issues

Notably, yield, for both 2D and 3D devices, is related primarily to defect density and total die size. Most 3D processes do not significantly alter the per square millimeter yields. Therefore, a 100 sqmm die in 2D has about the same yield as four stacked 25 sqmm dies.

Three applications that are obvious candidates for wafer-to-wafer 3D integration are memories, FPGAs and CMOS sensors. What these applications have in common is the ability to repair or ignore processing defects. Today, Tezzaron targets memory as its main application for wafer-to-wafer bonding. The vastly increased wire interconnect available using 3D integration enables a much higher degree of reparability. This provides a unique opportunity for memory exploitation.

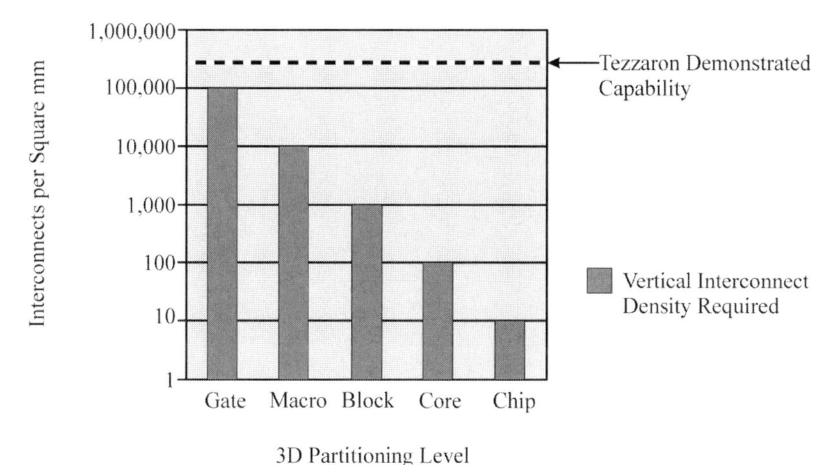

Figure 24.1 Interconnect density requirements by partition level.

24.4 Interconnect Density

In any 3D-IC design, the number of vertical interconnects needed – that is, the required interconnect density – depends on how the design is partitioned, as this determines how the 3D wiring is to be used. Figure 24.1 shows the interconnect densities required at five different levels of 3D partitioning.

Partitioning at the macro level isolates sections of circuitry as small as a 4-bit counter. Circuit elements can be separated onto different layers by their process requirements and then optimized by process. This level of partitioning enables true circuit level 3D integration. The required interconnect density for macro level partitioning falls within the range that can reasonably be built with today's techniques.

It can be argued that finer partitioning (gate or even transistor level) will not yield any additional benefit until transistors can be fabricated in layers on a single wafer. Today, adjacent transistors on a wafer are separated by perhaps only a few tenths of a μm. In contrast, transistors on two adjacent 3D layers, even in the most aggressive 3D stacking, are at least 3 or 4 μm apart. This suggests that practical 3D integration is currently limited to elements the size of small counters or larger.

Most 3D integration applications will be adequately served by a density of 10 000 interconnects per sqmm. There are some exceptions, of course. Among these exceptions, CMOS sensors represent a unique opportunity. The most often cited goal is to separate the sensor elements from the amplifiers. Sensor elements themselves might benefit from further separation, as they often have process requirements that would be well suited to separate wafer processing. It would be ideal to have one vertical connection per pixel with a pixel pitch of about 1.5 μm. This combination would drive the required interconnect density to nearly 400 000 per sqmm.

How much interconnect is needed for other applications? The answer is not a simple one. Each case must be examined as to its own requirements. The vertical

Table 24.1 Element sizes.

Element	Size (f^2)[a]	Comments
Standard cell gate	200–1000	Three connections
Standard cell flip-flop	5000	Five connections
16 bit sync-counter	125 000	Twenty connections
Opamp	300 000	Four connections
TSV (Tezzaron)	500	Includes spacing
Wafer-to-wafer bondpoint (Tezzaron)	350	Includes spacing
Chip-to-wafer bondpoint (Tezzaron)	35 000	Includes spacing

[a] "f" is the minimum feature size of a given process and "f^2" is the corresponding area.

interconnect does take silicon space and therefore costs transistors. Table 24.1 shows the area of various circuit elements, where "f" is the minimum feature size of a given process and "f^2" is the corresponding area.

Connecting a simple gate to other simple gates using 3D TSVs is very expensive. For example, using TSVs for all three I/Os of a NAND gate is a poor use of silicon area. Each TSV is 500f^2 while the gate itself is only 300f^2 – this translates to a TSV overhead of 80%! Vertically connecting a 16 bit sync-counter is much better, with less than 12% overhead for the TSV, and an Opamp would have less than 1% TSV overhead. Whether the goal of the 3D integration is density, cost, performance, power, or some combination thereof, thoughtful partitioning is crucial. 3D integration, just like 2D, requires prudent design choices and careful engineering.

24.5 Process Requirements for 3D DRAM

Tezzaron specifically sought to develop a process for high volume 3D memory. The requirements for high-performance 3D DRAM memory are as follows:

- DRAM is famous for low cost and tight margins; to compete, the 3D manufacturing flow must be very low cost.
- Maintaining or improving yield is paramount.
- The number of process steps must be minimized.
- Process requirements must be compatible with existing semiconductor processes.
- Wafer processing cannot change the transistor leakage profile.
- Partitioning for best process optimization requires a vertical interconnect density of about 1 million per 50 sqmm die.
- Interconnect is limited to the area between memory arrays; thus the pitch must be at most 4–5 μm, perhaps as low as 2 μm.
- Interconnect must be highly reliable, with less than 0.1 ppm failure rate before repair.
- An additional consideration is that Tezzaron, being fabless, needs the cooperation of its foundry partners for any process changes; hence any process changes must be "safe" and "minimal".

This set of requirements drove a solution that resulted in the copper-to-copper FaStack process, which addresses all of these issues.

24.6 FaStack Process Overview

FaStack uses only existing foundry processes and standard materials (copper and tungsten). Cost is kept low by using a wafer-to-wafer bond between bulk silicon wafers, SOI wafers, or a combination. (Stacking wafers of widely disparate temperature coefficient materials, like silicon to InP, is a bad fit for copper-to-copper bonding.) The equipment required is all standard and commercially available – EVG Bonders and Aligners, Okimoto Grinders and CMP. Processing is minimized by using a metal bond, so that one step simultaneously forms both the electrical and the mechanical bondpoints. The DRAM design is partitioned to place memory cells on wafers that are completely DRAM process and memory cell compatible; all other circuitry is on a separate wafer that is compatible with standard logic processing. Additional features of the process are described in the following sections.

24.7 Bonding Before Thinning

A notable distinction of the FaStack process is that bonding occurs before thinning. This process flow never requires a carrier wafer and avoids any handling of very thin wafers. Other processes that thin the wafers prior to attachment can cause stress in the thinned wafer to be either released or increased. These changes can manifest themselves in altered transistor characteristics, especially increased leakage. By bonding before thinning, these changes are eliminated and the wafer's leakage profile is maintained.

In addition, thinning before bonding leaves the thinned wafer vulnerable to deformation and nonlinear distortions. Bonding before thinning allows the extremely precise alignment necessary for high-density interconnect.

24.8 Tezzaron's TSVs

24.8.1 Via First TSVs

Tezzaron employs a via first technique, building the vertical interconnects into each wafer before bonding and thinning. This offers distinct advantages over via last process flows.

24.8.2
TSVs as Thinning Control

TSVs can be used for thinning control, serving as a CMP stop. When Tezzaron first demonstrated this process in 2001, copper vias were used in a BEOL flow. The tantalum barrier covering the copper plug was used as an *in situ* polish stop. Tezzaron's current FEOL flow normally uses a tungsten plug, but the nitride or oxide layer surrounding the plug performs the same function as a stop to the Si CMP processing. The ultimate result of this technique is to provide a substrate thickness control of $\pm 0.5\,\mu m$ across a 200 mm wafer.

Because the TSVs will be used in this manner, uniformity of depth is very important.

Three factors that could be expected to affect depth uniformity are the density of TSVs on a die, the diameter of the TSVs and the position of the die on the wafer. Figures 24.2–24.4 show the results of testing on these three factors. Within any lot, the typical 3σ depth variation is $\pm 2.7\%$. The total variation observed to date has displayed a 3σ of $\pm 5.0\%$.

24.8.3
TSVs as Alignment Markers

A second benefit of the via first technique is the ability to use the vias, which are clearly visible after thinning, to align the masks for additional backside process steps. A via last process flow requires either SOI (to see through the wafer) and/or a good guess. In either case, process tolerances tend to accumulate quickly, driving up the keepaway spacing and via landing sizes.

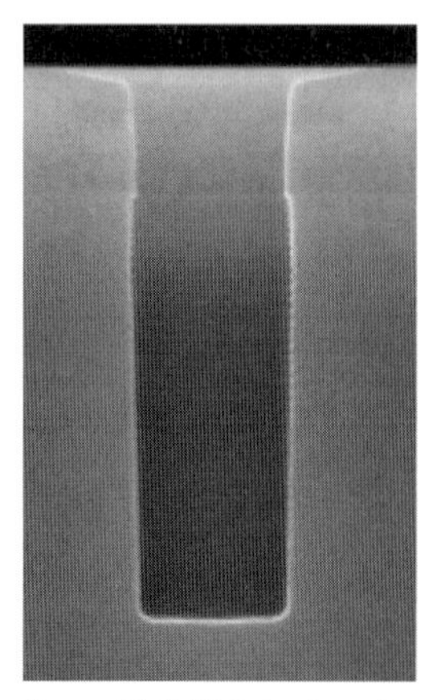

Figure 24.2 TSV depth variance by pattern density; left: TSV pattern density 0.2%, depth 8.192 μm; right: TSV pattern density 0.8%, depth 8.201 μm; within-die depth variance 0.01 μm.

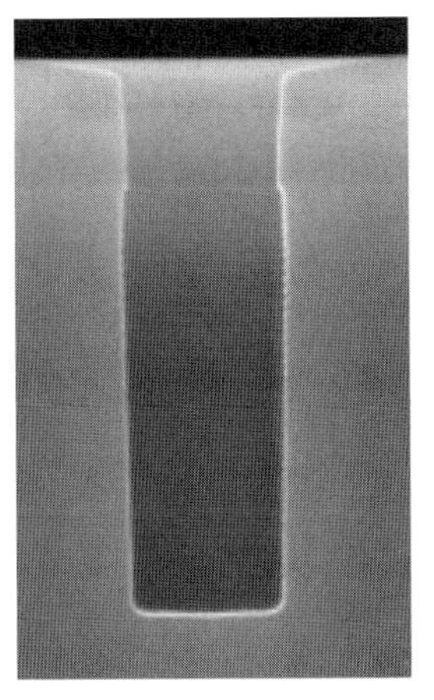
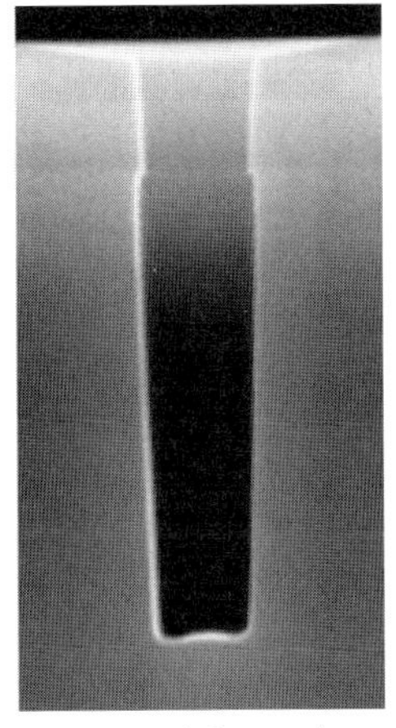

Figure 24.3 TSV depth variance by diameter; left: TSV diameter 3 μm, depth 8.192 μm; right: TSV diameter 2 μm, depth 8.300 μm; within-die depth variance 0.11 μm.

24.8.4 BEOL and FEOL

FaStack can use either BEOL or FEOL process flow. Two different types of vertical connect structures (SuperVia and SuperContact) are created, depending upon the flow: BEOL produces SuperVias and FEOL produces SuperContacts. Each offers unique benefits. Either technique can create either copper or tungsten interconnect; however, SuperVias are typically copper and SuperContacts are typically tungsten. This is the case because these are the metals typically used at that stage in processing.

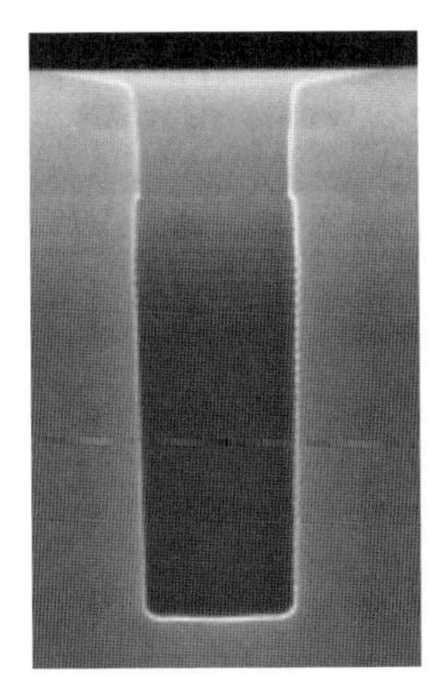
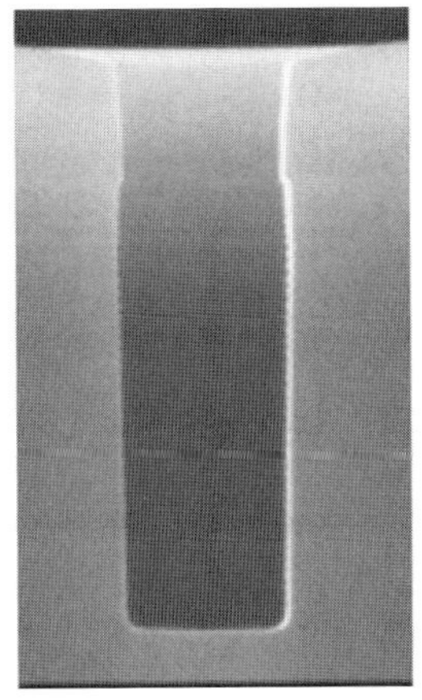
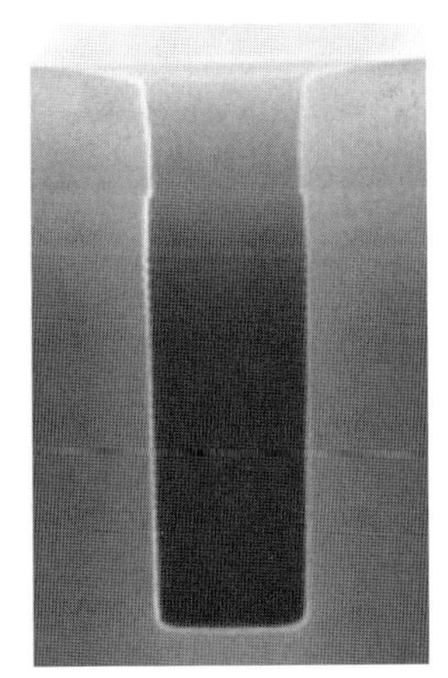

Figure 24.4 TSV depth variance by die placement; left: center of wafer, depth 8.192 μm; center: halfway to wafer edge, depth 8.255 μm; right: wafer edge, depth 8.340 μm; within-wafer depth variance 0.15 μm.

Table 24.2 SuperVia process flow (simplified).

	Process flow for SuperVia creation
1.	SuperVia mask, 4×4 μm
2.	ILD etch (through all layers of interconnect)
3.	SI etch, 4.5 μm
4.	ILD barrier deposition (SiO_2 or composite), 1000 Å
5.	Optional second ILD etch for connection to local metal wiring
6.	Ta/TaN deposition, 250 Å
7.	Copper seed deposition, 1500 Å
8.	Copper electroplate, 1.5 μm
9.	Room temperature anneal
10.	Copper CMP

Fabricating a tungsten SuperVia would require an additional copper layer to be put down for the bondpoints; using copper for SuperContacts would require special care and processing for the subsequent tungsten contacts.

24.8.5
SuperVia TSVs

Table 24.2 lists the steps to create SuperVia TSVs. The resulting architecture is illustrated in Figure 24.5, where the SuperVia employs a connection to local metal wiring. In any given design, not all SuperVias are likely to employ this type of connection.

SuperVias have the advantage of not requiring any foundry process changes, as the vertical interconnect is entirely post processed. Another advantage is that the via material is copper and extends from within the substrate to the top surface of the

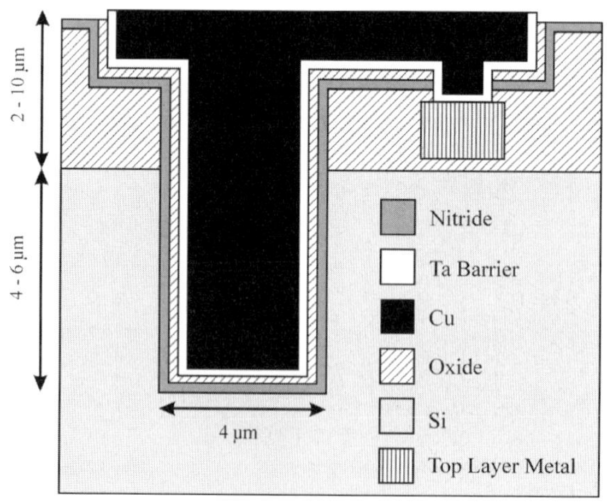

Figure 24.5 SuperVia architecture.

Table 24.3 SuperContact process flow (simplified).

	Process flow for SuperContact creation
1.	SuperContact mask, 1.2 × 1.2 μm
2.	ILD etch
3.	SI etch, 4.5 μm
4.	ILD barrier deposition (SiO_2 or Composite), 1000 Å
5.	Ti/TiN deposition, 250 Å/60 Å
6.	W deposition, 0.8 μm

wafer; this certainly helps with thermal transfer. However, the SuperVia also has significant disadvantages. Each SuperVia requires an open field area with no transistors and no interconnect. The SuperVia must also span the entire metal stack and extend further about 5 μm into the substrate, so it can easily be 12 μm long. To keep a conservative aspect ratio, the via quickly grows in diameter. This can constrain the density of the interconnect.

24.8.6 SuperContact TSVs

Table 24.3 lists the steps to create SuperContact TSVs; the resulting architecture is illustrated in Figure 24.6.

SuperContacts have their own advantages and disadvantages. The SuperContact is built during wafer fabrication, so it requires the foundry to perform a unique process step. While this step is relatively simple, it does represent a change and thus requires qualification. Pushing the vertical interconnect fabrication into the foundry

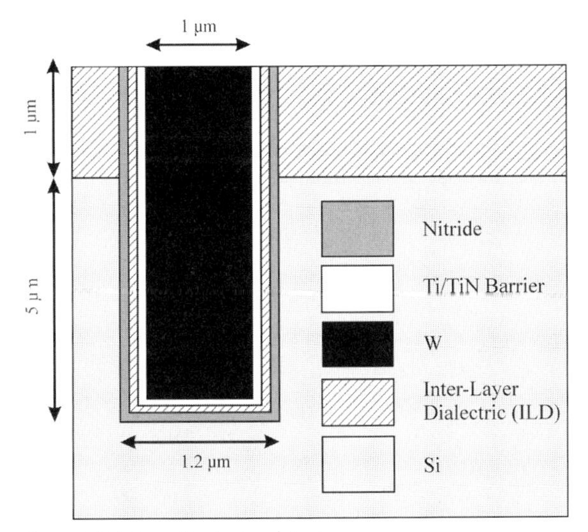

Figure 24.6 SuperContact architecture.

significantly reduces the process complexity and equipment requirement for the subsequent stacking operation. Another benefit of SuperContacts is that the diameter of the interconnect can be greatly reduced. The total length of a SuperContact is typically less than 6 μm; furthermore, because it is fabricated from tungsten, it tolerates a higher aspect ratio. The total cross sectional area of a SuperContact is less than 1/10 that of a SuperVia. The much smaller overall structure leads to a large reduction in the capacitance. Tungsten also provides a smaller heat expansion and thus creates less stress and permits closer placement to adjacent structures such as transistors; on the other hand, tungsten provides less heat transfer and higher resistance. So far, Tezzaron has not found heat transfer to be an issue; however, it always needs attention, and to date Tezzaron has not fabricated devices that need to rely on vertical interconnect for heat transfer. Tungsten's higher resistance is completely insignificant for signals, but power distribution may require tens to hundreds of vias, where resistance could become an issue. Overall, the FEOL SuperContact method is generally preferred.

24.8.7 TSV Characteristics and Scaling

Table 24.4 lists some salient characteristics of Tezzaron's interconnect. Figure 24.7 shows the relative sizes of various bondpoints compared to a buffered D flipflop design. TSVs such as the SuperVia and SuperContact will continue to scale. Tezzaron expects the SuperContact to reach the nanoTSV (sub-μm) scale within a year or so. Scaling down to via diameters of 0.5 μm or less are very likely over the next five to ten years.

24.9 Stacking Process Flow Details (with SuperContacts)

Figures 24.8–24.16 illustrate the stacking process with SuperContacts. Figure 24.17 shows a cross section of a two-wafer stack with one visible SuperContact. Once a 3D wafer stack is complete, it can be sawed and wire-bonded or flip-chipped, just as any normal 2D wafer would be.

Table 24.4 3D interconnect characteristics.

	SuperVia	SuperContact	Face-to-face bondpoint	Chip-to-wafer bondpoint
Size (μm)	1.2 × 1.2	1.2 × 1.2	1.7 × 1.7	10 × 10
Minimum Pitch (μm)	6.08	<4	2.4	25
Feedthrough capacitance (fF)	7	2–3	≪	<25
Series resistance (Ω)	<0.25	<0.35	<	<
Max interconnect (per mm^2)	25 000	100 000	170 000	1600

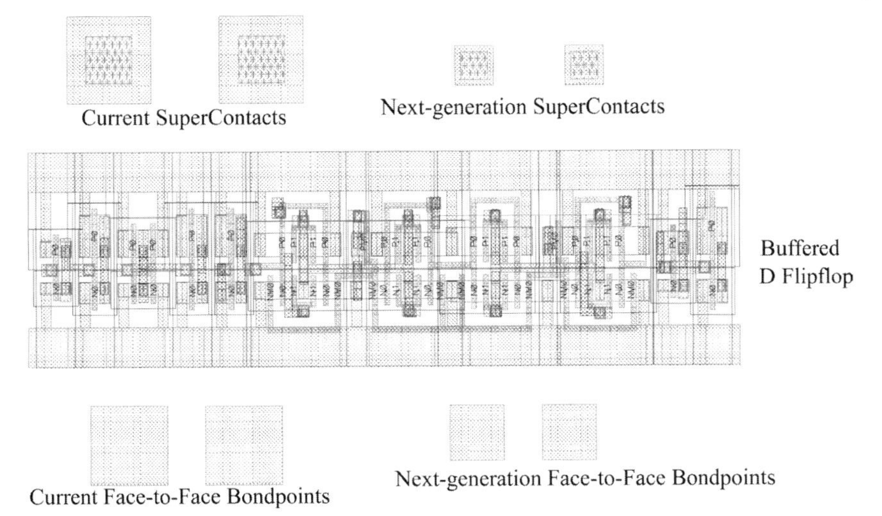

Figure 24.7 Bondpoints and a buffered D flipflop.

24.10 Stacking Process Flow with SuperVias

The stacking process with SuperVias is essentially identical to the process with SuperContacts except for the creation of the TSVs themselves (BEOL copper instead of FEOL tungsten). Figure 24.18 shows a cross section of a completed three-wafer stack with five visible SuperVias. None of these SuperVias employ connections to local metal wiring.

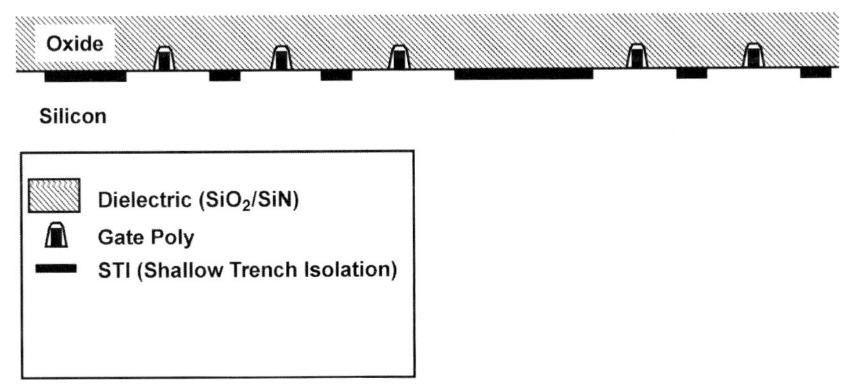

Figure 24.8 Cross-section of one wafer, immediately after transistors have been created, but before contact metal.

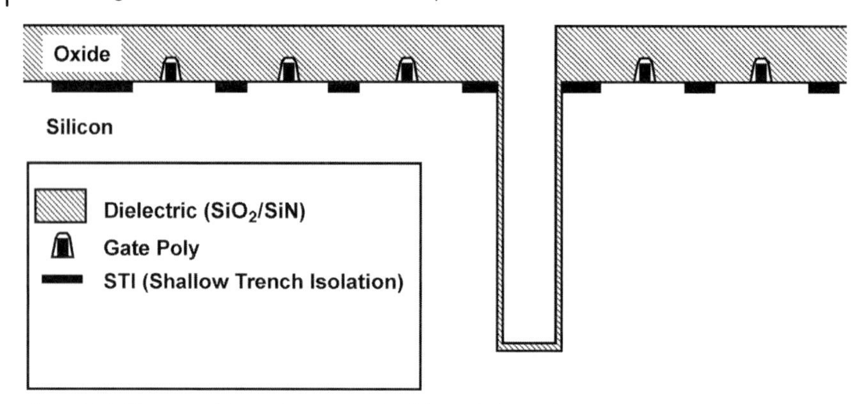

Figure 24.9 The vertical SuperContact is etched through the oxide and into the silicon substrate approximately 6 μm. The walls are lined with SiO_2/SiN.

24.11 Additional Stacking Process Issues

The following sections address some specific processing issues that are not apparent in the preceding series of figures.

24.11.1 Planarity

Wafers as they arrive from the foundry typically do not have very parallel surfaces. A pre-grind is done prior to any further processing, leaving the wafers with a typical total thickness variation (TTV) of <1 μm. This step eases the future backgrind and can greatly improve the overall substrate thickness control after back thinning.

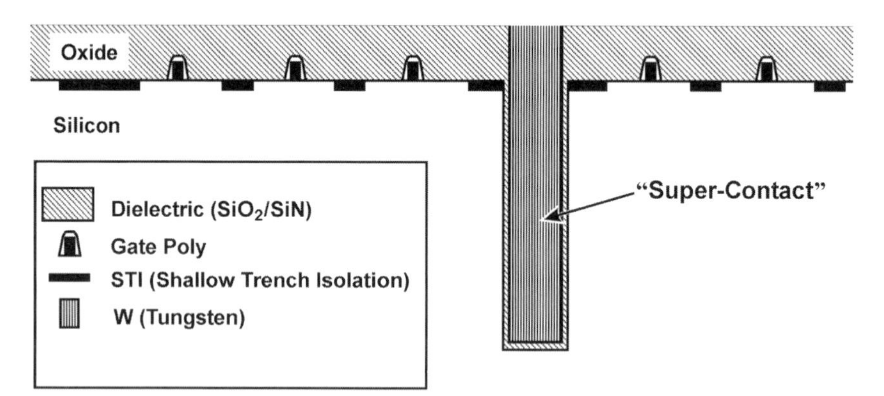

Figure 24.10 The SuperContact is filled with tungsten and finished with chemical-mechanical polishing (CMP). This completes the unique processing requirements at the wafer level.

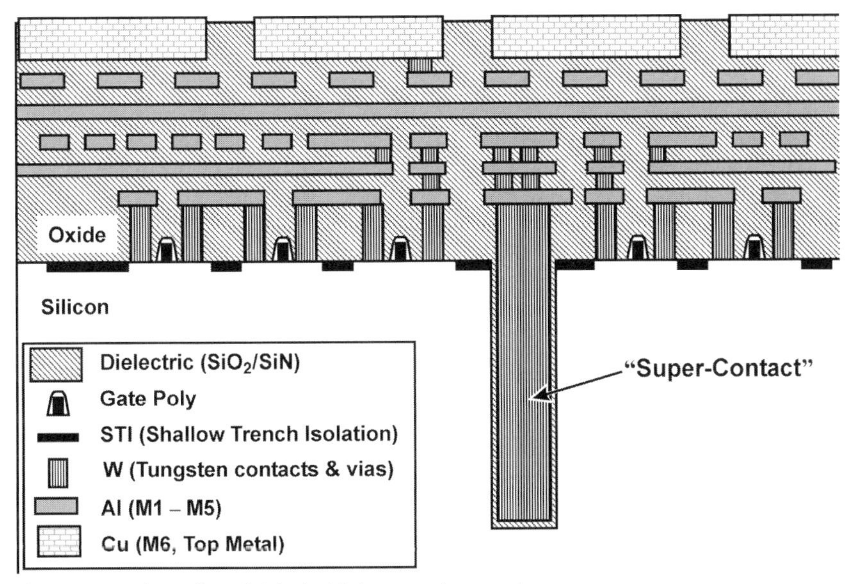

Figure 24.11 The wafer is finished with its normal processing, which can include a combination of aluminium and copper wiring layers. The last layer must be copper.

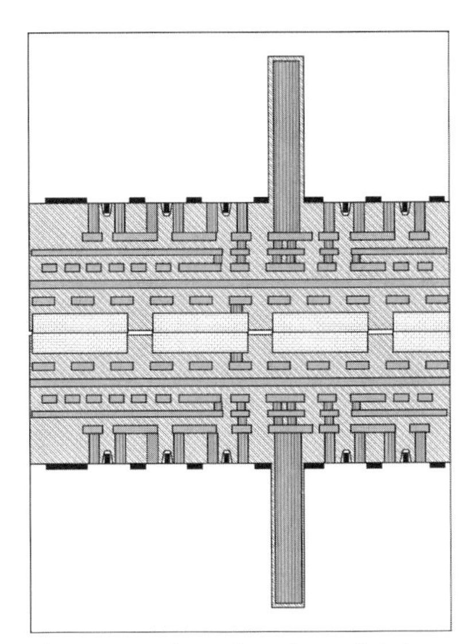

Figure 24.12 The oxide surface is slightly recessed on both wafers. They are then aligned and bonded in a copper thermal diffusion process that takes place in a vacuum at approximately 375 °C and 40 psi. Several minutes are required to form the bond. Typical cycle time within the bonder is 20 min.

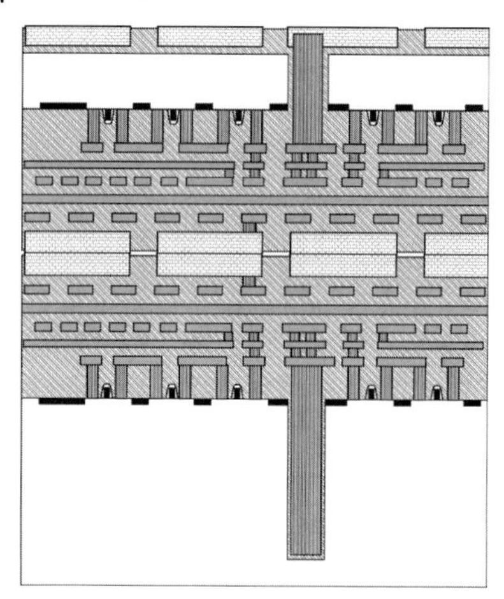

Figure 24.13 After bonding, the top wafer is thinned to the bottom of the Super-Contacts. This leaves a substrate thickness of about 4 μm. Thinning is done with a combination of wafer grinding, CMP, and etching. The backside of the thinned wafer is covered by an oxide, then a single damascene copper process creates bonding pads for subsequent stacking.

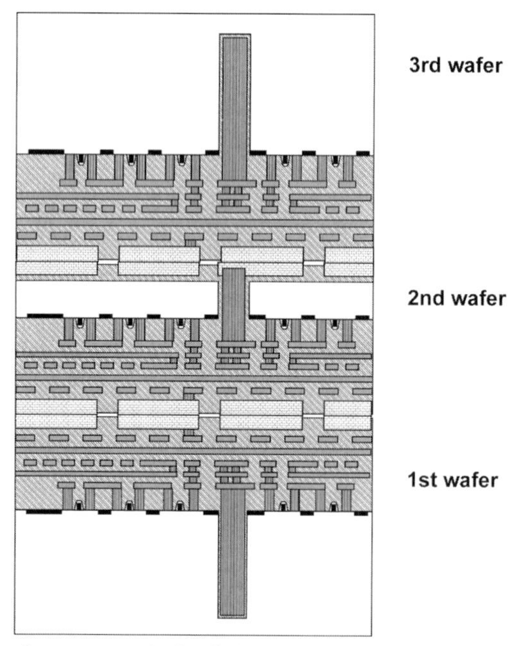

Figure 24.14 A third wafer has been added to the stack, using the same technique by which the second wafer was added.

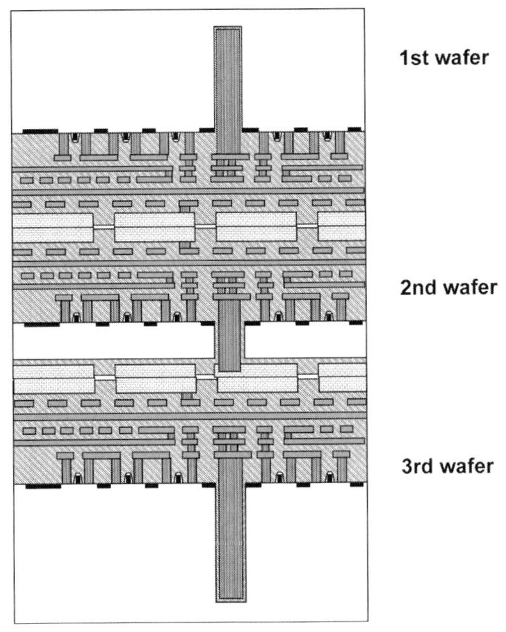

Figure 24.15 Now the stack is inverted. Final processing will be applied to the backside of the first wafer.

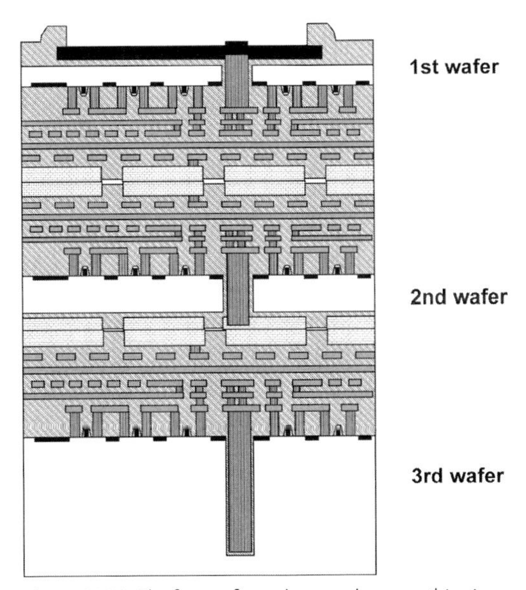

Figure 24.16 The first wafer undergoes the same thinning process used before, stopping on the tungsten Super-Contacts. Instead of a copper damascene process for bonding pads, an aluminium layer is deposited for normal wire bonding.

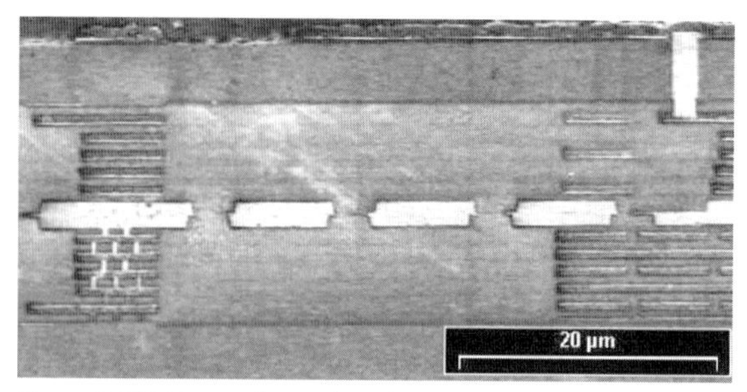

Figure 24.17 Two-wafer stack with SuperContact; Bright horizontal lines are bondpoints; vertical structure at upper right is a SuperContact.

24.11.2
Edge Grinding

When two wafers are bonded, the outer edge is left unattached due to the normal edge exclusions on the wafer. The largest contributor to the edge exclusion is the electrode ring used in copper plating. Even if full face copper were used, some small lip would remain. After stacking, when the top wafer is thinned, the excluded edge becomes thin, fragile, and hazardous. This thinned edge is prone to chipping and tearing, and the broken pieces pose contamination risks for future process steps. To correct this, Tezzaron performs edge grinding (Figure 24.19). The edge grinder that is normally used to bevel or form wafer edges can easily be programmed to remove just the upper wafer lip in a stack. At each successive layer of a stack, the grind is stepped slightly inward, producing a "wedding cake" structure.

24.11.3
Alignment

Precise wafer alignment is crucial to high density interconnect. Four factors affect the ultimate alignment that can be achieved:

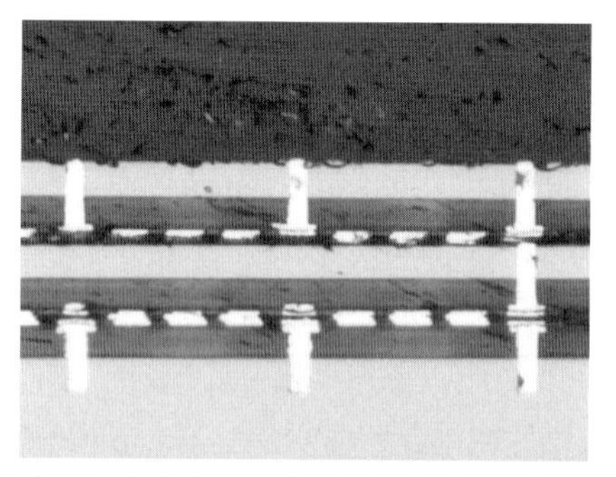

Figure 24.18 Three-wafer stack with SuperVias; two sets of bright horizontal lines are bondpoints; the five vertical structures are SuperVias.

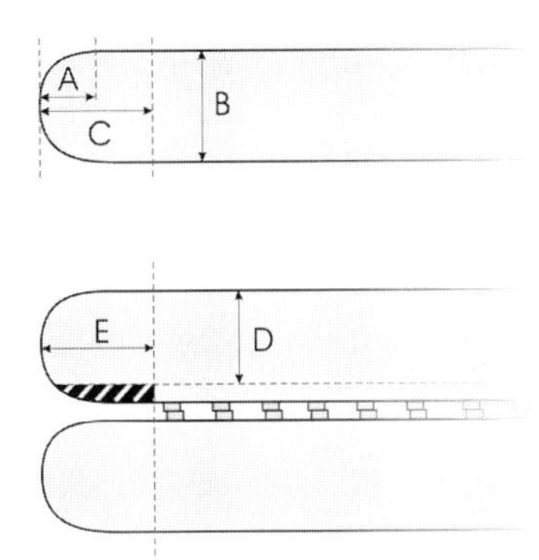

Figure 24.19 Wafer edges: Upper figure: cross-section of one wafer at an exaggerated scale. Bevel (A) is about 300 μm deep. Wafer thickness (B) is from 400 to 1000 μm. Edge exclusion (C) is from 3 to 5 mm. Lower figure: Cross-section of a bonded pair of wafers. The gap between the wafers is less than 0.1 μm. The bulk of the top wafer (D) will be removed, leaving from 5 to 15 μm. Excluded edge (E) , from 3 mm, will be ground away to remove the fragile unbonded edge (shaded).

1. Stepper runout: The wafer as originally patterned has an error of ~0.1 μm across it from the inaccuracy of the first patterns laid down on the wafer in foundry processing. This is true for both 200 and 300 mm wafers. This error is not related to the normal layer to layer alignments in semiconductor processing that produce considerably smaller errors.

2. Pre-bond alignment: This alignment can be done using IR through the wafer or using the EVG SmartView face to face aligner. Pre-bond alignment limitations are typically small, less than ~0.35 μm.

3. Wafer slip during bonding: This is a major error source. Many people have reported slips of 1–2 μm and slippage of 10 μm or more is possible. Tezzaron has done considerable work to reduce the problem; improvements in the wafer chuck have permitted almost a 10× reduction in slippage.

4. Wafer temperature mismatch at bonding: If one wafer is 1 °C warmer than the other, the error can be 0.4 μm on a 200 mm wafer. Manufacturers of newer bonders are going to great lengths to improve the thermal tracking of the wafers prior to bonding.

Tezzaron typically sees about a 0.3 μm (~1σ) misalignment on 200 mm wafers. One micron is a good 3σ number today. Over the next few years some improvement is expected; 0.5 μm 3σ on 300 mm wafers may be as good as it gets, given some of the basic issues that have been outlined. Also, following the previous discussion of required interconnect density with the current thinned wafer thicknesses, there may be little impetus to improve this value even to 0.5 μm. The SEM (scanning electron microscope) image in Figure 24.20 shows a pair of bonded

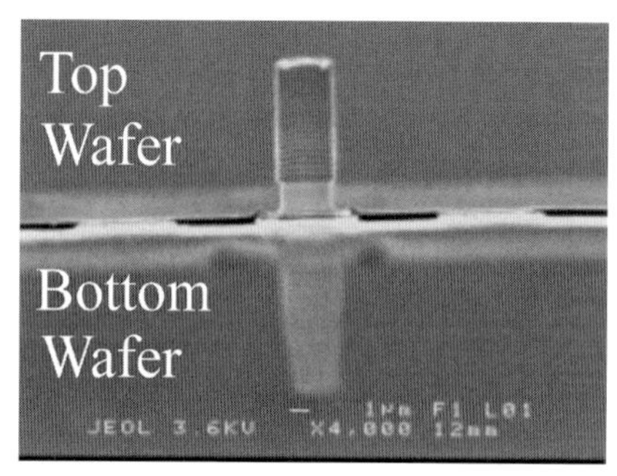

Figure 24.20 Wafer alignment; two wafers bonded face to face with aligned SuperVia TSVs.

wafers with two TSVs meeting at the bondpoint. The misalignment between the two TSVs is a mere 0.3 μm.

24.11.4 Bondpoint Area

Another critical consideration is the area consumed by bondpoints. Tezzaron has very strict rules regarding the type, shapes and patterns of bondpoints. The patterns must match on the two surfaces to be bonded. The copper surface density on any one bond face must be very uniform, but the density for a given bonded pair may be anywhere from 15 to 100% – a range that also accounts for the 3σ worst case alignments. Obviously, 100% bondpoint area would mean that there was only one electrical connection on the whole die – this would be useless. Most applications fall in the 35–45% realm. Mechanical issues surrounding the size, shape, pattern and density of the bondpoints are critical to good bonding. Table 24.5 lists typical dimensions of Tezzaron's bondpoints.

A bonded surface contains both mechanical and electrical bondpoints. Typically, only a small percentage of the total bondpoints are used to carry electrical signals. In Tezzaron's processor-memory stacked device, only ~1.5% of the bondpoints are electrical, but in the stacked CMOS sensor over 50% of the bondpoints are electrical. A process separated DRAM uses about 6% of its bondpoints to carry signals.

Table 24.5 Typical bondpoint dimensions.

Typical bondpoint dimensions		
Pitch (μm)	**Diameter (μm)**	**Spacing (μm)**
10.5	7.0	3.5
5.0	3.3	1.7
2.9	2.1	0.8
2.4	1.7	0.7

24.12
Working 3D Devices

Tezzaron has built various working test devices, including field programmable gate arrays (FPGAs), CMOS sensors, SRAMs, DRAMs and memory-processor combinations. These devices have demonstrated the dramatic performance impact that 3D integration can have.

The FPGA shows seamless integration of 3D interconnect with 2D layouts, with the vertical interconnect literally blending into the structure of the FPGA. Each of the hundreds of logic blocks contains 12 vertical connections to the next layer.

The CMOS sensor was created by separating the photo diodes from the amplifier circuitry. The performance of this first primitive 3D sensor suggests that significant cost and performance gains are possible. The test device also demonstrated good yields at pixel pitches down to 2.4 μm. In the test lot, most devices had 100% pixel yields, showing complete bond integrity over 25 000 vertical interconnects per device. CMOS sensors constructed in 3D like this have the advantages of 100% array area and backside illumination.

The memory test devices have proven the viability of separating memory cells from the rest of the circuitry, a key benefit to 3D integrated memories. A great concern was the effect of 3D processes, especially thinning, on device leakage. The physical separation of the sense amplifiers from the memory cells also caused some concern. The results of the test devices showed that neither of these concerns was warranted, and allowed complete vindication of the 3D memory concept.

Perhaps the most impressive of the test devices is the stacked memory-on-processor. This device demonstrated a 5× performance increase at equal power, or a 10× power decrease at equal performance. While this memory/CPU demonstration was unique in its scale of improvement, it shows the extreme merit of 3D development.

The various test devices showed the ability to pass signals to at least 1 GHz, although it is safe to assume that much higher frequencies would work as well. The highest power consumption device was only 3 W. Certainly this is a very low power level compared to a modern CPU; however, given the actual die size of the device used for this test, it is actually similar to a 100 W CPU in watts per sqmm.

Figure 24.21 shows the top of one of the test devices with its wire bonding. Notice there is no apparent circuitry; only the wire bond pads and some additional probe pads are visible. The top layer of silicon in this example is ~6 μm thick. A thin rectangular line of metal is visible outside the pads. This is the seal ring, which is present through the copper bond layer as well as through the top thinned layer.

24.13
Qualification Results

The qualification results to date have provided strong support to the belief that this process can be scaled to large volumes such as those associated with memories. Many

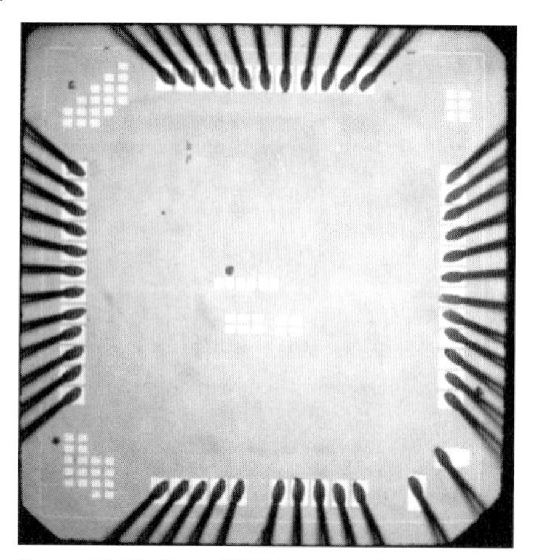

Figure 24.21 Top of finished test device.

typical qualification tests concentrate on the electrical performance of the semiconductor device, but for 3D devices the most significant tests are those that cause mechanical stress. As an example, testing of high voltage effects on the semiconductor life has very little to do with whether the device is 2D or 3D. Most assuredly, high mechanical stresses will ultimately cause electrical failures, and complete and thorough testing must be done on every commercial device, but the most significant 3D process issues can be measured with two basic tests: temperature cycling and high temperature storage. Tezzaron's copper-to-copper bonded components fabricated at 180 nm have been thoroughly tested for temperature cycling and high temperature storage. Over 100 000 device cycles (more than 100 devices in more than 1000 cycles) have produced no failures, proving the bonding to be reliable. Tezzaron used the JESD22-A104-B testing standard with test condition C (-65 to 150 °C) and soak mode 4 (15 min). Maximum temperature extremes create the most stress, and long soak times guarantee maximum "creep." The test devices were also exposed to 150 °C for 504 hours, a test that should expose any additional stress related failures. For the four test lots Tezzaron has processed so far, no failures have occurred. Additional tests and their results are described in the following sections.

24.13.1 Bonded Wafer Shear Testing

During back grinding and CMP the wafer stack is exposed to shear forces. Testing has shown the required force to shear the wafers apart is greater than 30 kiloton. This is almost an order of magnitude greater than required. The failure mode most often seen is copper bondpoints actually being ripped from the underlying glass. There have been no recorded instances of the copper to copper bond actually failing in properly bonded wafers.

24.13.2
Delamination: High Power Caused (Self-Forced)

A concern many have is whether hot spots, such as those on CPU components, will cause delamination due to the high thermal differentials. However, overclockers can rest easy; in hot-spot tests, the devices did not fail. A large on-chip resistor, about 1 sqmm, was used to create a 10 W $sqmm^{-1}$ power hot spot adjacent to a virtually zero power circuitry feature to create a very high temperature differential in a very small region. The device did stop correct operation due to the high on-die temperature, but when the thermal source was turned off and the temperature returned to a normal range, the devices returned to normal operation. This test was repeated on several devices, some of which were cycled more than 100 times. No failures were recorded, other than successfully burning out the resistor itself!

24.13.3
Transistor Performance Drift

Tezzaron tested 130 nm devices that were exposed to stacking processes. The results of the testing showed no significant statistical changes in transistor performance, as summarized in Tables 24.6–24.11

A second set of tests was performed on DRAM test devices. These were two-layer stacked devices with test memory cells on both a thinned layer and a standard non-thinned layer. DRAM is particularly sensitive to leakage; nominal DRAM memory cell leakage is often measured in femtoamperes. The tests showed no difference in leakage or performance between the thinned and the non-thinned cells; both sets of cells performed to specifications for leakage and access time.

Table 24.6 Processing effects – threshold voltage.

Threshold voltage (VTO) (V)						
	NMOS			PMOS		
Width/length (μm)	20/20	20/0.3	20/0.13	20/20	20/0.3	20/0.13
Pre-processing average	0.395	0.485	0.479	−0.055	−0.399	−0.398
Post-processing average	0.393	0.484	0.465	−0.357	−0.396	−0.404

Table 24.7 Processing effects – breakdown voltage.

Breakdown voltage BVDSS (V) (*Ids* = 2 μA)						
	NMOS			PMOS		
Width/length (μm)	20/20	20/0.3	20/0.13	20/20	20/0.3	20/0.13
Pre-processing average	3.380	3.220	3.220	4.100	4.000	2.780
Post-processing average	3.377	3.230	3.217	4.147	3.970	3.113

Table 24.8 Processing effects – subthreshold slope.

Subthreshold slope SUBSLP (mV dec^{-1})						
	NMOS			PMOS		
Width/length (μm)	20/20	20/0.3	20/0.13	20/20	20/0.3	20/0.13
Pre-processing average	75.840	76.820	79.380	–73.040	–76.960	–89.460
Post-processing average	74.367	76.100	78.567	–74.733	–76.833	–88.600

Table 24.9 Processing effects – saturation current.

Saturation current idsat (μA) ($Vd = Vg = 1.2$ V)						
	NMOS			PMOS		
Width/length (μm)	20/20	20/0.3	20/0.13	20/20	20/0.3	20/0.13
Pre-processing average	122.520	5152.000	9696.000	26.940	2061.800	5986.200
Post-processing average	121.500	5094.333	9840.333	26.897	1997.333	4473.000

Table 24.10 Processing effects – leakage current.

Leakage current ioff (pA) ($V = 1.4$ V)						
	NMOS			PMOS		
Width/length (μm)	20/20	20/0.3	20/0.13	20/20	20/0.3	20/0.13
Pre-processing average	151.820	638.900	3655.000	136.460	1285.120	2820.5000
Post-processing average	140.433	433.667	3237.667	211.333	910.333	1216.8000

Table 24.11 Processing effects – gate leakage current.

Gate leakage current GLEAK (nA)						
	NMOS			PMOS		
Width/length (μm)	20/20	20/0.3	20/0.13	20/20	20/0.3	20/0.13
Pre-processing average	1.200	1.172	1.190	0.909	0.883	0.886
Post-processing average	1.250	1.287	1.300	1.018	1.011	0.767

24.13.4
Life Testing

Life testing has been continuing on a few devices. These devices are run at nominal voltage and room temperature. They have continued to operate to the original tested performance for more than 10 000 hours.

24.13.5
Highly Accelerated Stress Testing (HAST)

One other mandatory test before production is HAST (Highly Accelerated Stress Testing). HAST testing verifies the ultimate integrity of the seal rings that protect the internal vertical copper connections. Tezzaron has not performed HAST testing to date, as all the test devices were all packaged in ceramic carriers that remove the usefulness of HAST testing. Tezzarons commercial parts will use standard plastic packaging, and HAST will be performed on all such devices.

24.14
FaStack Summary

Tezzaron's copper-to-copper process performs both chip-to-wafer and wafer-to-wafer bonding. TSVs are created using a via first process flow employing either FEOL or BEOL. SuperVias, usually made of copper, are produced in a BEOL flow and can be processed post wafer foundry. SuperContacts, usually made of tungsten, are produced in an FEOL flow at the original wafer foundry. SuperContacts are typically much smaller in diameter than SuperVias and do not interfere with routing metals as SuperVias do.

The copper bond is formed over several minutes at 375 °C and 40 psi. The bond can withstand shear forces of 30 kiloton. Wafer alignment is typically within 0.3 μm with a 1 μm 3σ alignment. Alignment tolerance is layer to layer and not cumulative in the stack. Vertical interconnect demonstrates high yield with less than 0.1 ppm failure. The stacking process is based on standard equipment. After stacking, wafers are thinned to a substrate thickness of ~3 μm. Costs are kept low by using native foundry materials and process flow and by creating simultaneous electrical and mechanical connection. The bond-before-thin flow requires no special wafer handing or wafer carriers.

All process steps and the entire flow have been demonstrated in the creation of three-wafer stacks. Four-wafer stacks have been accomplished and five-wafer stacks are doable in the near term. There is no fundamental limit to the number of wafers that can eventually be stacked.

3D integration is poised to become a mainstream process over the next several years because it promises cost improvements and performance benefits that can no longer be expected from scaling alone. At the 2007 3D-SIC conference, a participant showed photos from more than 100 different developers of TSVs. This shows a high level of confidence in the future of 3D ICs, as efforts on that scale are seldom performed just for the sake of science. Copper-to-copper bonding is just one of several methods being pursued by numerous entities. Copper-to-copper was perhaps the first bonding method to demonstrate various truly useful and functioning devices, but ultimately there are likely to be several winning processes and techniques suited for different applications. With any 3D process, as in 2D, success will depend heavily on the efforts of the designers who exploit the technology.

24.15 Abbreviations and Definitions

FEOL –	Front End Of Line: any process performed before the metal interconnect is laid down
BEOL –	Back End Of Line: any process performed during the creation of metal interconnect
Bondpoint –	a physical connection, mechanical and/or electrical, between two layers of a 3D-IC
TSV –	Through-Silicon Via: a vertical interconnect piercing the body of a die in a 3D-IC
NanoTSV –	any TSV measuring less than a μm in diameter
Via First –	building through-silicon vias into the wafer before bonding
Via Last –	creating through-silicon vias after wafers have been bonded
Wafer Scale –	any process performed across an entire wafer (not one die at a time)
FaStack –	Tezzaron's 3D stacking process.
SuperVia –	Tezzaron's BEOL TSV
SuperContact –	Tezzaron's FEOL TSV

25
3D Integration at Ziptronix, Inc.

Paul Enquist

25.1
Introduction

Semiconductor product requirements have been increasingly driven by form factor, in addition to the conventional metrics of cost and performance, to satisfy the demands of the wireless and handheld markets that currently dominate semiconductor consumption. 3D integration, consisting of a multilayer stack of, and interconnection between, conventional CMOS integrated circuits (ICs), has emerged as a solution to these requirements that exceeds the capabilities of what can be delivered in a single layer of CMOS technology. This 3D solution literally provides another dimension to continue the reduction of cost and increase in functionality and performance of semiconductor products, commonly referred to as Moore's Law [1].

3D integration technology fundamentally consists of three separate technology elements: bonding to stack the ICs, electrical interconnection between the separate ICs, and thinning to minimize the volume and facilitate electrical interconnection. There are several technology choices for each one of these technology elements, resulting in an even larger number of combinations competing for 3D market adoption and dominance.

These three separate technology elements have been evolving very rapidly to minimize costs while meeting increasing performance and form factor requirements. Early implementations of 3D technology consisted of adhesive bonding for stacking, backgrinding for thinning and wirebonding for electrical interconnection [2]. All three of these technology elements are very low cost and met the requirements of thinned stacks of ICs that drove the early adoption of 3D technology. A wide variety of combinations of IC stacks have been successful, including multilevel stacks of commodity memory, stacks of commodity memory on microprocessors, and stacks of analog and digital ICs for wireless and handheld devices [2].

Handbook of 3D Integration: Technology and Applications of 3D Integrated Circuits.
Edited by Philip Garrou, Christopher Bower and Peter Ramm

ISBN: 978-3-527-32034-9

Continued demand for increased performance and reduced form factor and cost led to the adoption of wafer bumping as a 3D technology, for example flip-chip or ball grid array (BGA), to both stack and electrically interconnect multiple layers of ICs [3]. For the number of 3D connections required by most products, the increase in bump bandwidth capability comes at a cost premium to wirebonding, which has limited its adoption and market share compared to 3D wirebond technology. This cost premium has been compensated by an improvement in supply chain management, commonly referred to as Package-on-Package (PoP) [4], where 3D stacks are assembled from burned in, packaged parts that greatly reduce the risk of including a bad part in the 3D stack that could otherwise scrap good parts in the 3D stack.

Simply replacing wirebonding with bumps is not sufficient for applications that require more than two IC layers in the 3D stack or package interconnectivity to a portion of a 3D stack consisting of more than one layer. These applications require a 3D via to form an electrical interconnection through ICs in the 3D stack. This via technology includes a silicon etch process to form the via through the transistor layer and into or through the silicon substrate and an isolated metallization process to form an electrical interconnection through the via without shorting to the silicon substrate or transistor layer. These 3D vias are typically referred to as through silicon vias (TSVs) [5] and can also be referred to as through die vias (TDVs) [6] if they extend not only through the silicon substrate but also through the CMOS foundry back-end-of-line (BEOL) multilevel interconnect stack of metal and inter-level metal dielectric. There are many different types of via technology, depending on when the vias are etched and filled in the 3D process flow. For example, TSVs or TDVs can be fabricated integral with the CMOS wafer fabrication, after CMOS fabrication but before 3D stacking, or after 3D stacking. An early preference to fabricate 3D vias after the CMOS fabrication can be attributed to a combination of a lower adoption barrier to implement 3D via technology outside the CMOS foundry than inside the CMOS foundry and applications that did not require a high density or a high number of 3D electrical interconnections within the 3D stack and were tolerant of a high exclusion volume of silicon near the 3D via.

The success of these early 3D technologies has been achieved with relatively large feature size as defined by 3D interconnection pitch and thickness of individual IC layers in the stack. The rapid development and adoption of these 3D technologies is indicative of the fundamental cost, form factor and performance benefits of 3D using feature sizes comparable to conventional packaging and assembly technologies. A further benefit of 3D can be expected with scaling of these feature sizes to dimensions comparable to conventional CMOS foundry Back End of Line (BEOL) integration technology.

Direct oxide bonding is a 3D integration technology that consists of unit process steps commonly used in the foundry BEOL for the manufacture of conventional 2D ICs. Moreover, direct oxide bonding results in a 3D integrated structure that is compatible with wafer scale BEOL and back-end fabrication, including via etching, metal via fill, oxide deposition, backgrinding and CMP. These attributes allow the fabrication of a 3D IC that exhibit interconnection pitch and thickness scaling that is more typical of that achieved in the integration of conventional 2D CMOS circuits than assembly and packaging of CMOS ICs.

This chapter begins with a description of the fundamentals of direct oxide bonding, with emphasis on a variation suitable for 3D IC fabrication that exhibits a high bond energy at low post-bond temperatures with resilience to void generation and is robust to subsequent BEOL and back-end fabrication. The use of this direct oxide bonding to achieve electrical interconnections across the bond interface integral to the bond process is then described. The relative advantages of these oxide bond technologies for 3D integration are discussed, including examples of their use in the fabrication of 3D ICs. Cost implications of these 3D integration processes and their integration into the supply chain are also summarized.

25.2 Direct Bonding

Direct bonding is a bond technology where atomic bonds are formed between two surfaces after placing these two surfaces into contact [7–11]. The process includes the preparation of an atomically smooth surface, typically <0.5 nm RMS, before the surfaces are placed into contact. The process has historically required a high temperature anneal after the surfaces are placed into contact to convert a weak, temporary van der Waals atomic bond into a much stronger, permanent covalent bond. An example of this conversion when direct bonding hydrophilic silicon surfaces is given in Equation 25.1 where weakly bonded hydroxol groups are converted into covalent oxygen bonding.

$$\mathrm{Si{-}OH} + \mathrm{Si{-}OH} = \mathrm{Si{-}O{-}Si} + \mathrm{H_2O} \tag{25.1}$$

The byproducts of this reaction can result in the generation of voids at the bond interface. This void generation can be exacerbated by the presence of hydrocarbons at the bond interface that act as nucleation sites for these byproducts [7, 8].

Direct bonding is distinct from other bond technologies in that a very high adhesive bond energy can be achieved between the two bonded surfaces without the insertion of adhesives, solders, alloys, metals and so on at the bond interface that require voltage, temperature, pressure, ultraviolet and so on to cure, reflow, melt, fuse and so on. Direct bonding is sometimes referred to as fusion bonding [8]; however, this is a misnomer because the application of heat, if required to increase the bond strength, does not melt or liquefy the bond interface.

A fundamental advantage of direct bonding is that the application of elevated temperature after initial bonding generally increases bond strength while many other bond technologies that require elevated temperature to form the initial bond are limited to post-bond temperature excursion near the bond temperature to avoid degradation (i.e. adhesive, solder and some types of metal and alloy bonding). This feature of direct bonding is critical for some applications, for example SOI bonded wafer fabrication where post-bond temperatures greater than 1000 °C may be required.

25.2.1
Direct Oxide Wafer Bonding

Direct oxide wafer bonding is a category of direct bonding where an oxide on the surface of the wafer, typically silicon oxide, is used as the direct bond interface. The oxide can be a grown oxide, for example a thermal silicon oxide on a silicon substrate, or a deposited oxide, by for example plasma enhanced chemical vapor deposition (PECVD), Tetraethyl Orthosilicate chemical vapor deposition (TEOS), high density plasma (HDP) or physical vapor deposition (PVD).

An advantage of a deposited oxide is that it is broadly applicable to a wide variety of wafers and, in particular, CMOS IC wafers where deposited oxides are typically used in BEOL CMOS IC manufacture. A deposited oxide can also be deposited to a thickness greater than the non-planarity of a CMOS IC wafer, facilitating subsequent surface planarization and smoothness to the specification required for direct bonding. This required planarization and smoothness can be readily achieved with chemo-mechanical polishing (CMP), which is also typically used in BEOL CMOS IC manufacture. The use of established deposited oxide and CMP manufacturing unit processes can thus be leveraged to implement a 3D integration direct oxide CMOS IC wafer bond.

Direct oxide wafer bonding can be used to 3D integrate CMOS ICs in three basic configurations; face-to-face (F2F), face-to-back (F2B) and back-to-back (B2B), where face refers to the CMOS surface and back refers to the silicon substrate surface. The deposited oxide and CMP planarization is applicable to both bonding surfaces in all three configurations. Alternatively, this planarization may be applied to only one of the bonding surfaces in the F2B or B2B configurations if the other bonding surface is a silicon substrate with a surface roughness suitable for direct bonding.

25.2.2
Low-Temperature Direct Oxide Wafer Bonding

Low-temperature direct oxide wafer bonding is a category of direct oxide wafer bonding where very high bond energies can be obtained at low or room temperature. The ability to obtain very high bond energies at low temperatures is required for applications where there is a large coefficient of thermal expansion (CTE) difference between materials being bonded or where materials have a limited thermal budget. These applications include heterogeneous substrates, encapsulation of CMOS IC or micro-electromechanical system (MEMS) devices, and 3D CMOS IC integration [12].

25.2.2.1 3D Integration Bond Strength Requirements

The bond strength requirements for a 3D integration process are given by the combination of requirements for the post-bond 3D integration process, packaging of the 3D device and 3D device operation and reliability.

3D CMOS IC Fabrication The fabrication of a 3D CMOS IC by the 3D integration of CMOS ICs fundamentally consists of three separate technology elements: bonding to stack the ICs, electrical interconnection between the separate ICs, and thinning to scale the vertical dimension and facilitate electrical interconnection. The non-conductive nature of direct oxide bonding allows capacitive [13] or inductive [14] coupling to achieve a 3D AC interconnect between stacked ICs immediately after bonding without further fabrication. However, a post-bond via etch and fill process after bonding is typically used to achieve a 3D AC or DC interconnect between the stacked ICs.

The 3D interconnection pitch capability of a via etch and fill process is limited by the aspect ratio of the via. Minimizing the thickness of the IC on top of the stack is thus advantageous to allow scaling of the via diameter, 3D interconnection pitch and 3D interconnect exclusion volume in the bonded IC. There is a substantial opportunity for thinning of an IC in a 3D stack because over 95% of the thickness of a typical CMOS IC wafer consists of a silicon substrate that is not required for device operation and is only used for wafer handling. The IC can be thinned before bonding; however, thinning after bonding typically allows thinner 3D IC layers to be achieved and avoids handling of thin die or wafers during the direct oxide bond process. Bonding of a temporary handle wafer to an IC wafer can be used to aggressively thin the IC wafer before direct oxide bonding, which is then removed after bonding [5, 12, 15, 16]. However, these additional handle wafer bonding and removal steps reduce yield and increase the cost of the 3D integration process.

A 3D direct oxide bond must thus be strong enough to support thinning and via etch/fill post-bond process steps. Moreover, this oxide bond strength must be obtained at a temperature within the thermal budget of the stacked ICs. The thermal budget is limited by pre-bond CMOS IC fabrication and testing. For example, 350–400 °C is compatible with many CMOS ICs, but lower temperatures will likely be required in future generations of CMOS ICs. In addition, 3D CMOS ICs stacked after 2D CMOS IC testing may be limited to 200 °C to sustain data retention.

Thinning after Bonding Thinning of a CMOS IC wafer after bonding typically includes backgrinding, CMP and etching. Thinning of Silicon-on-Insulator (SOI) CMOS IC wafers can be simplified by the Buried Oxide Layer (BOX) that exhibits very high etch selectivity to several wet and dry etches. The IC wafer can be thinned to about 2–20 μm, depending on the number of levels of interconnect metallization in the BEOL interconnect stack and the silicon thickness required for proper transistor operation. A high and uniform bond strength is required when thinning CMOS IC wafers to this extreme to avoid delamination induced by residual stresses in the thinned IC.

Thinning may also require thinning of die bonded to a wafer, for example when bonding Known Good Die (KGD) to improve yield or when stacking die of different sizes. Thinning of die may further increase the required bond strength and uniformity due to additional stresses exerted on die than on a wafer when thinning. For example, when thinning with an end-feed backgrinder, the periphery of all individual die on a die bonded wafer are subject to chipping while only the periphery of the wafer periphery is subject to chipping on a wafer bonded wafer.

Via Etch/Fill A via etch and fill process implemented after direct oxide bonding and thinning requires a via etched through the entire thickness of the upper bonded, thinned IC, the silicon oxide used to planarize the bonded ICs, including the direct oxide bond interface, and a portion of the lower bonded IC to expose a level of metallization for via filling and interconnect. The direct oxide bond needs to be strong enough to avoid lateral etching at the bond interface, which can complicate the via fill or compromise device functionality.

An advantage of direct oxide bond technology is its compatibility with BEOL via etch and fill technology. This compatibility is attributed to the use of silicon oxide in the BEOL or other inter-level metal dielectrics whose via etch and fill characteristics are compatible with silicon oxide.

Packaging The 3D IC packaging requirements are similar to those for 2D ICs and vary greatly by application. One notable exception is the singulation or dicing of 3D ICs from a 3D integrated wafer. Dicing through a bonded interface can generate significant stresses that can cause delamination if the bond is not strong enough. This problem can be exacerbated when dicing through a bonded interface near the surface, for example after a wafer has been aggressively thinned after bonding, or when there is a significant amount of metallization in the street. Die-to-wafer bonded 3D IC wafers typically do not have die bonded on the dicing street and thus do not have this dicing requirement.

An advantage of direct oxide bond technology is the lack of metal or other foreign material at the bond interface that can load the dicing blade and compromise kerf, throughput or blade wear.

Packaging does not require any additional temperature requirements as packaging temperatures are well within the CMOS IC thermal budget.

3D Device Operation and Reliability A 3D IC is subject to the same device operation and reliability requirements as a 2D IC. 3D integrated processes must meet several requirements, including temperature cycling, temperature storage and mechanical tests. Although the temperature extremes of these tests are lower than those in the CMOS BEOL or device packaging, the repetitive temperature cycling, combination of temperature and humidity, and other environmental criteria place additional requirements on the direct oxide bond strength that must be met.

Hermeticity is one example of an environmental criteria. An advantage of direct oxide bond technology is the resulting hermetic capability of a 3D IC, which can be comparable to a 2D IC because of the common use of silicon oxide in 2D CMOS fabrication.

25.2.2.2 3D Integration Bond Void Requirements

A 3D integration process that bonds 2D ICs needs to avoid generating voids at the bond interface that may compromise downstream 3D IC fabrication. There are generally two types of voids that can be generated with a direct bond process. The first type is generated when the bond surfaces are initially placed together and the second type is generated after the bond surfaces have been placed together, usually as a result of heating of the bonded surfaces.

Bond Voids Generated During Initial Placement A placement void is the generation of a void at a direct bond interface when two surfaces are placed together. The formation of placement voids is dependent upon the mechanical and surface properties of the 2D ICs that are placed together and the method in which the surfaces are placed together. For example, 2D ICs are typically composed of silicon with a Young's modulus of 150 GPa and a surface planarity characterized by bow and warp that describe their non-planarity. If two of these surfaces are arbitrarily placed together, they will first touch in locations with opposed high spots and then be drawn together by the direct bonding force, restrained by the Young's modulus and thickness of the ICs. The direct bond will thus be propagated, across the bond surface, from several sites across the bond surface. If the surfaces are bonded in vacuum, voids can be generated by different distances between the initial bond sites on the two surfaces that prohibit intimate contact between the two surfaces across the entire surface. If the surfaces are bonded in air or other ambient, voids can also be formed by the propagation of multiple bond fronts toward each other across the surface that trap the ambient and form a void.

The generation of these voids can be avoided by adjusting the surface properties and the method by which the surfaces are brought together. For example, if the surface bow exceeds the warp and extends away from the direct bond surface, initial surface contact will preferentially occur at a single location, resulting in a single bond front that propagates radially across the direct bond interface, simultaneously displacing ambient and avoiding voids.

Bond Voids Generated After Initial Placement Voids can also be generated at a direct bond interface after the initial placement of two direct bond surfaces results in the absence of placement voids. These voids can result from the generation of chemical reaction byproducts at the bond interface that are trapped at the bond interface. Chemical reactions at the bond interface and their activation energies are determined by the termination of the direct bond surfaces before they are placed together. Chemical reactions with higher activation energies require higher temperatures to generate void-generating byproducts. These byproduct voids are thus strongly dependent on the surface termination chemistry and the ability of the direct bond interfaces to absorb these byproducts.

For example, the surface termination described in Equation 25.1 has historically been a preferred direct bond surface termination. This chemistry terminates a silicon exposed surface with hydroxyl groups that can be converted into water, at elevated temperatures, that can form byproduct voids if not removed from the bond interface.

25.2.2.3 3D Integration Bond Format Requirements

A 3D integration process may require the stacking of 2D IC wafers and/or 2D IC die. The advantages of stacking wafers include bonding an entire wafer with a single bond and ease of maintaining a planar surface after bonding. The advantages of stacking die include accommodating different die sizes, different wafer sizes, and the use of Known Good Die (KGD) to improve yield [17]. The preference for die or wafer bonding depends on the application. In general, applications with high yielding die of the same size typically prefer wafer bonding while applications with low yielding die or different size die typically prefer die bonding. A 3D integration process that is well suited to either die or wafer bonding will thus be more broadly applicable.

25.2.2.4 Low-Temperature Direct Oxide Wafer Bonding Process

The primary 3D integration requirements for a direct oxide bond technology are thus high bond strength at low temperature, resilience to void generation at the bond interface and ability to bond either die or wafers. These requirements have been met with a comprehensive understanding of the mechanical, chemical and structural parameters that promote the formation of chemical or covalent bonds at low temperature [18–20]. The four key components of this technology are mechanical surface specification, chemical surface specification, bond surface alignment and placement, and post-placement bond strength increase.

Low-Temperature Direct Oxide Bond Mechanical Specification The mechanical specification for preparing the surface of a wafer for low temperature direct bonding includes ensuring that the thickness, bow, warp and surface roughness are within acceptable control limits. These limits depend on the Young's modulus of the wafer in that it is preferable for the wafer surfaces to be drawn into intimate contact by the attractive forces of the direct bond. For silicon wafers, the semi-standard values of thickness, bow and warp can be easily accommodated by the attractive forces of a direct oxide bond.

The use of direct oxide bonding to stack 2D CMOS ICs is complicated by the relative non-planarity of the IC surface compared to the planar silicon substrate surface. This non-planar surface is due to CMOS wafer fabrication steps, for example the pad cut through the top passivation layer for test and packaging. The non-planarity of the IC surface can vary significantly, depending upon the type of CMOS wafer fabrication process. For example, CMOS ICs built with an aluminium-based BEOL typically have a greater non-planarity than CMOS ICs built with a copper-based, dual damascene BEOL due to a CMP process step after the last metal deposition that improves planarity.

The IC surface non-planarity needs to be planarized to achieve direct bonding over the entire IC surface. This planarization is readily achieved with a combination of silicon oxide deposition and CMP, which are fabrication technologies used in CMOS BEOL fabrication. This CMP technology is also capable of achieving a <0.5 nm surface roughness specification for direct bonding with slight modification to typical production CMP processes that exhibit $\sim$1 nm surface roughness.

Low-Temperature Direct Oxide Bond Chemical Specification The chemical specification for preparing the surface of a wafer for low-temperature direct bonding includes activation and termination of the surface. Activation and termination can be accomplished in a single process step or in separate steps. The activation process includes the breaking of surface bonds to increase the surface reactivity to a desired chemical species termination and facilitate the removal of chemical byproducts after initiation of the direct bond. This process can result in the removal of a few monolayers of silicon oxide without increasing the surface roughness and can be implemented with a plasma or reactive ion etch using, for example oxygen, argon or nitrogen. The activation process may also serve to clean the surface from any residual hydrocarbons or other contaminants remaining after completion of the mechanical specification. Removal of this contamination avoids nucleation of chemical byproducts and generation of byproduct voids at the bond interface and allows diffusion of chemical byproducts away from the bond interface, as described below.

After or during activation, the surface is terminated with a desired chemical species. The chemical species is preferably reactive at low temperature and has byproducts that can readily diffuse away from the bond interface. Two examples of this termination are given below. Equation 25.2 describes an amine surface termination after, for example, exposure to an ammonia-based solution. Equation 25.3 describes a fluorinated surface termination after, for example, exposure to a HF-based solution.

$$\mathrm{Si{-}NH_2 + Si{-}NH_2 = Si{-}N{-}N{-}Si + 2H_2} \tag{25.2}$$

$$\mathrm{Si{-}HF + Si{-}HF - Si{-}F{-}F{-}Si + H_2} \tag{25.3}$$

After the direct oxide bond surfaces have been appropriately terminated, they are ready to be placed together and initiate the direct oxide bond chemical reaction, for example that given in Equations 25.2 or 25.3. Initiation of either of these chemical reactions results in the generation of a very strong atomic bond with hydrogen as a byproduct. The high diffusivity of hydrogen in silicon oxide facilitates removal of this byproduct from the direct oxide bond interface, driving the formation of additional strong atomic bonds that afford a high surface bond energy at low or room temperature.

The ability of silicon oxide at the direct bond interface to effectively remove byproducts can be accomplished with silicon oxide on either one or both sides of the direct bond interface. This allows 3D CMOS IC F2B bonding with silicon oxide on the face to either silicon oxide or silicon on the back without the subsequent formation of byproduct voids.

One of the requirements of a broadly applicable 3D integration technology is that it supports either wafer or die bonding. Low-temperature direct oxide bonding is compatible with die bonding with a slight modification of the chemical specification described above. After the mechanical specification is achieved with CMP, the wafer is singulated with dicing, etching, lasing, scribe/break and so on. The planarized surface may be protected with a protective coating, for example photoresist, that may be removed after the singulation process. It is necessary that removal of this protective coating not leave any residue that may facilitate byproduct void generation or increase the surface roughness beyond that acceptable for direct bonding. After singulation, KGDs may be selected and loaded into tooling compatible with the chemical specification described above. The design of this tooling at wafer scale allows a wafer of KGD to be simultaneously treated chemically without the need for individual die handling.

The use of this technology for 3D integration is not limited to planarized silicon oxide surfaces. A wide variety of dielectrics that are compatible with planarization of a CMOS IC surface or silicon substrate and the chemistry exemplified by Equations 25.2 and 25.3 may also be used. Examples include, low-k dielectrics, oxy-nitrides and other dielectrics synergistic with existing CMOS IC wafer manufacturing.

Low-Temperature Direct Oxide Bond Alignment and Placement After a surface of an IC die or wafer has been suitably planarized, activated and terminated, it is ready to be directly bonded to a surface of another suitably prepared IC die or wafer by simply placing these two surfaces together. The suitably prepared IC die or wafers are aligned prior to surface placement and then placed within an accuracy required by the application. An advantage of the low-temperature direct oxide bond technology for 3D integration is the inherent precise post-bond alignment accuracy that can be achieved. This precision is achieved by achieving a high bond strength after placement that does not require a post-placement bond process that can generate an alignment error greater than the pre-placement alignment and placement alignment errors. Alignment accuracies within $\pm 1\,\mu m$ over 3 standard deviations are readily achieved with low-temperature direct oxide bonding [21].

Post-Placement Bond Strength Increase The placement together of two suitably prepared low-temperature direct oxide bond CMOS IC surfaces spontaneously initiates a chemical reaction that yields a high density of covalent atomic bonds. The accumulation of these atomic bonds result in a high surface bond energy and shear strength at low or room temperature. For example, a surface bond energy in excess of $1\,J\,m^{-2}$ has been reported without requiring a post-bond temperature increase [12]. These bond energies are not obtained instantaneously upon placement but increase over time according to the low-temperature direct oxide bond kinetics given, for example, by the combination of Equations 25.2 or 25.3 and the removal of byproducts from the bond interface. These kinetics can be rapidly accelerated with moderate application of temperature after the CMOS IC surfaces are initially placed into contact. For example, a post-bond temperature of $100\,^{\circ}C$ has been reported to reduce the time required to achieve a bond energy in excess of $1\,J\,m^{-2}$ by an order of magnitude, and a moderate post-bond temperature increase of $150\,^{\circ}C$ has resulted in bond energies in excess of the silicon fracture strength [12].

These surface bond energies are adequate for post-bond 3D integration fabrication processes, including via etch and fill, die or wafer thinning with backgrind, CMP, or etch, and dicing. The temperatures at which these surface bond energies are obtained are well within the thermal budget of CMOS ICs. The lack of placement or byproduct void formation and the ability to bond either die or wafers further enable low-temperature direct oxide bonding as a suitable 3D integration process.

25.2.2.5 Low-Temperature Direct Wafer Bonded ICs

A scaleable 3D integration process requires the ability to have 2D IC layers in the 3D IC stack whose thickness is determined by the thickness of semiconductor required for transistor operation and thickness of the BEOL interconnect stack. This requires the removal of that portion of the semiconductor substrate that is only required for handling purposes, which is typically >95% of the substrate thickness. Removal of this substrate portion after bonding avoids issues associated with handling very thin die, but requires a uniformly high surface bond energy to avoid delamination.

This capability of low-temperature direct oxide bonding has been demonstrated with several applications, including the fabrication of 100 GHz symmetric intrinsic heterojunction bipolar transistor (SIHBT) devices using bonded indium phosphide-(InP) based devices [22] and 125 000, 7 μm pitch pixel staring focal plane array (FPA) devices with 100% interconnectivity, using bonded silicon-on-insulator based devices [12], where the entire substrate was removed after bonding, leaving a residual bonded layer of only 1–2 μm. The bonded IC die or wafer can be bonded with the IC side either "face down" or "face up" to a "face up" or "face down" lower die or wafer. 3D "face up" stacking can be accommodated by a temporary bond of the IC die or wafer to a handle wafer, followed by desired thinning of the IC substrate, low-temperature direct oxide bonding of the thinned substrate side to the lower IC die or wafer, and removal of the temporarily bonded handle wafer. For example, the detector silicon wafer of the 3D FPA device described above was bonded "face up" by use of a low-temperature direct oxide bonding handle wafer bond.

The ability to scale the thickness of a stacked IC by aggressive thinning of an IC after bonding further enables scaling of the 3D interconnection pitch. A 3D interconnection density approaching that typically achieved between upper levels of metal in conventional CMOS integrated circuits, $\sim 1\,\mu m^{-2}$, can be expected for a stacked, thinned IC layer that has a thickness comparable to these upper levels of metal. 3D interconnection densities of $\sim 0.1\,\mu m^{-2}$ have already been achieved for stacked ICs thinned to ~2 μm [16, 23], providing evidence of the scalability of a low-temperature direct oxide bond technology to 3D stack ICs followed by post-bond 3D IC integration processes of thinning, via etching and via filling.

25.3 Direct Bond Interconnect

Although the most aggressive 3D interconnection pitch and 3D stacked thickness scaling can be achieved with post-bond thinning, via etch and fill described above, some 3D structures are not able to be thinned adequately to achieve this pitch. For example, a 3D stacked IC may require a relatively thick silicon layer for transistor or detector operation and/or a BEOL multilevel interconnect stack of metal and inter-level metal dielectric. This minimum thickness correspondingly limits the scalability of the 3D interconnection pitch. Moreover, if 3D vias need to be etched and filled after 3D stacking, they necessarily disrupt the BEOL of the stacked die or wafer by requiring an exclusion volume for the via etch/fill.

The 3D interconnect BEOL disruption can be avoided with a 3D bonding technology that forms a vertical 3D electrical interconnection integral to the bond. 3D stacking with bumping is an example of this type of 3D stacking, but the non-planarity of this 3D technology limits the scaling to a 3D interconnection pitch ~50 μm. One approach to improve this scaling has been to modify the bump metallurgy to facilitate the formation of smaller bumps, pillars and so on [24, 25]. Another approach is to form a nominally planar surface that can form integral interconnects as part of the 3D stacking or bonding process.

Copper thermo-compression bonding [26, 27] is an example of a nominally planar 3D interconnect technology that forms integral interconnects as part of the bonding process. This process requires a bond cycle consisting of time, pressure and temperature after die or wafers have been aligned and stacked to effect the bond and electrical interconnections. This bond cycle typically requires considerable time in an expensive capital equipment tool that increases process cost compared to a low-temperature direct oxide bond process which only requires alignment and placement in a less expensive capital equipment tool.

Although the low-temperature direct oxide bond tool cycle time is significantly lower than copper thermo-compression bonding, it is not able to form an integral 3D interconnect with the bond or 3D stack. A technology opportunity thus existed for a low-temperature direct oxide bond technology that can form an integral 3D interconnect with the stack. This technology has been developed, patented [28, 29] and trademarked as Direct Bond Interconnect (DBI®) [30].

25.3.1 DBI® Process Flow

The DBI® 3D integration process is similar to the low-temperature direct oxide bond process with respect to the direct bond capability, but different with respect to the inclusion of DBI® metal contact structures that are preferably co-planarized with a silicon or dielectric surface and provide for an integral 3D interconnect with the bond. Several different methods for the formation of these DBI® contact structures are possible. Two such methods are described below.

25.3.1.1 DBI® Process – Plating Method

In this method, the DBI® metal contact structures are plated. One example of this process flow is given below, assuming the starting wafer is a conventional CMOS wafer with either a planarized tungsten or copper last via formed to the last metal, according to an aluminium or copper BEOL, respectively.

1. Wafer Start – planarized tungsten or copper last via formed to last metal
2. Seed Layer Deposition, for example by physical vapor deposition
3. Metal Contact Structure Plating, for example using a photoresist mask
4. Seed Layer Etch, for example with a metal dry etch with the metal contact as a mask
5. Oxide Deposition, for example by physical vapor deposition or PECVD
6. Planarization, for example by CMP, to direct oxide bond 0.5 nm spec.

25.3.1.2 DBI® Process – Damascene Method

In this method, the DBI® metal contact structures are plated or deposited by physical vapor deposition (PVD). One example of this process flow is given below, assuming the starting wafer is a conventional CMOS wafer with the last metal planarized:

1. Patterning and Via Etch to last metal
2. Metal Contact Structure Deposition, for example by PVD or electroplating
3. Planarization, for example by CMP, to direct bond 0.5 nm spec.

The inherent planarity of these methods allows a significant extension in the scalability of 3D interconnect pitch beyond the capability of other 3D technology. For example, the similarity of the DBI® Damascene Method with typical CMOS BEOL implies <1 μm^{-2} 3D contact structure density capability.

After two wafers surfaces are DBI® planarized, for example with one of the methods described above, two such die or wafer surfaces can be bonded by first activating and terminating the surface as described for the low-temperature direct oxide bond process. The surfaces are then aligned such that the metal contact structures are opposed and then placed together, spontaneously generating a mechanical low-temperature direct oxide bond between the two die or wafers. There are several methods to form a DBI® 3D electrical interconnection. For example, a 3D electrical interconnection can be formed spontaneously with the mechanical direct oxide bond if, for example, the contact metal extends nominally a few nm above the surface and is free of native oxide. Alternatively, a 3D electrical interconnection can be subsequently formed with heating, if, for example, the DBI® contact metal is a few nm below the surface and has a native oxide, due to a larger coefficient of thermal expansion of the DBI® contact metal to the surrounding material that results in a metallic bond and 3D electrical interconnection after heating. Other combinations of metal topography and surface condition are possible to achieve DBI® connections.

25.3.2 DBI® Physical and Electrical Data

DBI® 3D interconnections have been demonstrated in both die-to-wafer (D2W) and wafer-to-wafer (W2W) formats. Figure 25.1 shows a four interconnection cross-section of a 1 000 000 interconnect, 8 μm pitch DBI® daisy chain array fabricated on silicon using the Plating Method, thermal oxide isolation, aluminium seed metal

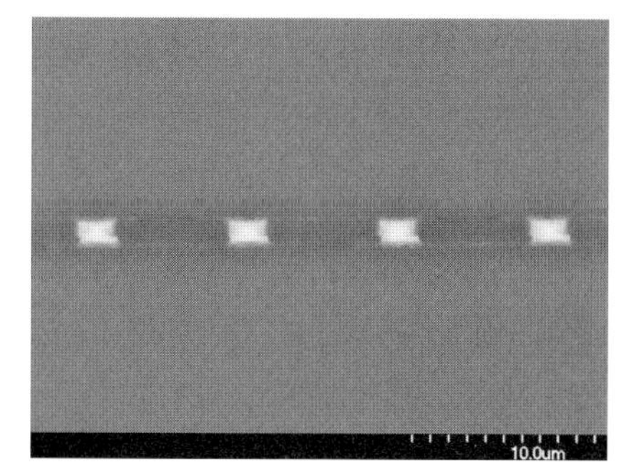

Figure 25.1 Scanning electron micrograph cross-section of 4 of 1 000 000, die-to-wafer, 8 μm pitch, 100% 3D interconnectivity DBI® electrical interconnections.

patterning and wet etching for lateral daisy chain routing, D2W format, and 350 °C heating cycle.

The typical resistance of these daisy chains is about 300 kΩ, indicating an average DBI® daisy chain link resistance of about 0.3 Ω. This resistance is comparable to that expected for the daisy chain link seed metal sheet resistance, indicating that the DBI® contact resistance is less than 50 mΩ.

The DBI® contact resistance is obtained after completion of the DBI® process and does not require an excess voltage to break down an interfacial oxide or other barrier. This has been verified by a high-resolution voltage source measurement performed before any other electrical testing of 1 000 000 element DBI® daisy chains. This measurement consists of a 10 nV voltage source, obtained with a combination of a 10 mV sensitivity HP 4140 B voltage source and a 1 000 000 : 1 voltage divider and a 0.1 pA sensitivity HP 3478A multimeter. The resistance of DBI® daisy chains initially measured with this 10 nV, 0.1 pA sensitivity agreed with the resistance measured at 100 mA, indicating that DBI® is "barrier-free" down to at least an average of ~50 fV.

Figure 25.2 shows a 200 mm W2W format 10 μm pitch DBI® daisy chain array fabricated on silicon using the Plating Method, thermal oxide isolation, aluminium seed metal patterning and wet etching for lateral daisy chain routing, and 350 °C heating cycle with about 150 000 000 3D interconnections arranged as approximately 330 independently testable daisy chain die with 460 000 connections each. After fabrication, the die are tested by removal of one of the silicon substrates and thermal oxide isolation to reveal one of the seed layers for test. The typical DBI® daisy chain resistance is about 150 kΩ or an average link resistance of about 0.3 Ω that is dominated by aluminium seed metal sheet resistance, providing additional indication that average DBI® contact resistance is less than 50 mΩ. Daisy chains that do not exhibit 100% 3D interconnectivity have had individual pairs of rows tested. Statistical analysis of this data indicates that these parts typically only have one or two 3D interconnections that are not operable and the daisy chain 3D interconnection defects are less than 1 ppm. Failure analysis of these defects reveals that point defects associated with patterning and wet etching of the seed layer are the dominant failure mode.

Figure 25.2 Photograph of approximately 150 000 000, wafer-to-wafer DBI® 3D interconnections from 330, 460 000 element, 10 μm pitch daisy chain die, after removal of silicon substrate.

25.3.3
DBI® Reliability Data

A preliminary evaluation of DBI® reliability indicates this 3D integration technology greatly exceeds JEDEC requirements. DBI® daisy chain bare die were subjected to 96 hour of HAST and 1000, −65 to 175 °C temperature cycles without any observable failures or significant change in resistance. These parts were further subjected to triple the HAST requirement (288 hour) and ten time the temperature cycling requirement (10 000 cycles) without observing any failures. These parts have also been quenched in liquid nitrogen (−196 °C) without observing any failures.

The DBI® daisy chain parts have also passed initial electro-migration tests; 1 000 000, 8 μm pitch DBI® daisy chains with 3 μm diameter contact area and 1 μm, 3σ alignment have not exhibited a change in resistance after 100 hour at 100 mA. Excellent electro-migration performance is expected for this 3D technology as the DBI® contact metal is not necessarily comprised of a solder or alloy.

25.4
Process Cost and Supply Chain Considerations

A significant portion of process cost is the capital equipment required to run the process. The DBI® capital equipment requirement for bonding is largely met by tooling that simply aligns die or wafers and places them into contact. This requirement can be inexpensively met with the use of conventional pick-and-place tools [21] as used for the fabrication of the parts shown in Figures 25.1 and 25.2. This simplicity provides a fundamental cost advantage for DBI® compared to other 3D integration bonding technologies. For example, copper thermo-compression technology is a 3D stacking technology that has demonstrated significant 3D capability [26, 27]. However, this technology requires a bond cycle after alignment and placement that requires the application of uniform temperature and pressure that considerably increases capital equipment cost per tool, and greatly increases cycle time, which greatly reduces throughput. Independent analysis of these effects has resulted in a projected W2W bond cost savings of as much as 75% for DBI® compared to other 3D bonding processes that require temperature and pressure bond cycles after alignment and placement [24].

Early generations of 3D technology were implemented in assembly, package, and test, or “back end” of the supply chain due to a combination of lowest cost, similarity of existing technology and minimum disruption of the supply chain. However, interestingly, the 3D technologies required to meet the next generation of 3D requirements are more similar to existing technologies in the foundry, or “front end,” than in the “back end.” For example, direct oxide bonding, DBI® and copper-thermo-compression bonding make use of chemo-mechanical polishing, which is a well-established “front end” BEOL technology. Another example is 3D via technology whose initial implementations were post-foundry but more recently has been implemented as a foundry *in situ* pre-BEOL process [26]. Advantages of this foundry implementation include a 3D via that does not disrupt the BEOL by requiring a BEOL 3D via exclusion, significantly shorter and denser 3D vias, and reduced downstream process CMOS degradation risk. Next generation 3D technology will thus find increasing synergy with established foundry processing that may result in at least a portion of this technology being adopted in the "front end." For example, assuming qualified 3D foundry *in situ* CMOS fabrication and DBI® CMP specification, wafers can exit the foundry "3D ready," reducing the investment that would otherwise be required in the highly cost-sensitive "back end" to achieve the improvement in next generation 3D IC performance.

References

1 Moore, G.E. (April 19 1965) Cramming more components onto integrated circuits. *Electronics*, **38**, 114–117.
2 Prismark – Stacked Die Report, (2002).
3 Walker, J. (June 2005) Market transition to 3-D integration and packaging: density, design, and decisions. 3-D Architectures for Semiconductor Integration and Packaging, Tempe, Arizona, USA.
4 Smith, L. (October 2006) 3-D package selection – key design, technical, business, and logistic factors. 3-D Architectures for Semiconductor Integration and Packaging, San Francisco, CA, USA.
5 Yole Development, 3-D IC Advanced Packaging Technologies, (February 2007), company report.
6 Enquist, P. (September 2006) High density direct bond interconnect (DBI™) technology for three dimensional integrated circuit applications. MRS Symposium Y, Boston, NH, USA.
7 Shimbo, M., Furukawa, K., Fukuda, K. and Tanzawa, K. (1986) *Journal of Applied Physics*, **60**, 2987.
8 Lasky, J.B., Stiffler, S.R., White, F.R. and Abernathy, J.R. (1985) *Proceedings of the IEEE IEDM*, **684**.
9 Lasky, J.B. (1986) *Applied Physics Letters*, **48**, 78.
10 Tong, Q.-Y. and Gosele, U. (1999) *Semiconductor Wafer Bonding*, Wiley.
11 Plosl, A. and Krauter, G. (1999) Wafer direct bonding: tailoring adhesion between brittle materials. *Materials Science and Engineering*, **R25**, 1–88.
12 Enquist, P. (September 2005) Room temperature direct wafer bonding for three dimensional integrated sensors. *Sensors and Materials*, **17** (6), 307–316.
13 Drost, R.J., Hopkins, R.D., HO, R. and Sutherland, I.E. (Sept 2004) Proximity communication. *IEEE Journal of Solid-State Circuits*, **39** (9), 1529–1535.
14 Franzon, P.D., Davis, R., Steer, M.B. *et al.* (September 2006) Contactless and via'd high-throughput 3D systems. MRS Symposium Y, Boston, NH, USA.
15 Garrou, P.E. and Vardaman, E.J. (March 2006) 3D integration at the wafer level. *TechSearch International*.
16 Topol, A.W., LaTulipe, D.C., Jr., Shi, L. *et al.* (1996) Three dimensional integrated circuits. *IBM Journal of Research and Development*, **50** (4/5), 491.
17 Enquist, P. (September 2006) Direct Bond Interconnect Slashes Large-Die SoC Manuracturing Costs, FSA Forum, pp. 12–13,45
18 Tong, Q.Y., Fountain, G.G. and Enquist, P.M. (June 7 2005) US Patent, 6,902,987.
19 Tong, Q.Y., Fountain, G.G. and Enquist, P.M. (May 9 2006) US Patent, 7,041,178.
20 Tong, Q.Y., Fountain, G.G. and Enquist, P.M. (September 19 2006) US Patent, 7,109,092.
21 Chou, H. (October 2006) Die and Wafer Level 3-D Packaging for Advanced ICs and Microelectronics. 3-D Architectures for

Semiconductor Integration and Packaging, San Francisco, CA, USA.
22 Enquist, P., Chow, D., Tong, Q.-Y. *et al.* (2000) Symmetric intrinsic HBT/RTD technology for functionally dense, LSI 100 GHz circuits. GOMAC.
23 Keast, C., Aull, B., Burns, J. *et al.* (September 2006) Three-dimensional integrated circuit fabrication technology for advanced focal planes. MRS Symposium, Boston, NH, USA.
24 Motoyoshi, M., Kamibayashi, K., Bonkohara, M. and Koyanagi, M. (2006) 3-D-LSI and its key supporting technologies. 3-D Architectures for Semiconductor Integration and Packaging, October 31–November 2, 2006.
25 Trezza, J. (November 2006) Multi-material system on Chi poC. 3-D Architectures for Semiconductor Integration and Packaging Conference, Burlingame, CA, USA.
26 Patti, R. (November 2004) The design and architecture of 3-D memory devices. 3-D Architectures for Semiconductor Integration and Packaging Conference, Burlingame, CA, USA.
27 Morrow, P. *et al.* (2004) Wafer level 3-D interconnects via Cu bonding. Advanced Metallization Conference.
28 Tong, Q.Y., Enquist, P.M. and Rose, A.S. (November 8 2005) US Patent, 6,962,835.
29 Tong, Q.Y. (2006) Room temperature metal direct bonding. *Applied Physics Letters*, **89**, 1.
30 www.ziptronix.com.

26
3D Integration ZyCube

Makoto Motoyoshi

26.1
Introduction

ZyCube was founded in 2002, based on the 3D technologies of ASET (Association of Super Electronics Technologies: Japanese Semiconductor Consortium) [1] and Tohoku University [2–8].

Figure 26.1 shows our expectation of 3D-LSI applications. The advantages of 3D-LSIs are less power, less cost, less space, less heat generation, less noise generation and the potential to increase circuit density per unit area. To transfer this into the main stream of LSI technology, it is necessary to suppress a rise in the cost of chip stack processes without degrading performance. Figure 26.2 shows the product road map. Starting from a sensor application, we will develop the 3D Super Chip, which consists of various functional and various size of thin chips with free combination. Following up on this technology we aim for bio-compatible LSIs, such as the artificial retina chip, and so on. Figure 26.3 shows the 3D-LSI process road map. In this figure, there are two technologies. The upper line is for the current 3D-LSI structure in which the TSV is formed under the peripheral bonding pads. In almost all LSI designs, there are no active devices under them, in order to avoid bonding damage. So the through hole design can be relaxed up to the bonding pad size. Another advantage is utilizing the current chips and designs with or without minor layout modification. The main merit of this technology is miniaturization. The lower line in Figure 26.3 is for next-generation 3D-LSIs. To facilitate the performance of 3D-LSI the circuit blocks in stacked chips need to connect directly with fine pitch TSVs and micro-bumps. Consequently, to avoid a chip area penalty, their pitches need to be shrunk to less than 5 μm.

26.2
Current 3D-LSI–New CSP Device for Sensors [9–11]

In an ideal image sensor, pinouts are preferably located on the opposite side of the die from the sensor array. For realizing this structure, 3D LSI technologies, such as

Handbook of 3D Integration: Technology and Applications of 3D Integrated Circuits.
Edited by Philip Garrou, Christopher Bower and Peter Ramm

ISBN: 978-3-527-32034-9

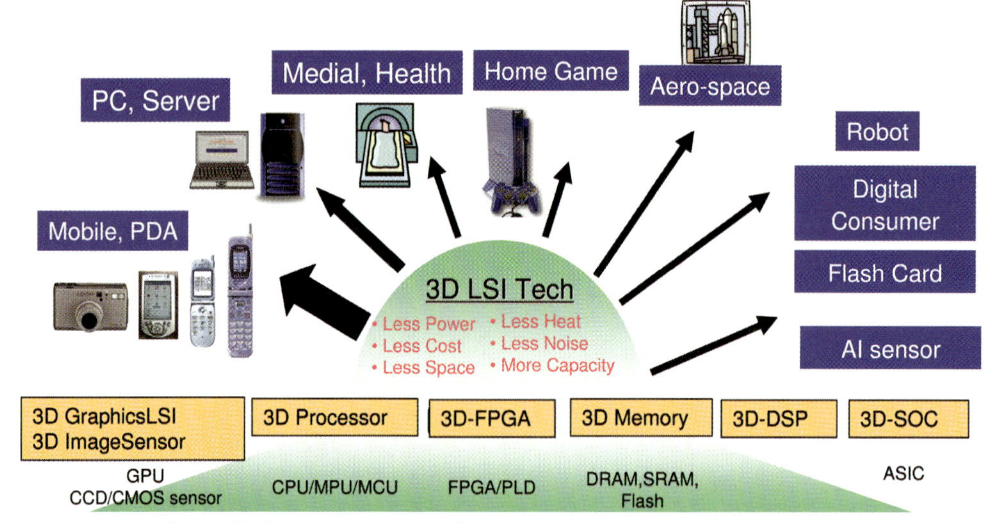

Figure 26.1 Potential application of 3D-LSI.

through silicon via (TSV), wafer thinning and bump formation are useful. Figure 26.4 shows the schematic view of a new chip size package (ZyCSP) for sensor applications. This structure does not have a real 3D structure, but applies the same technology as 3D integration. It is easy to expand the stacked structure, to include a sensor with DSP, or a sensor with memory and DSP. There are two issues in applying this technology to an image sensor device. One is maintaining a low temperature during

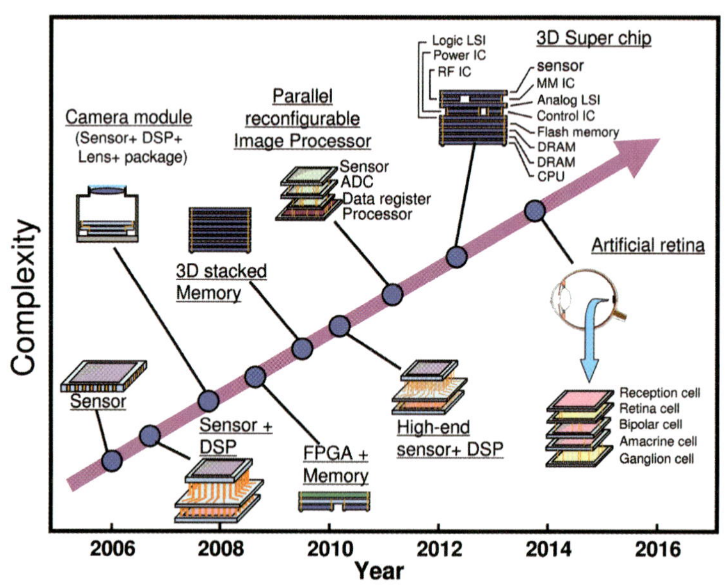

Figure 26.2 Product roadmap of 3D-LSI.

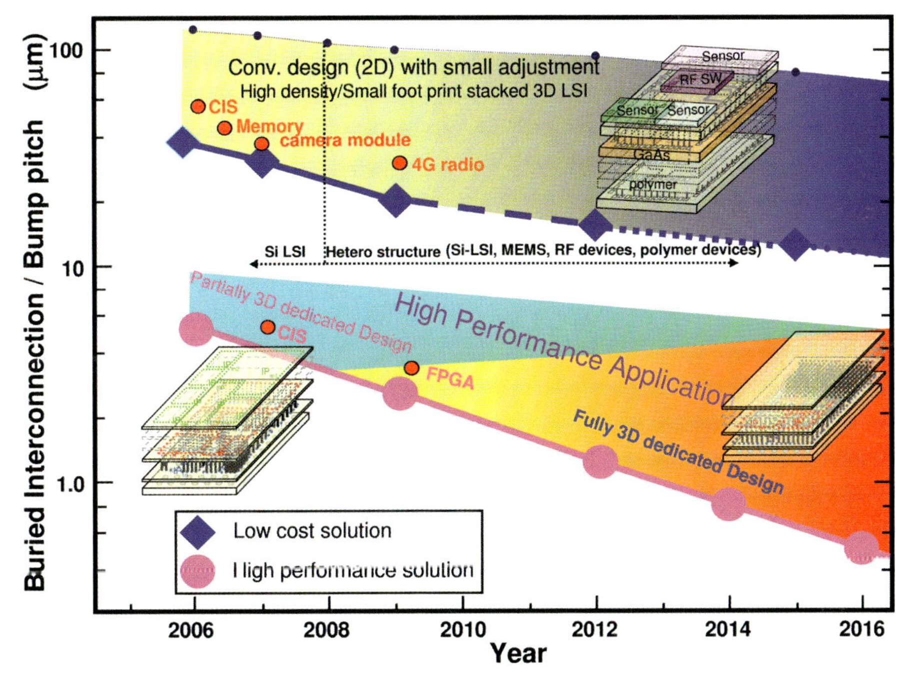

Figure 26.3 3D-LSI process road map. Upper line is for high density application and lower line is for high-performance application.

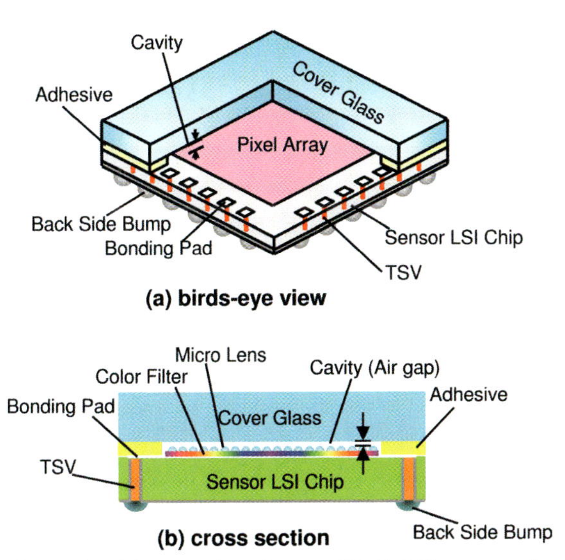

Figure 26.4 Schematic view of a new CSP (ZyCSP) structure for sensor application.

the fabrication process, because the thermal stability of the polymer micro lens and the color filter is less than 200 °C. This process is analogous to the CMOS-BEOL (back end of line) process. But the process temperature of most typical BEOL unit processes is around 350 °C. Therefore, it is necessary to lower the unit process temperature to 200 °C and optimize it, carefully without degradation of reliability. Another issue is avoiding degradation of the optical characteristics during the processes. At the finished wafer stage, the surface of the pixel array is not covered with a passivation layer, which is different from other LSI chips.

26.2.1 New Chip Size Package (ZyCSP) Process

This new chip size package process uses the dead area under bond pads and TSVs that links pads to backside interconnects under the pads. Figure 26.5 shows the process sequence for making the new chip size package (ZyCSP). After finishing the LSI process, a handle wafer is attached to the sensor LSI wafer by adhesive film (Figure 26.5a), followed by wafer thinning down to about 100 μm by back grinding and polishing (Figure 26.5b). Subsequent process steps for TSV formation are shown in Figure 26.5c and d. Back side deep Si etch and successive SiO_2 etch process with photoresist mask form through holes. Figure 26.6a shows a SEM image of a $60 \times 60\,\mu m$ through hole after deep Si etch. Figure 26.6b shows a FIB-prepared cross sectional SEM image of a 60 μm ϕ through hole with a depth of 100 μm. The sample consists of a sensor LSI wafer, adhesive layer and handle glass wafer. To avoid charging of the oxide layer and the handle glass wafer surface, the sample is coated with sputtered carbon. Optimizing Si/SiO_2 etch selectivity and Si etch rate, we can suppress the notch at the $Si–SiO_2$ interface. After making sidewall insulation, the

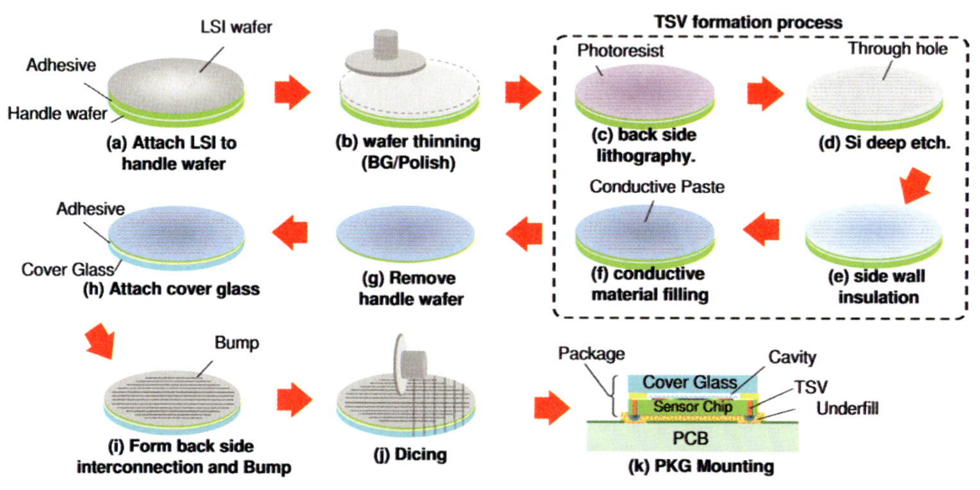

Figure 26.5 Process sequence of new CSP.

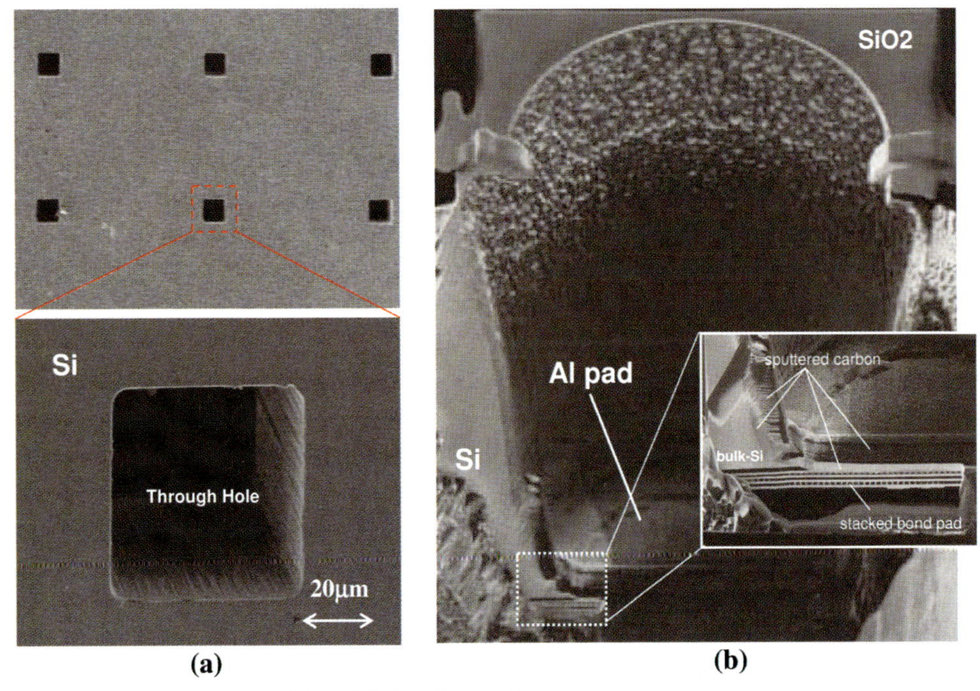

Figure 26.6 (a) A SEM image of through holes after Si etch. Through hole size is 60 × 60 μm and depth is about 100 μm. (b) A FIB-prepared cross sectional SEM image of a 60 μm-diameter through hole after Si/SiO_2 etch. This sample is coated by sputtered carbon for protection during the XeF_2 etch in the sample fabrication process. Sample structure: sensor LSI chip/adhesive/ cover glass.

through hole is filled with conductive material (Figure 26.5e and f). The figure shows one example which is filled with conductive paste by a printing technique. Then the handle wafer is replaced with a cover glass. This process is only for sensor applications (Figure 26.5g and h). Subsequently, after forming backside interconnection and bumps, the sensor wafer with cover glass is diced into the sensor chips (Figure 26.5i and j). Figure 26.7 shows a SEM image of the edge of the new CSP. Dicing is done from the front (cover glass) and back (Si) sides. The cover glass is chamfered by a 60° angled cut and subsequently diced fully.

In Figure 26.8, this process is compared with the conventional process. The main issue with the conventional process is low yield. As mentioned before, after finishing the LSI process, the polymer micro lenses are exposed without protecting material. Typically, assembly processes, such as dicing, die bonding and wire bonding, are performed in a cleanroom of lower classification than wafer fabrication. The lens surface is not flat and particle reduction is difficult. Moreover, the silicon dust during the dicing process has sharp edges, and it sticks into or stays on the lens surface. In contrast to the conventional process, in the new CSP process the lens surface is

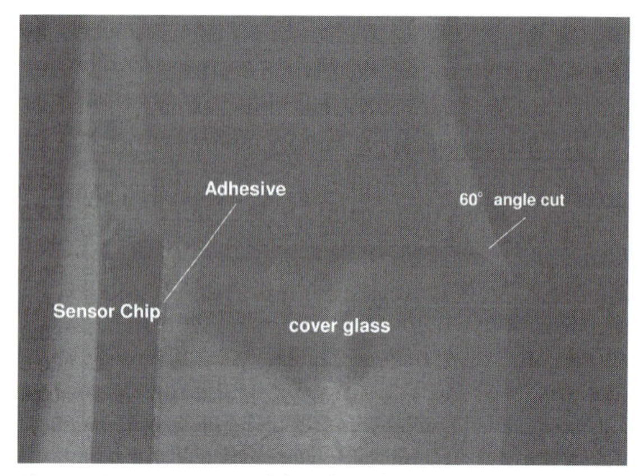

Figure 26.7 SEM image of the edge of ZyCSP. A 700 μm-thick cover glass is attached to a ~100 μm-thick sensor chip with adhesive.

protected by covering materials during these fabrication processes. Thus, yield loss with the dust during the CSP process does not occur and we can expect higher yield and quality.

26.2.2
TSV Filling Process

There are several candidates for the TSV filling process. Filling with conductive paste is a promising technology to reduce process cost. Figure 26.9 illustrates the process sequence of this technique. After thinning the LSI processed wafer, through holes are formed by deep Si etch and successive SiO_2reactive ion etch (RIE) with photoresist

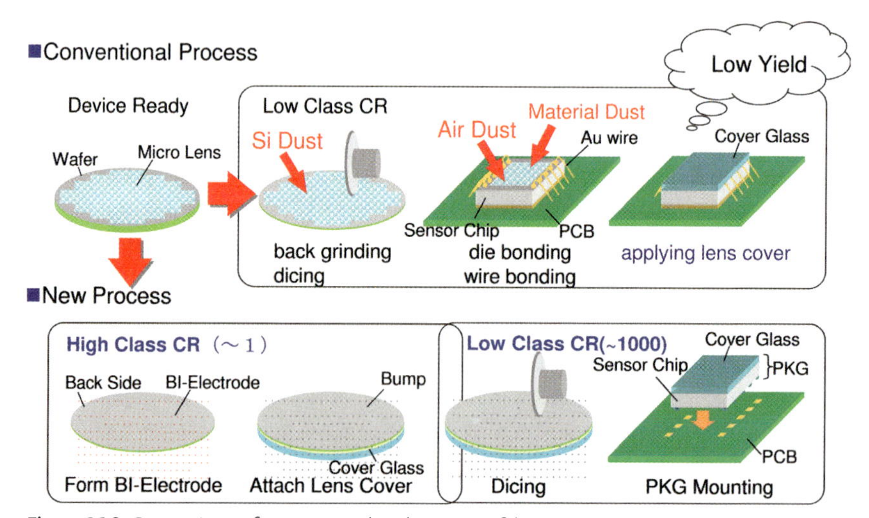

Figure 26.8 Comparison of conventional and new CSP fabrication processes.

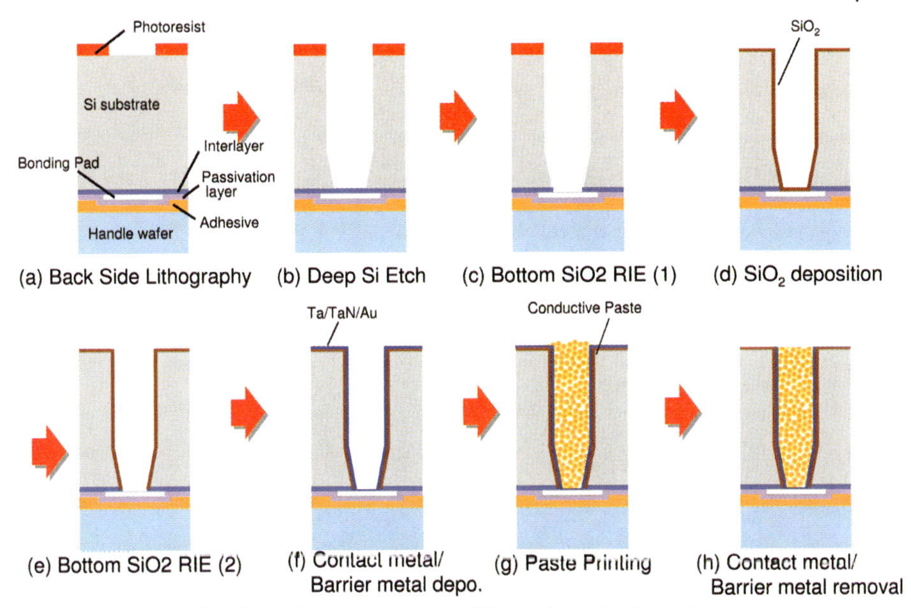

Figure 26.9 Process flow for buried interconnection filling with conductive paste.

mask to expose the bottom side of the bonding pad (Figure 26.9a–c). Then the side wall insulator is formed by using low-temperature plasma CVD-SiO_2 deposition and subsequent SiO_2 RIE (Figure 26.9d and e). Optimizing CVD and RIE conditions, only the bottom oxide of the through hole is completely removed to expose the backside of pad metal. After contact metal and diffusion barrier metal deposition, conductive paste is filled by a printing technique (Figure 26.9f and g). Finally, the contact metal and the diffusion barrier metal at the back side surface of the wafer are removed (Figure 26.9h). Figure 26.10 shows a cross sectional view of TSVs filled with Cu base paste material with a TSV diameter of 20 μm and depth of 270 μm. The paste showed

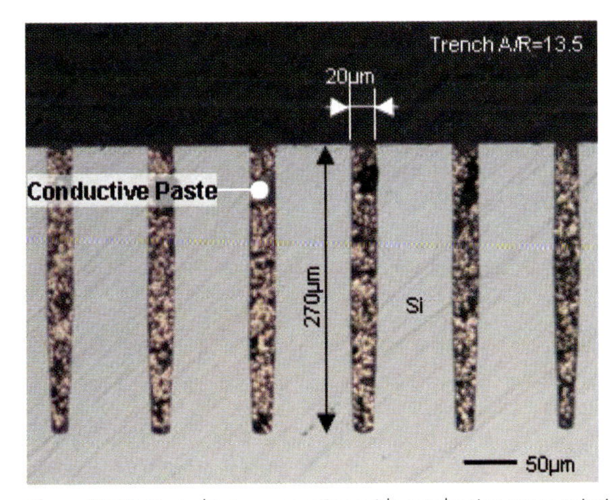

Figure 26.10 Buried interconnection with conductive paste printing.

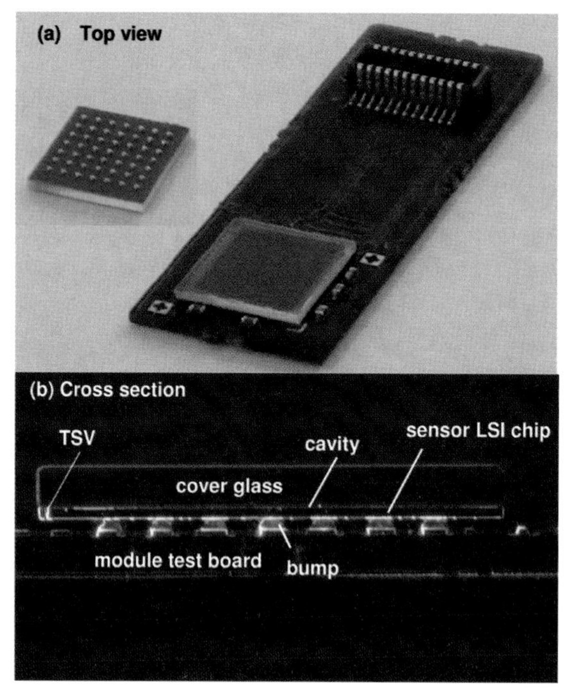

Figure 26.11 ZyCSP™ for sensor device (a) Top view image (b) Cross sectional image.

excellent filling characteristics. The merits of this method are the shorter turn around time and lower machine cost compared with other techniques, such as plating, metal CVD, and so on. The issues of this process are relatively high resistively and the contraction of paste material during the cure cycle. In this application, the cross section of the TSV is much larger than that of the interconnect in an LSI chip, so the parasitic resistance of the TSV does not affect device characteristics.

26.2.3 New Chip Size Package (ZyCSP)

Figure 26.11a shows the top view image of the New Chip size package mounted on an evaluation substrate. Figure 26.11b shows a cross sectional photo of the CSP. The cover glass is attached at the peripheral area of the chip with adhesive. An air gap is formed between the micro lens and the cover glass. The buried interconnection connects the pad electrode and the backside interconnection.

26.3 Future 3D-LSI Technology

There are five key technologies for realizing the future 3D-LSI: (1) wafer thinning, (2) adhesion of chips, (3) fine pitch buried interconnection, (4) micro-bump and (5)

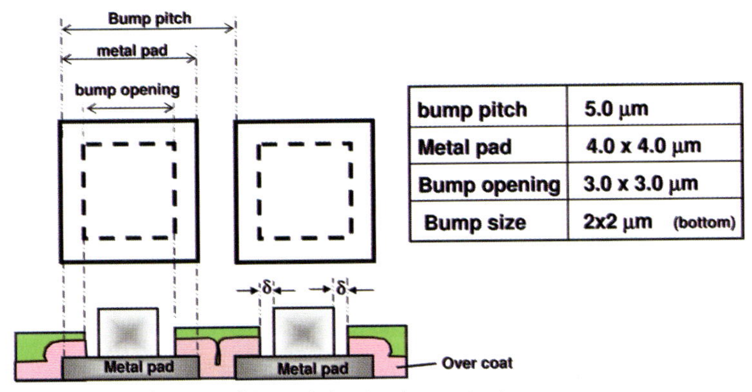

bump pitch	5.0 μm
Metal pad	4.0 x 4.0 μm
Bump opening	3.0 x 3.0 μm
Bump size	2x2 μm (bottom)

Figure 26.12 Schematic of 5 μm pitch micro-bump structure. Clearance grooves (δ) between bump opening and bump are formed by self-aligned process.

accurate and high speed chip alignment technology. In these technologies, micro-bump will be used for the face down architecture by itself, such as sensor and logic stacks, memory and CPU/GPU stacks, and so on. Figure 26.12 shows a schematic diagram of 5 μm pitch micro-bumps formed by a new micro-bump fabrication process. The bumps produced on the counter-face surface of LSIs are bonded thermo-compressively. The top of the bump from the LSI surface is less than 3 μm. Therefore, to obtain stable connections, in this fabrication process, the chip surface is planarized simultaneously. The clearance grooves between bump opening and bump, which absorb the deformation of the micro-bump and prevent a short-circuit between adjacent bumps, are formed by a self-aligned process. Figure 26.13 shows a photograph of the In/Au micro-bump chain TEG. The bump height is about 2.9 μm from the LSI surface and the size of the bump at the bottom is 2 × 2 μm. Figure 26.14 shows a cross sectional photo image of a 5 μm pitch micro-bump chain. The micro-bump has been developed as one part of fine pitch 3D buried interconnection. But the micro-bump itself can be used as the connection for a face down two chip

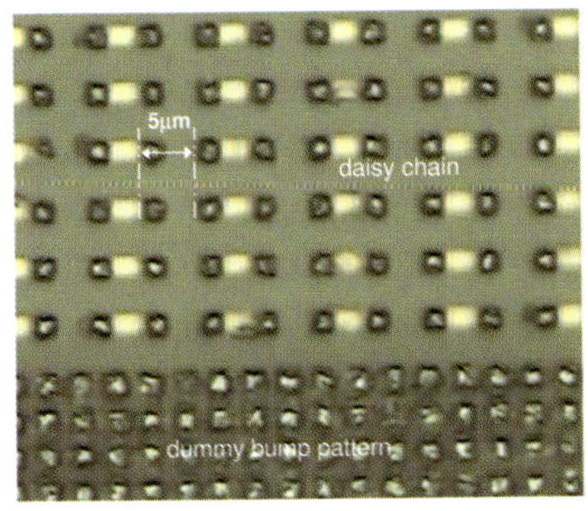

Figure 26.13 Top view of a 5 μm pitch daisy chain pattern before chip stacking.

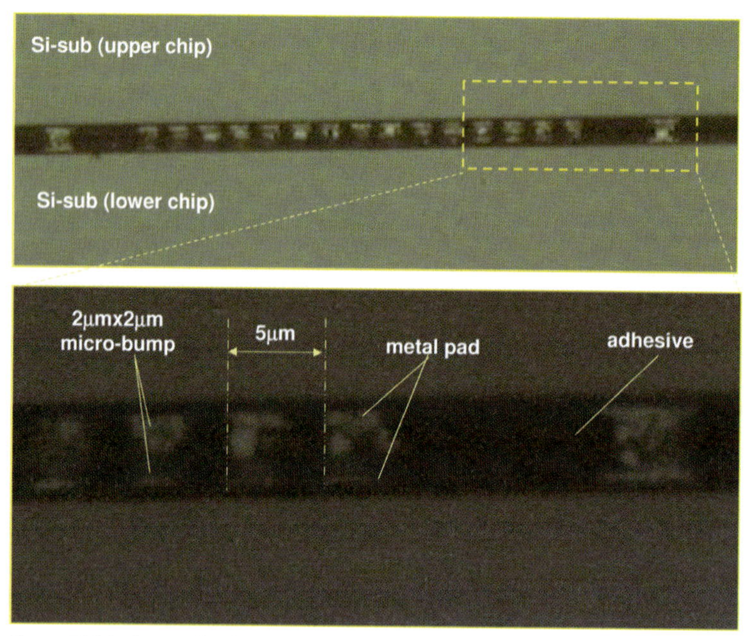

Figure 26.14 Chip to chip bonding with 5 μm pitch micro-bump.

architecture. Figure 26.15 shows an example of this application. By connecting the back illuminated type CMOS image sensor to Logic LSI, a high speed CMOS sensor system will be realizable. Other opportunities include CPU with memory, FPGA with memory, and graphic processor with memory.

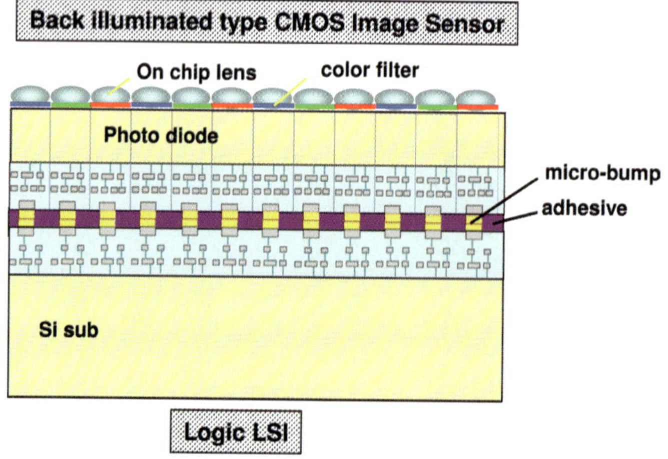

Figure 26.15 Face down architecture using micro-bump. This is an example of a high-performance CMOS image sensor.

References

1 Bonkohara, M. (1999) Proceedings 6th Annual KGD Industrial Workshop, session II-3.

2 Koyanagi, M. (1989) Roadblocks in Achieving Three-Dimensional LSI. Extended Abst. 8th Symposium on Future Electron Devices, pp. 50–60.

3 Takata, H., Nakano, T., Yokoyama, S. *et al.* (1991) A novel fabrication technology for optically interconnected three-dimensional LSI by wafer aligning and bonding technique. Extended Abststracts 1991 International Semiconductor Device Research Symposium, pp. 327–330.

4 Koyanagi, M., Kurino, H., Matsumoto, T. *et al.* (1998) New three dimensional integration technology for future system on-silicon LSIs. IEEE International Workshop on Chip Package Co-Design, 96–103.

5 Koyanagi, M., Kurino, H., Lee, K.W. *et al.* (1998) Future system-on-silicon LSI chips. *IEEE Micro*, 18 (4), 17–22.

6 Kurino, H., Lee, K.W., Nakamura, T. *et al.* (1999) Intelligent image sensor chip with three dimensional structure. *IEEE IEDM Technical Digest*, 879–882

7 Lee, K.W., Nakamura, T., Ono, T. *et al.* (2000) Three-dimensional shared memory fabricated using wafer stacking technology. *IEEE IEDM Technical Digest*, 165–168.

8 Ono, T., Mizukusa, T., Nakamura, T. *et al.* (2002) Three-dimensional processor system fabricating by wafer stacking technology. Proceedings International Symposium on Low-Power and High-Speed Chips, 186–193.

9 Motoyoshi, M., Kamibayashi, K., Bonkohara, M. and Koyanagi, M. (November 2006) 3D-LSI and its key supporting technologies. *Technical Digest on 3D Architecture for Semiconductor Integration and Packaging.*

10 Bonkohara, M., Motoyoshi, M., Kamibayashi, K. and Koyanagi, M. (2006) Three dimensional LSI integration technology by “chip on chip”, “chip on wafer” and “wafer on wafer” with system in a package. Extended Abstracts 2006 MRS Fall Meeting. p. 661.

11 Motoyoshi, M., Kamibayashi, K., Bonkohara, M. and Koyanagi, M. (2007) Current and future 3 dimensional LSI technologies. Technical Digest of the International 3D System Integration Conference, pp. 81–814.

IV
Design, Performance, and Thermal Management

Handbook of 3D Integration: Technology and Applications of 3D Integrated Circuits.
Edited by Philip Garrou, Christopher Bower and Peter Ramm

ISBN: 978-3-527-32034-9

27
Design for 3D Integration at North Carolina State University

Paul D. Franzon

27.1
Why 3D?

When is it advantageous to go vertical and when is it not? Stacking two wafers together and integrating them with vertical vias is not cheap. As a rough rule of thumb, the additional processing cost is about equivalent to that of adding two additional layers of metallization. This cost is even higher if individual die are stacked. This cost must be justified through performance gains or cost savings elsewhere in the system. This cost is much greater than simply that of even "high-end" sophisticated packaging. When might it possibly be justified?

Fortunately, there is a growing consensus that there are several, main-stream, circumstances which justify 3D integration. A list these of potential drivers for 3D integration is provided in Table 27.1. These will be discussed in turn.

The first, and most obvious, potential motivation is miniaturization. However, through-silicon 3D integration is rarely justified by the desire for miniaturization alone. For most circumstances, if volume reduction is the only goal, then it is much more cost effective to stack and wire-bond. This technology is already in wide-spread use in cell phones, and continues to grow in sophistication. However, one exception that is being widely explored is for memories. Wire-bonding can not be easily used to stack identical memory chips, as they are all the same size. In addition, there are system advantages to thinning and stacking multiple memory die such that the aggregate memory has the same end form factor as one memory package. For example, this technology could enable a credit card sized video storage and viewing device containing hundreds of hours of video.

The most explored advantage of 3D is to use it to reduce the interconnect distance between chip functions. Many researchers justify 3D from an interconnect delay, and interconnect power, perspective. Theoretically, the advantages can be substantial. Several studies have presented a Rent's rule style of analysis that presents significant advantages [3–5, 11, 12]. The basic argument relies on the fact that with each additional added layer of transistors there is a similar increase in the number of

Handbook of 3D Integration: Technology and Applications of 3D Integrated Circuits.
Edited by Philip Garrou, Christopher Bower and Peter Ramm

ISBN: 978-3-527-32034-9

Table 27.1 Issues that are potential drivers for 3D integration.

Driving issue	Case for 3D	Caveats
Miniaturization	Stacked memories. "Smart dust" sensors	For many cases, stacking and wire-bonding is sufficient
Interconnect delay	When delay in critical paths can be substantially reduced through 3D integration	Not all cases have a substantial advantage
Memory bandwidth	Logic on memory can dramatically improve memory bandwidth	While memory bandwidth can be improved dramatically, memory size can only be improved a little
Power consumption	In certain cases, a 3D architecture might have substantially lower power over a 2D one	Limited domain. In many cases, it does not
Mixed technology (heterogeneous) integration	Tightly integrated mixed technology (e.g. GaAs on silicon, or analog on digital) can bring many system advantages	Though might justify 3D integration, this driver might not justify vertical vias, except for the case of imaging arrays

circuit functions that can be interconnected within a fixed wire length. This leads to a 25% or more decrease in worst case wire length [3, 11], a similar decrease in interconnect power [5] and a decrease in chip area. However, experience shows that many designs do not realize this in practice. Fortunately, with careful choice appropriate design applications can be found. For example, FPGAs are very interconnect bound and can achieve substantial performance and power improvements when recast in 3D [12]. Other examples that have demonstrated significant advantages are discussed in Section 27.2.

Stacking memory die to create a new "super-memory" chip is not the only 3D application involving memory. An interesting and little explored area is logic-on-memory. That is creating a high bandwidth memory interface to the logic. For many end applications, the demand for memory bandwidth is growing rapidly. In many cases, this is due to the increased use of multi-core processors. With the addition of each processor, comes a similar requirement for increasing memory bandwidth. It is predicted that, by 2010, a 32-core CPU will require 1 TBps of off-chip memory bandwidth [6]. This, by itself, gives a fairly natural case for 3D, which has been only lightly explored, and then mainly in the context of general purpose computer architecture. For example, 3D caches can lead to 10–50% reductions in cache latency, depending on the benchmark used [9, 8]. Other applications that are likely to benefit from logic-on-memory include digital signal processing, graphics and networking. Some application-specific examples are given below. At least one company is exploring customized 3D memories for logic-on-memory applications [14].

The potential for power reduction comes from two directions. The prospective for interconnect power reduction was discussed above. An area that has been little explored is the potential for trading area for power. Given the relief provided by 3D integration on interconnect issues, this potential exists. Another obvious potential advantageous route to explore is to use 3D memory integration to reduce memory power. This is explored a little in one of the case studies below.

Finally, a compelling driver for 3D technology is mixed technology or heterogeneous integration. The main application explored to date has been imaging arrays. The advantage of using 3D is that no area has to be sacrificed on the imaging layer for circuitry (making it a more efficient photon collector), and considerable circuitry can be placed right underneath each pixel for pixel-level processing. For example, this approach has been used to produce a laser radar receiver array [1].

27.2 Interconnect-Driven Case Studies

At NC State University, the 3D Design group has developed several benchmarks establishing the potential for 3D, which are summarized here. All these benchmarks were investigated using the Lincoln Labs 3D SOI process [2]. This process permits three layers of three-metal, 0.18 μm SOI CMOS technology to be stacked and integrated using vias of approximately 1 μm foot-print. A cross-section of this process is shown in Figure 27.1. These benchmarks are as follows:

- A synthesized two-processor subsystem study [13] (MS thesis directed by W.R. Davis).
- A synthesized FFT subsystem study and analysis [7] (PhD thesis directed by W.R. Davis).

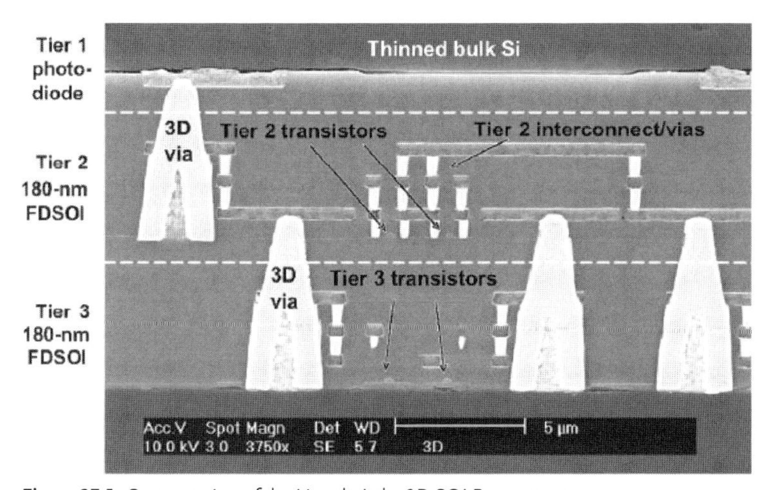

Figure 27.1 Cross section of the Lincoln Labs 3D SOI Process, as used in a photo-diode application. The applications described here-in were built using three 180 nm tiers. Taken from Reference [2].

- A full custom designed Ternary Content Addressable Memory [10]. (PhD thesis in the author's group).

All of the results presented here for these designs are based on extracted data from the detailed designs, not high level estimates. In the last two cases, the results have been implemented in silicon (though still in fabrication at the time of writing).

The design and analysis of a two-processor multi-CPU was executed in the Lincoln Labs 3D process and is presented in Reference [13]. The partitioned architecture is shown in Figure 27.2. Comparison with a 2D implementation in the identical base technology is given in Table 27.2. In this design the critical (slowest, most sensitive) delay path is through the "Wishbone traffic cop." By placing this module in the

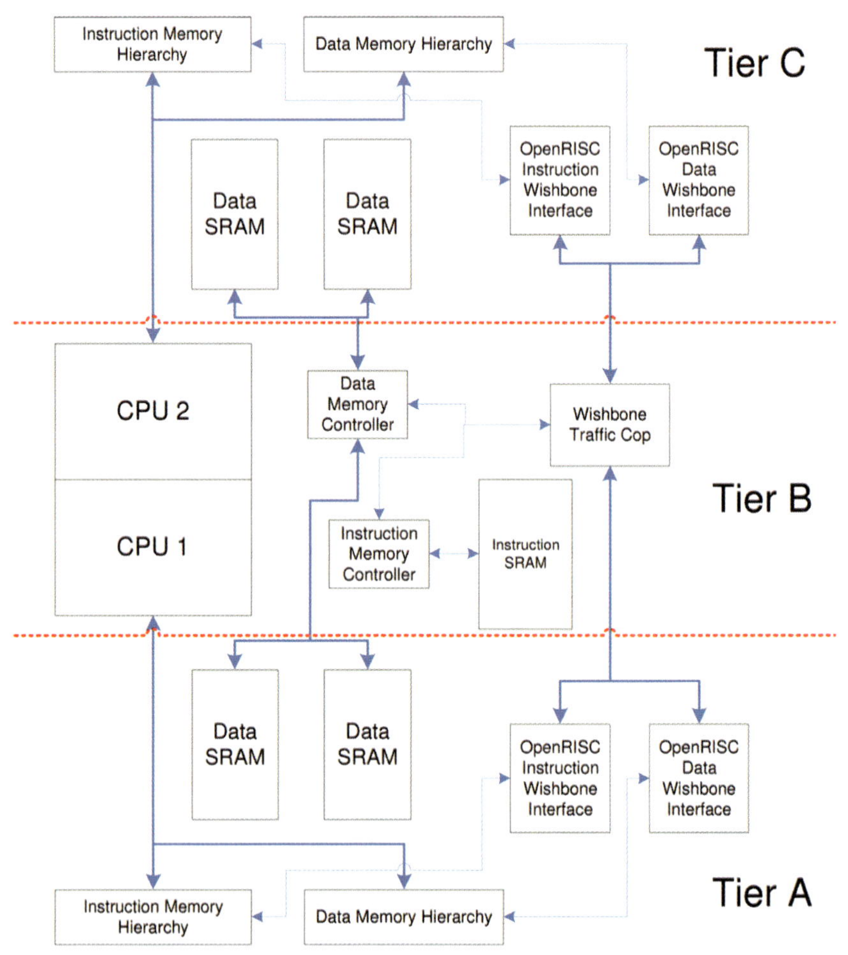

Figure 27.2 Architecture of a two-issue processor chip designed in the Lincoln Labs Process. Each tier refers to one tier of silicon in a 3-stack design. Taken from Reference [13].

Table 27.2 Improvement in interconnect delay for 3D two-CPU multiprocessor.

	2D	3D	Reduction
Critical path (ns)	25.13	17.87	28.9%
Data memory (ns)	36.6	24.86	32.07%
Instruction memory (ns)	37.82	24.01	36.51%

Taken from [13].

geometric center of the three-chip chip-stack, the savings in interconnect delay resulted in a 28.9% improvement in timing. Given that the path contains circuits as well as interconnect (and the circuits are unmodified), this is a very pleasing result and is in general agreement with theory. However, the overall decrease in area was about 1%, less than that predicted theoretically. The reduction in power was 3%.

In contrast, the results obtained in an implementation of a Fast Fourier Transform (FFT) Processor were not as compelling. The critical path was not as interconnect length dominated and a 3D implementation achieved only a 7% delay reduction [7]. However, Hua compared the performance for both designs with the improvements achievable by technology scaling, with the results presented in Figure 27.3. Adding two additional silicon layers is roughly equivalent to two generations of technology scaling.

Content Addressable Memories are used for associated lookups, for example much like a Thesaurus. You do not need to know the address of a word to find matches, all we have to do is provide the word. Using this process, a 3D Ternary Content Addressable Memory design was mapped onto 3D. At the cell level (Figure 27.4), the performance of the TCAM is dominated by the "Match Line" that indicates if a

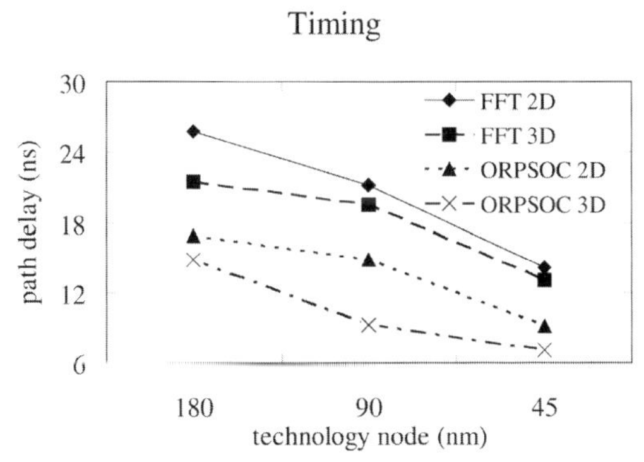

Figure 27.3 Improvement in path delay improvements achieved using a 3D 180 nm technology vs. technology scaling alone. Three metal layers were assumed for each silicon layer. FFT = Fast Fourier Transform. ORPSOC = Open core RisC Processor System on a Chip (the two core design presented in Figure 27.2). Adding two additional silicon layers is roughly equivalent to two generations of technology scaling. Taken from Reference [7].

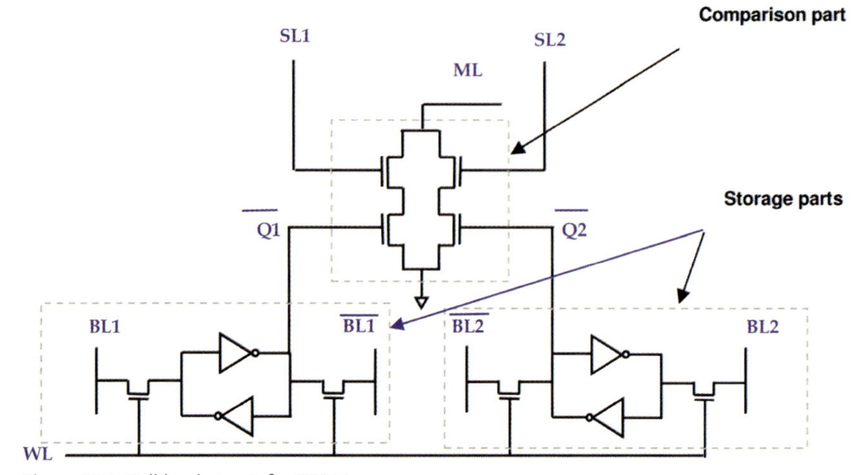

Figure 27.4 Cell-level circuit for TCAM.

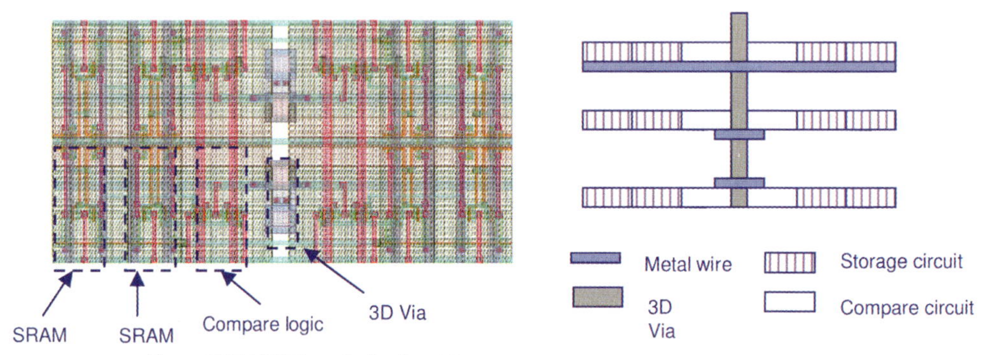

Figure 27.5 TCAM stacked cell architecture.

successful match is found. The match line is the highest capacitance line in the circuit. By stacking cells and careful optimization layout, its length can be reduced (Figure 27.5). This design was implemented and extracted using a 3D analysis tool. The results are shown in Table 27.3. With a total power savings of 23%, this provides a compelling case for 3D implementation.

Table 27.3

	2D Design	3D Design	Improvement
Match line power (μW)	44.6	32.3	28%
Total power (μW)	123	95.4	23%

27.3 Computer-Aided Design

In support of the 3D design community, the 3D Group at NC State University has developed a complete CAD flow, based on conventional tools provided by Cadence and PTC and with ongoing collaboration of PTC and the University of Minnesota. This flow is shown in Figure 27.6. Several changes have been implemented to enable 3D designs. Though a first release of the tools is available, the work in this tool flow is ongoing. In the first release, Cadence 2D tools were used, with the addition of a customized partitioning tool. PTCs Mechanica suite was used for thermal analysis, interfaced to the IC design tools via a macromodeling tool. In the second generation, 2D place and route will be replaced by a true 3D place and route tool.

For a designer, the main differences to consider in a 3D flow are as follows:

- 3D Floorplanning: True 3D floorplanning is needed to enable true 3D optimal designs. Careful consideration of the 3D spatial relationships of different modules in a design leads to short interconnect paths. The floorplan has a strong impact on thermal operation, as discussed below.
- Close integration with thermal analysis: This aspect is very important. Heat density, and thus temperature increases with each additional layer of transistor devices. For example, in all the designs above, the total heat density was almost tripled over that of a 2D design. Thermal analysis is essential at the floorplanning

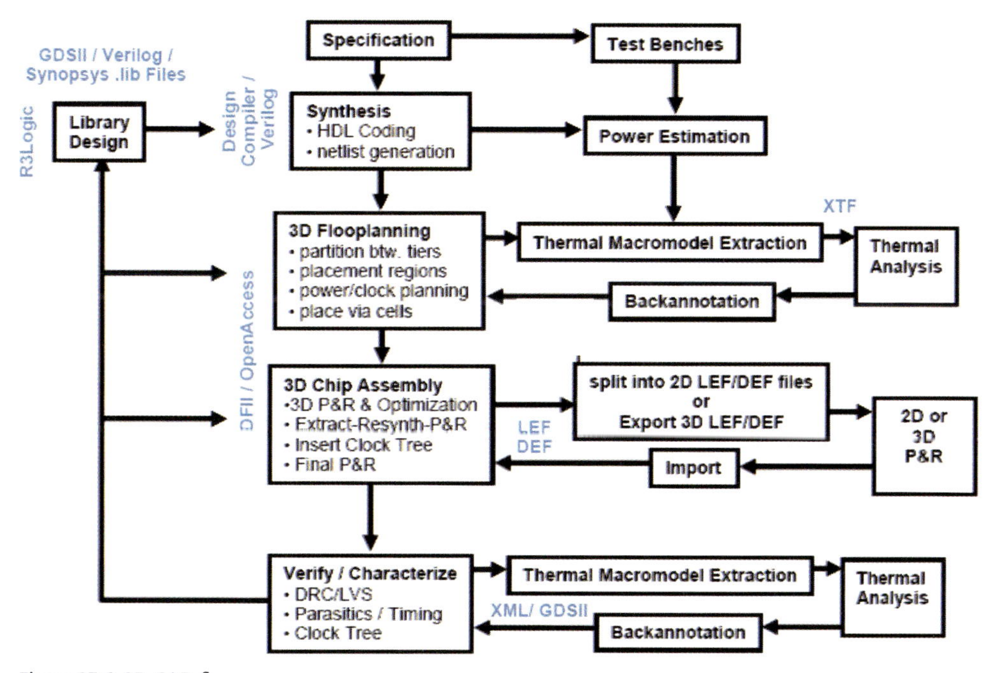

Figure 27.6 3D CAD flow.

level to determine if modules need to be rearranged in order to control temperature. It might also be used to determine if additional area should be allocated to thermal vias. However, notably, additional vertical vias only provide an incremental improvement when temperatures are too high. At this stage, temperature is better controlled through better choice of the design blocks implemented, and their arrangement in the 3D floorplan. For example, a "hot" module should be placed on a silicon layer closer to the heat sink. Thermal analysis is also important after detailed design is complete. For example, timing can not be predicted unless device temperature is well known. An important circuit to analyze is the clock circuit. The clock buffers not only produce a lot of heat (10–20% is typical) but require detailed timing analysis. Mis-prediction of clock skew can easily lead to chip failure. This is one circumstance where the addition of thermal vias around clock buffers can be used to provide fine-grain temperature control.

27.4 Discussion

With careful choice of application, and good design planning, integrated 3D designs can often offer a superior performance/cost point over their 2D equivalent. However, without both of these, 3D designs are not likely to offer a strong advantage, especially in designs where performance is dominated by the performance of multiple long-range interconnects.

3D design does bring some additional burdens though. The increased complexity and importance of thermal design was discussed above. Additional issues include the increased cost and risk of associated with prototyping, and yield management issues.

It is well known that the masks and wafer processing costs associated with the first prototype run in the multi-million dollar range, and increase rapidly with smaller technology nodes. This is true for one 2D design – imagine taping out multiple 2D designs simultaneously to create a 3D design. The cost and risk associated with a 3D ASIC could be quite daunting. There has not been much public discussion on this issue but possible approaches to alleviating this cost and risk include the following:

- Focusing on logic-on-memory. This separates the risks. Each can be approached independently.
- Make each chip in the stack identical at the physical layer. This means only one mask set and wafer run. Configure the connections using soft switches after stacking and integration. Alternatively, simply minimize the differences between each layer of silicon, for example limit the difference to one mask layer only.
- Multi-project runs.

Yield management is another issue that has been little discussed. If the yield of a single die at wafer level is 90%, then yield of a die from two stacked wafers would be 81%; from three wafers, 73%, and so on. This would create a clear financial disincentive for 3D. Fortunately, there are at least two ways to alleviate this issue.

One approach would be to use small die that have high yield. Another would be to use a die on wafer 3D integration approach, rather than wafer on wafer. If the separated die were even partially tested and sorted (as is commonly done before dicing), the yield would be that of undiced wafer die. If only good sites on the wafer were populated with mounted die, then the yield impact of 3D integration would be minimal.

References

1. Aull, B., Burns, J., Chen, C. *et al.* (Feb 2006) Laser radar imager based on 3D integration of Geiger-mode avalanche photodiodes and two SOI timing circuit laters. Proceedings of the IEEE ISSSC, pp. 1179–1188.
2. Burns, J.A., Aull, B.F., Chen, C.K. *et al.* (Oct. 2006) A wafer-scale 3 D circuit integration technology. *IEEE Transactions on Education*, **52** (10), 2507–2516.
3. Banerjee, K., Souri, S., Kapur, P. and Saraswat, K. (2001) 3-D ICs: A novel chip design for improving deep-submicrometer interconnect performance and systems-on-chip integration. *Proceedings of the IEEE*, **89** (5), pp. 602–633.
4. Davis, W., Wilson, J., Mick, S. *et al.* (Nov–Dec 2005) Demystifying 3D ICs: the pros and cons of going veritical. *IEEE Design and Test of Computers*, **222** (6), 498–510.
5. Das, S., Chandrakasan, A. and Reif, R. (2004) Timing, energy and thermal performance of three dimensional integrated circuits. Proceedings Great Lakes Symposium on VLSI, 338–343.
6. Hofstee, H.P. (May 2004) Future microprocessors and off-chip SOP interconnect. *IEEE Transactions on Advanced Packaging*, **27** (2), 301–303.
7. Hua, H. (2006) Design and Verification Methodology for Complex Three-Dimensional Digital Integrated Circuits, PhD Dissertation, NC State University, under the direction of W.R. Davis.
8. Kuhn, S.A., Kleiner, M.B., Ramm, P. and Weber, W. (Nov 1996) Performance modeling of the interconnect structure of a three-dimensional integrated RISC processor/cache system. *IEEE Transactions on CPMT, Part B*, **19** (4), 719–727.
9. Li, F., Nicopoulos, C., Richardson, T. *et al.* Design and management of 3D chip multiprocessors using network-in-memory, Proceedings ISCA'06, pp. 130–141.
10. Oh, E.C. and Franzon, P.D. (Oct. 2007) Design considerations and benefits of three-dimensional ternary content addressable memory. Proceedings IEEE CICC.
11. Rahman, A., Fan, A. and Reif, R. (2000) Comparison of key performance metrics in two and three dimensional interrgated circuits. Proceedings International Interconnect Technology Conference, pp. 18–20.
12. Rahman, A., Das, S., Chandrakasan, A. and Reif, R. (Feb 2003) Wiring requirement and three-dimensional integration technology for field programmable gate arrays. *IEEE Transactions on VLSI*, **11** (1), 44–54.
13. Schoenfliess, K. Performance Analysis of System-on-Chip Application of 3D Integrated Circuits, MS Thesis, NC State University, under the direction of W.R. Davis.
14. www.tezzaron.com.

28
Modeling Approaches and Design Methods for 3D System Design

Peter Schneider and Günter Elst

28.1
Introduction

3D Integration offers various capabilities for new system concepts and form factors. But, due to dense integration of different functional blocks, numerous physical interactions within the 3D systems exist. Especially, the influence of integration technology on system function has to be considered, preferably early in the design process.

Within the design process of integrated circuits, mostly on-chip interconnects are considered today. Using field solvers that are integrated in commercial design environments, parasitic RLC values are extracted. For digital circuits these values are provided using dedicated data formats like SPEF or DSPF and used for timing and cross talk investigations. Influences from packaging concerning thermal effects as well as EMC are usually considered after IC design. Thermal problems in ICs are examined in a considerable number of papers as well as at specific conferences and workshops like SEMITHERM and THERMINIC.

Currently, for 3D integration there is no established flow which considers the influence of packaging and integration technology on system behavior and reliability. However, the electrical behavior of interconnects, thermal management within the stack and thermo-mechanical issues are important problems that have to be solved. The following characteristics of 3D systems lead to new challenges for the design support:

- Functional layers are located closely on top of each other.
- Different materials and manufacturing technologies cause different properties of layers.
- There are many degrees of freedom for the system architecture.
- Design tools and design flows used for the functional layers might be very heterogeneous.

Currently, for modeling and simulation there are several methods and tools available that are also used in PCB or package design and can be adapted to 3D integration.

Handbook of 3D Integration: Technology and Applications of 3D Integrated Circuits.
Edited by Philip Garrou, Christopher Bower and Peter Ramm

ISBN: 978-3-527-32034-9

Published work is mainly focused on special topics like thermal management and modeling [1–4], detailed analysis of electromagnetic behavior of interconnects [5–9], circuit level modeling [10–17] and thermo-mechanical analysis [18–25]. Some papers cover the analysis of coupled physical problems [26, 27].

The main tasks within 3D design, which have to be supported by appropriate design methods and tools, are:

- System and architecture design, for example partitioning of system functions and their assignment to layers, design of data communication structures (e.g. busses) tailored on stacked layers and so on.
- Design of functional blocks, especially integration of the influence of neighboring layers.
- System verification, for example simulation studies, which allow the analysis of interactions between several layers.
- Reliability assessment.

Thus, these tasks are more or less strongly coupled. A real 3D design from scratch is a multi-criteria optimization task with a huge number of design parameters.

Since, currently, 3D integration mostly incorporates existing dies, there are a lot of constraints that decrease, on the one hand, the number of design parameters and, on the other hand, there are new problems that have to be solved.

The most important uncertainties for designers and the fields where information are desperately needed are:

- Thermal couplings between functional blocks that may influence the behavior semiconductor devices.
- Thermally induced stress that may influence system reliability and system function.
- Electrical characteristics of 3D interconnects and electrical coupling between layers, especially for power and signal integrity.

Depending on the kind of 3D integration, different effects are dominant. Figure 28.1 shows the basic approaches for 3D integration and the physical effects that are most important to these variants.

Reliability issues also have to be considered in the design process. They are strongly related to the interconnect technology and discussed in other chapters of this book in detail. In the following section we focus on the influence of 3D integration onto the system function. In the first part we describe how information mentioned above can be acquired by modeling and simulation and then can be provided to the designer. In the second part we focus on special aspects such as low power design and design for testability.

28.2 Modeling and Simulation

One basic problem within 3D design is the lack of information about effects that occur due to the dense integration of different components. Therefore, information

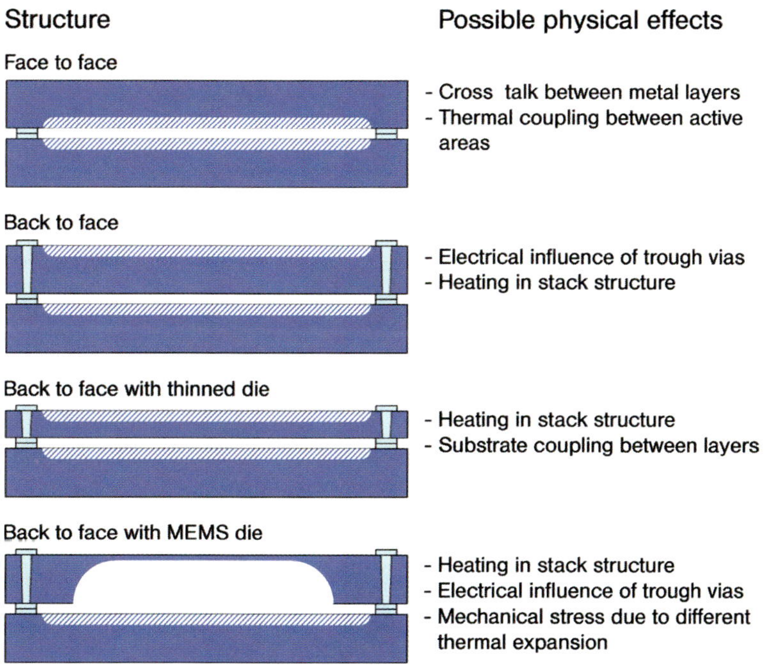

Structure | **Possible physical effects**

Face to face

- Cross talk between metal layers
- Thermal coupling between active areas

Back to face

- Electrical influence of trough vias
- Heating in stack structure

Back to face with thinned die

- Heating in stack structure
- Substrate coupling between layers

Back to face with MEMS die

- Heating in stack structure
- Electrical influence of trough vias
- Mechanical stress due to different thermal expansion

Figure 28.1 Basic approaches for 3D integration and possible physical effects.

from different physical domains has to be acquired and must be provided to designers. Due to the variety of structures and effects of different physical domains, efficient modeling approaches and simulation algorithms are key methods to solve several problems in this field. Therefore, an appropriate methodology for multi-level and multi-physics analysis of interconnect structures is required.

Depending on the design step, different models are needed and different types of simulations have to be carried out:

- Detailed analysis of physical effects based on the solution of partial differential equations (PDE) with algorithms like the Finite Element Method (FEM), Finite Difference Method (FDM) and others.
- Analysis of more complex systems based on less accurate but faster models. These models are usually based on differential-algebraic equations (DAE). This approach is widely used in analog circuit simulation, such as with SPICE and is also a basis for modeling languages like VHDL-AMS or Verilog-AMS.
- Simulation of digital circuits, including information about parasitics.
- System level simulations, which are usually carried out using signal flow oriented approaches and tools like MATLAB/SIMULINK.

Depending on the level of model abstraction, the information about physical effects has to be abstracted, too. Therefore, a hierarchical approach for modeling and simulation is implemented (Figure 28.2).

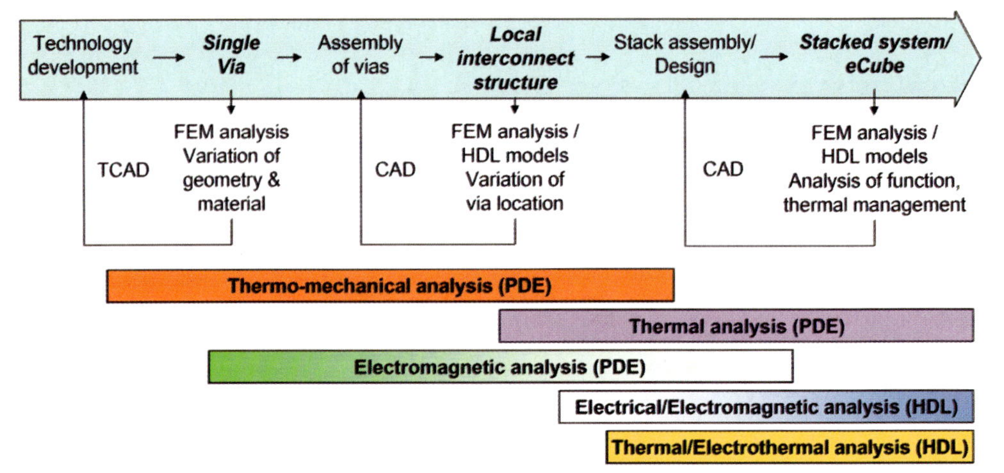

Figure 28.2 Hierarchical approach for modeling and simulation.

The main idea is to carry out analysis on one level of abstraction and use the output for optimizing on the same level as well as an input for modeling on the next level. For instance, thermo-mechanical analyses are used to investigate the reliability of single vias and small assemblies of such vias, respectively. Variations of geometry and material parameters can be analyzed. Performing electrostatic and electromagnetic calculations of single vias and local interconnect structures in the RF domain using PDE solvers, parasitic circuit elements like resistors, capacitors and inductors can be extracted and subsequently used to derive behavioral models for system level simulation. PDE solvers are also applied to investigate the thermal behavior of local interconnect structures and the stack assembly and to derive behavioral models. Electro-magnetic and thermal behavioral models can be combined with the electrical model of the entire system to enable complete system simulation considering the effects mentioned above.

28.2.1 Modular Modeling Approach

The geometry of a stacked structure is characterized by layers of a certain thickness and a considerable amount of interconnects between these layers. Assuming that layers can be connected using different integration technologies and that interconnections have the same shape between two layers, a modular modeling approach is chosen. Figure 28.3 shows the basic principle for combining geometrical and network models: the upper part shows subsections of the entire geometry and the lower part a simplified thermal network. Network elements represent the heat flow within the die, metal layers as well interchip vias. Usually, the thermal capacity also has to be considered, which is represented by capacitors between each network node and ground.

Since we have to deal with simulations within different physical domain, we aim at a tool-independent structural representation and a semi-automatic generation

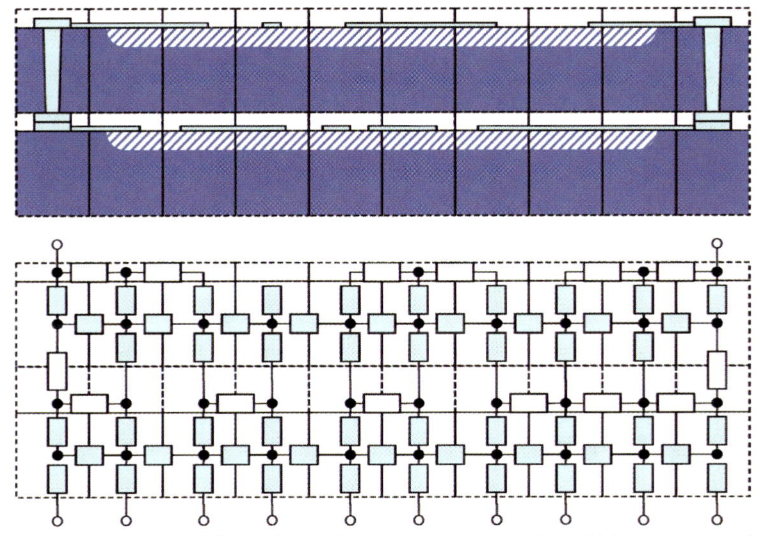

Figure 28.3 Principle of modular modeling – top: geometrical model, bottom: network model.

of models for PDE solvers. The basic idea is to generate parametric models of basic structures, which can be modified concerning geometrical (Figure 28.4) and material properties as well as the element type of the PDE simulator, which represents different physical effects. Complementarily, behavioral models for electrical and thermo-electrical simulations are also provided for these basic structures.

To build up models for stacked systems a library of basic structures must consist of models for normal die areas, areas with interchip interconnects, as well as areas without interconnects or PCB areas where the 3D system is mounted on. Figure 28.5 shows basic models for different interconnect technologies.

A tool-independent structural representation is carried out using XML. Figure 28.6 gives an example for an ICV-SLID.

For a combination of basic modules to form complex models within PDE solvers there are some restrictions concerning the interfaces between these modules:

- The size of contiguous geometrical entities has to be equal.
- Generally, materials of a layered structure should be identical in adjacent modules, but this is not absolutely necessary.
- The (finite element) mesh of contiguous interfaces should be identical or very similar to allow a mapping of the mesh by means of the FEM simulator.
- Boundary conditions and degrees of freedom at the interface areas have to match from module to module, such as heat flux and temperatures in the case of thermal investigations.

To allow the procedure of module combining at finite element level, uniform identifiers describing geometry, material, finite elements and so on have to be used. The values used for entity numbering, for example material type or element type numbers, have to be equal, too.

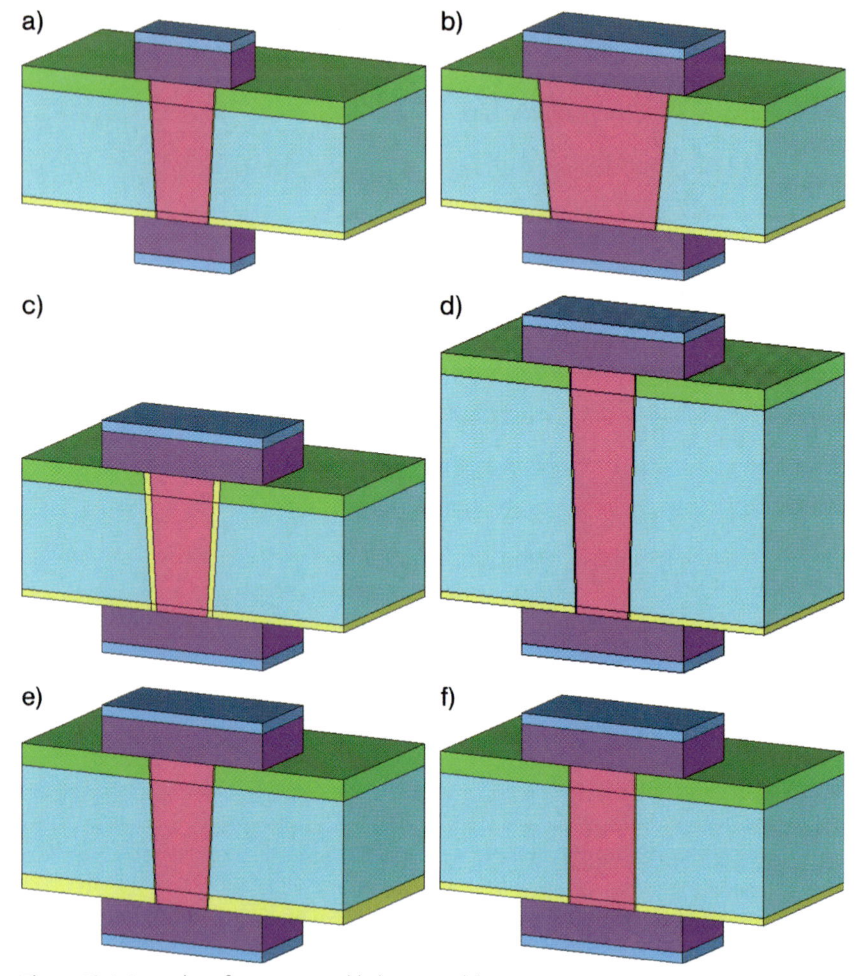

Figure 28.4 Examples of parameterizable basic models.

The modular approach described above allows one to perform detailed analysis of single basic modules for different physical effects, to generate complex models for PDE solvers, and, due to the correspondence between geometrical and HDL-models, also to build up complex system level models.

28.2.2 Simulation on Component Level

By performing simulations on the component level instead of the entire system, very detailed analyses of these limited structures (e.g. basic modules) are possible, for example using the finite element method. The main focus is on the determination of parameters that will be used in the entire system simulation or for first assessments:

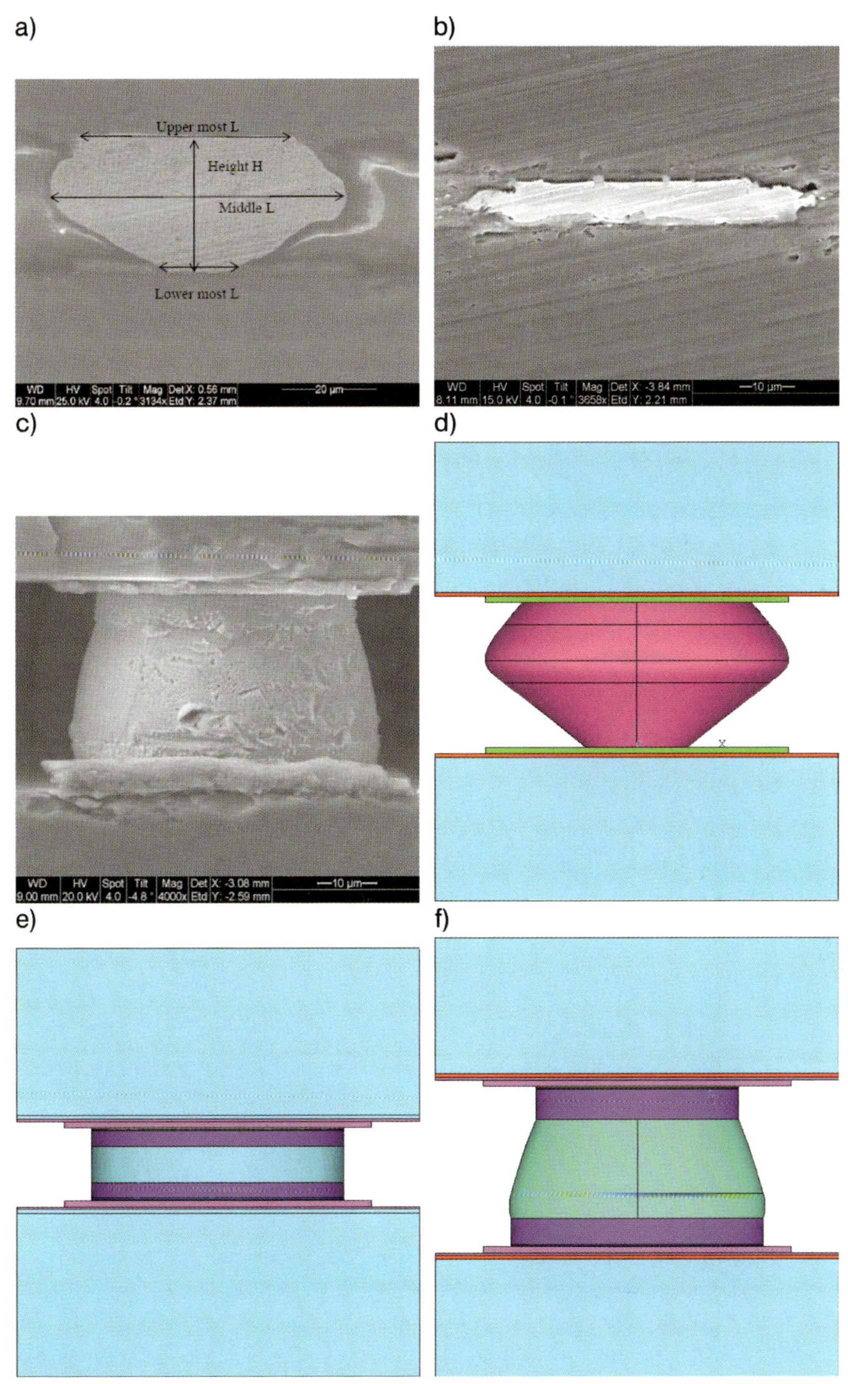

Figure 28.5 Basic models for different integration technologies: (a, d) Au stud bumps; (b, e) ICV-SLID; (c, f) µ-flip-chip (SEM figures by courtesy of SINTEF).

```xml
<Project>
  <Name>eCubes</Name>
  <Description>ICV cell geometry</Description>
  <Geometry>
    <Name>Dimension</Name>
    <Parameter>
      <Name>X</Name>
      <Value>50</Value>
    </Parameter>
    <Parameter>
      <Name>Y</Name>
      <Value>50</Value>
    </Parameter>
  </Geometry>
  <Geometry>
    <Name>Copper Pad</Name>
    <Parameter>
      <Name>X</Name>
      <Value>25</Value>
    </Parameter>
    <Parameter>
      <Name>Y</Name>
      <Value>25</Value>
    </Parameter>
    <Parameter>
      <Name>Height</Name>
      <Value>5</Value>
    </Parameter>
  </Geometry>
  .
  .
  .
</Project>
```

Figure 28.6 Tool-independent XML representation.

- Electrical parameters like resistance, capacitance, inductance for low and high frequencies.
- Thermal parameters like thermal resistance and thermal capacity.

28.2.2.1 Calculation of Circuit Parameters for Low Frequencies

For parameter estimation of parasitic circuit elements, field calculations using PDE solvers are an important method. Using the PDE solver itself or the results of electrostatic and electromagnetic calculations, the values of inductors, capacitors and resistors can be derived.

There are two main possibilities:

- Extraction of system matrices generated by the PDE solver and external derivation of the required parasitic element parameters.

- Electrostatic and electromagnetic calculations using the PDE solver.
  - Resistance: definition of current or current density – electrostatic simulation – derivation of resistance from resulting voltage drop.
  - Capacitance: applying voltages to the "electrodes" – electrostatic simulation – derivation of capacitance from resulting field energy; multiple "electrodes": multiple simulation runs always regarding two electrodes per run.
  - Inductance: electromagnetic simulation – derivation of inductance from resulting current-magnetic flow relation (inductivities and mutual inductivities); multiple "electrodes": multiple simulation runs always regarding two conductors per run.

The multiple simulation runs due to multiple "electrodes" are supported by special macros inside ANSYS.

Figure 28.7 illustrates the derivation of the resistance of an ICV-SLID structure and the capacitance of two adjacent structures.

Figure 28.8 shows results from field calculations and a basic network model for two adjacent SLID structures. Offsets in the manufacturing of single vias are taken into

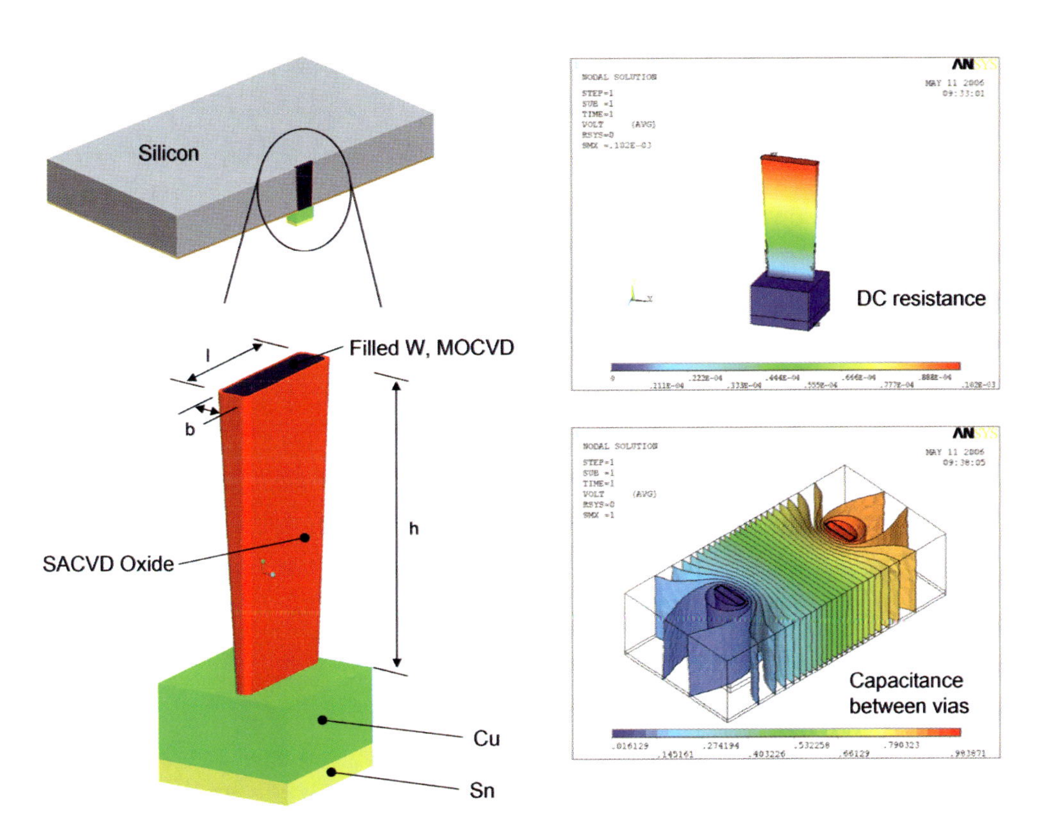

Figure 28.7 Simulation on component level – ICV-SLID and derivation of resistance and capacitance.

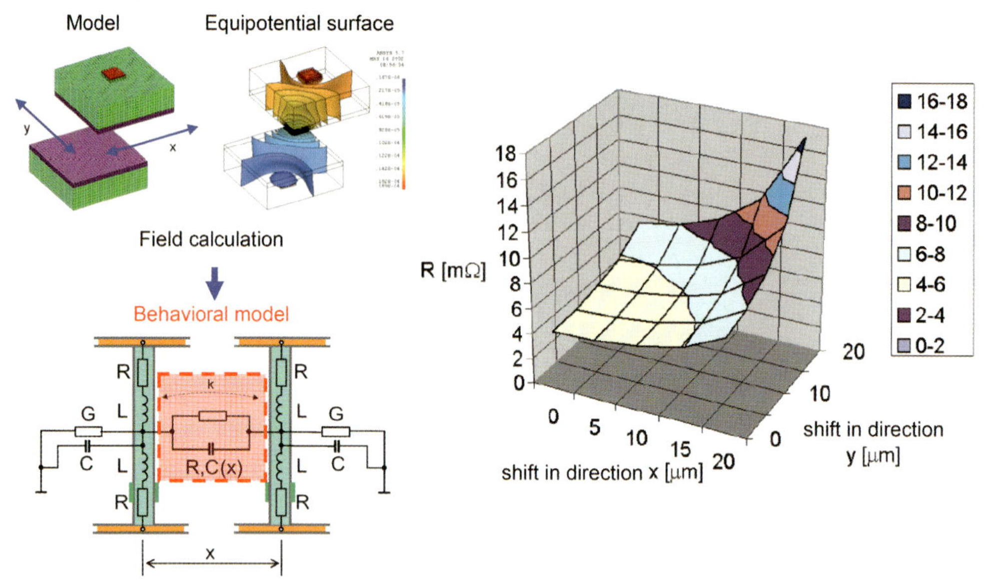

Figure 28.8 Basic equivalent network derived from field calculations.

account by the offset or shift parameters x and y. The coupling parameters $R(x)$ and $C(x)$ between the vias are determined as a function of the distance x. For multi-via structures a multitude of calculations has to be carried out to obtain the data for parameter estimation.

Not only distances or mounting tolerances are important parameters but also geometrical properties of basic models. Figure 28.9 for instance illustrates the influence of diameter and height of the interconnect structures already shown in Figure 28.5 on the resulting resistance. For technological reasons the height of the ICV-SLID structure is fixed at 13 μm.

28.2.2.2 **Electrical Behavior at High Frequencies**

Selected wires and interchip interconnects within a complex stacked system strongly influence the overall system performance. The most important issues, such as EMC and cross talk, are caused by electromagnetic coupling. Especially at high frequencies, the electrical behavior of inter-chip vias is important for the performance of the stacked system. Capacitive and inductive coupling between the vias and to the ground potential as well as the frequency dependence of the via resistance have to be considered.

The frequency dependence is based on the skin and proximity effect. The skin effect is a specific phenomenon in high frequencies, leading to a higher current density near the surface of a conductor than at its core. The electric current tends to flow at the skin because the energy flows in the dielectric around the conductor and permeates as a damped field into the conductor. Generally, the higher the frequency

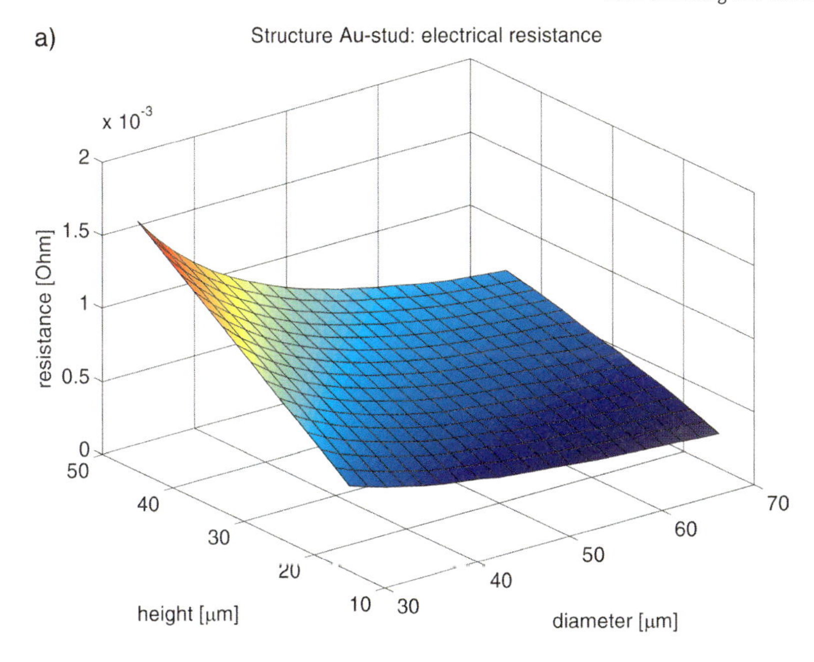

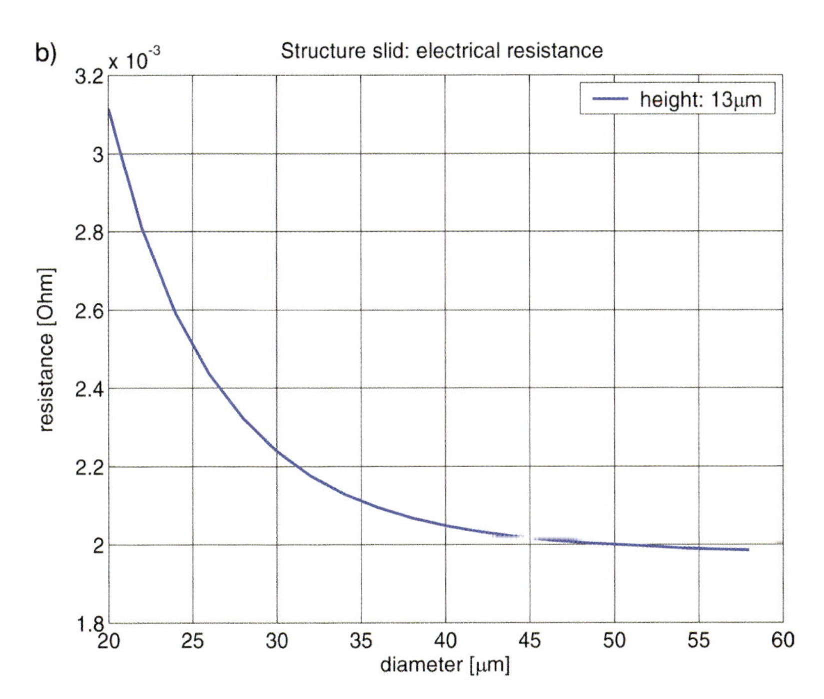

Figure 28.9 Parametric analysis of resistance of the interconnect structures shown in Figure 28.5 – (a) Au stud bumps; (b) ICV-SLID; and (c) μ-flip-chip.

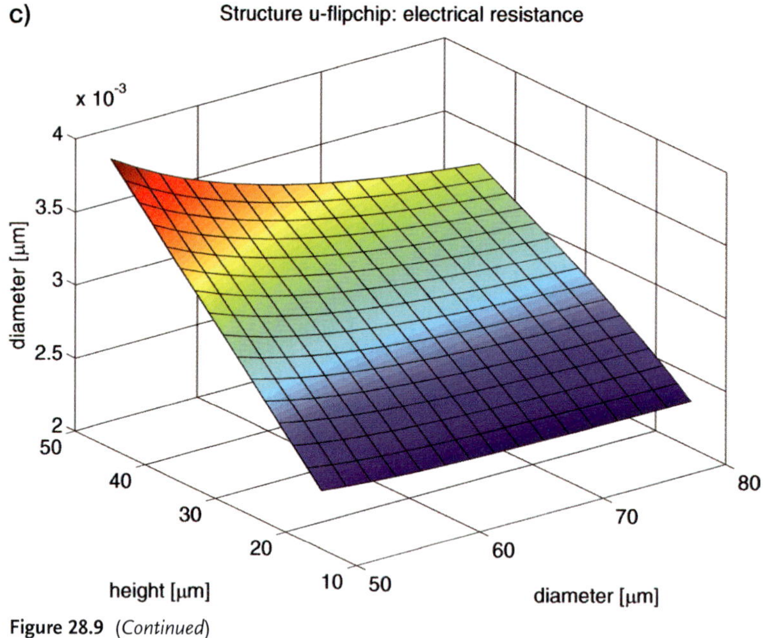

Figure 28.9 (*Continued*)

the stronger the skin effect. The decline of current density versus depth is specified by the skin depth δ, which is a measure of the distance over which the current falls to 1/e of its original value (Equation 28.1).

$$\delta = \sqrt{\frac{2}{\omega\mu\sigma}} \qquad \begin{aligned} \omega &= \text{angular frequency} \\ \mu &= \text{permeability} \\ \sigma &= \text{electrical conductivity} \end{aligned} \tag{28.1}$$

Assuming two adjacent conductors are close enough, the electromagnetic fields interact and modify the current distributions in the conductors. The resulting current crowding is called the proximity effect. If the currents flow in the same direction this effect has a repellent force on the current distributions and causes a displacement to the avert side of the conductors. Oppositely flowing currents result in displacement of the current to the facing sides. The strength of the proximity effect depends on the distance between the vias – the smaller the distance the stronger the proximity effect.

For illustration, the current density distribution referring to the DC value is shown in Figure 28.10 for a tungsten via with quadratic cross section ($5 \times 5 \mu m^2$) at different frequencies.

Figure 28.11a shows the resistance as function of frequency of a via with circular cross section with a diameter of 10 μm (continuous curve). A subdivision into four conductors with the same total area in a distance of 20 μm (see Figure 28.11b) leads to a minor improvement (dashed curve). Next, the frequency was fixed to 10 GHz and the distance of the vias was reduced from the starting value of 20 to 2.5 μm. This

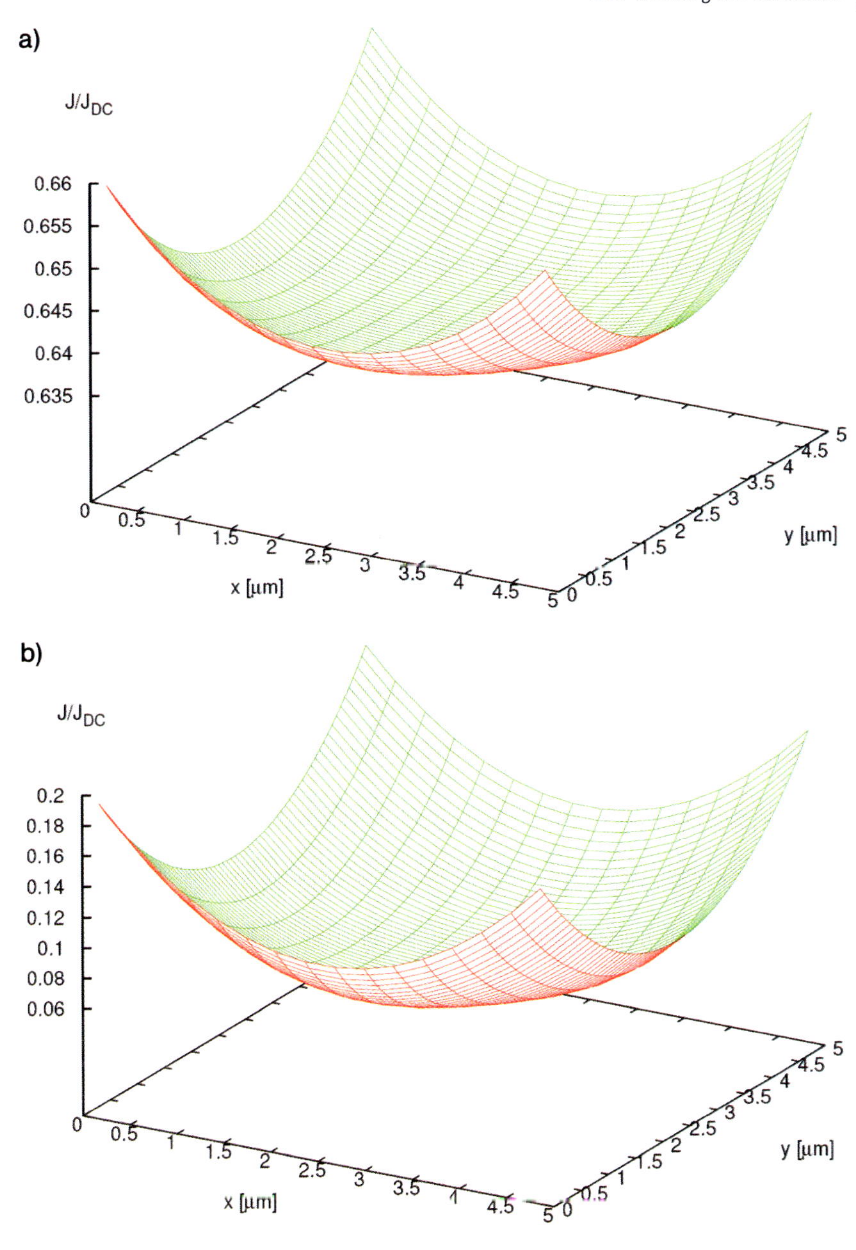

Figure 28.10 Current density distribution referring to the DC value with the frequency as parameter. (a) f = 1 GHz, (b) f = 10 GHz, (c) f = 100 GHz.

decrease in distance results in an increase in resistance (dotted curve) towards the value of the massive conductor.

Not only massive circular cross sections of vias are possible but also shapes such as those shown in Figure 28.12a–c. The required area is always the same ($7 \times 7\,\mu m^2$)

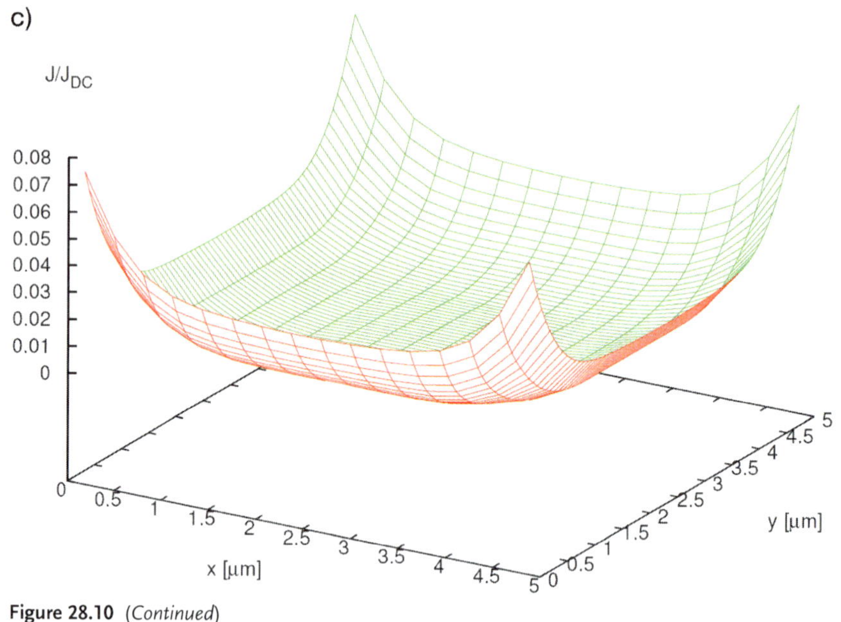

Figure 28.10 (*Continued*)

but the resistance as a function of frequency differs. Investigations of these cross sections are concept studies in which the manufacturability was disregarded in the first phase.

The structure in Figure 28.12a has a lower effective area than the structure in Figure 28.12b and, consequently, a higher DC value of the resistance; however, this disadvantage disappears at frequencies higher than approximately 3 GHz (dashed vs. continuous curve). Figure 28.12c has four outer parallel connected vias (1–4) and one inner separate via (5), which carries in the first case a current in the same direction as the outer vias. In the second case the inner via is not present and in the third case the current direction in via 5 is opposite to the outer vias. The resulting resistance distributions are shown in Figure 28.12d. For frequencies higher than about 5 GHz the resistance for case 3 (opposite current direction) becomes lower than the resistance of structures (a) and (b).

By providing the resistance distributions of the vias as behavioral models they can be used in detailed simulations of the electrical behavior of the system.

To describe the electrical behavior of linear networks at high frequencies, scattering parameters or so-called S-parameters are often used. This approach is also practicable to represent the electrical behavior of interconnect structures.

The S-parameter matrix describing an N-port network contains N^2 elements. Figure 28.13 depicts the most common two-port network. The relationship between the incident power waves (a_1, a_2), the reflected power waves (b_1, b_2) and the S-parameter matrix with the four parameters is shown in Equation 28.2.

$$\begin{pmatrix} b_1 \\ b_2 \end{pmatrix} = \begin{pmatrix} S_{11} & S_{12} \\ S_{21} & S_{22} \end{pmatrix} \cdot \begin{pmatrix} a_1 \\ a_2 \end{pmatrix} \qquad (28.2)$$

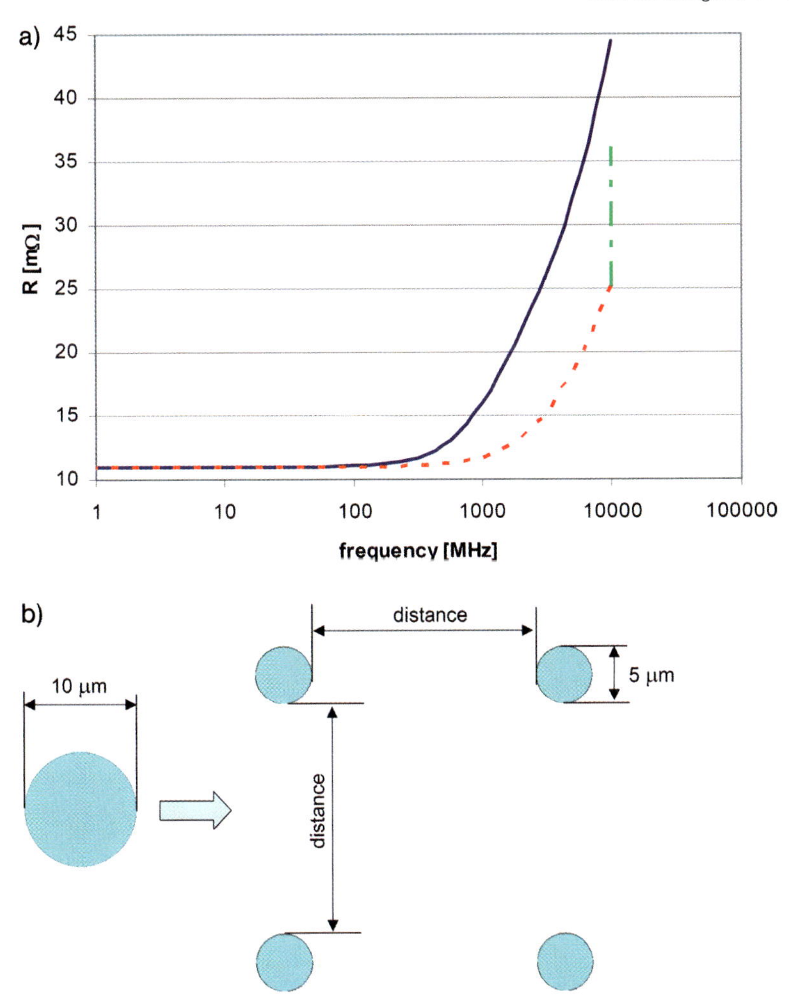

Figure 28.11 Resistance as a function of frequency. (a) (—) single via, (– – –) subdivided vias at fixed distance, (- - - -) subdivided vias with decreasing distance but at fixed frequency. (b) cross section of single via and subdivided vias.

The *S*-parameters of two-port networks are defined as follows: S_{21} and S_{12} describe the forward and reverse voltage gain, respectively, which is the transmission behavior of the structure. The reflexion at the ports is described by S_{11} and S_{22}, the input and output voltage reflection coefficients, respectively.

S-parameters are dimensionless complex numbers that can be described, for example, by magnitude and phase or real and imaginary part.

These parameters can be measured using real structures or can be determined with simulation tools, for example with the PDE solver Microwave Studio. The obtained characteristic curves of the S-parameters can be used to derive equivalent-networks of the investigated structures by parameter optimization.

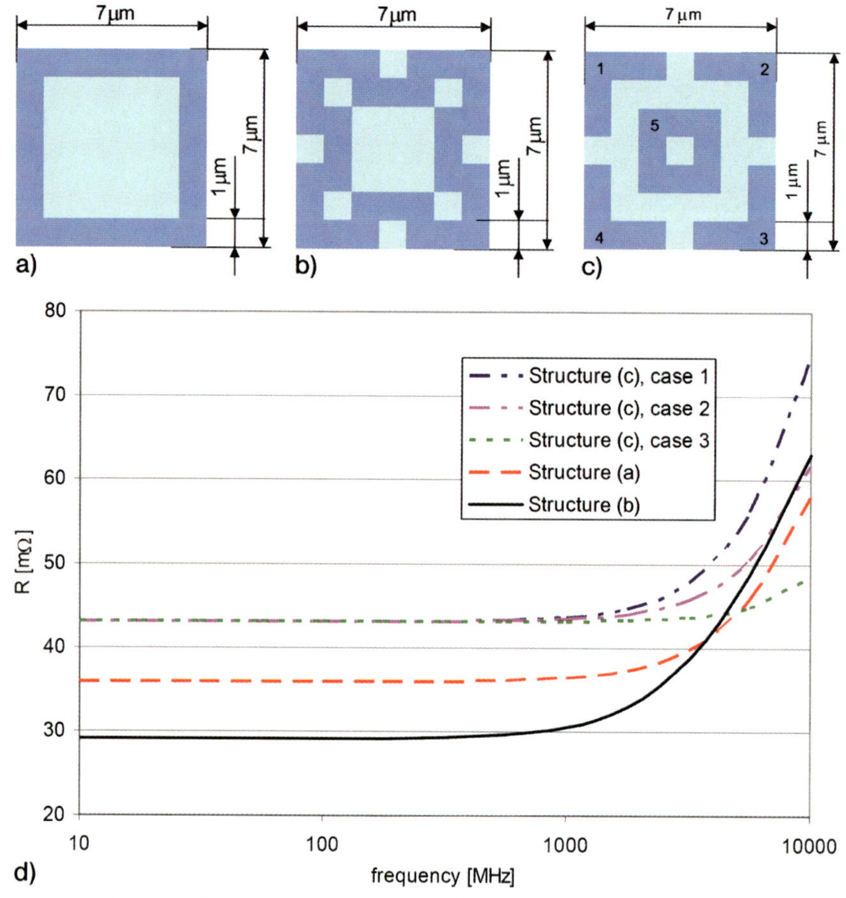

Figure 28.12 (a–c) Different cross sections of vias; (d) resulting resistance as a function of frequency.

Using given geometry and material parameters, a parametric 3D model of the ICV-SLID structure was prepared in the CST Microwave Studio (Figure 28.14).

Next, simulations up to a frequency of 100 GHz were carried out and the S-parameters S_{11}, S_{12}, S_{21} and S_{22} were derived. The magnitude distribution of

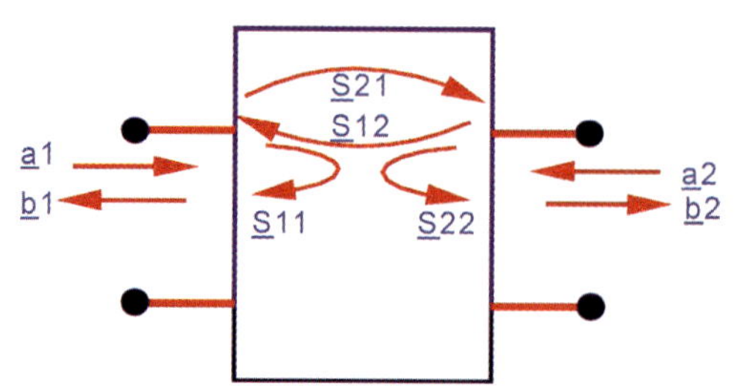

Figure 28.13 Two-port network with S-parameters [28]; the resulting S-parameter matrix is shown in Equation 28.2.

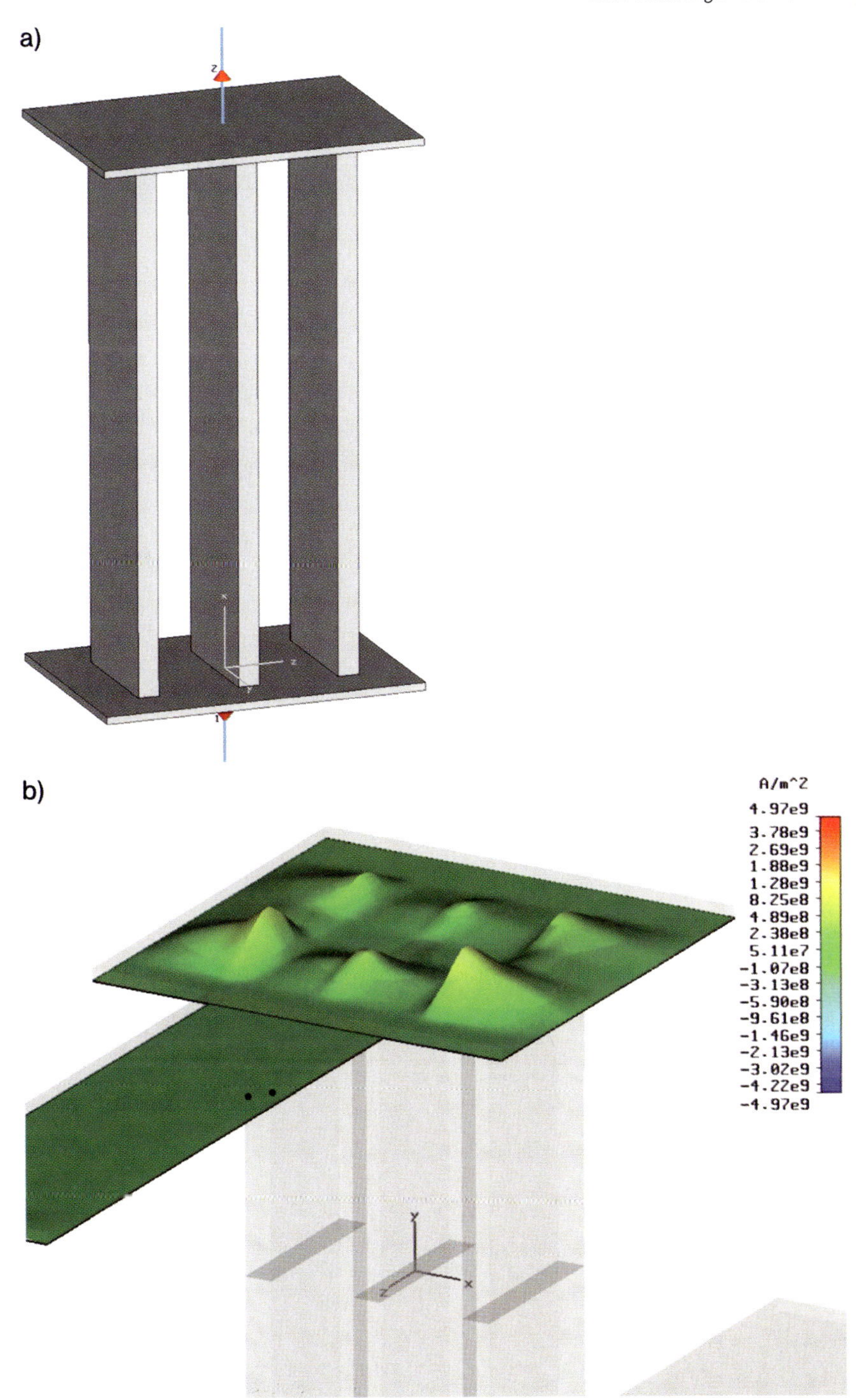

Figure 28.14 (a) 3D model of the ICV-SLID structure in CST Microwave Studio. Simulation result for current density in y-direction on top (b) and within the via structure (c).

c)

Figure 28.14 *(Continued)*

these quantities is shown in Figure 28.15 and the phase distribution is shown in Figure 28.16. Due to the symmetry of the structure the quantity S_{11} equals S_{22} and S_{12} equals S_{21}. Using this characteristic curves an equivalent-network of this structure will be derived in Section 28.2.4.

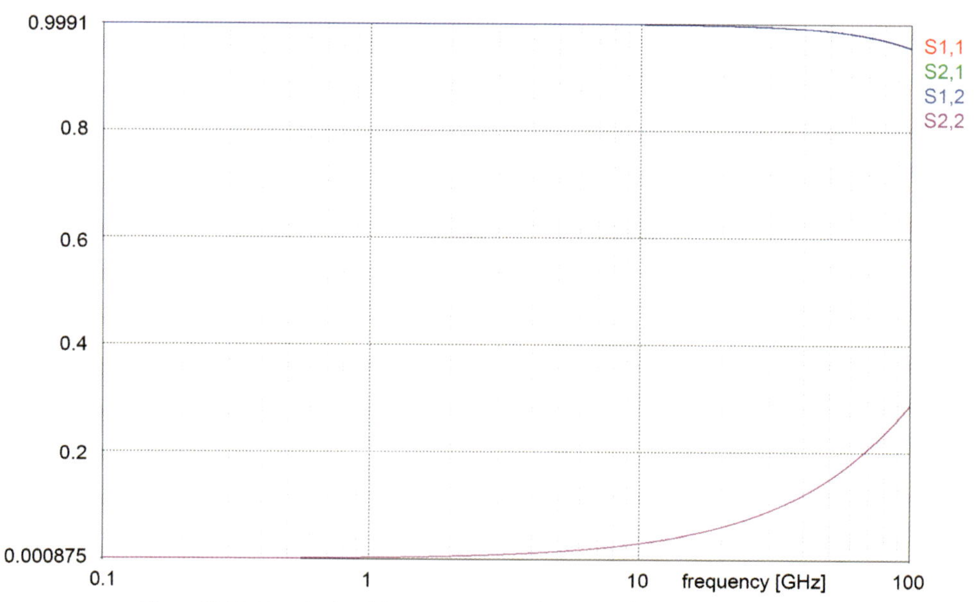

Figure 28.15 Magnitude of S-parameters S_{11}, S_{12}, S_{21} and S_{22}.

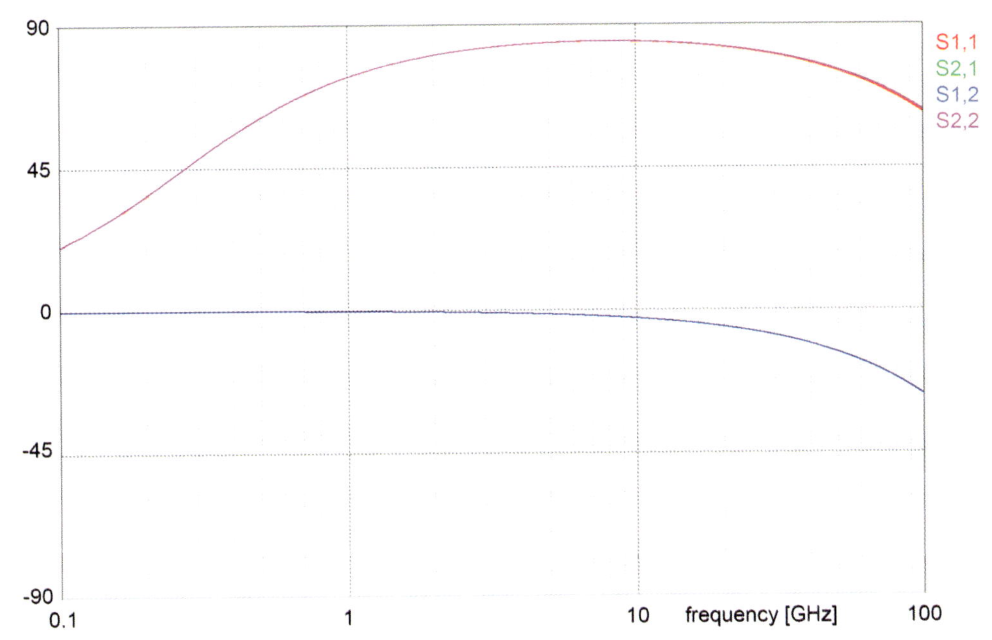

Figure 28.16 Phase of S-parameters S_{11}, S_{12}, S_{21} and S_{22}.

28.2.2.3 Cross Talk

The analysis of cross talk between wires at neighboring layers of a face-to-face stack and between neighboring wires on the same chip was performed with the layout given in Figure 28.17a (wire depth = 350 nm, wire width = 200 nm, wire length 100 μm, distance between wires 2 and 3 = 300 nm, distance between the chips = 8 μm). The termination resistors of wire 3 at input and output are 50 Ω, whereas wires 1 and 2 are only terminated at the end with 50 Ω. Because of the high distance d and the small dielectric constant between the layers (air: $\mu_r = 1$), the cross talk between wires on neighboring layers is approximately 1% of the cross talk between neighboring wires on the same layer (Figure 28.17b).

Evaluation of the results allows the following conclusions:

- The electrical parameters of ICV (see Figure 28.8) do not lead to a higher additional signal delay in comparison with the wire at on-chip level.
- The geometry of the ICV on the electrical behavior has no influence in the range up to 10 GHz (see Figures 28.11 and 28.12).
- In general, the cross talk between wires of neighboring layers is negligible in comparison with neighboring wires at one layer.

Probably, there will be no restrictions for high speed and RF applications in VSI through 3D wiring. On the contrary, the chance for shorter connections in 3D wiring will lead to lower delays.

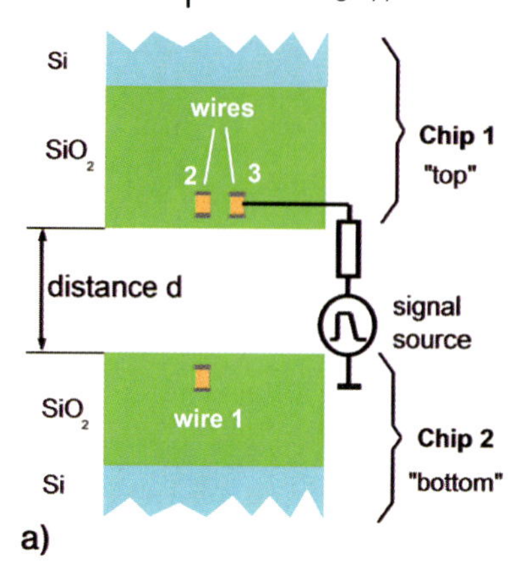

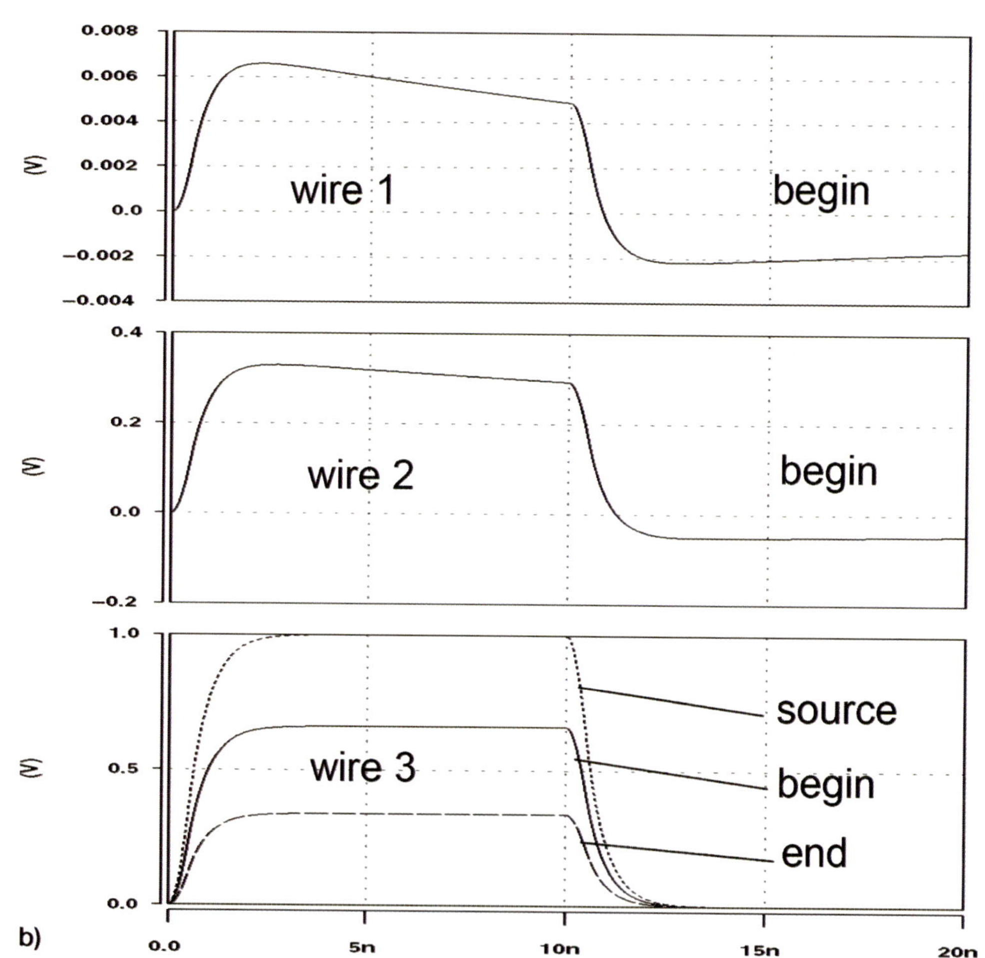

Figure 28.17 (a) Geometry for cross talk calculation; (b) some results of cross talk calculation.

28.2.2.4 **Thermal Behavior**

By performing thermal simulations of single interconnect structures (e.g. basic modules), parameters of these elements such as thermal resistance and capacitance can be derived. They are suitable for modeling thermal equivalent networks that will be used in the entire system simulation or for first assessments.

Similar to the investigations of electrical values, geometrical properties can be varied and used as parameters of thermal resistance and capacitance.

Figure 28.18 illustrates the influence of diameter and height of the interconnect structures already shown in Figure 28.5 on the thermal values. For technological reasons, the height of the ICV-SLID structure is fixed at 13 μm.

28.2.3 **Influence of Thermal Stress on MEMS**

Due to the different coefficients of thermal expansion of the layers, mechanical stress may be induced. On the one hand, this may cause reliability problems and, on the other hand, it might influence system behavior. Especially for the integration of sensors, problems may occur. Some effects are:

- Modified electrical conductivity caused by additional mechanical stress in piezoresistive elements, for example in pressure and force sensors.
- Modified suspension conditions for oscillating structures, for example for gyroscopes and accelerometers, which cause modified eigenfrequencies.

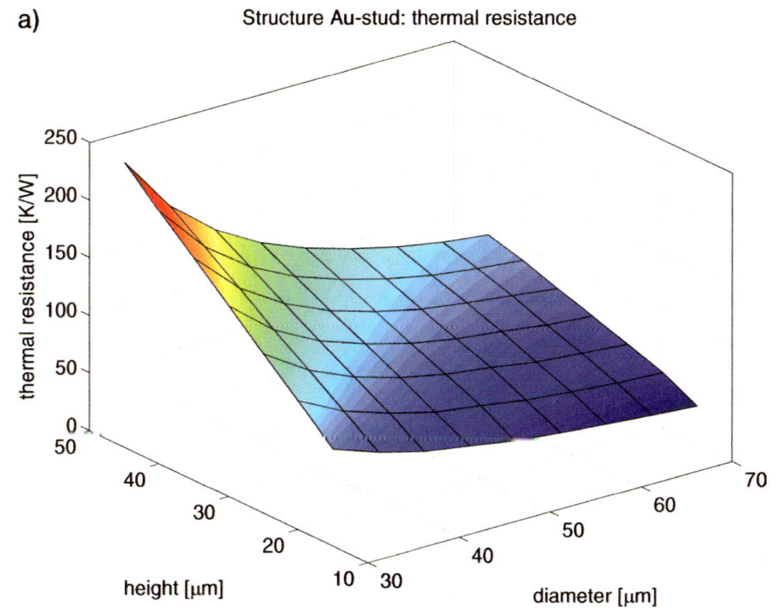

Figure 28.18 Parametric analysis of thermal resistance and capacitance of the interconnect structures shown in Figure 28.5 – (a, b) Au stud bumps; (c, d) ICV-SLID; and (e, f) μ-flip-chip.

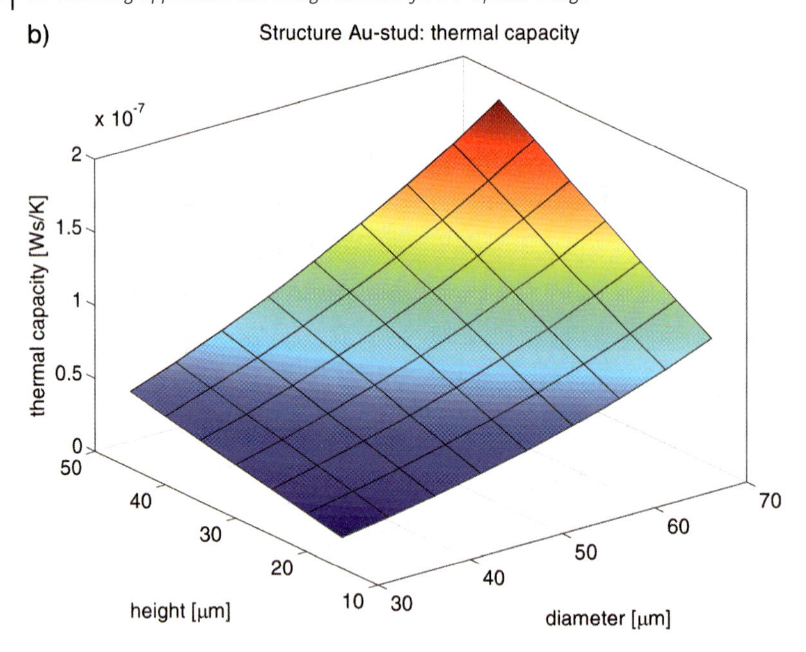

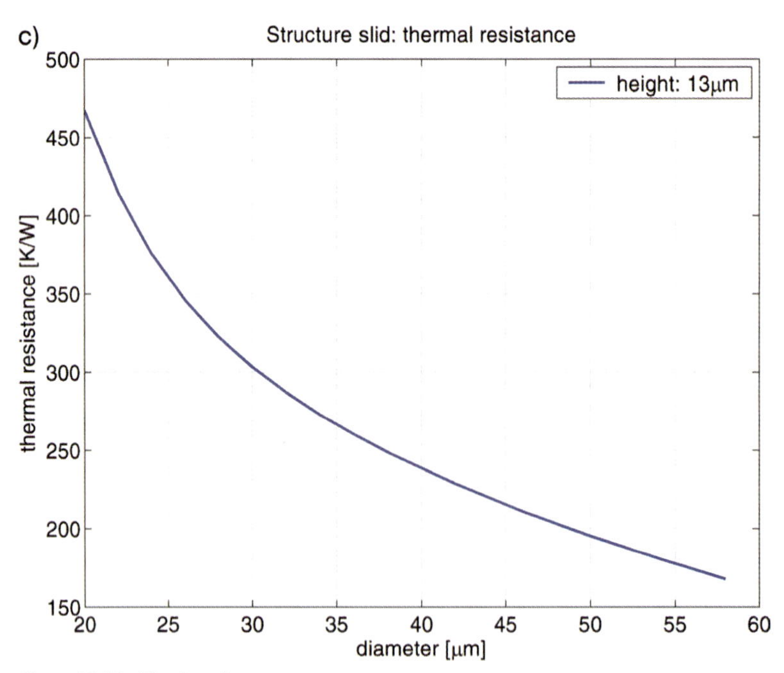

Figure 28.18 *(Continued)*

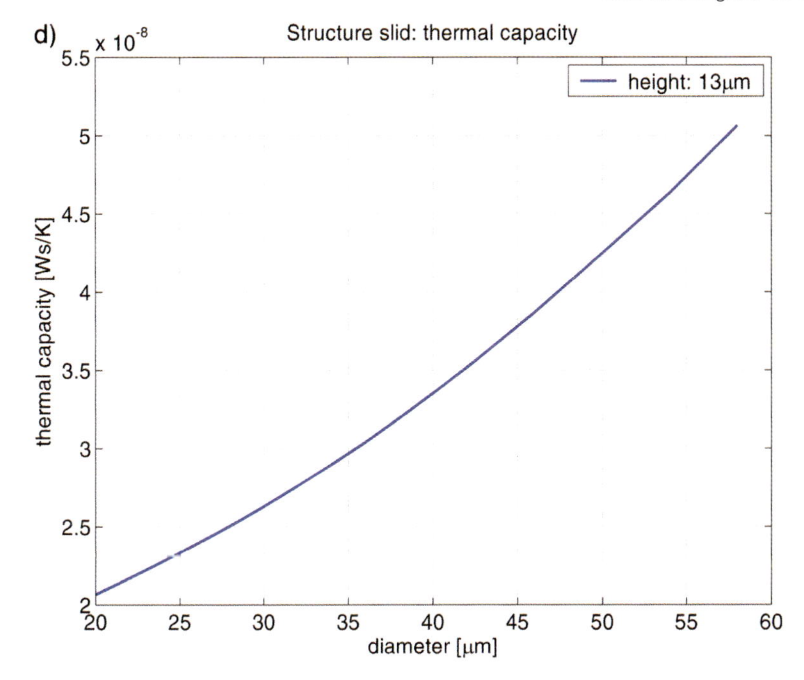

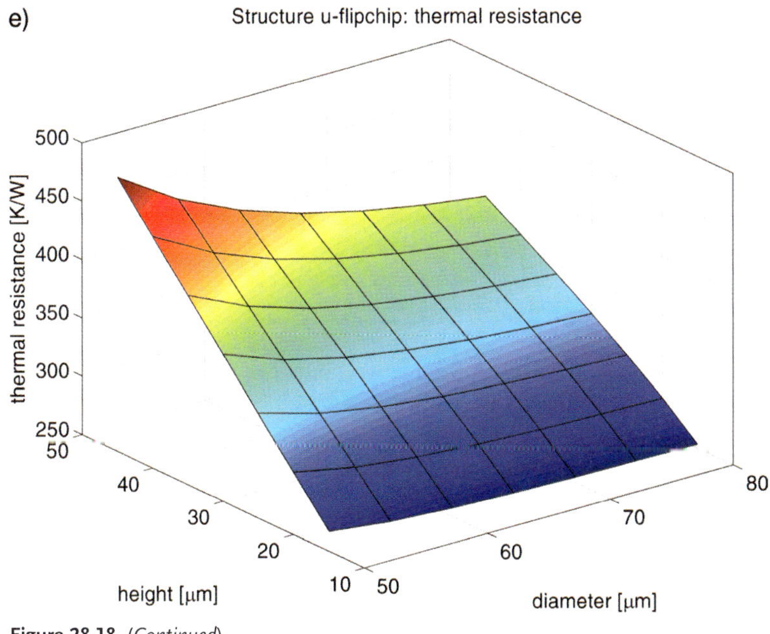

Figure 28.18 *(Continued)*

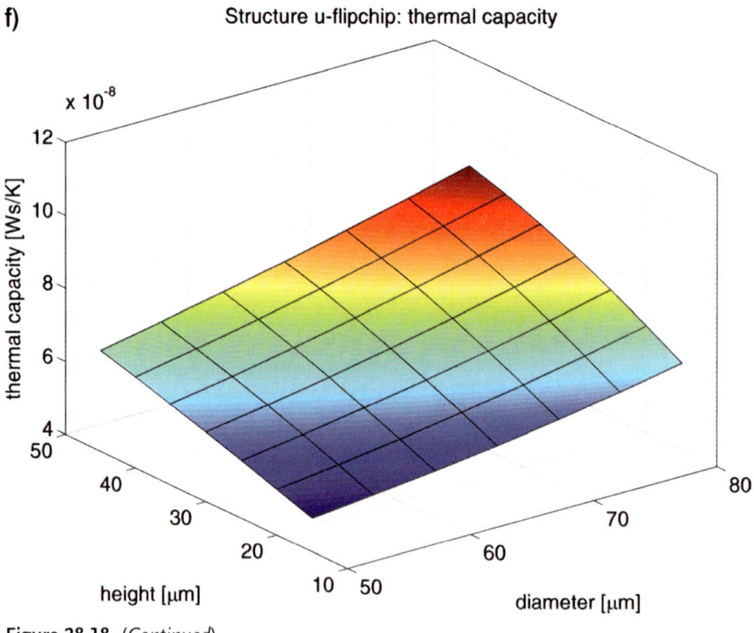

Figure 28.18 (*Continued*)

- Varying distances in comb drive structures for electrostatic actuation.
- Modified geometry of micro-fluidic channels.

These effects have to be both considered and compensated within the stack design.

28.2.4 Simulation of Complex Stack Structures

Thermal effects play an increasing role in micro systems and integrated circuits. Especially for the prevention of local hot spots, thermal simulations of the entire stack can be very helpful.

For a stack consisting of three layers (Figure 28.19), the results of a thermal analysis are shown in the following. For modeling, the modular approach described in Section 28.2.1 was applied.

Due to layout constraints and the request for short wiring between the layers a situation occurs where in one operation mode devices are active at one x-y-location in all three layers. This leads to a local hotspot like that shown in Figure 28.20 (left-hand side).

By introducing additional vias for heat transport (see Figure 28.20, right-hand side), the maximum temperature could be decreased. Presently, the investigations are being extended to analyses of additional operation modes and to combine of optimization criteria.

At this point we have shown the thermal simulation results of the stack structure using the PDE solver ANSYS.

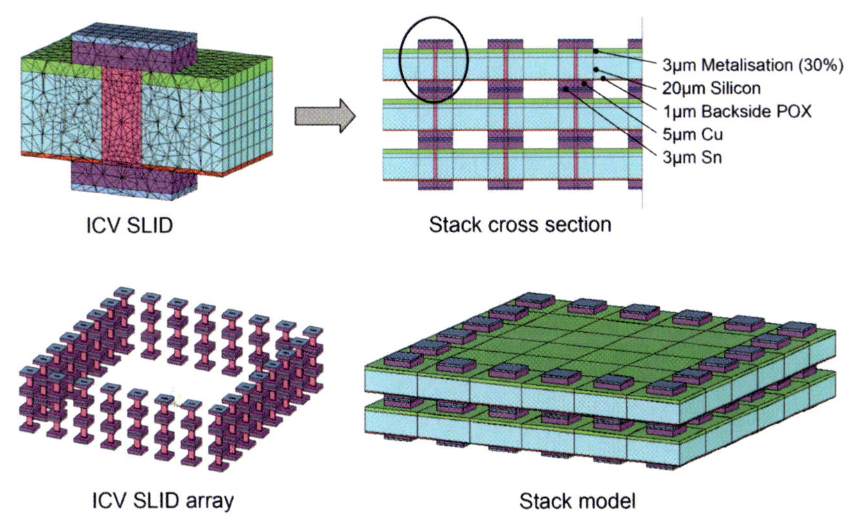

Figure 28.19 Stack structure built up of basic modules.

To examine electro-thermal interactions, the transistor models have to be extended to consider temperature changes. This includes the introduction of the local device temperature as a variable in all relevant electrical equations. Furthermore, it is necessary to calculate the power loss of the device depending on the electrical properties. Therefore, the circuit simulator has to fulfill some requirements, especially concerning the integration of new (or extended) device models.

Combined with a thermal network or behavioral model of the stacked structure, a coupled electro-thermal analysis of the entire system is possible (Figure 28.21).

To generate the thermal network of the stack, first the stack structure has to be modeled using, for example, the PDE solver ANSYS. The entire model is built up of basic modules that are generated for ANSYS using a common geometrical and material representation (see Figure 28.6). In a next step, the thermal network can be derived from the finite element model by order reduction methods directly (see Section 28.2.5) or by parameter optimization or approximation algorithms using the thermal simulation results of ANSYS.

Then, the devices of the network representation of the entire stack structure, which are temperature-dependent, and the devices with significant power loss have to be replaced by electro-thermal device models.

Finally, the resulting electrical circuit with thermal pins can be connected to the thermal network or to the reduced order behavioral model of this network for shorter simulation times on the DAE level.

28.2.5
Methods for Computer-Aided Model Generation for System Level

The process of model generation for system level simulation is supported by various methods: parameter optimization, model order reduction and approximation

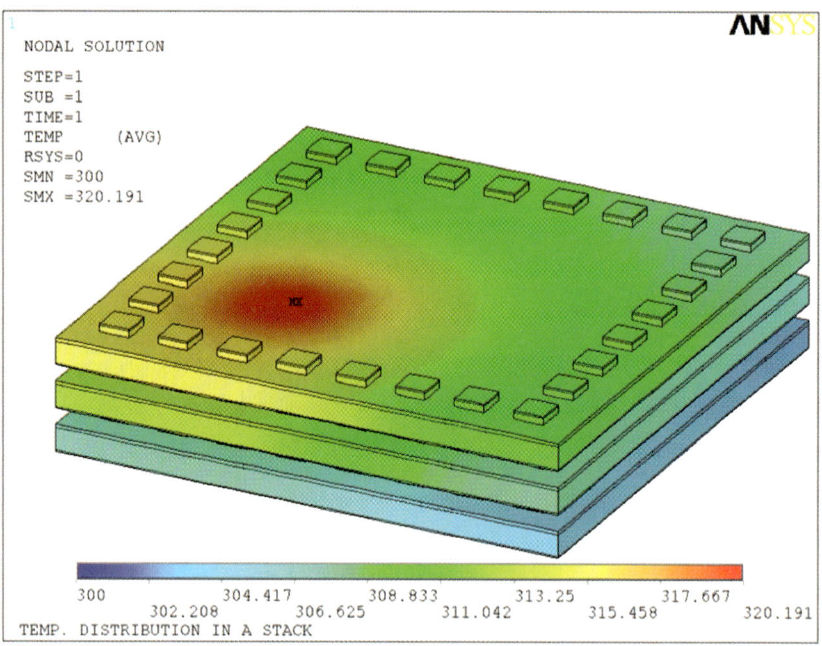

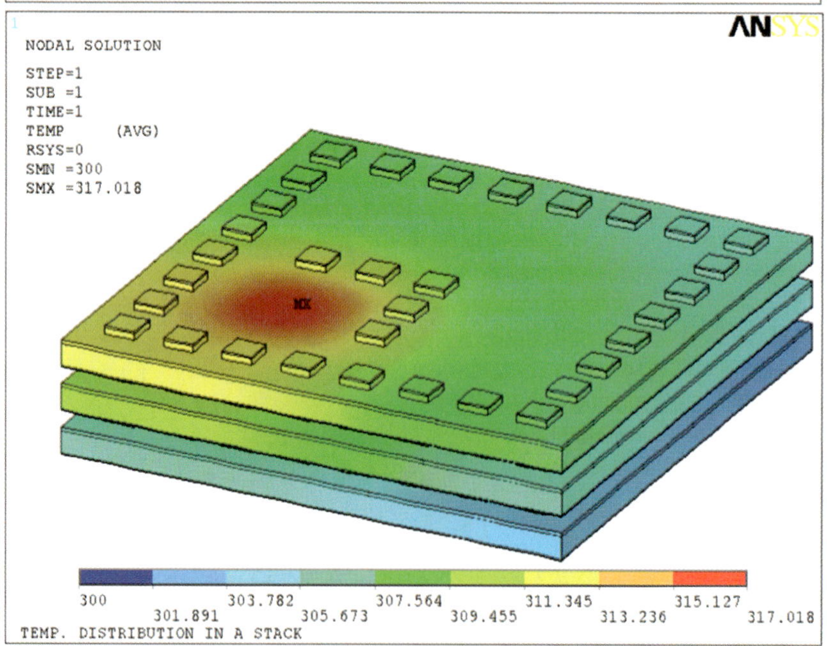

Figure 28.20 Left-hand side: stack structure with active devices in the lower left corner, $T_{max} = 320$ K; right-hand side: stack structure with additional vias, $T_{max} = 317$ K.

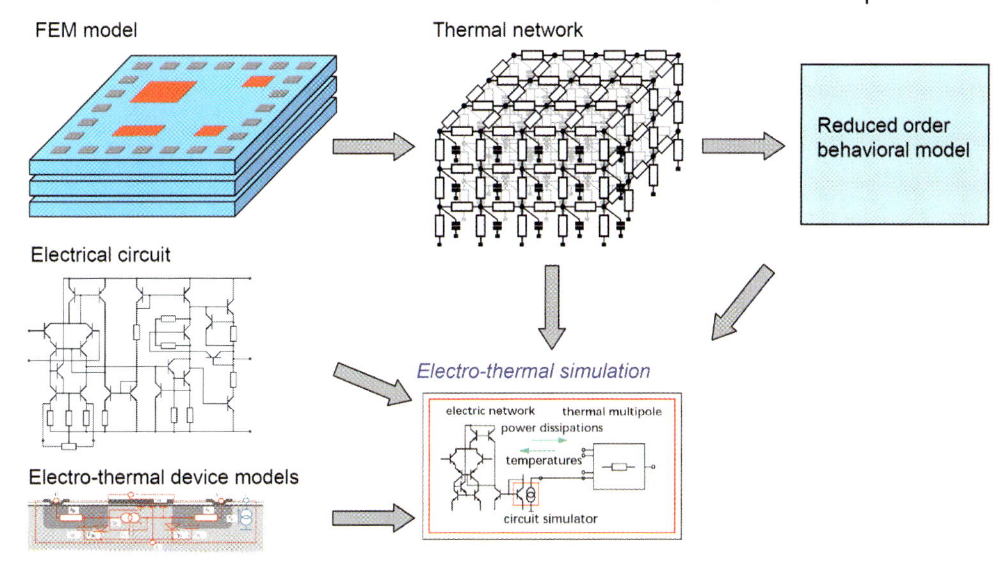

Figure 28.21 Electro-thermal simulation combining electrical circuit and thermal model of stack structure.

(Figure 28.22). These methods are combined with direct calculation of model parameters from PDE simulations.

28.2.5.1 **Optimization**

Using the previously calculated S-parameter distributions (see Figures 28.15 and 28.16), a parametric RLC network can be developed by numerical parameter optimization. For that purpose, generic network models are needed (Figure 28.23) for

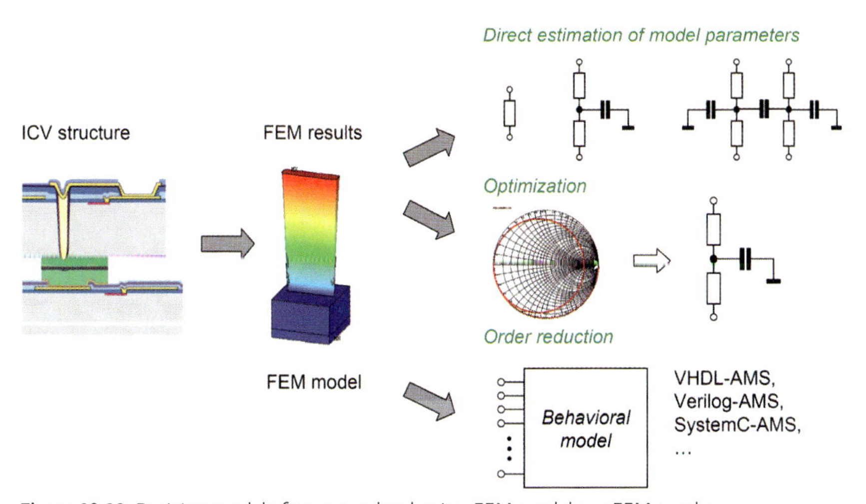

Figure 28.22 Deriving models for system level using FEM models or FEM results.

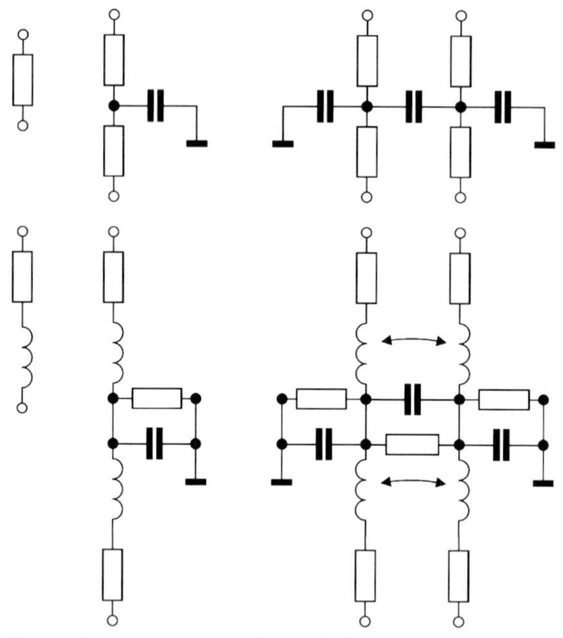

Figure 28.23 Generic network models for one and two vias.

which S-parameters are calculated with circuit simulators. Using an optimization tool, the parameters of these networks can be adapted. As objective function for optimization, the deviation of S-parameter curves from simulations in Microwave Studio and circuit simulation are used. The right choice of initial values of optimization parameters is important for good optimization results. This can be supported by a rough calculation of network parameters for a simplified structure.

Instead of deriving equivalent RLC networks, behavioral models in modeling languages like VHDL-AMS, Verilog-AMS, Modelica, and so on can also be parameterized [29].

28.2.5.2 **Model Order Reduction**

For the generation of behavioral models, for example, to investige thermal effects (see Section 28.2.4) on system level, model order reduction methods are applied. In the following, the mathematical background is described briefly.

The first step of deriving behavior models from partial differential equations (PDE) is semi-discretization of the PDE via the Finite Element Methods (FEM). This usually leads to linear time invariant (LTI) control systems with a large state space dimension as opposed to only a few input/output variables, as shown in Figure 28.24.

Now, the aim of model order reduction is to reduce the number of internal states while keeping the number of input/output variables, so that the original system can seamlessly be replaced by the reduced one.

As illustrated in Figure 28.25, for first order systems, the state space dimension can be reduced through projection of the system matrices by multiplication with a

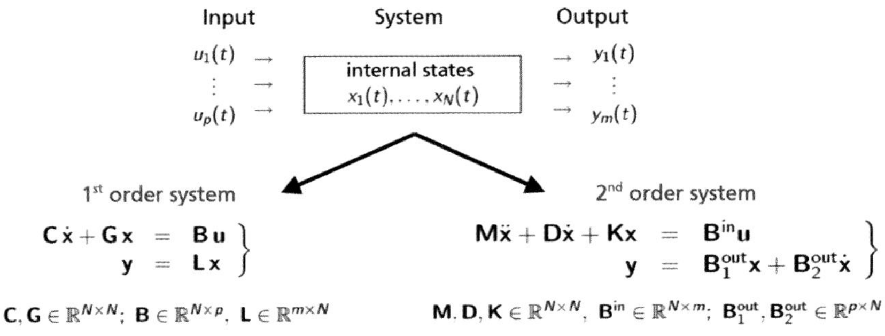

Figure 28.24 Large state space system of original system.

rectangular projection matrix $\mathbf{V}_n$. The resulting state space dimension equals the number of columns of $\mathbf{V}_n$. Existing approaches based on moment matching choose $\mathbf{V}_n$ so that structural properties of the original system like passivity and reciprocity are preserved. If the original system is of second order, which is typical for mechanical systems or RCL networks, the original system could be simply transformed into a mathematically equivalent first-order system. Projection will then lead to a first-order reduced system that cannot be transformed back into second order in general. However, there exists a projection scheme, similar to Figure 28.25, that maintains the second-order structure of the original system.

As an example, the stack structure with additional thermal vias already shown in the right-hand side of Figure 28.20 will be used. At the top layer of this three-layer structure we introduce a heat generating device in the front left corner and we want to investigate the temperature as a function of time at three points where heat sensitive devices are placed (see Figure 28.26). This leads to a system with one input and three outputs or four in/outputs.

First, the stack structure was modeled using the PDE solver ANSYS. The dimension of the resulting system matrices was about 95.000. Exporting the matrices and reducing the dimension using order reduction algorithms lead to a reduced system with a dimension of 40. Then a thermal behavioral model of the stack structure in the description language MAST was generated.

Finally, simulation runs using ANSYS and the system simulator SABER were carried out to compare the reduced order model with the original one. A power loss ideally changing from 0 to 100 mW at time 0 was applied to an appropriate

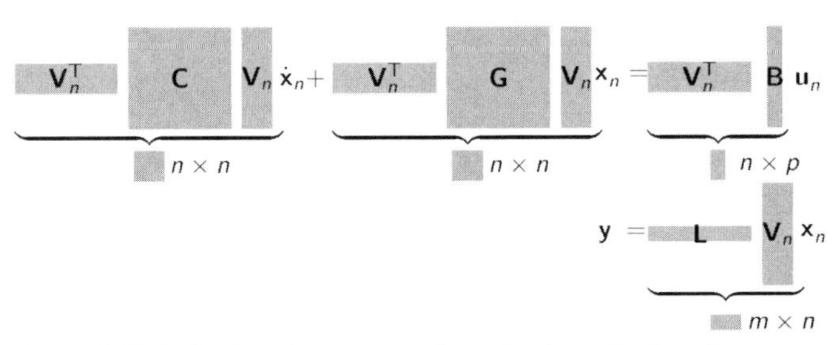

Figure 28.25 Reduction of state space dimension by projection of system matrices.

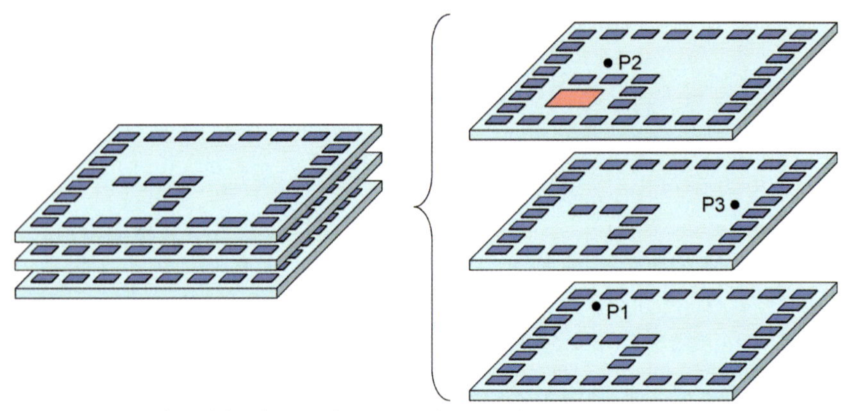

Figure 28.26 Stack model with power loss injected in upper layer (bottom left corner) and three points where heat sensitive elements are present.

area at the top layer and the temperature run at the three interesting points was calculated. Figure 28.27 illustrates the very good accordance between the results of the original system with a dimension of 95.000 and the reduced system with a dimension of 40.

Using the reduced behavioral model of the stack structure, electro-thermal simulations of the entire system are possible (see Figure 28.21). The electrical circuit

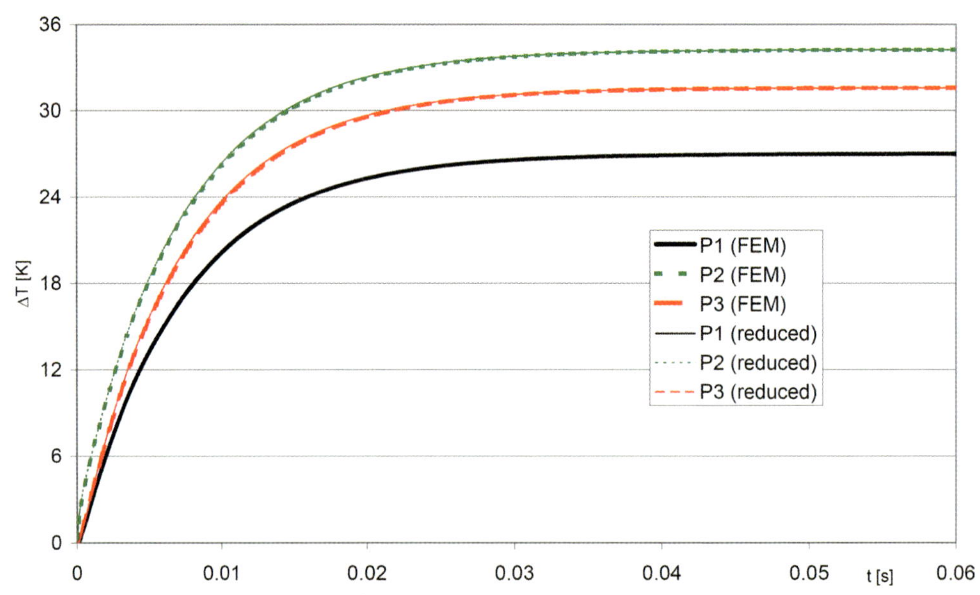

Figure 28.27 Increase of temperature at the three heat sensitive elements – comparing results using PDE solver (FEM) and reduced order model.

must contain electro-thermal device models for the heat loss generating as well as temperature sensitive elements, resulting in thermal pins of these elements where the thermal behavioral model of the stack can be connected to.

Furthermore, the thermal model can easily be connected with behavioral models regarding heat convection or radiation between, for example, stack and environment.

28.2.6 Model Validation

For model validation, the preparation and measurements of test structures and test circuits is mandatory. Furthermore, due to the dense integration, it is necessary to develop special measurement techniques, because devices of interest are often located inside the stack and not easily accessible for direct measurements.

28.2.6.1 Electrical Measurements

Figure 28.28 shows a sketch of a test structure for electrical measurements and the corresponding simulation model for the FEM simulator ANSYS. Furthermore, there are special structures which allow a measurement of the via resistance. Figure 28.29a shows a so-called Kelvin structure. Driving a current between port A and C and measuring the voltage drop between B and D the resistance of the via can be calculated easily by Ohms law. Another possibility for electrical measurements is the use of daisy chain structures, where several vias are situated in a serial connection.

28.2.6.2 Thermal Measurements

For thermal measurements in packaging, several methods can, more or less, be applied for stacked systems:

- direct measurement, for example, by measuring temperature dependent device parameters;
- thermal imaging, for example, lock-in thermography;
- liquid crystal microthermography;
- fiber optical thermometry probing.

A good overview of thermal measurement methods is given in Reference [30]. In addition, special techniques are available that allow a measurement within stacked structures; see, Reference [31].

28.2.7 Integration of Circuit or Behavioral Models into the Design Flow

Models for system level simulation are provided as SPICE macro-models or as behavioral models, formulated in description languages such as VHDL-AMS, Verilog-AMS, or MAST. This is equivalent to a mathematical description with algebraic differential equations (DAE).

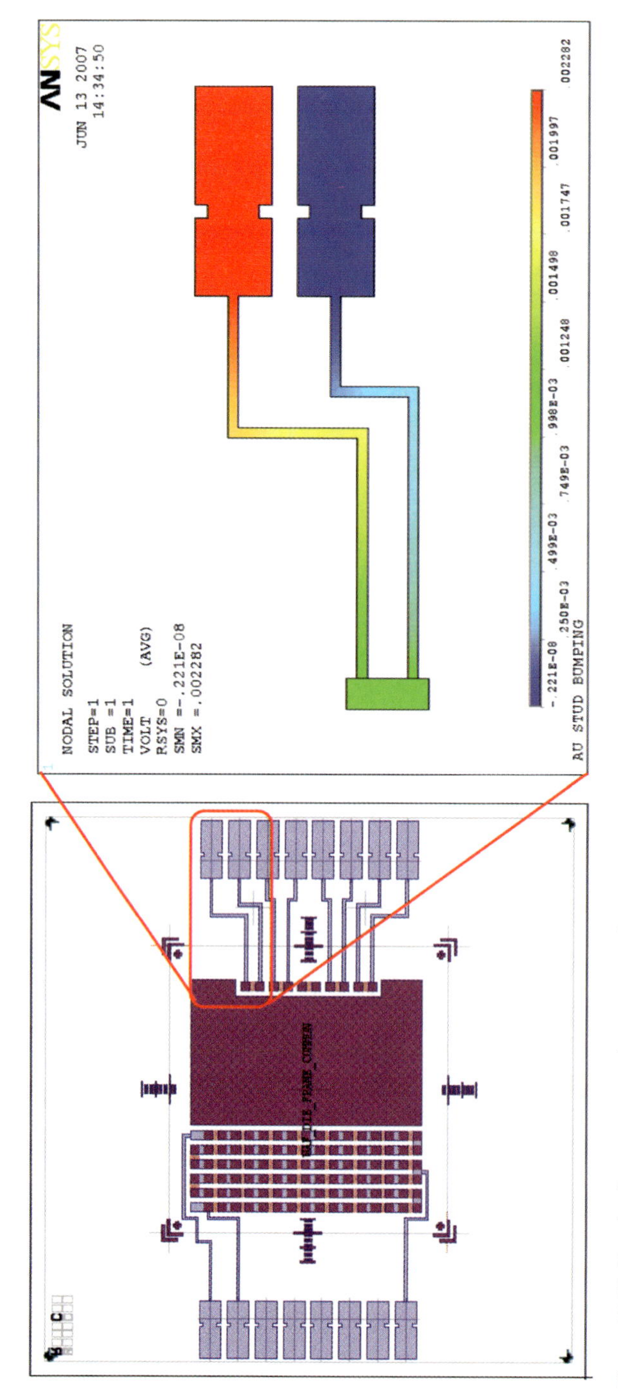

Figure 28.28 Sketch of test structure (by courtesy of SINTEF) and derived electrical model in ANSYS.

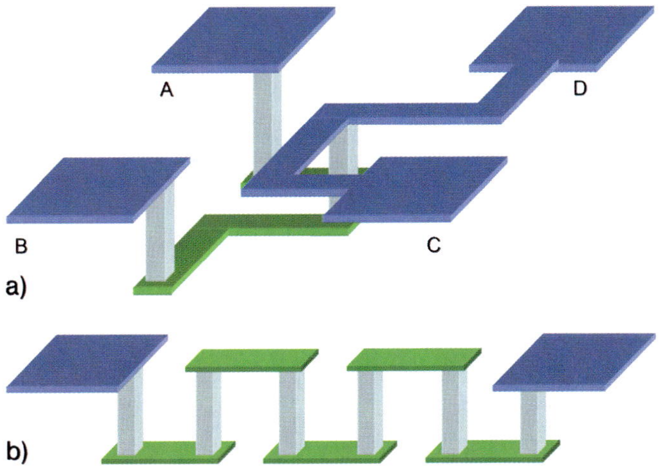

Figure 28.29 (a) Kelvin structure; (b) daisy chain.

28.2.7.1 SPICE-Model (Circuit Simulation)

Parameterized equivalent RLC networks (macro-models) can be used in analog circuit simulators like SPICE, HSPICE or PSpice. This kind of description is often preferred by designers because it can be interpreted and evaluated easily.

Figure 28.30 shows a parameterized SPICE sub circuit (.sbckt) of a generic RLC network suitable for the ICV-SLID structure shown in Figure 28.14.

```
.subckt SUB p1 p2
r1  p1  x1  0.5780297259945977
r2  p1  x4  0.05092364608820336
r3  p1  x7  0.5780297259945977
l1  x1  x2  1.385372709103731E-9
l2  x4  x5  2.849176465819848E-11
l3  x7  x8  1.385372709103731E-9
r4  x2  x5  1.209535880531847
c1  x2  x5  1.2324451720037E-5
r5  x5  x8  1.209535880531847
c2  x5  x8  1.2324451720037E-5
l4  x2  x3  1.385372709103731E-9
l5  x5  x6  2.849176465819848E-11
l6  x8  x9  1.385372709103731E-9
r6  x3  p2  0.5780297259945977
r7  x6  p2  0.05092364608820336
r8  x9  p2  0.5780297259945977
.ends
```

Figure 28.30 Parameterized SPICE sub circuit (.sbckt) of a generic RLC network suitable for the ICV-SLID structure in Figure 28.14.

28.2.7.2 HDL-Model (Mixed-Signal Simulation)

The behavior of a system or a component can be described at different abstraction levels:

- behavior at the boundaries, ignoring inner structure and physics (black box model);
- functional behavior, ignoring physics (structural model);
- physical behavior (physical model).

There are very different methods to describe this behavior, depending on the function of the system and abstraction level. Hardware Description Languages (HDLs) allow one to transform mathematical descriptions based on a Differential Algebraic Equation System (DAE) into a behavioral model. Thus, parameter dependencies and complex nonlinear relations can be described directly and very easily.

A selection of currently used HDLs with application fields are:

- MAST, analog and digital systems
- HDL-A, analog systems
- Spectre HDL, analog systems
- Verilog-A, analog systems
- Verilog-AMS, analog and digital systems
- VHDL, digital systems
- VHDL-AMS, analog and digital systems
- SystemC-AMS, digital and analog hard- and software systems.

Using these languages, complex models that represent multi-dimensional dependencies of circuit parameters from geometrical and material properties can be provided easily.

28.2.7.3 Digital Simulation

For digital systems, usually a full chip parasitic extraction is performed and the data is provided in SPEF (Standard Parasitics Exchange Format) files. Within these files, parasitic circuit elements – resistances, capacitances and inductances – of transmission lines are described in an ASCII format. It is defined as IEEE-Standard 1481-1999 – IEEE Standard for Integrated Circuit (IC) Delay and Power Calculation System.

Within SPEF, parasitics can be represented on different levels of abstraction, for example, as Pi-Models with an arbitrary number of segments, as reduced Pi-Models with a limited number of segments, or as a single capacitance.

Figures 28.31 and 28.32 show a small interconnect structure and the corresponding SPEF file.

Figure 28.33 gives a flow for integration of interchip via parasitics into a digital design flow.

28.2.7.4 Electro-thermal Simulation

The electro-thermal simulation has to be integrated into the circuit design flow. Figure 28.34 shows the correspondence of a electro-thermal simulation based on

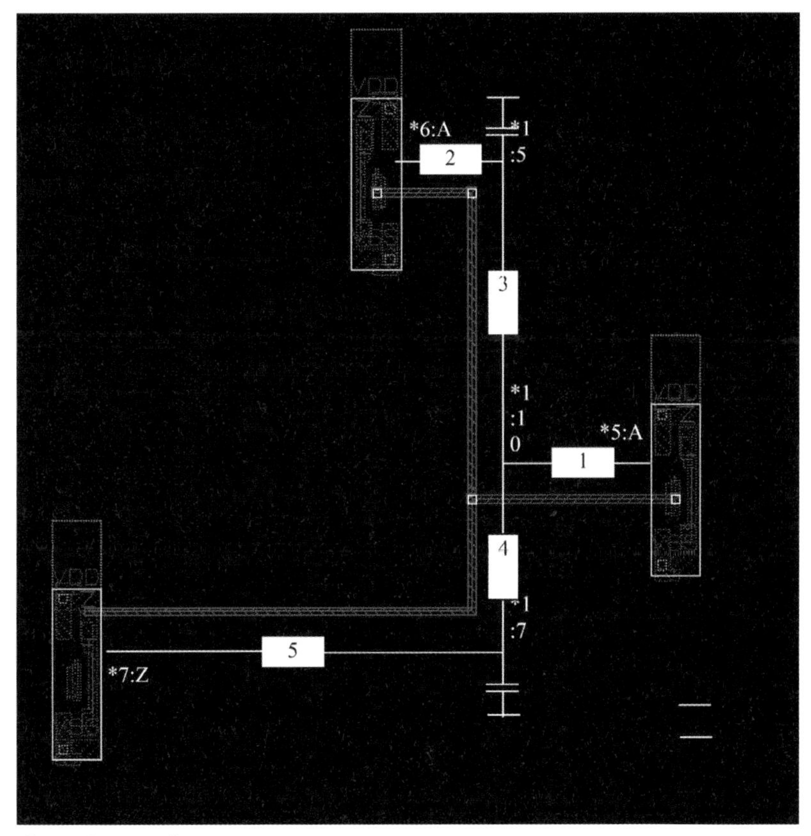

Figure 28.31 Small interconnect structure.

HDL-modeling and normal circuit simulation. For the coupled electro-thermal simulation there are two basic approaches:

- One-time electrical simulation, extraction of power losses, calculation of resulting temperatures using the thermal models, and using this temperature values as fixed device parameters during all subsequent electrical simulations.
- Fully coupled electrical-thermal simulation as described in Section 28.2.4: This requires replacement of all electrical devices that are temperature-dependent and of all devices with power loss by electro-thermal models.

Briefly summarized, based on the investigations mentioned above, the influence of integration technology can be considered within the design process. Minimization of the impact on system behavior requires:

- Methods for optimization and design improvement;
- New circuitry and new system architectures.

```
*SPEF "IEEE 1481-1999"
*DESIGN "o2"
*DATE "Mon Aug 20 12:23:31 2007"
*VENDOR "Synopsys"
*PROGRAM "Star-RCXT"
*VERSION "2006.06                    "
*DESIGN_FLOW "PIN_CAP NONE" "NAME_SCOPE LOCAL"
*DIVIDER /
*DELIMITER :
*BUS_DELIMITER []
*T_UNIT 1.00000 NS
*C_UNIT 1.00000 FF
*R_UNIT 1.00000 OHM
*L_UNIT 1.00000 HENRY

// COMMENTS

*NAME_MAP

*1 w1
*7 u1
*6 u2
*5 u3

*D_NET *1 1.64651

*CONN
*I *5:A I *C 65.3500 44.1800
*I *6:A I *C 60.3500 49.1800
*I *7:Z O *C 55.5200 42.3600 *D R_SIVX010

*CAP
1 *1:5  0.155816
2 *1:7  0.0210339

*RES
1 *5:A  *1:10 6.17814
2 *6:A  *1:5  1.85000
3 *1:5  *1:10 8.3476189
4 *1:10 *1:7  0.0691947
5 *7:Z  *1:7  7.25068
*END
```

Figure 28.32 SPEF file corresponding to Figure 28.31.

If system level models contain technology dependent parameters with given tolerance ranges, methods for parameter and tolerance optimization can be used; e.g. for improvement of signal integrity, thermal management and yield optimization. Furthermore approaches for low power design and design for manufacturability are based on such models and characteristics.

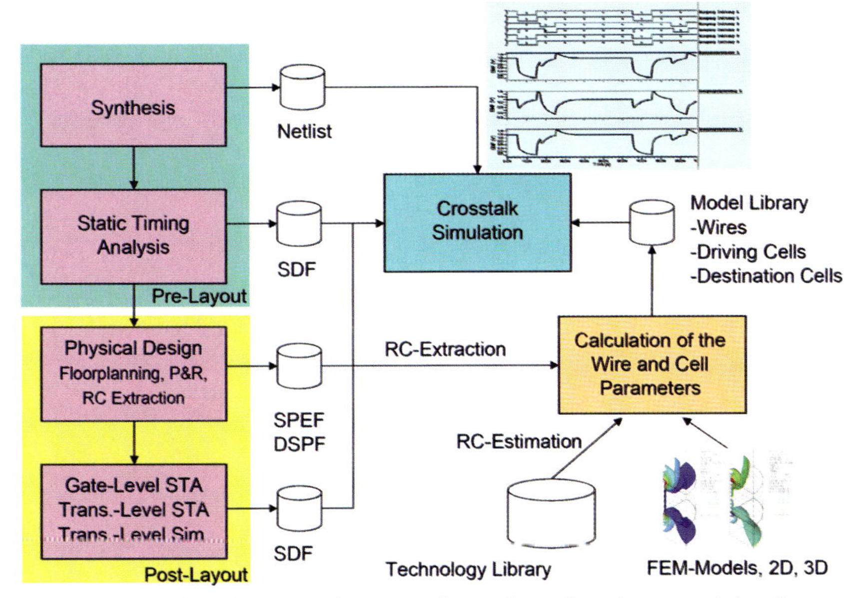

Figure 28.33 Digital simulation regarding cross talk according to floor planning and place & route.

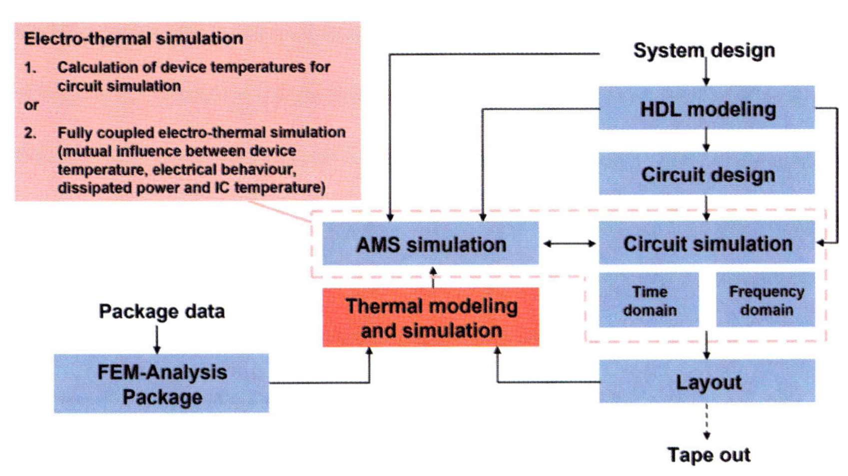

Figure 28.34 Electro-thermal simulation taking package and layout information into account.

28.3 Design Methods for 3D Integration

28.3.1 Low Power Design

Further optimization methods, generally included in synthesis steps, find dedicated circuit structures with the required properties. Examples are the minimization of the dynamic part of power consumption on logic level by reducing the switching activities in CMOS circuits. Because of the high density of transistors in 3D systems, electro-thermal coupling in active areas and a large heating in stacks are inevitable with a perceptible influence on the electrical behavior and reliability of devices.

Therefore, methods for extreme low power design are essential. The following means are highly promising for lower power loss:

- Reduction of the switching activity of registers and combinatory logic (for CMOS circuits).
- Switching off temporarily unneeded functional blocks (operating voltage).
- Isolation of unneeded sub circuits by prelogic – clamp on constant voltage level, operand isolation.
- New architectures and new circuit technology.

After estimation of the power consumption of the regarded circuit, the potential for power reduction is evaluated and suitable methods are selected. Dedicated tools, for example, from the company Synopsys (Design power, power compiler), can be used. To achieve realistic results, the gate net list must contain proper net load, which is determined by back annotation.

The power consumption is usually determined by simulation with a realistic input pattern rather than using less accurate statistical methods. Furthermore, a power-characterized target library must be available. This library must provide information about both:

- Internal power for minimization of dynamic power consumption;
- Leakage power for minimization of static power consumption.

Based on this information, power consumption and low power improvement can be determined.

Widely used methods are presented in the following.

28.3.1.1 Reduction of the Register Switching Activity

Changes of register states (switching activity) are determined during simulation. This process also consists of processing of data depending on the intended purpose (switching activity of individual, all or certain blocks of registers, profiling). Within the control part, usually specified by a Finite State Machine FSM, a potential for reduction of energy dissipation by minimization of the switching rate of status registers is given. Here the states can be coded in such a way that the most frequently occurring state transitions have the smallest Hamming distance. The frequency of the state transitions is derived from profiling the register switching activity (see Figure 28.35). State coding can be carried out, for example, by using Simulated Annealing.

Since power consumption of a circuit is usually proportional to the die area used, a main goal is to consider area reduction within state coding. This is possible by specifying minimization of logical functions as additional constraints for state coding.

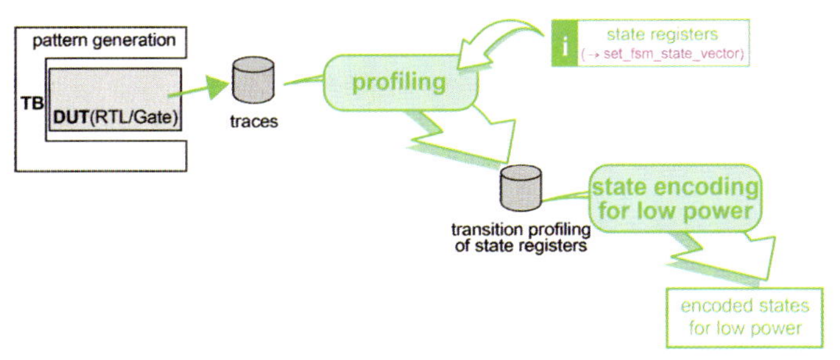

Figure 28.35 Methodology for estimation of power consumption.

28.3.1.2 Prelogic

In some designs (e.g. comparators) sub circuits can be completely deactivated after a definite decision by means of a suitable prelogic. Thus, the average switching activity can be lowered. Furthermore, for circuits with buffered inputs the number of input registers can be minimized by inclusion of a dedicated prelogic. For Finite State Machines (FSM), which remain for user-specific input sequences in the same state over many clock cycles, multiple computation of subsequent state and output can be prevented by special logic blocks. Accordingly, power dissipation of the circuit can be lowered. Reduction of switching activity also consists of removing peak and bumps, and the minimization of the length of combinational paths with high signal change rate.

28.3.1.3 Clock Gating

For conventional clock gating, usually the enable condition of the register/the register bank is used for controlling the switching logic. With the above-mentioned methods some options are set by the designer (clock gating style/conditions, resulting register width, and so on), while switching off logic is included in the elaboration step automatically. A very often used method is local clock gating, that is the condition for disabling registers is generated locally (see Figure 28.36). Without accessing further external information, the switching off condition is derived exclusively from a comparison of old and new register content (XOR function).

Additional switching off logic increases, firstly, the power consumption of the circuit. This depends on register switching rate, register width and the clock gating style. Therefore, it is necessary to control the replacement of registers considering the power characteristics of the cells. For this purpose, different simulations with a

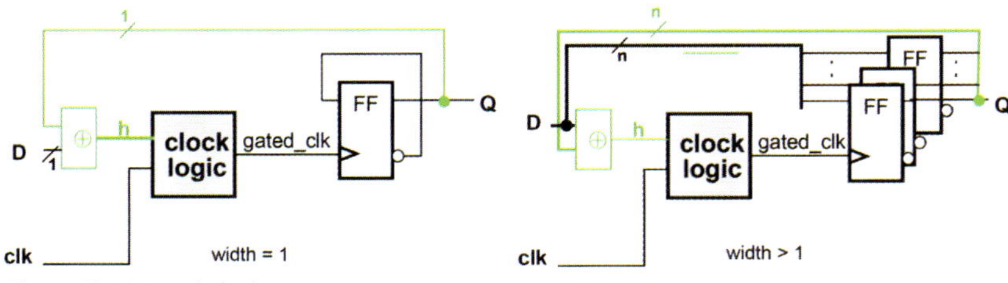

Figure 28.36 Local clock gating.

subsequent power analysis are performed. The results are then compared with the power consumption of the register cells without switching off logic. This information can be used for the conventional clock gating (e.g. for achieving minimum register width).

28.3.1.4 General Survey

The power dissipation of a synthesized circuit depends on several different parameters, for example the die area, number of registers, and choice of library elements. Optimization concerning an individual parameter usually does not lead necessarily to minimum power consumption. Low power design is a complex optimization task where multiple parameters must be considered.

28.3.1.5 Circuitry and System Architecture

Design improvements by local changes of the circuit are especially powerful optimization methods. However, special system architectures and circuit structures (circuitry) with low power consumption and local design improvements lead to the best circuit solution. With a systematic development of new system architectures, which use the possibilities of 3D integration concerning low power design, a high potential can be made available.

This is illustrated here by the example of a base band processor for mobile terminals with multi-standard and multi-band architecture and high computing performance. With realization as one chip, the system busses need about 30% of the chip area and nearly 50% of total power consumption. The system architecture allows 3D integration with much shorter vertical busses. Estimations with modified busses lead to a reduced power consumption of the system busses of only 30% of the one chip realization. The total power consumption of the stack is 65% of the total power of the single chip solution. Intel recognized this potential and has worked on design and fabrication of 3D microprocessors – see Reference [32]. The anticipated results of this work are a reduction of bus power around 60% and a decrease of logic power of about 15%. Ultimately, this leads to an overall decrease of power, delay and costs!

Last but not least dedicated design strategies can help to find an optimal or partially optimal solution in several design steps. The power consumption and temperature level caused by inner thermal loads are estimated by simulation based methods, the circuit parts with high power losses are located, and floor planning is performed with the objective of a nearly uniform distribution of the thermal sources on the chip area or in the stack volume.

28.3.2 Design for Testability

For the test, two have to be solved:

- First, test of every layer before stacking, to achieve a high yield of the stack.
- Second, test of the complete stack with its complex functionality in an acceptable time and with sufficient large fault coverage.

The design of a tailored hierarchical clock system in synchronous logic ensures the sufficient correctness of the timing in SoC or SiP of high complexity. A well chosen hierarchy concerning the layers is the best premise for the separate testing of the layers.

To test the vertical connections (ICV) and the complete stack, a design for testability is necessary.

Methods and principles of design for testability, such as scan methods, self-test approaches, and integration of complete test functions on the chip, help to obtain efficient test sequences with the necessary fault coverage for highly complex SoC or SiP.

A suitable method is the built-in self-test (BIST) with additional hardware for the self-test processing of some hardware components and for the output of the test results. In addition, this method makes it more difficult to access the inside of the circuit after stacking. This security characteristic is particularly important for several chips, for example, chip cards and wireless networks.

In practice the test-per-scan-BISTs are used mostly. This method is well supported by design systems. The essential disadvantage of the long testing time can be reduced by several parallel scan paths. The functional self-test can be shortened by the structural BIST substantially.

Using the BIST, the following subtasks have to be solved:

- Work out a BIST framework for the test-per-scan-BIST;
- Work out the BIST components;
- Improvement of the fault coverage;
- Test of the vertical connections (interchip vias);
- Consideration of low power requirements of the test hardware;
- Installation of the BIST framework and the BIST components in selected circuits;
- Integration of these methods into the used design flow.

Bidirectional lines are a problem for scan design. It can only be solved by increased costs of hardware and with a higher design effort. Accordingly, circuit structures with bidirectional lines should not be included in the BIST.

A typical BIST framework for the rather conventional partitioning according to functional modules is shown in Figure 28.37. The separation from other functional modules happens with the help of multiplexers. The BIST frameworks need the following components: pseudo random pattern generator (PRPG), multiple input shift register (MISR), phase shifter, compactor and BIST control.

Reasonably, adjacent scan chains receive, mostly from each other, independent random patterns generated by the phase shifter. The central BIST control receives the information, by instruction, which modules (partitions) are to be tested at the same time. Figure 28.38 shows the interaction between the central BIST control and the BIST framework. After execution of BIST, signaled by "bist_done", the content of signature registers (MISR) are read and the signatures are compared either on or outside of the chip.

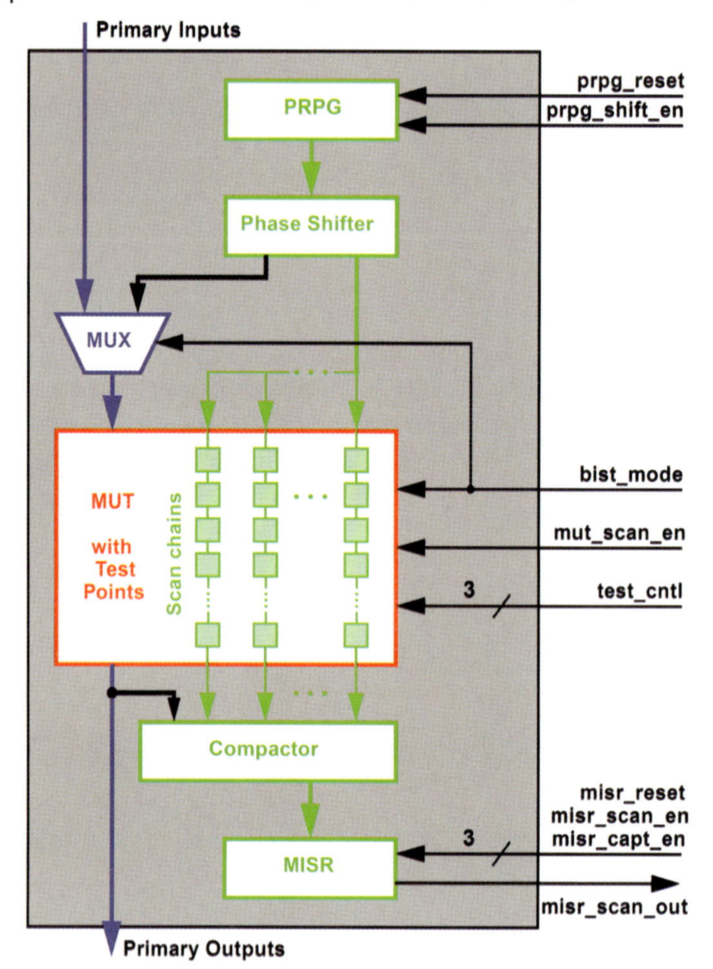

Figure 28.37 A typical BIST framework for selected module (MUT).

Circuits with complete scan pathes can be tested with a deterministic test pattern to nearly 100%. With a random test pattern, how they are used by BIST, one achieves frequently only a clearly smaller fault coverage. This characteristic is called "random pattern resistance". A set of methods are available to improve fault coverage (e.g. reseeding, weighted random pattern testing, bit fixing and bit flipping). One usual method is represented by inserting additional test points to improve the controllability on the one hand (e.g. outputs of a multiple AND) and to improve observability on the other hand; see Reference [33]. The control points are activated by control signals at different times during the test (multi-phase test point insertion). Points of observation are summarized by XOR gates and stored into additional scan-flip-flops.

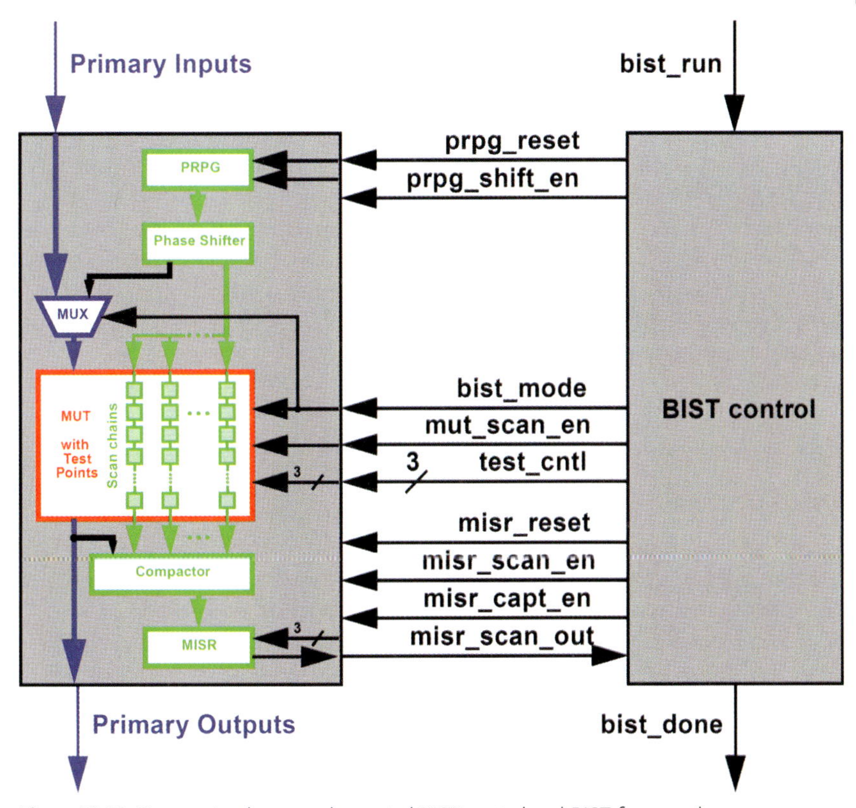

Figure 28.38 Cooperation between the central BIST control and BIST framework.

28.4
Conclusions

The goal of all investigations mentioned above is the design support for 3D micro systems. Therefore, the adaptation of design flows and the integration of results within these design flows are important tasks, which are also covered by the described methodology. All results are provided using data formats and languages of the dedicated design tools.

Depending on the design task, models on different levels of abstraction have be used. For analog and mixed-signal systems, these are usually SPICE net lists or behavioral models in VHDL-AMS or Verilog-AMS. Digital system design is supported by models for cross talk and signal-dependent delays that are derived from more detailed models and analysis.

Due to the dense integration, an important task for the entire stack design is thermal management. This includes the localization of hot spots, adjustment of the stack layout to decrease critical temperatures, as well as reliability issues due to thermally induced mechanical stresses. Furthermore, the influence of thermal effects on the behavior of semiconductor devices and micro structures, for example sensor elements, has to be investigated. These tasks are supported either by detailed analysis using PDE solvers or by system level simulation, for example for electro-thermal interactions, using appropriate behavioral models [34–36].

The design of 3D systems demands methods and tools mainly for extreme low power design and design for testability. Some optimization methods, generally included in synthesis steps, find dedicated circuit structures with the required properties. Examples are the minimization of the dynamic part of power consumption on logic level by reducing the switching activities in CMOS circuits. Known methods of design for testability, such as scan methods, self-test approaches and integration of complete test functions on the chip, help to obtain efficient test sequences with the necessary fault coverage.

Acknowledgments

We gratefully acknowledge the work of our colleagues at Fraunhofer IIS/EAS, which is the basis for this chapter. Especially, we thank Sven Reitz, Andreas Wilde, Jörn Stolle, Jörg Becker, Roland Martin, Jürgen Haufe, Andreas Köhler Matthias Gulbins, Horst Süße and Andy Heinig for their willing cooperation over several years.

Furthermore, we express our sincere appreciation to our colleagues at Fraunhofer IZM in Munich and Berlin, especially Josef Weber, Peter Ramm and Eberhard Kaulfersch, and our colleagues at Sintef, Infineon, Philips and NXP for their highly effective cooperation.

This chapter is partly based on the project e-CUBES, supported by the European Commission under support no. IST-026461.

References

1 Leduc, Patrick, de Crecy, Francois, Fayolle, Murielle *et al.* (2007) Challenges for 3D IC integration: bonding quality and thermal management. International Interconnect Technology Conference, IEEE 2007, 4–6 June 2007, pp. 210–212.

2 Dongkeun, Oh, Chen, Charlie Chung Ping and Hu, Yu Hen (2007) 3DFFT: thermal analysis of non-homogeneous IC using 3D FFT Green function method. 8th International Symposium on Quality Electronic Design, 2007. ISQED'07, 26–28 March 2007, pp. 567–572.

3 Xue, Lei, Liu, C.C., Kim, Hong-Seung *et al.* (March 2003) Three-dimensional integration: technology, use, and issues for mixed-signal applications. *IEEE Transactions on Electron Devices*, **50** (3), 601–609.

4 Huang, Wei, Stan, M.R. and Skadron, K. (Dec. 2005) Parameterized physical compact thermal modeling. *IEEE Transactions on Components and Packaging Technologies*, **28** (4), 615–622.

5 Munteanu, I. *Robust Analog Design through 3D EM Simulation*, Proceedings Analog '05, March 2005, Hannover, p. 37–46.

6 Maeda, S., Kashiwa, T. and Fukai, I. (Dec 1991) Full wave analysis of propagation characteristics of a through hole using the finite-difference time-domain method. *IEEE Transactions on Microwave Theory and Techniques*, **39** (12), 2154–2159.

7 Chtchekatourov, V., Coccetti, F. and Russer, P. (2001) Full-wave analysis and model-based parameter estimation approaches for Y-matrix computation of microwave distributed RF circuits. IEEE MTT-S International Microwave Symposium Digest, 2001, 20–25 May 2001, **2**, pp. 1037–1040.

8 Sabelka, R. (February 2001) Dreidimensionale Finite Elemente Simulation von Verdrahtungsstrukturen auf Integrierten Schaltungen. Dissertation, Technische Universität Wien.

9 Zhai, Xiaoshe, Song, Zhengxiang, Geng, Yingsan *et al.* (2006) Hybridized 3D-FDTD and circuit simulator for analysis of PCB via's signal integrity. 17th International Zurich Symposium on Electromagnetic Compatibility, EMC-Zurich 2006, 27 Feb.–3 March 2006, pp. 89–92.

10 Mei, S. and Ismail, Y.I. (April 2004) Modeling skin and proximity effects with reduced realizable RL circuits. *IEEE Transactions on Very Large Scale Integration (VLSI) Systems*, **12** (4), 437–447.

11 Wollenberg, G. and Kochetov, S.V. (February 2003) Modeling the skin effect in wire-like 3D interconnection structures with arbitrary cross section by a new modification of the PEEC method. 15th international Zurich Symposium and Technical Exhibition on Electromagnetic Compatibility, pp. 609–614.

12 Ruehli, A.E. (1974) Equivalent circuit models for three-dimensional multiconductor systems. *IEEE Transactions on Microwave Theory and Techniques*, **22** (3), 216–221.

13 Kamon, M., Marques, N., Silveira, L. and White, J. (1997) Generating reduced order models via PEEC for capturing skin and proximity effects. Proceedings IEEE 6th Topical Meeting in Electrical Performance of Electronic Packaging, Monterey, CA, San Jose, California, October 1997.

14 Arona, N.D. (2003) Modeling and characterization of copper interconnects for SoC design. International Conference on Simulation of Semiconductor Processes and Devices, Boston Marriott Cambridge, September 3–5 2003.

15 Antonini, G., Scogna, A.C. and Orlandi, A. (April–June 2003) S-parameters characterization of through, blind, and buried via holes. *IEEE Transactions on Mobile Computing*, **2** (2), 174–184.

16 Ryu, Chunghyun, Lee, Jiwang, Lee, Hyein *et al.* (2006) High frequency electrical model of through wafer via for 3-D stacked chip packaging. 1st Electronics System Integration Technology Conference, September 2006, 1, pp. 215–220.

17 Ghouz, H.H.M. and El-Sharawy, E.-B. (Dec. 1996) An accurate equivalent circuit model of flip chip and via interconnects. *IEEE Transactions on Microwave Theory and Techniques*, **44** (12), Part 2, 2543–2554.

18 Lau, J.H. and Pao, Y. (1997) *Solder Joint Reliability of BGA, CSP, Flip Chip, and Fine Pitch SMT Assemblies*, McGraw-Hill, New York.

19 Auersperg, J., Schubert, A., Vogel, D. *et al.* (1997) Fracture and damage evaluation in chip scale packages and flip-chip assemblies by FEA and MicroDAC. *Application of Fracture Mechanics in Electronic Packaging*, **20**, 133–138.

20 Schubert, A., Dudek, R., Auersperg, J. *et al.* (1997) Thermo-mechnical reliability analysis of flip-chip assemblies by combined microdac and finite element method. Conference Proceedings Interpack'97, Hawaii, USA, pp. 1647–1654.

21 Dudek, R. Schubert, A. and Michel, B. (2000) Thermo-mechanical reliability of microcomponents. Proceedings 3. International Conference on Micromaterials, Berlin, Germany, April 17–19, 2000, pp. 206–213.

22 Schubert, A., Dudek, R., Michel, B. and Reichl, H. (2000) Package reliability studies by experimental and numerical analysis. Proceedings 3. International

Conference on Micromaterials Materials, Berlin, Germany, April 17–19, 2000, pp. 110–118.

23 Wittler, O., Sprafke, P. and Michel, B. (2003) Elastic and viscoelastic fracture analysis of cracks in polymer encapsulations. in Fracture of Polymers, Composites and Adhesives II, 3rd ESIS TC4 Conference (Hrsg.), (eds J. Williams, A. Pavan and B. Blackman), ESIS Publication, Elsevier, Amsterdam.

24 Jansen, K.M.B., Wang, L., Yang, D.G. *et al.* (2004) Constitutive modeling of moulding compounds. Proceedings of the 54th Electronic Components and Technology Conference (ECTC2004), Las Vegas, June 1–4 2004, IEEE Catalog number 04CH37546C, ISSN: 0569-5503, ISBN: 0-7803-8366-4, IEEE pp. 890–894.

25 Auersperg, J., Seiler, B., Cadalen, E. *et al.* (2005) Fracture mechanics based crack and delamination risk evaluation and RSM/DOE concepts for advanced microelectronics applications. Proceedings of 6th IEEE EuroSimE Conference and Exhibition, Berlin, Germany, April 18–20 2005, pp. 197–200.

26 Pillai, E., Rostan, F. and Wiesbeck, W. (27 May 1993) Derivation of equivalent circuits for via holes from full wave models. *Electronics Letters*, **29** (11), 1026–1028.

27 Sommer, J.-P., Michel, B. and Ostmann, A. (2005) Numerical characterization of electronic packaging solutions based on hidden dies. 6th International Conference on Electronic Packaging Technology, 2005, 30 Aug.–2 Sept. 2005, pp. 300–306.

28 Sischka, F. (2002) *S-Parameter Basics for Modeling Engineers*, Agilent Techn.

29 Schneider, P., Schneider, A. and Schwarz, P. (2005) A modular approach for simulation-based optimization of technical systems. Proceedings, 3rd MIT Conference on Computational Fluid and Solid Mechanics, June 14–17 2005, Cambridge, Massachusetts, pp. 1288–1291.

30 Harper, C.A. (ed.), (2000) *Electronic Packaging and Interconnection Handbook*, 3rd edn, McGraw-Hill, New York.

31 Wunderle, B., May, D., Braun, T. *et al.* (2007) Non-destructive failure analysis and modeling of encapsulated miniature smd ceramic chip capacitors using thermal and mechanical loading. Proceedings, 13th International Workshop on THERMAL INVESTIGATIONS of ICs and SYSTEMS, 17–19 September 2007, Budapest, Hungary.

32 Morrow, P. *et al.* (2006) Design and fabrication of 3D microprocessors. Paper Y3.2, Applications of 3D Integration, MSR Fall Meeting, Boston, November 26–30.

33 Tamarapalli, N. and Rajski, J. (1996) Constructive multi-phase test point insertion for scan-based BIST. International Test Conference, Washington, D.C., October 1996, pp. 649–658.

34 Ramm, P., Klumpp, A., Merkel, R. *et al.* (2003) 3D system integration by chip-to-wafer stacking technologies. Extended Abstracts of the International Conference on Solid State Devices and Materials, Tokyo, pp. 376–377.

35 Schneider, P., Reitz, S., Wilde, A. *et al.* (25–27 April 2007) Towards a methodology for analysis of interconnect structures for 3D integration of micro systems. Proceedings DTIP 2007, Stresa, pp. 162–168.

36 Schneider, P., Reitz, S., Wilde, A. *et al.* (22–24 October 2007) Design support for 3D Integration by physical oriented modeling of interconnect structures. Proceedings 3D Architectures for Semiconductor Integration and Packaging, Burlingame.

29
Multiproject Circuit Design and Layout in Lincoln Laboratory's 3D Technology

James Burns, Robert Berger, Nisha Checka, Craig Keast, Brian Tyrrell, and Bruce Wheeler

29.1
Introduction

The implementation of Lincoln Laboratory's three-dimensional (3D) circuit design techniques closely followed the development of the 3D fabrication technology. DARPA funded a 3D Multiproject program to make Lincoln's 3D circuit integration technology available to designers in universities and commercial institutions so that experience gained by all participants through circuit design and testing of their 3D chips would lead to greater understanding of the advantages of 3D integrated circuits (ICs). 3D circuit designs were submitted by these institutions and were integrated into a 3D Multiproject chip along with Lincoln Laboratory designs. Initially, 3D circuits and test devices were designed and laid out as if a 3D circuit were distinctly different from its 2D counterpart. However, experience gained in 3D circuit layout and 3D technology development made clear that a more direct design approach was required to simplify the design and layout process, particularly for researchers unfamiliar with the details of 3D technology. This required the development of 3D design rules, 3D design and layout enhancements to Mentor [1] and Cadence [2] computer-aided design (CAD) tools, and circuit design submission rules to ensure data compatibility when the designs were incorporated in the 3D Multiproject chip.

29.2
3D Design and Layout Practice

Three steps were adopted to aid the design and layout of 3D chips. The first was to view a 3D chip as a 2D chip but with multiple SOI, polysilicon, and interconnect layers. The scanning electron micrograph (SEM) of the 3D chip of Figure 20.2 in Chapter 20 (Lincoln Laboratory's 3D Circuit Integration Technology), was obtained from a 3D ring oscillator and can be viewed as a 2D chip with three silicon-on-insulator (SOI) levels, two buried oxide layers, three polysilicon layers and ten

Handbook of 3D Integration: Technology and Applications of 3D Integrated Circuits.
Edited by Philip Garrou, Christopher Bower and Peter Ramm

ISBN: 978-3-527-32034-9

metallization levels. The second step was to introduce a new term, tier, to avoid confusion with the various terms used to describe design and fabrication layers. A tier is the section of a completed wafer that contains the active silicon, polysilcon and interconnect and if it is a SOI wafer it also contains the buried oxide (BOX). As discussed in Chapter 20, the 3D integration process bonds, transfers and interconnects tiers to build a 3D chip. The third step was to design and lay out a 3D chip as it would be viewed in its completed state with the tiers numbered from 1 to 3, with tier 1 at the bottom and the bond pads located tier 3, the top tier, as shown in Figure 20.1 of Chapter 20.

3D design rules were developed and embedded into the existing 2D design rules for the laboratory's FDSOI technology [3], which increased the number of rules by 25%, a value consistent with the additional SOI, polysilicon, and interconnect layers. The current design rules are the third revision and reflect technology and design advancements. Four features defined by the rules are related to the 3D via shown in Figure 20.5 of Chapter 20. The first is the size of the 3D cut, which defines the resist window to etch the oxide to the metal annulus. The second is the size of the metal annulus in the upper tier. The third is the 3D land in the lower tiers, which is contacted by the tungsten plug. The fourth is the spacing between 3D vias and between 3D vias and other layers to avoid shorting of the tungsten plug to them. The present-scaled 3D via design consists of a 3-μm square landing pad, a 1.5-μm square annulus opening and a 1.75-μm 3D cut. The total etched oxide depth is 8 μm. The design of these inter-tier connections significantly reduced the pitch of 3D vias to 6 μm from 26 μm used in the 3D imager first reported [4]. In addition, metal fill densities were modified to ensure that the planarity achieved after chemical mechanical polishing (CMP) satisfies the requirements for low-temperature bonding.

Experience from three previous 3D chip designs indicated that the CAD tools used for design and layout needed refinement to reduce design errors and reticle compilation time. Some major problems in the past were misalignment of the tiers, which led to nonfunctional 3D vias, 3D vias that lacked metal lands and faulty designs due to misidentification of layers. To relieve some of these difficulties, Lincoln Laboratory incorporated the 3D design rules into a Mentor Graphics-based 3D design kit. DARPA also funded North Carolina State University to develop a Cadence-based 3D design kit (courtesy of Rhett Davis, North Carolina State University). Modifications to these design kits included layer displays that were unique to each tier. For example, the polysilicon layers in each tier are the same color but have different shading.

A major challenge in 3D design is placement of sub-circuits, such as an arithmetic logic unit (ALU) or a ring oscillator, within a tier and its placement among the three tiers; this topic has been the subject of a great deal of research [5]. An experimental architectural and topological design tool [6] was developed to optimize the 3D architecture of a circuit as a function of the 3D via design rules, the number of tiers and the circuit area. A cost function is defined, which could be performance, circuit area or processing complexity. Then, a set of simulations is run to determine what process and design factors would minimize the function. The regularity of an imager design makes it a natural application for such a tool, and the tool was applied

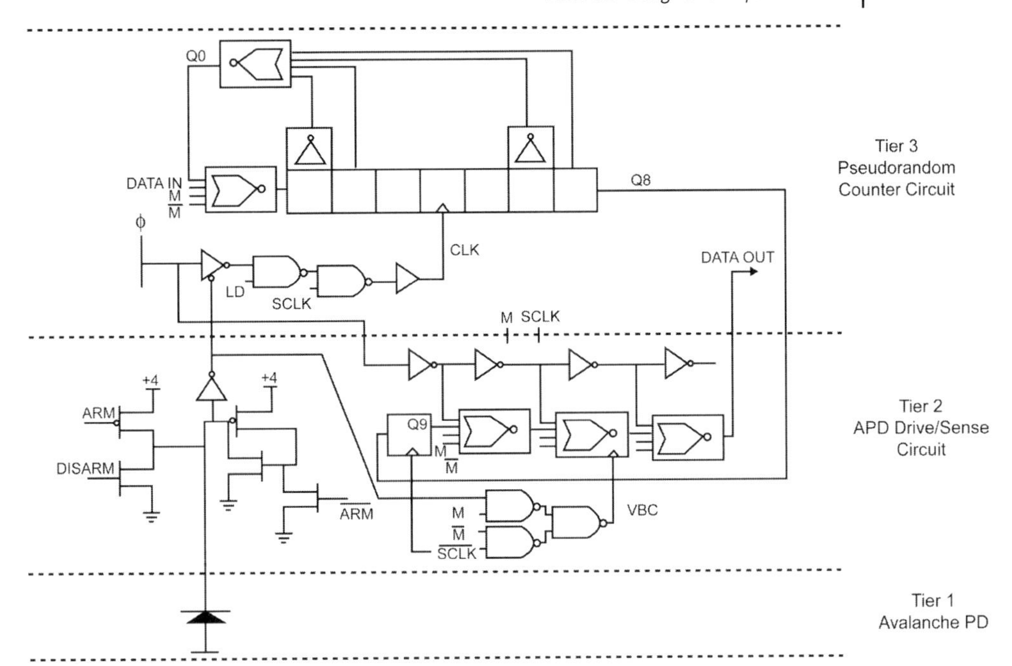

Figure 29.1 Optimization analysis studied the pixel size of the 3D-LADAR chip as a function of portioning the sub-circuits shown above among seven tiers. There are 256 transistors per pixel.

to the laser radar (LADAR) imager discussed in Chapter 20 and shown in Figure 29.1. For this analysis the cost function was the pixel size, and the principal variables were the number of tiers for the design and the placement of the various sub-circuits among seven tiers. The results of the analysis to determine the minimum pixel size

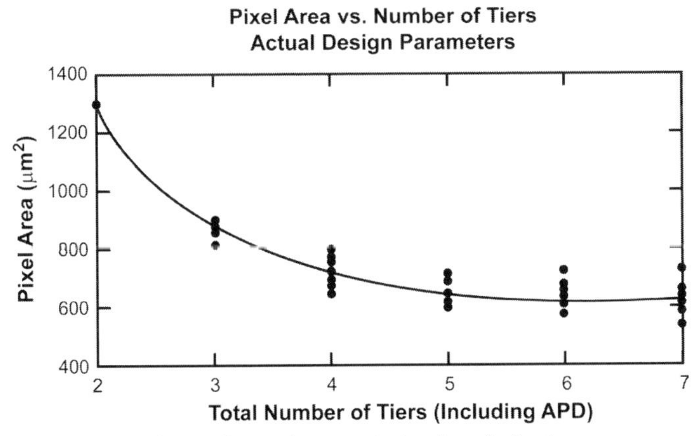

Figure 29.2 Analysis indicates that partitioning the sub-circuits among four tiers results in a optimum 3D architecture when pixel size is used as the criterion.

are shown in Figure 29.2 and indicate the optimum number of tiers is four but also point out that a three-tier design is superior to a two-tier design. The actual 3D chip consisted of three tiers, which was a compromise among processing complexity, fabrication throughput and pixel size. This analysis demonstrated the utility of a 3D-specific architecture tool that would be very valuable in the early stages of design to optimize 3D circuit architecture.

29.3 Design and Submission Procedures

The designer is required to adhere to specific layout and submission requirements since the design will be integrated into a 3D Multiproject chip. As previously discussed, a chip is designed as viewed in its completed state with the bond pads on the topmost tier. A layer numbering procedure is used that appends the conventional layer numbers for 2D design to the tier number. For example, the numbers for polysilicon layers on tiers 1 and 3 are 209 and 409, respectively, while the metal 1 numbers for layers on tiers 2 and 3 are 327 and 427, respectively. These procedures ensure that the tiers have a common origin and that rotation of tiers 2 and 3 about the x-axis, required for reticle compilation, can be performed automatically at Lincoln Laboratory. 3D designs must be submitted as a gds file with the top cell of the design identified as 3DP_org_Design.gds where 3DP is the multiproject program name, org is the institution's name, and Design is the design name chosen by the designer. The bond pads must be placed in the top cell. The extent of the cell is defined as X = ###, Y = ### in mm units, and deliberate design rule violations, if permitted by Lincoln Laboratory, must be identified.

Upon receipt at Lincoln Laboratory, a design is reviewed for obvious errors using a gds viewer and then run through a post-processing procedure to add implant layers, bloat specific features and add fill layers. Then design rule checks (DRCs) are run to verify that 3D-specific rules are satisfied. For example, a 3D land must exist for each 3D cut, the metal surround rules for a 3D cut in tier 2 must be satisfied in metal 3 of tier 1, and the metal surround rules for back metal 1 in tier 2 must be satisfied for a 3D cut in tier 3. A layout vs. schematic (LVS) extraction is also run as part of the post-processing procedure. The DRC report, LVS netlist and the post-processed files are then sent to the designer for review and correction if necessary. After a final review, all the 3D designs, test devices, alignment targets and process control structures are integrated to create the master gds file. The master file is separated into individual photolayers for reticle manufacturing at a mask vendor.

The design practices discussed in the previous section were used by 20 institutions to design 3D chips that were included in DARPAs first 3D Multiproject chip, shown in Figure 29.3. Various applications were investigated to study 3D design and performance issues. 3D-SRAM memories were designed by researchers at RPI (courtesy of J. McDonald, Rensselaer Polytechnic Institute). The devices are the first 3-tier memories fabricated and promise that 3D will lead to greater design and

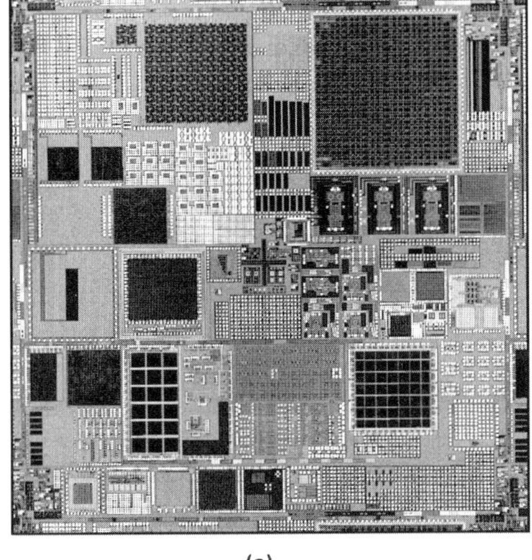

(a)

3D FPGAs, digital, and digital/mixed-signal/RF
ASICs exploiting parallelism of 3D interconnects
3D analog continuous-time processor
3D-integrated S-band digital beamformer
Stacked memory (SRAM, Flash, and CAM)
Self-powered CMOS logic (scavenging)
Integrated 3D nano-radio and RF tags
Intelligent 3D-interconnect evaluation circuits
DC and RF-coupled interconnect devices
Low-power multi-gigabit 3D data links
Noise coupling/cross-talk test structures and circuits
Thermal 3D test structures and circuits

(b)

Figure 29.3 (a) Photograph of the first 3D Multiproject chip, which contained integrated 3D chips designed by 21 institutions in addition to circuits designed internally and (b) list of the 3D circuits and applications.

programming flexibility. 3D designs from Cornell, shown in Figure 29.4, included field-programmable gate arrays (FPGAs) that promise higher-bandwidth performance due to reduced interconnect delay. Also included in the 3D chip were low-voltage adaptive analog circuits with backgated transistors that exhibited reduced RF cross-talk through 3D-integrated ground planes. A 3D analog-digital converter for imaging applications was designed at Stanford (courtesy of S-M Lee, Stanford University) that achieved a high frame rate with a 50-μm pitch. A 3D design by UCLA [7] investigated capacitive-coupled interconnects and demonstrated a base-band impulse shaping interconnect and self-synchronized RF interconnect at >11 GHz with a bit error rate less than 1×10^{-14}. A 3D imager [8] demonstrated the feasibility of a three-color imager. A collaborative effort by the Naval Research Laboratory, BAE Systems and Cornell University designed 3D-integrated CMOS circuitry with micromechanical resonators that achieved a $Q = 4700$ at 34 MHz (Figure 29.5).

A second 3D Multiproject program that includes 20 participants has begun. Additional topics to be studied are optimum 3D layout to control heat dissipation, heat extraction from 3D circuits, performance enhancements with scaled 3D vias, and the utility of an interconnect tier to provide signal distribution and redundancy (Figure 29.6). Test data are not available as of this writing, but the results are sure to lead to further understanding of the promise of 3D design.

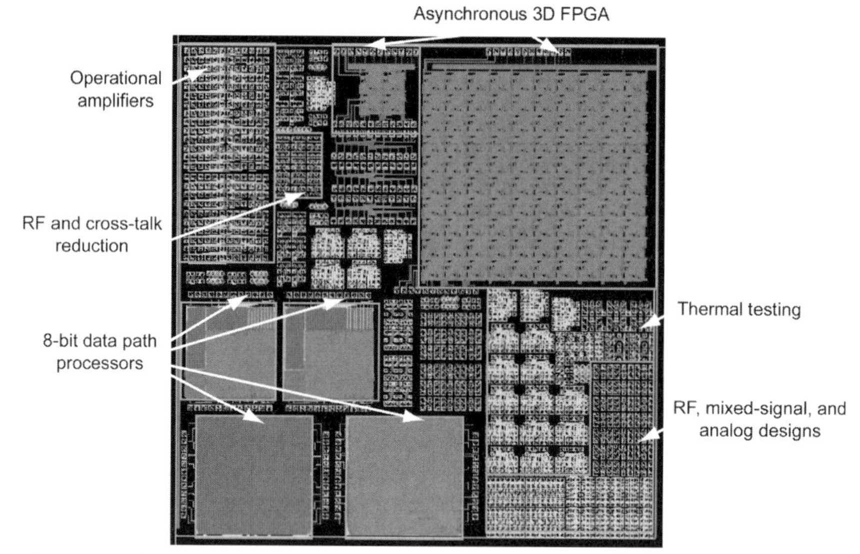

Figure 29.4 Chip design incorporating a series of functional asynchronous 3D FPGAs and low-voltage adaptive analog circuits, with backgating and RF cross-talk reduction through 3D-integrated ground planes. Courtesy of S. Tiwari of Cornell University.

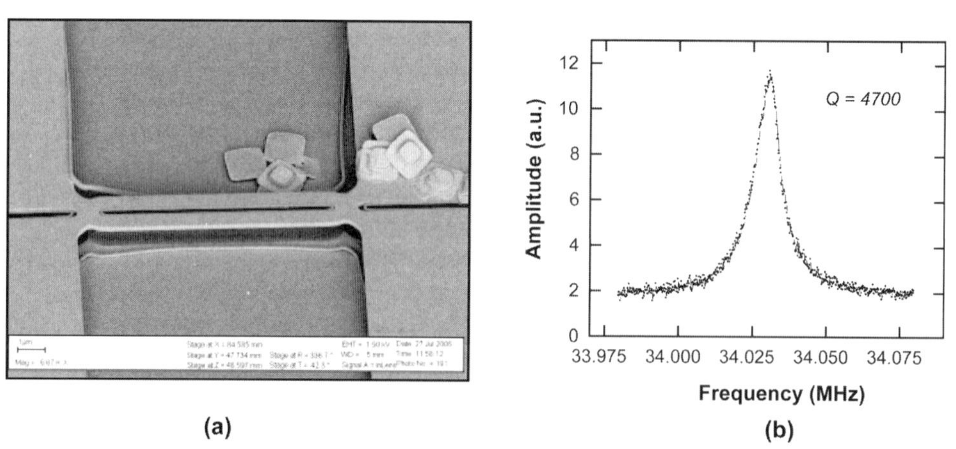

Figure 29.5 (a) SEM of a micromechanical resonator and (b) the frequency response. The resonator was 3D integrated with CMOS. Courtesy of M. Zalalutdinov of the Naval Research Laboratory.

Acknowledgments

This work was sponsored by the Defence Advanced Research Projects Agency under Air Force contract #FA8721-05-C-0002. Opinions, interpretations, conclusions, and

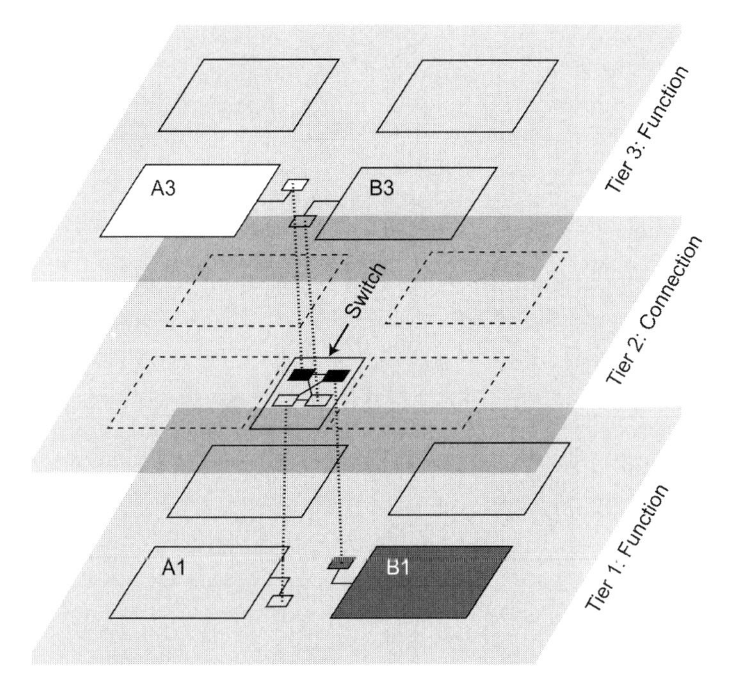

Figure 29.6 A design concept being tested on the second 3D Multiproject chip is the use of tier 2 as a smart interconnection tier to achieve reduced signal delays and power supply droop. The tier can be part of a redundancy strategy.

recommendations are those of the authors and are not necessarily endorsed by the United States Government. The authors wish to acknowledge the editorial assistance of Karen Challberg.

References

1 Mentor Graphics, IC Nanometer Design Tool Suite.
2 Cadence Virtuoso Design Tool.
3 MITLL Low-Power FDSOI CMOS Process Design Guide, Revision 2006:7, Advanced Silicon Technology Group, MIT Lincoln Laboratory, 244 Wood St., Lexington, MA 02420.
4 Burns, J., McIlrath, L., Keast, C. *et al.* (2001) Three-dimensional integrated circuits for low-power, high-bandwidth systems on a chip. Digest Tech. Papers IEEE International Solid-State Circuits Conference, pp. 268–269.
5 Reber, M. and Tielert, R. (1996) Benefits of vertically stacked integrated circuits for sequential logic. *Proceedings of IEEE International Symposium on Circuits and Systems*, **4**, 121–124.
6 Tyrrell, B. (June 2004) Development of an architectural design tool for 3-D VLSI sensors, MS Thesis, Massachusetts Institute of Technology, Department of Electrical Engineering.

7 Gu, Q., Xu, Z., Ko, J. and Chang, M.C.F. (2007) Two 10Gb/s/pin low power interconnect methods for 3D ICs. Digest Tech. Papers IEEE International Solid-State Circuits Conference, pp. 364–365.

8 Culurciello, E. and Weerakoon, P. (2007) Three-dimensional photodetectors in 3D silicon-on-insulator technology. *IEEE Electron Device Letters*, **28** (2), 117–119.

30
Computer-Aided Design for 3D Circuits at the University of Minnesota

Sachin S. Sapatnekar

30.1
Introduction

3D technology opens up an entirely new dimension, literally and figuratively, for circuit design, making new sets of design structures and architectures available and feasible. For a chip designer at the leading-edge, this presents a range of opportunities, and also a significant set of challenges. Even in the conventional 2D world, chip design is an extremely laborious and time-intensive task; with the new horizons opened up by 3D, the range of choices increases, as does the level of complexity. Computer-aided design (CAD) tools, which are indispensable in 2D design, become even more significant in addressing the challenges of 3D. This chapter presents an overview of critical CAD technologies that play in facilitating 3D design.

One of the primary motivators for 3D technologies is related to the dominant effects of interconnects in nanoscale technologies, and the addition of a third dimension provides significant relief in this respect. For example, for a 2D chip with a footprint of $2L \times 2L$, the longest wire can be $4L$ units long. However, the same circuitry can be accommodated within a four-layer 3D chip of footprint $L \times L$, where the longest wire, neglecting z dimension distances that are typically $\ll L$, is $2L$. The reduction in interconnect lengths does not just affect the longest wire: in general, 3D can effect reductions in the average interconnect lengths (in comparison with 2D implementations, for the same circuit size), lower wire congestion, as well as by denser integration, which results in the replacement of chip-to-chip interconnections by intra-chip connections. In addition, the increased packing density improves the computation per unit volume.

The move from 2D to 3D is inherently a topological change, and therefore, several of the problems that are unique to 3D circuits are related to physical design optimizations that determine the circuit layout. This applies to layout optimizations such as placement and routing, which will be discussed in this chapter.

Handbook of 3D Integration: Technology and Applications of 3D Integrated Circuits.
Edited by Philip Garrou, Christopher Bower and Peter Ramm

ISBN: 978-3-527-32034-9

Another issue is related to the fact that circuitry can be packed more densely in 3D than in 2D. While this is clearly a major advantage, it also brings about new limitations and challenges to the designer, in terms of how the chip interacts with the environment. A k-tier 3D chip could use k times as much current as a single 2D chip of the same footprint; however, the packaging technology is not appreciably different. This has major implications from the point of view of packaging limitations:

- First, the 3D chip generates k times the power of the 2D chip, which implies that the corresponding heat generated must be sent out to the environment using a substantially similar package, or that 3D chips must face higher temperatures. The latter is highly undesirable, since elevated temperatures can hurt performance and reliability, in addition to introducing variabilities in the performance of the chip. Therefore, on-chip thermal management is a critical issue in 3D design.
- Second, the package must be capable of supplying k times the current through the power supply (V_{dd} and ground) pins, as compared to the 2D chip. Given that reliable power grid design is a major bottleneck even for 2D designs, this implies that significant resources have to be invested in building a bulletproof power grid for the 3D chip.

30.2 Thermal Analysis of 3D Designs

Thermal issues are a key factor in 3D design, and therefore we begin this chapter with an exposition of some key techniques for performing thermal analysis in 3D circuits. At the full-chip level, the ideas of heat transfer can be applied to determine the on-chip temperature. A typical design consists of multiple tiers of active devices stacked on each other, each dissipating power and generating heat. This heat is then transferred through the silicon and dielectric structure to the heat sink, and the strength of the heat conduction paths determines the corresponding rise in temperature. Briefly, if the effective thermal conductivities from the dominant heat sources to the heat sink are small, then the rise in temperature will be small; if not, not.

Heat conduction at the macro scale implied by such systems is governed by Fourier's Law, given by:

$$\nabla^2 T(\mathbf{r}) + \frac{g(\mathbf{r})}{k_\mathbf{r}} = \frac{\rho c}{k_\mathbf{r}} \frac{\partial T}{\partial t}$$

where k is the thermal conductivity at the particular location, ρ is the density of the material and c is the specific heat capacity; g is the volume power density, which is also location dependent. Usually, the problem is formulated in three-dimensional space, therefore $\mathbf{r}$ is a three-dimensional array $\mathbf{r} = (x, y, z)$. Since the time constant of on-chip temperature change is usually in the order of milliseconds, while the operating frequency of electric signal is in the range of picoseconds, for the purposes of analysis, it is sufficient to solve the steady-state problem.

The heat diffusion equation can then be simplified as Poisson's equation (Equation 30.1):

$$\nabla^2 T(\mathbf{r}) = -\frac{g(\mathbf{r})}{k_r} \tag{30.1}$$

The boundary conditions for this equation reflect the heat sinking environment. For multilayer systems that have piecewise constant thermal conductivities, a continuity equation is applied for the temperature and heat flux at the boundaries of each piecewise constant region. From the sides of the chip, the boundary conditions are taken to be adiabatic, since no heat can escape. For the heat sink, it is reasonable to coarsely assume an isothermal condition, which states that the heat sink is at a constant temperature, the ambient temperature (a constant value that has the function of the ground node in an electrical circuit). Alternatively, a conductive boundary condition could use a macromodel for the heat sink, connected to a node representing the ambient temperature.

We will discuss two techniques for on-chip thermal analysis of 3D ICs, using the finite difference method (FDM) and finite element analysis (FEA), respectively. The FDM discretizes the space into regions, and the temperature of each region is represented by the temperature at its center. By approximating the derivative as a finite difference, the partial differential equation is written as a system of linear algebraic equations of the form shown in Equation 30.2:

$$G\mathbf{T} = \mathbf{P} \tag{30.2}$$

where G is the thermal conductance matrix, $\mathbf{T}$ is the vector of unknown temperatures, and $\mathbf{P}$ is the vector of power dissipation over all regions. The classical theory of heat transfer draws an analogy between the relationships between nodes representing adjacent finite difference regions and the electrical resistance between two nodes in a circuit [1]. This thermal-electrical analogy allows the representation of the heat flow network in the form of a grid of thermal resistors. This grid can be solved using routine electrical circuit analysis techniques, by thinking of thermal resistors as electrical resistors, voltages as temperatures and currents as power.

Like the FDM, FEA also discretizes the design space into regions known as elements. The temperature at any point within each element is a polynomial interpolation of the temperatures at its vertices, and the task of FEA is to find the vertex temperatures for the discretized partial differential equation.

For rectangular structures of the type encountered in integrated circuits, a rectangular cuboidal element is a useful structure to enable the simulation of heat conduction in the lateral directions without aberrations in the prime directions. The FEA results in a matrix of the type shown in Equation 30.3:

$$K\mathbf{T} = \mathbf{P} \tag{30.3}$$

The left-hand side matrix, K, known as the global stiffness matrix, can be constructed using stamps for the finite elements and the boundary conditions. In comparison with the FDM matrix G, the K matrix is typically more dense, but the level of discretization for comparable accuracy in FEA is typically smaller, implying that the dimension of K can be smaller than that of G.

Both the FDM and the FEA equations are solved rapidly using a standard linear solvers. Direct Cholesky factorization based solvers can be very efficient for sparse FDM matrices, while preconditioned iterative methods (with good preconditioners) can deliver excellent results [2]. For iterative solvers, clever adjustments of the convergence criteria may be employed to achieve accuracy–runtime tradeoffs. A third category of stochastic solvers, based on random walks [3], has recently been shown to be extremely effective for solving systems of this type, particularly for incremental or partial analysis, but also for solving full systems.

30.3 Thermally-Driven Placement and Routing of 3D Designs

For standard cell-based 3D designs, we describe a flow, illustrated in Figure 30.1, for performing placement and routing with built-in techniques for thermal mitigation. The input to the system is a technology-mapped netlist and a description of the library, and the physical design process consists of several steps. Temperature is treated as a first-class citizen during this optimization, in addition to other conventional metrics, and inter-tier via reduction is also considered to be a desirable goal. In the placement step, the standard cells are arranged in rows within the tiers of the three-dimensional circuit. Since thermal considerations are particularly important in 3D ASIC-like circuits, this procedure must spread the cells to achieve a reasonable temperature distribution, while also capturing traditional placement requirements. In the second step, the temperature distribution is made more uniform by the judicious positioning of thermal vias within the placement, which achieves improved

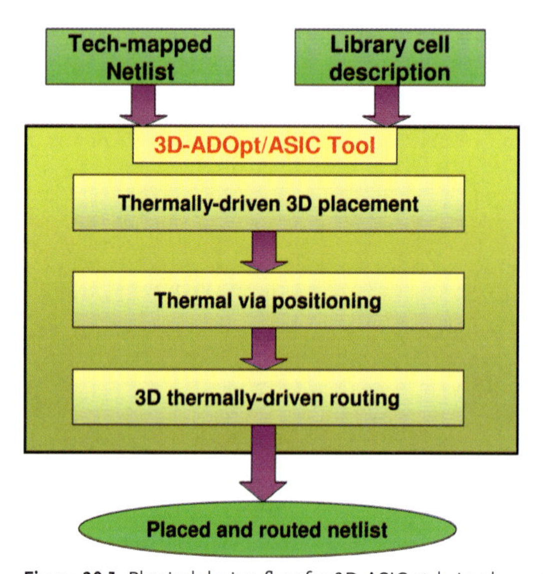

Figure 30.1 Physical design flow for 3D ASIC-style implementations [4].

heat removal. These vias correspond to inter-tier metal connections that have no electrical function, but, instead, constitute a passive cooling technology that draws heat from the problem areas to the heat sink. Finally, the placement goes through a routing step to obtain a completed layout. During routing, several objectives and constraints must be taken into consideration, including avoiding blockages due to areas occupied by thermal vias, incorporating the effect of temperature on the delays of the routed wires, and, of course, traditional objectives such as wire length, timing, congestion and routing completion. We now describe each of these steps in further detail.

30.3.1 Thermally-Driven 3D Placement

This section describes two generations of the 3D-ADOpt placement tool developed at the University of Minnesota.

The first generation placer [5] used a force-directed placement paradigm, where an analogy to Hooke's law for springs is used between cells. Cells that are connected to each other by a net are deemed to have an attractive force (with the idea that this will bring them closer together, and shorten the overall net lengths). Connections to the boundaries of the chip help to keep the cells distributed.

In each iteration, each cell is treated as a point object represented by its center, and the minimum energy state of the system is computed, corresponding to a specific location for each cell. However, this solution has a few limitations. First, it does not incorporate thermal considerations. Second, the relaxation of an inherently discrete placement problem to this continuous space neglects the fact that cells are not point objects but have nonzero areas. The solution above may result in illegal placements where cells overlap. Third, the problem is solved in the continuous space, and the solution is snapped on to the set of allowable discrete locations.

To overcome some of these issues, iterations are introduced, with new repulsive forces added in. The first set of repulsive forces is based on an embedded fast FEA-based thermal analysis, and moves cells away from hot regions. Note that in early iterations this analysis could be more approximate and fast; as we come closer to the final placement it is sped up. The second set of repulsive forces relates to cell overlaps: regions with large overlaps introduce repulsive forces that move the cells to sparser regions.

The second generation placer [7] observes that since 3D layouts have very limited flexibility in the third dimension (with a small number of layers and a fixed set of discrete locations), the assumption of near-continuity used in force-directed placers is inherently limited in the z direction, where the number of layers is typically in the range of 3–10 in foreseeable technologies. Given this limitation, partitioning works better than a force-directed method. Accordingly, this work performs global placement using recursive bisectioning. To speed up the placement, instead of an embedded thermal analysis engine, thermal effects are incorporated through thermal resistance reduction nets, which are attractive forces that provide incentives for high power nets to remain close to the heat sink.

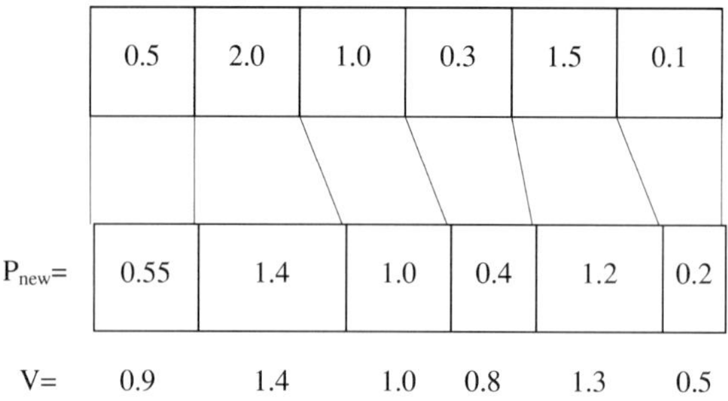

Figure 30.2 Pressure equalization towards 1.0 with damping [6].

The global placement step is followed by coarse legalization, in which a novel cell-shifting approach is proposed. This generalizes the methods in FastPlace [8] by allowing shift moves to adjust the boundaries of both sparsely and densely populated cells using a computationally simple method. The process uses an analogy to a gas law problem in which there is a series of containers with movable walls and different pressures in each container. According to the gas laws, the shared walls of the containers would move so that the pressures equalize. For this problem, the initial pressure corresponds to the initial occupation density of each cell, and the goal is not to equalize the pressures, but to bring them to a nominal unit value. Accordingly, this corresponds to the pressure equalization problem, but with a limit that the pressure of an overcongested region (with occupancy >1.0) should be brought down to 1.0 units and not lower. An example 1D configuration is shown in Figure 30.2. In practice, a damping procedure is used during this expansion to help preserve cell ordering during these moves, for stability. Cells are then moved to these new boundaries (cell-shifting), and the process is repeated iteratively. Next, global moves to an optimal region, or local moves in the vicinity of the current location of the cells, are used to change cell locations. These moves may be cell swaps or unilateral moves, and are chosen so that they reduce the placement cost function.

Finally, detailed legalization generates a final nonoverlapping layout. The approach is shown to provide excellent tradeoffs between parameters such as the number of interlayer vias, wire length, temperature. Figure 30.3 shows an example tradeoff between the number of inter-tier vias and the total wirelength for the benchmark circuit ibm01 [6]. In general, if the number of inter-tier vias is not restricted, these may be freely used; however, as this number becomes tighter, there is severe contention for these resources. To overcome this contention, it is essential for wires to detour to reach an appropriate via, resulting in increased wirelengths. While the total number of inter-tier vias is fixed by technological constraints, these may be split up in various ways between the power grid, the thermal via network and signal nets. Therefore, the number of available inter-tier vias available to signal lines is a design choice, and this tradeoff curve permits the designer to make appropriate choices.

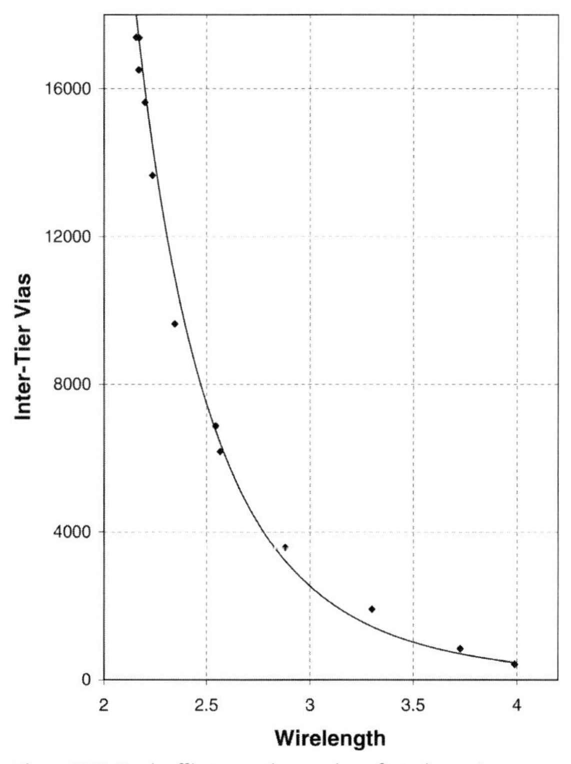

Figure 30.3 Tradeoff between the number of interlayer vias versus the total wire length for the benchmark ibm01 [6].

30.3.2 Automated Thermal Via Insertion for Heat Removal

While silicon is a good thermal conductor, with half or more of the conductivity of typical metals, many of the materials used in 3D technologies are strong insulators that place severe restrictions on the amount of heat that can be removed, even under the best placement solution. The materials include epoxy bonding materials used to attach 3D tiers, or field oxide, or the insulator in an SOI technology. Therefore, the use of deliberate metal lines that serve as heat removing channels, called "thermal vias," are an important ingredient of the total thermal solution. The second step in the flow determines the optimal positions of thermal vias in the placement that provides an overall improvement in the temperature distribution. In realistic 3D technologies, the dimensions of these inter-tier vias are of the order of $5 \times 5\,\mu m$.

In principle, the problem of placing thermal vias can be viewed as one of determining one of two conductivities (corresponding to the presence or absence of metal) at every candidate point where a thermal via may be placed in the chip. However, in practice, it is easy to see that such an approach could lead to an extremely large search space that is exponential in the number of possible positions; note that the set of possible positions in itself is extremely large. Quite apart from the size of the search space, such an approach is unrealistic for several other reasons. First, the wanton addition of thermal vias in any arbitrary region of the layout would lead to nightmares for a router, which would have to navigate around these blockages. Second, from a practical standpoint, it is unreasonable to perform full-chip thermal analysis, particularly in the inner loop of an optimizer, at the granularity of individual thermal vias. At this level of detail, individual elements would have to correspond to the size of a thermal via, and the size of the FEA stiffness matrix would become extremely large.

Fortunately, there are reasonable ways to overcome each of these issues. The blockage problem may be controlled by enforcing discipline within the design, designating a specific set of areas within the chip as potential thermal via sites. These could be chosen as specific inter-row regions in the cell-based layout, and the optimizer would determine the density with which these are filled with thermal vias. The advantage to the router is obvious, since only these regions are potential blockages, which is much easier to handle. To control the FEA stiffness matrix size, one could work with a two-level scheme with relatively large elements, where the average thermal conductivity of each region is a design variable. Once this average conductivity is chosen, it could be translated back into a precise distribution of thermal vias within the element that achieves that average conductivity.

An algorithm for solving this problem is described in Reference [9]. The technique has been applied to a range of benchmark circuits, with over 158 000 cells, and the insertion of thermal vias shows an improvement in the average temperature of about 30% [9], with runtimes of a couple of minutes. Therefore, thermal via addition has a more dramatic effect on temperature reduction than thermal placement.

Figures 30.4 and 30.5 show the 3D layout of the benchmark struct, before and after the addition of thermal vias, respectively. As before, red and blue regions in the thermal map represent hot and cool regions, respectively. Remarkably, the greatest concentration of thermal vias is *not* in the hottest regions, as one might expect at first. The intuition behind this is as follows: if we consider the center of the uppermost tier, a major reason why it is hot is because the tier below it is at an elevated temperature. Adding thermal vias to remove heat from the second tier, therefore, effectively also reduces significantly the temperature of the top tier. For this reason, the regions where the insertion of thermal vias is most effective are those that have high thermal gradients.

30.3.3 Thermally-Driven 3D Routing

Once the cells have been placed and the locations of the thermal vias determined, the routing stage finds the optimal interconnections between the wires. As in 2D routing, it is important to optimize the wire length, the delay and the congestion. In addition, several 3D-specific issues come into play. First, the delay of a wire increases with its temperature, so that more critical wires should avoid the hottest regions, as far as possible. Second, inter-tier vias are a valuable resource that must be optimally allocated among the nets. Third, congestion management and blockage avoidance

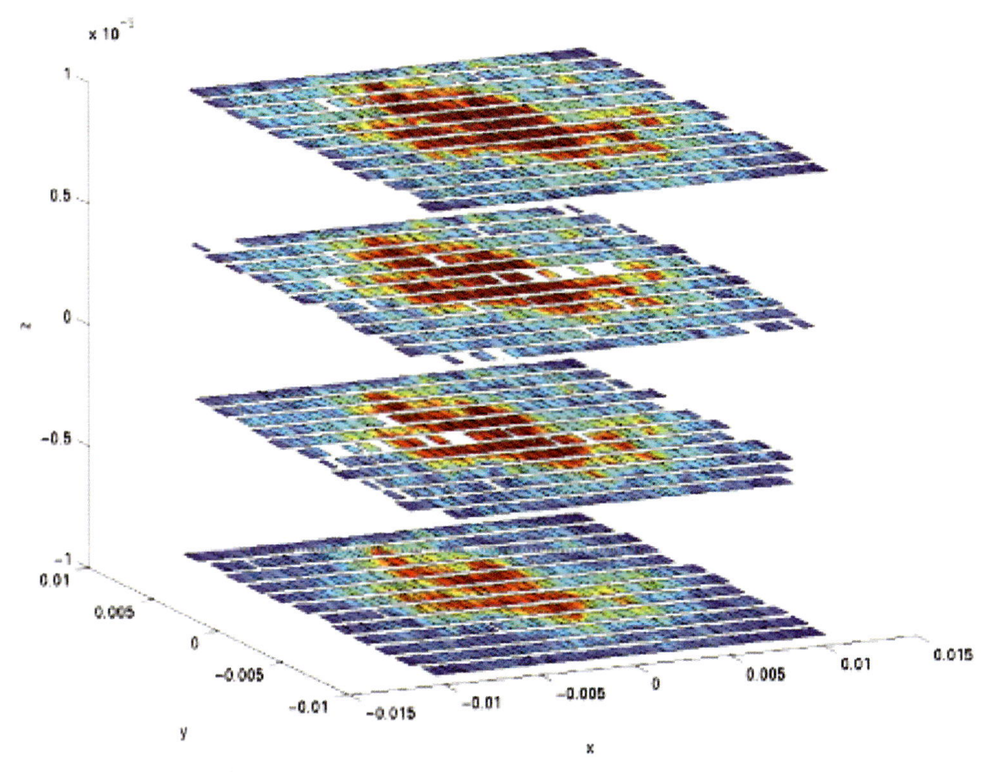

Figure 30.4 Thermal profile of struct without thermal vias [4].

is more complex with the addition of a third dimension. For instance, a signal via or thermal via that spans two or more tiers constitutes a blockage that wires must navigate around.

Each of the above issues can be managed through exploiting the flexibilities available in determining the precise route within the bounding box of a net, or perhaps even considering slight detours outside the bounding box, when an increase in the wire length may improve the delay or congestion or may provide further flexibility for inter-tier via assignment.

Consider the problem of routing in a three-tier technology (Figure 30.6). The layout is gridded into rectangular tiles, each with a horizontal and vertical capacity that determines the number of wires that can traverse the tile, and an inter-tier via capacity that determines the number of free vias available in that tile. These capacities account for the resources allocated for non-signal wires (e.g. power and clock wires) as well as the resources used by thermal vias. For a single net, as shown in the figure, the degrees of freedom that are available are in choosing the locations of the inter-tier vias, and selecting the precise routes within each tier. The locations of inter-tier vias will depend on the resource contention for vias within each grid. Moreover, critical wires should avoid the high-temperature tiles, as far as possible.

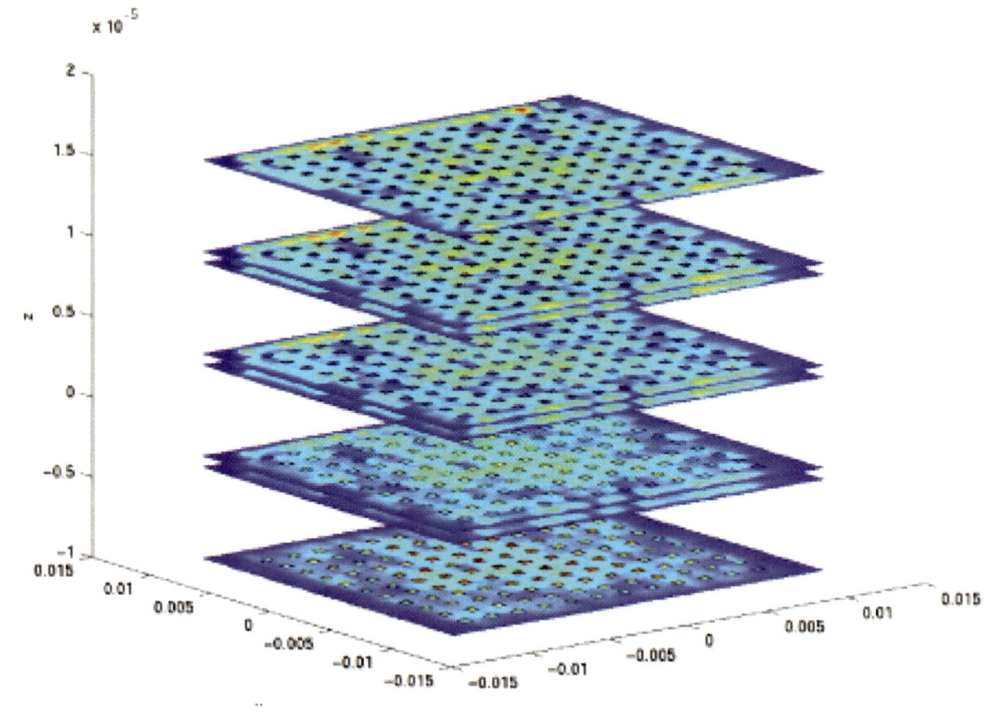

Figure 30.5 Thermal profile of struct after thermal via insertion [4].

Just as thermal vias enhance vertical heat conduction, we can also introduce thermal wires to improve lateral heat conduction and help vertical thermal vias to reduce hot spots temperature efficiently: for those hot spots where only a restricted number of thermal vias can be added, we can use thermal wires to conduct heat laterally, and then remove heat through thermal vias in adjoining grids. Thermal wires have the subsidiary benefit of aiding manufacturability: in unevenly packed regions of the layout, thermal wires can act as metal fill that aids the evenness of the topography after a Chemical-Mechanical Planarization (CMP) step. The thermal vias and thermal wires collectively form a heat conduction grid that removes heat to the sink: thermal vias perform the bulk of the conduction to the heat sink, while

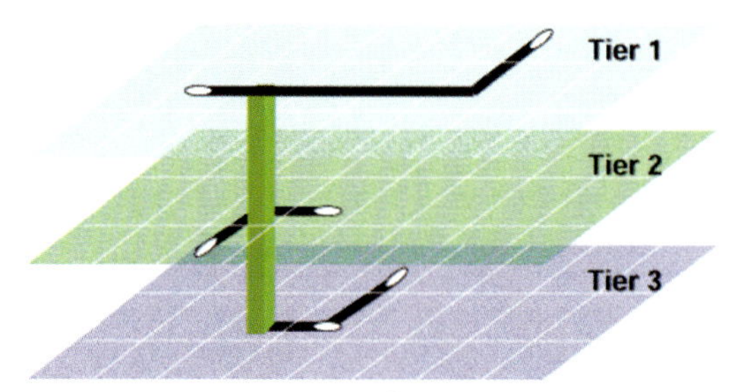

Figure 30.6 An example route for a net in a three-tier 3D technology [4].

thermal wires help distribute the heat paths over multiple thermal vias. Since thermal vias and thermal wires contend for, respectively, vertical and lateral routing resources with signal wires, they should be well planned to satisfy the temperature and routability requirements.

In Reference [10], a method for effectively reducing on-chip temperatures is proposed, through the appropriate insertion of thermal vias and thermal wires. A routing solution that obeys thermal and routing capacity constraints is then presented.

The routing scheme begins with building minimum spanning tree (MST) estimators for each net, for the purpose of estimating routing congestion. Next, signal inter-tier via assignment is performed using a hierarchical mincost network flow formulation. Signal interlayer via assignment is then performed at the boundaries of group pairs at each level in a top-down way following the hierarchy: the assignment is conducted for the topmost level group boundary first, and then at the boundaries of group pairs of lower levels in the hierarchy and so on.

Once the locations of the inter-tier vias have been determined, the problem reduces to a 2D problem on each layer, of connecting each net to the pins and inter-tier via locations on that layer. However, this has to be performed in a thermally conscious manner, and a thermally-driven 2D maze routing approach is employed, incorporating a thermal penalty term in the cost function.

This is used as an initial solution that is now iteratively improved. Adjoint sensitivity analysis is employed to determine the sensitivity of a hot spot to the thermal via density. These sensitivities are used to set up a linear programming (LP) based thermal via/wire insertion procedure to reduce temperature: in each step, the thermal via densities are adjusted under the local linear model provided by the sensitivities. After this, if the layout violates any congestion constraints, a rip-up-and-reroute step is employed to satisfy them again. The procedure repeats iteratively until the thermal requirements are met, or until no further improvement is possible. Experimental results show that the scheme can effectively resolve the contentions between thermal via/wire and routing, generating a solution satisfying both congestion and temperature requirements. The overall flow of the method is outlined in Figure 30.7.

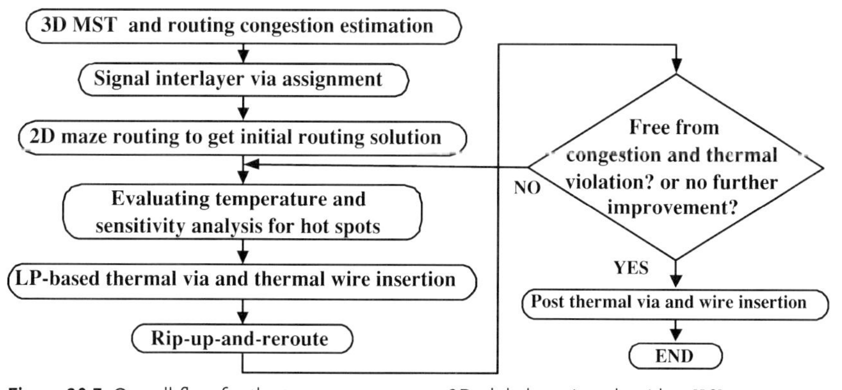

Figure 30.7 Overall flow for the temperature-aware 3D global routing algorithm [10].

30.4 Power Grid Design in 3D

3D integrated circuits are extremely limited in terms of their ability to deliver power, due to I/O pin limitations. I/O pins fall into two categories: those used to deliver signals and those that deliver power (V_{dd} or ground). While increased on-chip functionalities lead to a demand for more signal I/O pins, the demand for power is stronger. In contemporary and future technologies, one-half to two-thirds of all I/O pins must be dedicated to power delivery so as to reduce the worst case IR-drop and L (di/dt) noise in the power grids, while the total number of package pins does not increase significantly. As a result, as the total current consumption of the chip goes up due to the increasing circuit complexity and higher switching frequency, each power pin must deliver a larger amount of current to the chip. This trend can be clearly seen in Figure 30.8, which is created based on the data from ITRS [11]. In other words, as the IC technology advances, the number of pins for each unit of current delivered is actually reduced.

The pin limitation problem is exacerbated in 3D ICs: for a 3D chip with k tiers, the amount of current to be supplied is k times as much as for a 2D chip with the same footprint, but the number of available pins is essentially the same. Viewing this another way, if we were to transform a 2D IC into a k-tier 3D IC implementing the same functionality, the number of pins accessible to the circuit will be reduced to $1/k$ of the original value because of the much smaller footprint area. Figure 30.9 shows an example of transforming a 2D IC into an equivalent 3-tier 3D IC, in which the number of pins that can be placed on the 3D chip is only one-third of that for the corresponding 2D chip.

The idea of stacked V_{dd} levels is particularly useful in this context. In Reference [12], a high-tension power delivery scheme was proposed to reduce power grid noise and the effect of electromigration. In this new circuit paradigm, logic blocks are stacked several levels high and power is delivered to the circuit as multiples of the regular

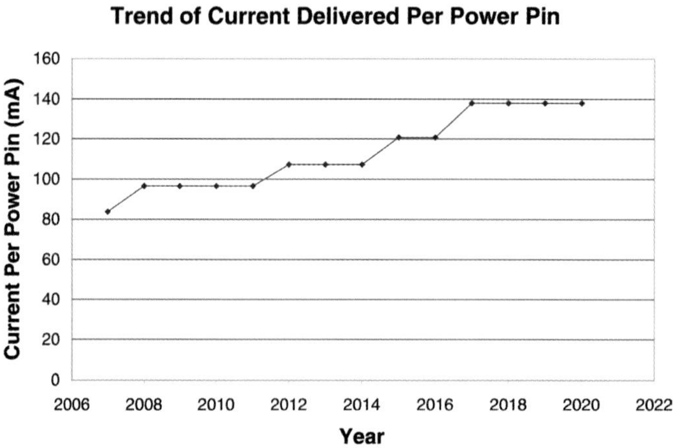

Figure 30.8 Trend of current delivered per power pin based on the data from ITRS.

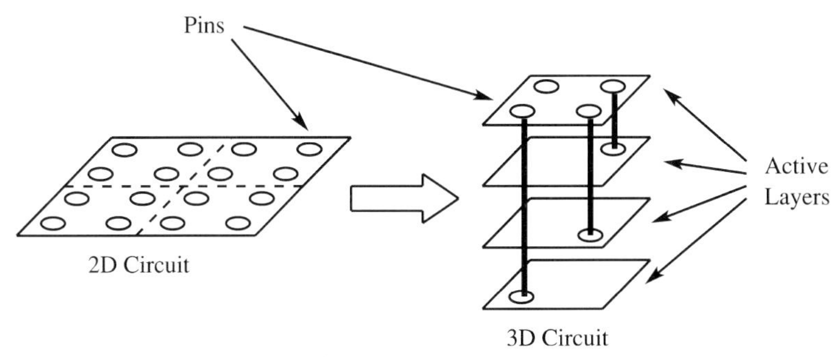

Figure 30.9 Power pin limitations for a 3D chip, as compared to a 2D chip with the same amount of circuitry.

supply voltage V_{dd}. Next, the delivered high-tension supply voltage is divided into several V_{dd} domains, each of which has a range of V_{dd}, and circuit blocks are distributed to different V_{dd} domains. Voltage regulators are used to control the voltage levels of internal supply rails. This is illustrated in Figure 30.10.

The advantage of this new circuit structure is that the current can be "recycled" between stacks. When logic blocks are stacked n levels high and the current requirements between logic blocks operating in different V_{dd} domains are balanced, the current flowing through each external power grid would be reduced to $1/n$ of the original value, where "external power grid" refer to a power grid that is connected to power pins, that is nV_{dd} and GND rails in an n-level stacked-V_{dd} circuit. Therefore, voltage drop, noise and electromigration issues can be significantly alleviated.

The key point here is to build partitions in such a way that the current can indeed be recycled through successive stacks of V_{dd} layers; if it cannot, it flows through the voltage regulators and is wasted. A forthcoming publication [14] presents a novel partitioning method for solving this problem. The chip is partitioned into regions, each of which has one voltage regulator. A Voronoi tessellation of the space

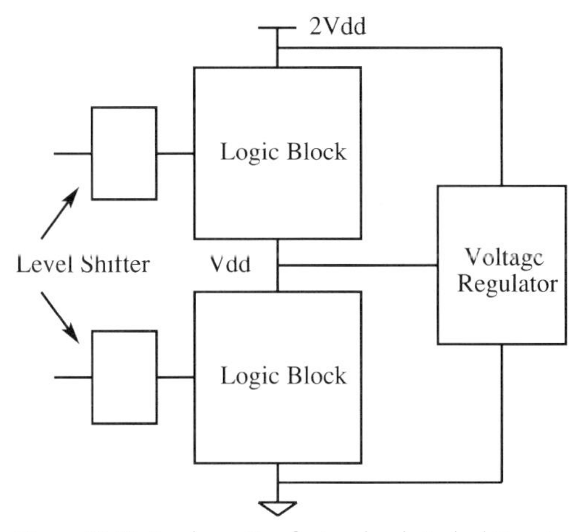

Figure 30.10 A schematic of a two-level stacked-V_{dd} circuit structure [13].

determines the fraction of current drawn by each block from each regulator. Next, the partitioner allocates blocks to regions so as to minimize wasted power. Results on a 3D benchmark circuit show that 95% of the power is usefully recycled, and only 5% is wasted, through the use of an intelligent partitioning scheme.

30.5 Conclusion

This chapter presents a set of critical problems to be solved in building useful CAD tools for 3D design. CAD tools are already indispensable to manage the complexities of 2D design; they will be even more so for 3D technologies. While some tools can be leveraged from 2D technologies, physical design tools, such as placement and routing, must be rethought. Moreover, thermal and power grid considerations become more critical in 3D, since the connections to the outside world, for both power supply and heat removal, are very limited. Currently research in 3D CAD is in its infancy, and as these circuit technologies mature and enter the mainstream this area will correspondingly expand.

References

1 Ozisik, M.N. (1968) *Boundary Value Problems of Heat Conduction*, Dover, New York.

2 Saad, Y. (2003) *Iterative Methods for Sparse Linear Systems*, 2nd edn., SIAM, New York.

3 Qian, H. (2005) Stochastic and Hybrid Linear Equation Solvers and their Applications in VLSI Design Automation, PhD thesis, University of Minnesota, Minneapolis, MN.

4 Ababei, C., Feng, Y., Goplen, B. *et al.* (November–December 2005) Placement and routing in 3D integrated circuits. *IEEE Design & Test*, **22** (6), 520–531.

5 Goplen, B. and Sapatnekar, S.S. (2003) Efficient thermal placement of standard cells in 3D ICs using a force directed approach. in Proceedings of the IEEE/ ACM International Conference on Computer-Aided Design, pp. 86–89.

6 Goplen, B. (2006) Advanced placement techniques for future VLSI circuits, PhD thesis, University of Minnesota, Minneapolis, MN.

7 Goplen, B. and Sapatnekar, S.S. (2007) Placement of 3D ICs with thermal and interlayer via considerations. in Proceedings of the ACM/IEEE Design Automation Conference, pp. 626–631.

8 Viswanathan, N. and Chu, C.C.-N. (2004) FastPlace: efficient analytical placement using cell shifting, iterative local refinement and a hybrid net model. Proceedings of the ACM International Symposium on Physical Design, pp. 26–33.

9 Goplen, B. and Sapatnekar, S.S. (2005) Thermal via placement in 3D ICs. Proceedings of the ACM International Symposium on Physical Design, pp. 167–174.

10 Zhang, T., Zhan, Y. and Sapatnekar, S.S. (2006) *Temperature-aware routing in 3D ICs*. in Proceedings of the Asia-South Pacific Design Automation Conference, pp. 309–314.

11 Semiconductor Industry Association . (2006) International technology

roadmap for semiconductors, Available at http://public.itrs.net/Links/2006Update/2006UpdateFinal.htm. (Nov. 29, 2007).

12 Rajapandian, S., Shepard, K., Hazucha, P. and Karnik, T. (2005) *High-tension power delivery: Operating 0.18 μm CMOS digital logic at 5.4 V.* Proceedings of the IEEE International Solid-State Circuits Conference, pp. 298–299.

13 Zhan, Y. (2007) High efficiency analysis and optimization algorithms in electronic design automation, PhD thesis, University of Minnesota, Minneapolis, MN.

14 Zhan, Y., Zhang, T. and Sapatnekar, S.S. (2007) *Module assignment for pin-limited designs under the stacked-*V_{dd} paradigm. in Proceedings of the IEEE/ACM International Conference on Computer-Aided Design, pp. 656–659.

31
Electrical Performance of 3D Circuits

Arne Heittmann and Ulrich Ramacher

31.1
Introduction

In recent decades, progress in semiconductor technology was mainly driven by the reduction of minimum feature sizes and, as a consequence, the ability to realize low-cost components. To fulfill special demands on the component level even different technologies emerged that are well suited for realizing memory (DRAM, Flash), sensors, low-power digital circuits, high speed digital circuits, RF circuits, analogue circuits or high voltage components, for instance. However, the concurrent technology roadmap for micro- and nanoelectronics only marginally accounts for technology demands in the area of system integration, which are given by applications like battery supplied high-performance computing, automotive computing and integrated microsystems containing sensors, components for analogue signal conditioning as well as digital processors for information processing. Such systems are composed of several components to realize a complete application.

31.1.1
Example 1: Baseband Processors in Mobile Phones

Obtaining flexible high-performance chip architectures for baseband processing is one of the most challenging problems in realizing battery-driven wireless terminals. These systems must not only support multiple communication standards like UMTS FDD, CMDA2000, GSM/SPRS/EDGE, IEEE 802.11b, IEEE 802.11g, Bluetooth, DAB and GPS, some of them must be executable in a concurrent way (Figure 31.1). Since for battery-driven devices low power consumption is indispensable, past as well as recent approaches for system implementation relied on the preparation of special purpose macros for signal-processing parts of the physical layer that were controlled by a programmable DSP, for instance. Nevertheless, nowadays the number of

Handbook of 3D Integration: Technology and Applications of 3D Integrated Circuits.
Edited by Philip Garrou, Christopher Bower and Peter Ramm

ISBN: 978-3-527-32034-9

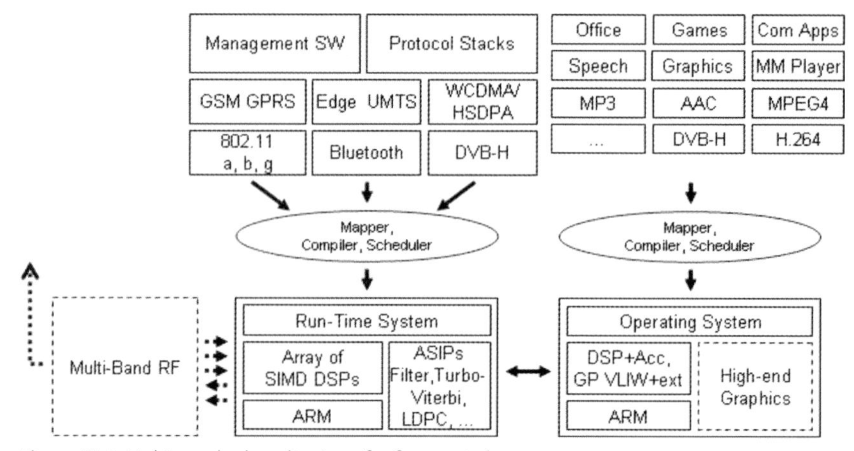

Figure 31.1 Multi-standard applications for future wireless terminals call for multi-processor systems.

standards is steadily increasing, which requires enhanced flexibility in the baseband processing that is hard to achieve using a macro-based design style. Indeed, a macro-based architecture sticks to the specification that was given in the beginning. Changes in specification as well as the addition of new standards require (possibly a complete) redesign of the chip architecture. To continue with the conventional design style would imply increasing the number of macros, with a concomitant increase in chip area and increased design cycle. Consequently, the time to market is prolonged and chip development is more and more within the critical path.

In contrast to the well-known approaches sketched above, the necessity for flexibility as well as high-performance computing with low power consumption can be fulfilled satisfactorily by a Software-Defined-Radio (SDR) platform [1]. The underlying hardware architecture is given by a multi-processor cluster of single-instruction-multiple-data (SIMD) DSP cores (Figure 31.2). A SIMD core is particularly suited for the algorithms used in communication systems, which exhibit a considerably large data parallelism. To provide the performance required for the widespread standards UMTS FDD, CDMA2000, GSM/GPRS/EDGE class 12, IEEE 802.11b, IEEE 802.11g, Bluetooth, DAB and GPS, several clusters of SIMD cores have to be implemented. In addition, filtering processors for channel encoding as well as decoding are integrated in the architecture. Last but not least, a general-purpose processor of ARMxx type is used for protocol stack software. As shown recently [2], up to 25% of power consumption is caused by accessing the memory system of the given multiprocessor architecture. Hence, optimizing accesses and introducing hierarchy to the memory system is the most promising approach to reducing power consumption. However, this optimization requires a detailed view of the principles of how functions are mapped on the processor cluster.

Communication standards that are executed on a multiprocessor architecture have to be divided into single functions. These functions have to be distributed on different processors and will be executed in parallel to achieve maximum performance gain.

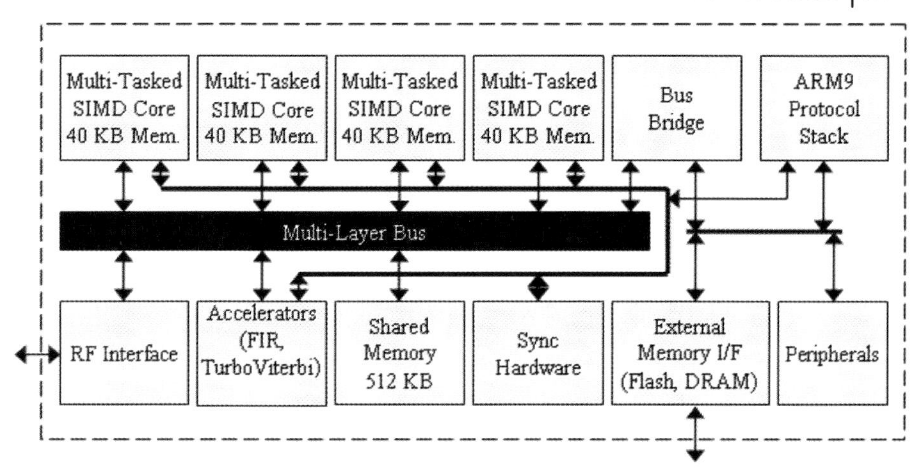

Figure 31.2 Multiprocessors architecture from Infineon.

Within SDR, one design task is to assign the execution of particular functions to dedicated processors.

To avoid idle cycles the mapping of functions to processors has to happen such that the throughput of each processor is maximal and the hand over of data between processors is minimized. As data are to be fetched and stored from and in memory and the memory can not always be placed next to the processor, communication paths in a 2D chip can be long and, hence, consume considerable power.

All processors have access to a common memory system via a multilayer bus architecture. Since this memory has to have connections to all processors global wiring needs to be implemented that includes data lines, synchronization lines as well as address lines and control signals.

The implementation of long communication lines, which affords large system busses, causes two problems concerning the electrical performance. Usually, large busses are realized using metal lines with narrow space without any shielding in-between due to aspects concerning area occupation. However, for technologies with minimum features sizes below 100 nm cross talk caused by line-to-line capacity between neighboring data lines increases the latency for signal transmission. Diminished performance of the bus system is inevitable.

In addition, long metal lines cause a large capacitive load that has to be charged by large drivers. The charging and discharging of long interconnect lines consumes a considerable amount of power.

For instance, a interconnect line 6 mm long (which represents a typical global metal line) with a drawn width of 760 nm will give a capacitive load of about 1.5 pF (90 nm technology node, line-to-line space 500 nm) and a RC time constant of 700 ps. By supplying the memory interface with a typical voltage of $V = 1.2$ V an energy of about 1.1 pJ per bit has to be considered for each state change of the line. With four parallel busses (in our case 400 bit for each bus, including data lines, address lines and control lines) to the memory system running at a moderate frequency of 40 MHz an average power consumption of about 70 mW has to be regarded due to switching power only.

When dealing with a total power budget of 250 mW for a typical wireless device it becomes obvious that a reduction of switching power for global busses is mandatory. By reducing the switching power from 70 down to 7 mW either the performance of the system could be increased by 25% (by keeping the total power budget at 250 mW) or the power consumption can be reduced by the same amount. In both cases, a clear advantage is brought to the customer.

One solution to tackle the reduction of parasitic capacitances for global interconnections is the concept of 3D stacking. For instance, if we break up the proposed multi-processor architecture into five layers, using four layers for SIMD processors and one layer for the memory system, the global interconnection length from processors to the shared memory system will be reduced from approximately 6 mm down to 120 μm (four layers, 30 μm each; see Figure 31.3). As we will see later, a complicated characteristics for vertical interconnection has to be taken into account, but a parasitic capacitance of <50 fF per vertical contact and layer can be achieved easily. A vertical wire embedded in a four-layer stack (from top to bottom) has an expected parasitic capacitance of <200 fF, which represents an 86% reduction of capacitive load compared to a single horizontal wire. In addition, the resistance in series can be reduced from several 100 Ω down to <10 Ω, reducing the RC time constant considerably. Since vertical contacts, usually, are arranged in a wide-line-to-line space (due to technology constraints), cross talk of neighboring lines can be neglected. A performance gain for the memory access will be observed.

Nevertheless, to realize a 3D multilayer stack, vertical contacts with a pitch of <10 μm and a diameter of <5 μm have to be provided by an appropriate 3D technology

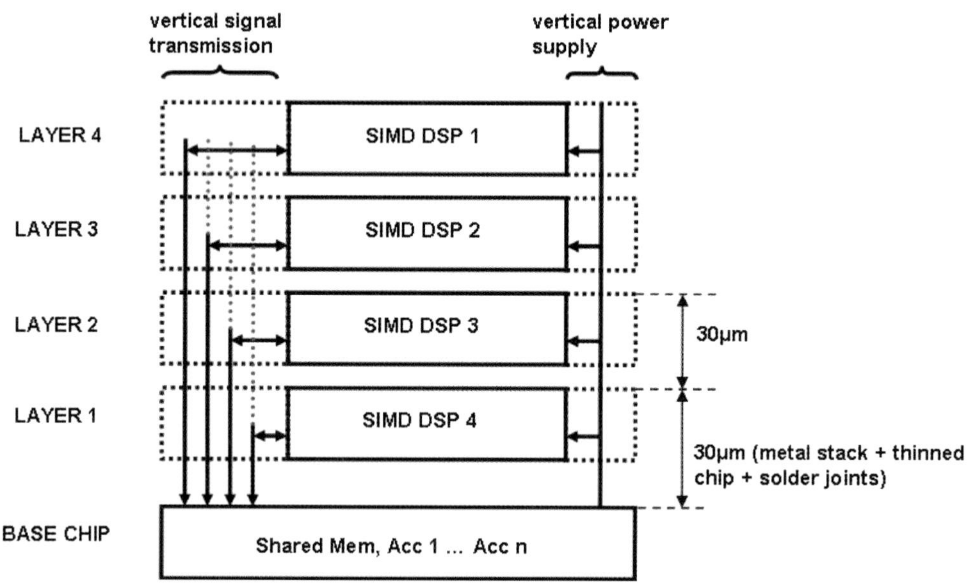

Figure 31.3 3D stacked multiprocessor architecture with minimized global bus length to the shared memory system.

to connect 1600 signals down to the memory system. Precise chip-to-chip alignment, precise creation of via holes as well as contacts and a robust and reliable wafer handling are mandatory. Here, a dedicated 3D technology is to be developed that will be described later.

31.1.2 Example 2: Advanced Man–Machine Interface for Cell Phones

In the future most operations on cell phones will rely on a robust man–machine interface based upon robust speech processing as well as image processing. User identification and user tracking for an advanced videophone are some examples of the many applications that will be available on high-end cell phones within the next decade. However, image processing architectures that operate robustly under arbitrary conditions are still lacking. This is especially true if one considers a low-power hardware architecture with real time capabilities for image processing.

Recently, it turned out that architectures based on pulsed neural networks are well-suited for basic image processing tasks. Image segmentation and feature detection are examples of functions frequently used in image processing that can be realized very efficiently with respect to power consumption as well as speed if artificial neural networks based on pulse processing are used [3–5].

However, aspects of low power consumption as well as real time capabilities require special hardware architectures that both retain flexibility and take into account the special demands of pulse processing. Information processing using neural networks typically is based on few different types of dynamic elements like synapses and neurons. Networks of neurons connected by synapses realize simple functions that can be used to compose filter functions, detectors and object recognizer. Dedicated functions rather emerge by defining network architecture in terms of connections than by properly defining the dynamic elements used in these architectures [6, 7].

Nevertheless, some constraints of the dynamic elements affect characteristic properties of the architecture, which has led to certain implementations. The most important element of a network architecture is the so-called integrate-and-fire (IAF) neuron [8].

Basically, an IAF neuron has two states: receive and send. In the receive-state the so-called synaptic input current I_{syn} (Equation 31.1) is continuously integrated on the neuron's membrane to the membrane potential a_{K} (Equation 31.2):

$$I_{\mathrm{syn,K}}(t) = i_{0,\mathrm{K}} \cdot W_{\mathrm{K0}} + \sum_{L \in N_{\mathrm{K}}} W_{\mathrm{KL}}(t) \cdot X_{\mathrm{L}}(t) \quad (31.1)$$

$$a_{\mathrm{K}} = \int_{T_0}^{T_0+t} I_{\mathrm{syn,K}}(t')\mathrm{d}t' \quad (31.2)$$

W_{KL} represents the dynamic connection weight (a synapse) from neuron L to neuron K. N_{K} is the set of neurons that are connected to neuron K where K is a postsynaptic neuron. In addition, the current $i_{0,\mathrm{K}}$ is a continuous signal that may be used to couple

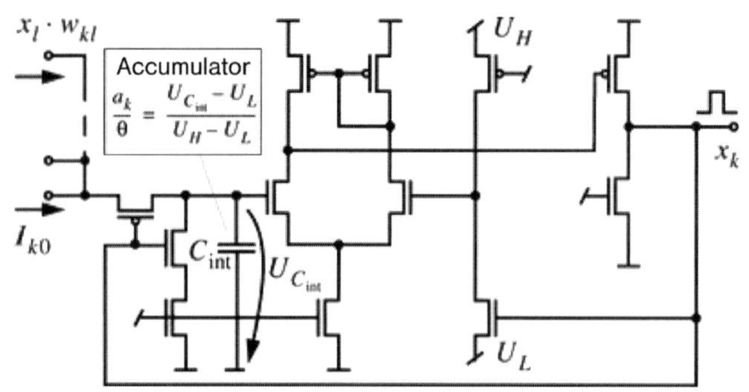

Figure 31.4 Schematic view of an IAF neuron. C_{int} represents the membrane capacitance on which the synaptic current will be integrated. The output pulse X_K is used to discriminate the send-state from the receive-state and connects C_{int} either to the synaptic current or a current source for the purpose of discharging C_{int}.

sensor signals (e.g. pixel sensors from an image sensor) directly to a particular neuron K.

If the membrane potential reaches a given threshold (theta) the neuron changes its state from receive to send while sending a pulse of fixed duration t_d (pulse width). To simplify the design of artificial neurons the shape of a pulse is given by a rectangular characteristic. This rectangular output signal is formally described by $X_K(t)$.

If the pulse is completed the neuron resets its membrane to an initial value [below the threshold (theta)], changes its state to receive and the integration of the synaptic current starts again.

An IAF neuron can simply be designed by using a transconductance amplifier, a capacitor, a current source for discharging the capacitor and a switching network. Figure 31.4 shows a simple circuit for a compact IAF neuron. The capacitor C_{int} represents the membrane that will be either charged by synaptic current [if the pulse $X_K(t)$ is off] or discharged by the current source [if the pulse $X_K(t)$ is on]. The pulse width is given by the reset level of the membrane, the threshold, the capacitance of C_{int} as well as the current provided by the current source.

In a conventional environment (planar chips) several neurons are embedded in a regular array structure to implement a large number of neurons. Programmable routing resources have to be added as well as several synapses to design networks, suited for feature extraction, for instance.

However, by analyzing a particular test chip we observed that by continuing with a conventional design style, on one hand limitations with respect to the number of integrated neurons clearly emerge while, on the other hand, low power consumption for the system is hard to achieve, even by exploiting low power circuit techniques.

In the test chip given by the block structure (Figure 31.5) a system consisting of 128×128 neurons as well as 65.000 synapses used as a segmentation device was implemented [5]. Pixel information is conveyed via a D/A interface to individual neurons, which start pulsing after receiving pixel information.

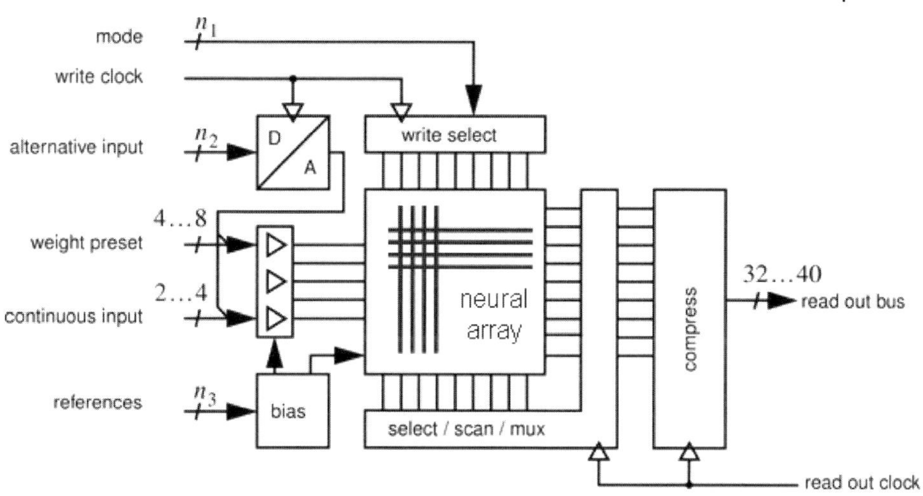

Figure 31.5 Block architecture of a neural test chip. The neural array, consisting of neurons as well as synapses, is connected to the encoding circuit (select/scan/mux), which enables monitoring pulse events via the read out bus. The D/A converter is either used to transfer pixel information from an imager to the neural array or to preset internal nodes of the neural array.

To monitor the pulse activity of all neurons a so-called address-event-encoder was implemented. This encoder asynchronously detects pulse events of individual neurons, encodes the neurons that were recently firing using an address-event-representation (AER) and transmits the encoded information via an interface to the periphery (read-out bus). If system complexity increases (i.e. number of neurons rises), such an interface is not only used to monitor activity but will be still required to couple particular networks. The performance of such an interface has to be designed with respect to the dynamic range of the pulse rate of particular neurons.

For the given technology (130 nm CMOS) currents can be made as small as 0.1 nA using a subthreshold circuit technique. Then, the offset compared to currents caused by noise is large enough (>6 dB) to realize reliable artificial neurons.

For different sizes of the capacitor C_{int} the required performance with respect to the number of events that have to be encoded can be calculated. Table 31.1 gives an estimation for the maximum pulse rate, and derived from that the number of expected events per second for different network sizes. On one hand, the network size is clearly restricted to the ability of the encoder to achieve an appropriate throughput. On the other hand, by keeping a reasonable throughput of the encoder, the area increases due to overhead caused by a large capacity C_{int}.

In Table 31.1 a worst case scenario was assumed – given by the assumption that all neurons fire with the maximum pulse rate. The values for area occupation are based on a layout for a single neuron (Figure 31.6). For an implementation of $n = 16384$ neurons and a capacitance of 400 fF we have measured a power consumption of 3 mW for the neural array (full rate) while the circuit that encodes the pulse events has

Table 31.1 Pulse rates as well as area occupation for an IAF neuron comprising different values for C_{int} (130 nm CMOS technology).[a]

C_{int} (fF)	Max. rate (kHz)	$n=1024$ (MHz)	$n=16384$ (MHz)	$n=65536$ (MHz)	Area/neuron (MHz)
800	2.5	2.56	40.9	163.84	608
400	5.0	5.1	81.9	327.68	336
200	10.0	10.2	163.84	655.36	200
100	20.0	20.4	327.68	1310.72	132
75	26.7	27.6	436.8	1747	115

[a] For different network sizes n (number of implemented neurons pulsing at their maximum frequency) the required encoding frequencies for the AER encoder are depicted.

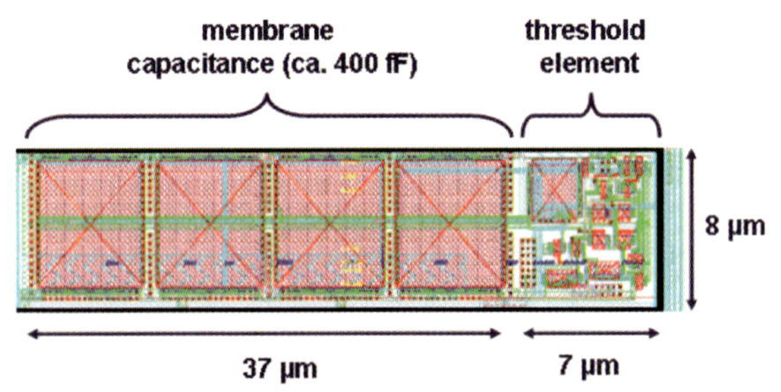

Figure 31.6 Layout of an IAF neuron realized in a 130 nm CMOS technology. Most of the area is occupied by the membrane capacitor.

a power consumption of up to 250 mW (clock frequency $f_{CLK} = 200$ MHz). Clearly, power consumption becomes the limiting factor for such a concept of implementing a neural architecture.

Instead, for a direct transmission of pulse activity to adjacent layers of a neural architecture using a 3D-concept an AER circuit would not be necessary. Here, a 3D stacking approach definitely supports an interface consisting of high bandwidth at the lowest computational effort and, hence, power consumption.

However, the scenario sketched above will be valid only at the early stages of the image processing architecture. After low-level feature extraction, pulse activity becomes more and more sparsely distributed, relaxing the performance requirements of an encoding circuit considerably. At early stages of image processing (low level feature extraction) bandwidth requirements from layer to layer are comparably high but network structure is regular. Networks for feature detectors can be hardwired. Layers of subsequent processing steps have relaxed bandwidth requirements but are less regular in structure with regard to active connections (e.g. associative memories). Here, it becomes more important to have a programmable routing architecture, which can be implemented based on a transmission of encoded pulse events using a 3D interconnect architecture.

31.2
3D Chip Stack Technology

In the past, several methods for 3D integration have been developed to exploit the third dimension for device integration [9]. In the stacked silicon-on-isolator (SOI) technology [10] silicon dioxide layers and active silicon layer alternate in a multilayer fashion. Since SOI technology is not a standard technology this approach is only well suited for circuit concepts that are already based on SOI.

To embed standard technologies like CMOS in a 3D process, different concepts have to be used. A very simple approach is to embed the active silicon in specific support materials. These support materials carry the required vertical interconnect structures and compensate for mechanical stress [11, 12]. Since the active silicon is bonded to the support layer, vertical interconnections are realized more or less outside of the chip border with a low contact density.

The highest density in terms of vertical contacts per unit area can only be achieved if contacts are placed on the surface of the active chips. Then, the chips are directly stacked on top of each other in a back-to-face fashion without additional support layers. Contacts are realized through solder joints, which on one hand realize the electrical connections, and on the other hand realize a solid mechanical support between adjacent layers.

There are two different ways well suited for realizing a concept consisting of high contact densities. In one approach the well-known standard processes (e.g. CMOS) run on a standard wafer from front-end processes to back-end metallization. Then, so-called interchip vias are etched and metallized from the front side of the wafer [13, 14]. After via metallization a support carrier is attached to the processed wafer, which helps in stabilizing the wafer during the thinning process. The thinning process is stopped if the through-chip vias are visible at the rear side of the wafer. Finally, the chips are soldered into a 3D stack. Critical steps include the alignment of the vias and fabricating reliable solder joints.

The required deep front-side etching of through-holes comes with different problems. The most unfavourable prerequisite is the required aspect ratio for via metallization, since the through-holes have a depth of at least 30–40 μm. The through-holes must run through the whole metal stack, the active silicon and the residual depth of the silicon after wafer thinning.

This process can be optimized by considering the function of the through-holes. Since the through-holes are used especially to realize the rear-end contacts of the wafer, those holes have to be realized from the rear-end of the chip terminating on the first regular metallization of the standard process only [15, 16]. Metallization within the metal stack of the chip to the surface could be obtained by using the given regular metal layers of the standard process.

If rear-end holes are favored, the process concept has to be extended. Such a concept is clearly based on additional back-side processes and, in addition, cannot start at regular CMOS processing since masks required for structuring the rear-end of the wafer need to be aligned on specific structures to hit landing pads in the first metal layer during back-side etching of the vias.

31.2.1
3D Process with Self-Adjusting Back-Side Contacts

The process we propose here [17–20] requires preprocessed wafers to enable mask alignment from the back-side. At the same time, a thickness control mechanism for wafer thinning has to be prepared.

First, an epitaxial etch-stop layer is grown on a raw silicon wafer. This etch-stop layer is used to realize a precise residual wafer thickness after thinning, which results in precise alignment of back-side vias to the landing pads in the first metal layer of the active CMOS. Via diameters can be outlined in a very easily while the pitch between neighboring vias can be made as small as possible, resulting in high connection densities between adjacent layers of the 3D stack.

We use a copper-tin soldering process based on the solid-liquid interdiffusion (SOLID) process [21] to create the electrical and mechanical connection between the single chip layers. This allows for subsequent stacking without the degeneration of previous solder joints.

Using this process, we created true multilayer stacks and tested them with respect to the static electrical properties of ohmic contacts and interchip vias. We then characterized parameters such as through resistance, substrate leakage and via-via leakage currents. We directly incorporated these results in the design of test circuits that will create test for stuck-at failures of the interchip connections after stack assembly.

To characterize electrical properties of the 3D interconnect vias for the proposed 3D process a brief introduction of particular processing steps is given in the next subsections. Figure 31.7 illustrates the main steps for the complete 3D stacking process.

31.2.2
Wafer-Preparation

The raw wafers are prepared to enable mask alignment as well as to implement structures used for controlling the thinning procedure.

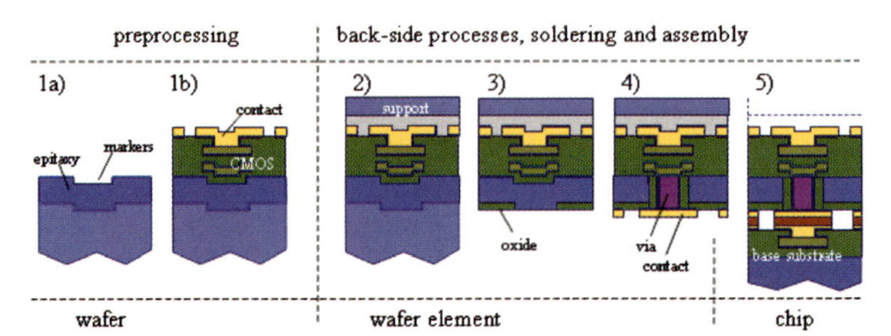

Figure 31.7 Process flow of 3D stacking. (1a) Etch-marker and epitaxy; (1b) CMOS processing; (2) glue support material to wafer; (3) thinning and back-side oxidation; (4) via etching, via metallization, back-side contact; (5) assembly and soldering.

First, small markers are etched on the wafer's surface, which realize structures sufficient to align masks up to a precision of <1 μm using optical equipment. An important feature of the proposed concept is a self-limiting thinning process with specified alignment markers. These markers help in precisely aligning the first front-side mask of the standard CMOS process with the first back-side mask, creating the back-side contacts.

For this purpose, the blank surface of a wafer is patterned with markers and overgrown by a silicon epitaxy layer. The crystal growth area consists of two layers: the first layer is a thin p+ layer (highly doped with boron). The second layer is doped according to the requirements of standard CMOS having an appropriate size, which defines the residual thickness of the thinned layer. The p+ layer simplifies the thinning process compared to other concepts because its properties let it act as a selective etch stop.

Two masks have to be aligned: the first mask of the regular CMOS process and the mask used to define the via holes that, later, will be etched from the back-side. After etching the markers an epitaxial silicon layer is grown on the wafer. At the border between the raw wafer and the epitaxial layer a p+-etch-stop layer is drawn-in. The size of the epitaxial layer has to be in the range of about 10 μm to separate the CMOS layer and p+ layer sufficiently.

Since the epitaxial growth of silicon keeps the structure of the initial surface to the concurrent surface (especially sharp edges) at any time, the etch markers will be seen on the surface of the epitaxial layer (Figures 31.8–31.10).

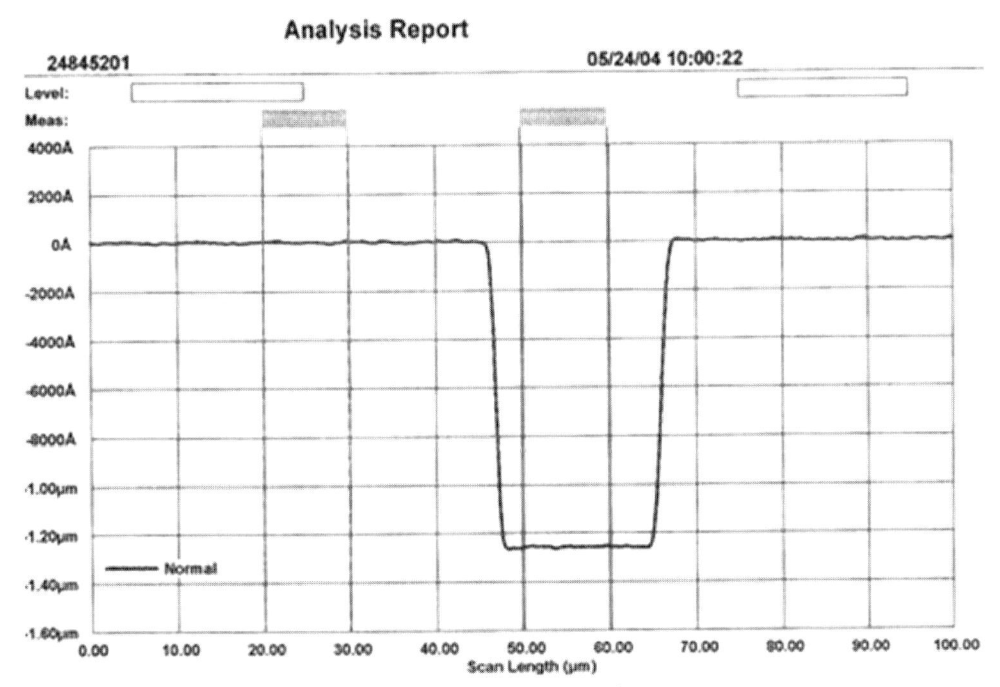

Figure 31.8 Profile of etch marker etched off the surface of a raw wafer.

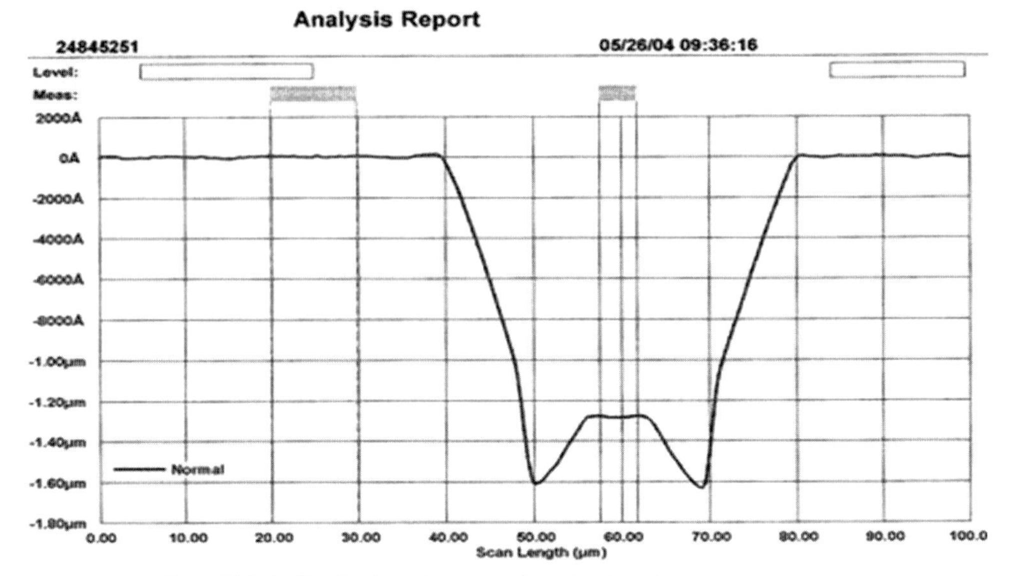

Figure 31.9 Profile of etch marker grown through epitaxy layer.

31.2.3
CMOS Processing and Front Side Metallization

The first mask of the CMOS process has to be aligned to the etch markers seen on the surface of the epitaxial layer. If this is achieved all active as well as passive structures from CMOS process will be aligned with respect to these etch markers.

For the vertical interconnects a copper-tin soldering process based on the solid-liquid interdiffusion (SOLID) process is used to create the electrical and mechanical connections between particular layers of the stack. Here, on the front side of the wafer the chip passivation is opened at specific positions where the vertical connections will be created.

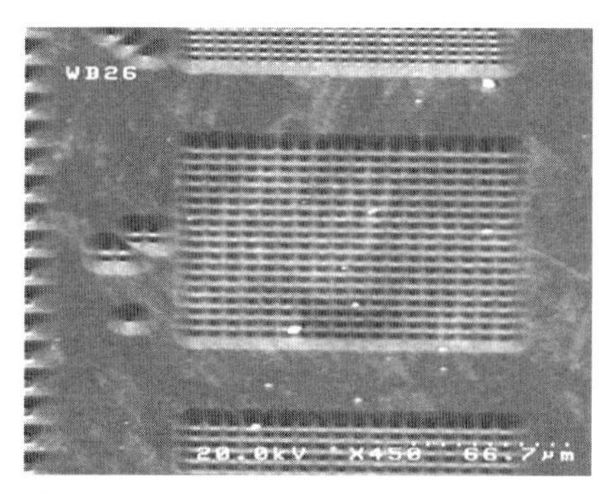

Figure 31.10 SEM view: Etch markers seen on epitaxy layer.

To maintain a low thermal budget for the whole process to prevent circuit degradation, deposition of metals by CVD has to be avoided since here the thermal budget usually exceed 300 °C. Therefore, a copper layer is applied using an electroplating process (which runs at room temperature) using sputter deposition forming the Cu seed. This layer forms one side of each 3D contact.

31.2.4 Wafer Thinning

A carrier is glued to the front side of the wafer to improve the mechanical stability that is required during the thinning process as well as for the chip assembly. The first step of the thinning process is realized using a rapid mechanical thinning step (lapping, grinding), resulting in a wafer thickness of approximately 50 μm. The mechanical thinning is followed by wet chemical etching process up to the epitaxial etch stop layer using a KOH solution. Since the selectivity (ratio of etch rate) of KOH is about 100 : 1 for undoped silicon and silicon, respectively, the wet etching process results in a very planar surface of the wafer's back side. In addition, the alignment markers implemented at the beginning are now visible on the rear side of the wafer.

31.2.5 Via Etching and Sidewall Isolation

The vias are etched with vertical sidewalls using an anisotropic silicon etching process that has to be aligned to the etch markers using a lithography step. By our experimental setup the via diameter is limited to about 5 × 5 μm. A sidewall oxidation is realized before via metallization to isolate the conducting via from the substrate. Sidewall oxidation is achieved using a low temperature PECVD process followed by an anodic oxidation process that closes so-called pin holes. An anodic oxidation is used because most low-temperature dielectric deposition processes will result in low-quality isolation layers. Here, conductive connections to the substrate through pin holes cause a high leakage current. For the proposed 3D concept a self-healing process closing the pin holes was found and implemented. Anodic oxidation will selectively oxidize free Si surface areas up to a thickness determined by the applied voltage. During the anodic oxidation process, the chip is placed in an electrolyte consisting of ethylene glycol. A DC voltage is applied with the chip forming the anode and a platinum wire as the cathode. This process is driven by the electric field across the oxide layer. Hence, the growth rate of SiO_2 is high in areas of thin oxide layer while for areas of bulky SiO_2 the additional growth rate of SiO_2 can be neglected. A self-healing behaviour will be observed. As an experimental result, a typical oxide thickness of anodic oxide is in the range of about 50 nm.

As preparation for the copper electroplating step, we deposit a titanium-tungsten diffusion barrier and the copper seed layer. This metallization step fills the vias with copper and prepares the solder joints with the deposition of tin. Figure 31.11 shows a through-chip via with copper metallization.

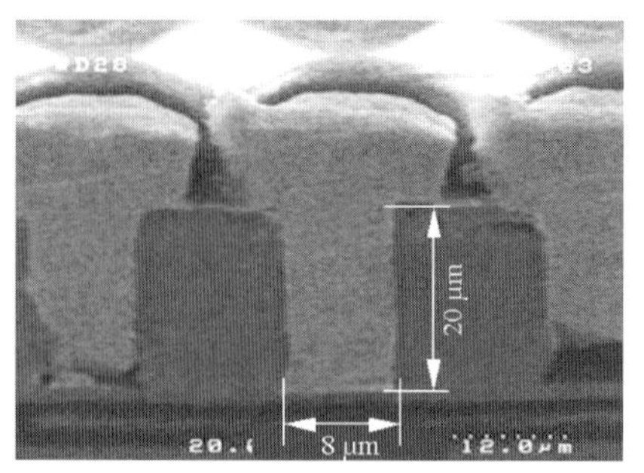

Figure 31.11 Cross section of via metallization.

31.2.6 Test and Soldering

To test single chips, we can use via metallization from the rear-side to evaluate the functionality of the chip before soldering. This selection of known-good dies requires separating the elements into single chips (including the carrier for stabilization) and will significantly increase yield.

After test, only known-good-dies (KGD) will be soldered in a chip level solder process. For this process the rear-side of one chip regarded to be known-good is aligned to the front side of the previously soldered chip. After alignment, the top chip is positioned on top of the bottom chip followed by a heating process. Removing the carrier by dissolving the glue and cleaning the surface are the last steps to prepare the top surface for the next layer to be soldered.

We perform the following solder step with the SOLID process at 300 °C, which involves a high-temperature melting phase. This phase is mechanically stable above 600 °C and enables the stacking of further layers on top without melting the already-soldered joints below. Figure 31.12 shows a view of a solder joint.

In the three-layer stack shown in Figure 31.13, a displacement of the first solder level is not visible. This proves the high temperature stability of all solder joints above

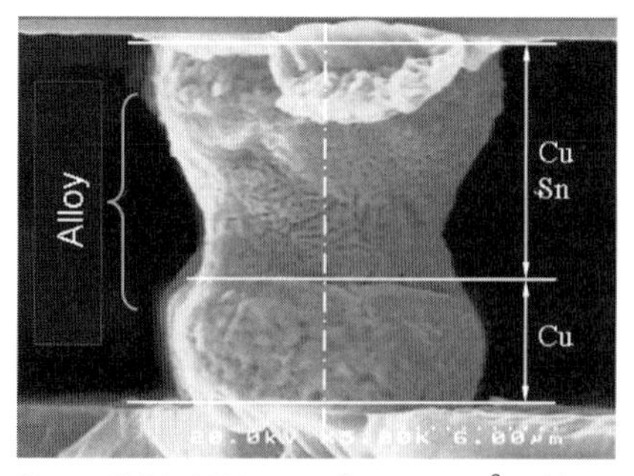

Figure 31.12 SEM view of a $9 \times 9\,\mu m^2$ solder joint.

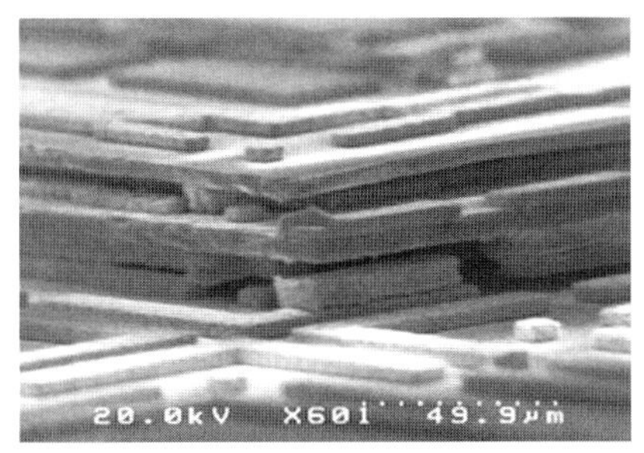

Figure 31.13 SEM image of a chip stack with two thinned chips and bottom substrate soldered in a back-to-face technology.

the solder temperature of 300 °C and, in turn, reveals the formation of the high-temperature melting copper-tin phase.

31.3 Electrical Performance of 3D Contacts

In terms of design, the electrical performance of 3D contacts is of interest.

For a 3D power supply system the ohmic characteristic of contacts has to be evaluated with respect to contact sizes and number of contacts used for a specific power net.

Digital signal transmission from layer to layer is influenced by capacitive coupling from neighboring contacts as well as coupling to the substrate. Here, low RC-constants for vertical signal transmission lines are of interest. High contact densities enabling vertical bus systems with a large number of data lines on the other hand lead directly to a trade-off between resistivity (contact size) and bus width.

Low frequency analogue signal transmission requires a linear IV-characteristic with minimum leakage into the substrate to avoid signal distortion.

To address the different aspects of electrical characteristics different measurement setups were evaluated.

31.3.1 Isolation, Cross Resistance and Via-Metal Resistance

Within our proposed 3D stacking concept vias are etched from the backside through the thinned silicon wafer and terminate on the first metal layer. To characterize the path solder pad–via–first metal a measurement setup was prepared. After thinning a wafer down to 10 µm the front side of the thinned wafer was prepared with aluminium while the back side was prepared with silicon dioxide. Vias were etched from the back side and metallized. Two types of vias were created: conducting vias that contact directly the aluminium and isolating vias that were separated from the aluminium by a silicon dioxide layer. Figure 31.14 illustrates the VI-characteristics

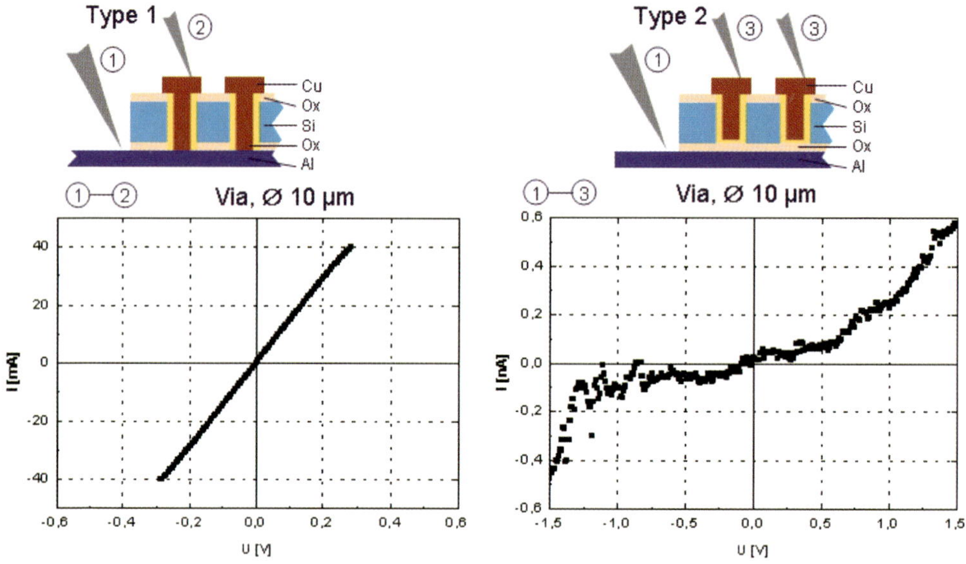

Figure 31.14 IV-characteristic of a Al-terminated via (left) and an isolated via (right). The diode-like IV-characteristic is caused by pin holes.

and the measurement setup. The through current using the conducting path was in the range 10–100 mA while for the SiO_2-terminated via a maximum leakage current of 600 pA was measured.

31.3.2
Solder Connection and Cu Wires

For this setup two chips structured with solder connections were soldered. One chip acts as a base substrate having front-side metallization while the second chip is structured with back-side metallization. Within this setup it is possible to characterize the solder connection itself and the copper lines that may act as wiring resource on front side as well as on the back side. Figure 31.15 shows the resistivity of copper wiring for different line lengths. In addition, the solder resistance of a $10 \times 10\,\mu m$ solder joint was measured.

31.3.3
Via and Solder Joint

Within this setup a bottom chip was structured with Cu wires and solder joints. Again, a top chip was realized using a thinned silicon wafer that has metallized through-vias, front side contacts as well as back-side pads for soldering. Both chips were soldered. The resistance measured in this test is typically $R = 2.57\,\Omega$. This result includes serial resistance, such as the resistance of the copper line and the measure-

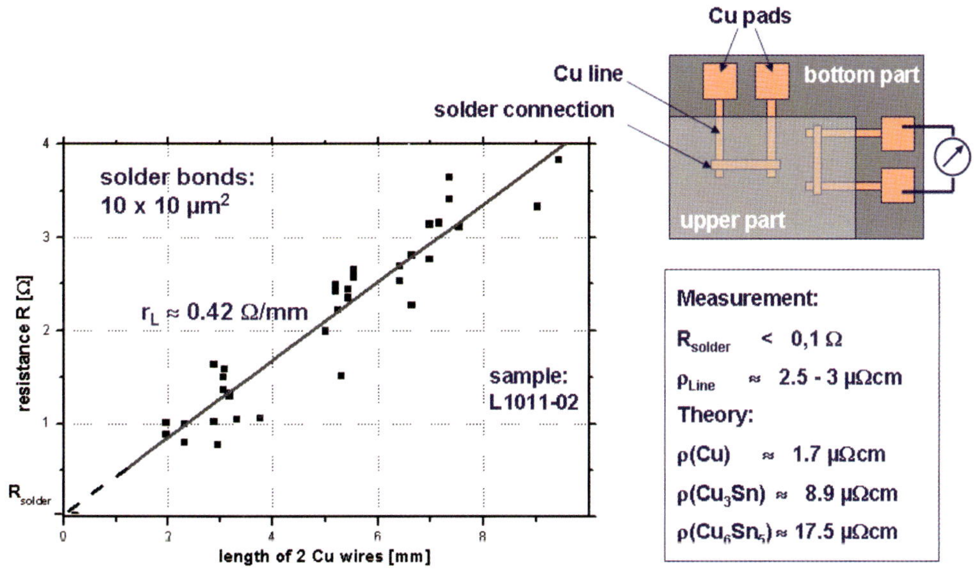

Figure 31.15 Experimental setup for measuring of copper resistivity acting as wiring resource. Two chips structured with conducting lines were soldered back-to-face.

ment setup. The copper line accounts for the main fraction of the resistance value with approximately 1.2 Ω. The via connection and the solder joint as well as the measurement setup generate the remaining resistance of 1.4 Ω in total. Here, we conclude that the resistance of a solder joint is less than 0.5 Ω (Figure 31.16).

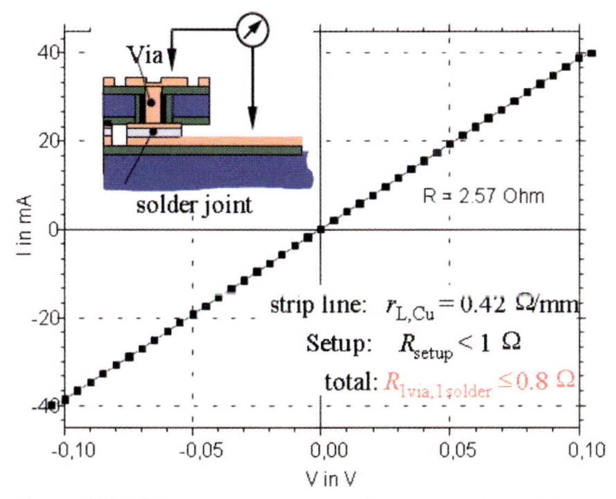

Figure 31.16 Measurement setup to characterize the resistance of a via that appears to be switched in series to the resistor formed by the solder joint.

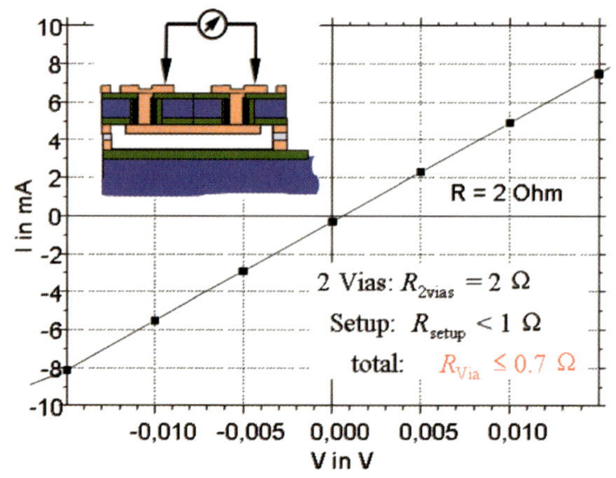

Figure 31.17 Measurement of a bridge connection with minimum distance to characterize the resistance of via interconnects.

31.3.4
Via Bridge

The via bridge is a structure containing a front-to-back and a back-to-front connection. Considering the resistances of the measurement setup, the copper wiring as well as the vias, it can be concluded that the resistance of one via is less than 1 Ω, see Figure 31.17.

31.3.5
Via Leakage

Pin holes caused by incomplete sidewall oxidation of the rear-side vias lead to a leakage current into the substrate. Figure 31.18 shows the diode characteristics obtained after low-temperature sidewall oxidation using CVD (skipping the anodic

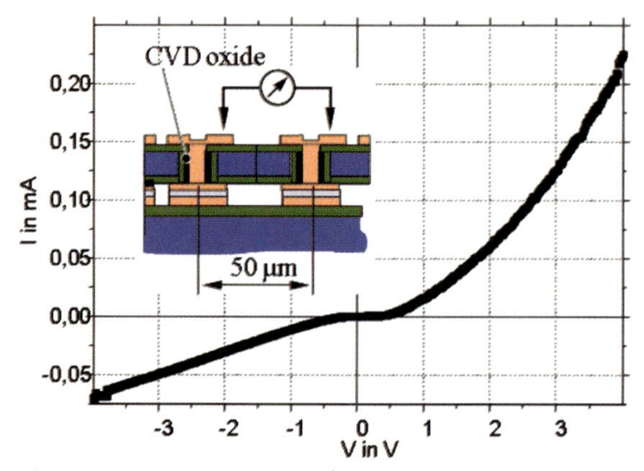

Figure 31.18 IV-measurement of diode characteristics to characterize pin holes.

oxidation) and via metallization. This leakage current can cause galvanic via-to-via coupling and probably increases the bulk potential. To avoid this leakage current all pin holes have to be closed in a reliable way. As already mentioned, pin holes can be closed using a self-healing anodic oxidation process. Figure shows the current characteristics obtained after an anodic oxidation of the via sidewalls. Here, a via sidewall leakage current below 10^{-5} A cm^{-2} was measured at a cross-potential of 10 V (Figure 31.18). This leads to an average leakage current per via of less the 10 pA, which is sufficient even for low power applications.

31.3.6
Equivalent Circuit for Simulation

By considering the different materials, processing steps and sizes of structures, an equivalent circuit of a rear-end via was generated that can be used for circuit simulations (Figure 31.19). To enhance reliability of the soldering process vias and solder pads are spatially displaced (Figure 31.20). Hence, mechanical stress carried out during the soldering process is compensated in the area beside the vias and releases mechanical stress from the via holes.

Using the results from measurements described in previous sections the resistive elements can be estimated directly. In addition, for dynamic simulations capacitive elements are of interest. In a first assumption all capacitors can be approximated using a plane-parallel capacitor. The oxide thickness of the bottom capacitor C_{Sub} is about 500 nm, resulting in a specific capacitance of 0.064 fF µm^{-2}. The total parasitic capacitance of the bottom wire into the substrate is about 23 fF. The oxide of the via sidewalls is about 200 nm thick, and hence a specific capacitance of 0.16 fF µm^{-2} is expected. In total, a C_{Si} of about 7.9 fF has to be taken into account. Effectively, a large resistor R_{Si} has to be considered to be in series to this capacitor; consequently, C_{Si} can be neglected. The via-to-via coupling capacitance C_{via} is equal to $C_{Si}/2$. Even for vias

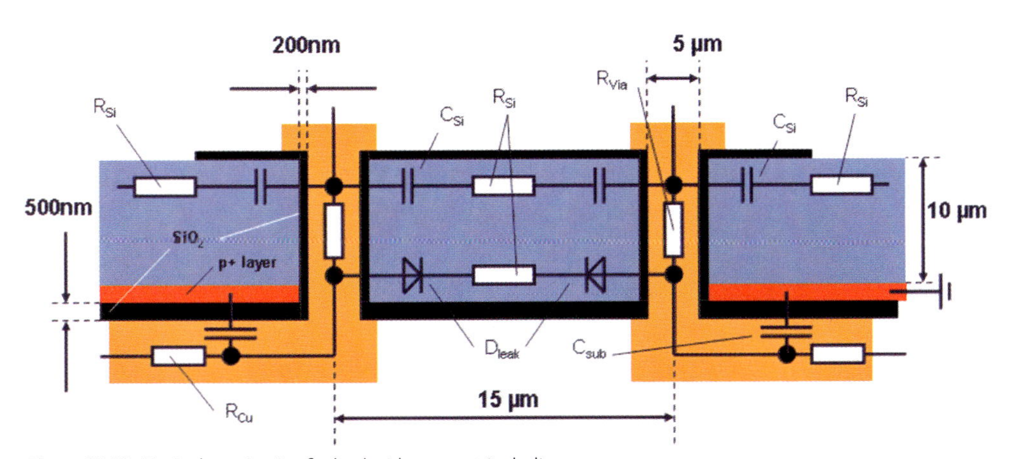

Figure 31.19 Equivalent circuit of a back-side contact including leakage diodes caused by pin holes.

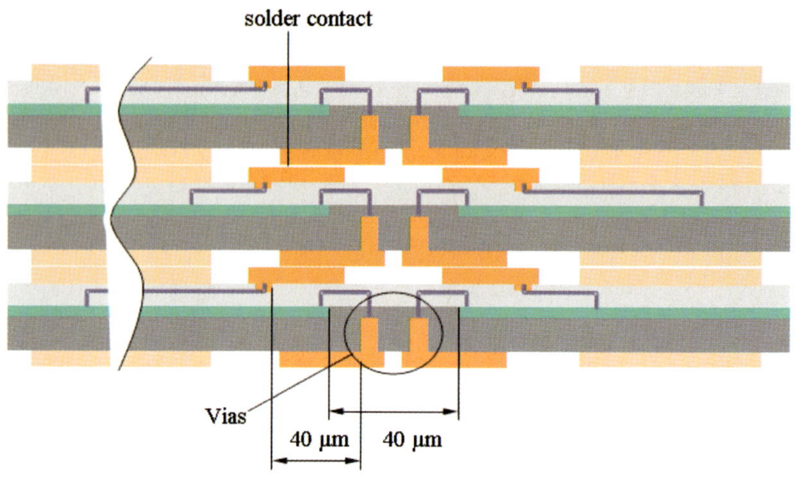

Figure 31.20 Cross section of a three-layer stack. Vias and solder contacts are spatially displaced to release mechanical stress from the vias during the soldering process.

placed below 1 µm this capacitance remains constant. Only the coupling resistor formed by the silicon substrate is reduced, resulting in a more effective via-to-via coupling.

31.4 Summary and Conclusion

So far, we have realized a mechanical stack with seven layers (Figure 31.21).

In the present test design, we were able to implement a maximum via density of more than 4 400 vias per square millimetre. The density is limited due to the

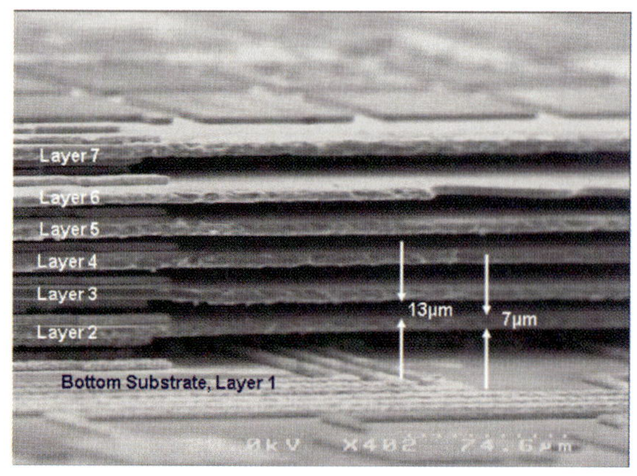

Figure 31.21 SEM image of a mechanical seven-layer stack, including bottom substrate and interchip via connections.

alignment accuracy of the test equipment and allows for a minimum distance of 10 μm between two vias of 5 μm width. Nevertheless, a trade-off between high contact density and high circuit density has to be considered since vertical vias occupy active chip area.

After the complete multilayer stack has been processed, some via connections might fail. Furthermore, problems during soldering can cause open contacts of short circuited signal lines. Detecting theses errors requires the implementation of additional test circuits to the functional CMOS circuitry. With these results, it is possible to design a suitable circuit for a systematic via test to sort out non-functional vias. A short test run at start-up may detect these errors and registers the interchip connections found to be good or bad [22]. This information can be used to reroute vertical signal paths through redundant via holes.

31.4.1 The Vision Cube

Once we have a cost-effective 3D stack technology at hand and have made sufficient progress in designing artificial intelligent systems, systems will be needed that combine speech and vision processing with communication processing in a small volume. Our vision is depicted in Figure 31.22 – The Dialogue modem cube.

The upper layers of the stack contain the vision and speech system accompanied by an extra layer for the memory. They are followed by the higher functions concerning scene interpretation and understanding as well as its encoding. The remaining layers contain the wireless modem and service agents. Such a small dialogue modem, as we would like to call it, could be used in PDAs and robots for the home (as a main building block), and so on.

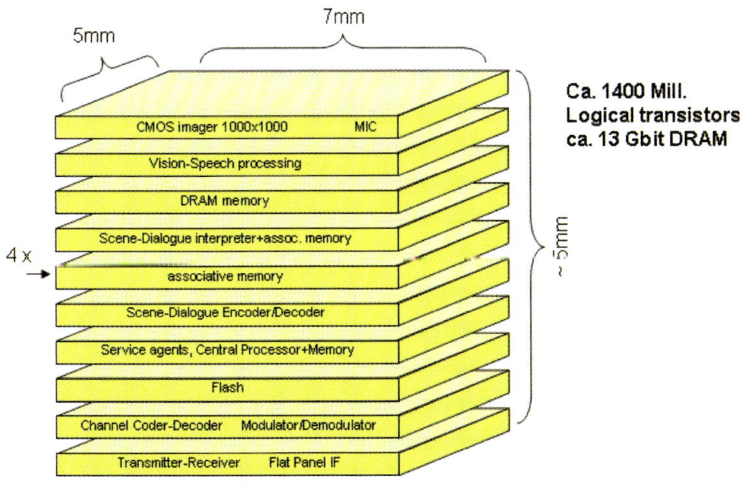

Figure 31.22 The Vision cube.

Acknowledgment

Some of the work reported here was performed in the VisionIC project, which was funded by the German Federal Ministry of Education and Research (BMBF).

The contributions of the following colleagues and partners are gratefully acknowledged: Peter Benkart, Andreas Munding, Alexander Kaiser, Markus Bschorr, Prof. Erhard Kohn, Prof. Hans-Jörg Pfleiderer, Holger Hübner, Wolfgang Raab, Hans-Martin Bluethgen, Jörg Schreiter, Jens-Uwe Schlüssler, Daniel Matolin, Christian Mayr and Prof. Rene Schüffny.

References

1 Ramacher, U. (Oct 2007) *Software-Defined Radio Prospects for Multistandard Mobile Phones. IEEE Computer Magazine.*

2 Raab, W., Bluethgen, H.-M. and Ramacher, U. (2004) A Low-power Memory Hierarchy for a Fully Programmable Baseband Processor. WMPI, pp. 102–106.

3 Matolin, D., Schreiter, J., Schueffny, R. *et al.* (2004) Simulation and implementation of an analog VLSI pulse-coupled neural network for image segmentation. The 47th IEEE International Midwest Symposium on Circuits and Systems, pp. 397–400.

4 Heittmann, A., Matolin, D., Schreiter, J. *et al.* (2002) An analog VLSI pulsed neural network for image segmentation using adaptive connection weights. International Conference on Artificial Neural Networks, pp. 1293–1298.

5 Schreiter, J., Ramacher, U., Heittmann, A. *et al.* (2004) Cellular pulse coupled neural network with adaptive weights for image segmentation and its VLSI implementation. *SPIE*, **5298**, 290–296.

6 Heittmann, A. and Ramacher, U. (2004) An architecture for feature detection utilizing dynamic synapses. The 47th IEEE International Midwest Symposium on Circuits and Systems, pp. 373–376.

7 Mayr, C., Heittmann, A. and Schueffny, R. (2007) Gabor-like image filtering using a neural microcircuit. *IEEE Transactions on Neural Networks*, **18**, 955–958.

8 Gerstner, W. and Kistler, W. (2002) *Spiking Neuron Models*, Cambridge University Press, Cambridge.

9 Al-Sarawi, S.F., Abbott, D. and Franzon, P.D. (1998) A review of 3-D packaging technology. *IEEE Transactions on Components, Packaging, and Manufacturing Technology, Part B*, **21**, 2–14.

10 Ohtake, K. *et al.* (1991) Four-story structured character recognition sensor image with 3D integration. *Microelectronic Engineering*, **15**, 179–182.

11 Becker, K.F. *et al.* (2004) Stackable system-on-packages with integrated components. *IEEE Transactions of Advanced Packaging*, **27**, 268–277.

12 Lin, C.W.C., Chiang, S.C.L. and Yang, T.K.A. (2003) 3D stackable packages with bumpless interconnect technology. Proceedings 5th Electronics Packaging Technology Conference (EPTC 03), IEEE Press, pp. 8–12.

13 Ramm, P. *et al.* (2001) InterChip via technology for vertical system integration. Proceedings IEEE 2001 Int'l Interconnect Technology Conference, IEEE Press, pp. 160–163.

14 Ok, S.J., Kim, C. and Baldwin, D.F. (2003) High density, high aspect ratio through-wafer electrical interconnect vias for MEMS packaging. *IEEE Transactions on Advanced Packaging*, **26**, 302–309.

15 Burkett, S.L. *et al.* (2004) Advanced processing techniques for through-wafer

interconnects. *Journal of Vacuum Science and Technology B*, **22**, 248–256.

16 Das, S. *et al.* (2004) Technology, performance, and computer-aided design of three-dimensional integration. Proceedings 2004 Electrochemical Society Meeting, ACM Press, pp. 108–115.

17 Kaiser, A., Munding, A., Benkart, B. *et al.* (2005) 3D chip integration technology for microsystems. 207th ECS Meeting.

18 Munding, A., Kaiser, A., Benkart, P. *et al.* (2004) Chip stacking technology for 3D-integration of sensor systems. Proceedings HeTech.

19 Benkart, P., Kaiser, A., Munding, A. *et al.* (2005) 3D chip stack technology using through chip interconnects. *IEEE Design & Test of Computers*, **22**, 512–518.

20 Benkart, P., Kaiser, A., Munding, A. *et al.* (2005) 3D chip stack technology using through-chip interconnects. *IEEE Design & Test of Computers*, **22**, 512–518.

21 Huebner, H. *et al.* (2002) Face-to-face chip integration with full metal interface. Proceedings Advanced Metallization Conference (AMC 02), Materials Research Society, pp. 53–58.

22 Bschorr, M., Pfleiderer, H.-J., Benkart, P. *et al.* (2005) Eine test- und ansteuerschaltung für eine neuartige 3D-verbindungstechnologie (in German). *Advances in Radio Science*, **3**, 305–310.

32
Testing of 3D Circuits

T.M. Mak

32.1
Introduction

3D integration is a technology that promises many possibilities. It can allow a higher level of much needed, integration of disparate process/circuit technologies (such as SiGe RF circuit on CMOS baseband) [1]. It also solves bandwidth issues [2] with high-performance multiple-cores microprocessors. In addition, in the case where device scaling may stall, it still allows scaling in a different dimension. However, just like device scaling, the new process must be manufacturable and high yielding. Otherwise, it will remain as a niche product.

One aspect of manufacturing is that of Test. Traditionally, wafers are probed and individual dies are tested/sorted (generally called "wafer sort") before they are packaged (plastic or ceramic housing for the die). The packages are then tested again for specification and their respective functionality. Additionally, some products may be burn-in by the component manufacturer to reduce infant mortality. Wafer sort is essential for the obvious reason that defects are present even in the best semiconductor fabrication processes. It is a shame to throw away a good (and often expensive) package with a die that contains defects (Figure 32.1). In addition, the more dies you integrate, the higher is the required quality of the dies that you want to integrate. Therefore, modern semiconductor manufacturing includes wafer sort to collect as much information on individual dies before they are sent on to assembly.

While testing of 3D integrated systems will probably follow a path similar to conventional testing, the way the wafers or dies are fabricated and integrated will lead to a somewhat different wafer or die level testing. Do we sort/test the individual wafers first before they are bonded together? Do we want to sort the dies first and stack the matched dies onto each other? We examine the details in this chapter.

Handbook of 3D Integration: Technology and Applications of 3D Integrated Circuits.
Edited by Philip Garrou, Christopher Bower and Peter Ramm

ISBN: 978-3-527-32034-9

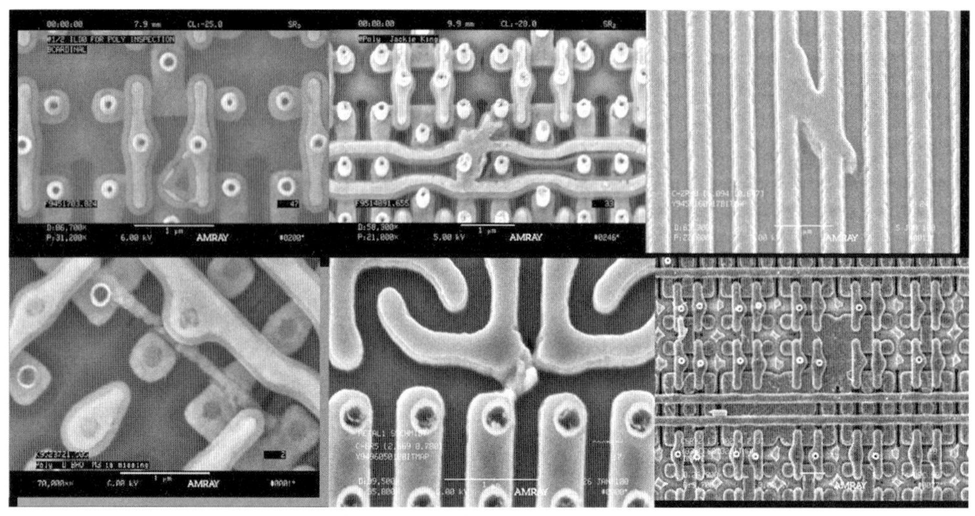

Figure 32.1 Example defects in manufacturing.

32.2
Yield and 3D Integration

Yield is a very significant term in manufacturing. It is the ratio of the manufactured "good" items over that of the total manufactured items. Each process step, whether it is adding material (like metal deposition) to the base wafer or subtracting material (like etching), will introduce its specific types of defects and hence lower the yield. Transistor level defects may include leaky junctions or weak transistors. Wiring defects may include bridges or opens (especially with copper Damascene processing). While most manufacturing processes are perfected to a large degree before they are ramped in volume, there is no such thing as defect free or perfect manufacturing. Even a 99.9% yield for each individual step will bring forth a 67% of yield for a 400-step process.

Of course, a lowered yield will increase the cost of the product sold as a smaller number of good products are sold with more or less the same production cost (the cost of making wafers is the same with or without the defects). All manufacturing costs will have to be amortized over a smaller number of saleable units.

Defects are measured and reported in terms of defect density. It is the number of defects over a unit area (usually a square cm). Fabrication plants have machines that actually can count the defects and track their locations after various process steps [3]. Of course, individually observed defects at any process layer may or may not result into bad chips since not all of the surface area of the chip is utilized in any given layer. There are also process steps that may produce defects that are not observable (e.g. ion-implantation), but nonetheless impact the fabricated circuits.

Therefore, the yield or probability of a good dies is proportional to the area of an individual die. The larger the die, the lower is the yield (Figure 32.2). A smaller die

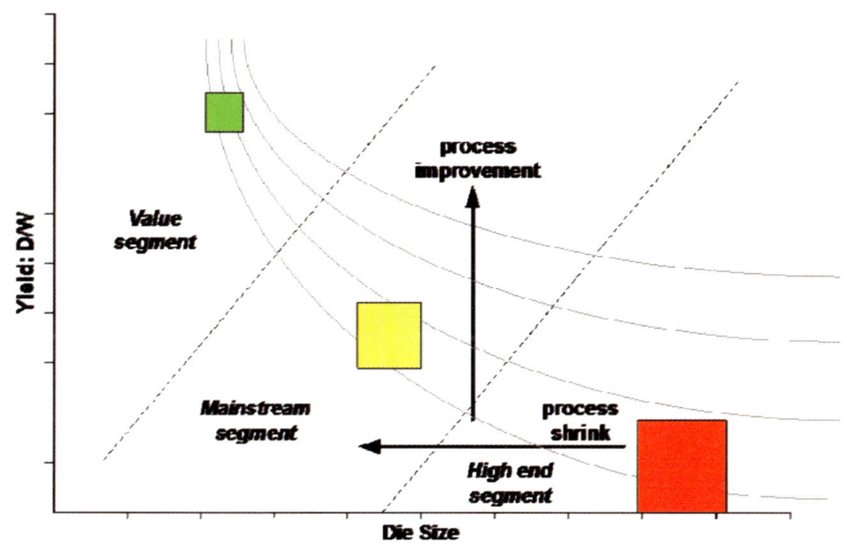

Figure 32.2 Yield–die size relationship.

is obviously better in terms of yield, but then you have to factor in less functionality per die, multiple packages, the higher interconnect capacitance (hence more power and lower performance) and the larger board space and system footprint. Consequently, it is a complicated accounting. There are several good reasons to pursue 3D integration to improve system cost, but we must keep an eye on the price we have to pay.

Even though 3D integration does not make the overall die footprint bigger (the dies are stacked onto each other). It essentially doubles the die area used for a 2-die stack and triples the area for 3-die stacking. There is a small gain in that a smaller die will result in more yield-able dies per wafer [4].

In another words, yield is also proportional to the number of layers. The more layers the lower is the yield. To be fair, 3D integration may potentially result in shorter interconnects [since the wires may run vertically via the short Through Silicon Vias (TSVs) to another circuit on another layer] and potentially fewer layers. For some proposed 3D structures, the layer saving may not be obvious (e.g. memory over processor).

With wafer stacking (which is the more common stacking method), there is simply no choice in matching up of good versus bad die. Whatever you have on the same die location, good or bad dies are stacked. So, for this kind of stacking, yield is a product of the individual die yield (and multiplied by the yield with the TSV and bonding process, too). In other words, with this kind of stacking, you would have the equivalent of a large die with twice the area of each die, but on the footprint of a single die. The footprint may be smaller, performance may be higher and power may be lower but yield is not on your side and the cost will be much higher.

A more subtle point in this wafer level integration is that we also have to consider lot to lot, wafer to wafer and even across-die process variations (see Figure 32.3).

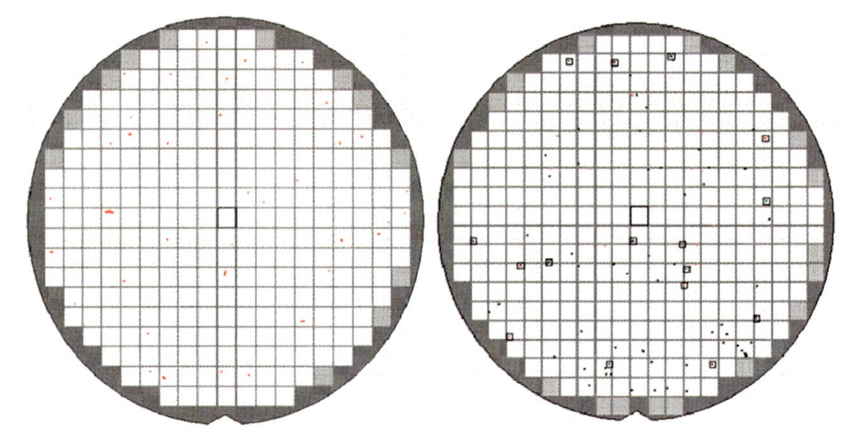

Figure 32.3 Defects on either wafer differs in location.

Wafer stacking would essentially bring two unknown wafers together. There is little correlation between the process parameter distributions across the two wafers shown in Figure 32.3. Consequently, even if the stack consists of defect-free dies, the resulting performance will be limited by the slower chip in the 2-die stack – this may be enough to undermine the performance gain from lowered wire capacitance.

In Figure 32.4 the colored circle represents the wafer and each square represents an individual die. The color denotes the speed of the transistors or the gates. The color change from blue (fast) to red (slow) spans a delay about twice as long. From this picture, we can observe not only wafer level variation but also on-die process variation. Even though these wafers probably are picked to illustrate their diverse

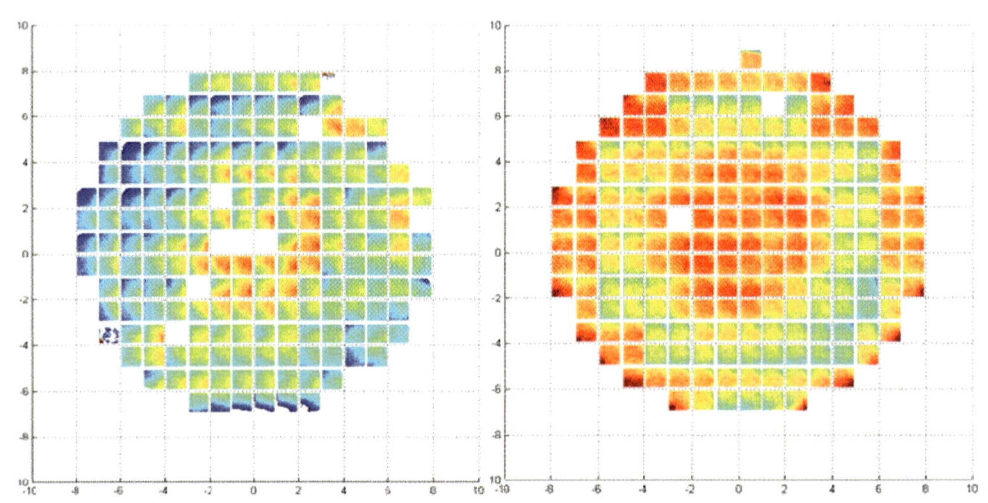

Figure 32.4 Process distributions across wafer (and across die).

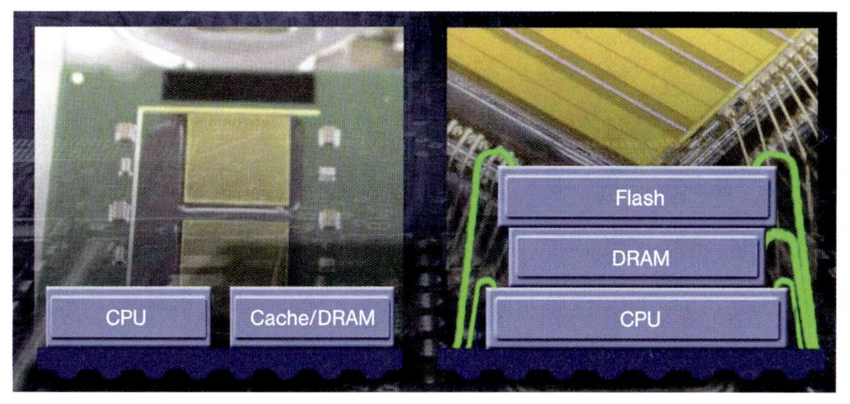

Figure 32.5 MCM versus SiP.

process variation, the reality can be just as bad, if not worse. Future scaling is probably going to introduce more process variation as dimensions keep getting smaller and we do not have the shorter wavelength lithography to match. The smaller structures are also subjected to molecular irregularities, which are beyond the scope of this chapter.

This is in stark contrast to package level integration, such as Multi-Chip Module (MCM), System In Package (SiP) and System On Package (SoP). Figure 32.5 shows MCM and SiP; SoP is similar to MCM/SiP with the exception that discrete components (capacitors or inductors) are also integrated into the package itself as thin film components or embedded discrete components. For these integration technologies, good dies are selected through wafer sort and only the matching dies (such as speed or power consumption level) are packaged together in the same package. Although there are still test limitations with wafer sort (see next section on KGD), this does improve the yield significantly.

32.3 Known Good Die (KGD)

Since yield is so important in bringing a product into the mainstream market, how do we improve the yield? In fact, this is not a new problem and has been around with various forms of package level integration. Before 3D die stacking technology was pursued, people were integrating more dies on the same substrate or in the same package. These systems may be called MCM, SiP or SoP. Some may simply consist of two or more dies on the same substrate (MCM). Some may even have die stacking (just glued together without the TSV) and be inter-connected with bond wires to the substrate. Some may even have sophisticated passive components built into the package layers (SoP).

They are all impacted by the same dreaded yield formula – multiplication of individual die yield. The problem is summarily called Known Good Dies (KGD).

That is, to have a high yielding package level yield, the quality of the die must be known before they are integrated. Since, in all these situations, the quality of the die must be known at the wafer level, this essentially asks for complete testing to be performed at the wafer sort process. If a defect or performance deficiency is discovered on any of the dies after the dies are assembled onto the substrate or package, either re-work has to be done or the whole package has to be discarded. The yield loss will add to the cost of shippable packages. Re-work would require that the bad die be identified (diagnosis aids needed), the bad die be removed, the mounting site be cleaned for remounting and, eventually, a new die be put in place. Many of these operations are delicate, tricky and labor consuming. Re-work is not a recipe for high-volume mainstream applications.

However, even though a high yield is required, getting KGD is not easy. Owing to the lack of a package, the connection from the tester or automatic test equipment (ATE) to the wafer/die is through a hardware interface called a probe card. Owing to the tight bond pad pitch and high density pads, tungsten alloy needles are used for probe cards (Figure 32.6). Transitioning from wire bond to area bonding technology like Controlled Collapse Chip Connection (C4) made the situation even worse. Probe cards today have more than a thousand and they all have to register themselves to individual pads. If debris is collected on the probe tips over extended use, test quality may suffer. Over the years, there have been numerous advances made in probing technology but it has yet to see main stream usage in the industry. The probing needles are definitely the performance limiter. They limit the instant power drawn (creating $-L\ \mathrm{d}i/\mathrm{d}t$ noise). It also creates an impedance discontinuity for high frequency signals. Traditionally, only low frequency tests can be conducted at the wafer level and, more often than not, these tests should not demand any power draw that the needles cannot provide. It is a double whammy.

Figure 32.6 A high density C4 probe card.

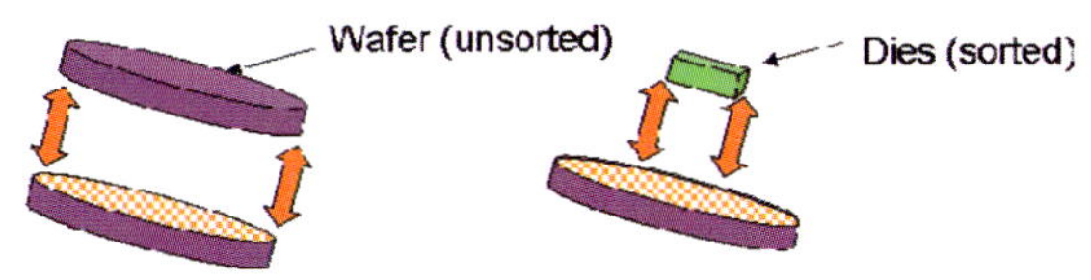

Figure 32.7 Wafer level stacking versus die level wafer stacking.

How do we obtain KGD with tests performed at only low frequency and with limited power? Enter the world of structural testing and design for test (DFT). Structural testing is essentially what its name implies: testing the individual structures of the chip and not necessarily its combined functionality nor performance. Scan and Built-in-self-test (BIST) [5] are common for logic, while Memory BIST (MBIST) or Array BIST (ABIST) are common for memory. With DFT, the relevant circuits are tested for integrity and the tests are not necessarily run at speed since design efforts for DFT should be minimized. For example, scan divides up the complex sequential state machines into logic paths between scanned nodes, which are directly controllable and observable as a serial scan chain. Hence testing only consists of testing these logic paths and the whole sequential complexity gets broken down into combinational complexity only. Also, test circuits typically run at slow speeds, so they are not design intensive and, very often, these DFT circuits are synthesized using common CAD tools.

Better KGD knowledge through more comprehensive structural tests does not necessarily solve the problem of wafer stacking. Simply knowing where the bad dies are does not prevent these bad dies being stacked on top of good die with a wafer stack (Figure 32.7). With the defective die located exactly where they are, there is simply no maneuvering possible with wafer stacking and new methods of design and test have to be figured out. However, it does help with die stacking where a pre-tested KGD can be selected to stack on top of another pre-tested KGD, whether the bottom die is still in the wafer form or singulated. This is especially important when stacking more than two dies. Low per-layer yields simply will make this economically unfeasible.

32.4 Wafer Stacking Versus Die Stacking

Here we mention different implications for wafer level stacking and die stacking. We have to clarify the various possible options so that a clearer picture is presented.

Wafer level stacking is the most straightforward type of stacking, with one wafer stacked on top of another. However, there is the option of a face to face bonding or back to face bonding. With face to face bonding (Figure 32.8), the active layers (where the transistors are) can be very close to each other (with the separation of their respective metal layers). The connections between the wafers are copper pads, which are directly bonded together. With this configuration, one side of the sandwich will be connected to a heat dissipating element such as a heat spreader or a heatsink. The

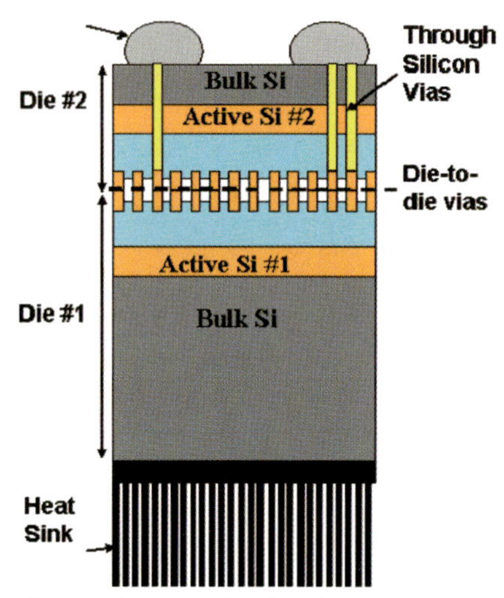

Figure 32.8 Face to face bonding.

other side of the sandwich has to connect to pins/pads/balls for signal and power. These must then be brought out with through silicon vias (TSVs), and C4 pads must be constructed at the other side of the upper layer after the TSV is exposed.

One example process flow is shown schematically in Figure 32.9 (with the progression from left to right). TSVs are pre-fabricated and are initially buried. They are not exposed until the wafers are bonded and the upper layer is thinned. The final C4 pads will then be added to the combined stack.

With this kind of stacking configuration, the worst case situation occurs. The bottom die has tens of thousands to hundreds of thousands of copper pads, but their small size and sheer number make probing the signals impossible. For the same reason, the top wafer is not probe-able from the copper-pad side. Moreover, the TSVs are buried and C4 pads are not yet fabricated, preventing them from being accessed

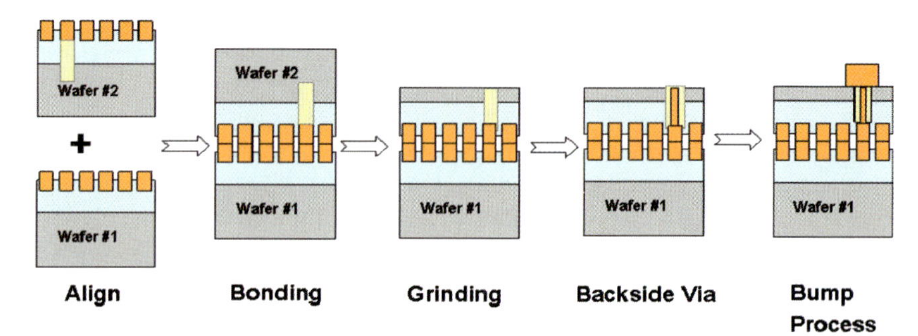

Figure 32.9 Build process of face to face bonding.

Figure 32.10 Pre-thinned wafers are extremely fragile.

from the other side. Consequently, the whole stack is not testable in any way and we just have to live with the yield consequence.

The other option of wafer level stacking is front to back stacking [12]. This will enable a more testable setup for the top die as the C4 can be pre-fabricated on the top layer. However, the top wafer has to be pre-thinned before bonding to the bottom. Moreover, additional copper pads have to be fabricated to provide the bonding surface. This is certainly a much more delicate process than face to face wafer stacking as thinned wafers are less than the thickness of a sheet of paper with little mechanical rigidity (Figure 32.10). It is not clear if this is a favored option for wafer stacking.

This brings us back to the die stacking option. Die stacking would seem to be more desirable due to the possibility of pre-selecting die for stacking. However, the same processing limitations exist. Do we use front to front stacking? Do we use front to back stacking? Do we fully process (with C4 and thinning) the top wafer before sorting it, thereby allowing the maximum test access? How do we probe a paper thin wafer without cracking it? If we sort the wafer without the full processing then it, again, will limit the access and test coverage (buried TSVs, tiny pads). How do we handle singulated, thinned dies? When do we add the C4 solder bumps? (wafer level processing or die level processing?) How do we thin the die stack afterward? If we carry out face to back die stacking, how do sort the bottom die?

After some mental gymnastics in dealing with different configurations of die orientations, we may have to ask ourselves: is there any configuration that will lend itself to conventional wafer probing? The answer is "No" (although we do not want to preclude any breakthrough in this area). One option is to devise ever smaller and denser probe arrays so that the micro-copper pads of each half of the stack can be probed independently. Another option is to design in extra test pads that are of more conventional dimension to allow conventional probing. However, since these will take up much more area and reduce the area for the micro-pads, this type of hybrid padding is popular neither among the testers (who want more power and test data channels) nor designers (who want maximum micro-pads for signaling).

32.5 Defect Tolerant and Fault Tolerant 3D Stacks

Given the above, why not live with the fact that defects are where they are and just work around them? Once we accept defects, we can proceed to examine defect tolerance and fault tolerance.

Defect tolerance is not entirely new. High density DRAM has long had repair capability that allows extra rows or columns or even blocks of cells to be included with every chip. Upon detection, bad cells (or rows or columns) are fused out and the extra rows/columns are fused as replacement. DRAM will not be at the price point where they are without repair capability. Other storage systems have similar capabilities [6]. Hard disk surfaces are never defect free. Again, testing will reveal the bad sectors and they are simply remapped with a spare sector elsewhere on the disks.

These are all manufacturing techniques that are used to improve yield. But what about failures that show up in the field? Here the concept of fault tolerance comes into play. A very high level of reliability from the system is expected in certain computing applications, such as aerospace/transportation, financial transactions and so on. Errors are not acceptable in such applications. To prevent system corruption or downtime, high reliability systems have fault tolerant features built in. This may include error detection with information redundancy, duplicate modular redundancy and triple modular redundancy and recovery system such as check-pointing and rollback.

Of course, many of these schemes are very expensive in terms of resources – in the form of die area, performance or power and it goes against the idea of high level integration. The authors are not suggesting a direct deployment of these techniques but rather the concept itself. If we have to live with defects, we have to be able to tolerate them. Test mechanisms must be put in place to ease the testing and even facilitate diagnosis so that the exact failing structure can be identified. Once they are identified, a reconfiguration mechanism will remove or remap spare resources. Not only do these test mechanisms have to work with the manufacturing test equipment, it is much preferred that they can be periodically called into service for identifying other reliability issues after the chip is deployed in the field. The architecture of the chip will also allow certain levels of redundancy so that spares are available. In the domain of microprocessors, this is already set with the trend toward multi-core processing. With multiple cores on the same piece of silicon [7, 8], core sparing or core disabling is possible. Mesh type on-chip network also lends itself to rerouting and easy faulty unit isolation [11]. A chip with fewer functional cores will work fine with a minor reduction in performance. With the system performance level as the main marketing attributes, fewer cores and lower frequencies will result in a slightly lower price point for the product line. In a lower level of granularity than a full processing core, features that are not critical (speculative execution, extra processing units or special processing units) can simply be disabled as graceful degradation [9]. Again, performance may suffer, but not the overall system level functionality.

Another trend that justifies fault tolerance is the accelerated device degradation and infant mortality rate of future highly scaled devices [10]. Fault tolerant and reconfiguration not only improve yield for integration, it also guards against any future degradation and wearout issues that may arise when the product is already deployed in the field.

References

1 Mallik, D. *et al.* (Nov 9 2005) Advanced package technologies for high-performance systems. *Intel Technology Journal*, **9** (4), 259–271.

2 Polka, A.P. *et al.* (August 22 2007) Package technology to address the memory bandwidth challenge for tera-scale computing. *Intel Technology Journal*, **11** (03), http://www.intel.com/technology/itj/index.com, Design for fault-tolerance in system ES model 900.

3 Mittal, S. *et al.* (Nov 2005) Line defect control to maximize yields. *Intel Technology Journal*, **2** (4), http://www.intel.com/technology/itj/index.com.

4 Young, I. (2007) 3D design opportunities and challenges for microprocessors. presented at the Advanced Circuit Forum of ISSCC.

5 Wang, L.-T., Wu, C.-W. and Wen, X. (eds) (2006) *VLSI Test Principles and Architectures: Design for Testability*, Morgan Kaufmann, San Francisco, CA.

6 Hampson, C. (1997) Redundancy and high-volume manufacturing methods. *Intel Technology Journal*, **1** (02), 4th quarter, http://www.intel.com/technology/itj/index.com.

7 Held, J. *et al.* (2006) *From a few cores to many: A tera-scale computing research overview, Research at Intel White Paper*, http://download.intel.com/research/platform/terascale/terascale_overview_paper.pdf.

8 Vangal, S. *et al.* (Feb 12 2007) An 80-Tile 1.28 TFLOPS Network-on-Chip in 65nm CMOS. Proceedings of ISSCC 2007 (IEEE International Solid-State Circuits Conference), 98–589.

9 Spainhower, L., Isenberg, J., Chillarege, R. and Berding, J. (8–10 July 1992) Twenty-Second International Symposium on Fault-Tolerant Computing, 1992. FTCS-22. Digest of Papers. pp. 38–47.

10 Borkar, S. (Nov–Dec 2005) Challenges in reliable system design in the presence of transistor variability and degradation. *IEEE Micro*, **25** (6), 10–16.

11 Azimi, M. *et al.* (August 22 2007) Integration challenges and tradeoffs for tera-scale architectures. *Intel Technology Journal*, **11** (3), http://www.intel.com/technology/itj/index.com.

12 Topol, A.W. *et al.* (2006) Three-dimensional integrated circuits. *IBM Journal of Research and Development*, **50** (4/5), 491–506.

33
Thermal Management of Vertically Integrated Packages

Thomas Brunschwiler and Bruno Michel

33.1
Introduction

For the second time in the history of electronic components heat removal from integrated circuits (ICs) has become a limiting factor. It already forced the industry once, in the 1990s, into revolutionary technology changes, namely, achieving the transition from the bipolar era to the new era of CMOS technology (Figure 33.1). Today we observe again such a marked transition: from the evolutionary scaling of clock frequencies to multi-core microprocessor architectures and the advent of 3D vertical integration, with the benefit of reduced power dissipation for the computing performance needed.

Vertically integrated packages will especially challenge the thermal management community, and will require new innovative ways to remove the dissipated heat from a volume and not just a single plane.

33.1.1
Power Dissipation in Electronic Components

The dissipated power in a CMOS microprocessor is the sum of the power to drive the logic gates, signal interconnects, clock networks and the cache memory. The power dissipated by field-effect transistors can be separated into switching, short-circuit and leakage power. At present the switching power accounts for 70% of the total power dissipated [1]. It is a second-order function of the supply voltage (V_{dd}) and proportional to the gate capacitance (C) and the clock frequency (f):

$$P_{sw} = \frac{1}{2} C f V_{dd}^2 \tag{33.1}$$

Increasing resistances and current density limits in electrical wires prevented a proportional reduction of V_{dd} during dimensional scaling. This and the increasing clock frequency are the main reason for the CMOS power density explosion. Despite that, the absolute power dissipated per switching event is still reduced. At gate lengths

Handbook of 3D Integration: Technology and Applications of 3D Integrated Circuits.
Edited by Philip Garrou, Christopher Bower and Peter Ramm

ISBN: 978-3-527-32034-9

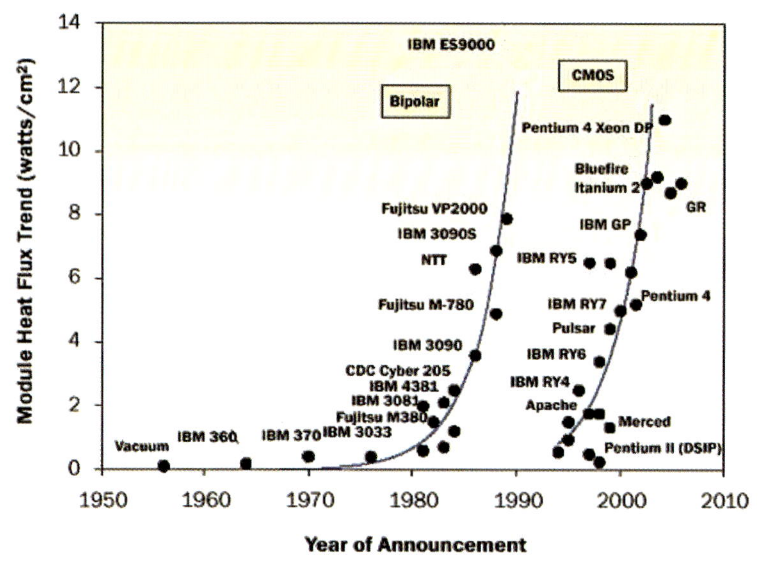

Figure 33.1 Module heat flux explosion at bipolar times – and about ten years later in CMOS technology. (From Roger Schmidt. With permission).

below 90 nm, the leakage power due to electron tunneling through the nanometre-scale gate dielectric film starts to exceed the switching power. However, by introducing high-k gate oxides with their increased dielectric thicknesses, the parasitic power dissipation can be reduced for one to two generations. By dimensional scaling the cross sections of on-chip electrical wires are reduced while at the same time the chip size increased. This is called the wiring-crisis and results in increasing ohmic power dissipation in the metal wires. In 3D packages, the wire length and the number of signal repeaters can be decreased, reducing total power dissipation. Low-k dielectrics or air gaps between the individual wires reduce the wire capacitance but, because of nano-porosity, are poor thermal conductors.

33.1.2 Motivation for Thermal Management

33.1.2.1 Temperature Influence on Performance

High temperatures reduce the mobility of semiconductor carriers and increase metal resistivity, causing an increase in time constants. To keep the clock frequency constant, the supply voltage is raised, which in turn results in additional power dissipation and further temperature increases. As leakage currents scale also exponentially with temperature, operation of ICs at sub-ambient temperatures was proposed. For a coefficient of performance (COP) of thermal compression loops between 3 and 5, the power dissipated by the cooling loop exceeds the power savings that can be realized by reducing the device temperature, rendering the system economically and ecologically inefficient [2].

33.1.2.2 **Temperature-Driven Reliability**

Mechanical stress can lead to catastrophic failure due to thermal expansion mismatch of package material combinations. Component life time (L) depends on aging effects, such as electromigration, diffusion, relaxation, delamination and voiding, and is exponentially reduced with temperature (T) according to the Arrhenius law:

$$L(T) = A\left(e^{\frac{E_a}{kT}} - 1\right) \tag{33.2}$$

with a system-specific constant (A), the activation energy (E_a) and Boltzmann's constant (k). For a temperature increase of 20 °C, the lifetime of the system is reduced by a factor of two. Temperature cycling due to switch events or varying computational loads is the source of thermal stress fatigue and voiding in thermal interface materials [3].

33.1.2.3 **Thermal Limits**

The maximum tolerated junction temperature varies, depending on device technology and product needs such as performance and targeted lifetime. High-performance logic is most sensitive to temperature. Low-performance logic, memory and handheld devices as well as automotive electronics are operated at moderate clock frequencies and consist of robust transistor designs with larger gate lengths and gate dielectric thicknesses (Table 33.1).

33.2 Fundamentals of Heat Transfer

Heat transfer is caused by thermal gradients or mass transfer. To reach equilibrium, the system relaxes by conduction, convection or radiation.

33.2.1 Conduction

Heat conduction is the transfer of kinetic and vibrational energy from atom to atom or molecule to molecule thorough a solid, non-moving fluid or gas. Fourier's law describes the heat diffusion in continuum space:

Table 33.1 IC junction temperature limit for various applications [4].

Application	Temperature limit (°C)
High-performance logic	95
Low-performance logic	125
Memory devices	125
Handheld devices	125
Automotive electronics	175

$$\dot{q} = -k\frac{\mathrm{d}T}{\mathrm{d}x} \tag{33.3}$$

where $\dot{q}$ is the heat flux, $\mathrm{d}T/\mathrm{d}x$ the spatial thermal gradient and k the thermal conductivity. For steady-state conditions and one-dimensional heat flux through a plane wall with thickness L the thermal impedance is:

$$R_{\mathrm{cond}} = \frac{\Delta T}{\dot{q}} = \frac{L}{k} \tag{33.4}$$

33.2.2 Convection

The transfer of heat from a solid to a moving fluid/gas or vice versa is called convection. Close to the interface, it is dominated by heat conduction as the relative fluid velocity is low because of the non-slip condition of the fluid at the wall. With increasing distance from the solid surface, the velocity increases and heat removal is dominated by mass transport. The solid–fluid temperature gradient and the heat transfer coefficient (h) define the heat flux:

$$\dot{q} = h(T_{\mathrm{wall}} - T_{\mathrm{fluid}}) \tag{33.5}$$

Correspondingly the convective thermal impedance is:

$$R_{\mathrm{conv}} = \frac{\Delta T}{\dot{q}} = \frac{1}{h} \tag{33.6}$$

The heat transfer coefficient can be derived from the non-dimensional Nusselt (Nu) number:

$$Nu(x) = \frac{h(x) \cdot L_H}{k} \tag{33.7}$$

It depends on the heat transfer geometry with the characteristic length (L_H), the local position (x), the Reynolds (Re) and Prandtl number (Pr). Re describes the ratio of inertial to viscous forces in the fluid. In straight channels, the flow is laminar at $Re < 2300$. Pr describes the ratio of the momentum diffusivity to the thermal diffusivity and is defined by the coolant properties. Semi-empirical correlations are defined (Table 33.2) [5] to calculate the non-dimensional Nusselt (Nu) number. In most cases they are of the form:

Table 33.2 Coefficients for the local Nusselt number Nu_0 in the stagnation (center) zone of a jet impinging normal to a surface.

Flow regime	*a*	*b*	*c*	Reference
Laminar	1.648	1/2	1/3	[6]
Turbulent	0.93	1/2	0.4	[7]

$$Nu \propto a \cdot Re^b \cdot Pr^c \qquad (33.8)$$

In the case of laminar flow and fully developed thermal and hydrodynamic boundary layers, b and c are equal to zero and Nu becomes a constant.

33.3 Thermal-Packaging Modeling

33.3.1 Temperature and Power Map Prediction During IC Design

Signal delay and power dissipation depend on the local circuit temperature, and require the modeling of power maps and thermal behavior already in the chip design phase. Especially, high-performance ICs having gate lengths <90 nm are sensitive to existing temperature variations, which can be as large as 40 K. Thermal analysis tools compute full-chip, 3D thermal profiles from the electrical layout and are used for timing and electromigration analysis.

33.3.2 Design and Optimization of Thermal Packages

33.3.2.1 Analytical Method

The partial differential equation (PDE) of heat conduction can analytically only be solved for simple geometries. The result is a trigonometric series, which can be approximated in a closed-form equation. Lee *et al.* [8] have derived such results to define the spreading resistance of heat sinks.

33.3.2.2 Numerical Approaches

Arbitrary heat conduction problems are solved numerically by finite-difference and finite-element methods (FEM). The PDE is solved numerically starting from initial conditions, and the procedure is repeated until the energy balance converges. Finally, temperature and heat flux can be extracted for any point in the control volume by interpolation.

Convective heat and mass transfer is modeled by finite-volume methods that solve the Navier–Stokes equation. These computational fluid dynamics (CFD) methods are computationally demanding because they co-solve energy, mass and momentum equations. Using temperature-independent fluid properties decouples the energy equations and reduces the computation cost to some extent.

33.3.2.3 Efficient Modeling

Given the complexity and length scale range (from transistor to computer chassis) in a typical package, the problem is solved in sub-systems using a bottom-up approach: The result from the detailed study serves as the boundary condition for the next-higher-level model. Depending on the level of optimization, several approximation methods have been developed, and will be discussed below.

Equivalent R/C Networks For steady-state and transient chip-level power map and temperature prediction, the analogy of an electrical R/C network is used to represent individual thermal components. Electrical network analyzers then solve the differential equations.

Board Level Design with Compact Modeling In the two-resistor compact model approach [9], an individual chip package is represented by two resistors from junction to board and junction to lid. These values are derived from detailed package-level models and are specified by the supplier. The heat conduction in the heat sink is then solved in detail by FEM. The heat convection can be computed by CFD or by simplified fluid network modeling and using heat transfer correlations for standard geometries.

33.4 Metrology in Thermal Packaging

33.4.1 Characterization of Thermal Components

Static thermal gradient and heat flux measurements define the thermal resistance of a package and its components with temperature sensors such as thermocouples (TC) or resistance temperature detectors (RTD) between components of interest. To analyze the thermal conductivity of thermal interface materials (TIM), one-dimensional heat flux and bondline thickness measurements are used. Transient methods measure the temperature response of an applied power step or pulse as a function of time with a single temperature sensor at the source [10]. With subsequent mathematical transformation the thermal resistance and capacitance values of the components are determined. Fluid properties are measured in radial heat conduction mode, the so-called hot-wire method. Solid materials are characterized in one-dimensional heat flux condition using the laser flash or resistive heater method. Here, the R/C values of individual layers in a multilayer system, such as a vertically-integrated package, can be defined by the structure function analysis [11]. To capture the relevant thermal response of a package with component time constants ranging from microseconds to minutes, logarithmic time sampling is used. The method can also accelerate the thermal qualification of packages.

33.4.2 Power Map Measurement

Spatially resolved imaging of microprocessor temperatures [12] provides the processor power map at different load conditions (Figure 33.2). The heat conduction matrix

a)

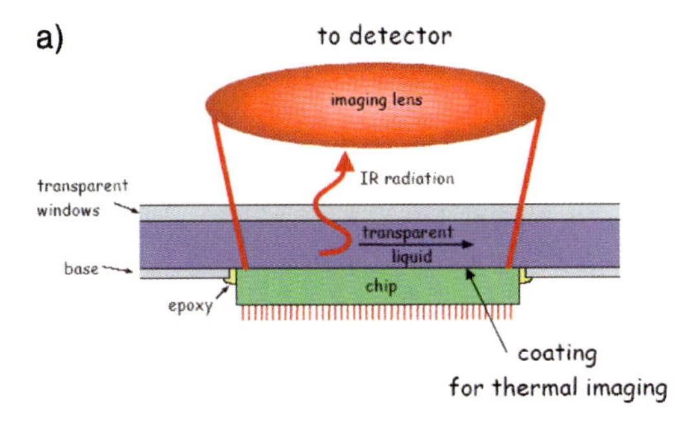

b)

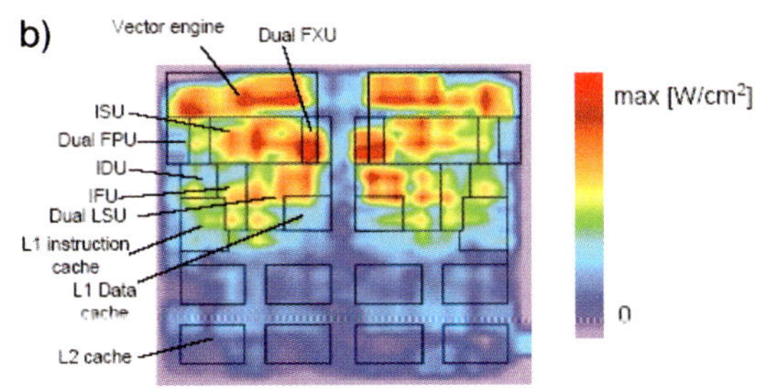

Figure 33.2 Cross section through test setup (a) and measured power map (b) of a dual-core microprocessor. (From Reference [12] with permission.)

representing the processor is measured by stepwise local heating at the chip frontside by means of a laser beam. During operation the back-side temperature is measured by infrared imaging. The heat removal is carried out with forced liquid back-side cooling using an infrared-transparent coolant. With this method, it is possible to analyze hotspots during operation.

33.5 Thermal Packaging Components

33.5.1 Thermal Interface Materials

The minimum distance between two solid surfaces is defined by the roughness and the warp. Because of the low thermal conductivity of air ($0.0245\,\mathrm{W\,m^{-1}\,K^{-1}}$), a 1-μm-wide gap results in a thermal impedance of $40.8\,\mathrm{K\,mm^2\,W^{-1}}$. Particle-laden polymers or solder materials are used as thermal interface materials (TIM) to reduce the thermal contact resistance (i.e. indium thermal conductivity $= 81\,\mathrm{W\,m^{-1}\,K^{-1}}$). Owing to the high stiffness of solder materials, thermo-mechanical stress caused by thermal expansion mismatch and thermal gradients in the package are transferred to the chip and can cause catastrophic failure or low cycle fatigue. This limits the use of solders to small chips and material sets having a low expansion-coefficient mismatch. Thermal greases consist of an oil matrix laden with high-thermally-conductive metallic or ceramic particles with sizes <50 μm. The thermal conductivity increases exponentially above the percolation threshold (30–50 vol.% fill factor) [13]. Formulations with fill factors above 65% reach thermal conductivities of >4 $W m^{-1} K^{-1}$. At higher particle loading, the high viscosity can cause chip fracture or solder-ball crushing during assembly.

33.5.1.1 **Hierarchical Nested Channels**

Hierarchical nested channels allow higher fill factors in greases [14]. The pressure applied to reach a certain bondline thickness (BLT) in a given time in the presence of a viscous medium is proportional to the area. By subdividing a surface, the squeeze dynamics is increased because the paste is evacuated by channels separating the posts. Channel diameters are maximized as the pressure drop in the channels scales with the inverse of the fourth power on the hydraulic diameter. For heat transfer reasons, the post fill factor is maximized. These contradictory requirements can be only fulfilled by hierarchical channel networks (Figure 33.3).

33.5.1.2 **Control of Particle Stacking**

After the squeeze process, particle stacks with high thermal conductivity along the diagonals of the flat chip surface can be observed (Figure 33.4). They are caused by drag force balancing during squeeze flow along the die diagonal and define the ultimate bondline thickness. By employing stack control channels, the fill factor of these areas is increased and the bondline thickness reduced [15].

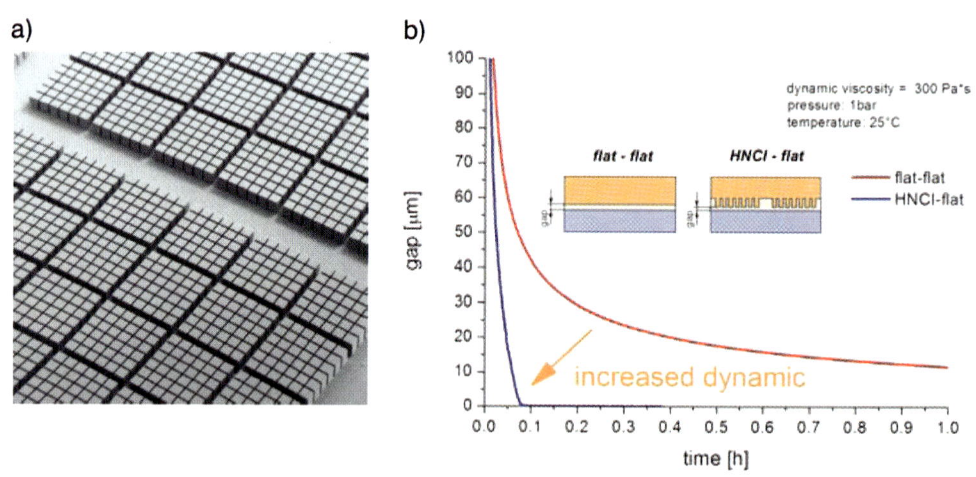

Figure 33.3 (a) Scanning electron micrograph (SEM) of a hierarchical nested channel network etched into silicon and (b) improved bondline formation dynamics. (From Reference [14] with permission.)

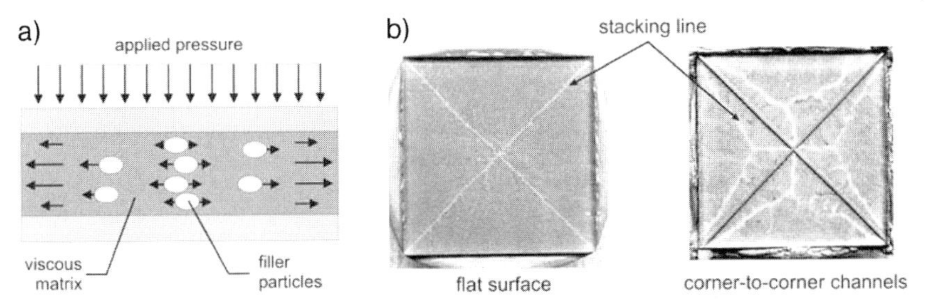

Figure 33.4 (a) Force balancing and (b) particle stacking with and without particle control channels during and after bondline formation. (From Reference [15] with permission.)

33.5.2
Advanced Air Heat Sinks

The ubiquitous availability and compatibility of air with electronics makes air cooling the heat removal technique of choice. Air fins increase the total surface area and reduce the convective thermal resistance by a factor of 100 so that chip heat flux values of up to 70 W cm^{-2} can be handled. For higher heat fluxes, water- or methanol-filled heat pipes or vapor chambers are used to increase the heat spreading in the heat sink base. Fluid evaporates on hot areas, condenses in cooler areas and the fluid is transported back to the heat source by capillary forces in the wicking structures. A limiting factor is the critical effective heat flux at around 140 W cm^{-2} at which the liquid film on the heat source dries out, resulting in an abrupt increase in thermal resistance.

33.5.3
Forced Convective Liquid Cold Plates

Forced convective liquid cooling handles higher heat flux levels because of the 3000-fold increased specific thermal capacity and sixfold increased thermal conductivity of water compared with air. In micro-scale liquid cold plates, the flow is typically laminar (Reynolds number < 2300). Two basic heat transfer architectures are possible: micro-channels with increased wetted surface area but mostly developed thermal boundary layers, and impinging fluid jets with a highly undeveloped boundary layer in the stagnation zone. Microscale coolers require parametric optimization as both the heat transfer (h) and the pressure drop are inversely proportionally to the channel or jet hydraulic diameter (d_h) – the ideal cold plate consumes a minimal pumping power for the heat transfer specified.

High-performance microchannel cold plates for cooling of 400 W cm^{-2} are composed of parallel flow sections to reduce the total pressure drop and the junction temperature non-uniformity due to fluid temperature increase from inlet to outlet. The convective heat transfer is increased further by a staggered fin arrangement to break up the thermal boundary layer (Figure 33.5) [16].

Microscale jet impingement cold plates with >40 000 nozzles use a distributed return architecture for uniform heat removal over the chip surface. One nozzle and its

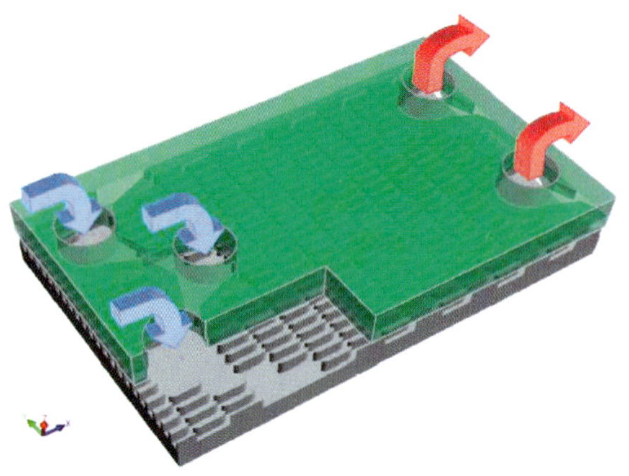

Figure 33.5 3D rendering of assembled staggered fin microchannel cold plate. (From Reference [16] with permission.)

drainage hole form an independent, scalable unit cell. A tree-like double branching fluid network delivers the coolant to the nozzle and drains it with minimal pressure drop and uniform cell flow rates. Heat transfer coefficients up to 10 W cm^{-2} K^{-1} can be realized, allowing a heat flux limit of 350 W cm^{-2} at a temperature increase of 65 K from the fluid inlet to the junction temperature (Figure 33.6) [17].

33.6 Heat Removal in Vertically-Integrated Packages

33.6.1 Main Challenges for Traditional Back-Side Heat Removal

When stacking multiple active layers, the heat flux and thermal resistance accumulate in the package. The back-side heat removal limit is estimated using an analogous

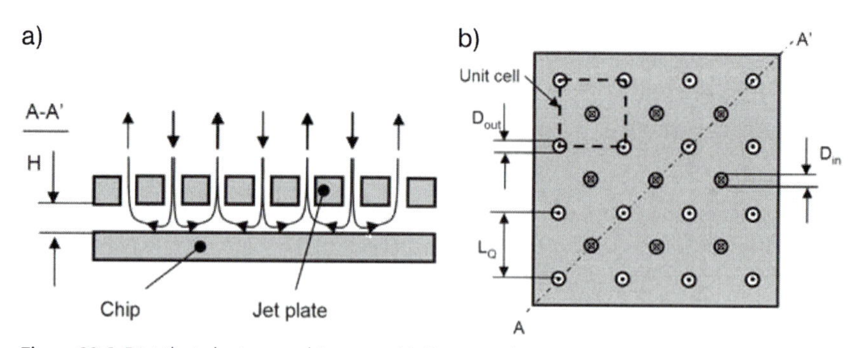

Figure 33.6 Distributed return architecture. (a) Cross section showing inlet jets with neighboring drainage holes. (b) Top view of the face-centered-cubic jet and return array. (From Reference [17] with permission.)

Table 33.3 Parameters used for back-side heat removal limit estimation.

Component	Thermal resistance ($K\,mm^2\,W^{-1}$)	Component	Power dissipation ($W\,cm^{-2}$)
Cold plate	7	MPU cache	60
Thermal interface	6	MPU hotspot	240
Dielectric + wires	3.5	MPU uniform	100
Dielectric + wires memory	1.4	Memory	10
Die thickness:			
Uppermost	500 μm		
Others	100 μm		

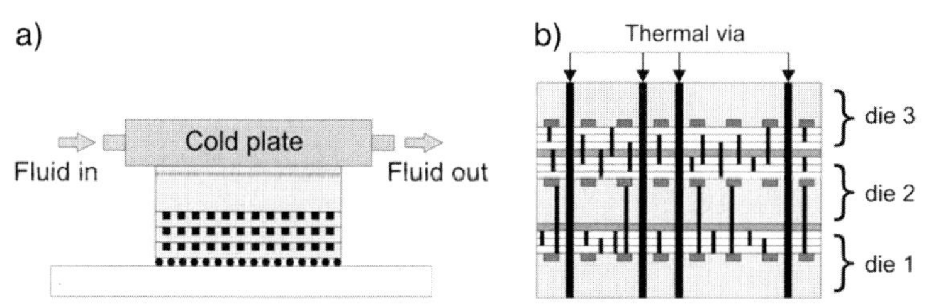

Figure 33.7 (a) Back-side-cooled 3D package and (b) thermal vias in a 3D chip stack.

resistor network, assuming one-dimensional heat flux (Figure 33.7a). The thermal resistance of the dielectric layers with metal wires is calculated using predicted 45-nm node dimensions [4]. We use best-case values for the cold plate and thermal interface (Table 33.3). The test cases are performed at a standard and a conservative maximal thermal budget of 65 and 45 K, respectively (Table 33.4). The maximum memory junction temperature is 15 K higher than in the microprocessor unit (MPU). A high-performance MPU with peak heat flux of four times the cache heat flux is compared with a MPU with uniform power dissipation (Table 33.4).

Table 33.4 Back-side cooling limits: temperature gradient $T_{junction\ MPU} - T_{fluid\ inlet}$ for different chip stacks at thermal budgets of 65 and 45 K.[a]

ΔT_{max}	65 K	ΔT_{max}	45 K
Acceptable temperature	ΔT (K)	**Acceptable temperature**	ΔT (K)
12× mem/MPU hs/cp	60.6	2× mem/MPUhs/cp	43.8
MPU hs/3× mem/cp	61.8	MPU hs/mem/cp	46
2× mem/MPU hs/MPU cache/cp	65	2× MPU uniform/cp	38
3× MPU uniform/cp	63.3	**Exceeds temperature limit**	
Exceeds temperature limit		MPU hs/MPU cache	55.7
2× MPU hs/cp	111.1	2× MPU hs	111.1

[a] mem: memory, cp: cold plate incl. TIM, hs: hotspot.

It is advantageous to place the MPU close to the cold plate and to avoid or offset hotspots. Memory below and on a single MPU can be handled by back-side cooling. Two MPUs with non-overlapping hotspots can be cooled at a budget of 65 K, but with overlapping hotspots they exceed the limit for both cases, which reduces the freedom for electrical design optimization. Therefore, new innovative heat-removal concepts have to be investigated.

33.6.2 Heat Conduction Improvement with Thermal Vias (TV)

Low-k dielectric materials or air gaps reduce the parasitic capacitance, signal flight time and crosstalk of electrical wires. Unfortunately, the heat conduction of these materials is also reduced because of their nano-porosity [18]. Worst case and ultimate goal is the air gap technology with 62 times lower heat conduction than SiO_2. One way to keep the thermal conduction through the dielectric wire layers acceptable is to implement thermal vias (Figure 33.7b). Owing to the 1000-fold increased thermal conductivity of copper compared to the dielectrics the via area fill factor can stay low and the loss in silicon real estate is marginal. Different temperature-aware routing algorithms with thermal via insertion are proposed to minimize wiring blockage [19, 20]. With further scaling, the thickness of the metal wires and the silicon-on-insulator film is below the mean free path of phonons. The lateral thermal conductivity is reduced owing to dominant phonon scattering at the layer interface and the thermal-via density needs to be increased.

33.6.3 Interlayer Thermal Management

33.6.3.1 Lateral Heat Conduction

Conductive spacers between the chip layers smooth out heat flux peaks or conduct the heat laterally out of the package (Figure 33.8) [21]. To be effective, a spacer thickness of more than 1 mm is needed and scales with chip size. This limits the technology to peripherally-routed or low vertical interconnect density 3D packages and small die sizes. Alternatively, thin form-factor vapor chambers may be used as heat spreaders [22]. So far, no vertical-interconnect implementations have been realized.

33.6.3.2 Forced Convective Liquid Interlayer Heat Removal

In interlayer convective cooling, the chip stack is placed in a fluid containment and the coolant is forced between the individual dies to remove heat from adjacent

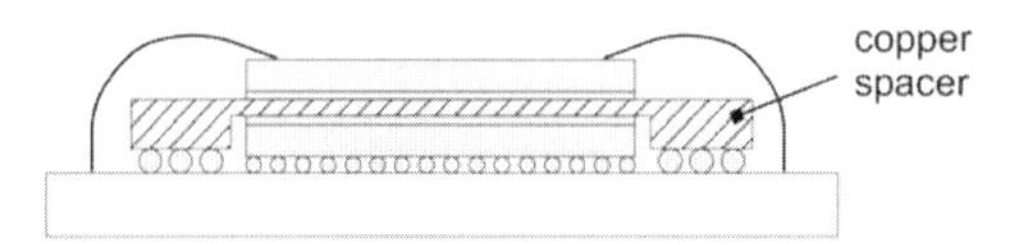

Figure 33.8 Lateral heat removal with a copper spacer.

Table 33.5 Performance of interlayer convective heat removal.

Chip size (mm)	Fluid gap (μm)	Interconnect/pitch (μm)	Coolant	Power density ($\mathrm{W\,cm^{-2}}$)	Reference
10	100	Area array/200	Water	394	[23]
10	10	Peripheral	Water	25	[20]
14.14	300	Area array/800	Two-phase water	140	[25]
10	50	Peripheral	FC-77	50	[24]

layers (Table 33.5 and Figure 33.9a). Vertical through vias can be integrated, increase the wetted surface area and induce mixing of the fluid. If water is used, then the electrical interconnects have to be hermetically sealed. For dielectric coolants no electrical insulation is needed, but material compatibility with polymers has to be verified.

The junction temperature response from fluid inlet to outlet is shown qualitatively in Figure 33.9b. Parallel heat conduction through the silicon and wire layers causes a small and constant temperature increase ($\Delta T_{\text{conduction}}$). The convective heat transfer coefficient is maximum at the fluid inlet, reduces with developing boundary layer thickness and reaches an asymptotic value for fully-developed conditions. Consequently the convective thermal gradient ($\Delta T_{\text{convection}}$) increases. The accumulated sensible heat in the fluid causes a linear increase in ΔT_{heat} along the chip surface. The maximum junction temperature is found at the fluid exit. For non-uniform heat flux, the hotspots should be arranged close to the fluid inlet to reduce the temperature maxima and gradients along the die.

To implement true area array vertical interconnects, heat transfer structures such as microchannels, staggered or inline pin fins have to be implemented. Pressure drop limits the fin pitch to values >50 μm [23]. At such high fluid resistances and with the resulting low flow rate, the temperature increase on the junction is mainly dominated by the fluid temperature increase, favoring a heat transfer structure with minimal friction. At pitches >200 μm, the flow rate is increased and the convective thermal gradient dominates, requiring improved heat transfer structures such as pin fins with a high grade of fluid mixing. Because of the increased viscosity and reduced thermal capacity of dielectric fluids, the flow rates are reduced and the fluid temperature

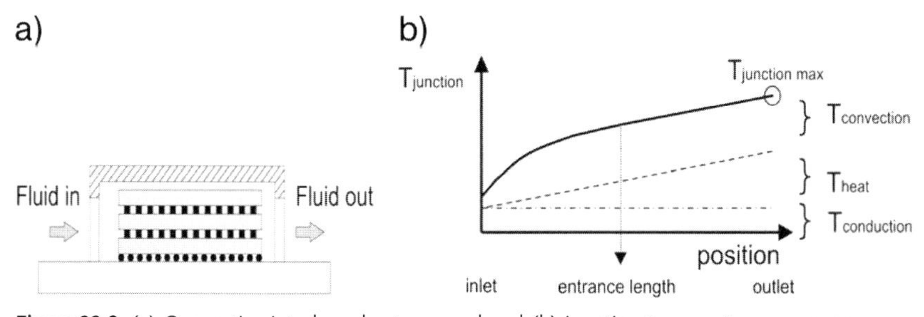

Figure 33.9 (a) Convective interlayer heat removal and (b) junction temperature response.

increase is dominant. They can handle moderate heat flux levels of 50 W cm^{-2} at reduced package complexity [24].

33.6.4 Conclusion

Heat removal in 3D stacks is a major challenge and might limit progress in the area of high-performance packages. Stacks with a single MPU layer but multiple memory dies can be back-side-cooled in the traditional way with high-performance air heat sinks or liquid cold plates. At higher numbers of stack levels and with nano-porous low-k dielectric materials, efficient algorithms are required to minimize wire blockage and the loss of active silicon real estate caused by thermal via-insertion. True scalability is provided only by convective interlayer heat removal at moderate interconnect pitches >50 μm but at the price of added complexity.

References

1 Rahman, A. and Reif, R. (2001) Thermal analysis of three-dimensional (3-D) integrated circutis (ICs). Proceedings IEEE 2001 International Interconnect Technology Conference, 4–6 June 2001, pp. 157–159.

2 Jain, A. and Ramanathan, S. (2006) Theoretical investigation of sub-ambient on-chip microprocessor cooling. Proceedings 10th Intersociety Conference on Thermal and Thermomechanical Phenomena in Electronics Systems ITHERM 06, 30 May–2 June 2006, pp. 765–770.

3 Kraus, A. and Bar-Cohen, A. (1983) *Thermal Analysis and Control of Electronic Equipment*, Hemisphere Publishing Corporation, New York, Chapter 2.

4 International Technology Roadmap for Semiconductors (ITRS), http://public.itrs.net (2007).

5 Verein Deutscher Ingenieure VDI-Wärmeatlas (2006) Springer-Verlag, Berlin, Heidelberg, Germany.

6 Scholtz, M. and Trass, O. (1970) Mass transfer in a nonuniform impinging jet. *The AlChE Journal*, **16**, 82–96.

7 Pan, Y., Stevens, J. and Webb, B. (1970) Effect of nozzle configuration on transport in the stagnation zone of axisymmetric impinging free-surface jets: Part 2. Local heat transfer. *Journal of Heat Transfer*, **114**, 880–886.

8 Lee, S., Song, S., Au, V. and Moran, K. (1995) Constriction/spreading resistance model for electroncs packaging. Proceedings ASME/JSME Thermal Engineering Conference, Volume 4, pp. 199–206.

9 Stiver, D. and Shidore, S. (2002) The extraction of a two-resistor/two-capacitor model for common IC packages and their implementation in CFD, Proceedings IMAPS 2002, Denver, CO.

10 Smith, B., Brunschwiler, T. and Michel, B. (Sept. 17–19 2007) Utility of transient testing to characterize thermal interface materials, Proceedings Thermal Investigations of ICs and Systems Therminic 2007, Budapest, Hungary, pp. 6–11.

11 Rencz, M. (March 2005) Thermal issues in stacked die packages. Proceedings 21st Annual IEEE Semiconductor Thermal Measurement and Management Symposium SEMI-THERM 2005, pp. 307–312.

12 Hamann, H., Lacey, J., Cohen, E. and Atherton, C. (2006) Power distribution

measurement of the Dual Core Power PCTM 970^{MP} Microprocessor, Proceedings IEEE International Solid-State Circuits Conference, Session 29.

13 Devpura, A., Phelan, P. and Prasher, R. (2001) Size effects on the thermal conductivity of polymers laden with highly conductive filler particles. *Microscale Thermophysical Engineering*, **5**, 177–189.

14 Brunschwiler, T., Kloter, U., Linderman, R. *et al.* (2007) Hierarchically nested channels for fast squeezing interfaces with reduced thermal resistance. *IEEE Transactions on Components and Packaging Technologies*, **30** (2), 226–234.

15 Linderman, R., Brunschwiler, T., Kloter, U. *et al.* (2007) Hierarchical nested surface channels for reduced particle stacking and low-resistance thermal interfaces. Proceedings 23rd Annual IEEE Semiconductor Thermal Measurement and Management Symp, San Jose, CA, March 2007, pp. 87–94.

16 Colgan, E.G. *et al.* (June 2007) A Practical implementation of silicon microchannel coolers for high power chips. *IEEE Transactions on Components and Packaging Technologies*, **30** (2), 218–225.

17 Brunschwiler, T., Rothuizen, H., Fabbri, M. *et al.* Direct liquid-jet impingement cooling with micron-sized nozzle array and distributed return architecture. Proceedings ITHERM 2006, San Diego, CA, pp. 196–203.

18 Im, S., Srivastava, N., Banerjee, K. and Goodson, K. (2005) Scaling analysis of multilevel interconnect temperatures for high-performance ICs. *IEEE Transactions on Electron Devices*, **52** (12), 2710–2719.

19 Wong E. and Lim, S. (2006) 3D floor-planning with thermal vias. Proceedings Design, Automation and Test in Europe DATE 06, March 2006, Vol. 1, pp. 1–6.

20 Takahashi, K. *et al.* (2004) Process integration of 3D chip stack with vertical interconnection. Proceedings 54th IEEE Electronic Components and Technology Conference, Vol. 1, pp. 601–609.

21 Lee, H. *et al.* (March 2005) Thermal characterization of high performance MCP with silicon spacer having low thermal impedance. Proceedings 21st Annual IEEE Semiconductor Thermal Measurement and Management Symp. SEMI-THERM 2005, pp. 322–326.

22 Popova, N., Schaeffer, C., Sarno, C. *et al.* (2005) Thermal management for stacked 3D microelectronic packages. IEEE 36th Power Electronics Specialists Conference PESC 05, pp. 1761–1766.

23 Brunschwiler, T., Michel, B., Rothuizen, H., Kloter, U., Wunderle, B., Oppermann, H. and Reichl, H. (2008) Forced convective interlayer cooling in vertically integrated packages. Proceedings ITHERM 2008, Orlando.

24 Chen, X., Toh, K. and Chai, J. (2002) Direct liquid cooling of a stacked multichip module. Proceedings Electronics Packaging Technology Conference, pp. 380–384.

25 Koo, J., Im, S., Jiang, L. and Goodson, K. (2005) Integrated microchannel cooling for three-dimensional electronic circuit architectures. *Journal of Heat Transfer*, **127**, 49–58.

V
Applications

Handbook of 3D Integration: Technology and Applications of 3D Integrated Circuits.
Edited by Philip Garrou, Christopher Bower and Peter Ramm

ISBN: 978-3-527-32034-9

34
3D and Microprocessors

Pat Morrow and Sriram Muthukumar

34.1
Introduction

In this chapter we examine the advantages of 3D stacking applied to microprocessors and related integrated microprocessor systems. In general, microprocessors are driving towards lower power consumption, increased performance, reduced form factor and increased integration. 3D technology can enable improvements in all of these areas. For conventional process scaling, the signal delay time (RC) is expected to increase with technology node mostly from the increasing resistance of the wires. In addition, the aggregated interconnect length increases since more wires are being used in each layer and more metallization layers are being added. Hence, for microprocessor systems, it is most important to focus primarily on using 3D to reduce wiring [1–3]. Available 3D integration processing schemes can be roughly grouped by their inter-strata connection pitch capability: tight pitch ($\leq 10\,\mu m$) and loose pitch ($\geq 100\,\mu m$). These two groups are not meant to be all inclusive, but this definition is useful for illustrative purposes. Similarly, for 3D microprocessor applications, we can divide the wire reduction tasks into two basic categories based on the inter-strata connection pitch. The first we call "logic + memory" stacking and generically includes stacking cache, main memory or strata with similar functions onto a high-performance logic device. These applications typically do not require tight inter-strata pitch 3D processing. The second category, which we call "logic + logic" stacking, involves splitting a logic area between two or more strata and requires much tighter pitches than "logic + memory" stacking. The inter-strata pitch therefore is a primary determinant of whether a 3D microprocessor application will work with a 3D integration process and vice versa.

Two primary challenges are faced when integrating 3D stacking with microprocessors. The first is integrating 3D stacking process steps with modern high-performance microprocessor fabrication processes. Areas of concern are 3D processing with strain-enhanced Si devices and low-K dielectrics, both of which are sensitive to stresses. 3D stacking typically involves thinning device layers to less than

Handbook of 3D Integration: Technology and Applications of 3D Integrated Circuits.
Edited by Philip Garrou, Christopher Bower and Peter Ramm

ISBN: 978-3-527-32034-9

100 μm, making devices more prone to stress effects. In addition, through-silicon vias (TSV) as well as the bonding structures are areas where stress effects could potentially be introduced. For example, thermo-mechanical stress may arise from mismatch in the coefficient of thermal expansion (CTE) during 3D stacking. Thus, to ensure minimal impact to device characteristics, the processing approach needs to be considered carefully. The second challenge involves thermal ramifications of the 3D stacked system as stacking effectively increases the transistor density. Thermal concerns are especially important in microprocessors since they have a higher power density than other applications and heat dissipating paths are limited. Although thermals are *potentially* worse in a stacked microprocessor configuration, they have been shown to be manageable through intelligent design [4]. The rest of this chapter reviews some design and fabrication options that best integrate (i.e. least disruptive) 3D stacking concepts with microprocessor systems and address the associated challenges.

34.2
Design of 3D Microprocessor Systems

34.2.1
Introduction

As stated earlier, the main goal is to reduce interconnect length since more than 30% of the power in a microprocessor can be consumed in backend interconnect wire. With 3D integration, multiple strata of different types can be stacked with a high bandwidth, low latency and low power interface. Additionally, wire reduction using 3D provides new microarchitecture opportunities to trade off performance, power and area. One design consideration in 3D microprocessors is to determine the number of active strata that will be integrated. Simple benchmarking analysis can provide some estimation of the relative benefits with increasing number of strata. For example, List *et al.* [5] generically analyzed interconnect performance assuming the average wire length followed a $L_{2D} \times n^{-1/2}$ trend, where n is the number of strata and a L_{2D} is the average interconnect length in the 2D planar layout. More complicated analysis using Rent's rule type models [1, 6, 7] and simulations of 3D circuit blocks across multiple strata [8, 9] show similar trends where it is clear that the largest benefit occurs from stacking the first two layers. In addition to diminishing benefits with increasing number of strata, there are also processing issues that may practically limit the number of strata. For these reasons and for simplicity, the cases presented below show some real situations of implementing 3D layout in a microprocessor across two strata. These go beyond simple estimation and can demonstrate the specific advantages and challenges of eliminating wire with 3D stacking. Figure 34.1 illustrates an example of a two strata 3D stack. This schematic shows the case where two strata are connected face-to-face using metallic bonding structures, but the analysis in the following sections is not dependent on this specific integration scheme.

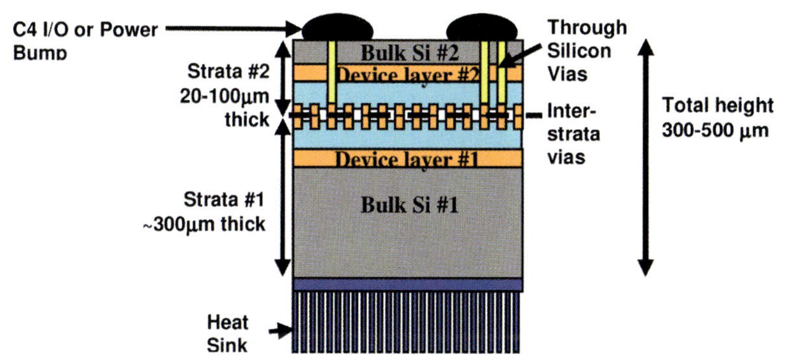

Figure 34.1 Sample 3D structure. The example here shows the case of a bulk-silicon two-strata stack that has been face-to-face stacked with metal bonding.

Case 1: Shortening wires dedicated to off-die interfaces to memory by stacking memory onto logic. This case belongs to the "logic + memory" category. As an example, Section 34.2.2 shows that stacking a large SRAM or DRAM cache on the Intel Core 2 Duo results in higher bandwidth (BW) and shorter latency access to large amounts of storage.

Case 2: Decreasing the length of wires connecting blocks by vertically stacking the blocks. Section 34.2.3.1 shows an example where the microarchitecture of an Intel Pentium 4 family product is split across two die to construct a 3D floorplan. This is in the "logic + logic" stacking category and takes advantage of increased transistor density eliminating wire between blocks of microarchitecture, resulting in shorter latencies between blocks yielding higher performance and reduced power consumption.

Case 3: Decreasing the length of wires within a functional unit block by block splitting. This is also in the "logic + logic" stacking category; Section 34.2.3.2 shows a simple example where a data cache is split into two strata, which also results in reduced latency and reduced power consumption.

34.2.2
Example of Logic + Memory Stacking: Stacked Cache

Some of the advantages of increasing the on-die cache capacity by stacking are increased performance by capturing larger working sets, reduction of off-die bandwidth requirements by accessing more data on die instead of externally, and reduction of system power by reducing bus activity through fewer main memory accesses. Figure 34.2 shows an illustration of the considered structure. Simulations were run to compare the baseline performance of the Intel Core 2 Duo to cases where a large SRAM or DRAM cache was stacked on top. Overall, this resulted in higher bandwidth (BW) and shorter latency access to large amounts of storage. Figure 34.3a shows the baseline microprocessor floorplan overlayed with its power map and Figure 34.3b shows the resulting thermal map. RMS benchmarks were calculated

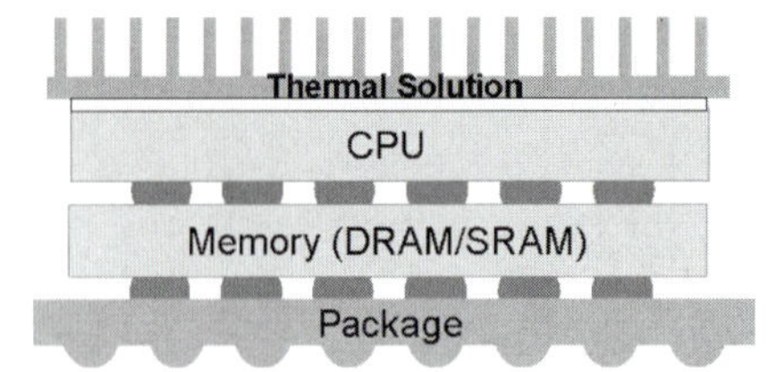

Figure 34.2 DRAM stacked with CPU. The CPU side of the stack is closer to the heat sink.

from a suite of workloads and are described in Reference [2]. Figure 34.4a shows Cycles per Memory Access (CPMA) for average results of 12 RMS benchmarks, comparing the baseline (first bar) to the three stacking options of 12 MB SRAM, 32 MB DRAM and 64 MB DRAM (second to fourth bars). The secondary Y-axis on the graph along with the solid line shows the resulting off-die bandwidth. With stacked cache, the off-die bandwidth is reduced 3× while simultaneously decreasing the CPMA by 13% and yielding a 66% average power reduction in average bus power due to reduced bus activity. A simplifying aspect of stacking DRAM memory on logic is that the thermals are not problematic to manage, primarily because DRAMs consume little power and the hotter die (processor) is close to the heat sink (Figure 34.2). An in-depth thermal calculation of a 92 W microprocessor stacked with a DRAM showed the peak temperature only increased by 1.92 °C (see Figure 34.4b). A more in-depth description of this performance and thermal analysis is found in References [2, 4].

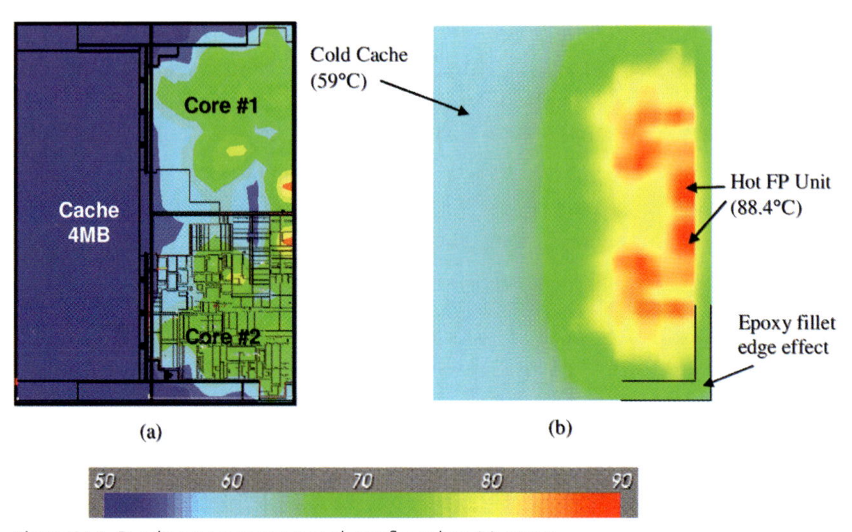

Figure 34.3 Baseline microprocessor planar floorplan: (a) power map; (b) thermal map. (Adapted from Reference [10] and used with permission.)

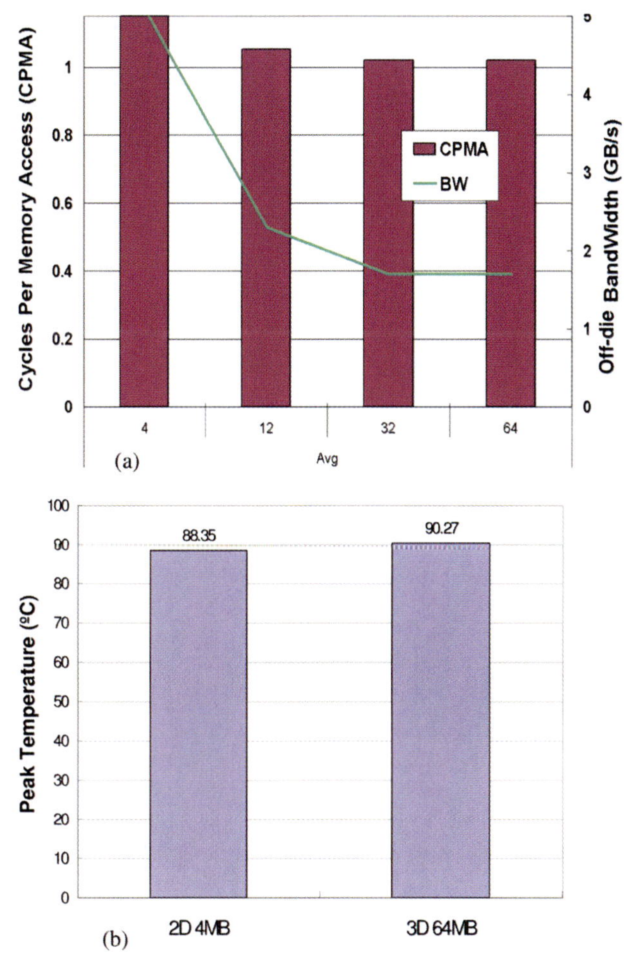

Figure 34.4 (a) Average performance for two thread RMS as cache capacity increases from 4 to 64 MB; (b) peak temperature of 3D DRAM compared to 2D and no DRAM. (Adapted from Reference [10] and used with permission.)

34.2.3
"Logic + Logic" Stacking: Examples of Partitioning a Microprocessor into Two Strata

The scope and flexibility of options for microarchitecture improvements are directly dependent on the inter-strata via pitch and are determined by the process technology. Obviously, via pitches that are too loose will require additional intra-die routing to accommodate around them. A low via density also limits the granularity at which a system can be divided between multiple die. When we partition a microprocessor into strata to reduce interconnect length, it can be considered at two basic levels – splitting at a block level and splitting within a block. The following two sections show examples of this "logic + logic" stacking.

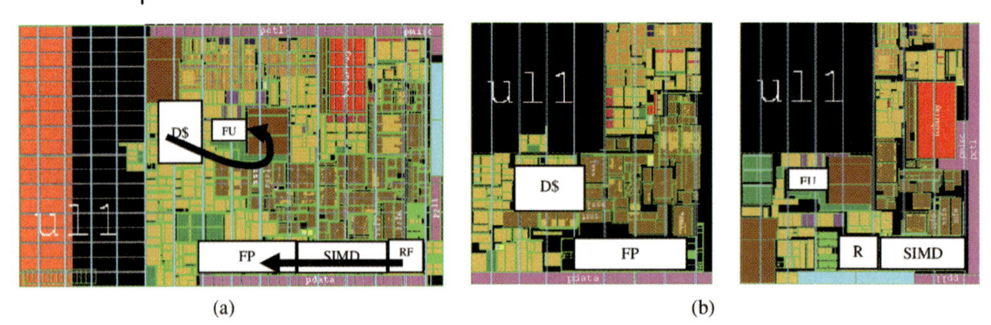

Figure 34.5 (a) Baseline Intel Pentium 4 Family Product 4; (b) 3D floorplan of (a). (Adapted from Reference [10] and used with permission.)

34.2.3.1 "Logic + Logic" Stacking Example 1: Rearranging Functional Blocks into Two Strata

The simplest way to explore the benefits of "logic + logic" stacking is to rearrange the relatively large functional blocks from a planar layout to a stacked layout. An example of how this can be done is shown in Figure 34.5a where an Intel Pentium 4 family processor was split into two strata. The new 3D floorplan, which reduces the footprint by 50%, is shown in Figure 34.5b. Overall, this level of "logic + logic" stacking enabled blocks to be moved closer in proximity, thus reducing the inter-block latency and power. When doing this, it is important to focus much of the layout arrangement effort on known performance sensitive pipelines. One example is load-to-use delay, which is critical for the overall performance of most benchmarks. Figure 34.5 illustrates the path between the first level data cache (D$) and the data input to the functional units (FU). As is seen in Figure 34.5a the worst case path occurs when load data must travel from the far edge of the data cache, across the data cache to the farthest functional unit. Due entirely to planar floorplan limitations, this will yield at least one clock cycle of wire delay. Figure 34.5b shows that a 3D floorplan can overlap the D$ and functional units. In the 3D floorplan, the load data only travels to the center of the D$, at which point it is routed to the other die to the center of the functional units. Now, that same worst case path contains only half the routing distance because the data is only traversing half of the data cache and half of the functional units. This particular case removes one clock cycle of delay in the load-to-use delay. This situation is also favorable thermally because the D$ is a relatively low power block and the new floorplan with the D$ on top of the functional units is in lower power density than the planar floorplan's hottest area over the instruction scheduler. Many other pipestages can be removed using a similar approach, eventually resulting in removal of approximately 25% of all pipestages with a 15% performance improvement from reduction of instruction execution latency. Table 34.1 shows the complete list of pipestages removed from the microarchitecture.

As pointed out earlier, a concern of 3D stacking is the potential doubling of power density and the thermal consequences. This 3D floorplan increases power density by 1.3× and the temperature by 14 °C. This temperature increase may be mitigated

Table 34.1 "Logic + Logic" 3D stacking performance improvement and pipeline changes.

Functionality	% of Stages eliminated	Performance gain (%)
Front-end pipeline	12.5	~0.2
Trace cache read	20	~0.33
Rename allocation	25	~0.66
FP inst latency	Variable	~4.0
Int register file read	25	~0.5
Data cache read	25	~1.5
Instruction loop	17	~1.0
Retire to de-allocation	20	~1.0
FP load latency	35	~2.0
Store lifetime	30	~3.0
Total	~25	~15

since there is a performance gain of 15% along with the 15% power reduction. It is possible to voltage and frequency scale the final results to reach a neutral peak temperature for the 3D floorplan, which results in a 34% power reduction and 8% performance improvement. There are also other frequency and voltage scaling options (Table 34.2). For example, scaling to neutral performance yields a 54% power reduction and scaling for maximum performance yields an almost 30% increase in performance. This example of "logic + logic" stacking in a microprocessor shows how significant advantages can be obtained using only simple block level layout modifications. More in-depth details of this analysis can be found in Reference [2] and additional design optimizations are expected to improve these results further.

Table 34.2 Frequency and voltage scaling the "Logic + Logic" stacked 3D floorplan; conversion equations for temperature, power, Vcc, and frequency are included.[a–c]

	Pwr	Pwr (%)	Temp (°C)	Perf (%)	Vcc	Freq
Baseline	147	100	99	100	1	1
Same Pwr	147	100	127	129	1	1.18
Same freq	125	85	113	115	1	1
Same temp	97.28	66	99	108	0.92	0.92
Same perf	68.2	46	77	100	0.82	0.82

[a] Perf versus frequency: 0.82% perf for 1% frequency.
[b] Freq versus Vcc: 1% for 1% in Vcc.
[c] 3D provides 15% added perf; 15% pwr savings at same frequency.

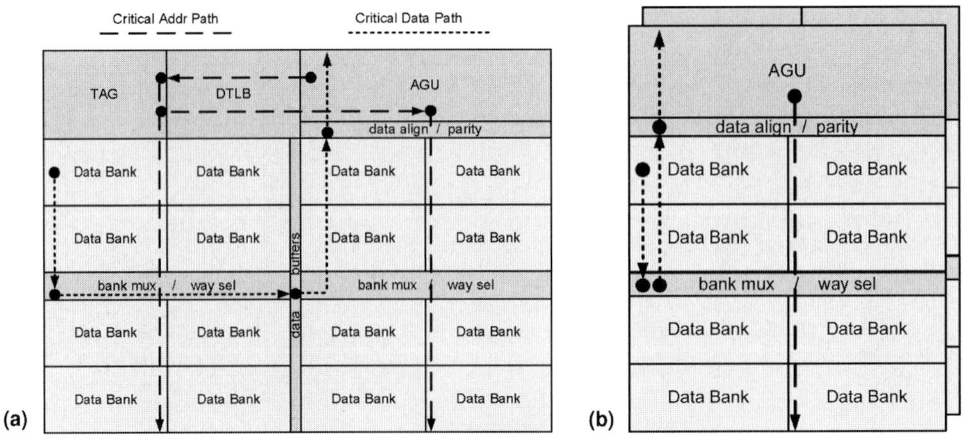

Figure 34.6 First level data cache. (a) Baseline 2D; (b) 3D implementation. (Adapted from Reference [10] and used with permission.)

34.2.3.2 "Logic + Logic" Stacking Example 2: Partitioning a Functional Block into Two Strata

The next level in complexity is to split the microarchitecture of the individual blocks into two strata. A simple example that demonstrates how this can be done is splitting a first level data cache. Figure 34.6a shows the layout of a 32 kB SRAM first level data cache of an IA32 microprocessor. This particular cache is 2 way set and 16 bank data with parity and includes an Address Generation Unit (AGU), Data Translation Lookaside Buffer (DTLB), Tag Address memory (TAG), 16 Data Banks with data bank muxes, way selects, alignment and parity logic and buffering for the long metal data busses. Again, the focus is to reduce the critical timing path, which in this case is from address generation to data result. The critical timing path has several long interconnect routes, as shown by the dashed and dotted lines. When the cache was redesigned in 3D (Figure 34.6b), the new floorplan resulted in approximately 50% of the original footprint and reduced the critical address path significantly by sharing the address path between the top and bottom die. In fact, the horizontal data bus routing and its buffering are completely eliminated. In addition, almost 20% total silicon area is saved because the bank muxes and way select logic in the center of the unit can share devices, primarily from sharing of sense and write circuits between the banks. When the cache was simulated using Intel's 65 nm CMOS technology, the results showed an overall 10% reduction in read latency. Even though there is a small (and expected) increase in the sense time due to increased sense amp sharing, it is more than compensated by the horizontal metal reduction. A third benefit is seen with 25% power reduction coming from the shorter address, select, data wires, reduced clock loading and sharing of clock distribution between the two strata. An in-depth description of these 3D data cache results can be found in Reference [3]. Simultaneous area reduction, power reduction and latency reduction have significant final benefits to a microprocessor. Similar benefits were found when more complicated blocks were redesigned in 3D, and are expected throughout the blocks in a microprocessor design.

34.2.3.3 Concluding Remarks Regarding Design of 3D Microprocessors

These simple cases demonstrate how 3D can be used to improve microprocessors, both in power, performance and area. It is of course possible to simultaneously apply the methodology of all of the above examples (integrating appropriate block arrangement, block redesign, and stacked cache or even stacked main memory) to improve the bottom line value of the microprocessor system. Although the inter-strata via pitch (restricted by strata alignment) is the primary limitation to 3D microarchitecture options, today's available manufacturing equipment and processes already enable reduction in global and semi-global wiring, which typically occupies the top one or two metallization layers.

34.3 Fabrication of 3D Microprocessor Systems

34.3.1 Introduction

As we have seen in earlier chapters of this book, there are many methods and processes to choose from when adapting 3D stacking for microprocessors. Specifically for this application, we desire to choose a process that is simplest to integrate (i.e. least disruptive) with the standard manufacturing flow and enables high inter-strata interconnect density and bandwidth. The first item to choose involves what type of bonding to employ between strata. Bonding techniques can be organized generally into "dielectric bonding" and "metallic bonding" (see, for example, References [11–15]). Multiple options for dielectric bonding processes exist [12–14], but a major disadvantage of these techniques is that each inter-wafer connection requires a through-silicon-via (TSV). Figure 34.7a shows the dilemma where active device area is consumed by each inter-strata connection made by a TSV, ultimately resulting in an expensive sacrifice of device density for inter-wafer bandwidth [2]. In contrast, for the case of metal bonding (Figure 34.7b), the bonded inter-wafer connections do not consume active device area, enabling a significantly higher signal bandwidth between device layers. Of course, TSV are still needed for power delivery, chip I/O and signal routing through the device strata, as well as for inter-wafer connections for multi-strata stacking if desired. Even after taking this into account, metal bonding integration still consumes less device area than dielectric bonding and hence has a via density and bandwidth advantage over dielectric bonding. The diameter of the TSV is another variable that impacts the amount of consumed device layer. Although smaller diameter TSV reduces stacked device layer consumption the minimum diameter is fabrication limited, as determined by the minimum thickness of the strata layer due to via-filling aspect ratio. If the process flow includes handling of thin wafers or thin die, the minimum strata layer thickness can in turn be defined by these handling

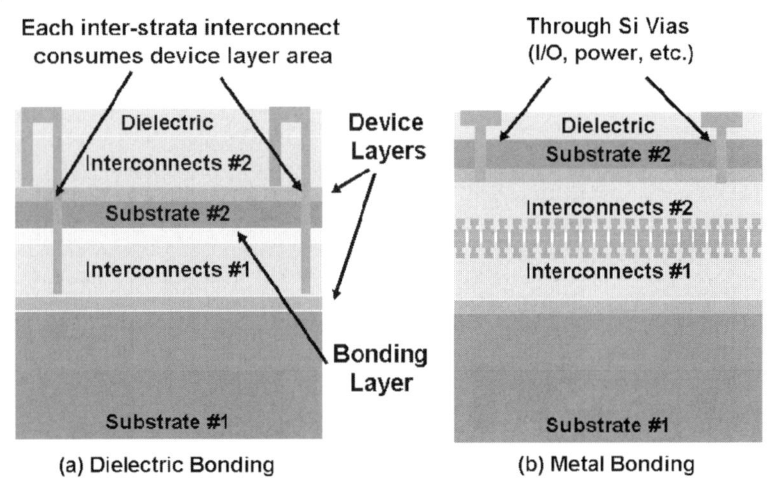

Figure 34.7 General schematics comparing dielectric bonding (a) with metal bonding (b). Each inter-strata connection for dielectric bonding requires a TSV. Each TSV consumes active area, which limits the device density. Inter-strata vias for metal bonding do not require TSV. Metal bonding only requires TSV for power, I/O, and providing connections for multi-strata stacking.

requirements, which becomes more problematic for very thin strata. Finally, important design parameters such as power delivery and reliability requirements also impose design rules on minimum TSV sizes. Thus, balancing device area consumed by TSV with these other requirements will ultimately define their size.

The decision on whether to do the final bonding at a die to die level, a die to wafer level or a wafer to wafer level depends on which microprocessor application is targeted. Die level stacking is probably the best choice for "logic + memory" applications for the following reasons: First, die stacking enables stacking of different size die, which makes it flexible when selecting an SRAM or DRAM size to be matched to a microprocessor. Second, memory stacking does not require very tight inter-strata pitches and hence alignment throughput will be less of an issue. Third, pre-testing both the microprocessor and the memory die is relatively straightforward, and one should be able take advantage of the known good die (KGD) yield benefit provided by die stacking. In contrast, wafer stacking may be a better choice for "logic + logic" applications because of its throughput advantages, especially when very precise alignment (tight via pitch) is required, as is the case when splitting a microprocessor. One requirement for wafer stacking is that the dies need to be the same size, and, as seen from the examples in Section 34.2.3, that is usually the case for microarchitecture splitting. Furthermore, when the logic is split, it may be not be as straightforward to test the individual pieces and select them based on KGD for yield improvement, which is another reason for die stacking to be less compelling with this application. However, notably, our definition of "logic + logic" stacking here does not refer to system-on-chip (SOC) type architectures, which have

Table 34.3 Via processing options for 3D stacking and microprocessors.

TSV attributes impacting microprocessor	Via first approach	Via last approach
Via filling materials	W. Cu not a very viable option due to high-temperature annealing steps in front-end CMOS	Cu, W. Cu is preferred from a cost perspective
Inter-strata pitch	Both tight pitch ($\leq$10 µm) and loose pitch options ($\geq$100 µm)	Both tight pitch ($\leq$10 µm) and loose pitch options ($\geq$100 µm)
TSV size options	~5 µm diameter preferred. Larger via diameters will impact downstream processing	~5 µm diameter for tight pitch options and 15–60 µm diameters for loose pitch options
Dielectric materials	CMOS based	CMOS based
Ease of integration to microprocessor manufacturing flow	Difficult. Contamination to line and impact to transistor scaling	Easy. Feasible using existing back-end process technology options
Cost of manufacturing	Higher	Potentially lower
3D stacking options	Wafer stacking only. Ultrathin die (<25 µm) handling in assembly is most expensive	Both wafer-stacking and die-stacking feasible
Multi-chip/multi-device 3D integration	More complicated. Design rules need to be consistent across chips/devices	Simpler. TSV design rules mostly independent of device design

multi-functional device integration, but the required interconnection bandwidth belongs to the loose pitch category. Finally, wafer stacking enables the benefits of thinner strata and reduces the burden of handling unsupported ultrathin die or wafer during assembly/packaging.

Another process choice is to decide whether the TSV will be made using a "Via-First Approach" or a "Via-Last Approach." Table 34.3 compares TSV attributes based on these approaches. "Via-First Approach" has the TSVs fabricated along with the fabrication of active circuitry and prior to thinning, dicing and assembly while "Via-Last Approach" has the TSVs fabricated after the fabrication of active circuitry and back-end layers but prior to dicing and assembly. One advantage of the "Via-First Approach" is that it minimizes the thin wafer handling and processing steps [16–18] However, its major disadvantage is that this type of integration disrupts the standard process flow, typically requiring significant process reengineering and adding constraints on design rules of transistor scaling. In contrast, fabricating TSVs after completion of the active circuitry, as in the "Via-Last Approach," avoids conflicts with the standard process flow and allows for more design options in defining TSV shape as well as how they contact to the circuitry. Since the standard fabrication process for microprocessors is long, very complicated and expensive, there is a major advantage in using the "Via-Last Approach" and avoiding all the issues associated with having to reengineer parts of the standard process. The "Via-Last Approach" also does not require use of the expensive equipment and clean room protocols used in fabricating active circuitry, thus reducing overall manufacturing costs.

The next two sections describe two process flows which were targeted specifically for 3D microprocessor applications. Section 3.2 describes a wafer stacking process designed for logic splitting and Section 3.3 describes a die stacking process targeted for stacking memory on logic. Both processes use "Via-Last Approach" and fabricate the TSV at a wafer level for cost and throughput. As expected, the TSV sizes are very different: TSV for die stacking are on the order of 15–60 μm and on the order of 5 μm for the wafer stacking process. Similarly, the inter-strata via pitch are tighter for wafer stacking (<8 μm) and about typical flip-chip bump layer-size pitch (>100 μm) for the die stacking process.

34.3.2 Wafer Stacking Using Copper Bonding

34.3.2.1 Copper Bonding Process

For the microprocessor logic splitting application that requires a very high density of inter-strata vias, it becomes paramount that the quality and yield of the connections be good. Wafer level copper bonding is one option that has been shown to provide high yield and good quality connections with no detectable interface resistance [11]. Experiments with this process were completed using 300 mm wafers to demonstrate compatibility with today's microprocessor fabrication processes. The basic process flow for making a two-layer structure with wafer level copper bonding is shown in Figure 34.8. As this is a via-last process flow, the wafers are finished up to a final metal layer that will become the bonding layer between strata. The bonding structures are finalized by a slight recess of the oxide to ensure contact with the counterparts. The wafers are then bonded face to face, the top one is thinned down to a final pre-defined thickness (typically in the range of 5–30 μm) and TSV are fabricated. The thinning

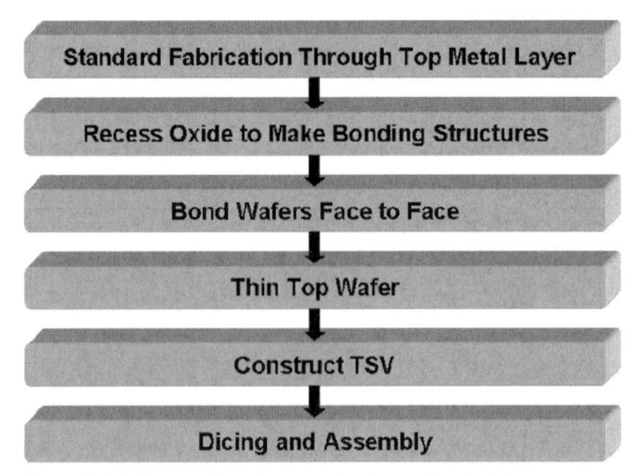

Figure 34.8 Process flow diagram for the wafer stacking process example. The flow is for two-strata.

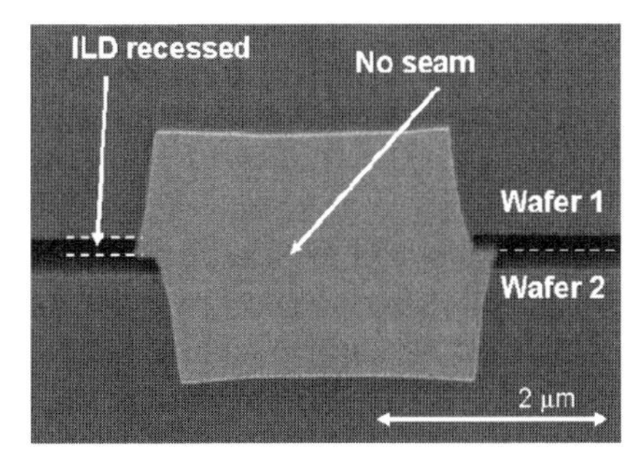

Figure 34.9 Cross section of a bonded structure showing no interface seam between the two pieces of the structure. The ILD is recessed slightly to ensure contact between the two pieces of metal.

process also includes a sub-threshold damage removal step after grinding. Additional discussions on sub-threshold damage are provided in References [19, 20]. Since significant power needs to be delivered to the microprocessor, copper was chosen for its low resistance as the TSV conducting material. Figure 34.9 shows a cross section of a structure bonded under optimum conditions and indicates good bonding by the lack of an observed seam or gap layer. Figure 34.10 shows a CSAM (C-Mode Scanning Acoustic Microscopy) image, demonstrating that good bonding can be achieved across a 300 mm wafer. Figure 34.11 shows a sample distribution of measured inter-wafer chain resistances. In the graph, each point is a chain of 4096 inter-wafer links where the bonding structure pitch was about 9 μm. In addition to the measured data, the graph shows the predicted resistance calculated from measured interconnect dimensions, as well as the resistance values calculated using measured dimensions with a ±5% Cu height variation in one of the layers. The distributions are tight and are within the expected range, indicating no detectable interface resistance. Various sizes and pitches of bonding structures were studied and it is concluded that the

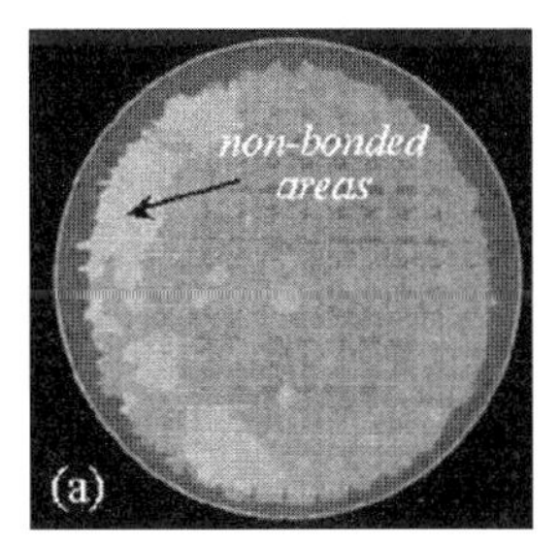

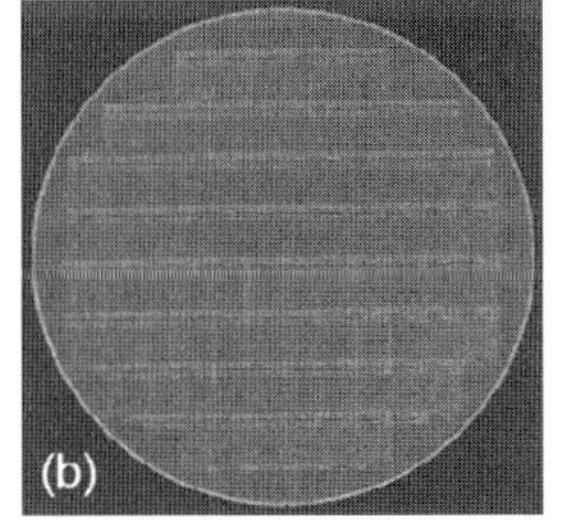

Figure 34.10 CSAM images of 300 mm copper bonded wafers. Processed under (a) non-optimal conditions, showing areas that failed to bond, and (b) optimal conditions, showing good bonding across the 300 mm wafer. (Adapted from Reference [11] and used with permission.)

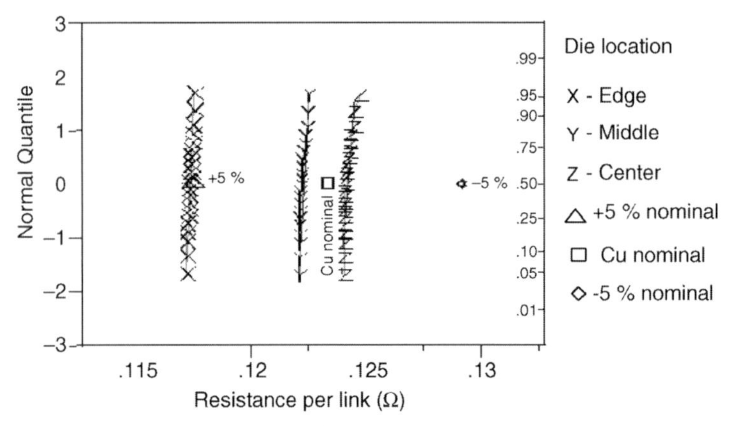

Figure 34.11 Sample distribution of the resistance per link of chains with 4906 links each, from three die on a bonded pair of 300 mm wafers. The die are labeled Center, Middle and Edge, which refers to their position with respect to the wafer center. The nominal resistance is calculated from structure geometries and two calculated reference points are shown where thickness is varied by ±5%. From this, it is concluded that the bonding interface resistance is negligible. (Adapted from Reference [11] and used with permission.)

minimum pitch of the inter-strata vias is limited by the consistent cross-300 mm wafer alignment capability of the equipment, which is expected to improve as tooling advances.

34.3.2.2 **Device Wafers Stacked with Copper Bonding Technology**

Since wafer level copper bonding looks attractive from a connection point of view, the next step is to look at its compatibility with circuitry. To do this, device wafers were fabricated on the same process technology used in Intel's 65 nm products and then used to test the effect of stacking. This process technology includes strained-silicon devices built on bulk-silicon and low-K dielectrics. A cross sectional image of a fabricated stack and a schematic of the final structure are shown in Figure 34.12. Device and circuit tests showed there was no degradation found in individual N- and P-channel MOSFETS [21] or simple ring oscillator circuits [4]. Figure 34.13 shows some representative $I_{\mathrm{off}} - I_{\mathrm{on}}$ and $I_{\mathrm{off}} - V_{\mathrm{t}}$ data comparing stacked and thinned (14, 19 μm) wafers with non-bonded baseline wafers. The outliers were due to non-optimal backside patterning of some of the testing pads. Larger functional blocks in the form of 4 Mb SRAMs were also tested in the Cu-stacked configuration and also found to function with similar performance as those in non-stacked wafers. The SRAM data were confirmed for wafers thinned down to a thickness of 5 μm [21].

34.3.2.3 **Wafer Stacking with Copper Bonding Conclusions**

From these data it was concluded that this wafer level stacking process with copper bonding can provide the tight pitches required for logic splitting and is compatible with modern microprocessor fabrication processes. More details on the specific pitch requirements for the processor are described in References [2, 4]. Finally, dicing of

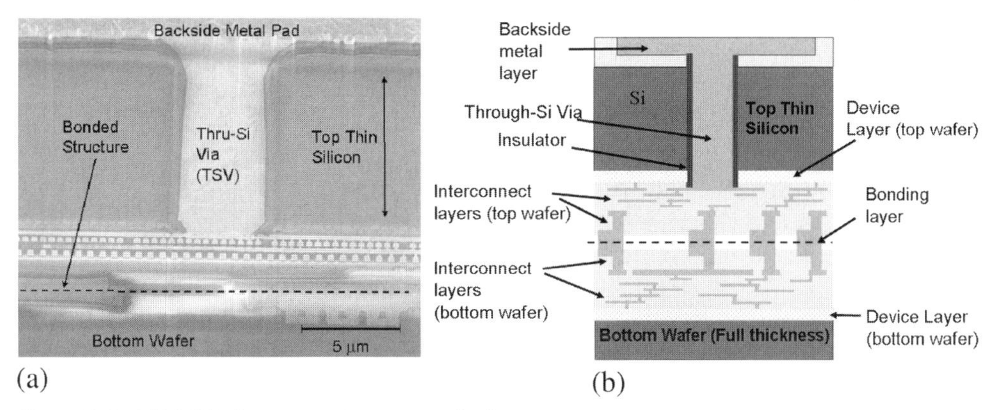

Figure 34.12 (a) XSEM of representative structure (wafer stacking); (b) schematic. (Adapted from References [10, 21] and used with permission.)

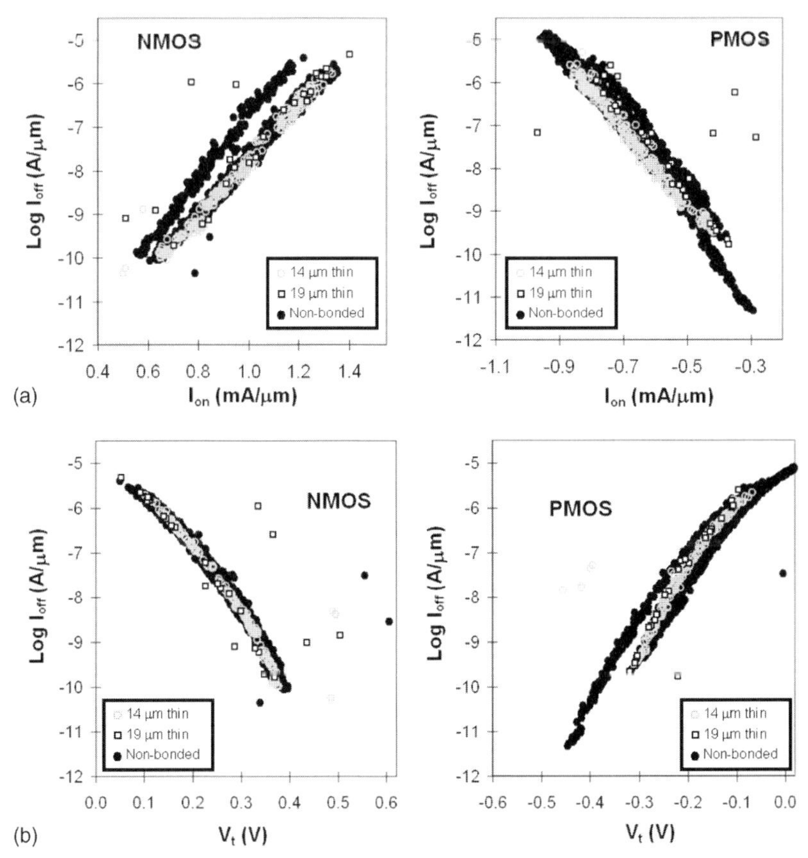

Figure 34.13 $I_{\rm off} - I_{\rm on}$ (a) and $V_{\rm t} - I_{\rm off}$ (b), comparing devices in thin stacked wafers and non-bonded wafers. (Adapted from References [10, 21] and used with permission.)

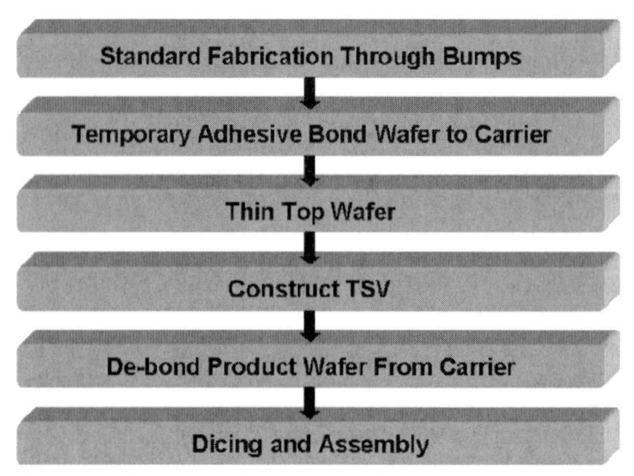

Figure 34.14 Process flow diagram for the die stacking process example.

the two-strata wafer stack was also demonstrated and produced an edge quality comparable with dicing of non-bonded wafers.

34.3.3 Die Stacking via Metal Bonding

A die stacking process also using via-last TSV fabrication was developed with many similar steps as the wafer stacking process. Figure 34.14 shows an example of a flow. Conceptually, the biggest difference in this flow compared to wafer stacking is the temporary bonding of the device wafer to a wafer support system (WSS) instead of a permanent bond between two device wafers. The WSS enables handling of thin wafers (<100 μm) through the TSV fabrication steps, which are similar to the TSV processing steps used in the wafer bonding process. The WSS includes a carrier, which can be either glass or silicon and a strippable adhesive for carrier separation after TSV fabrication and prior to dicing and assembly. The adhesive needs to have high thermal stability, high modulus for post grind quality and an ability to be stripped at end-of-line TSV processing. Figure 34.15 shows an example of post grind

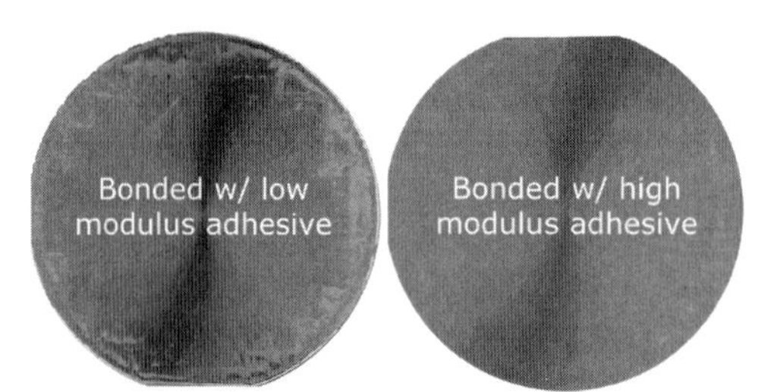

Figure 34.15 Silicon grind quality using low modulus and high modulus adhesives. (Adapted from References [10, 23] and used with permission.)

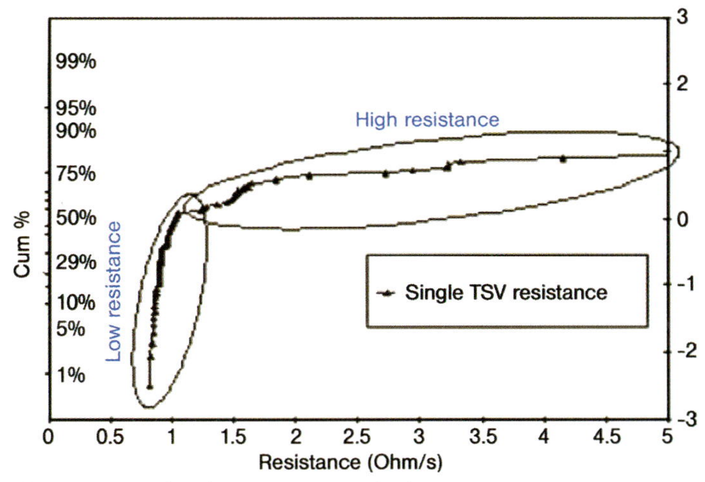

Figure 34.16 Single Kelvin TSV resistance landing on tungsten contacts. The low resistance group agrees with calculations and the high resistance group arises from known issues with a non-optimized wafer level process. (Adapted from References [10, 23] and used with permission.)

quality, comparing low and high modulus adhesives. Additional details related to WSS and adhesives can be found in Reference [22]. Since this die stacking process targets looser inter-strata pitch applications such as stacking memory on logic, the TSV can be made significantly larger ($\sim$15 μm) than those used in the wafer stacking process ($<\sim$5 μm). The inter-strata pitch limited by typical flip-chip bump sizes >100 μm relaxes some of the TSV processing requirements on alignment tolerances.

Although it is straightforward to land TSV on copper metallization, as was done in the wafer stacking example in Section 34.3.2, it is also possible for the TSV to land on an array of tungsten contacts, which for some applications has layout advantages. Figure 34.16 shows a typical wafer level distribution of individual TSV resistance measured using a Kelvin structure where the TSV landed on an array of tungsten contacts. The measured low resistance distribution agrees with theoretical calculations and the high resistance portion resulted from non-uniform wafer processing issues associated with the current non-optimized process. Capacitance from a line load of ten TSVs was measured to be less than 1 pF and was limited by the resolution of the instrumentation.

Since these TSV used in the die stacking process are relatively large, it is important to understand their effect on devices as a function of proximity to establish "keep out zones." TSVs were fabricated on an Intel 90 nm silicon device wafer at increasing distances to paired planar NMOS discrete transistors (Figure 34.17). Saturation current (I_{DSAT}) and threshold voltage (V_T) data were collected prior to TSV processing and compared with post processing data. Figure 34.18 shows a sample wafer level normalized plot of shift in I_{DSAT} and V_T for both close and far proximity from the TSV.

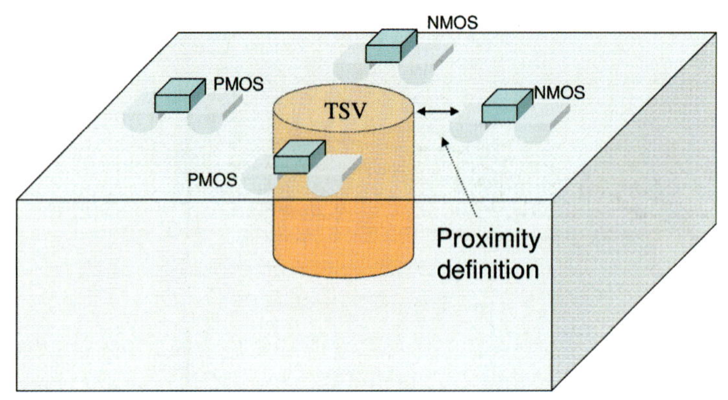

Figure 34.17 "Keep-out" zone (KOZ) requirements were determined by placing TSVs within arrays of discrete planar transistors.

In general, there may be a slightly stronger TSV proximity effect on short channel devices compared to long channel devices but, overall, the shifts in device parameters were negligible in all cases. The tail distributions for the measurements are again due to non-optimized process uniformity issues. Notably, negligible shifts are observed even without a subsurface damage removal step. Based on this, it is concluded that keep out zones will have an insignificant impact on die area.

Multi-strata stacks with TSV were assembled using this die stacking approach, integrating traditional copper-solder joints and copper-tin-copper thermo-compression interconnects. Figure 34.19a shows a representative cross-sectional view of a seven-die stack assembled in a front to back configuration. Figure 34.19b shows the completed stack including underfill. In this example, each die was about 75 μm thick. More details of the die stacking process can be found in Reference [23]. Finally, Figure 34.20 shows a 3D stacked Microprocessor on a SRAM die prototype fabricated using the die stacking process and currently in testing at Intel [24]. The image shows the thin SRAM die in the center of image with the many core processor die stacked on top and the package connection at the bottom. In this case the two active die are stacked face to face and the interconnections are made with typical C4-type processing, including epoxy underfill and solder.

34.4 Conclusions

3D is an exciting new process technology that can provide substantial power and performance benefits to microprocessors, even with simple design modifications. This chapter shows some specific examples of how to implement 3D in microprocessors but more investigation is needed to explore fully what is capable from microarchitecture, affordability and HVM manufacturing scalability perspectives. As shown in several earlier chapters in this book, there are many 3D processes available

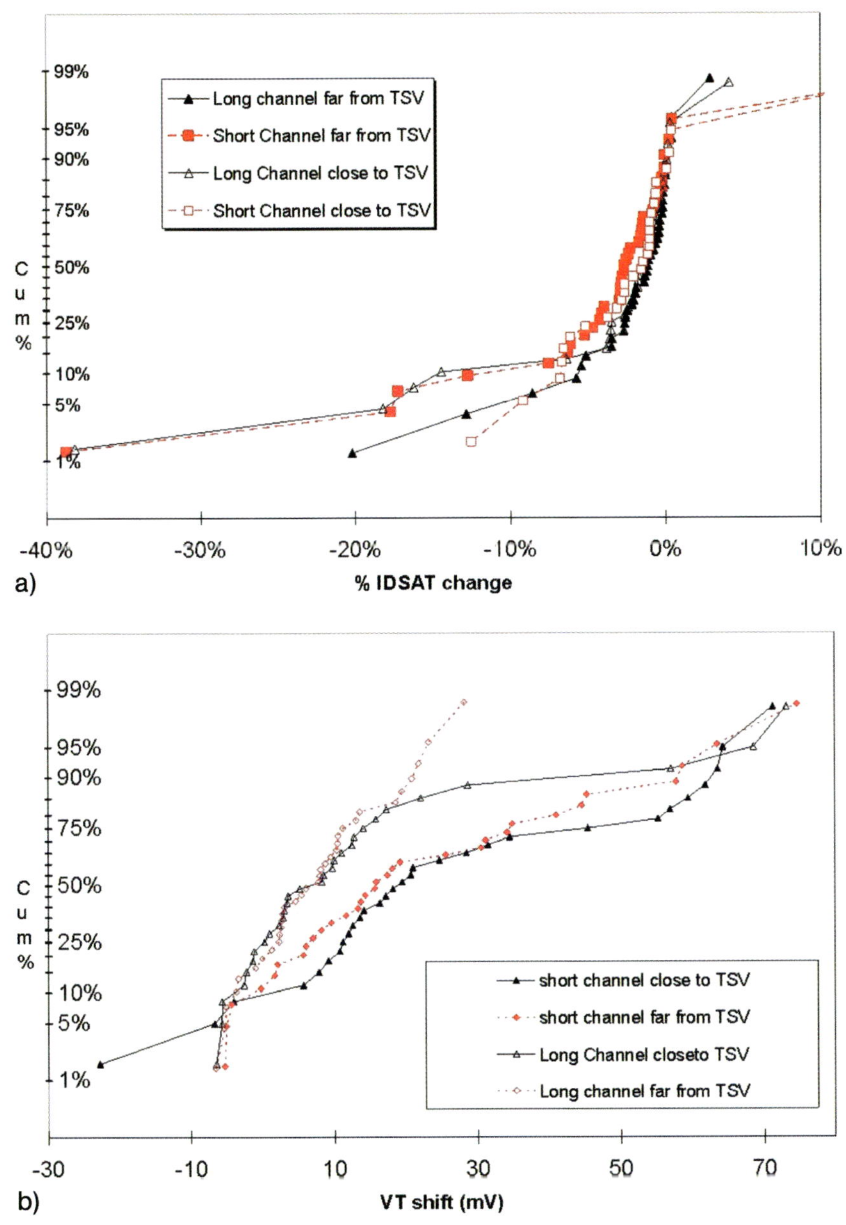

Figure 34.18 I_{DSAT} change (a) and V_T shifts (b) for short and long channel n-MOS relative to proximity of TSV. (Adapted from References [10, 23] and used with permission.)

to choose from, but it has been determined that the two process technology examples reviewed here are suited to integration with Intel's microprocessor technology. Immediate applications integrating 3D stacking onto microprocessors are targeted at "logic + memory" applications [24]. Looking towards the future, tooling used for

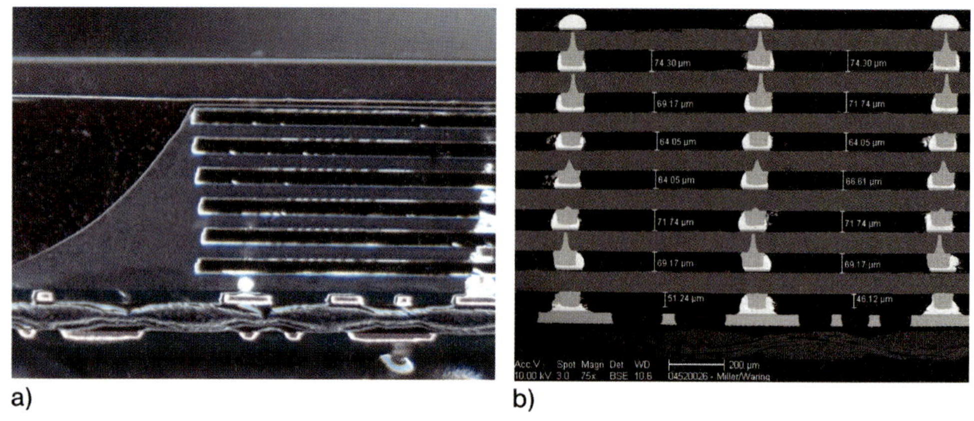

Figure 34.19 Cross-section of a seven-die assembly stacking. (Adapted from References [10, 23] and used with permission.)

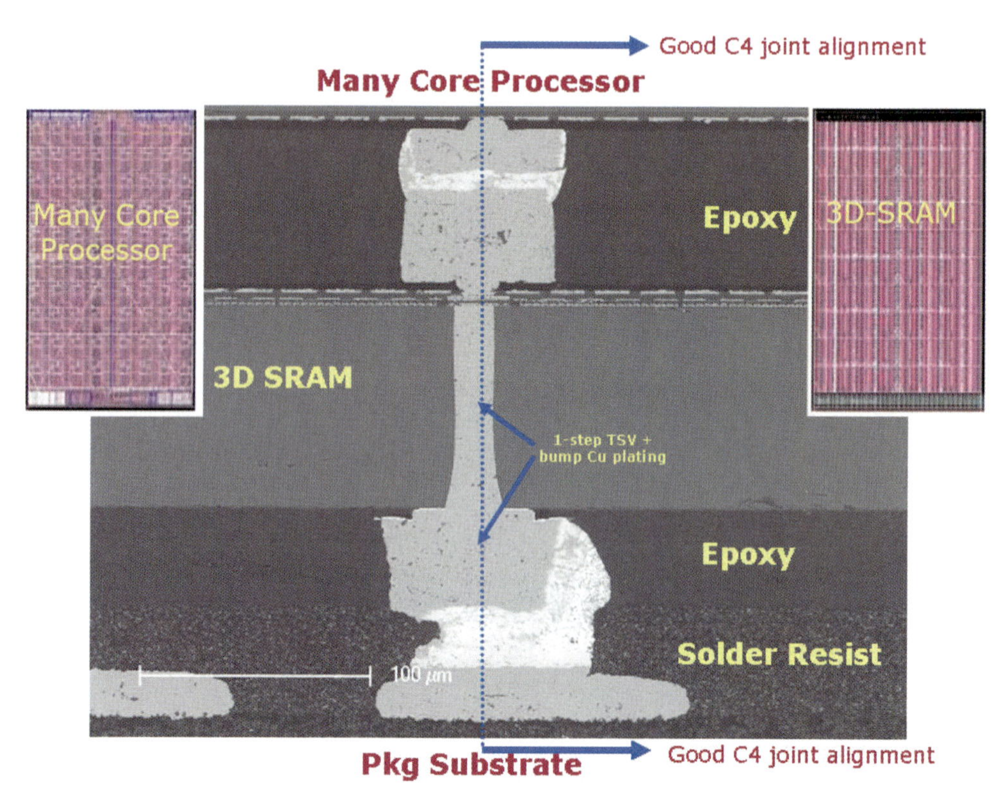

Figure 34.20 3D stacking prototype of Many Core Processor on SRAM. The prototype uses Intel's 80 core processor die face-to-face stacked with a SRAM die that is connected to the package substrates through TSVs. 3D stacking here helps connect the cores on die to its own memory partitions. (Source: Intel.)

alignment is expected to improve significantly, enabling an increase in density of inter-strata vias. As this capability increases, even more microarchitecture improvements will become possible.

References

1 Nelson, D.W., Webb, C., McCauley, D. *et al.* (September 2004) A 3D interconnect methodology applied to iA32-class architectures for performance improvement through RC mitigation. Proceedings of the VMIC, pp. 78–83.

2 Annavaram, M., Black, B., Brekelbaum, N. *et al.* (December 2006) Die stacking (3D) microarchitecture. Proceedings of the 39th Annual International Symposium on Microarchitecture.

3 Reed, P., Yeung, G. and Black, B. (2005) Design aspects of a microprocessor data cache using 3D die interconnect technology. *Proceedings of the ICICDT*, 15–18.

4 Black, B., Brekelbaum, N., DeVale, J. *et al.* (February 2007) 3D design challenges, Proceedings of the International Solid State Circuits Conference. (in press).

5 List, S., Bamal, M., Stucchi, M. and Maex, K. (2006) A global view of interconnects. *Microelectronics Engineering*, **83** (11–12), pp. 2200–2207.

6 Rahman, A. and Reif, R. (Dec. 2000) System-level performance evaluation of three-dimensional integrated circuits. *IEEE Transactions on very Large Scale Integration. Special Issue System-Level Interconnect Prediction*, **8**, 671–678.

7 Das, S., Chandrakasan, A.P. and Reif, R. (2004) Calibration of Rent's Rule models for three-dimensional integrated circuits. *IEEE Transactions on Very Large Scale Integration Systems*, **12**, 359–366.

8 Puttaswamy, K. and Loh, G.H. (March 2006) Implementing register files for high-performance microprocessors in a die-stacked (3D) technology. IEEE Annual Symposium on Emerging VLSI Technologies and Architectures, 2006, Vol 00.

9 Puttaswamy, K. and Loh, G.H. (2006) Dynamic instruction schedulers in a 3-dimensional integration technology. Proceedings of the 16th ACM Great Lakes Symposium on VLSI, pp. 153–158.

10 Morrow, P., Black, B., Kobrinsky, M.J. *et al.* (2007) Design and fabrication of 3D microprocessors. MRS Proceedings, Volume 970, Enabling Technologies for 3D Integration (eds C. Bower, P. Garrou, P. Ramm and K. Takahashi), Materials Research Society, pp. 91–103.

11 Morrow, P.R., Kobrinsky, M.J., Ramanathan, S. *et al.* (2004) Wafer-level 3D interconnects via Cu bonding. Proceedings of the Advanced Metallization Conference, pp. 125–130.

12 Kwon, Y., Yu, J., McMahon, J.J. *et al.* (2004) Evaluation of thin dielectric-glue wafer-bonding for three-dimensional integrated circuit applications. *Materials Research Society Symposium Proceedings*, **812**, F6.16.1.

13 Gutmann, R.J., Lu, J.-Q., Pozder, S. *et al.* (October 2003) A wafer-level 3-D IC technology platform. Proceedings Advanced Metallization Conference, pp. 19–26.

14 Topol, A.W., Furman, B.K., Guarini, K.W. *et al.* (2004) Enabling technologies for wafer-level bonding of 3D MEMS and integrated circuit structures. Proceedings 54th Electronic Components and Technology Conference, vol. 1. pp. 931–938.

15 Tan, C.S. and Reif, R. (2005) Silicon multilayer stacking based on copper wafer bonding. *Electrochemical and Solid-State Letters*, **8**, G147–G149.

16 Tanida, K., Umemoto, M., Tomita, Y. and Tago, M. (2003) Ultra-high-density 3D chip stacking technology. Proceedings 53rd

Electronic Components and Technology Conference, pp. 1084–1089.

17 Hara, K., Kurashima, Y., Hashimoto, N. *et al.* (Aug. 2005) Optimization for chip stack in 3-D packaging. *IEEE Transactions on Advanced Packaging*, **28** (3), 367–376.

18 Andry, P.S., Tsang, C., Sprogis, E. *et al.* (May 2006) A CMOS-compatible process for fabricating electrical through-vias in silicon. 2006 Proceedings 56th Electronic Components & Technology Conference.

19 Pei, Z.J., Billingsley, S.R. and Miura, S. (Jul. 1999) Grinding induced subsurface cracks in silicon wafers. *International Journal of Machine Tools & Manufacture*, **39** (7), 1103–1116.

20 Sandireddy, S. and Jiang, T. (April 2005) Advanced wafer thinning technologies to enable multichip packages. *WMED*, 24–27.

21 Morrow, P.R., Park, C.-M., Ramanathan, S. *et al.* (2006) Three-dimensional wafer stacking via Cu-Cu bonding integrated with 65 nm strained-Si/low-k CMOS technology. *Electron Device Letters*, **2** (5), 335–337.

22 Kulkarni, S., Prack, E., Arana, L. and Bai, Y. (March 2006) Evaluation of adhesive wafer bonding and processes for 3D die stacking using TSV technologies. International Conference on Device Packaging, Scottsdale, AZ, USA.

23 Newman, M. *et al.* (May 2006) Fabrication and electrical characterization of 3D vertical interconnects. 2006 Proceedings 56th Electronic Components & Technology Conference.

24 http://cache-www.intel.com/cd/00/00/33/04/330426_330426.pdf (accessed july 2007).

35
3D Memories

Mark Tuttle

35.1
Introduction

Over the past 25 years, engineers have labored to build the smallest, and thereby, cheapest integrated circuit (IC) memories possible. The wafer processing cost for a given technology node is relatively fixed; therefore, obtaining more ICs per wafer ensures the lowest cost. To accomplish this year after year, the main approach has been to aggressively shrink the minimum cell area at a break-neck pace on the order of 30% per year, along with a commensurate vertical scaling of devices (Figure 35.1). For many years, the dynamic random access memory (DRAM) was the minimum feature technology driver; however, as recently as 2005, the new focus has become the Not-AND (NAND) flash memory product family. This turn in attention is due to density benefits and the simpler cell layout realized by the smaller NAND memory element. Because a memory IC may have billions of memory cells, there is an extreme need to optimize the minimum feature size and layout of each memory cell. With the level of engineering and innovation required to remain competitive in the memory market, it is easy to understand that using 3D integration to build memory structures in the 3rd dimension and thus attain large gains in memory density is an attractive proposition.

3D integration for memory can be as straightforward as physically stacking and electrically connecting the leads from individually packaged parts together, or as complicated as building memory cell elements on different layers of single crystal silicon interconnected at the memory cell level. Between those extremes, there are many approaches for 3D integration, with associated advantages and disadvantages for each.

35.2
Applications

The ultimate die size achieved for a memory IC is related to several factors, including the creativity of the circuit and layout designers, the minimum feature

Handbook of 3D Integration: Technology and Applications of 3D Integrated Circuits.
Edited by Philip Garrou, Christopher Bower and Peter Ramm

ISBN: 978-3-527-32034-9

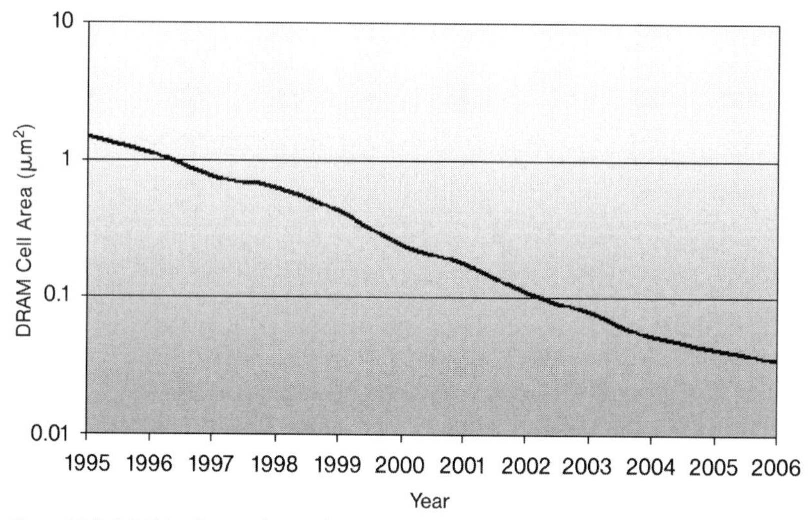

Figure 35.1 DRAM cell area change from 1995 to 2006.

size available for processing, and the technology used for packaging. The pricing pressure on memories is intense, and is the driving force behind exacting the most dice from each wafer, ultimately producing a small, packaged part for the customer. In addition to low cost, however, customers have various needs for smaller size, higher performance and high reliability. From a DRAM memory perspective, performance needs will drive the demand for compact 3D structures. As DRAM data rates climb above 1 Gbit s^{-1}, this becomes much more important due to the electrical parasitics, which tend to be substantial for conventional stacking structures using wire bonding. For NAND memory, the main market pressure for 3D comes from size rather than data rate performance, which is about an order of magnitude lower than DRAMs. To fit eight (and more) NAND ICs in one package for the latest mobile products it is critical to minimize stacking height, as well as overall length and width.

There are two levels of interconnectivity, or stacking, that can be employed to improve both the performance and size of the final memory package. The first approach is the use of intra-circuit connections, where one layer of device structures (possibly including transistors) is disposed over the original IC substrate. This approach requires very small vias, on the order of the minimum feature size, to take full advantage of this technique and make interconnections on array pitch. Larger vias could be used, but would need to occur much less frequently to prevent the die size from becoming overly large. The purpose of this type of stacking would be to optimize each layer for its unique characteristics. An example would be to build a DRAM memory access transistor on one substrate, and build the capacitor layer on a separate substrate that is electrically connected with a via, and to bond the two substrates together. Another example would be to build row and column decoders, and other peripheral devices, on an additional layer, shrinking the overall external area required for the final device. Neither layer would have a fully-functioning IC, but together they would form the completed IC. This concept is particularly interesting for bringing together widely different technologies, for example, pairing a silicon-based controller chip with a sensor made out of a non-silicon substrate, such as a II–VI compound or other exotic material. The vias could be as small as required for memory pitched cells, or larger for the connection of whole circuits built on multiple levels.

The second, more imminent, approach is the use of inter-chip connections, but on a much smaller scale than wire bonding or stacking packages will allow. The use of relatively large through wafer interconnects (TWIs), also called through silicon vias (TSVs), on the order of 40 μm in diameter, to facilitate the connection of the bond pads vertically, produces a very small interconnection length and a very small final package thickness. Short leads improve the electrical parasitics associated with the vias, such as resistance, capacitance and inductance. In addition, incorporating a thinned silicon substrate will substantially reduce the final package height, as well as keep the TWI structure even shorter.

Although individual DRAM memory ICs are commonly configured in widths of X4, X8 and X16, the final system memory density in many applications is increased by connecting multiple ICs together on a printed circuit board (PCB) called a memory module, thus increasing the total memory available (Figure 35.2). A typical DRAM memory module for a work station would have 18 or 36 ICs configured as X4 to achieve a 72 bit word (64 bit wide data, plus error correction). This connection can be done by placing 18 individually packaged ICs on a PCB. However, placing 36 chips for the higher density increases the profile of the module and often requires a stacked memory structure.

There are numerous IC stacking options, including package on package (PoP) stacking and multiple chip package (MCP) stacking, where the IC chips are stacked prior to encapsulation. The increase in memory density for a given area or volume is a clear advantage for mobile products and others that are space constrained. The MCP can be additionally broken down into (a) those chips that are electrically connected with wire bonding (Figures 35.3 and 35.4) or external surface interconnection techniques and (b) the TWI approach (Figure 35.5).

Since memory is not a stand-alone component and requires other integrated circuits for interfacing and output there is a natural inclination to build multiple types of electrical function on one chip. Although successful at lower levels of integration, integrating a large microprocessor with a large amount of memory poses major

Figure 35.2 Micron modules.

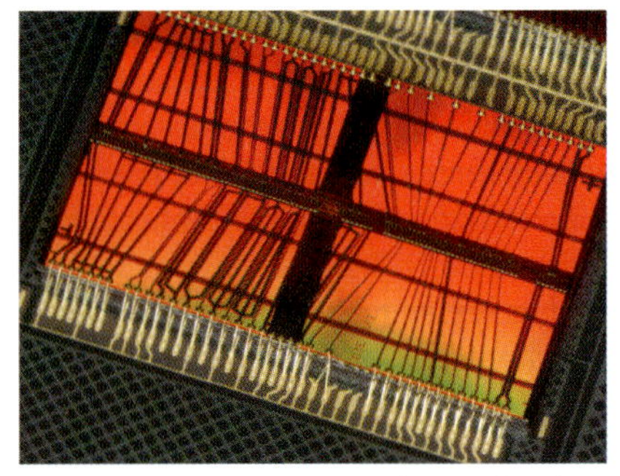

Figure 35.3 Micron COB two-dice stack with RDL to edge.

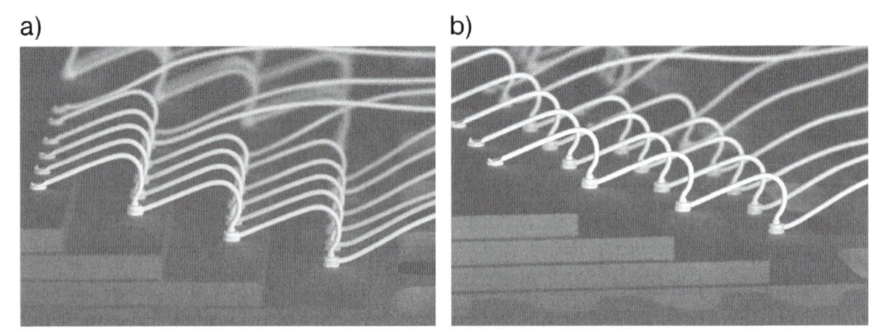

Figure 35.4 Micron NAND QDP.

difficulties as the optimum process technology for fast logic transistors and for low leakage memory access transistors is very different. Consequently, it does not make functional or economic sense to build them on the same two-dimensional piece of silicon. System in package (SiP) is the combination of multiple technologies in one package – a package that can also include non-IC components, such as resistors, capacitors, quartz crystals and even lenses for optical systems.

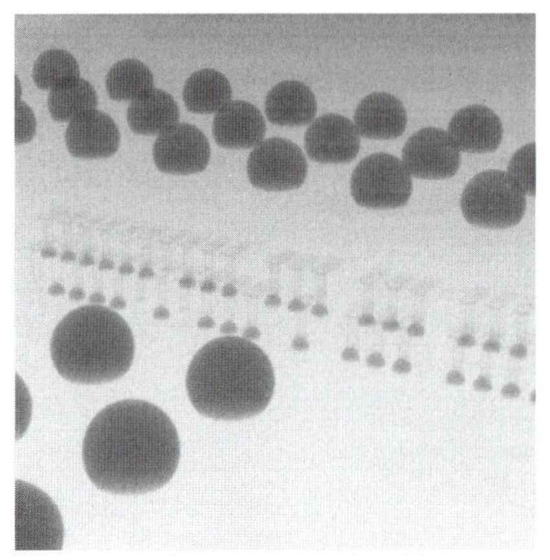

Figure 35.5 X-ray of a Micron TWI stack.

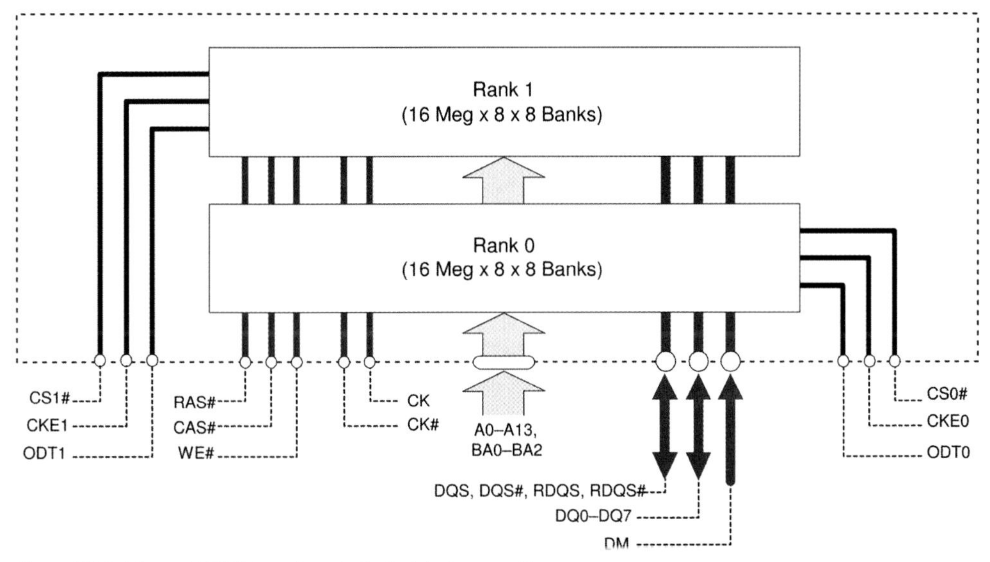

Figure 35.6 Micron DRAM two-dice stack electrical connection.

To route signals to the correct chip in a two-dice stack, and to avoid multiplying the parasitic problem by connecting chips in parallel at the bond pad, it may be necessary to use more vertical interconnections than there are standard bond pads. This matter can be solved by generating smaller vias, so that multiple vias are run through a single bond pad. It is also possible to utilize "spare" bond pads that are used for probe or debug, but are not normally bonded out in some memory configurations.

Figure 35.6 shows an example of the electrical connections required for a two-dice stacked configuration using two, 1 Gbit dice in one package to make a 2 Gbit device. The connections that must be unique for each IC in the two-dice stack are the clock enable (CKE), on-die termination (ODT) and chip select (CS). The rest of the terminations may be made in parallel to each IC. The concept is the same for a larger stack, although a more complicated connection structure is required to uniquely communicate with each IC while minimizing connections.

35.3 Redistribution Layer

A Redistribution Layer (RDL) is a connection system designed to move the bond pads to any location on the front die surface, allowing wire bonding or direct chip attach without the constraints of the original pad location and pitch. After the front-end fabrication and electrical probe, the wafers receive a layer of photosensitive polymer, such as polyimide (PI), benzocyclobutene (BCB) or polybenzoxazole (PBO) to provide a buffer layer under the future RDL bond pads. The polymer is patterned to open the original probe pads, and a metal is then deposited and patterned. The metal typically contains a thin barrier and adhesion layer, covered with a full thickness aluminium metallization, which is photolithographically patterned and etched, or alternatively covered with a Cu seed layer, which is masked with a photoresist pattern and plated with Cu to produce a full thickness Cu layer. This structure is then protected by another level of a photosensitive polymer, which is patterned to open the RDL bond pads. It requires three masking levels, unless the first polymer layer is combined with the passivation masking process in the front-end fab. The RDL process can be used over conventional Al metallization or Cu metallization. If Cu is used in either the lower levels or the RDL level, barriers may be needed to prevent Cu migration. The RDL bond pads may be finished with a plated Ni/Au to provide an under bump metallization (UBM) for a solder or plated bump, or a very hard landing pad for the wire bond process. Figures 35.3 and 35.7 show a Micron DRAM RDL structure. The RDL pads may be electrically probed again, if necessary, to verify yield or performance.

The RDL bond pads will have a substantially increased pitch, which allows more options in attaching the chip, such as flip chip on module (FCOM) or flip chip in package (FCIP), in addition to rearranging the location of pads to optimize the distribution of electrical signals as well as power and ground. The buffer layer of polymer under the RDL pads also allows the use of wire bonding directly over active circuitry. Wire bonding directly over standard silicon dioxide/silicon nitride passivation is normally avoided because of the reliability risk of cracking the brittle dielectrics.

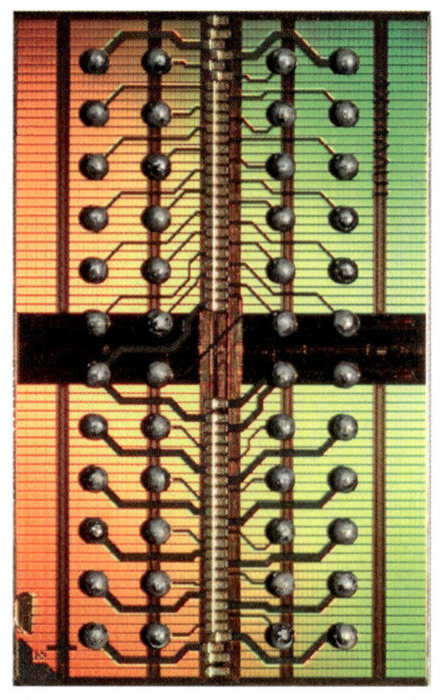

Figure 35.7 Micron DRAM RDL structure with solder balls.

An additional benefit of the RDL is its use for 3D stacking. The TWI cannot be placed in the RDL pads unless there are no conductors or active devices below them. However, the RDL structures can be used to advantage on the bottom die in the stack to accommodate a loose pitch or special ball out on the final module. Although it will increase the electrical parasitics, RDL can be used on the intermediate chips in the stack, but would normally require both front and backside RDL.

The conventional front side RDL process can also be implemented on the wafer backside in conjunction with a TWI structure, to make connections. Depending upon how it is used in the stack, a backside RDL may be used with or without a front side RDL. The backside RDL construction may be a little different, since there are no problems with potentially damaging active devices under this RDL. However, the basic concept of insulator, metal, insulator is appropriate to prevent shorting or reliability issues. The conductors can be built of the same materials and methods as the front side RDL, although some method of alignment to the front side, or the TWI structures, must be performed. This can be accomplished with lithography tools which reference the front side of the wafer while exposing resist on the backside, or lithography tools which reference the TWI structures that are visible on the backside. A major complication, however, is that the backside RDL is typically carried out on a wafer that has been thinned to its final thickness. Handling a wafer less than 300 μm thick, using standard semiconductor tools, can be very difficult. Typically, a handle or carrier substrate would be temporarily attached to the thinned, live wafer in order to transport the wafer through all of the process steps without breakage. A glass or silicon substrate is commonly used as the carrier, although special consideration needs to be placed on the issues of the coefficient of thermal expansion (CTE) mismatch, thermal conductivity and the ability to see through the substrate for alignment purposes.

35.4 Through Wafer Interconnect

The TWI structure may be built at many points during and after front-end wafer fabrication. Clearly, the structure that is built needs to be compatible with the materials present, but the earlier that the TWI structure is implemented into the process the more difficult the processing compatibility issue becomes. There are many complications in building a TWI structure that is stable during high temperature diffusion steps, does not cause any contamination or deleterious effects on the wafer, is compatible with future processing and has the necessary electrical properties in the finished TWI. These considerations can severely limit the choices of TWI materials that can be used, especially the conductor. Once the wafer has completed the alloying and passivation steps, the choice of materials is limited by the need to minimize process temperatures to prevent any change in electrical parameters or mechanical damage to the wafer, such as warping or cracking. This implies the temperatures need to be kept below the alloy and passivation deposition temperatures (typically in the range of 400 °C), and the CTE and stress of the materials must be carefully considered.

The TWI used for inter-chip connection technology will ultimately have four basic characteristics:

1. A conductive portion from one surface of the wafer to the other.
2. A dielectric portion isolating the conductor from the silicon substrate.
3. A method for connecting one end of the conductor to the IC, or passing through without connecting.
4. A method for connecting the ends of the conductor to another IC or substrate.

There are many ways to achieve these characteristics. In the case of connection of bond pads, the holes can be made from the front side of the wafer (passing through the bond pads, and then uncover the TWI by grinding and polishing away the wafer backside) or from the backside of the wafer (by mounting a temporary carrier on the front side, grinding and polishing the wafer backside, and then passing through the wafer and bond pads). Each approach has its own processing issues to solve, but the end is essentially the same.

Forming the TWI from the front side means that the processing typically occurs on a full thickness wafer, and alignments are straightforward because device structures, as well as the wafer alignment keys, are in plain view for standard lithography tools. High powered pulsed lasers can be used to ablate a hole fairly rapidly, and do not require photolithography. However, the heat-affected zone may need to be chemically etched away, enlarging the hole, and making it difficult to control the penetration depth precisely. It is also difficult to control the diameter, achieve smooth sidewalls and manage the resulting slag. For a small number of TWIs per die on a small diameter wafer, the laser throughput can be very reasonable. However, since they are made one at a time, a large number of TWIs per die, or a large number of dice per wafer, translates to a relatively slow throughput. Alternatively, standard photo/etch processes can be used to generate a reproducible and clean hole through the bond pad, interlevel dielectric (ILD) and into the silicon substrate. The throughput using photolithography/etching is independent of the number of TWIs per die, or the number of dice per wafer, and the depth of each hole can be controlled very uniformly across the wafer. Although dry etching the silicon substrate is a relatively fast process (over 15 $\mu m\ min^{-1}$, depending on width and depth), the depth of the holes must be kept as short as possible to minimize the total time. Tapering of the hole can also be reproducibly controlled to facilitate subsequent coating and filling. For most TWI connections, the conformal dielectric deposited in the hole must be thick and have a low dielectric constant to reduce the coupling to the silicon substrate. The conductor can be formed by many different methods, including plating, physical vapor deposition (PVD), chemical vapor deposition (CVD), molten metals and combinations of these methods. Since the primary conduction, particularly at high frequency, occurs on the sidewalls of the conductor, it is not necessary in large TWI structures for the conductor to fill the entire TWI volume. The TWI core can be filled with other metals, and even with an insulator such as a filled polymer, and retain the conduction required. Figure 35.8 shows a simulation at 100 MHz of the current flow in a TWI with a copper jacket surrounding a solder core. Clearly, at high frequencies the skin effect will minimize the advantage of having a solid plug of a low resistivity conductor.

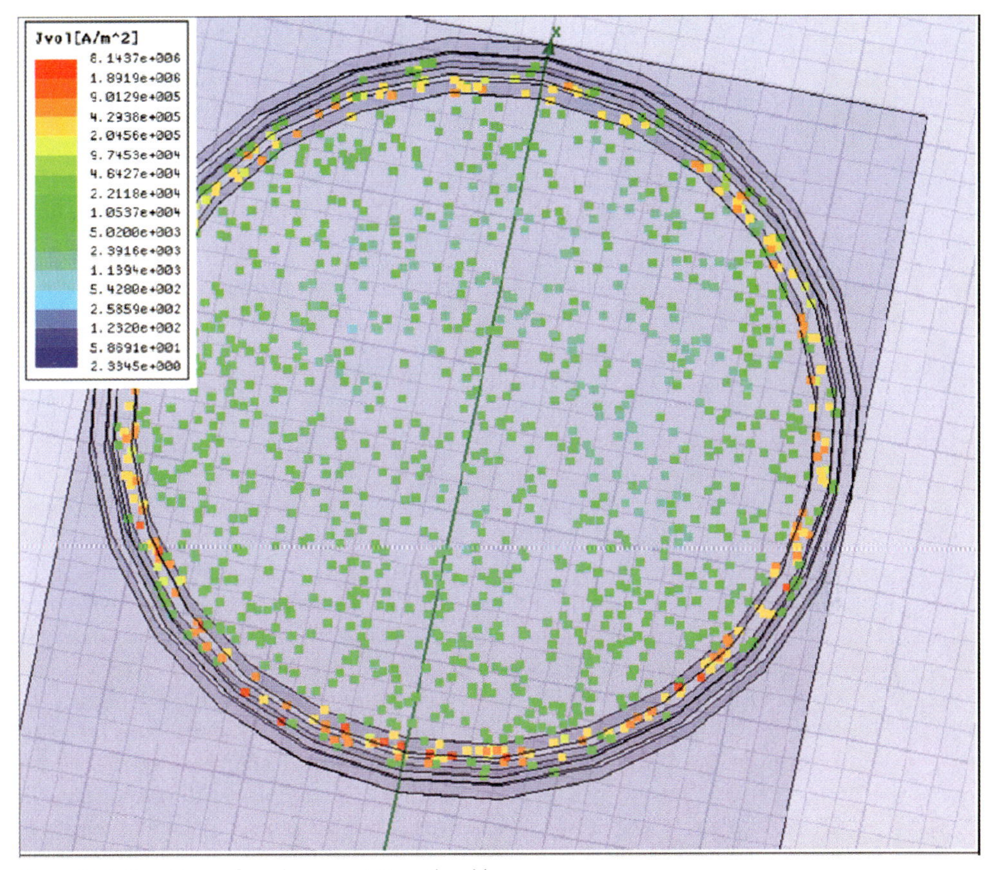

Figure 35.8 Simulation of conduction in TWI with solder core.

Although the TWI core could be left unfilled in a large TWI, there may be structural or reliability problems created with leaving the void.

A fundamental aspect of the construction is how to connect the TWI conductor with the bond pad where needed, and how to avoid connecting the TWI conductor in those situations where the TWI needs to pass through an IC without making a connection. Making the connection to the pad is done by opening the passivation layer over the bond pad, and allowing the TWI conductor to make contact to the pad when the connection is made to the other components in the stack. If the passivation is not removed from the bond pad, the TWI sidewall dielectric will prevent contact to the bond pad, but will not prevent connection of the TWI to a substrate or another IC in the stack.

Forming the TWI from the backside means that the processing for the TWI hole will occur after the wafer is thinned and a carrier is attached to the wafer front side (to allow easy handling through the processing equipment). Alignment to the front side is required, so the use of an exposure system that puts an alignment key on the

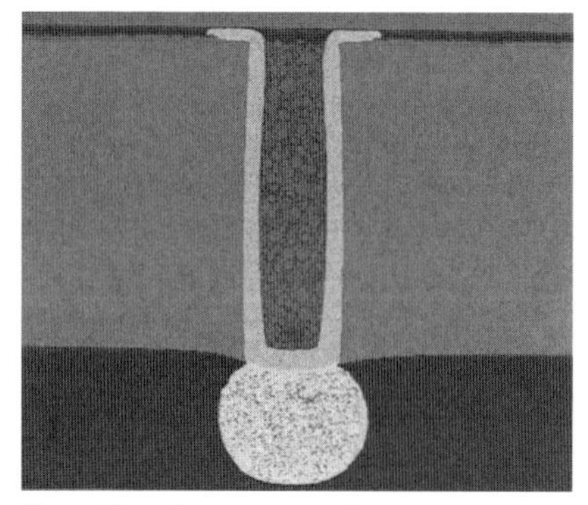

Figure 35.9 Cross-section of a TWI structure.

backside (after thinning), or a tool that allows each photolithographic level to be referenced to the front side, is required. If the TWI is to be used for stacking with some pads connected to the TWI, and some not, then the TWI hole needs to eventually penetrate the bond pad and connect using a method such as in the front side example above. However, if all pads can be connected to the TWI, then the hole can stop on the underside of the bond pad without penetrating. Construction of the dielectric, conductor and fill would continue as in the front side TWI case. A connection to the backside TWI will be necessary, and this can be done using a backside RDL process, or simply a dielectric deposition with photolithography and etch to open the TWI conductor. A silicon etch may be used first, to recess the silicon backside and allow the TWI conductors to protrude. Placing under bump metallization (UBM) on the ends of the TWI allows the formation of a plated bump or solder ball to be used to make the connection to the substrate or IC. Figure 35.9 shows a cross-section of a TWI constructed after the bond pads were formed, and shows a polymer core, along with a backside solder ball attached to the TWI.

35.5 Stacking

The number of dice in a stack is determined by the application. However, there are usually physical constraints based on the maximum allowable package size, electrical constraints based on the parasitics, and space constraints for making the electrical connections to select and use each IC. For TWI stacking, it is worth noting that the top die in a stack does not normally need to have a TWI, unless it needs to be contacted from the backside. All other dice in the stack will need TWI structures. An important consideration for stacking is to have very flat wafers and/or dice in their relaxed, unchucked state. Stacking dice with significant curvature can be particularly difficult with larger dice.

The connection of one to another must be made with a method that assures contact, and provides insulation between connections. Underfill is used to isolate the conductors and, depending on the process, it may be possible to use a no-flow underfill.

There are three basic approaches to stacking dice with TWI:

1. Wafer to wafer (WTW)
2. Die to wafer (DTW)
3. Die to die (DTD).

An inherently appealing aspect of WTW stacking is that the processing is done completely at wafer level, and thereby allows many dice to be handled and connected at one time. For high density memories, unfortunately, this same characteristic has some major drawbacks. Because the yield is lowest on the newest, highest density, most complicated and largest parts, and because these new generations are the very parts most likely to benefit by stacking to extend to the highest memory densities, it is very unlikely the yield will be high enough to have a high probability that two good dice will be stacked together. This clearly gets exponentially worse with each layer of stacking. In addition, memories need to be graded for performance, so that all of the dice in a stack will have matching characteristics. The random stacking of dice of variable performance, along with a significant number of defective dice, overrides the benefits of WTW for memories.

DTW stacking provides an advantage in registration for the bottom layer, since the wafer is intact, as well as not requiring an additional tool and process step for arranging the bottom dice in a reconstructed array, and the wafer can be utilized for handling. This also has a benefit during dicing in that there is no significant variability in the die to die spacing for X, Y and theta variation in bottom dice placement, as there is in a reconstruct situation. If encapsulation is done prior to singulation, this bottom die registration accuracy also translates to more uniform sidewall encapsulation thickness. Although matching good die to good die can be done easily with mapping, it is much more difficult to match the performance of each die, since this varies within a wafer and wafer to wafer. A major downside to DTW is that the defective dice require the pick and place of another defective die or dummy die on top, or else the subsequent process has to be flexible enough to allow singulation with some locations missing a full stack. This problem is significantly exacerbated with stacking higher than two.

DTD stacking is achieved by mounting a bottom die on a reconstructed array, or on a substrate/lead frame. The dice to be stacked will have been selected for yield and performance and, in most cases, will be thinned prior to pick and place. Although this does not have the registration advantage of using an intact wafer as the substrate, the substrate/lead frame approach is more compatible with conventional assembly tools, such as encapsulation and trim and form.

The orientation of the dice after stacking, whether Front to Front (F2F) or Front to Back (F2B), has implications for IC design and layout, as well as assembly. By stacking two ICs with the same bond pad layout F2B, the bond pad orientation will line up vertically automatically, and is the more straightforward. However, it may be desirable to reduce the total distance between the circuits of two stacked dice, and this can be done by stacking F2F. For F2F bonding, either the bond pads must have axial symmetry about the rotated axis (which can often occur in DRAMs with a single row of bond pads down the center of the long axis of the IC) or one or both of the ICs need an RDL layer to redistribute the bond pads so that they line up vertically for connection. This produces a fan out, fan in structure for each die, which may add unacceptable levels of electrical parasitics.

Stacked Chip on Board (COB) is a method of forming 3D connections using conventional assembly infrastructure. Although size and performance can be issues for some applications, the cost is low and the infrastructure exists. A stacked DRAM COB with two or four ICs is commonly built by placing the bottom die on a substrate, and then wire bonding to the substrate. A low loop wire bond technique is used, and a spacer is placed or formed on top of the bottom die to allow the space necessary for the bonds on the IC pads, then the next die in the stack is place on top. This continues until all the parts are stacked and wire bonded. The bond pads typically must be routed to the edge of the die in the original pad layout, or by using RDL (Figure 35.3). For NAND, with fewer bond pads than DRAM, the bond pads are often designed on one end of the IC to allow the stacked parts to be shingled to expose the bond pads on the end, without requiring spacers or RDL, which minimizes height and cost. The wire bonds may be cascaded from each IC to the IC below, or may all connect directly to the substrate from each IC (Figure 35.5). NAND is also commonly stacked in a directly vertical structure.

Another method of stacking utilizes the concept of package on package (PoP). This is the most straightforward, since it connects the leads of conventionally packaged parts; however, the large size and increased electrical parasitics do not make it attractive for compact, high-performance applications.

The different approaches to stacking produce different electrical results. A common method for looking at the electrical effect of the package is to use the eye diagram, a plot of voltage versus time, to measure jitter. Jitter is a variation in the switching transitions and provides a measure of the size of the switching window, thereby giving an indication of the electrical switching margin. The more jitter, the more the eye diagram closes and margin is reduced. Jitter becomes a major problem as data rates increase. Comparing the simulated eye diagrams of a stack of two, 2 Gbit, DDR3 DRAMs in a COB configuration and a TWI configuration, there is a wider timing window and less jitter over the whole simulated range, from 0.8 up to 1.6 Gbits s^{-1}, with about a 36% improvement in both jitter and timing window margin at 1.6 Gbits s^{-1} (Figure 35.10).

35.6
Additional Issues

One unique issue for DRAMs is a problem associated with thinning the wafers or individual ICs for stacking to achieve the minimum package height. The DRAM cell structure is a capacitor storage element with an access transistor. The capacitor and access device have an expected amount of charge leakage that is compensated for by periodically refreshing the data stored on the capacitor. The refresh specification is

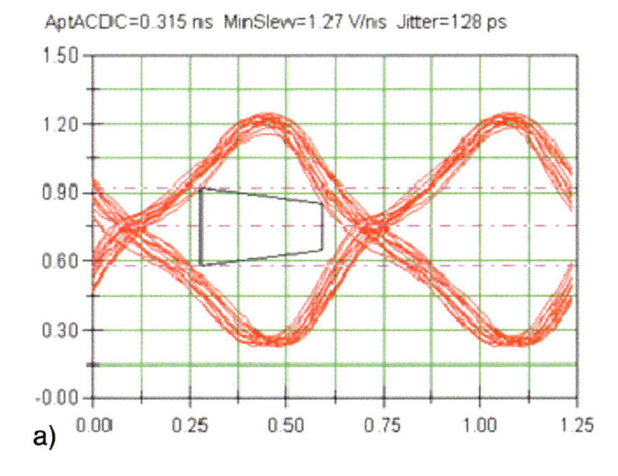

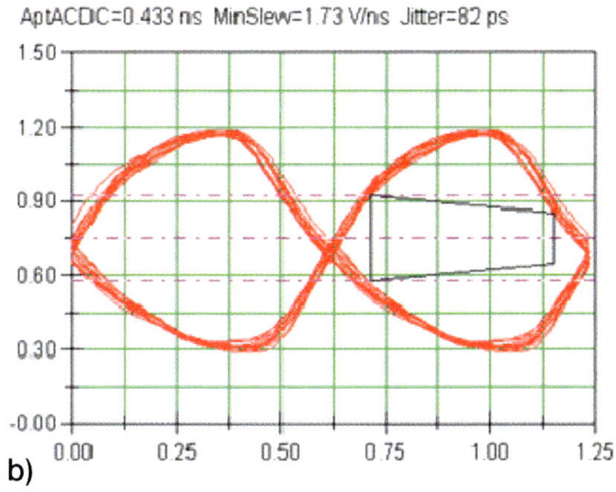

Figure 35.10 Eye diagram simulation of a DDR3 stack of two, 2 Gbit DRAMs in a COB (a) and TWI (b) configuration.

64 ms in most applications. However, if the leakage exceeds a threshold level prior to refresh, the data is lost from that bit, and the error correction must repair it. This retention time has a distribution over the billions of bits on the chip and it is the weakest bits (after redundancy has been implemented) that govern the minimum refresh time required. In addition, there is a very small number of variable retention time (VRT) bits in which the retention time is good one clock cycle, and bad the next, and then good again in a statistical process. The number of VRT bits and bits with a constant short retention time may be adversely affected by thinning the wafer. The fabrication process used, the design, the method that was used to thin the wafer, as well as the wafer's final thickness can affect the number of VRTs and the bits with short refresh time. The thinner the wafer, the more likely a problem will be seen. This is particularly a problem below 100 μm and has a large implication for stacking, since the original wafer thickness for a standard 300 mm wafer is 775 μm, and thinning the wafer below 100 μm is very desirable to minimize electrical parasitics, as well as overall package height.

Wafer Level Packaging (WLP) describes the concept of extending the wafer format processing from the front-end fabrication, through electrical probing and into assembly. Complete extension of the concept would include wafer level encapsulation, burn in and test, with singulation occurring just prior to shipment. This minimizes the individual handling of each die at an early point in the back end processing, and can potentially increase the throughput and lower the cost. Since these are wafer level processes that are being applied to all of the dice on a wafer at once, the smaller the die size, the lower the cost for each die. Micron Technology, Inc. has developed Osmium™ technology, which is a WLP capability including RDL, TWI and wafer level encapsulation (WLE). 3D stacking does complicate WLE, but there are methods being developed to accomplish it.

35.7 Future of 3D Memories

Smaller, cheaper, faster and more reliable. These attributes continue to drive the production of IC memory. A 3D approach that improves any one of these attributes has the potential to influence the market substantially. In the near term, we can expect to see the commercialization of the connection of memories into stacks utilizing large, bond pad type connections. The bond pad connections provide a large available area without devices below them. Current bond pad sizes exist primarily to allow production probing at the wafer level to provide fast feedback to the fabrication area of yield and performance. As probe and interconnect technology improves, much smaller diameter interconnects between chips will be used. These smaller interconnects allow the TWI location to be placed outside of bond pad regions, and are appropriate for intra-circuit connections.

It is conceivable to build the front end of line transistors and the back end of line interconnects separately, but simultaneously, and then join them together to lower the overall fabrication cycle time. This will require the development of extremely fine alignment control to join wafers together with device level alignment.

As optical, inductive, capacitive and RF interconnect technology improve, those interconnects which can tolerate the space necessary for encoding, transmitting, receiving and decoding signals, could make TWI structures unnecessary in 3D structures.

Acknowledgment

I would like to thank Jeff Janzen, Mark Hiatt, and Chad Cobbley for their invaluable help and advice.

36
3D Read-Out Integrated Circuits for Advanced Sensor Arrays

Christopher Bower

36.1
Introduction

This chapter focuses on the advantages and implementation of 3D ICs when applied to advanced sensor arrays. There are several efforts to commercialize 3D CMOS sensors that utilize through silicon vias (TSVs) along the perimeter of the device chip. These efforts are covered in Chapters 15 and 26. Here, the focus is on applications that require at least one TSV per pixel. Particular emphasis is placed on the class of pixelated detectors that utilize multiplexing silicon CMOS read-out ICs (ROICs). Figure 36.1a and b show the configuration that is currently used on many state-of-the-art sensor arrays. The device consists of a detector layer that is hybridized (i.e. interconnected) with an underlying ROIC chip. The detector layer can be fabricated using a wide range of materials [1], but HgCdTe and InSb are prominent materials for infrared imaging, which is the application that drives this technology. Infrared imagers are often referred to as infrared focal plane arrays, or simply focal plane arrays (FPAs), and are the subject of several reviews [1, 2]. Most commonly, the detector chip is hybridized with the ROIC using indium or indium alloy bump bonding techniques [3]. However, some of the wafer and die bonding technology currently being developed for 3D stacking of heterogeneous materials [4, 5] will likely compete with indium bonding for some future applications.

The continuous shrinkage of IC design rules has allowed ROIC designers to pack increasing amounts of circuitry within the footprint of the pixel unit cell. However, the push towards smaller pixels has somewhat countered the amount of new pixel-level functionality that can be achieved. In current 2D ROICs, the unit cell circuitry is often limited to a preamplifier and a storage capacitor. Read-out circuitry is now considered a technological barrier for many applications, including third-generation night vision sensors [6] and adaptive microsystems [7]. Another application currently limited by readout technology is laser-radar (LADAR) sensors that require pixel-level time-stamping circuitry [8]. This circuitry is difficult to implement within the confines of a two-dimensional unit cell.

Handbook of 3D Integration: Technology and Applications of 3D Integrated Circuits.
Edited by Philip Garrou, Christopher Bower and Peter Ramm

ISBN: 978-3-527-32034-9

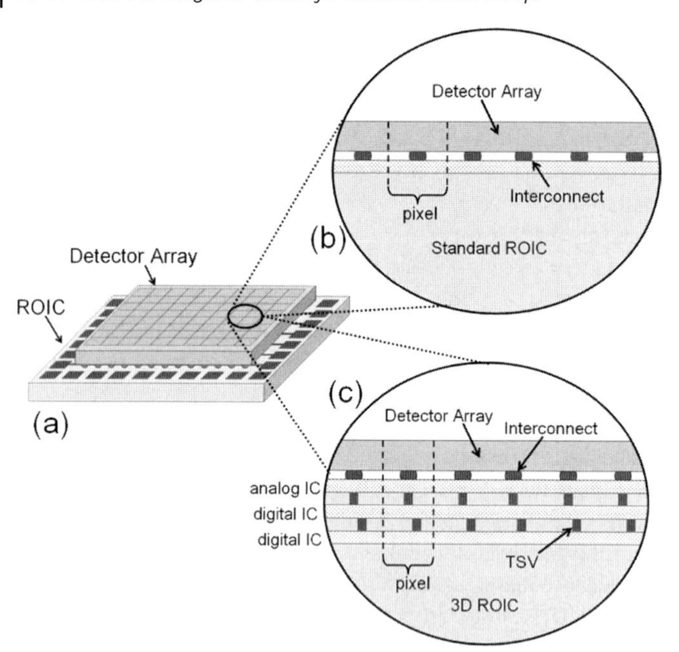

Figure 36.1 (a) Schematic drawing of a detector array hybridized to a ROIC. The cross-section insets show (b) a standard two-layer sensor (detector and ROIC) that does not require TSVs and (c) an example of a 3D sensor consisting of four tiers of electronics, the detector plus a 3D ROIC with two digital and one analog tier.

3D IC integration eliminates the restriction of the 2D unit cell. Using 3D concepts, designers can shrink pixel sizes while adding more circuitry to underlying ICs. This is why pixelated architectures have been referred to as the "low hanging fruit" for 3D integration. Figure 36.1c is a cartoon cross section illustrating a 3D ROIC that has three distinct IC layers. In this case, assuming the same pixel size, the designer would have three times more real estate for circuitry. Note that some of the additional design space will be consumed by the vertical interconnects. Additionally, the individual IC layers (or tiers) can be fabricated using optimized process technology. In this example (Figure 36.1c), the bottom two tiers could be fabricated using a digital IC foundry and the third tier could be fabricated using an analog foundry process.

36.2 Current Activity in 3D ROICs

36.2.1 The DARPA VISA Program

DARPA has been one of the major sources of funding in the area of 3D integration. Clearly, the military has a long standing and keen interest in advanced sensors, with particular emphasis on advanced imaging systems. The VISA (Vertically Interconnected Sensor Arrays) program [8–10] is focused on developing 3D ROICs to enable FPAs with high dynamic range (>20 bit), high readout rates (>10 kHz) and small pixels (less than 25 × 25 μm). Companies and institutes currently participating in VISA include DRS Infrared Technologies, RTI International, Raytheon, Ziptronix, Rockwell Scientific and Lincoln Labs. To date, much of the VISA activities remain undisclosed. The following section reviews the results published by the DRS Infrared Technologies/RTI International team. The Raytheon/Ziptronix team recently issued a press release highlighting some of their VISA results [11].

Horn *et al.* [8] have described several applications that may be enabled by the 3D ROICs being developed within the VISA program. Here, we give a brief summary of these VISA applications.

36.2.1.1 **Active Imaging**

Active imaging refers to laser-radar (LADAR) imaging that will allow next generation imagers to extend target identification range. Figure 36.2 illustrates how LADAR is envisioned to operate. An eye-safe short-wave infrared (SWIR) laser pulse illuminates a target and then the reflected photons are captured by the SWIR FPA. The "time-of-flight" of the photon can be determined using high speed digital timing circuitry built into the FPA. It is anticipated that 3D integration will allow designers to add the necessary high speed timing circuitry within the confines of the unit cell. The VISA program envisions sensor arrays that combine active imaging with passive LWIR FPAs.

36.2.1.2 **Improved Sensitivity**

The 3D integration of analog and digital wafers allows the implementation of wide dynamic range circuits [8]. These circuit concepts lead to vastly improved dynamic range and charge storage capacity in the unit cell. This increased charge storage translates into a predicted 10× sensitivity improvement for long wave infrared (LWIR) FPAs.

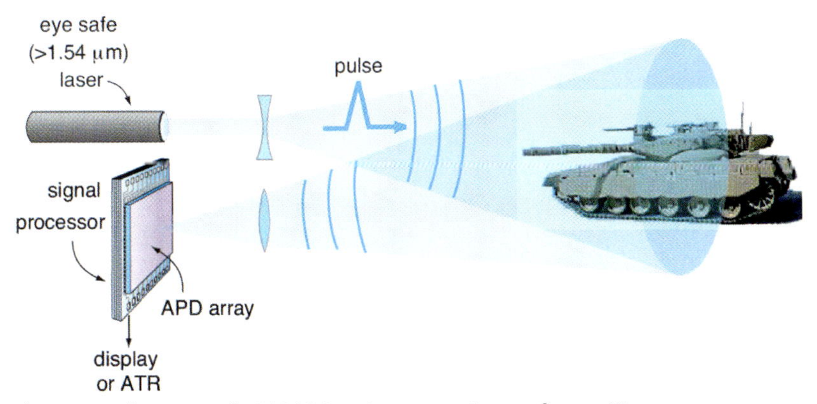

Figure 36.2 Illustration of a LADAR imaging system. From reference [8].

36.2.1.3 Camera-on-a-Chip

The VISA technology will ultimately allow many board level functions to move on chip (onto one of the IC tiers). It is envisioned that this could reduce the number of electronic boards required for a state-of-the-art camera. Horn *et al.* [8] analyze the benefit for a specific camera based on a 640 × 480 VO_x microbolometer array. They found that the number of electronic boards could be reduced from three to one, and they determined that the total system power would reduce from 3.1 to 0.78 watts. Balcerak *et al.* [12] have discussed how the expected power reduction and miniaturization enabled by VISA will help enable unattended sensor networks.

36.2.1.4 Resistance to Laser Jamming

The VISA 3D ROICs are expected to allow readout frame rates of 10 kHz. It is anticipated that a VISA FPA would defeat pulsed laser jamming by operating between the laser pulses.

36.2.2 The DRS/RTI Infrared Focal Plane Array

DRS Infrared Technologies and RTI International have been jointly developing a high-performance infrared focal plane array that is enabled by a 3D ROIC [13–16]. The objective FPA consists of three electronic tiers: a digital IC (Tier 1), an analog IC (Tier 2) and a HgCdTe photodiode array (Tier 3). The circuit diagram for the ROIC unit cell is shown in Figure 36.3. To achieve high dynamic range, the circuit combines an analog charge removal circuit (Tier 2) with a digital counting circuit (Tier 1) [8]. The

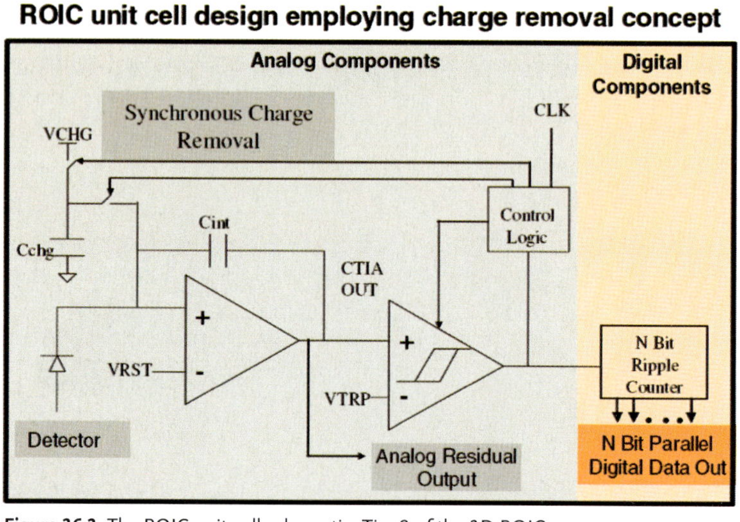

Figure 36.3 The ROIC unit cell schematic. Tier 2 of the 3D ROIC contains the analog components and the TSVs. The digital components are in Tier 1 of the 3D stack. From reference [13].

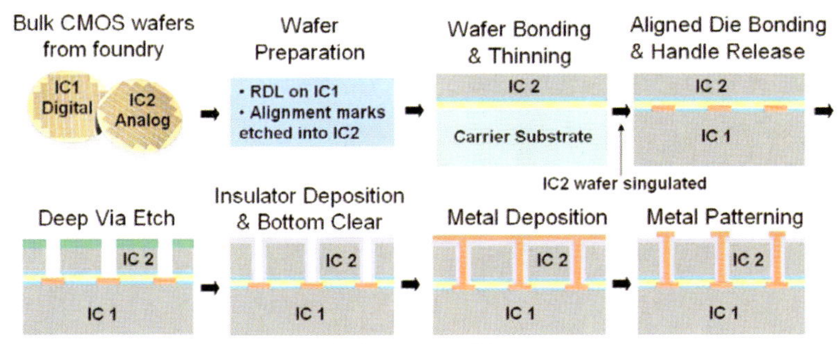

Figure 36.4 DRS/RTI process flow for fabrication of the 3D ROIC. From reference [14] (© 2006 IEEE).

3D design allows this circuitry to fit within a 30 μm unit cell. It is estimated that the same circuit would require at least a 50 μm unit cell in a traditional 2D design [13]. Another benefit of 3D integration is the possibility to achieve large power savings. In this case, a three-fold decrease in power consumption is expected because the Tier 1 digital circuit (0.18 μm design rules) was designed to operate at a lower voltage than the analog circuit (0.25 μm design rules).

Figure 36.4 shows the process flow used to fabricate the 3D ROIC [14]. Here, the Tier 2 wafer is thinned and bonded to the 3D stack before the TSVs are formed – therefore it is a "vias last" process. The analog wafer was fabricated using a 0.25 μm analog foundry process and the digital wafer was fabricated using a 0.18 μm digital foundry process. In this case, the analog wafer was designed to have exclusion zones to accommodate the TSV. The size of the exclusion zone was determined by experiments studying the impact of the spacing between the TSV and transistors (n and p channel MOSFETs) [14]. TSV to transistor distances as small as 1.5 μm showed no degradation in the transistor IV characteristics. First, a metal redistribution layer was added to the digital IC (Tier 1). Alignment marks were deep etched into the analog IC (Tier 2) to facilitate wafer singulation and aligned bonding. The analog wafer was bonded face-down onto a temporary handle wafer and thinned by backgrinding and CMP to approximately 30 μm. After thinning, the dicing and bonding alignment marks are now visible from the backside of the analog wafer. Next, the analog wafer is singulated into die while still attached to the handle wafer. The analog IC die are then aligned and bonded to the digital IC wafer using a high accuracy "flip chip" die bonder with split prism optics. A polymer adhesive is used for the die attach. The reported post-bond alignment accuracy between the layers is better than 2 μm. The handle die is subsequently released, leaving the 30 μm thick analog IC face-up on the digital wafer. Photolithography is used to pattern the etch mask for the vertical interconnects. The high aspect ratio vias are etched using a combination of oxide and deep silicon etching. The vias land on the metal redistribution layer previously added to the digital IC (Tier 1). A conformal insulator is deposited and selectively cleared from the bottom of the via. The insulated via is then filled with copper metallization, and finally the copper on the surface is patterned to accommodate the detector layer (Tier 3). Figure 36.5 shows a

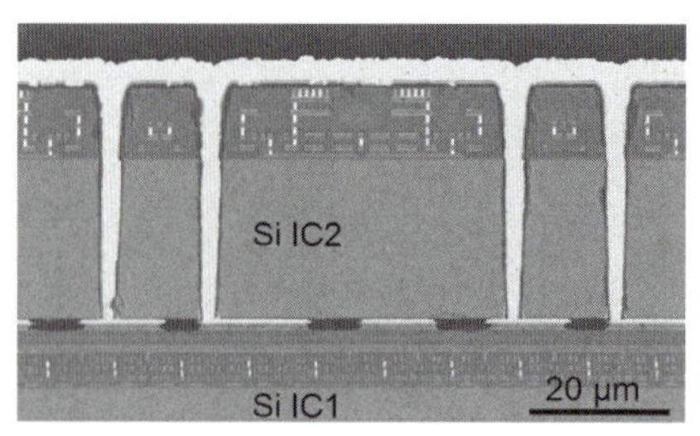

Figure 36.5 Cross-sectional electron micrograph of the two-tier IC stack following the 3D integration process. From reference [14] (© 2006 IEEE).

cross-sectional scanning electron micrograph of the vertically interconnected analog and digital wafers following deposition of the copper. This picture shows how the TSV goes through both the top oxide and the bulk silicon of the analog IC. In this case, the TSV exclusion zone includes both the silicon layer and the wiring levels in the analog IC.

In theory, various detector materials could now be hybridized with the two-layer 3D ROIC. In this case, an HgCdTe photodiode array is hybridized with the 3D ROIC using the high-density vertically integrated photodiode (HDVIP) process developed by DRS Infrared Technologies [17]. Interestingly, the HDVIP essentially constitutes a 3D integration process for HgCdTe, and has already been applied to multiple levels of HgCdTe in two-color detectors [18]. Figure 36.6 shows a schematic of the HDVIP structure. Figure 36.7 shows a bi-directional cross section micrograph of the fully fabricated three tier FPA.

As of this writing, the measured performance of the fully integrated 3D focal plane array has not been disclosed. Earlier, the team reported on a focal plane array where the analog IC was replaced with a surrogate passive silicon layer [14]. A cross-sectional micrograph of this structure is shown in Figure 36.8a. In this case,

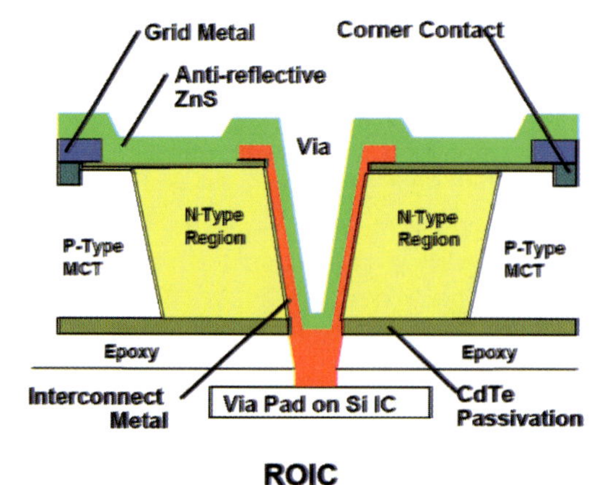

Figure 36.6 Schematic of the HDVIP™ structure. From reference [18].

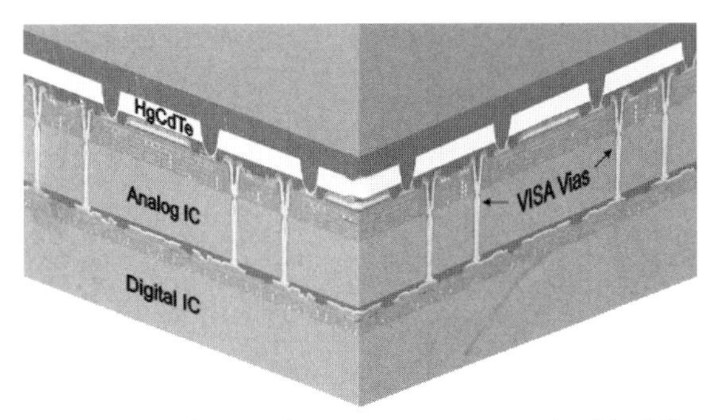

Figure 36.7 Bi-directional cross-section micrograph of the fully fabricated three-tier IR FPA. From reference [13].

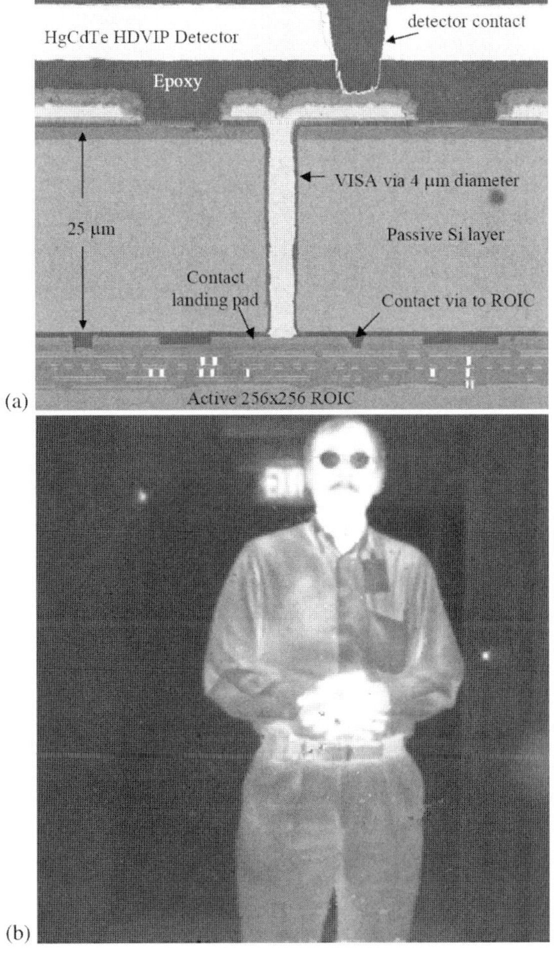

Figure 36.8 (a) Cross-section electron micrograph of the 3D FPA test structure with the surrogate passive silicon tier 2. (b) Thermal image from the 3D FPA test structure. From reference [14] (© 2006 IEEE).

the ROIC (Tier 1) allowed the team to measure pixel operability and to determine if the TSV added any additional noise to the detector signal. The noise performance of the 3D FPA containing the TSV was essentially identical to the noise performance of a standard FPA. Figure 36.8b shows a thermal image (MWIR) taken with the 256 × 256 3D FPA. The device was packaged and operated at 77 K within a laboratory dewar. No defects or artifacts were observed that could be traced to the 3D integration process. To study robustness of the TSVs, the device was cycled 1000 times from 77 K to room temperature. No change in detector performance or pixel operability was observed following the thermal cycles. To study the TSV operability, the HgCdTe layer was replaced with a top common copper electrode [14]. The multiplexing ROIC (Tier 1) allowed for convenient testing of every TSV (65 536) in the 256 × 256 array. The best TSV operability was 99.98%, which

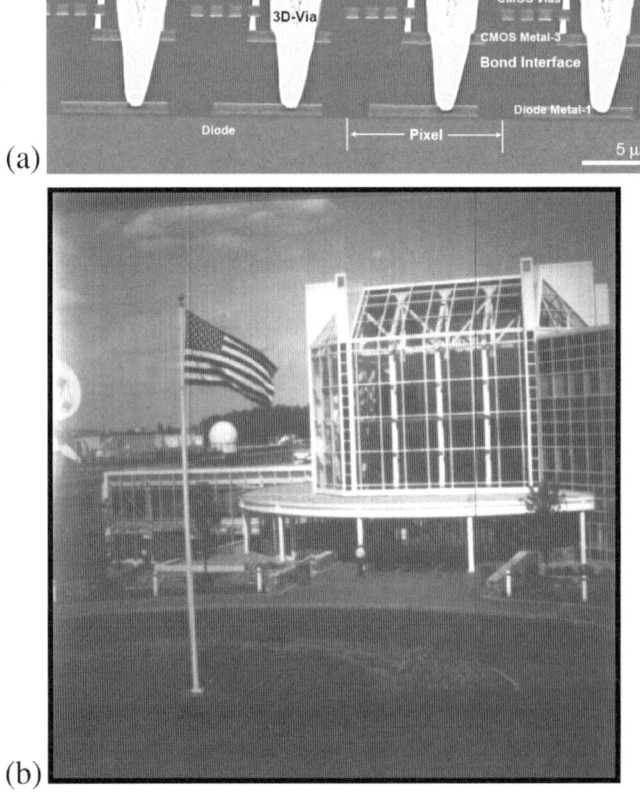

Figure 36.9 (a) SEM cross-section of the MIT Lincoln Laboratory 3D Megapixel CMOS imager and (b) an image captured with the device. From reference [19] (© 2005 IEEE).

corresponded to 14 non-functional TSVs out of 65 536. The team also fabricated 4-wire Ohm test structures to measure the resistance of TSVs. These "vias last" copper TSVs (4 μm diameter) were found to have an average resistance of 140 mΩ. The specific contact resistivity between the TSV and the landing metal pad was estimated to be $1 \times 10^{-8}\,\Omega\text{cm}^2$.

36.2.3
MIT Lincoln Laboratory's 3D Imagers

MIT Lincoln Laboratory has developed a well-known 3D integration process based on direct oxide bonding of FDSOI CMOS wafers. The MIT Lincoln Laboratory 3D integration process and imager demonstrations were described in detail in Chapter 20. Here, we will briefly summarize the two imagers that have been designed and fabricated.

A megapixel CMOS visible image sensor with 100% fill factor was enabled by 3D integration [19]. Figure 36.9a shows a cross-section of the 3D integrated imager. The imager consisted of a 1024 × 1024 array of 8 μm pixels. The photodiode tier was fabricated in high resistivity (>3000 Ω-cm) bulk silicon wafers. The SOI-CMOS tier was fabricated using a 0.35 μm FDSOI-CMOS process. The 3D integration enabled the detector tier (p + n photodiodes) to achieve 100% fill factor. Each sensor array contained over one million vertical interconnects and the pixel operability was measured to be greater than 99.9%. Figure 36.9b is an example image captured with the imager; see also Figure 20.10 (Chapter 20).

MIT Lincoln Laboratory has also developed a laser-radar (LADAR) chip [20] that combines an array of Geiger-mode avalanche photodiodes (APDs) with underlying

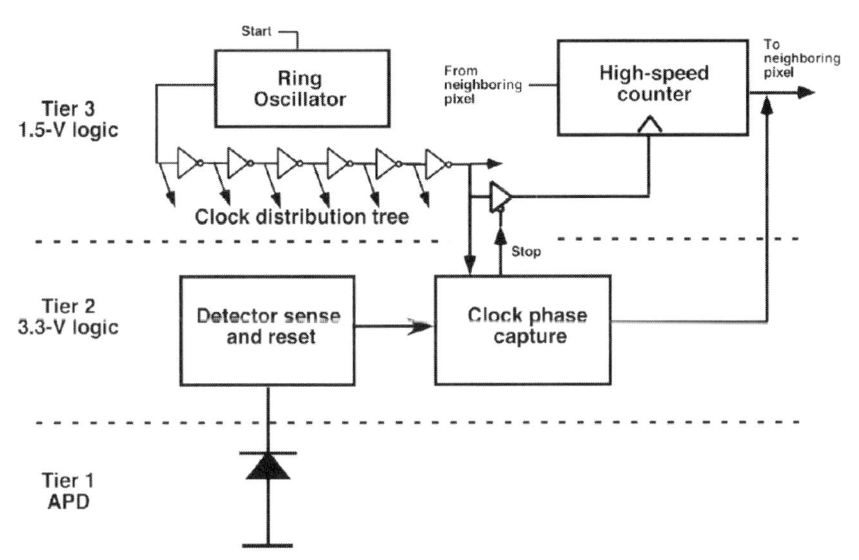

Figure 36.10 Block diagram of the 3D LADAR unit cell. From reference [20] (© 2006 IEEE).

high speed timing circuitry fabricated using FDSOI-CMOS. The chip consisted of a 64 × 64 array of 30 μm pixels. Figure 36.10 shows a block diagram of the three-tier unit cell. Figure 20.9 shows both a CAD rendering and cross-sectional SEM micrograph of the 3D LADAR chip. The APDs (Tier 1) were fabricated on a bulk silicon wafer. Tier 2 was fabricated using a 3.3 V, 0.35 μm FDSOI-CMOS process and Tier 3 was fabricated using a 1.5 V, 0.18 μm FDSOI-CMOS process. Each unit cell (pixel) contained six vertical interconnects between the layers (one between the APD and Tier 2, and five between Tiers 2 and 3).

36.2.4
Tohoku University's Neuromorphic Vision Chip

The group at Tohoku University have designed and fabricated a 3D imager that was inspired by the structure and functionality of the human retina [21, 22]. Figure 36.11a shows a simplified cross-section of the human retina and Figure 36.11b shows the

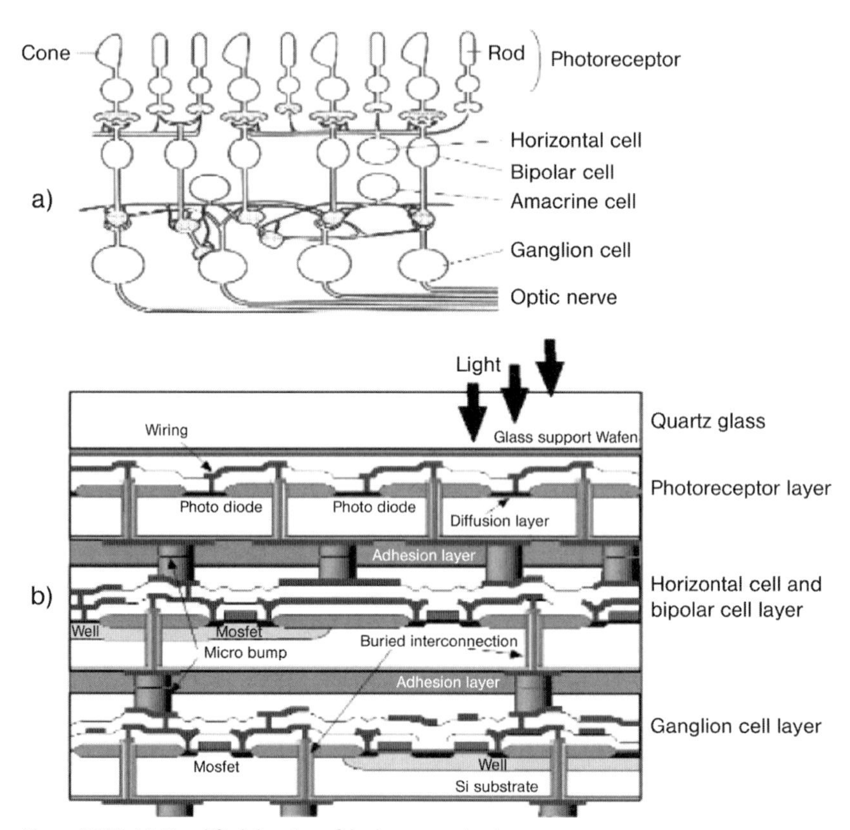

Figure 36.11 (a) Simplified drawing of the human retina in cross-section and (b) a schematic of the 3D IC fabricated by Tohoku University to mimic the human retina. From reference [22] (© 2001 IEEE).

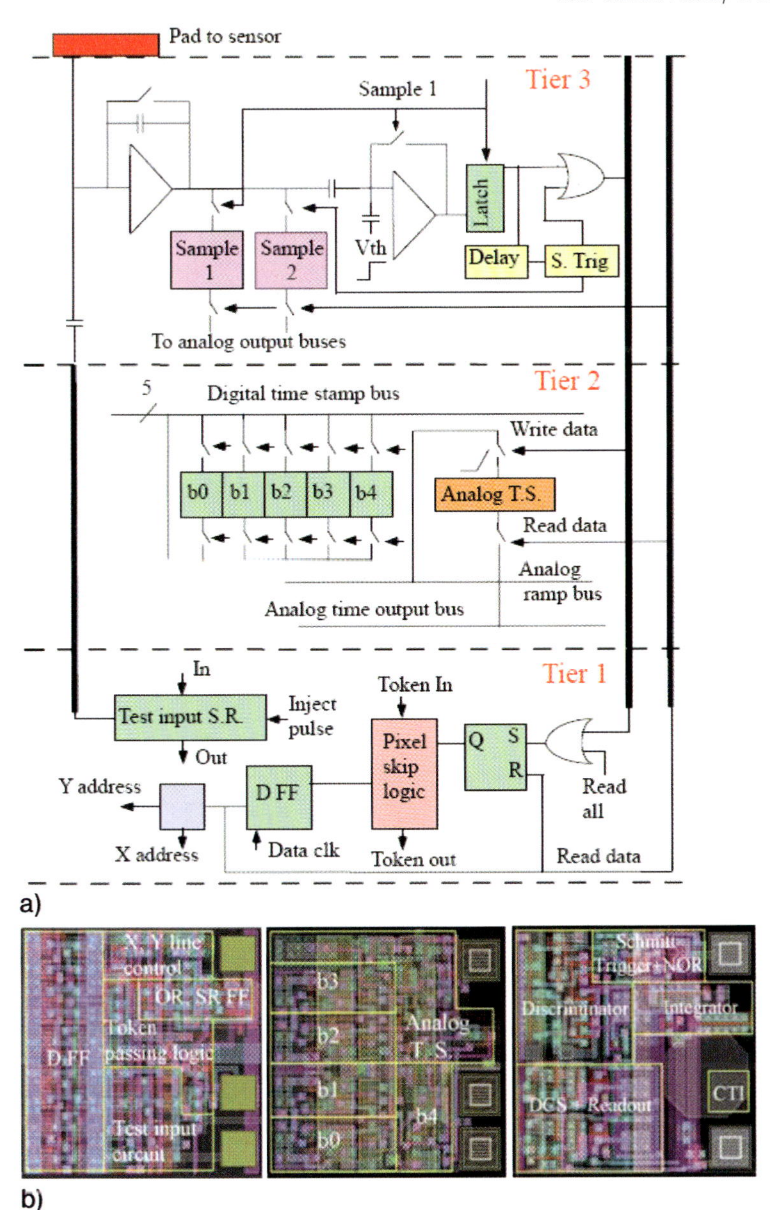

Figure 36.12 (a) Schematic illustrating how the Fermilab chip is partitioned into three tiers. The bold vertical lines are the vertical interconnects between the tiers. (b) Screenshots of the three separate tier layouts [24].

schematic of the 3D IC that was fabricated. The 3D integration process used buried polysilicon TSVs. After formation of the TSVs, the wafers were thinned and In-Au microbumps were used to bond and interconnect the IC tiers.

36.2.5 3D ROICs for High Energy Physics

The high-energy physics community has also started some activities in the area of 3D ROICs for advanced detector arrays [23, 24]. The group at Fermilab has designed a 3D ASIC that could potentially find application in the International Linear Collider (ILC) [24]. A 64×64 demonstration version of this 3D ASIC chip is currently submitted to the MIT Lincoln Laboratories multi-project run [Chapter 29]. Figure 36.12a is a schematic illustrating how the circuit is partitioned into three tiers. Tier 1 consists of digital logic (65 transistors), Tier 2 consists of time-stamping circuitry (72 transistors) and Tier 3 consists of analog circuitry and storage capacitors (38 transistors). The detector will be hybridized to Tier 3 following the 3D integration process. Each pixel contains 175 transistors and the footprint of the pixel is only $20 \times 20\,\mu m$. Figure 36.12b shows the pixel layout of the three tiers.

36.3 Conclusions

Research and development in 3D integration is leading to the realization of vertically stacked ICs that are interconnected with high density vertical interconnects. Pixelated devices, including both sensor and actuator arrays, are expected to benefit tremendously from 3D integration. Essentially, the pixel unit cell electronics will no longer be confined by the pixel 2D footprint. This will allow designers to shrink pixel footprints while also adding more complex electronics into each pixel.

References

1 Rogalski, A. (2003) Infrared detectors: status and trends. *Progress in Quantum Electronics*, **27**, 59–210.

2 Scribner, D.A., Kruer, M.R. and Killiany, J.M. (1991) Infrared focal plane arrays. *Proceedings of the IEEE*, **79**, 66–85.

3 John, J., Zimmerman, L., De Moor, P. and Van Hoof, C. (2004) High-density hybrid interconnect methodologies. *Nuclear Instruments and Methods in Physics Research A*, **531**, 202–208.

4 Tong, Q.-Y. (2006) Room temperature metal direct bonding. *Applied Physics Letters*, **89**, 182101.

5 Warner, K. *et al.* (May 2006) Layer transfer of FDSOI CMOS to 150 mm InP substrates for mixed-material integration. International Conference on Indium Phosphide and Related Materials, Princeton, NJ, pp. 226–228.

6 Norton, P. (2006) Third-generation sensors for night vision. *Opto-Electronics Review*, **14** (1), 1–10.

7 Zolper, J.C. and Biercuk, M.J. (2006) The path to adaptive microsystems, intelligent integrated microsystems. Proceedings SPIE, (eds R.A. Athale and J.C. Zolper), SPIE, 6232, pp. 1–14.
8 Horn, S., Norton, P., Carson, K., Eden, R. and Clement, R. (2004) Vertically-integrated sensor arrays – VISA, Infrared technology and applications XXX. in: Proceedings SPIE (eds B.F. Andresen and G.F. Fulop), SPIE, 5406, pp. 332–340.
9 http://www.darpa.mil/MTO/Programs/visa. (Oct. 2007)
10 Balcerak, R. and Horn, S. (2005) Progress in the development of vertically integrated sensor arrays, infrared technology and applications XXXI. in: Proceedings SPIE, (eds B.F. Andresen and G.F. Fulop), SPIE, 5783, pp. 384–391.
11 http://www.ziptronix.com/news/apr05_2007.html. (Oct. 2007)
12 Balcerak, R., Thurston, J. and Breedlove, J. (2005) Vertically integrated sensor array technology for unattended sensor networks, unattended ground sensor technologies and applications VII. in: Proceedings SPIE (ed. E.M. Carapezza), SPIE, 5796, pp. 1–6.
13 Temple, D., Bower, C.A., Malta, D., Robinson, J.E. *et al.* (2006) 3-D integration technology for high performance detector arrays, MRS Proceedings Volume 970, enabling technologies for 3-D integration. in: Proceedings MRS, (eds C. Bower, P. Garrou, P. Ramm and K. Takahashi), Material Research Society V 970, pp. 115–121.
14 Bower, C., Malta, D., Temple, D., Robinson, J.E. *et al.* (May 2006) High density vertical interconnects for 3-D integration of silicon integrated circuits. 2006 Proceedings 56th Electronic Components & Technology Conference, pp. 399–403.
15 Temple, D., Bower, C.A., Malta, D. *et al.* (2006) High density 3-D integration technology for massively parallel signal processing in advanced infrared focal plane array sensors. *Proceedings of IEDM*, San Francisco, CA.
16 Robinson, J., Coffman, P., Skokan, M. *et al.* (2006) Vertically integrated sensor arrays (VISA) for enhanced performance HgCdTe FPAs. Proceedings of Military Sensing Symposia, Orlando, FL.
17 Kinch, M.A. (2001) HDVIP™ FPA technology at DRS. *Proceedings of SPIE*, **4369**, 566–579.
18 Dreiske, P.D. (2005) Development of two-color focal-plane arrays based on HDVIP™. *Proceedings of SPIE*, **5783**, 325.
19 Suntharalingam, V., Berger, R., Burns, J.A. *et al.* (2005) Megapixel CMOS image sensor fabricated in three-dimensional integrated circuit technology. Digest Tech. Papers IEEE International Solid-State Circuits Conference, pp. 356–357.
20 Aull, B., Burns, J., Chen, C. *et al.* (2006) Laser radar imager based on three-dimensional integration of Geiger-mode avalance photodiodes with two SOI timing-circuit layers. Digest Tech. Papers IEEE International Solid-State Circuits Conference, pp. 304–305.
21 Kurino, H., Lee, K.W., Nakamura, T. *et al.* (1999) Intelligent image sensor chip with three dimensional structure. *Proceedings IEDM*, 879–882, Washington, DC.
22 Koyanagi, M., Nakagawa, Y., Lee, K.-W. *et al.* (2001) Neuromorphic vision chip fabricated using three-dimensional integration technology. *Proceedings ISSCC*, pp. 270–271, San Francisco, CA.
23 Yarema, R. (2006) Development of 3D Integrated Circuits for HEP, 12th LHC Electronics Workshop, Valencia, Spain, http://epp.fnal.gov/. (Oct. 2007)
24 Yarema, R. Development of 3D Integrated Circuits for HEP, FERMILAB-PUB-06-343-E.

37
Power Devices

Marc de Samber, Eric van Grunsven, and David Heyes

37.1
Introduction

In portable devices such as cellular phones and PDAs it can be advantageous to apply miniaturized components. An important driver for miniaturized components is the need for increased functionality, leading to optimal volume and area use. Miniaturization not only results in a reduced foot print of the components on the printed board but it can also have a positive effect on the device performance. For example, shorter interconnections lead to better high frequency behavior and shorter thermal pathway. The ultimate miniaturization is reached when packaging the component into a Chip Size Package (CSP). For single ICs this can be achieved by simple wafer level back end rerouting of the original bond pads into a pattern compatible with second level interconnect. This allows mounting the component directly onto the printed board in the application. Finishing the ICs with, for example, solder balls results in a Ball Grid Array (BGA) configuration. Such basic rerouting is not sufficient for vertical discrete components. Vertical discrete components have active interconnect on the front side and on the back side of the wafer and require addressing therefore both top side and bottom side of the device. In a plastic package (SO type) this is achieved by first mounting the die onto a leadframe using an electrically conductive die attach material such as solder, AuGe bonding or conductive glue. Wires are used to connect the top side of the die towards leads on the lead frame. This results in an electronic package with a planar interconnect scheme ready for mounting onto a printed board. In a flip chip CSP also all connectons (I/Os) have to be brought to one side of the final CSP product. If this is to be done on the wafer level then through wafer interconnect (TWI), using through wafer vias (TWV), is the method to address the back side. TWV is also called through silicon vias (TSV). Although TWI requires rather complicated technologies, wafer level processing, resulting in simultaneous fabrication of a large number of packages, limits the additional packaging cost. Limiting the additional total product cost will be of utmost importance if one considers using wafer level CSP packaging for low cost discrete semiconductor products.

Handbook of 3D Integration: Technology and Applications of 3D Integrated Circuits.
Edited by Philip Garrou, Christopher Bower and Peter Ramm

ISBN: 978-3-527-32034-9

37.2
Wafer Level Packaging for Discrete Semiconductor Devices

The specific boundary conditions when pursuing wafer level packaging for vertical discrete components are not limited to the sole but very challenging need for a through wafer interconnection. Discrete components, offering a single electrical functionality such as a diode or transistor or some small integrated functionality such as like resistor-equipped transistor or dual transistor, are typically very small dies. This means that the die area can easily be below 1 mm^2. There is an ongoing cost drive to further miniaturize the discrete dies. This has an immediate effect on the through wafer via formation method of choice and on the available area for realizing such via. Diode and transistor fabrication can be performed with rather simple processes and with a limited number of lithographic steps, yielding low cost wafers with huge numbers of dies per wafer. The packaging methods need to be very high volume and low cost so that die free package costs are very low. Concerning the package footprint, discrete packages are certainly not driving printed board assembly development. This means that the I/Os of discrete CSPs will have to match other CSP products such as mini-BGA. This limits the possibilities for I/O pitch reduction associated with small die use.

Consequently, the additional wafer level back end process steps to achieve a wafer level CSP have to be in line with boundary conditions such as low cost, high yield, meeting standard assembly rules.

37.3
Packaging for PowerMOSFET Devices

In powerMOSFET devices there is a trend towards miniaturized packages. The reasons for pursuing small size packages are not only related to the occupied board area but also aim at achieving the maximal die-to-package ratio and hence the maximal switching capacity per occupied area. Figure 37.1 shows this miniaturization driver, resulting in the final wish to achieve a 100% "filling" of the package.

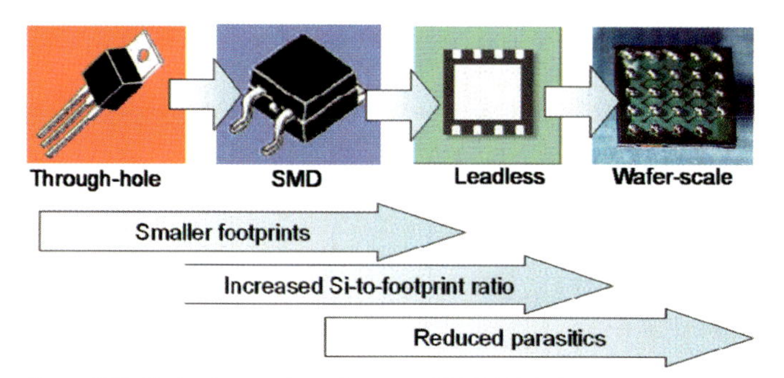

Figure 37.1 Packaging roadmap for powerMOSFET devices.

Table 37.1 Silicon to printed board ratio for various power package types.

	Max chip size (mm^2)	pcb Footprint (mm^2)	Si-pcb ratio (%)
SOT404	25	150	17
SOT428	13	66	20
SO8	9	30	30
LF-Pak	12	30	40
WSP			100

Table 37.1 gives the filling factor (active area versus package footprint this is footprint) for various package types. The table shows the max chip size, printed board footprint and Si-to-printed board ratiom for various packages and the roadmap towards the largest Si to printed board ratio. With a CSP the maximal Si to printed board ratio can be achieved, so the wished 100% "functional filling" of the package.

When packaging vertical MOSFET devices very specific requirements show up. Such devices are high power type, putting stringent requirements on the electrical and thermal connections. Electrical connections should be very low ohmic to reduce the internal dissipated energy. The generated temperature is to be conducted away from the die efficiently, through the package into the printed board or into additional heat sinking parts. In a standard plastic package these requirements are tackled by the use of optimized die-to-leadframe bonding, the use of multiple or thicker bonding wires, applying exposed die pads or additional thermal connections. Figure 37.2 shows the basic elements of a "standard" plastic package with a vertical power MOSFET.

The picture in Figure 37.2 shows a metal lead frame, on which the die is mounted, the MOSFETdie and the various bonding wires. The thinner wire on the left contacts the gate bonding pad and the three thicker wires connect the source contact of the MOSFET.

When introducing CSPs, as fabricated on wafer level, the specifications achieved with the above-mentioned packaging methods should at least be safeguarded.

Figure 37.2 Plastic packaged MOSFET (without top molding).

As mentioned already in general terms for all discrete semiconductor devices, powerMOSFET dies are also being miniaturized. Next generation MOSFET processes allow the transistor functionality to be realized with ever decreasing die sizes, resulting in tougher packaging challenges. For the realization of a wafer level package this has various effects. The available area for the via formation is decreasing while the connections have to keep meeting the electrical specifications. Smaller die sizes result in worse thermal pathways. Furthermore, the interconnect to the printed board has to fit on the die while still fullfilling the boundary conditions of the I/Os. This means that at least three I/Os have to be accomodated on the CSP for a transistor and that sufficient cross-section is required in the I/Os to allow sufficient dissipation towards the printed board.

The MOSFETs used in the development of our wafer level CSP packaging process are vertical trench MOSFETs. Figure 37.3 depicts a typical basic build-up of such MOSFET die.

As shown in Figure 37.2, the device has, on the front side, bond pads for source and gate of the transistor. Typical for such a device is that it is optimally designed for the main function of the transistor. This means that as much area as possible is used for the source area. The gate structure is optimized in switching performance and the area reserved for addressing of the gate in the package is minimized. The third electrical connection to the MOSFET transistor is the drain contact, which is almost always formed as a non-patterned back wafer contact, based on high conductive bulk Si combined with a thin film metal stack. This metal stack forms the ohmic connection towards that doped Si bulk material. When a flip chip package type is envisaged the drain connection is to be formed in some way on the front side of the die. As mentioned a maximal area of the die is used for forming the source. Reserving part of the active area of the front side for addressing the drain contact would limit the performance of the transistor unless the die is enlarged for that purpose. This is in conflict with the wish to increase MOSFET performance per die area. Therefore, the TWI should not consume active area. Concerning the interconnect, all required I/Os have to fit on the front side of the die, whatever the layout in the silicon.

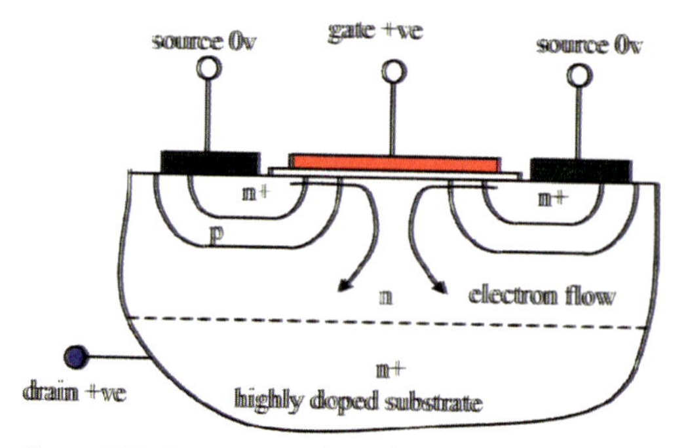

Figure 37.3 Cross section through vertical MOSFET.

37.4 Chip Size Packaging of Vertical MOSFETs

There are various ways to achieve a miniturized vertical MOSFET package. If one does not include the smallest size plastic packages that form the large market volume solution, two types can be considered, namely near- and real-CSP, which are discussed in turn below.

The first type are the so-called near-CSP products. Such packages are larger than the naked dies. Typically, these package types are leadless, which means that they do not have extended connections sticking out of the package body. Certainly, therefore, these near-CSP products can be considered highly miniaturized.

In the market there are, amongst others, two important MOSFET near-CSP packages worth consideration and comparison with real-CSP devices.

A first example of a near-CSP and leadless power device package is the Fairchild MOSFET BGA package [1]. This package is basically formed by mounting the die drainside-down into a metal part, forming in this way the connection to the drain of the MOSFET. By putting solder balls directly onto the die for gate and source contact and onto the top side of the metal block for the drain contact the package I/Os are formed. A drawing of this package type is shown in Figure 37.4.

Another near-CSP package is the DirectFET MOSFET package from International Rectifier [2]. This package type is comparable in build-up to the MOSFET BGA, but the metal part is now a metal can type part instead of a metal block. The electrical I/Os for gate and source are formed with LGA contacts on modified or rerouted bond pads on the original die and for the drain contact the clip-like contacts on the metal can are used. Figure 37.5 shows an artists' impression of the product.

The two near-CSP package types just discussed are not real CSPs (depending on the definition of CSP) and are certainly not fabricated on wafer level. Fabrication

Figure 37.4 MOSFET BGA from Fairchild.

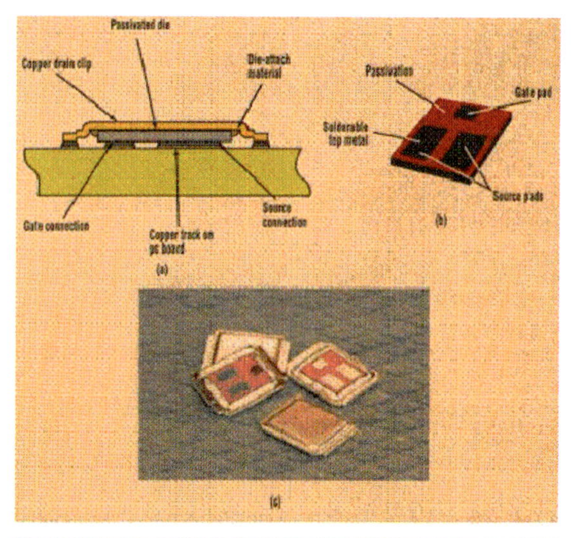

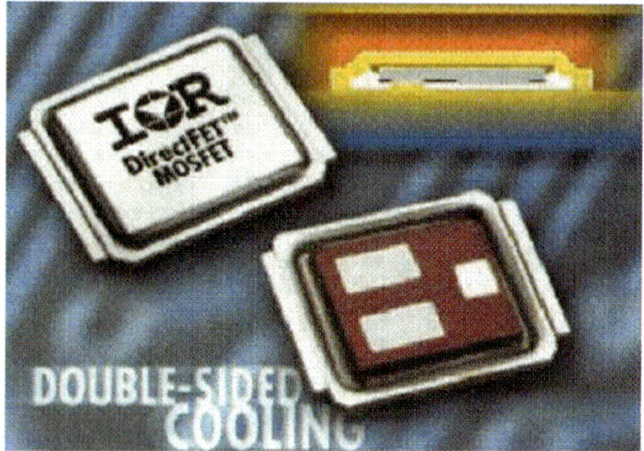

Figure 37.5 DirectFET package from International Rectifier.

on wafer level or on another type of multiple parallel processing (e.g. on matrix lead frames) is, however, essential to cope with the added back end processing cost.

The so-called real-CSP packages are typically fabricated on wafer level. In such CSPs the drain contact of the vertical MOSFET has thus to be brought to the other side of the wafer with the dies still being part of the total wafer. This is done with a TWI. This TWI might in principle be formed at any point in the wafer processing but front end integration would require other TWI processes than back end ones. In case of implementation of front end TWI the thermal budget of the back end process should be compatible with the TWI processes.

A straightforward contact method is to form a diffused plug through the epi-layer into the drain bulk silicon. This is a front end method and is, for example, used by International Rectifier. Figure 37.6 shows such an IR FC product [2].

The drain contacting method is based on forming a deep diffusion plug through the active layer and into the bulk substrate, which is on drain potential. This approach

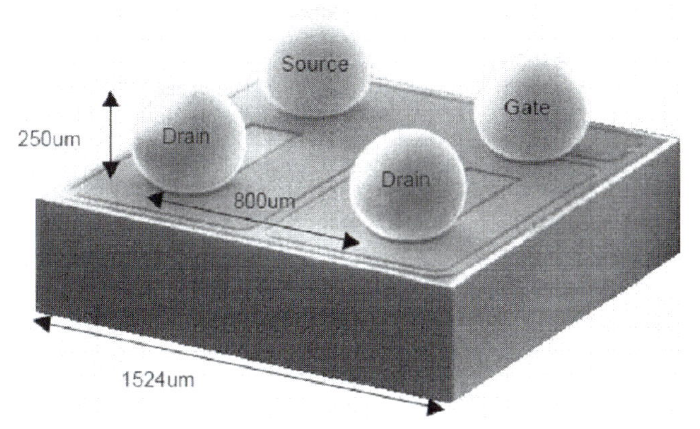

Figure 37.6 IR FC type MOSFET with diffused plug drain contact.

requires using part of the active source area of the MOSFET for forming that plug area. In addition, there is a limitation in the conductivity of such diffused front-to-back connection due to the limited solubility of dopants in the silicon. The final resistivity will in this case be a trade-off with consuming active area for the plug.

The theoretically best electrical connection can be achieved with a through wafer metal contact. Therefore, it would be optimal to use a real metal plug for the drain contact through the wafer.

At Philips a back end TWI was studied based on such a real-metal interconnect. This technology of choice is a BEOL type with non-specific IC design [3]. It can be compared in that way with certain 3D TWI SiP concepts. The TWI process starts after fully finalizing the IC wafer process. In principle this would then even allow a pre-testing of the transistors on wafer level (as far as is relevant, as the back end wafer level packaging does not allow rejection of failing dies from further processing).

Our approach is based on the fabrication of a TWI in the sawing/separation lane. Such an approach yields a product-independent TWI. Figure 37.7 shows the layout of the approach.

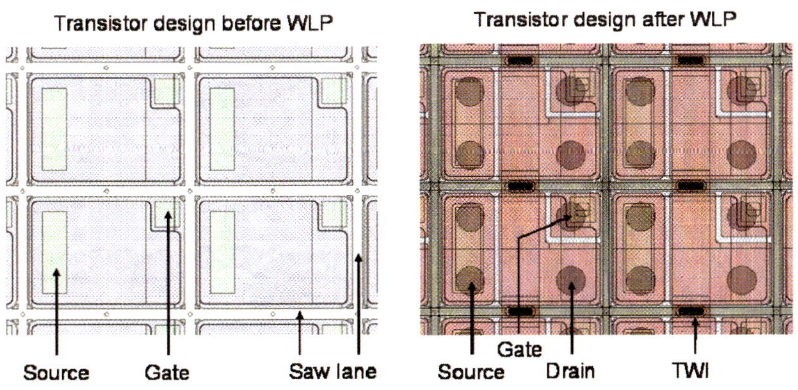

Figure 37.7 Layout of the Philips/NXP TWI for MOSFETs.

The left-hand drawing in Figure 37.7 shows a group of four MOSFET dies. Source and gate areas are depicted. In the right-hand drawing in Figure 37.7 additional process layers are visualized. The metal rerouting is shown in purple and the solder ball UBM can be seen as grey circles. The TWI via is drawn in black and is located in the saw lane. This is because obviously a metal plug also consumes active area. Owing to the intrinsic low resistivity of the thick Cu metal which was used, the via area can already be limited. To overcome the consumption of silicon area to the maximum a method was developed to locate the vias in loose area, more precisely in the saw lanes. By doing so, a via is shared between two neighboring dies. This physical via "disappears" during the final separation of the wafer into CSP products, and the only feature left is a half-via (which is basically a top-to-bottom metal strip on the two sides of the CSP).

In the next section, details of the technology and the decision factors for certain choices are further elucidated.

Here we compare the only two wafer-level real-CSP technologies, that is, either with a diffusion plug or with a metal plug. We have shown that for certain conditions a metal top-bottom connection always has better conductance performance than a diffused contact. Figure 37.8 compares RDSon as a function of % chip area that is used for the via for a diffused-plug type and a TWI type CSP (naked die).

Figure 37.8 shows two measured RDSon values for a Philips/NXP TWI based power CSP (green dot) and a commercial diffused-plug product (red diamond). The simulated curve shows that for the given die area and MOSFET type a diffusion plug can never result in as low RDSon numbers as with a metal TWI. This is because of the trade-off of the competing effects on resistance and consumed area in the diffused plug type device. For the metal TWI in the sawing lane area there is in theory no effect on the device size.

Obviously, there are physical limitations in the approach of making the TWI in the sawing lanes. First of all, the physical dimension, that is the width of the via, has to fit in that sawing lane. Therefore, minimal widths were used, in combination with an

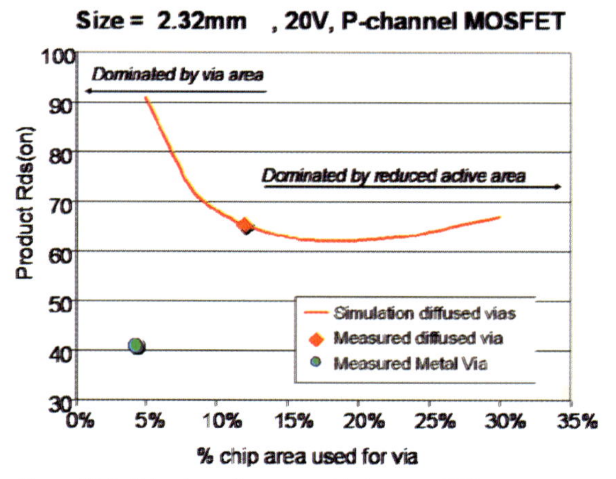

Figure 37.8 Side view of an assembled power CSP.

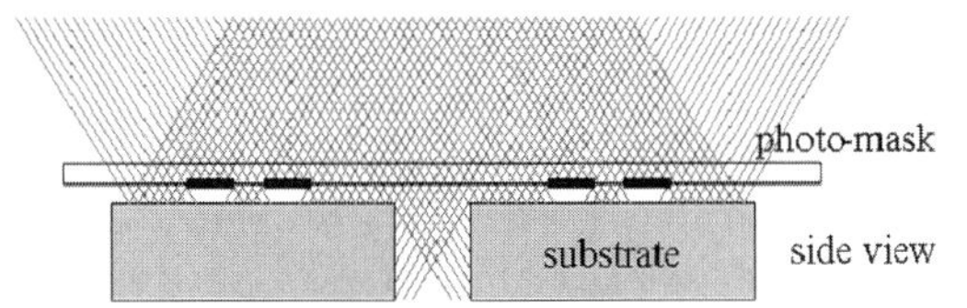

Figure 37.9 Picture of a metal TWI in a test device.

elongated shape to achieve low RDson. By using a shared via the via width has no effect on the resistance of that part. Only the width of the metal track, which is determined by the length of the via, defines, together with the metal thickness and the wafer thickness, the resistance of the TWI. This is portrayed in Figure 37.9.

One can see in Figure 37.9 and on the layout in Figure 37.7 that a via is shared between two dies. The via width in Figure 37.9 is 200 μm; clearly, there is only metal on the long edge of the via and not on the short edges. The processing of such TWI is further explained in the next section, and considerations and limitations with respect to the via width are discussed.

37.5 Metal TWI Process for Vertical MOSFETs

The wafer level packaging process starts from standard vertical D-MOS wafers that have finished the back end process [4]. At this stage the wafer has aluminium bond pads and a silicon nitride passivation. As discussed earlier, when packaging such a device in a plastic package the backside drain is soldered to the lead frame. The source and gate are connected to the lead frame by means of wire bonding (Figure 37.2). Also mentioned earlier is the fact that a wafer level packaged device requires all connections to be addressable on one side of the device. This has a strong impact on the design. Firstly, the real estate of devices needs to be large enough to allocate the solder balls. Secondly, a through wafer interconnect to reroute the back side drain to the front side is required. Thirdly, the new drain interconnect has to be formed on top of originally source area.

As a first processing step, with the purpose to allow rerouting on top of active source area, an additional BCB passivation layer is applied. This BCB layer establishes extra electrical separation between the rerouted drain connection and the source area and moreover some mechanical decoupling between the device and solder ball is achieved. Next, a through wafer via is created. There are various via forming processes available, such as Deep Reactive Ion Etching (DRIE), laser drilling, laser ablation, powder blasting, and so on. Each technology has its benefits and drawbacks [5]. Although laser processing is a single via process the relatively low number of vias in our application results in a lower cost solution compared to DRIE. Using DRIE etching, all vias on the wafer are made simultaneously. However the etching process is rather slow and requires expensive masking steps.

Therefore, laser processing was chosen as via method for the CSP fabrication. A triple YAG laser is used and the through silicon via is formed by laser ablation.

To prevent redeposition of silicon during this ablation process a PVA protection layer is applied. After rinsing off this protection layer no residues are formed on the device except at the very edge of the via. These residues, called burr, are a consequence of the "silicon melt" coagulate at the edge of the vias. This effect can be minimized by tuning the laser process. The potential risk of wafer weakening due to laser vias, as compared to the "stressless" DRIE method, was quantified by performing bending tests [5]. Another potential risk is the formation of "defects" in the active parts of the devices – called the affected zone effect. This effect starts from the edge of the fabricated via and can be quantified as a distance from that via edge. Using a sensitive npn bipolar transistor test die it was shown that the affected zone is limited to well below 20 μm.

As discussed earlier there is a wish to have little or preferably no sacrifice of silicon area for the via. The minimal via size, for an elongated via this is the via width, is determined by several factors. First, there is the physical limitation in size defined by the laser ablation process. For 150 μm wafer thickness a via width as low as 20 μm could be achieved. However, such vias show a large spread in geometry and shape. In addition, post-processing becomes impossible as there is no straightforward back-end technology available for depositing thick metal layers in such small vias. Another limitation in the via width is related to alignment. There is some spread in via position, defined by the laser system and the x-y table. This affects not only the required alignment of subsequent lithography steps, such as for the TWI metal patterning. The via should also allow later separation of the wafer into individual parts, using a straight cutting process. This means that all vias should be on a virtual line within the specifications of the separation process, allowing a single pass separation process for each separation lane.

It was found that 100 μm wide vias are very compatible with the available processes and the 150 μm wafer thickness. Although the technology has been developed for this via geometry the 100 μm via width is in conflict with the wish not to sacrifice silicon area. Typical separation lanes on the MOSFET wafer are 50 μm, so the implementation of a 100 μm wide via requires reservation of some extra area on top of the separation area and, therefore, a die layout change. Consequently, a 50 μm wide via was also considered, as this geometry is in line with the 50 μm separation lane and also with some available "clear" area between the separation area and active device. As a result, both 100 and 50 μm wide vias have been studied and the post-processing evaluated.

The formation of a conductive path from one side to the other of the wafer is the next process step. This needs some consideration of the functional specifications and the device build up to arrive at the best choice of materials and processes [6]. For this specific MOSFET device it was chosen not to apply an isolation layer in the via and to use a copper plated pattern for the top-to-bottom interconnect. Via isolation is not necessary because the area in the sawing lane is all through the wafer at drain potential.

The TWI and the rerouting layer on the front and back side of the wafer are formed by a Cu layer. To reduce the contribution of this redistribution pattern to the total RDSon of the power device this Cu pattern must have a thickness of 12 μm. Having a

Cu layer this thick in the via would mean that during later singulation of the CSP devices by sawing this Cu layer would be cut too. This is not acceptable as the Cu would pollute the sawing process. One way to avoid this is to pattern the Cu inside the via and thus clear the sawing lane. Thus, the redistribution pattern process requires simultaneous patterning on the front side, the backside and inside the via. This requires a specific three-dimensional patterning technique, consisting of conformal application of photoresist, exposure on front and back side and angle exposure inside the via. In other words a three-dimensional (3D) photolithography process is used [3]. This 3D photolithography process consits of the next process steps:

1. Front and bottom application of a TiCu plating base.
2. Enforcing the copper thickness by electrolytic plating.
3. Conformal application of photoresist (electro depositable photoresist, negative tone).
4. Inclined illumination on front side of the wafer, perpendicular illumination on backside and spray development.
5. Etching of thick copper.
6. Photoresist removal.

The plating base is sputtered using DC-magnetron sputtering. In two separate runs, 100 nm Ti and 500 nm Cu are applied on the front side and the back side of the wafer. Despite the quite vertical walls inside the via, which strongly reduces metal deposition stronly on the walls, the two-sided sputtering yields sufficient metal on the side walls to allow Cu plating. Copper plating is used to increase the metal thickness to 12 μm. This is carried out in a double-sided tank-plater using $CuSO_4$ based plating chemicals (Enthone LP1). For optimal conformal plating a dedicated ring holder is used that also mechanically supports the thin fragile wafer in the highly agitated solution. The CuTi layer is patterned by photolithography and etching. A conformal photosensitive layer is applied by using a electro depositable photoresist from Shipley. The use of this negative tone photoresist in combination with Cu etching allows the use of a lithography process with inclined exposure. Figure 37.10 shows the principle of inclined exposure [7].

Owing to shadow effect during the exposure only one-directional inclination is possible. This is why vias are all placed in one orientation (in x or y). This method has been used for the mainstream Cu patterning process.

A more sophisticated way to achieve local inclined exposure is the use of diffractive elements on the photomask [9, 10]. Tailoring these diffractive elements to the

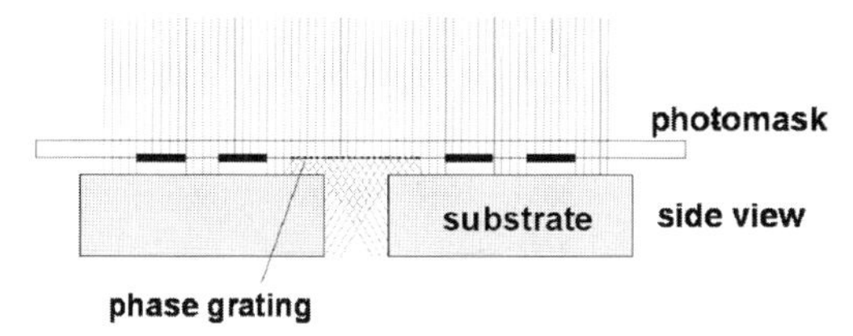

Figure 37.10 Schematic representation of inclined exposure method.

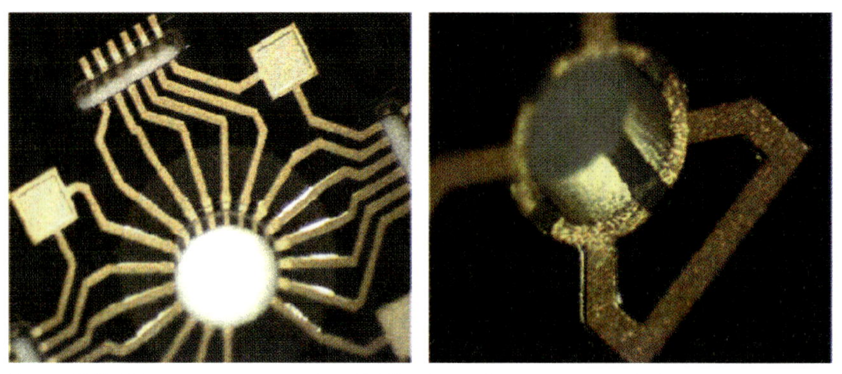

Figure 37.11 Schematic representation of diffractive optics exposure method.

exposure wavelength, the 0-order diffraction can be decreased while enhancing the first-order of diffraction. This is portrayed in Figure 37.10.

Patterning with this diffractive optics principle has several advantages. First, the patterning accuracy on the lateral surface is much better. In addition, standard exposure tools can be used and it allows varying the inclination angle locally on the mask. Especially, the latter advantage is very useful for patterning on all sides of a via. This enables multiple leads through a single round via. Test structures that suggest this are shown in Figure 37.11 [9].

The main drawback of this process is the high cost of the photomasks. However, this method would allow a single via to be shared between four, rather than two, TWI CSP devices, improving even further the optimal use of silicon area.

As mentioned earlier, the "simple" inclined exposure method was further used to make TWI powerCSP devices.

The next step after resist exposure is developing the photoresist, etching the copper and removing the photoresist. This results in a thick Cu redistribution layer on the front side, backside and inside the via. The result is shown in Figure 37.12.

Figure 37.13 shows, on the left-hand side, a picture of a device type based on a 100 μm wide via at the stage of Cu structuring. To allow a good view on the via metallization and the metal inside the via a 200 μm via type pattern is shown in the right-hand picture of Figure 37.12.

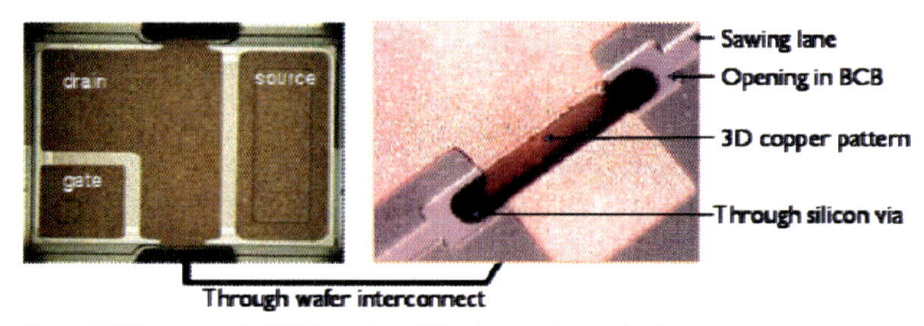

Figure 37.12 Any-angle TWI by using diffractive optics method.

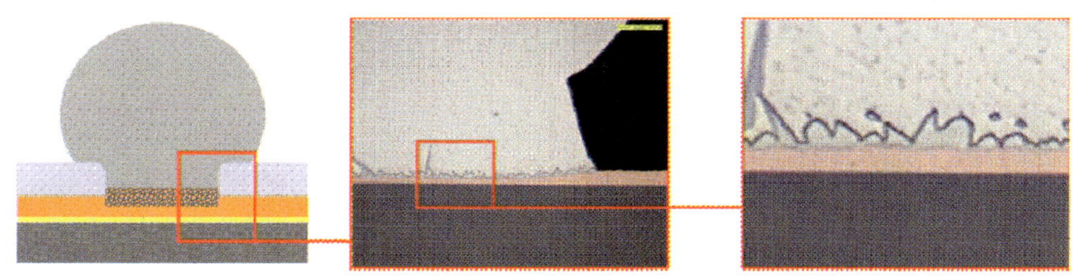

Figure 37.13 Pictures of fabrication status after Cu etching (see text for details).

The thickness of the Cu layer has an additional benefit. It allows placing solder balls directly on the copper redistribution layer without diffusion barrier. Obviously, part of the Cu will dissolve into the solder; however, only to the extent that the maximum solubility for that specific temperature and volume is reached. Calculations and experiments have proven that after five times reflow process approximately 6 μm of copper will be dissolved into the solder ball. This means that a residual 6 μm of copper is left at the interface with the device [6]. This is shown in Figure 37.13.

In this situation there is no need for finishing the Cu layer with a diffusion barrier.

To localize the solder balls and to prevent solder flowing over the Cu patterns a solder resist is required. Various processes are available to apply a solder resist. However, the presence of TW vias and the fragility of the wafer limits the choices. Two processes have been evaluated: lamination of solder resist and dip coating of BCB. Both processes give good results but both have drawbacks. Applying a laminate solder resist exerces force on the wafer such that any irregularity or particle on the wafer will cause wafer breakage. Dip coating of BCB, on the other hand, requires specific tooling, not only for the dip coating process but also for drying the wafer. Lamination and dip coating processes have given some issues in process automation.

The interconnect of the MOSFET CSP towards the printed board is established with a BGA interconnect. For the first developed power CSPs a rather straightforward BGA pitch of 0.8 mm could be used [6]. However, for the smaller dies, with last developments having resulted in dies <1 mm^2, 0.5 mm pitch BGA and even 0.4 mm pitch BGAs patterns are required [7]. There are several established techniques for wafer bumping, each having its own strengths and weaknesses. The two best known methods for applying solder balls are electroplating of solder or solder paste printing. However, for the power MOSFET CSPs it is critical that the bumps have a large cross-sectional area to keep the resistance low. As mentioned earlier, the RDSon of the transistor is its major electrical parameter. Our technology partners DEK and TUB/FhG Berlin have developed for this a stencil printing technique for applying preshaped SnAgCu solder balls [8]. Printing preformed solder balls allows deposition of a high volume of solder to maximize the mentioned interconnect cross section within the physical limits of bump pitch and solder behavior. The challenge was not only in achieving JEDEC size solder balls with high yield but also to be able to process on fragile 6-inch wafers that are 145 μm thick and which are

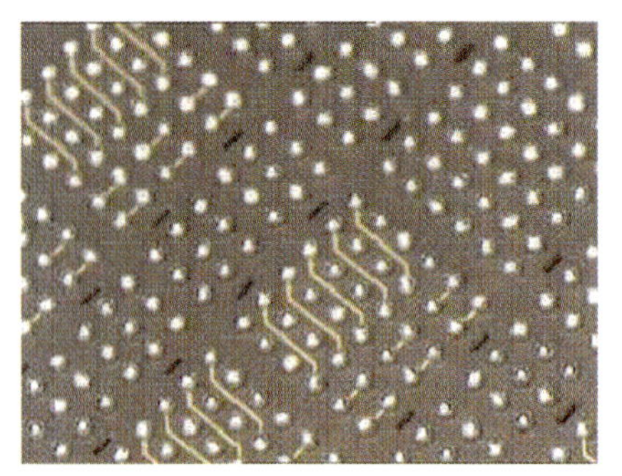

Figure 37.14 Pictures showing evidence of Cu consumption during multiple reflows.

weakened by the TWI vias. In Figure 37.14 a photograph is shown as an example of these printing results.

On the picture in Figure 37.14 a test patterns for mimicing the real wafer level processing conditions can be seen. The main challenging factors for solder bump printing are in the TWI vias and the thin wafer. The 145 μm wafer might bow during the printing process or get damaged during the ball remelting process. The vias, on the other hand, can disturb the printing action due to loss of vacuum in the holding tool.

The tests done by DEK in the framework of the Blue Whale EU project have demonstrated that a highly repeatable bump height of 260 μm could be obtained for the 0.5 mm pitch BGA pattern [8]. The bump yield for this process on this type of wafers was established at over 99.9%.

The final process step is singulation. For the vias 100 μm wide this is done by sawing through the sawing lanes. As discussed, the metal interconnect in the 100 μm vias is patterned in such way that the sawing can be done without cutting through metal, allowing standard dicing to be used. In addition, the width of the vias and the position of the vias along the sawing lane is within the tolerances of the sawing process with respect to sawing blade width (20 μm) and sawing accuracy.

As already noted, for the 50 μm via type CSPs patterning of the metal in the via is no longer possible, and so standard dicing becomes difficult. These devices have been separated by using a cutting laser. Figure 37.15 shows the results of such separation.

The top two pictures in Figure 37.15 are views of a wafer while the bottom two pictures show separated devices seen from top and bottom side.

The tolerance of the laser cutting process and the small laser spot used allows separation with minor yield loss. Obviously, for the 50 μm vias tolerances are tight and process optimization of the laser via formation process and the laser separation process are required.

Using cross sectioning, the via and other features of the power CSP are shown in Figure 37.16.

In the left-hand picture of Figure 37.16 the Cu wrapped around the device and through the via can be seen. The solder ball attached to the thick Cu is also shown.

Figure 37.15 A test wafer with preformed solder balls.

The layer build-up of the device can be seen in Figure 37.17. In this picture a complete power MOSFET CSP is represented. The place of cross sectioning is shown by the yellow line on the right-hand image.

The patterned Cu metallization and solder ball attached to the source area (left-hand side) and gate area (right-hand side) are clearly visible. In addition, the back side patterning of the Cu, including etched marking, can be seen.

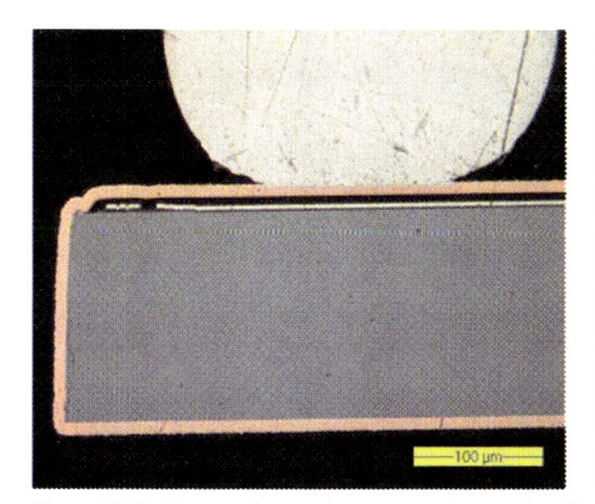

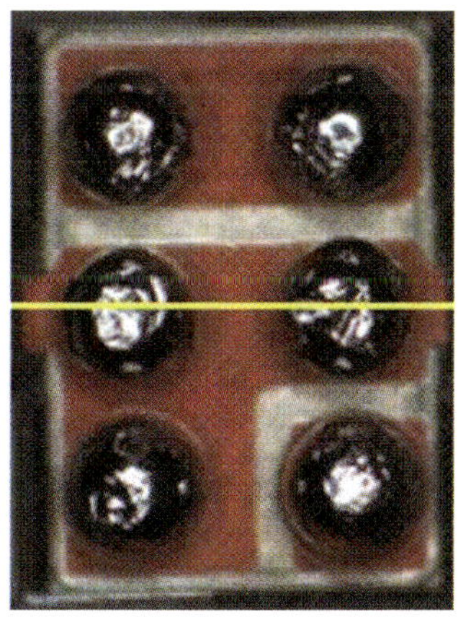

Figure 37.16 Pictures showing laser separation results of power CSP devices with 50 μm via.

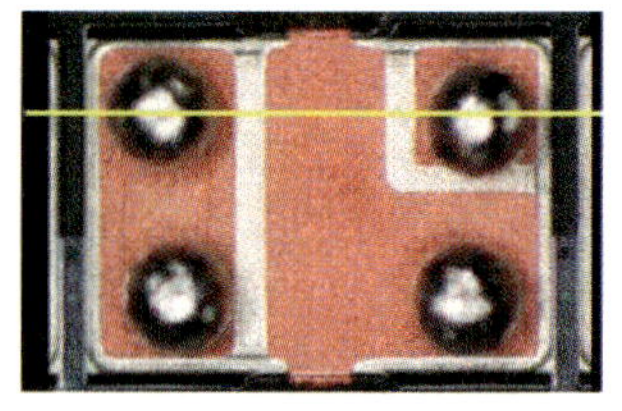

Figure 37.17 Cross section through power CSP; cutting area is shown by the yellow line in the right-hand picture.

37.6 Further Evaluation of the TWI MOSFET CSPs

The powerCSP products have been electrically and thermally evaluated and next subjected to assembly trials and reliability testing [11, 12].

As mentioned earlier, the on-resistance of the packaged MOSFET is of prime importance. This criterium is therefore a main monitor of the electrical quality. Next to that the naked die package has a potential risk in leakage currents (e.g. gate-to-source leak). This parameter is also checked.

To allow electrical evaluation the devices were mounted on printed boards using standard assembly techniques (based on solder printing, mounting and reflow soldering). Automotive test conditions were chosen for device evaluation and, therefore, polyimide type printed boards were required. In Figure 37.18 an assembled CSP product is shown as an example.

The side view picture in Figure 37.19 shows the attach by the solder balls to the printed board.

The device in Figure 37.19 is a type with a 100 μm wide via TWI. In this image the cut-through via and the patterned Cu in that via are clearly visible. The picture shows the envisaged large solder cross section (for achieving a low RDSon) and a high stand-off (for thermo-mechanical reliability), which is the result of using preformed solder balls.

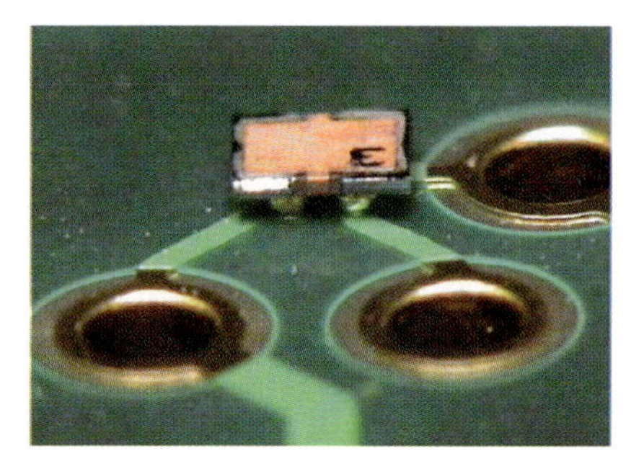

Figure 37.18 Cross section showing complete device (see text for details).

Figure 37.19 Metal TWI CSP mounted on evaluation board.

Table 37.2 gives experimental RDSon values for our metal TWI CSP devices. These numbers are to be compared with the product requirements from existing plastic-packaged similar devices.

Transistor type numbers listed in the table are the various types under study. Type 1 has a 100 μm wide via and four solder balls at a pitch of 0.8/0.5 mm (in x and y direction). Type 2 has the same BGA layout and pitch but is based on a 50 μm via. Types 3 and 4 are again 100 μm via based but have deviating BGA patterns (four balls on 0.5 mm pitch for type 3 and six balls on pitch 0.4/0.5 mm for type 4).

The data in Table 37.2 show that for the various powerCSP types the chosen TWI type resulted in within-specification RDSon values, even for the type with 50 μm via (type 2) (Figure 37.20).

A possible issue with miniaturized devices in general and CSPs in particular is the thermal performance. To verify this the thermal properties of a specific type power MOSFET CSP were compared with traditional packages with the same die inside [6]. The thermal resistance from junction to ambient was measured. The CSP component is a 0.8 mm pitch BGA type with 24 solderballs. Its Rth_{j-a} is $7\,KW^{-1}$. A SOT404 packaged device with an internal heat sink has a Rth_{j-a} of $0.5\,KW^{-1}$ but an often used device without heatsink such as a SO8 packaged device has a Rth_{j-a} of $25\,KW^{-1}$. So the thermal resistance of a CSP packaged product is in-between the two conventional packages. However, thermal modeling has revealed that in real applications the net heat transfer is largely limited by the printed board, and the CSP fullfills the thermal requirements.

The electrical and thermal evaluations are clear decision elements that the approach with TWI based on metal and fitting fully in the sawing lanes

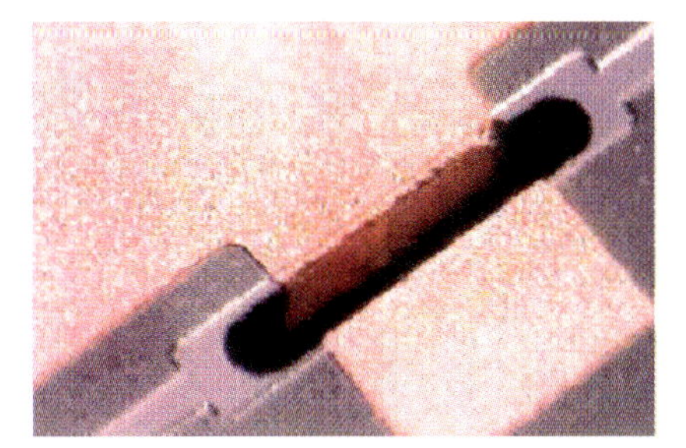

Figure 37.20 Comparison diffused versus metallic contact for RDSON for various die sizes.

Table 37.2 Mean RSDon values for various powerCSP transistors and specifications as a reference.

	Rsdon (mΩ) at Vgs = 2.5 V and ID = 1.5 A		
Transistor type	**Mean value**	**Spec min**	**Spec max**
1	38.0	36.0	43.0
2	39.1	36.0	43.0
3	38.6	36.0	43.0
4	39.0	36.0	43.0

(so without sacrificing silicon area) is fully valid for the considered power transistor devices.

Other important electrical specifications such as gate-to-source leakage, breakdown voltages and so on are also not negatively affected by the conversion of the standard plastic packaged devices into a naked CSP device. This proves the legitimity of the choice of TWI concept, materials and the processes.

37.7 Outlook

Although the number of fabricated wafers during the study was rather limited, we conclude that the process is feasible. Device yield losses are limited and detectable and the initial high risk of wafer loss (breaking of wafers during thin film processing) has been solved by proper post-processing of the wafers after thinning. A next step into further development towards industrialization would be to obtain more evidence of the reliability of the parts. The studied parts yielded good reliability results (temperature cycling, temperature shock, HAST and pressure cooker) but the number of tested devices is too small to draw final conclusions.

Industrialization potential obviously not only depends on technological possibilities but is largely determined by other factors. For example, the decision as where to install the process, either as a back-end in a wafer fab facility or in an assembly factory, is a difficult one. For a wafer fab certain process steps are not standard, so equipment is not available (e.g. laser system) or materials introduce risk of contamination in that specific fab (e.g. Cu metallization). In addition, the CSP process can create a non-balance in the equipment utilization, depending on fabrication volumes. This can have a large negative effect on the final cost price of the devices.

In addition to this there is still hesitation over market acceptance of such devices. This holds for leadless packages in general as it is difficult to check the assembly quality. For miniature power devices in particular there is also the possible customer perception that a small device might not be capable of handling the high currents and power. These market restrictions are certainly not related to the discussed concept with metal TWI only, but are valid for all (near)CSP discrete devices and power devices.

References

1 www.fairchildsemi.com (2008).
2 www.irf.com (2008).
3 Nellissen, A. *et al.* US patent 6240621B1.
4 Bloos, H. *et al.* US patent 6420755B1.
5 Polyakov, A. *et al.* (2004) Comparison of Via-fabrication techniques for through-wafer electrical interconnect applications. Electronic Components and Technology Conference.
6 van Grunsven, E. *et al.* (2003) Wafer level chip size packaging technology for power devices using low ohmic through hole vias. 14th European Microelectronics and Packaging Conference and exhibition, Friedrichshafen, Germany, June 23–25 2003.
7 de Samber, M. *et al.* (2004) Through wafer vias for power transistors. 3rd European Microelectronics and Packaging Symposium, Prague, Czech Republic, June 16–18 2004
8 Various authors from the EU Blue Whale consortium, special session at the 3rd European Microelectronics and Packaging Symposium, (2004) Prague, Czech Republic, June 16–18 2004
9 Nellissen, T. *et al.* (2003) A novel photolithographic method for realizing 3-D interconnection patterns on electronic modules. 14th European Microelectronics and Packaging Conference and Exhibition, Friedrichshafen, Germany, June 23–25 2003.
10 Nellissen, T. *et al.* (2004) Development of an advanced three-dimensional MCM-D substrate level patterning technique. 3rd European Microelectronics and Packaging Symposium, Prague, Czech Republic, June 16–18 2004.
11 de Samber, M. *et al.* (2004) Through wafer interconnection technologies for advanced electronic devices. EPTC Conference, Singapore.
12 de Samber, M. *et al.* (2005) Fabrication and evaluation of miniaturized CSP power transistors, PROC EMPC, Brugge, June 12–15 2005.

38
Wireless Sensor Systems – The e-CUBES Project

Adrian M. Ionescu, Eric Beyne, Tierry Hilt, Thomas Herndl, Pierre Nicole, Mihai Sanduleanu, Anton Sauer, Herbert Shea, Maaike Taklo, Co Van Veen, Josef Weber, Werner Weber, Jürgen M. Wolf, and Peter Ramm

38.1
Introduction

This chapter reports on the application of 3D integration technologies in wireless sensor systems. It essentially focuses on the research approach proposed by e-CUBES (http://ecubes.org), a European integrated project to investigate and develop wireless small sensor cubes that communicate among each other. e-CUBES addresses various multi-disciplinary applications in the field of wireless sensor networks, with emphasis on: (i) distributed smart monitoring for aeronautics and space applications, (ii) wireless sensor networks for health and fitness and (iii) distributed intelligent automotive control. 3D integration is considered a key enabling technology for e-CUBES. In general, the goal of e-CUBES is to advance the micro-system technologies to allow for the cost effective realization of highly miniaturized, truly autonomous systems for ambient intelligence. A possible roadmap of for e-CUBES is also discussed at the end of the chapter.

Pioneering work in the field of autonomous systems for ambient intelligence was carried out within the Smart Dust project imagined by a group of the University of California at Berkeley [1, 2]. The goal of the Smart Dust project was to build a self-contained, millimeter-scale sensing and communication platform (Figure 38.1) for a massively distributed sensor network. Their device was proposed to be the size of a grain of sand, containing sensors, computational ability, bi-directional wireless communications and power supply, while being inexpensive enough to deploy by the hundreds. The science and engineering goal of the project was to use state-of-the art technologies (as opposed to futuristic technologies). In contrast, e-CUBES project proposes to take advantage of most advanced 3D integration technology [3–6] to realize tiny communicating objects.

Intelligent sensing of the human environment is one of the key future applications of highly miniaturized sensor nodes that build up a wireless ad-hoc sensor network

Handbook of 3D Integration: Technology and Applications of 3D Integrated Circuits.
Edited by Philip Garrou, Christopher Bower and Peter Ramm

ISBN: 978-3-527-32034-9

Figure 38.1 Smart Dust multi-chip node displaced featuring in a 4 mm^3 volume a solar cell power source, temperature, light and acceleration sensors, an 8-bit ADC, and bidirectional optical communication. (Reprinted with permission, IEEE copyright.)

with access to other global networks. These autonomous sensor systems will be, among others, used in logistics, traffic control, households, as well as for functional monitoring and stimulation inside the human body. Such multi-functional sensor systems will make contributions in new fields like ambient intelligence, security [e.g. unambiguous identification of people (biometrics) and objects as well as situation appraisal], food quality controlling, health monitoring, production supervision, and so on.

In Germany, Fraunhofer IZM proposed the "e-Grain" concept ("electronic grains") for highly tiny, miniaturized autonomous sensor nodes that are able to build a wireless network to communicate with each other (Figure 38.2a). They are freely programmable and to a certain degree modular. At the same time, they are universal

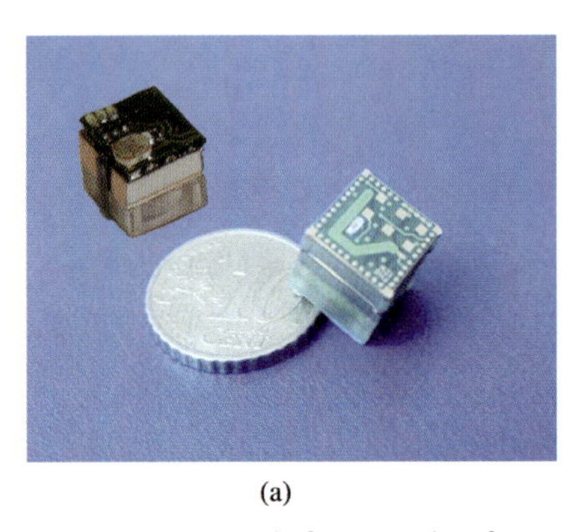

(a)

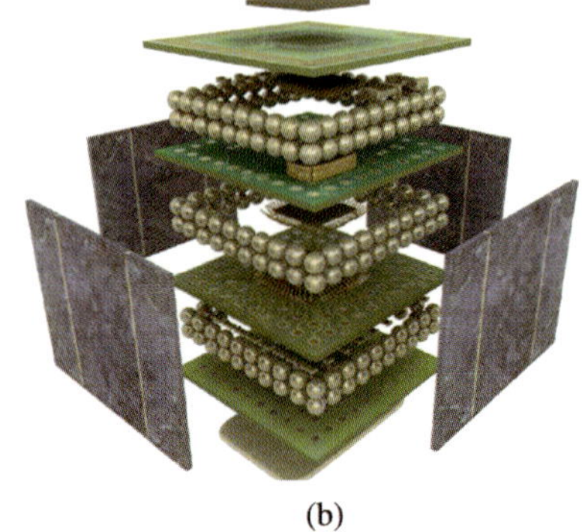

(b)

Figure 38.2 (a) Fraunhofer's examples of wireless sensor nodes for temperature and light measurement realized with advanced Chip on Board (COB) technologies using stacked printed circuit boards and folded multilyer flex substrate. The technical parameters of the sensor node are: operating frequency, 2.4 GHz; operating range, 1 m; repetition rate, 1 s^{-1}; power supply, 2 × 1.5 V batteries; operating time, >500 hour; size, 10 × 10 × 10 mm (with power supply); (b) IMECs 1 cm^3 EEG/ECG SIP demonstrator.

and partly specialized, through the integration of specific sensors. Fraunhofer IZM works on different integration levels to realize these wireless sensor nodes. One is focused on advanced board level integration using flip chip and wire bond technology as well as embedding of devices in printed circuit boards.

Another example in Europe concerns IMECs efforts on 3D-SiP (System in Package); in their approach, individual layers are processed in parallel, tested and, finally, stacked as "known-good devices." This goal is likely to be reached first for 3D-SiP, followed by 3D-SoC (System on Chip). For the realization of a 3D-SiP (Figure 38.2b), a relatively low interconnect density in the third dimension is used and each layer should, therefore, consist of a well-defined sub-circuit block. These components, which may be individual SiP packages, can be pre-tested ("known-good-SiP") to ensure a high final assembly yield.

At a European scale the concept that unites effort from different industries, research and academic groups from different countries is called e-CUBES (electronic cubes); in this chapter we make systematic use of this concept (described in Section 38.2).

In terms of design, hardware, technology and software, such concepts represent an entirely challenge and require a synergy of individual integration technologies at an exceptionally early stage. The realization of such extremely miniaturized and highly complex sensor systems requires the development of specific integration technologies. The Heterogeneous Integration of devices and components originating from different technologies for sensing, electrical signal and data processing, wireless communication, power conversion and storage is a key enabler for the realization of such sensor nodes. Solutions have to be found for media access, sensor signal processing and storage, data communication and power management. The Hetero System Integration concept is particularly gaining importance due to the shorter time to market cycle and a high degree of flexibility at lower cost and risk assessment compared to SoC solutions.

The technology poses particular challenges with regard to the desirable sizes (a few cubic mm), the need to achieve continuous operation through an integrated or external wireless power supply and the necessity of allowing multiple nodes to communicate in the network.

38.2 e-CUBES Concept

e-CUBES is a large multidisciplinary integrated project demanding a strong interaction between the consortium partners and complex coordination. Owing to the challenges simultaneously raised in e-CUBES by the technology platform for 3D integration, 3D system architectures, communication interfaces and protocols and driving applications, a consistent analysis of the 3D functionality, design and reliability, with scientific advice from both academia and industry is need for success.

The general concept of e-CUBES (Figures 38.3 and 38.4) clearly identifies the needed functional sub-blocks that should be 3D-integrated in a single functional

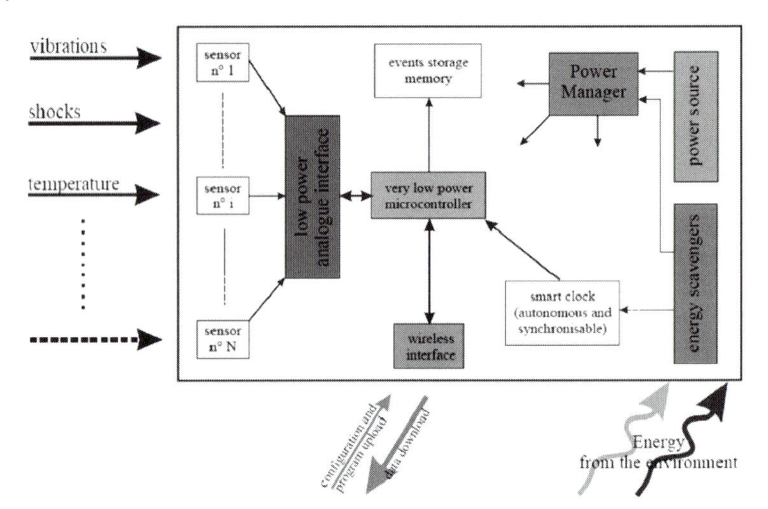

Figure 38.3 General architecture of e-CUBES, including functional sub-blocks.

object called the electronic cube (e-CUBE). An e-CUBE object has some key capabilities in a very limited volume: sensing, analog IC interface, microcontroller, memory, wireless interface and power management (possibly including energy scavengers). Clearly, this configuration can be very different from one application to another and not all layers are useful or needed in all applications. The sensing and communication are essential for practically all types of demonstration and applications. It is very important from the conceptual stage to evaluate and realize the need for cost effectiveness of such a highly miniaturized multi-function system, eventually pointing to the use of 3D integration technologies.

Another key feature of the e-CUBES is their ability to support the implementation of wireless sensor networks (Figure 38.5), a field that today is considered of strategic

Figure 38.4 Drawing of a 3D object (e-CUBE) that integrates all the functional layers described inside the e-CUBES box from Figure 38.3.

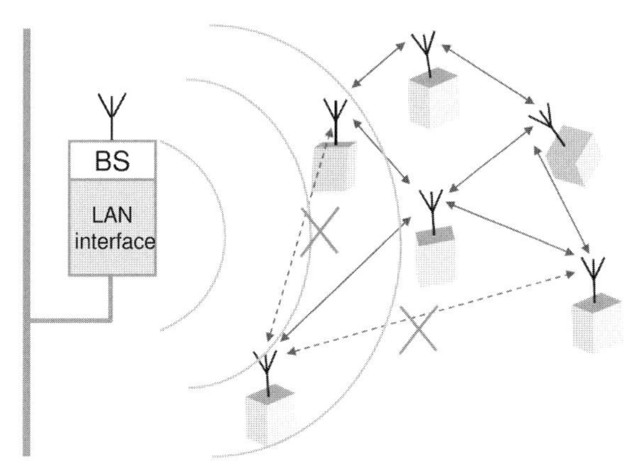

Figure 38.5 Wireless sensor network based on e-CUBES.

importance for Ambient Intelligence in Europe. The communication interfaces and specific communication protocols should be developed together with the hardware technology to define the operation of the whole system, according to the application specification.

e-CUBES strategy and vision particularly fits with the position of European IC industry, which, contrarily to some big US companies, is mostly interested in bringing system and application oriented solutions, targeting an increased functionality per unit of volume.

38.3 Enabling 3D Integration Technologies

For future applications, wireless sensor systems will be highly miniaturized. 3D integration technologies have to be applied because of their relevant benefits: extreme system volume reduction, reduction of power consumption (for lifetime enhancement), reliability improvement and low cost fabrication for meeting mass-market requirements.

The application-specific benefits of the 3D approach are illustrated by:

- Extreme miniaturization – (extreme system volume reduction) for implantation (health), for tire-vulcanization (automotive), for weight reduction (aeronautics).
- Reduction of power consumption (up to 30%) – for life time enhancement reduction of wiring length by vertical through chip interconnects.
- Significant enhancement of reliability – as required by aeronautics and automotive by reduction of about 50% of interconnects.
- Flexible and scalable module kits – as required by aeronautics and automotive compared to lateral system integration.
- Yield improvement – by sub-modules approach versus embedded (SoC); avoidance of monolithic technology mix.

- Improved thermal behavior by inter-chip vias – important for extreme temperature ranges of space, aeronautics and automotive environment.
- Safety due to affordable redundancy.
- 3D configured antennas – to acquire omni-directional pattern for communication important for higher frequencies caused by extreme miniaturization important in environments with limited line-of-sight.
- Dramatic system cost reduction for meeting mass-market requirements – by wafer level 3D integration of Known Good Dies (chip-to-wafer technologies).

In general, no one 3D integration technology is suitable for the fabrication of the large variety of 3D integrated systems. Moreover, even one single product may need several different technologies for a cost-effective fabrication. Wireless sensor systems represent an excellent example of the need of a suitable mixture. Consisting of MEMS, ASICs, memories, antennas and power modules they can only be fabricated in a cost-efficient way by application of specific optimized 3D technologies for the integration of the different sub-modules. Regarding the performance of the enabling technologies, two relevant 3D integration concept families are preferentially chosen in the e-CUBES project:

1. Wafer-level die stacking (without and with through-Si vias).
2. 3D Assembly of sub-modules.

3D technologies of the first category are described in "Wafer-Level 3D System Integration" (Chapter 16). Within e-CUBES the ICV-SLID technology [3] represents the mainstream concept for 3D integration of integrated circuits, as controllers and memories, by chip-to-wafer stacking with very high interconnect densities (10^4–10^6 cm^{-2}). Moreover, the fabrication of wireless sensor systems needs technologies optimized for integration of sensors and 3D assembly of sub-modules, as radio- and power-modules. In consequence, main objectives of the e-CUBES project are the development of specific integration technologies, namely for sensors, energy sources and antennas.

Integration of MEMS sensors for wireless sensor systems requires careful attention to the special features of the devices. A MEMS sensor has movable parts and the functionality of the device typically relies on a well-controlled mechanical situation. Some MEMS devices are quite bulky and need a certain volume, several MEMS devices cannot tolerate large packaging stresses, and many MEMS sensors (i.e. pressure sensors) need a window to the environment. Such requirements give limitations as to where and how a MEMS sensor can be integrated in a wireless sensor node, and the limitations are application specific.

Despite limitations, examples exist of successful stacking of MEMS and other devices, see for instance the ASIC stacked on top of the 3-axis accelerometer in Figure 38.6. However, all the interconnections are realized with wire bonds in the given example. To achieve extreme miniaturization of a wireless sensor node, and to benefit from all items listed above for a complete 3D approach, all wiring should be realized through the stacked devices. Wire bonds should be replaced with through-Si vias combined with electrical and mechanical interconnection points between the

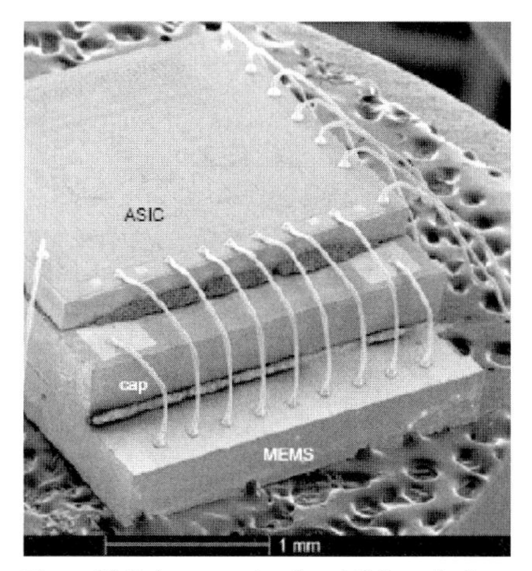

Figure 38.6 An example of an ASIC stacked on top of a MEMS sensor, where all interconnections are realized using wire bonds. [Source: Kionix KXM52 3-Axis Accelerometer (www.chipworks.com).]

devices. The removal of long wire bonds is especially beneficial for sensors based on a capacitive sensing principle (rather than, that is, a piezo-resistive sensing principle), since parasitic effects will be reduced.

The challenge is to find through-Si via technologies and interconnection technologies suitable for a MEMS sensor. In general, thinning of wafers down to around 50 μm and below makes it easy to realize reliable through-Si vias, but thinning is not always a viable solution for MEMS. Many MEMS sensors require mechanical stability and strength, or a certain volume or mass. Thin MEMS devices exist, but a typical thickness of a MEMS wafer is still of the order of 200–400 μm. The smallest holes that can realistically be made through such a wafer thickness have a diameter of about 10–20 μm. PlanOptik (www.quarzglas-heinrich.de/html/planoptik.html) produces such thick wafers with vias by structuring a silicon wafer by DRIE and filling the cavities with a borofloat glass. After filling, the wafer is ground and polished, leaving a wafer containing Si pins isolated by glass trenches. Silex (www.silex.com.au) proposes an alternative technology for equally thick wafers where the silicon wafers are etched by DRIE. A dielectric that fills the trenches is deposited, and finally the wafers are ground on the back-side to expose the filled trenches. SINTEF has presented a solution for 300 μm thick silicon wafers where the walls of hollow vias with aspect ratio (AR) 15 are covered with poly silicon [6]. By keeping the vias hollow, the costly and time-consuming process of completely filling 10–20 μm diameter holes is avoided. Hollow vias also eliminate the reliability concerns related to the large mismatch in coefficient of thermal expansion between the substrate silicon and filling materials, like for instance Cu. Reliability concerns are application specific, but several MEMS sensors are quite sensible for any thermo-mechanically induced stress. For the three mentioned through-Si via technologies the typical via pitch is $\sim$100 μm and the through-Si via resistance is in the range of a few ohms.

For the electrical and mechanical interconnection between a MEMS sensor and other devices in a 3D integrated system, flip chip bonding is the most relevant technology if wire bonding is to be avoided. Flip chip bonding is a mature process and there are numerous commercial suppliers of flip chip services, but again the special features of MEMS must be considered. A MEMS wafer is typically smaller and has a lower yield than an ASIC wafer. Chip-to-wafer bonding is therefore in general expected to be a more economical and flexible solution than wafer-to-wafer bonding. The chips to be bonded will often all be silicon, so global stress issues are negligible (local stress issues may still be relevant since the interconnecting material is not necessarily thermally matched to the silicon). As a consequence, small bumps and a small stand-off height are desired, in contrast to the large bump size used for flip chip bonding a silicon device onto a ceramic or plastic substrate. To comply with possible post-processing of the 3D stack, the bonding should be able to withstand temperatures in the range of 200–300 °C without significant deterioration. Finally, environmental concerns demand the use of lead-free materials. Examples of technologies that fulfill the mentioned requirements (chip-to-wafer bonding, a low gap size, high temperature tolerance and lead-free) and are suitable for several MEMS sensors are Au stud bump bonding, lead-free electroplated solder microbumps (like SnAg and AuSn microbumps) and the SLID technology using Cu/Sn intermetallic bonding [3]. Figure 38.7 shows an example of a possible 3D integration of a MEMS pressure sensor and an ASIC.

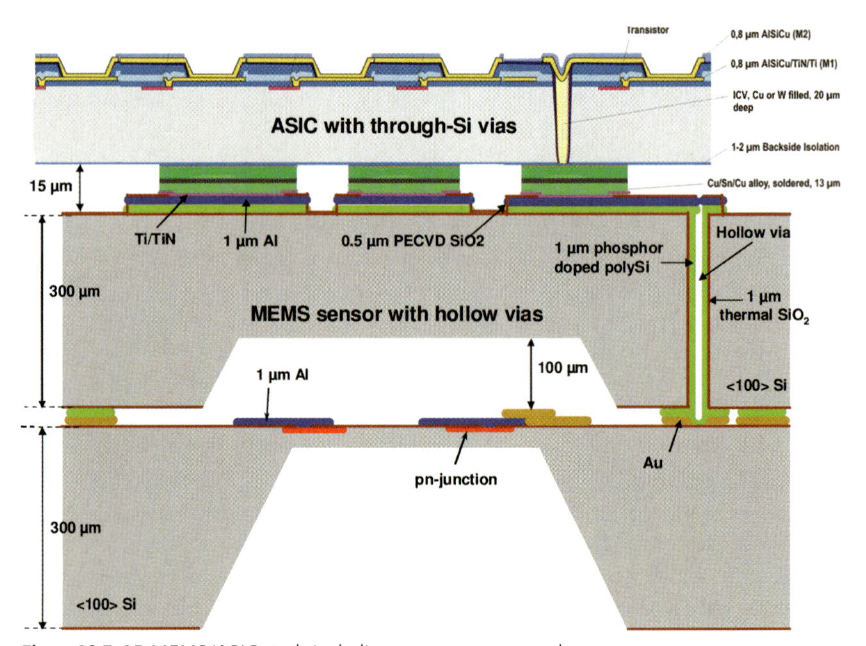

Figure 38.7 3D MEMS/ASIC stack, including pressure sensor and corresponding signal conditioning ASIC. The 3D integration is realized by use of deep hollow vias [6] and ICV-SLID [3] technology.

Several of the technologies presented here as compatible with the requirements of a typical MEMS sensor have been exploited in the e-CUBES project for the automotive demonstrator, which is described in more detail below.

38.4 e-CUBES GHz Radios

The e-CUBES radio is the communication block and interface of a wireless sensor network node; it has to answer challenging requirements in terms of 3D integration, power consumption, communication distance, standards and versatility. One of the first successful results concerning a fully integrated low power RF radio, realized by 3D-SIP stacking of an RF front-end CSP and a digital base band CSP, was reported in Reference [5] (Figure 38.8).

In e-CUBES, the pursed solutions are: (i) 2.45 GHz integrated radio for the automotive demonstrator and (ii) a 17 GHz ultra-low power solution for the health and fitness and aeronautical demonstrator. Both are designed using BAW resonators but MEMS-based resonators solution can be an alternative.

38.4.1 2.4 GHz Radio for Automotive Applications

The main rationale behind a 2.4 GHz radio for automotive applications is to reduce the size of the whole transceiver, allowing a very compact 3D integration of the radio system and the reduction of power consumption of the wireless node.

Figure 38.8 $7 \times 7 \times 2.5\,mm^3$ 3D-SIP stacking of an RF front-end CSP and a digital base band CSP for low power radio; after Reference [5].

The radio transceiver chip is embedded in a system composed of a BAR (Bulk Acoustic Resonator) device, a MEMS pressure sensor stack and a signal conditioning ASIC. This subsystem, assembled by means of a set of sophisticated interconnect technologies (inter-chip vias, redistribution layers and interposer) constitutes the major part of a TPMS (Tire Pressure Monitoring System).

The e-Cubes radio and its embedding into a wireless-node are illustrated in Figure 38.9. The core of the radio is the oscillator, which makes use of a BAR device. This approach replaces a crystal oscillator with PLL (Phase Locked Loop) and enables a higher degree of integration and minimization of radio turn on times due to the superior power-on start up times in the hundreds nanoseconds range. Furthermore, it reduces the complexity due to the lack of a PLL and RF-mixer in the transmitter part, which exhibits direct modulation and offers ultra-low power operation. In the receiver a high selectivity BAR device is used as RF input filter for channel selection, while an image-reject architecture introduces high sensitivity. The antenna is attached to the radio via an RF-interface and a matching network, which can be integrated into the RF front-end section of the transceiver ASIC.

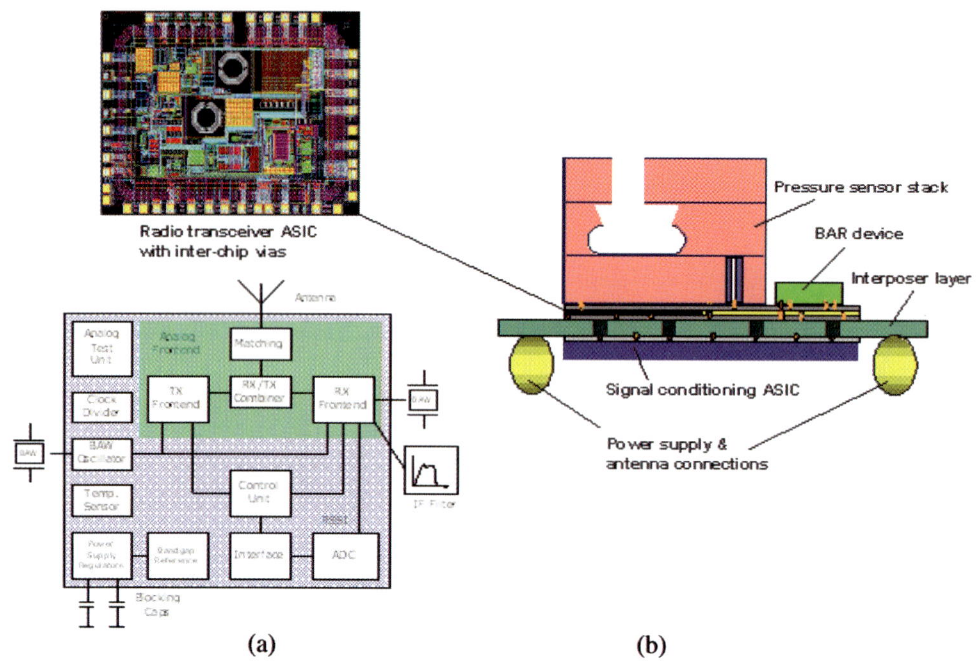

Figure 38.9 (a) Block diagram of e-CUBES radio for automotive applications and layout; (b) automotive radio subsystem and 3D implementation using Inter-Chip-Vias (ICVs) for RF interconnections.

38.4.2
17 GHz Ultra-Low Power e-Cube Radio for Wireless Body Area Network

For Wireless Sensor Networks (WSN) applied to body area networks, two needs have been identified in e-CUBES: first, concerning a technology capable of providing multi-Gbit s^{-1} data links for an in-room wireless personal area network (WPAN) system, and, second, a solution for ultra-low power wireless links for various sensor applications. The use of the 17 GHz unlicensed band (in Europe) together with an ultra-low-power radio architecture that does not require a standard PLL are proposed in e-CUBES to minimize the consumed energy per bit for longer sensor node lifetime. The 17 GHz unlicensed band is chosen due to the considerations of the available 200 MHz bandwidth, fewer potential interferers, small antenna size and good indoor propagation properties. These attributes enable relatively high data rate communication (10 Mb s^{-1}), thereby lowering the energy per bit to less than 2 nJ bit^{-1} for a WSN.

The 17 GHz radio system is designed so that the transceiver does not require a PLL. This reduces the turn-on time. The local oscillator may be derived from a bulk acoustic wave (BAW) device or cavity-type resonator. In this way, a low phase noise reference is achieved. However, the issue of frequency accuracy must be considered and handled at the system level, in an asymmetric master-slave network. BAW resonators below 10 GHz will be available in the near future. Therefore, an 8.6 GHz BAW resonator with absolute frequency accuracy of $\pm 0.3\%$ is realizable for this application. The proposed transceiver architecture is shown in Figure 38.10, and was reported extensively by Philips Research in [7, 8]. This is the slave part of an asymmetrical system employing a master device and ultra-low-power nodes. The receiver is based on direct down-conversion architecture. The LO signal is generated by a VCO with a BAW device at approximately 8.6 GHz, or one-half of the RF carrier frequency. Sub-harmonic mixers are employed to produce I/Q baseband signals. OOK and OFSK demodulation are supported in the receiver chain and the A/D

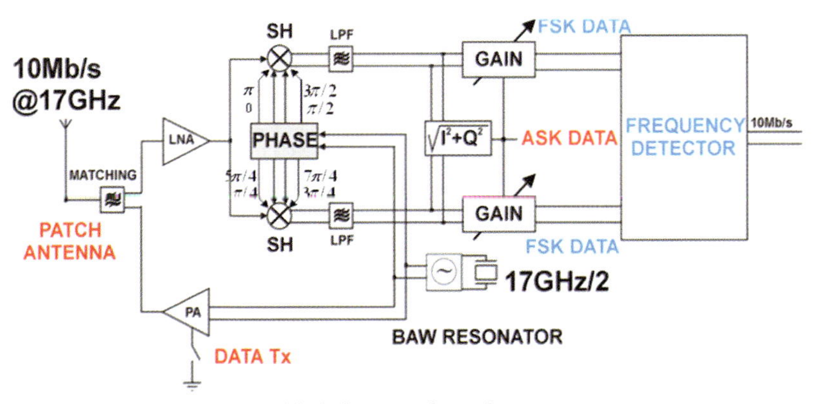

Figure 38.10 Tx/Rx Transmitter block diagram; after References [7, 8].

converter is eliminated to simplify the receiver. The OOK transmitter is also advantageous as it offers a very low-power implementation. In this architecture, all of the demodulation functions are performed in the analog domain, thereby eliminating the A/D converter, which reduces power consumption. OOK demodulation is achieved using a square root detector, which compares the square root of the amplitude of I/Q signals with a threshold voltage to decide whether the baseband signals are "1" or "0." A frequency detector is used for OFSK demodulation.

The transmitter shares the oscillator with the receiver and is turned-on/off by the data to provide an OOK modulation. As the transmitter is switched ON/OFF during data transmission, it may reduce the transmitter power consumption compared to that consumed by an FSK transmitter. In addition, as OOK signals only contain "0" and "1" PA linearity is not important. The problem of absolute frequency accuracy for the different Tx/Rx modulation formats can be solved using a master-slave (asymmetrical) system. The slave transmits first an OOK signal at frequency fRF1. The master device is located in the same room and locks onto this transmission. It then re-transmits an FSK signal with the required data on a carrier at fRF1. The ability of the master to continuously listen and search for the slave's transmission in both the time and frequency spaces is enabled by the fact that it can be mains powered and, consequently, has sufficient processing power to execute this search. Similar algorithms have been demonstrated in software-based GPS receivers.

38.4.3 The Role of RF MEMS in e-CUBES

The 3D integration of some emerging technologies such as RF MEMS (Radio Frequency Micro-Electro-Mechanical-Systems) is another key challenge for the e-CUBES; RF MEMS technology can provide substantial power savings while preserving or improving the high frequency performance. In the previous section, the key role of BAW resonators in designing low power PLL-free radio architectures has been demonstrated. While BAW solution seems the most adapted for few GHz frequencies, the MEM (Micro-Electro-Mechanical) resonator offers very flexible solutions for multi-frequency design in the tens to hundreds of MHz based on the control of lithographic dimension and not by the layer thickness. Open challenges related to high quality factors, low motional resistances requiring nano-gap technology for MEM resonators aiming at oscillator design, their temperature drift and 3D packaging are under exploration by many groups. Other future challenges concern the reliable 3D integration of BAW and/or MEM resonators.

In the past, the simplest examples of 3D-integrated RF MEMS were the above-IC high-Q passives: inductors [9] (Figure 38.11a) and capacitors (Figure 38.11b). In e-CUBES, the realization of phase shifters based on MEM switches for antenna beam steering [10] (Figure 38.12) is addressed. RF MEMS switches are employed in periodically loaded CPWs to achieve distributed MEMS transmission lines (DTMLs) phase shifter. The idea is to load periodically voltage-control variable capacitors tuning the distributed capacitance, the phase velocity and the propagation delay through the transmission line changes. The rationale behind this choice relates to the

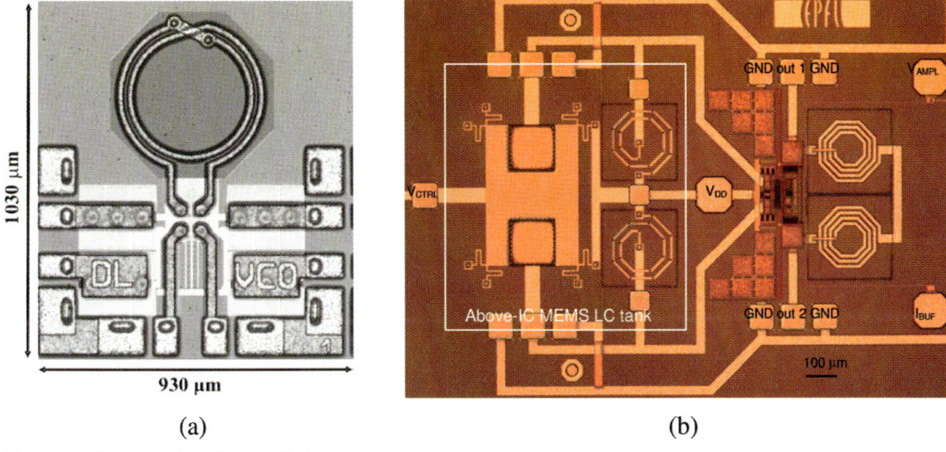

Figure 38.11 Examples of RF MEMS devices with 3D integration capability: (a) *Above-IC* high-Q inductor, by IMEC; (b) *Above-IC* double air-gap MEMS capacitor and suspended inductors (called MEMS LC tank) in 2.4 GHz CMOS VCO, by EPFL.

fact that RF MEMS capacitors overcome pin diodes limitations at very high frequency, offering an increased linearity, lower insertion losses and improved isolation at RF levels, in the context of a negligible DC power consumption (due to electrostatic actuation). The success of the RF MEMS switches is still conditioned by an appropriate packaging and solving reliability problems.

38.5 e-CUBES Applications and Roadmap

One key objective of e-CUBES is to elaborate a mid- and long-term technological and application roadmap to draw guidelines and priorities both for the research and the industrial directions. The e-CUBES roadmap is essentially structured at three levels:

1. Limits of miniaturization (from cm^3 to sub-mm^3, see Figure 38.13) and complexity (from two up to tens of functional interconnected layers in future 3D SoC).
2. Needed 3D technologies to answer to the application specifications (power, RF communication, sensing, and so on, in limited volume), and possible actions for standardization. In this perspective, technology bottlenecks should be predicted as early as possible and ways to overpass them at mid- and long-term proposed.
3. Visionary demonstrators and their impact on human life but also on the European markets and business models. Notably, reliability issues associated with 3D SOC design and fabrication are also of high importance, the reliability studies being distributed at various levels, from new materials and devices to the 3D system level.

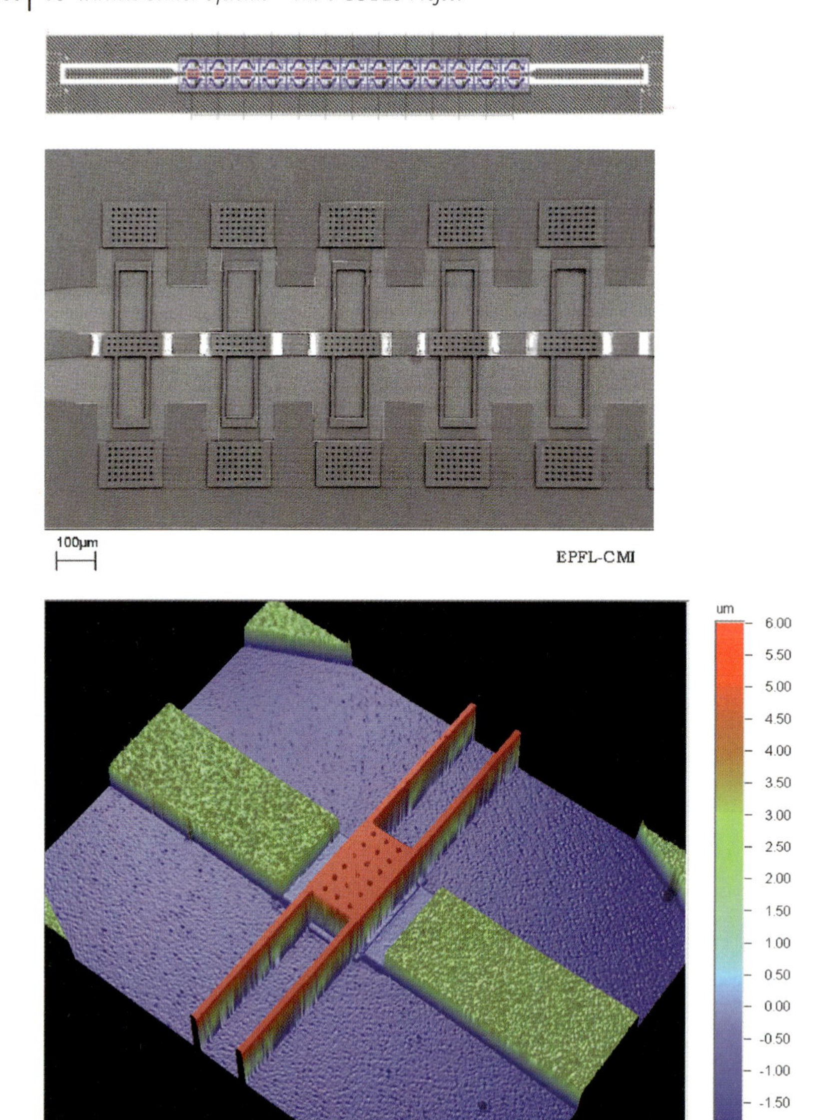

Figure 38.12 (a) Cell design and corresponding SEM picture of *above-IC* 10–20 GHz DMTL phase shifter based on MEM switches fabricated with a low temperature process (using polyimide as sacrificial layer), by EPFL [10]; (b) 2D profile of the capacitive switch used in the core DMTL cell, demonstrating excellent flatness (quasi-zero stress membrane) after release of sacrificial layer.

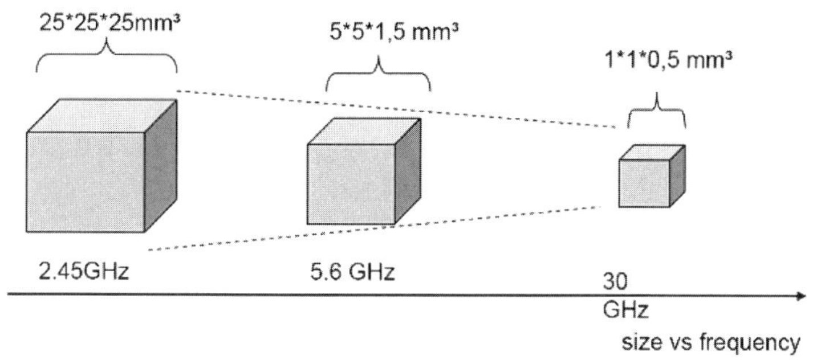

Figure 38.13 e-CUBES size versus frequency diagram.

On the other hand, 3D integration is expected to generate a performance evolution that cannot be rivaled by traditional 2D approaches, with a large range of applications (Figure 38.14). In the following, airborne and space, automotive and health and fitness application domains of e-CUBES are briefly discussed in terms of challenges and requirements. Table 38.1 briefly describes the four main demonstrators of e-CUBES projects in terms of some defined operational parameters: range, sensor movement and medium for wireless communications.

38.5.1 Airborne and Space Demonstrator

The roadmap for the introduction of e-CUBES in distributed smart monitoring and particularly in the aeronautical industry, as part of the onboard equipment, should be first compliant with the general trends in the aerospace domain, and, second, in line with the general aeronautical standards, particularly as far as reliability and safety are

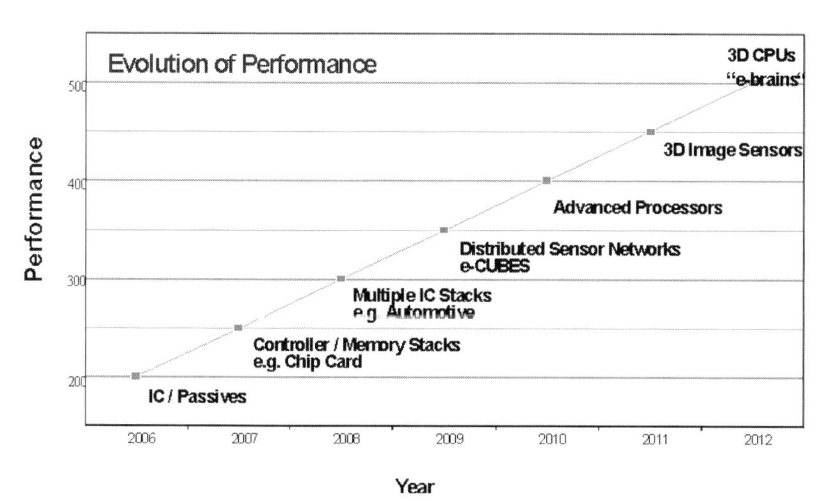

Figure 38.14 Evolution of performance due to 3D integration in various applications domains versus time. (Courtesy of FhG-IZM, Germany.)

Table 38.1 e-CUBES demonstrators and their main operational parameters.

Demostrator Operational parameter (units)	Aeronautics	Space	Health	Automotive
Range (=distance between sensors)	0.5–5 m	1 km to 10s of km	10s of cm to 30 m	10s cm to a few m
Sensor movement (In the network)	Fixed	Fixed or mobile	Mobile	Fixed or mobile
Medium for wireless communications	Metal + composite + free space	Free space	Human tissue (composite and conductive)	Metal + composite + free space
Volume allowed (cm^3)	~1	1–10	<A few mm^3	~1
Lifetime	20 yr min	From a few hours to 10s of years	From a few minutes to 10 yr	10–20 yr
Frequency range for communications (GHz)	>20	<2.4	From 100s kHz up to 40 GHz	>10
Type of data transfer	Burst mode + continuous mode	Burst mode + continuous mode	Continuous mode	Continuous mode
Data rates per sensor ($kB\,s^{-1}$)	Burst mode: 64 MB in 15 s; continuous mode: a few $kB\,s^{-1}$	Burst mode: 16 kB in 1 s; continuous mode: 10s of $kB\,s^{-1}$	Continuous: a few $kB\,s^{-1}$?	Continuous: 10s of $kB\,s^{-1}$?
Temperature range (°C)	−55 to +125	−200 to +250	+15 to +40	−30 to +200
Main available energy from the environment	Vibrations	Solar	Mechanical energy	Vibrations heat flux
Necessity for a self-organizing network	Yes (optimizing the connection to a master station)	Yes (increasing communications links efficiency)	Yes (minimizing the radio power through the body)	Yes (optimizing the connection to a master station)

considered as key factors in this domain. A separate (or dedicated) space roadmap is not presented; it is simply assumed that many of the requirements made by the aeronautical industry are indirectly paving the way for the 3D technology needed in space applications. However, e-CUBES partners are very aware that some of the space scenarios will require operation of WSNs with different distances between the e-CUBES and more extreme environmental conditions.

38.5.1.1 Airborne Applications of e-CUBES

The general trends in the mainstream of the aeronautical domain, specifically for passenger and cargo transportation, are:

- Safety: There is a continuous need to provide an increased level of flight safety by using new methods of early failure detection. The "black box" does not provide such an objective, as it is a mean of enquiry after an accident. Most of the sensors used today in the plane structures are dedicated to the flight control itself, and some are also used for maintenance management. But ATM is today the main solution where efforts go to maintain or increase safety levels whereas the flight traffic and the number of platforms increase, 4–10% per year, depending on the country and type of flight (short to long range).
- Flight operation cost reduction: Competition between companies demands a decreasing ticket cost. Fuel is today the largest cost contributor (now around 50% of the ticket cost – variable from country to country, airport from airport), which is leading to a general move towards using new flight routes to save fuel (for instance going over the North Pole to link Asia and USA or using different hub level using platforms such as the A380 for the long range primary hubs). Fuel cost increase is becoming the bigger problem in the Aeronautical domain today, with a possible consequence being sudden downturn in Aeronautics. Among the different factors contributing to the costs, maintenance accounts for around 30% and new ways of air fleet operations and maintenance management are much sought after by the companies. The reduction of the labor cost of the maintenance operations would ask to reduce the number of general periodic maintenance operations. These would be replaced by maintenance operations more focused on the necessities drawn from finer and smarter platform automated monitoring.
- Security: Examples from past years (not only since the terrorist actions on the Twin Towers) have shown that a civil plane could be a powerful terrorist weapon against which civil and military authorities have few solutions. In particular, there is a need of situational information onboard for ground concerned people.
- New fabrication technologies: and related techniques for the plane structure, with the purpose of reducing the structural weight of the plane (including the engines), while increasing the limits to rupture and the plane survivability to hazardous situations.
- More silent engines: necessary to have some compatibility between a huge traffic increase (between 4 and 8% per year depending on the geographical location) and

restricted areas availability near cities. Cabin noise reduction is also considered for increases passenger comfort.

- Progression towards the electrical plane: meaning the development of electrical functions for power actuations, replacing more and more mechanical, pneumatic and hydraulic functions. This context is becoming favorable towards electrical interfaces directly provided by wireless sensors and actuators, such as studied in the e-CUBES project.

In general, WSN can bring added value to each of the all above requirements and, in particular, e-CUBES is contributing to the following issues:

- Earlier detection: of mechanical and electrical failures for safer flights thanks to the smart local processing in each sensor combined to the huge number of sensors distributed in the structure.
- Reduction of maintenance cost: thanks to the continuous, distributed and decentralized monitoring of the whole platform (and not just only a few points in central areas) all along the flight.
- Contribution to security issues: by providing later on wireless cameras distributed in key areas of the structure to monitor the presence of people in critical situations on the ground or during the flight. The possibility of having non wire-connected sensors is a key factor to reducing sabotaging by terrorists.
- Drastic reduction of the weight: due the cabling; it was clear from the beginning of the project that, thanks to e-CUBES technology, there will be a great impact, even compared to optical cables. Moreover, the installation of the WSN on the platform is easier during the fabrication process itself, compared to cable solutions, and its upgradeability will practically eliminate the costs due to time spent on cable installation or reinstallation, and integrity checking.
- Cabin noise reduction: using active techniques based on a distribution of acoustic sensors combined to vibrating actuators is seen as a key point for increased passengers comfort.

Finally, if the "electrification" of the airplane of the future is confirmed as a big push towards more reliability and simplifications to the airplane infrastructure, then it paves the way to the introduction of e-cubes serving as a general interface between the Flight Control Unit and each electrical power actuator.

Figure 38.15 depicts the roadmap for technology block needs in the research and development and industrial phases of e-CUBES aeronautical demonstrator.

38.5.1.2 Space Applications of e-CUBES

One can contemplate the building of a satellite in a cube, the bricks of which (propulsion, navigation, communication, etc.) would be around mm^3 to cm^3 and would be then assembled through some standard 3-D technique (Bus Metal). An artists view of a future complex space demonstrator with multi sensors, communication, computing and storage is given in Figure 38.16. In the following, a few space

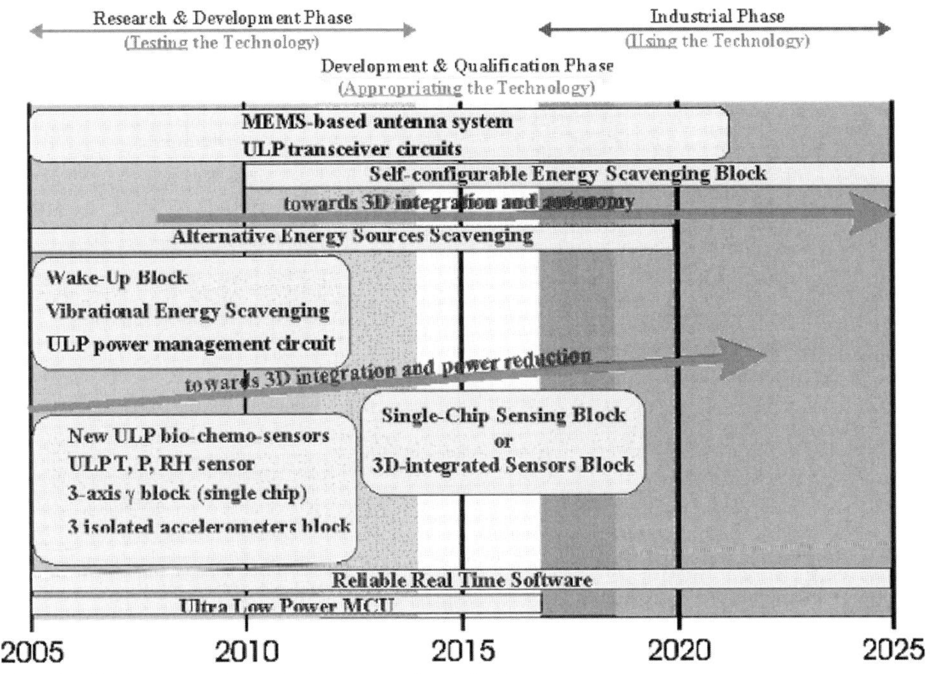

Figure 38.15 Roadmap of needs in R&D and industrialization phases of aeronautical e-CUBES demonstrator.

applications scenarios of wireless sensor networks based on e-CUBES are analyzed. Depending on the exploration mission objectives, sensing nodes made of e-CUBES would have to face very different constrains [12]. Sensing nodes can be fixed either on the ground of a planet or asteroid or moving relative to each other. The nodes

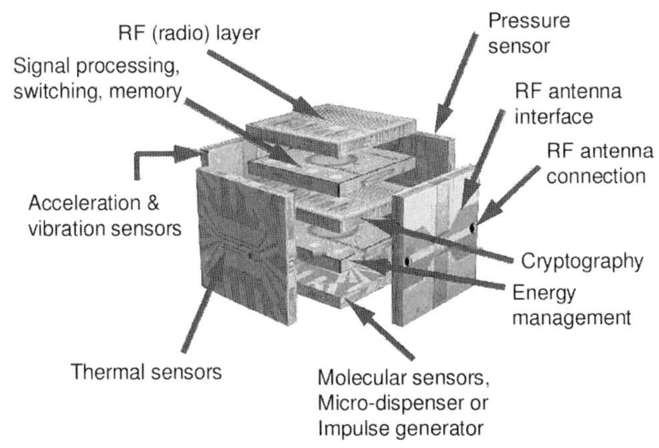

Figure 38.16 e-CUBES concept for space application (sensor node or satellite) with multi-functional layers.

deployment technique would have a strong impact on the node network. The nodes would conceptually have component needs similar to the airborne ones: MEMS sensor (analog or digital), A/D converter (if analog sensors), microcontroller (for signal conditioning, communication protocol and power management), DSP layer, memory, RF transceiver, antenna and power supply.

In a mission where nodes are falling through the atmosphere of a planet while taking measurements (Figure 38.17a), the relay for large distance transmission would fall among the nodes at the same speed. The duration of such a mission could be relatively short (up to a few hours) and the sampling rate quite high (for instance a data acquisition per second). In a scenario where the sensing nodes land on the ground of a planet or moon (Figure 38.17b), a relay would transmit the data collected by the nodes to an orbiter or directly to earth. The mission could last up to a few years, while the sampling rate would be very low (for instance one data acquisition per hour). An alternative to this scenario for low mass solar system objects, that is, asteroids, is to anchor the nodes into the ground, for instance to conduct seismic measurements (Figure 38.17c). More advanced scenarios would concern moving nodes over a long period of time. For instance actively moving nodes (micro robot) on the ground of planets or a moon could collect data while they would receive directions from earth regarding the surface area to be explored. In this perspective, different node design can be considered. For instance, the nodes could be constituted of a central heavy part that contains the electronics package into an inner sphere that is displaced by electroactive polymer actuators into an external hull. Slow motion of the actuators would induce a rolling behavior by continuously translating the center of gravity, while a quick motion of the actuators would make the node jump over obstacles (Figure 38.17d). A more complex scenario could concern a cloud of nodes that would rebound on the surface of a low mass object, that is, an asteroid (Figure 38.17e). The low attraction and the absence of gas would enable the nodes to rebound at high altitude (a few kilometers) with relatively low impact speed. Acceleration sensors could provide data on the surface nature, for instance to locate a good landing place.

Three essential needs of WSN for space exploration are not yet met by commercial WSN: (i) communication at distances up to 10 km, (ii) localization functionalities (preferably without relying on GPS/Galileo-type technologies or on RSS) and (iii) dynamic rapid self-organization. Today commercial WSN products can be used to demonstrate on earth the major aspects of an exploration mission but they are not yet mature enough to be used in space to collect reliable scientific data. To develop WSN for space exploration an effort is made in the e-CUBES project towards increasing the communication range and implementing localization functionalities. This research is combined with understanding radiation and environmental constraints and deriving criteria for the associated packaging and shielding issues. The packaging of the node must be robust and hermetic to allow for reliable operation on the asteroid, planet or moon, as well as safe transit from earth. The package serves as a mechanical support, and can provide electrical routing of signals and power. For cm^3 packages, ceramic chip carriers are an appealing solution, due to their robustness and also for instance due to the ability to build a patch antenna

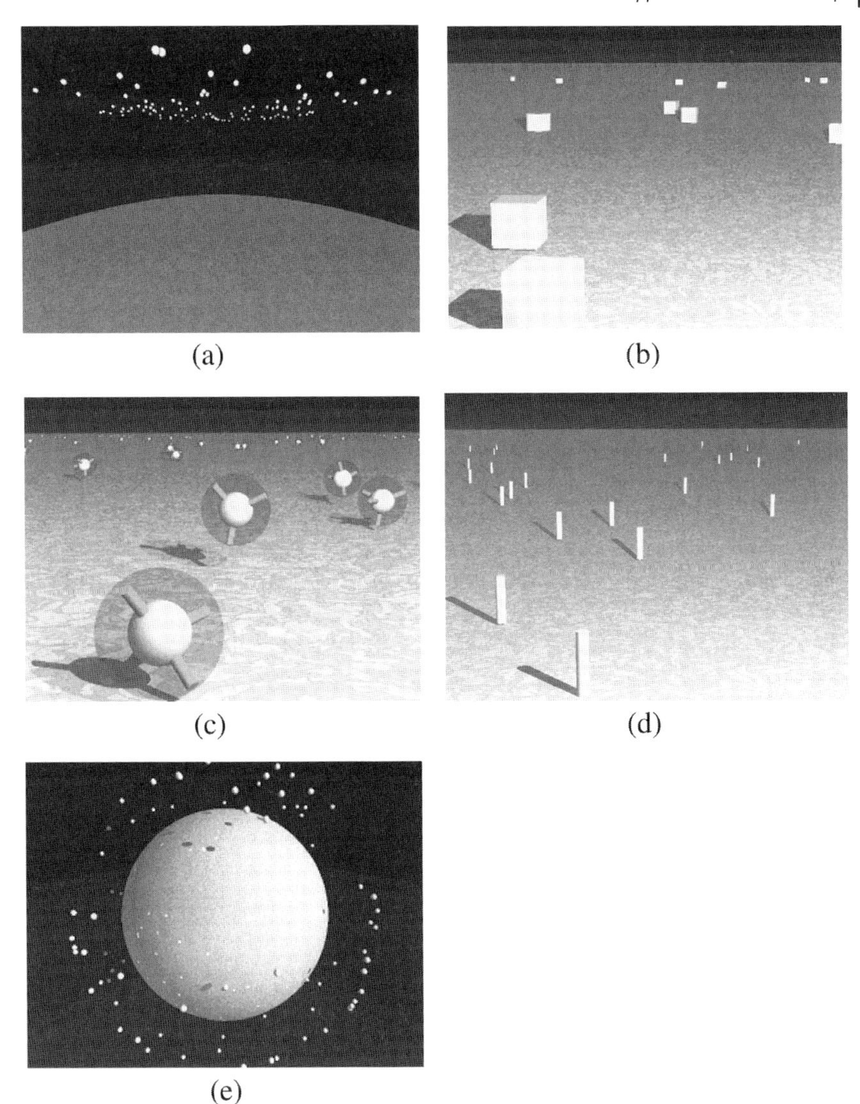

Figure 38.17 Various scenarios of WSN based on e-CUBES for space exploration: (a) atmospheric measurements, where the sensing nodes are falling through the atmosphere of a planet; (b) sensing nodes are located on the ground of a planet or moon; (c) nodes are anchored into the ground of an asteroid; (d) mobile nodes on the ground of a planet or moon; and (e) rebounding nodes around an asteroid (low mass object).

directly into the lid. The e-CUBES aims to integrate all the node elements into few mm^3 using direct chip to chip stacking and bonding. Radiation shielding (most likely in the form of a few mm of Al) may be required and may play a large role in node size and mass.

38.5.2 Automotive Demonstrator

Automotive applications of e-CUBES have their own opportunities and specific constraints; in general, automotive electronics use technologies different from other electronic applications and the technological processes and innovation cycles are equally different. While areas such as memory, computing and communication are closely linked to the semiconductor technology roadmap, the automotive field also uses very special technologies (such as power or MEMS) and has very strict quality requirements (high/low temperatures, acceleration, resistance against chemicals and so on). On the other hand, the degree of integration in automotive electronics is often much lower than in other applications, and this area is therefore often considered less demanding. However, the opposite is true, as automotive applications follow parameter sets other than integration density alone. These other parameters are in fact extremely demanding and very specific, depending on the application. Indeed, while the integration roadmap (Moore's law) is not irrelevant, automotive electronics lags the integration roadmap for other applications by about two generations.

Owing to the great variety and specificity of technological approaches of automotive applications, the variety of back-end integration approaches (of which 3D integration is one) is also large. Notably, for each field of application the technical solution may be different. No standardization or limitation to one particular technology is possible, but the demand for perfectly suited low-cost solutions is always high.

Sensor networks are researched in various automotive applications, in particular in Tire Pressure Monitoring Systems (TPMS) installed as autonomous sensor nodes. A tire-mounted TPMS should have a maximum weight 5 gram for a tiny volume (<0.5 cm^3) that includes package, power supply unit and antenna. An ultra-low power ASIC (reducing power consumption by factor of 10 compared to existing solutions) and the combined use of a battery and of an energy harvester are foreseen to enable a 0-year lifetime for the TPMS. Moreover, the TPMS should be very robust and, especially, withstand large values of acceleration (up to 1000–3000 *g*). The e-CUBES automotive system demonstrator depicted in Figure 38.18 exploits various 3D integration technologies and is expected to fulfill all these challenging requirements at a low cost (estimated at 2–5 € per unit).

38.5.3 Health and Fitness Demonstrator

People are increasingly taking responsibility for their health and physical condition: they want to manage their own health and feel good. Solutions that help them improve their personal well-being have high added value, for which they are, presumably, prepared to invest considerably. These types of solutions typically consist of three technology elements:

1. Unobtrusive sensors – small, wireless, virtually invisible, low weight – that measure relevant body parameters;

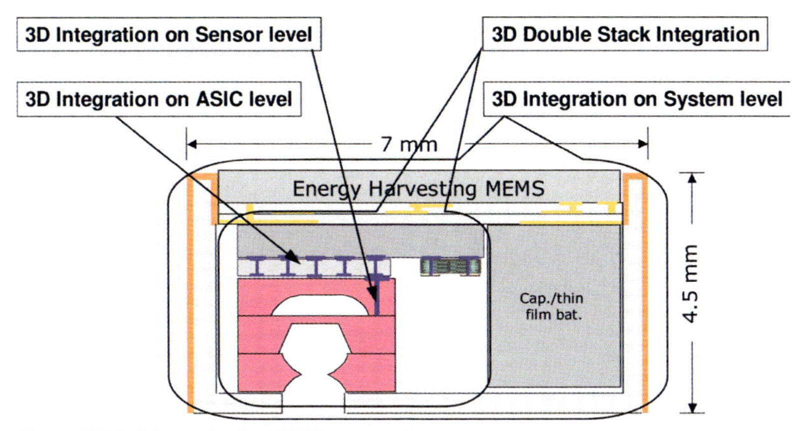

Figure 38.18 Schematic of e-CUBES automotive system demonstrator with energy harvesting power supply.

2. Algorithms that interpret these measurements; and coaching algorithms that, based on the current status, provide the user with feedback and guidance to reach the desired goals;
3. Wireless connectivity to transfer the data to a body network for further processing.

Within e-CUBES concept, an unobtrusive measurement of relevant body parameter (temperature, respiration, heart beat, etc.) becomes possible. When e-CUBES is small enough, implantations offer a unique possibility to medicate at certain positions in the body at certain times. For instance, the insulin level can be measured and the release of hormones can be stimulated. Using wireless sensors (e-CUBES) can greatly help in the diagnosis of neurological disorders such as Parkinson's or Alzheimer's disease. Parkinson's can so far be diagnosed only through behavioral changes, principally changes in gait.

In the field of health and fitness (intelligent human aid systems, diagnostics) health monitoring systems already exist. They are bulky and, due to the cables, the free action radius is very restricted. A wireless low-weight and easy-to-place "medical e-CUBE" would return a quality of life in a way never previously thought of (monitoring blood glucose level and release of necessary amount of insulin).

In the future, the following evolution is foreseen for health and fitness e-CUBE features: more sensor functionalities, more power-saving after application study, potentially wirelessly re-chargeable, 17 GHz radio with or without BWA resonators and packaging smaller than cm^3. The software could be also wirelessly upgrade-able.

38.6 Conclusion

The status and challenges related to the applications of emerging 3D integration technologies in wireless sensor systems have been reported and discussed. While recent technical progress in the frame of the e-CUBES integrated project readily points to the feasibility of various demonstrators in airborne, automotive and health and fitness fields, the cost effectiveness appears as a key supplementary criterion to drive the technical choices for future market success.

Acknowledgments

The authors acknowledge the funding of the integrated project e-CUBES by the European Commission under support-no. IST-026461

References

1 Warneke, B., Last, M., Liebowitz, B. and Pister, K.S.J. (Jan. 2001) Smart Dust: communicating with a cubic-millimeter computer. *Computer,* **34** (1), 44–51.
2 Cook, B.W., Lanzisera, S. and Pister, K.S.J. (June 2006) SoC issues for RF smart dust. *Proceedings of the IEEE,* **94** (6), 1177–1196.
3 Ramm, P., Klumpp, A., Merkel, R. *et al.* (2004) Vertical system integration by using inter-chip vias and solid-liquid-interdiffusion bonding. *Japanese Journal of Applied Physics,* **43** (7A), 829–830.
4 Ramm, P., Bonfert, D., Gieser, H. *et al.* (June 2001) InterChip via technology for vertical system integration. Proceedings of the IEEE International Interconnect Technology Conference, pp. 160–162.
5 Beyne, E. (2004) 3D Interconnection and Packaging: Impending Reality or Still a Dream? Proceedings of the IEEE International Solid-State Circuits Conference, ISSCC 2004, 15–19 February 2004; San Francisco, CA, USA, IEEE, pp. 138–145.
6 Lietaer, N., Storas, P., Breivik, L. and Moe, S. (2006) *Journal of Micromechanics and Microengineering,* **16** (6), S29–S34.
7 Sanduleanu, M.A.T. (2006) 17 GHz Ultra Low Power Radio-The 1 nJ/bit paradigm, Low power radio Workshop, ESSCIRC 2006, Montreux, September.
8 Sanduleanu, M.A.T. *et al.* (2007) 17 GHz RF front-ends for low-power wireless sensor networks. Proceedings of BCTM, Boston, October.
9 Sun, X., Dupuis, O., Linten, D. *et al.* (Nov. 2006) High-Q above-IC inductors using thin-film wafer-level packaging technology demonstrated on 90-nm RF-CMOS 5-GHz VCO and 24-GHz LNA. *IEEE Transactions on Advanced Packaging,* **29** (4), 810–817.
10 Fernandez-Bolanos, M. Badia *et al.* (2007) RF MEMS capacitive switch on semi-suspended CPW using low-loss HRS. Proceeding of MNE, Copenhagen, pp. 171–174.
11 Dubois, P., Botteron, C., Mitev, V. *et al.* (2008) Ad-hoc wireless sensor networks for exploration of solar-system bodies. *Acta Astronautica,* (in press).
12 Weber, W. Three-dimensional integration of silicon chips for automotive applications (eds C.A. Bower, P. Garou, K. Takahashi and P. Ramm), MRS Proceedings, Enabling Technologies for 3-D Integration, Volume 970, Material Research Society, ISBN: 978-1-55899-927-5.

Conclusions

Phil Garrou, Christopher Bower, and Peter Ramm

As 2007 drew to a close, 3D integration technology had become a very "hot" topic in the semiconductor industry. 3D integration articles are showing up monthly in almost all of the well-known IC trade magazines and research journals. "3D" is on the verge of competing with "nano" as a new technology buzz word. Is all the "buzz" about 3D integration justified? Researchers are well aware that "buzz" does not always translate into commercial success. However, in this case, as pointed out by several chapter authors, 3D integration is a very natural progression for the semiconductor industry.

With everyone in agreement that Moore's Law will encounter the "red brick wall" somewhere in the 32–22 nm nodes, the ITRS roadmap now shows 3D integration as a key technique for achieving higher transistor integration densities. While we are awaiting the ultimate replacement for CMOS technology, 3D approaches promise to improve performance (shorter interconnects), improve yield (individual device layers made on optimized process), decrease footprint and add functionality (non-silicon functionalities) without major changes in materials or technology. "Going Vertical" appears to be the only logical way to go.

The chapters in this handbook show that many institutes and companies have demonstrated full 3D integration processes. Several universities and institutes have also made great strides on improving individual unit operations and design and modeling routines. Figure 1 shows a map of some of the global activity in 3D integration.

A large percentage of the process flows that are demonstrated early on in a technological evolution are usually feasible (i.e., work in the laboratory) but ultimately are not commercially viable, thus many of the process sequences shown in this handbook may ultimately never reach commercialization. We leave it up to you to pick the winners. Certainly recent history teaches us that manufacturing processes based on wafer-level fabrication have a comparatively favorable cost structure and performance. For most applications wafer yield and chip area issues will be problematic for full wafer stacking approaches to 3D integration. In consequence, chip-to-wafer stacking concepts utilizing known dies (KGD) only are advantageous

Handbook of 3D Integration: Technology and Applications of 3D Integrated Circuits.
Edited by Philip Garrou, Christopher Bower and Peter Ramm

ISBN: 978-3-527-32034-9

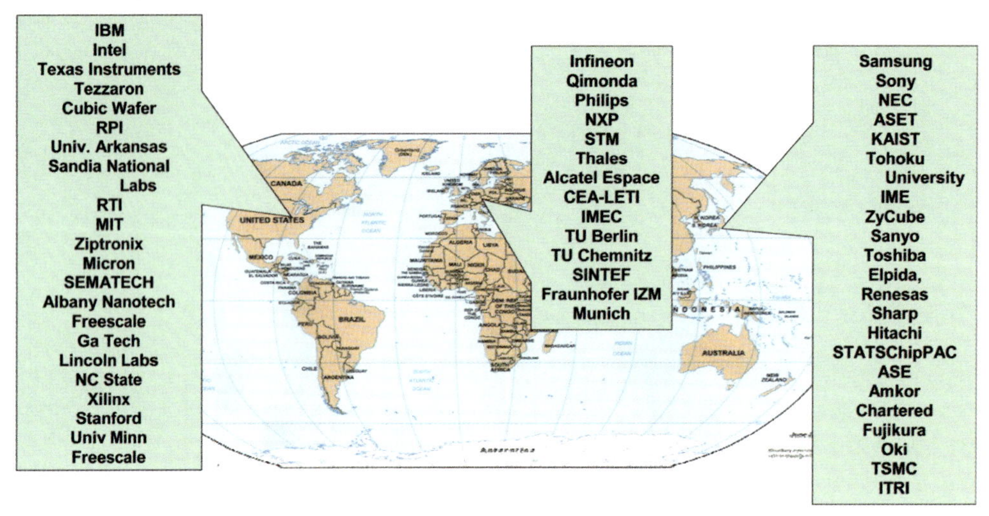

Figure 1 A sampling of global activities in 3D integration.

for 3D integration of devices with non-identical chip sizes or/and low wafer yield. The ultimate winning process is probably yet to be seen.

There certainly is a need for reliable and cost-effective solutions for 3D specific processes such as:

- bonding (temperature, force, alignment accuracy, defect density, etc.);
- thinning and dicing (uniformity, stress, die strength and so on);
- handling of thinned wafers and dies (productivity, process integration, etc.);
- integration of vertical interconnects (size, dead-zone, resistivity, reliability, etc.).

There are significant challenges and concerns in the areas of:

- electrical performance issues as, for example, cross talk and parasitics;
- need of 3D specific product development tools (simulation and modeling, design, process and test);
- thermal issues (e.g., increased power densities for stacked chips);
- reliability.

In conclusion, we propose that the future technology platform for improved integration density, performance, functionality and productivity of microelectronic systems over the next decade will be 3D integration. Although there are significant challenges, no real "show stopper" has been identified. One can envision that further improvements in 3D processes and design technology will lead to new 3D architectures (in analogy to human brains) and completely new products, such as, for example, ultra-miniaturized ambient sensor systems (in analogy to human sense organs). The future promises to be an exciting time for players in the 3D microelectronics arena!

Index

Handbook of 3D Integration: Technology and Applications of 3D Integrated Circuits.
Edited by Philip Garrou, Christopher Bower and Peter Ramm

ISBN: 978-3-527-32034-9

e

f

g

h

j

k

l

q

r

s

u